Transistor and Integrated Electronics

Transistor and Integrated Electronics

MILTON S. KIVER

Publisher and Editor
Electronic Packaging and Production Magazine

Fourth Edition

McGRAW-HILL BOOK COMPANY

New York ■ St. Louis ■ San Francisco ■ Düsseldorf ■ Johannesburg
Kuala Lumpur ■ London ■ Mexico ■ Montreal ■ New Delhi
Panama ■ Rio de Janeiro ■ Singapore ■ Sydney ■ Toronto

TRANSISTOR AND INTEGRATED ELECTRONICS

Library of Congress Catalog Card Number 76-37102

07-034942-8

1 2 3 4 5 6 7 8 9 0 KPKP 7 9 8 7 6 5 4 3 2

This book was set in Linotype News Gothic by The Maple Press Company, and printed and bound by Kingsport Press, Inc.
The designer was Judith Michael; the drawings were done by Vantage Art, Inc. The editors were Alan W. Lowe, M. E. Margolies, and Catherine A. Kerr. James E. Lee supervised production.

Contents

Preface

We are now well into the era of semiconductor electronics. No longer does any doubt remain about whether this technology, started in 1946, will be able to supplant vacuum tubes. The struggle has already been decided; and vacuum tubes are now in the terminal stage of their existence, gradually disappearing as the equipment in which they are employed reaches total obsolescence and is discarded.

Semiconductor electronics is today a young and vital technology; new developments are spilling out of the research laboratories almost on a daily basis. Some of these developments represent entirely new concepts in fabrication or usage; others are simply refinements in existing devices; and all have a bearing on the forward thrust of this new technology, although obviously some will be of far greater importance than others.

This fourth edition of Transistor and Integrated Electronics represents a complete revision of the third edition which was entitled simply Transistors. The revision is extensive—in many instances, chapters have been rewritten. All material deemed no longer applicable to present devices has been deleted; all characteristics and circuits have been updated; and a considerable amount of new material, including three new chapters, has been added.

This book is designed for electronics technicians as well as all other technical workers who desire to gain a working knowledge of transistors and semiconductor devices and their applications. The discussion starts with modern electron theory and progresses to the operation of diodes and transistors. Chapter 3 examines the characteristics exhibited by transistors under a variety of normal operating conditions and illustrates what factors place limitations on the range of transistor applications with respect to power, frequency, signal level, and temperature. A typical manufacturer's data sheet is also presented and analyzed for the information it presents.

Chapter 4 discusses the historical progression of a variety of transistor types, from the early point-contact units to recent transistors capable of operating far into the megahertz region. As this story unfolds, the reader is able to perceive clearly the complete dependence of product development on fabricating techniques and to see how laboriously obstacles have had to be surmounted step by step. It is perhaps one of the most important chapters in the book, because the processing techniques discussed therein are so vital to the manufacture of the integrated circuits that are now beginning to dominate the entire field of electronics.

The rising importance of field-effect transistors made it desirable to examine these devices in a separate chapter (Chapter 5). Thus, a full discussion of FETs in their most common forms is presented, including their fabrication and operation.

With the first five chapters serving as a foundation, the reader is then presented with a wide range of transistor applications in Chapters 6 (amplifiers), 7 (oscillators), 9 (radio, both AM and FM), 10 (computers), and 11 (television). All circuit explanations employ the well-proven step-by-step approach, starting with the simplest facts and proceeding gradually to the more complex. No mathematics of any significant difficulty is used in the text.

Chapter 8, an entirely new chapter in this revision, explores the integrated circuit, showing how it is basically related to the transistor not only in terms of operation but also in terms of similarity in fabrication. The integrated circuit is a natural development in the growth of semiconductor electronics; its form and function can be readily mastered once a solid, basic understanding of transistors has been achieved. Chapter 8 was positioned where it is in the text because radios, computers, and television receivers are each making increasing use of integrated circuits—and in the case of computers, integrated circuits are by far the overriding basic element employed.

Chapter 12 focuses attention on the growing number of other solid-state devices that have been developed and the mechanisms that were put to work to make them practical. These include the Gunn diode, light-emitting diodes (LEDs), photo diodes and transistors, zener diodes, varactor diodes, and thyristors and silicon controlled rectifiers, among others. It is interesting to note that the majority of these devices are the result of research work done on electronic-behavior mechanisms associated with PN junctions and their reaction to external stimuli.

In Chapter 13, the reader is introduced to transistor circuit design. The presentation should make the behavior of transistors in various circuits and at different frequencies better understood. For readers who wish to become generally familiar with circuit design, this chapter will provide all the information required; for those who intend to engage in circuit design, it will serve as a gateway to the more advanced texts available.

A discussion of the various precautions to observe when servicing transistor circuits and transistor devices as well as a method of approach to the servicing of all such units is given in Chapter 14. Finally, in Chapter 15, a series of simple and easily worked transistor experiments are included for readers who may wish to learn of transistor operation firsthand. To the practical man, such experiments offer an avenue for acquiring valuable experiences. It is strongly urged that this chapter not be slighted, since semiconductor electronics is an eminently practical subject and no full appreciation of its capabilities can be achieved without exposure to the actual devices themselves.

The book can be used in technical institutes, electronics schools, vocational-technical schools, armed-service schools, and industrial training programs in electronics, as well as for home study. Questions are included for each chapter so that an instructor may test the progress of a student or as a form of self-test for those studying alone.

The author wishes to extend his thanks and appreciation to the many firms in the field for the data, technical assistance, and photographs which they so graciously provided. The author is also indebted to Peter Henry of Fair Haven, N. J., for his technical assistance, particularly in the discussions relating to the fabrication and processing of transistors and integrated circuits. Mr. Henry's many years of experience in this segment of the electronics industry was particularly helpful in illuminating some of the reasons why certain techniques are employed.

Milton S. Kiver

Transistor and Integrated Electronics

Chapter One

Introduction to Modern Electron Theory

The story of the transistor is in large measure the story of matter and of how the scientists at the Bell Telephone laboratories were able to make that matter amplify electric currents. Even if the transistor were not responsible for today's huge semiconductor industry, it would still be important for its contribution to our understanding of solid-state matter. It demonstrated, for the first time in history, amplification in solids. Prior to the transistor, amplification was achieved only with vacuum tubes.

The invention of the transistor is officially credited to John Bardeen, William Shockley, and W. H. Brattain, three scientists working for the Bell Telephone laboratories. The first public announcement of the transistor was made in June, 1948. Since then, advances in design and processing techniques have been enormous. Consider that on a single silicon chip smaller than a fingernail we can now have hundreds of transistors, resistors, and diodes—in short, a complete circuit. In addition, the complexity and efficiency of these circuits continues to grow at an unbelievable rate.

The invention of the transistor in 1948 thus led not only to the establishment of an exciting new industry, one that is having a profound effect on every facet of our daily life, but also to the realization of a centuries-old dream, the landing of a man on the moon in 1969. When we stop to consider that all this has been accomplished with an industry barely in its infancy, we can well appreciate the even more remarkable developments that must lie hidden in the undiscovered depths of this astounding technology.

In order that we may develop an understanding of semiconductor circuitry, let us review what we know concerning the structure of matter and the role that the electron plays in that structure.

ATOMS AND MOLECULES

Every substance or material that we come in contact with can be divided into particles known as molecules. These are the smallest segments into which a substance can be divided and still retain all its individual characteristics. Molecular units are so minute that we have not been able to devise instruments that will enable us to see them. In order to see something that is extremely small in size, we must design an instrument, such as an optical microscope or electron microscope, that will detect this "something" and then enlarge it so that we can see it. How, then, can we distinguish between the molecules of the substance we are checking and the molecules of the instrument? Although it does not seem likely that we shall ever see a direct picture of a molecule, highly refined indirect methods have been developed for determining molecular structure so that we know quite a bit about it.

We know, for example, that the molecule of a substance may consist of a single element or a complex association of a number of elements. Chemists and physicists have discovered over a period of hundreds of years that there are more than 100 different elements that either singly or in combination make up all the matter on this planet. This figure has not remained a static one but has gradually risen as man's knowledge and scientific understanding have deepened. It is possible that many more elements will be discovered. Some of the more familiar elements are hydrogen, oxygen, gold, silver, copper, and iron.

A molecule, then, is the smallest portion of a substance that retains all the physical and chemical properties of that substance. But what about the elements? If we were to consider the smallest portion of an element that is still identifiable as that element, we would be dealing with the atom. Each different element is represented by a different atom.

ATOMIC STRUCTURE

As we work our way down the size scale, we come first to molecules, then to the elements that compose the molecules, and finally to the atoms that represent the elements. When we investigate atoms, we find that they consist of a centrally situated nucleus with a net positive charge surrounded by a number of electrons that revolve about the nucleus. The central positive charge is said to be due to protons. Each of the electrons has a negative electrical charge. In a stable atom, the positive charge of the nucleus is exactly counterbalanced by the total negative charge of the externally revolving electrons. The net overall electrical charge is zero, which is the normal state of most atoms.

The atom possessing the simplest structure is hydrogen. It consists of a positive nucleus containing a single proton. Revolving around this proton is a single electron. The illustration most commonly employed for the hydrogen atom is shown in Fig. 1.1. Actually, we have learned enough about atomic structure to know that Fig. 1.1 is a highly simplified picture of the hydrogen atom. But we would not gain any greater understanding of transistor action

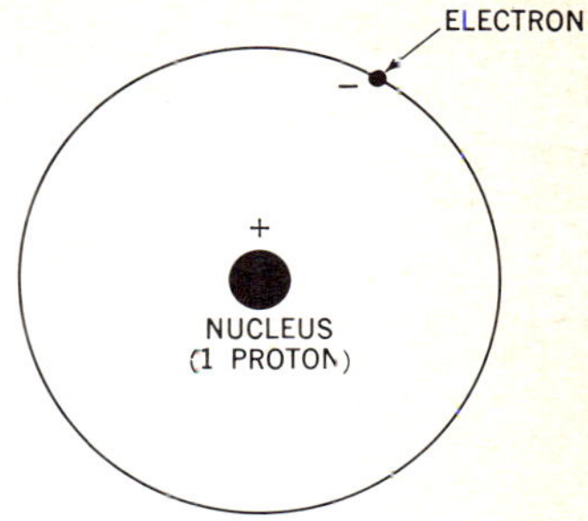

FIGURE 1.1
The structure of a hydrogen atom.

by modifying this illustration to conform to a more modern theory; and its simplicity does impart an understanding that might not otherwise be obtained. Hence, we shall remain with this method of representation.

Helium follows next in order of complexity; its atomic structure is indicated in Fig. 1.2. The central positive charge is offset by two electrons rotating about the nucleus. Each element then follows in numerical turn, with the central positive charge increasing in steps of one and being electrically counterbalanced by additional electrons revolving in paths, or orbits, about the nucleus. It should be noted that the nucleus and the associated electrons soon form fairly complex structures.

The orbiting electrons do not follow random paths; they are instead confined to definite energy levels. These levels can be visualized as shells, each successive shell being spaced at a greater distance from the nucleus. The closest energy level, or shell, to the nucleus carries one electron (as in the hydrogen atom) or two electrons (as in helium and all other atoms). These electrons may rotate at any angle about the nucleus, but they are more or less bound to remain within the confines of the shell.

The nucleus of the third element, lithium, has a positive charge of 3, which is electrically counterbalanced by three negative electrons revolving around the nucleus (Fig. 1.3). Two of the electrons are confined to the first

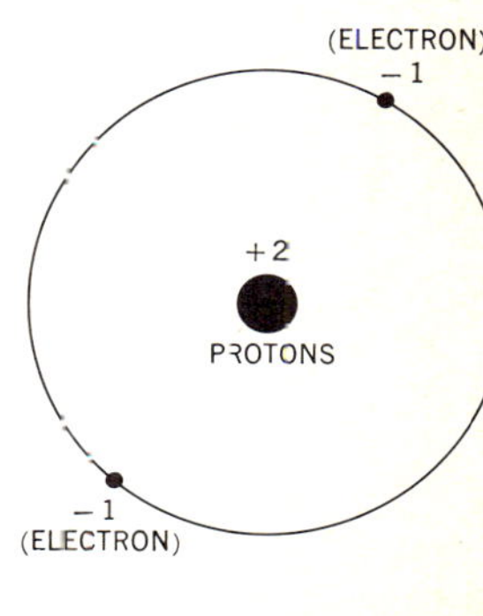

FIGURE 1.2
The atomic structure of helium.

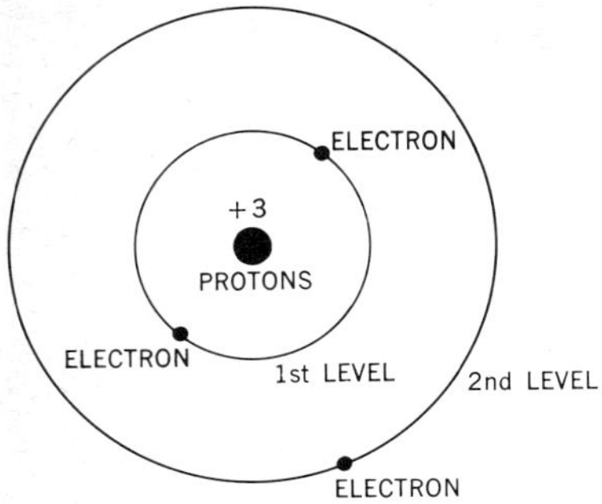

FIGURE 1.3
The atomic structure of lithium.

energy level (shell), just mentioned for hydrogen and helium. The orbit of the third electron, however, is much farther removed from the nucleus and is in a new energy level. This new second energy level is capable of holding eight electrons, a condition which is achieved by the element neon. Neon, with a positive charge of 10, has two electrons in the first shell and eight electrons in the second shell (Fig. 1.4). The third, or next, energy level (or shell) can hold eighteen electrons; the fourth level can hold thirty-two electrons. Beyond this, there are two additional shells, but these are never entirely filled, since there are only 103 known elements. In fact, we are not certain just how many they could hold.

To visualize the relative size of the nucleus and the distance to the first energy level, let us enlarge a helium atom so that the nucleus gets as big as a golf ball. If we could do this, the electrons in the first shell would be 20 mi away and be as large as marbles. From this, it is not hard to understand how two atoms could pass through each other's system without coming into physical contact with each other.

COMPOSITION OF THE NUCLEUS

Only in the element hydrogen does the nucleus consist solely of protons. As the element's atomic number increases, it is found that the positively charged

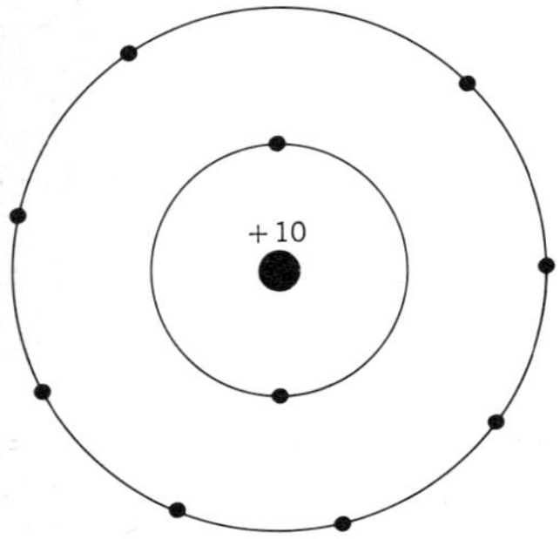

FIGURE 1.4
Neon, with a positive nuclear charge of 10, has two electrons in the first shell and eight electrons in the second shell.

nucleus contains protons and neutrons. A neutron has no charge and a mass slightly greater than that of a proton. Hence, the neutron does not alter the electrical charge of the nucleus.

In recent years a number of additional particles have been discovered in the nucleus: the meson can have a positive or a negative charge, whereas the neutrino has no charge. The mass of each particle is quite small, scarcely large enough to be a significant factor in the overall atomic weight of the element.

The foregoing discussion has been given in some detail so that the reader will better understand transistor operation. The transistor is a solid-state device composed essentially of atoms of silicon arranged in a definite geometric pattern, or lattice. The flow of current through this material depends upon our ability to dislodge electrons from the outer shell of the various silicon atoms. Hence, it was first necessary to understand how the electrons are arranged around the nucleus of the atom. With this understanding, we can now turn our attention to additional information concerning electron behavior in atoms.

ELECTRON BEHAVIOR IN ATOMS

It was noted in the preceding discussion that the electrons filled the first shell first, then the second shell, the third shell, etc. If an atom is to be stable, no electrons will be found in an outer shell unless the inner shells are completely filled. (This is true of the first three shells; in the fourth and higher shells, there is less tendency to follow this rigorous pattern, probably because with increasing distance, the influence of the nucleus on the outer electrons decreases rapidly.) A shell is in its most stable state when it carries a full complement of electrons. For the innermost level, this means two electrons; for the next shell, eight electrons; etc. Elements whose outer ring is not complete are more chemically active than are elements whose outer ring is complete. For example, sodium is a very active element. Its atomic number is 11, which means that the two inner energy levels (shells) are filled to capacity (i.e., two and eight electrons, respectively). The third ring contains one electron; and since sodium would be more stable without any electrons in this third ring, we find that sodium is chemically active because it readily loses or gives up this sole electron. As a matter of fact, sodium is so anxious to give up its extra electron, that it is never found by itself in nature. A suitable element that combines readily with sodium is fluorine. These elements react well together because fluorine (atomic number 9) has seven electrons in its second ring, and the addition of one electron completes this ring. One way of looking at this combination is to consider the sodium atom as giving up its lone third-ring electron to form a stable atom having two complete rings, or shells, while fluorine takes this extra electron and forms two complete shells.

A better explanation is to say that sodium and fluorine share these additional eight electrons, permitting each to have a completed outer ring. That is, fluorine uses the one sodium electron to complete its second shell and sodium uses the seven fluorine electrons to complete its third shell. Since the sodium and fluorine atoms do not leave one another but coexist in the crystal structure, this latter view is the more realistic. This view also provides a better insight into transistor action, since electron-sharing is an integral part of the silicon crystal.

Atoms in which the outer shell possesses a complete compliment of electrons are stable and well satisfied and will not enter readily into chemical combination with other elements. There are six such elements: helium, neon, argon, krypton, xenon, and radon. In helium, the first ring is complete with two electrons. In neon, the next heaviest of these atoms, there are two electrons in the first shell and eight in the second shell. Argon, the third inert element, has eighteen electrons divided into three shells. The first shell is complete with two electrons, the next shell has its full quota of eight, and the third shell has eight electrons. Since the maximum capacity of the third shell was previously indicated to be eighteen, it would appear that in argon this shell is not complete. This is not true, however, because it was found that all shells beyond the first one can be divided into subshells (Fig. 1.5). For the second main shell we have two subshells, one holding a maximum of two electrons and one a total of six electrons. Note that the total previously given for this second shell, eight, still holds, but the eight electrons are divided into subshells of two and six.

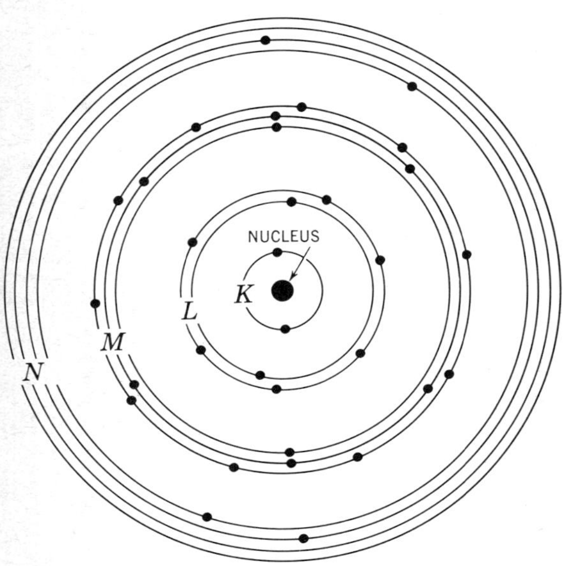

FIGURE 1.5
All energy levels, or shells, beyond the first, the *K* shell, are divided into subshells. Each group of shells is designated by a different letter beginning with *K* and followed by *L* to *Q*. The first subshell of any group is given the letter *s*; the second subshell, *p*; the third subshell, *d*; and the fourth subshell, *f*. It is thought that no known element possesses more than four subshells. The arrangement shown is for the element germanium.

For the third main shell, the maximum number of electrons is eighteen, divided into three subshells of two, six, and ten. In argon, the first two subshells are complete, and this has the same stabilizing effect as though we had completely filled all three subshells. This explains why sodium is "satisfied" when it shares its electron with fluorine's seven electrons in its (sodium) third shell. What happens is that the first and second sublevels of the sodium third shell are filled.

The atomic number of inert element krypton is 36, and here all sublevels of the first three main shells are filled with electrons. In addition, two sublevels of the fourth main shell are filled with electrons.

The chemical behavior of any element, then, is directly related to the number of electrons contained in its outermost shell and how close this shell comes to being filled. If the shell (or one or more of its subshells) is filled, the atom needs no additional electrons and therefore has no tendency to enter into chemical combination with other elements. If the shell is not filled, the atom requires electrons and is considered to be chemically active. These important outermost electrons are called valence electrons, and their energy levels, or shells, are called *valence levels.*

Electrical conductivity can be related to chemical activity because an atom whose outer ring is filled shows no tendency to part with any of its electrons. Since free electrons are needed to obtain an electron current, the inert (satisfied outer-shell) elements are insulators. On the other hand, atoms that part easily with an electron in order to end up with a complete shell are good conductors. Copper, for example, has an atomic number 29; this means that all subshells of the first three main shells are completely filled (i.e., 2, 8, 18), leaving one electron for the first subshell of the fourth ring. Copper is an excellent conductor of electricity because it will give up this one electron readily.

A considerable amount of energy would be required to remove a second electron from this copper atom, because the second electron would have to come from a complete ring and since this is a particularly stable condition, the atom would strongly resist the removal. However, if we applied enough energy, the electron could be removed.

ELECTRON REMOVAL

If one or more electrons are removed from an atom, the atom is no longer electrically neutral in charge. If one electron is removed, the nucleus has one more positive charge than the outer electrons, and the overall charge is $+1$. The atom has now become what is known as an ion; in this case, a positive ion.

It is also possible for an atom to gain an electron producing an overall electric charge of -1. The atom now becomes a negative ion. This situation occurs when an atom needs one or more electrons to complete a ring and attempts to obtain this additional electron from some other element, particularly one having a lone electron in its outermost ring.

Energy is required to remove an electron from an atom. Electrons have to be forced loose from atoms, and one way to pry them loose is to provide enough energy to allow them to escape the attractive force of the positive nucleus. Common forms of energy, particularly in electronic devices, are electric fields, heat, light, and bombardment by some other particle. In wires, for example, we force the copper atoms to give up an electron each by applying an emf across the ends of the conductor (Fig. 1.6*a*). In a vacuum tube, we heat a cathode until the outer-ring electrons have absorbed enough energy to escape from their respective atoms and leap into the interelectrode space (Fig. 1.6*b*). There they are attracted by an emf on the plate; and we thus get a flow of current through the tube and the outer circuitry system. Light, as an activating agent, is employed in photoelectric tubes; the energy that light rays bring to the atoms of the photoelectric cell enables some of the

FIGURE 1.6
Several common methods of separating electrons from their atoms.

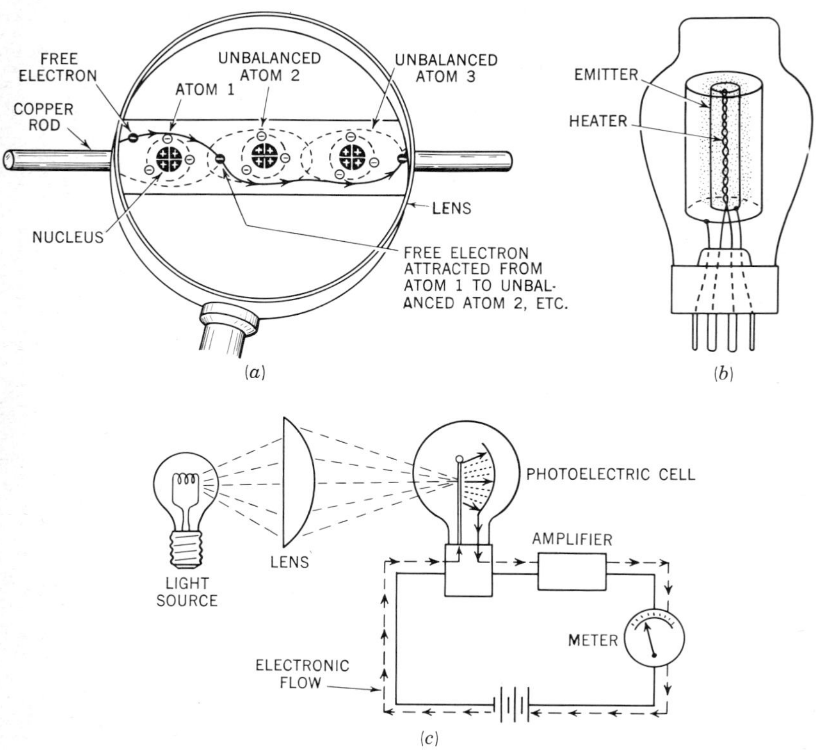

electrons to escape from their atoms and again reach a positively charged anode (Fig. 1.6*c*). This removal, or addition of electrons from atoms by energy transfer, is called *ionization,* and is extremely important to the electronics industry.

THE QUANTUM THEORY

The mechanism by which bound electrons are freed has been the subject of considerable investigation, and certain facts that are important to us in our study of transistors have been discovered. For example, it has been revealed that when energy is supplied to an electron held in an atom, a definite amount of energy has to be supplied in order for it to have an effect on the electron. The various shells in an atom represent definite energy levels, and in order to move an electron from a lower shell (or subshell) to a higher shell, a definite amount of energy is required. Failure to provide enough energy to the electron will cause it to remain at its present level. This is true even if the energy provided is just barely shy of the required amount.

If more than enough energy is provided for the electron to leave its orbit and move to the next higher level, then the excess will be to no avail unless enough extra is provided to enable the electron to move to a still higher shell. In other words, energy is required in definite discrete amounts, called *quanta,* and the electrons can receive these quanta only in whole numbers, such as 1, 2, or 3 quanta.

Electrons can lose energy as well as receive it, and when an electron in an atom loses energy, it moves to a level that is closer to the nucleus. This lost energy may appear as heat, as in a conductor when current is passed through it, or as visible light.

Different elements have different energy levels for their electrons, and the amount of energy absorbed or released consequently varies as the electrons move from level to level. This accounts for the different colors of light emitted by various substances when they are excited.

Heating sodium over an open flame produces a characteristic yellowish light; neon gas, when activated in electric signs, emits an orange-red glow. The energy required to produce red light is less than the energy to produce blue light. This is because the energy in a quanta bundle depends on frequency and red has a lower frequency than blue.

In gases, electrons of one atom tend to act independently of electrons of other atoms. In solids, however, the forces that bind atoms together greatly modify the behavior of the associated electrons; and we are here dealing with the aggregate action of many electrons rather than with individual electrons. One direct consequence of the close proximity of atoms in a solid is to cause the individual energy levels that exist for an isolated atom (depicted in Fig. 1.5) to break up to form bands of energy levels. Within the bands, discrete

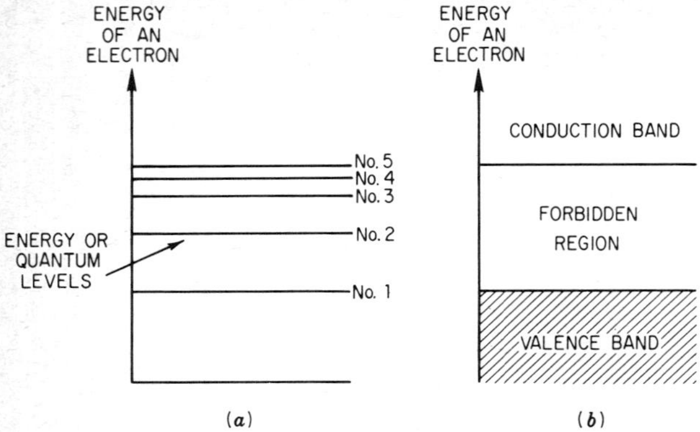

FIGURE 1.7
(*a*) Energy levels in an isolated atom. (*b*) Energy bands in a solid.

permissible energy levels still exist, but the act of bringing many atoms close together has produced many more permissible energy levels. It has also caused some energy levels to disappear.

Thus, where before in an isolated atom we had an energy-level arrangement such as that shown in Fig. 1.7*a*, in a solid material we find bands of energy as shown in Fig. 1.7*b*. As a matter of fact, only the three upper bands are shown in Fig. 1.7*b*. Additional energy bands exist below the valence band, but these are not important to an understanding of semiconductor behavior and will hence not be discussed.

The uppermost band of Fig. 1.7*b* is the conduction band. When electrons are found there, they can be easily removed by the application of external electric fields. When a material has many electrons in the conduction band, it acts as a good conductor of electricity.

Below the conduction band is a series of energy levels that collectively form the forbidden band. Electrons are never found in this band. Electrons may jump back and forth from the bottom valence band to the top conduction band, but they never come to rest in the forbidden band.

The valence band is formed by a series of energy levels containing the valence electrons. These electrons are more or less bound to the individual atoms, restricting their range of movement far more so than the range of movement of electrons in the conduction band. Electrons can be moved from the valence band to the conduction band by application of energy, generally thermal energy. This movement happens to a certain extent in semiconductors. (We shall cover this point more extensively in subsequent discussions.) Of more immediate interest is the fact that the extent of the forbidden band, or the separation between the conduction and valence bands, will determine whether a substance is an insulator, semiconductor, or conductor. Figure 1.8 shows the difference between insulators, semiconductors, and conductors in terms of their three bands. Figure 1.8*a* is called an *insulator* because of the wide extent of the forbidden band. The wider this band, the greater the amount of energy that must be fed to any electron in the valence band in order to raise it to the conduction band where it can become a carrier of electricity. Obviously, in an insulator, a large amount of energy is required to get a minute amount of current flow.

In a semiconductor, the forbidden band is smaller (Fig. 1.8*b*), which means that less energy is needed to bring an electron from the valence band through the forbidden gap to the conduction band. Hence, in semiconductors, more current will flow for a certain applied voltage, but this current will not be as large as we would obtain in a conductor.

The third illustration, Fig. 1.8*c*, is for a conductor. Here, the valence and conduction bands overlap. Now, we need only a very small amount of energy to move electrons into the conduction band; consequently, electricity is readily passed by conductors. All this, of course, is common knowledge; what is not so well known is how the various energy levels within a molecule cause the molecule to act as an insulator, semiconductor, or conductor. That is where the quantum theory so admirably fills in the gaps.

The side axis of each of the three illustrations in Fig. 1.8 is labeled simply as "energy." It is the generally accepted practice for physicists to use electronvolts as a convenient measure of energy.

An electronvolt is the energy acquired by an electron in falling through a potential difference of one volt. If we use this method of measuring energy,

FIGURE 1.8
The difference between insulators, semiconductors, and conductors in terms of their valence bands, forbidden bands, and conduction bands. (*a*) Insulator; (*b*) semiconductor; (*c*) conductor.

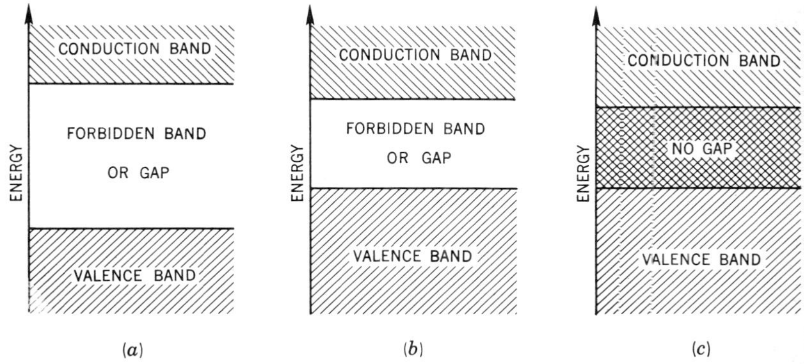

then the width of the forbidden band in an insulator is 1 eV (electronvolt) or more.

For the semiconductor silicon, the width of this band is 1.1 eV; for germanium, another semiconductor, the band width is 0.7 eV. Where the forbidden band is absent, i.e., with conductors, we need only 0.01 eV to bring an electron into the conduction band.

It should be understood, of course, that each of the three illustrations in Fig. 1.8 is representative of a class of materials and that there exist many substances whose characteristics fall somewhere in between those of insulators and semiconductors on the one hand and semiconductors and conductors on the other. Silicon in the highly purified state, for example, is a very poor conductor of electricity. If certain controlled impurities are added to silicon, we find that its conduction increases significantly, indicating that while a substance may be basically an insulator, its properties can be altered. This alteration is actually what happens in silicon and germanium transistors and is responsible for the ability of these units to function usefully. In Chap. 2 we shall see what happens to silicon and germanium when selected impurities are added.

RADIATION AND SEMICONDUCTORS

Space exploration and nuclear radiation have led to the creation of a large new market for radiation-hardened (resistant) semiconductor devices. These environments where semiconductors encounter radiation are those of nuclear weapon radiation, radiation from nuclear reactors, and the radiation in the Van Allen space belts.

Semiconductor material is subject to two fundamental radiation effects: displacement and ionization. Displacement occurs when an atom is knocked from its original crystal-lattice position to a new position. Ionization forms ionized atoms and free electrons when the energy levels of orbital electrons are increased enough to break their attraction for the parent atom.

Transistor operation is dependent on the introduction of a few parts per million of impurities into the semiconductor material. If even a small percentage of these added impurities are knocked out of position, the transistor will not function—which happens in certain radiation environments.

There are four types of radiation: neutron, proton, gamma-ray, and electron. Neutron radiation causes displacement-type damage and is generated by nuclear explosions that also produce gamma rays that cause ionization damage to semiconductors. Ionization damage is transient; and with time, a device will recover its useful electrical characteristics. Displacement damage is permanent. Proton, gamma-ray, and electron radiations cause ionization-type damage and are found in nuclear reactor and space environments.

We will discuss in Chap. 4 some processing techniques that harden semiconductor devices for radiation environments.

QUESTIONS

1.1 What advantages that transistors possess make them especially attractive for communications applications?

1.2. Differentiate atoms, molecules, and elements. Name 15 elements that you have personally come in contact with, either singly or in combination.

1.3. Describe the structure of an atom in general terms. Compare this structure with that of the solar system.

1.4. How are the electrons arranged within an atom? Consider first a simple element, then a fairly complex element.

1.5. What differences exist between the nucleus of a simple atom, such as hydrogen, and the nucleus of a complex atom?

1.6. What causes an element to be chemically active? Stable?

1.7. Is there any apparent relationship between the chemical activity of an atom and its electrical conductivity? Explain.

1.8. What happens when an atom gains an additional electron? Loses an electron? What is the altered atom called?

1.9. What methods may be used to remove electrons from an atom? Describe one or two methods in detail.

1.10. What is the quantum theory with respect to electron removal?

1.11. Define the valence band, conduction band, and forbidden band.

1.12. Explain the difference between conductors, insulators, and semiconductors in terms of energy bands.

1.13. Name two radiation effects in semiconductors.

Chapter Two

PN Junctions, Diodes, and Transistors

It was noted in the preceding chapter that the chemical activity of an atom is determined primarily by the number of electrons contained in the outermost ring of the atom. When this ring is filled, the atom is stable and shows little inclination to combine with an atom of any other element. The activity increases, however, when the number of electrons is less than the full number to complete a ring. Because of the importance of these electrons, they are given the special name of valence electrons. Furthermore, it is common practice in illustrations of atoms to show only the valence electrons (Fig. 2.1). The carbon atom, with an atomic number 6, is shown here. The $+6$ at the center represents the nuclear charge. Since there are four valence electrons, we know that these electrons are in the second ring; the two electrons not shown would be in the first ring.

Of immediate interest in transistors is silicon; this element also contains four electrons in its outermost ring. The atomic number of silicon is 14, giving us two completed shells of two and eight electrons each and four electrons in the fourth shell. The latter electrons are the valence electrons, represented in Fig. 2.2.

LATTICE STRUCTURE AND CRYSTALS

Silicon in the solid state possesses a crystalline structure in which a group of silicon atoms combine, through their valence electrons, to form a repeated structure having a number of basic cubical lattices, such as shown in Fig. 2.3. Each of the "balls" in the illustration represents a silicon atom; the rods between the balls represent the electronic forces binding each atom to its neighbors.

The cubical configuration, known as a diamond structure, is characteristic of the solid state of a number of elements, among them carbon, silicon, and germanium. A large, visible silicon crystal would be composed of millions upon millions of these basic cubical lattices.

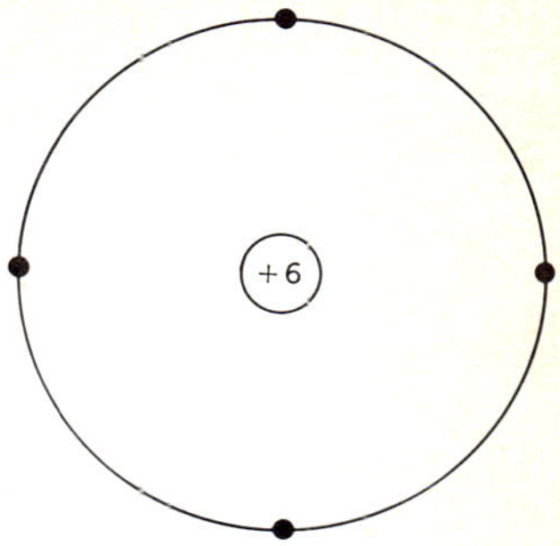

FIGURE 2.1
A simplified illustration of the carbon atom. Only the valence electrons are shown.

A two-dimensional illustration of the manner in which the silicon atoms are bound to one another is given in Fig. 2.4. Focusing our attention on any one of the central atoms, we see that each of its four electrons is shared by four other silicon atoms. This gives the central atom a total of eight electrons in its outermost ring; four of these electrons are its own, and the other four it "borrows" from the surrounding atoms. The eight electrons in this subring make the silicon crystal a stable element.

What is true of a central atom is true of all its neighbors: each shares its four outer electrons with four other silicon atoms. This, too, is shown in Fig. 2.4 for a limited number of atoms. All the valence electrons are tightly held together. Consequently, pure silicon is not a very good conductor of electricity.

The sharing of the valence electrons between two or more atoms produces a shared, or covalent, bond between the atoms. It is this bond which is largely responsible for the cohesion which the crystal structure possesses and which actually keeps the crystal structure intact.

Note that by this sharing to form complete valence rings, silicon in its pure form is a fairly good insulator. The reason silicon is not a complete insulator but rather a semiconductor stems from the fact that thermal agitation, arising from the energy imparted to the electrons by the heat of its surround-

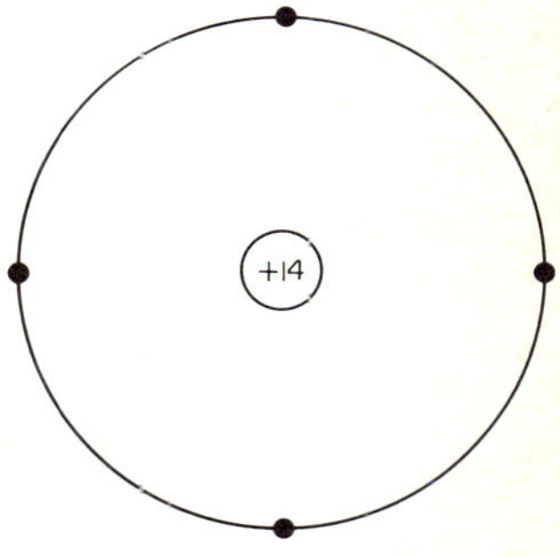

FIGURE 2.2
The silicon atom, using the simplified method of presentation.

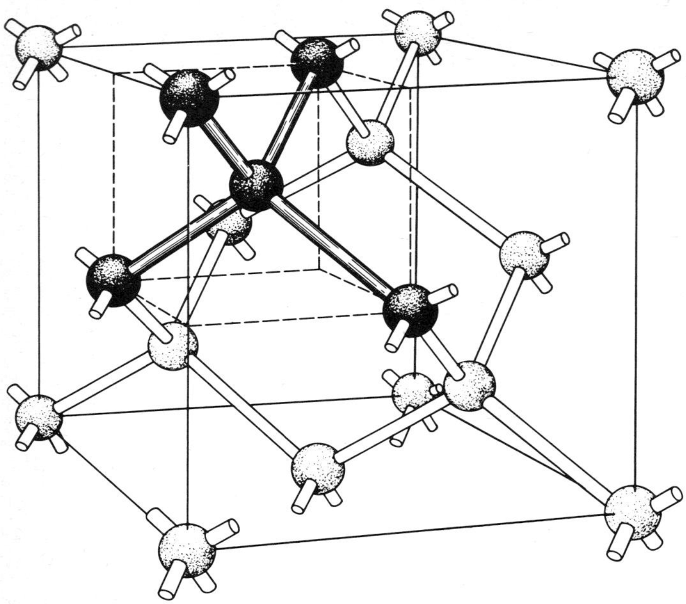

FIGURE 2.3
A typical crystal structure. Each sphere represents a nucleus surrounded by its inner shells. The spokes that join the atoms and support the structure represent the covalent bonds. *(After W. Shockley, "Electrons and Holes in Semiconductors," D. Van Nostrand Company, Inc., Princeton, N.J., 1950.)*

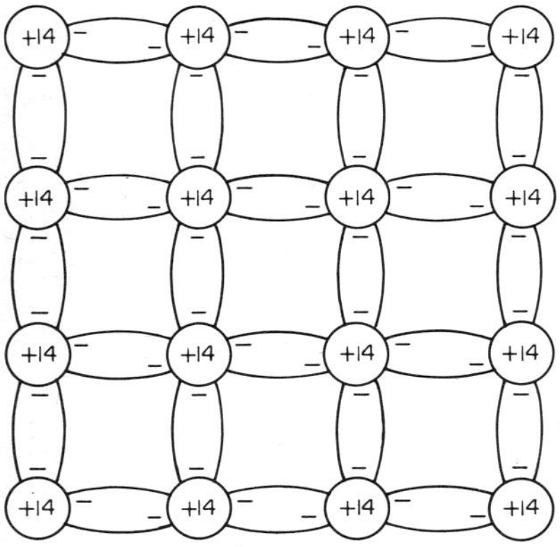

FIGURE 2.4
Representation of electron-paired bonds in silicon.

ings, causes an electron here and there to break away from its bond, move up into its conduction band, and wander through the crystal-lattice structure in a more or less aimless manner.

ELECTRONS AND HOLES

The bond from which an electron escapes is left with a deficiency, and this deficiency has been given the rather descriptive name of "hole," as though a physical hole had actually been left by the removal of the electron.

From our previous training we accept the fact that electrons are quite mobile and may be moved from point to point. Many experiments have been run to prove this fact. In our own experience with electric circuits we have never observed any action that would cause us to think otherwise.

However, in this case, it is also valid to state that the hole left by the electron will wander about within the crystal. This point is not so easy to accept, and yet it has been conclusively demonstrated that holes do travel through silicon crystals. This fact warrants a more detailed description, since understanding transistor action requires the acceptance of the hole-travel concept.

When a bound electron departs, the charge deficiency, or hole, that it leaves behind is confined to the valence ring of the atom. If, now, a nearby electron held in a covalent bond acquires enough energy to leave its bond and jump into the waiting hole, then in essence what we have had is a shift in position of the positively charged hole from its first position to this new position (Fig. 2.5). This same action can occur a number of times, with successive changes in hole position, so that we can say that a hole drifts about in a random manner the same as the electron that originally left the hole.

The foregoing discussion has dealt with a single electron and a single hole, but in actual crystals there would be many such pairs. And with many negative electrons and positive holes present, a considerable number of recombinations will be taking place all the time. Actually, in pure silicon at room temperature, one bond in ten is broken at any given time. Generation is the term used to describe the phenomenon of bond-breaking, by light or by heat. The reverse process, when an electron drops into a hole, is called recombination. A dynamic balance between generation and recombination exists in a crystal at equilibrium.

If the energy supplied to the crystal is an electric field developed by the application of an emf across the silicon crystal, then the motion of the electrons and the holes will be less random and more directed, in a direction determined by the voltage. Electrons will move toward the positive terminal of the battery, while the holes will drift toward the negative terminal of the

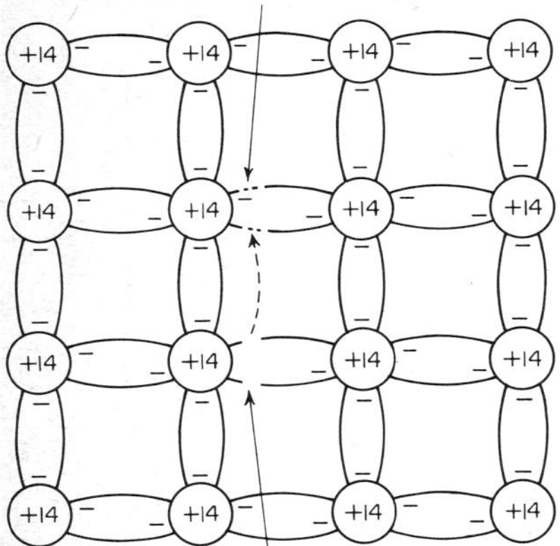

FIGURE 2.5
Method by which holes travel through a silicon lattice structure.

battery. The opposite movements of these two charges do not cancel each other, as one might suppose. Rather they aid each other. This was demonstrated in an experiment performed in 1889 by the physicist H. A. Rowland. He placed negative charges of static electricity on an ebonite disk separated by raised portions of the disk. When the disk was rotated at high speed, a magnetic field was produced, and the field was identical with what would have been expected if a flow of electrons had taken place in a loop of wire in the same direction of rotation.

Rowland then removed the negative charges and replaced them by an equivalent number of positive charges. The disk was now rotated in the opposite direction, and the resulting magnetic field had exactly the same direction as its predecessor. Thus, we obtain the same electrical effect whether we have negative charges (i.e., electrons) moving in one direction or equivalent positive charges (i.e., holes) traveling in the opposite direction. Ohm's law, or any other electrical law we know, would yield identical results in either case. This is a significant fact to remember because we have become used to thinking only of electron flow and the idea of mobile positive charges comes somewhat as a surprise. Both types of current flow have to be appreciated, since both occur in transistors.

It is interesting to note the rates at which electrons and holes drift (carrier drift mobilities) through silicon and germanium. For electrons: silicon 1,600 centimeters per volt-seconds (cm/V-s) and germanium 3,600 cm/V-s; for holes: silicon 400 cm/V-s and germanium 1,800 cm/V-s. These rates are at room temperature with no applied emf. When a voltage is applied, the rates increase considerably.

N-TYPE SILICON

Externally applied heat and light may be used to produce electrons and holes in a silicon crystal, but a much more efficient method of achieving the same results is to add very small amounts of selected impurities, generally to an extent no greater than a few parts per million. This means that each impurity atom can be regarded as isolated in the crystal and unaffected by neighboring impurity atoms, since the impurity spacing is relatively large. The impurity enters the crystalline structure of the silicon and takes the place of single silicon atoms at various points throughout the crystal-lattice structure. This situation is represented in Fig. 2.6. Phosphorus has five valence electrons, four of which enter into covalent bonds with four surrounding silicon atoms. This is in accordance with the structural arrangement in a silicon atom. The fifth electron has a large orbit and is bound rather loosely to the phosphorus atom. Thus, the bond between this electron and the phosphorus atom is easily broken. What we have actually done by the addition of minute quantities of phosphorus to the silicon structure is to provide the silicon with a source of free electrons. Elements such as phosphorus or antimony which serve as sources of electrons are called *donor impurities*. Furthermore, the silicon crystal containing these impurities is known as N-type silicon. The N, of course, refers to the fact that the electrical conduction through the crystal is done by electrons, which possess a negative charge.

P-TYPE SILICON

It is also possible to add impurities that possess three rather than five electrons in their outer orbit. Boron, gallium, and indium are examples of such substances. As with phosphorus, the new trivalent atom will replace a silicon atom in the structure (Fig. 2.7). However, in this case, instead of having

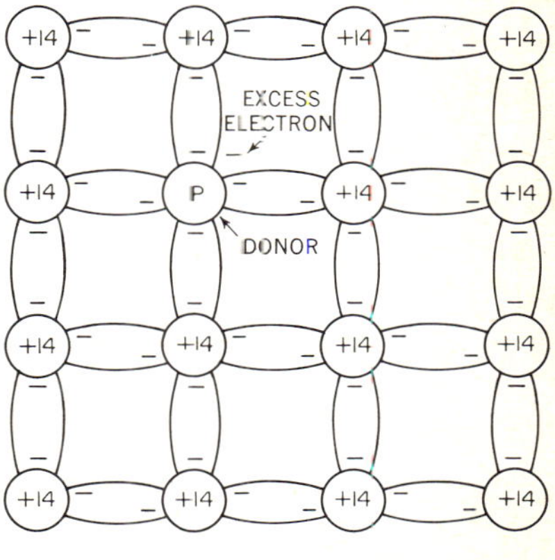

FIGURE 2.6
The effect of an impurity atom replacing a silicon atom. In this illustration, the impurity is phosphorus.

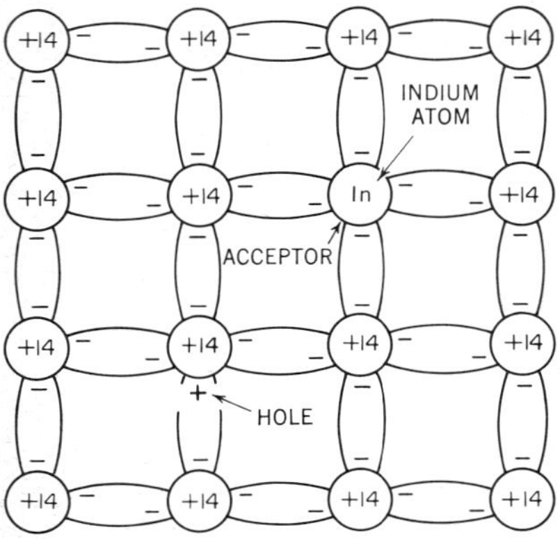

FIGURE 2.7
A hole is produced when an atom of a trivalent impurity such as indium replaces a silicon atom.

an excess of one electron, we now find ourselves with a deficiency of one electron. In order to complete the four electron-pair bonds, the trivalent atom "accepts" an electron from a nearby silicon bond. Actually, the extra electron accepted by the boron atom is excited out of the valence band or, frankly, stolen from a nearby valence bond. The net result of this "robbery" is to leave a hole in the neighboring electron-pair bond. With all this, trivalent atoms such as boron are known as acceptors, not robbers.

Thus, when certain impurities that have only three valence electrons are added to the silicon-crystal structure, a series of holes are produced. Under the stress of an applied emf, electrons from other nearby bonds will be attracted to these holes, thereby filling the gaps but creating a similar number of holes in their former bonds. Thus, we have the equivalent of a movement of holes through crystalline structure, and conduction is said to take place by holes.

The silicon crystals that contain these acceptor atoms are known as P-type silicon. By the careful selection of the impurity to be added, we can make a silicon crystal either N type or P type. Both are employed in transistors, and it is important that the reader understand the differences between them and how electrical conduction occurs through each.

It should be noted that a number of holes are present in N-type silicon because of the normal breaking of bonds arising from the absorption of heat or light in pure silicon. However, the electrons released because of the addition of phosphorus or other pentavalent impurity atoms are by far the principal conductors of electricity in N-type silicon. By the same token, free electrons exist in P-type silicon; but again, it is the holes created by the addition of trivalent impurity atoms that account for the major portion of the electrical conduction that takes place here. The holes in N-type silicon and the electrons in P-type silicon are called minority carriers.

The impurities must be added in carefully controlled amounts; otherwise the silicon-crystal structure is modified to such an extent that transistor action is not obtained. Impurities are added to the silicon crystal at precisely controlled temperatures in furnaces that can hold temperatures to $\pm 1/4$°C. This process is known as diffusion and will be described in Chap. 4.

The reader may wonder what would happen if both acceptor and donor impurities were added to a wafer of silicon. The holes created by the acceptor atoms would be promptly filled by electrons of the donor atoms. If both impurities were present in equal amounts, the excess electrons would merely fill the excess holes and the silicon would act as pure silicon containing no impurities. When a donor state has been canceled by an acceptor, that is, with respect to its ability to conduct, this cancelation process is known as compensation. However, it is important to remember that if one impurity were present in greater amounts, the electrons or holes it provided would become the principal carriers of electricity.

We might pause here for a moment and note what effect the addition of impurities has on the energy-band distribution in the atomic structure of crystal lattice. Before the impurity addition, the energy bands were depicted as shown in Fig. 2.8*a*. These were the valence band, the forbidden band, and the top conduction band. If we now diffuse a minute amount of donor impurity into a silicon wafer, we find that extra energy levels exist that were not present in the pure semiconductor. The extra electron added now establishes an energy level in the forbidden band just below the conduction band (Fig. 2.8*b*). The position would appear to be logical, since this extra electron is held only loosely by its parent atom and should be relatively easy to remove from that atom. This would bring the position of the extra electron close to the conduction band. It would not be completely within the conduction band, since that would assume that the electron were completely free of any influence exerted by its parent atom.

The amount of energy required to raise the electrons provided by the impurity atom to the conduction band is on the order of 0.05 electronvolts (eV). This is far less than the 1.1 eV needed to bring an electron from the

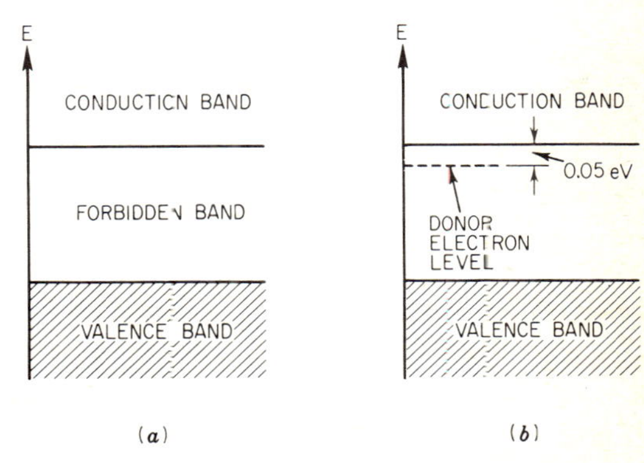

FIGURE 2.8
(*a*) Energy levels in a pure semiconductor. (*b*) Energy levels in a semiconductor with a donor impurity.

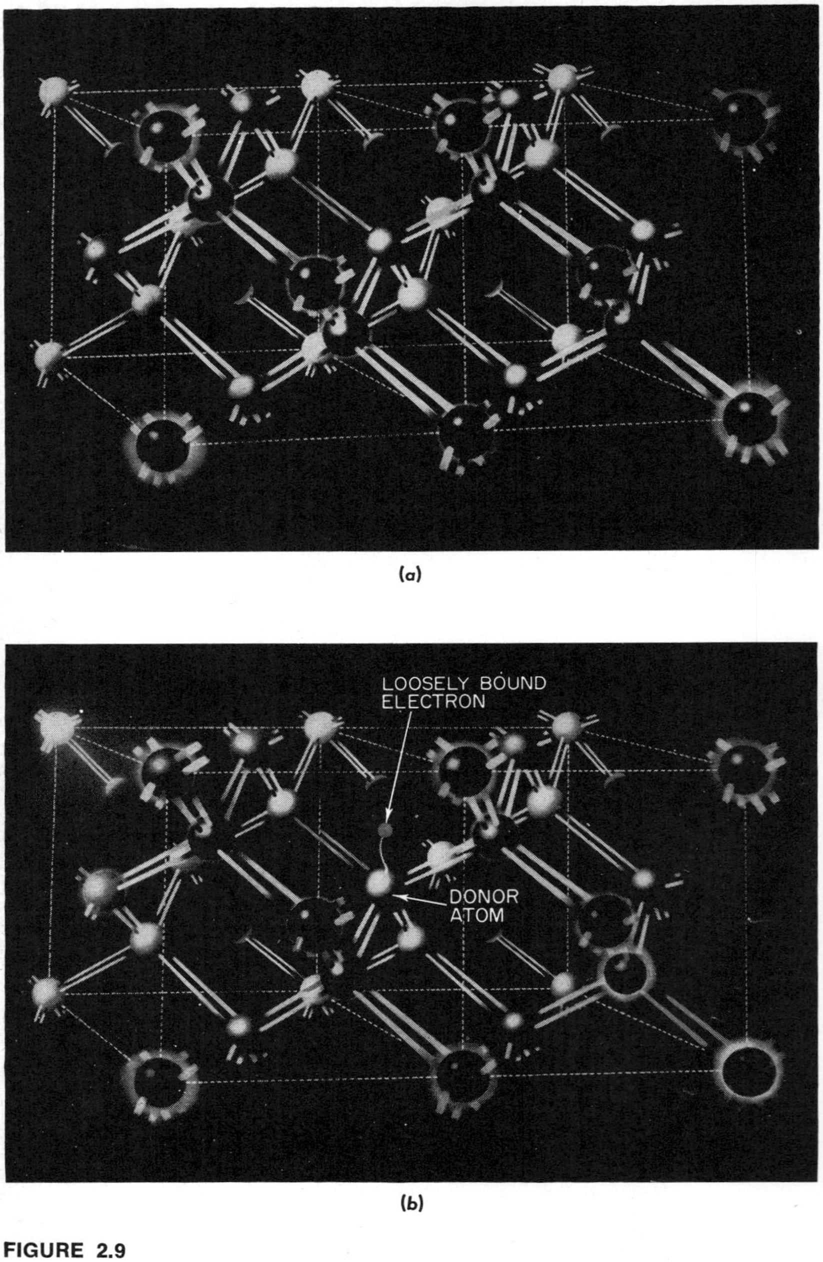

FIGURE 2.9
A pictorial representation of silicon lattice showing shared electron bonds and the manner in which holes and electrons travel from point to point within the crystal structure. (*a*) Pure-silicon crystal. (*b*) Donor atom substituted in silicon lattice. Note loosely bound extra electron.

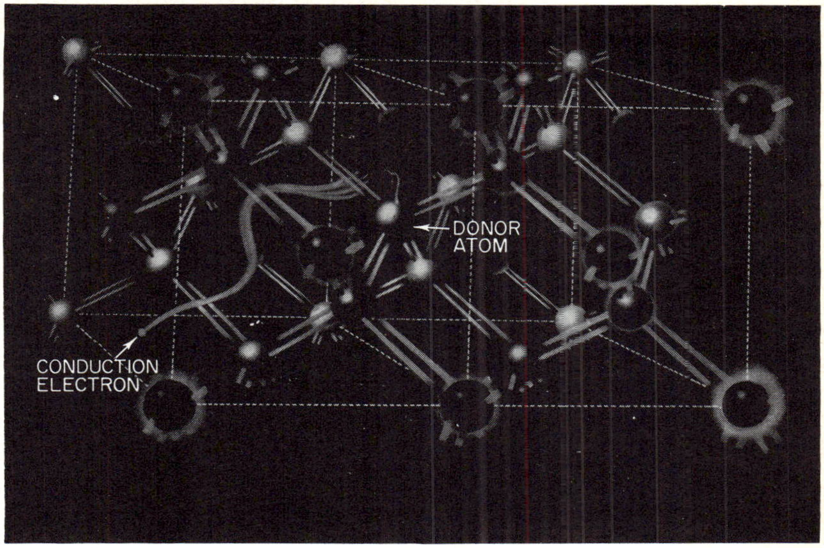

(c)

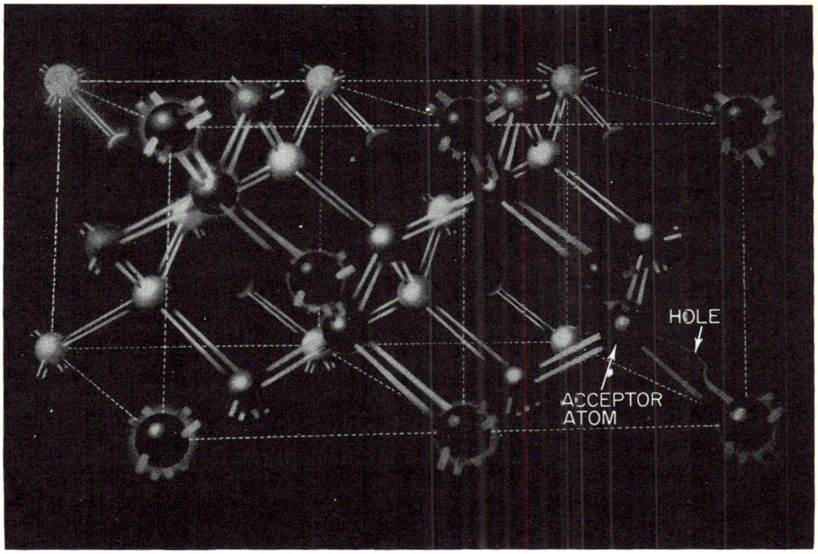

(d)

(c) N-type conduction; electron free of donor atom. This may occur because of heat or electric energy. (d) Acceptor atom substituted in silicon lattice. Note unfilled band in lower right-hand corner of illustration.

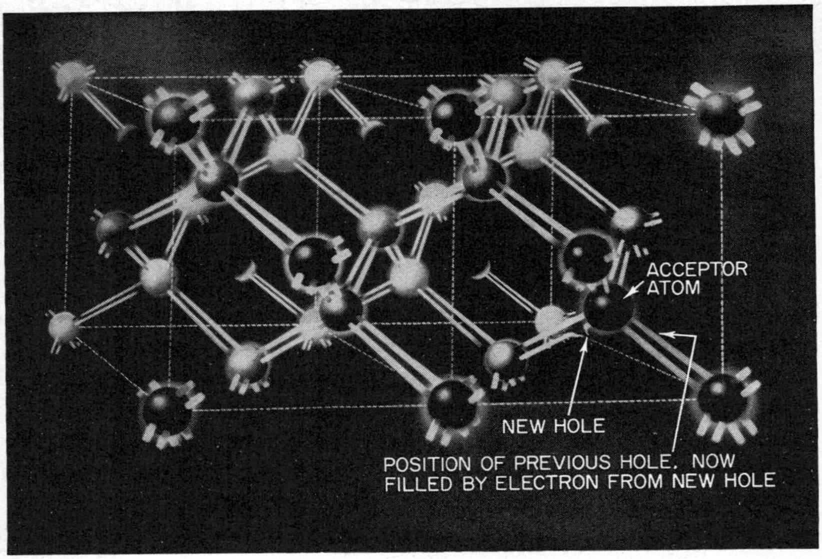

(e)

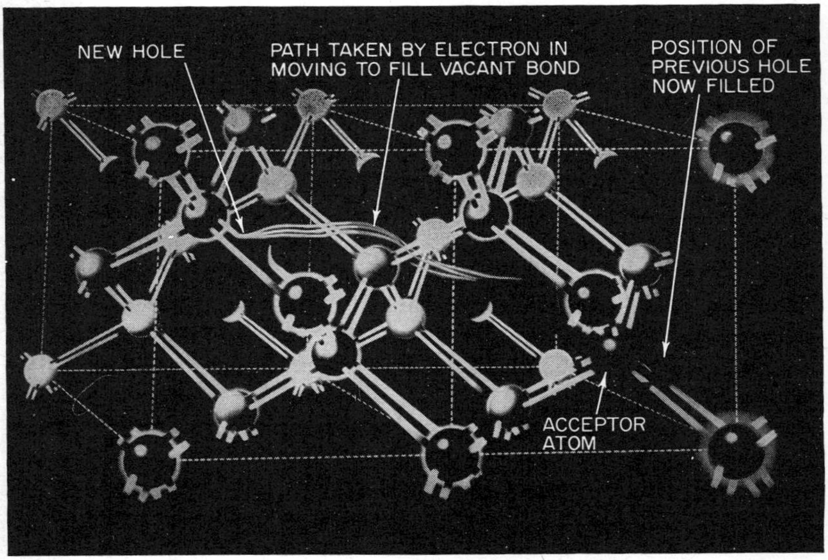

(*f*)

(*e*) Filling vacant bond by thermal excitation of electron from nearby bond. (*f*) P-type conduction. Hole migrating through lattice by excitation of electron from bond to bond. (*General Electric Company and Electrical Engineering.*)

valence band of a silicon atom to the conduction band. The difference is a little less in pure germanium, where the required amount of energy is 0.7 eV.

Note that the forbidden band of Fig. 2.8*a* exists only in a pure semiconductor. When an impurity is added, the extent of this band is modified as indicated in Fig. 2.8*b*.

Through a similar process, addition of a P-type impurity will also modify the energy-level distribution of a semiconductor, again making it easier to bring about conduction. Thus, with impurities, the conduction provided by a semiconductor is overwhelmingly taken care of by the extra electrons or holes provided by the impurities. Very little current is contributed by the breaking of covalent bonds. When a semiconductor does not contain any impurities, it is said to be in the intrinsic state, as shown in Fig. 2.9*a*.

PN JUNCTIONS

If we take a section of N-type silicon and a similar section of P-type silicon and join the two together, as in Fig. 2.10*a*, we obtain a device that is known as a silicon diode. The N-type silicon is at the right, and the P-type is at the left. The circles at the right with a positive sign represent the donor atoms. They have positive signs because their fifth electrons have been removed, leaving each donor atom with a $+1$ charge. The free electron is indicated by the negative sign just below each circle.

By the same reasoning, an acceptor atom in the P-type silicon is represented by a circle with a negative sign, the latter being due to the presence of the additional electron that was robbed from a neighboring electron-pair bond. The hole left by this electron is represented by a small plus sign.

When these two silicon sections are joined together, one might suppose that all the excess electrons on the right would immediately cross the junction and combine with the excess holes on the left. This action actually starts to occur; but before it can progress very far, it is brought to a halt. The reason is that when the sections are first brought together, some of the electrons in the N-type silicon cross the junction and combine with a corresponding number of holes in the P section. Since the N section was initially electrically neutral, loss of some of its electrons leaves it with a net positive charge. This charge increases, and a point at which no additional electrons are capable of overcoming this positive force is quickly reached.

By the same token, the P section itself was also electrically neutral at the start. When it loses some of its positive holes through combination with electrons from the N section, it develops a net negative charge. As additional electrons from the N section attempt to approach the junction, not only are they held back by the net positive charge existing in their own section but they are also repelled by the net negative charge of the P section. Thus, the migratory action is halted.

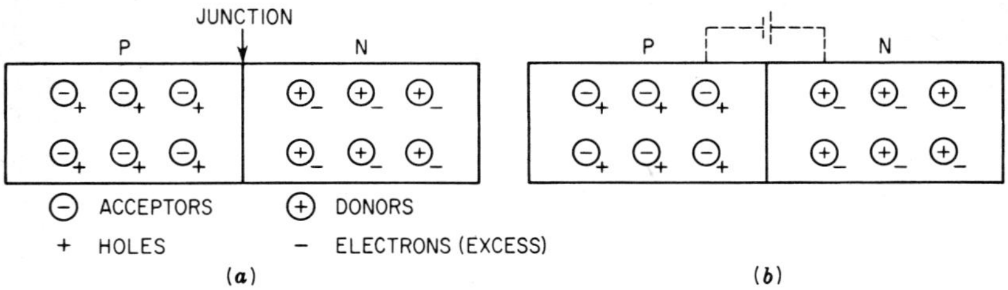

FIGURE 2.10
(*a*) A PN junction forming a familiar silicon diode. (*b*) The battery drawn across the PN junction represents the restraining forces present at the junction.

The region in the immediate vicinity of the junction is called the depletion region because of the absence of any mobile charges there. In the N section, the free electrons initially present were lost, as described above, by combination with the free holes just on the other side of the junction in the P section. Beyond the depletion region, free electrons are present in the N section. These free carriers, however, are restrained from approaching the junction, as explained above.

Several pictorial methods have been employed to indicate the restraining forces present at a junction. In one illustration, a small battery is placed across the junction in the manner shown in Fig. 2.10*b*. The negative terminal of the battery connects to the P-type silicon, while the positive terminal of the battery attaches to the N-type silicon side. Electrons attempting to travel from the N silicon side to the P side encounter the negative field of the battery and are repelled. By the same token, holes attempting to move from the P side to the N side see the positive battery terminal and they too are repelled.

A second method of representation utilizes what are called potential "hills." The electrons on the N side have to climb a negative potential hill in order to reach the P side (Fig. 2.11*b*). The hill is of course the repelling force of the acceptor atoms. On the other side of the junction, the holes have to climb a positive potential hill in order to move to the right (Fig. 2.11*a*).

In order to produce a flow of current across the junction, we must reduce the potential hill that exists there. This can be done by applying a negative voltage to the N-type material (Fig. 2.12) and a positive voltage to the P-type material. The free electrons in the N section are repelled by the negative battery field and move toward the PN junction. At the same time, the positive holes in the P section are forced toward the junction by the repelling force of the positive battery field. If the battery potential is strong enough, it will lower the potential hill at the junction and enable the carriers to move across to the opposite side.

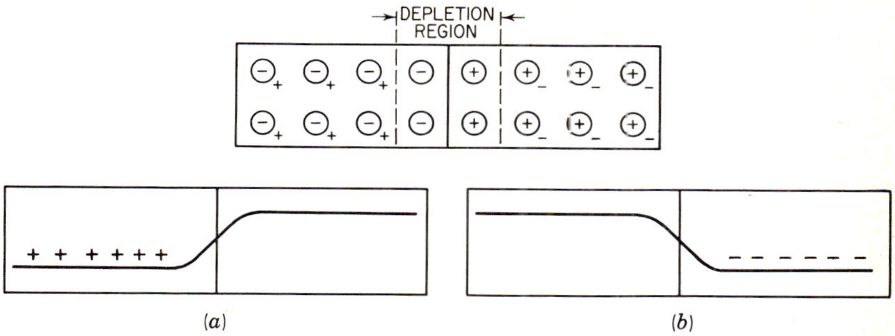

FIGURE 2.11
A second method of indicating the forces present at the PN junction that prevent the combination of the holes at the left with the electrons at the right.

Once the junction is crossed, a number of electrons and holes will combine. For each hole that combines with an electron from the N-type silicon, an electron from an electron-pair bond in the crystal near the positive terminal of the battery leaves the crystal and enters the positive terminal of the battery. This creates a new hole that moves to the junction under the force of the applied emf. For each electron that combines with a hole, an electron enters the crystal from the negative terminal of the battery. In this way we maintain the continuity of current flow. Stoppage at any point immediately breaks the entire circuit, just as it does in an ordinary electric circuit. If this were not so, electrons would pile up at some point and result in a gradually increasing charge or potential at that point. Since this does not happen, we must treat the circuit operation in the manner just indicated.

Note that current flow in the N region is by electrons; in the P region, the current is carried by holes (Fig. 2.13). As we approach the junction, we find both types of carriers. The overall value of current, however, remains constant and is a function of the applied voltage.

FIGURE 2.12
The effect on the PN junction of application of forward bias.

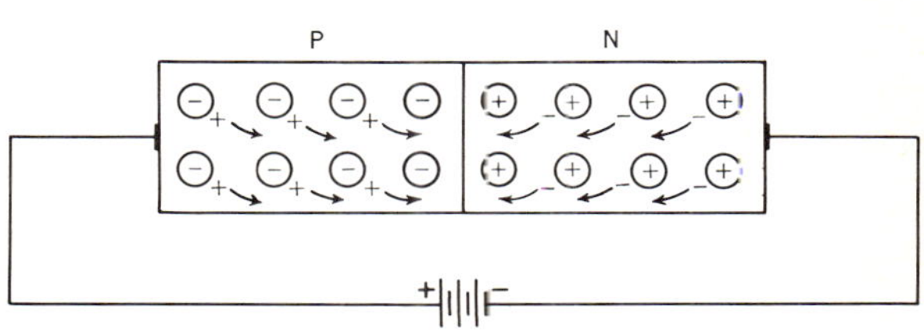

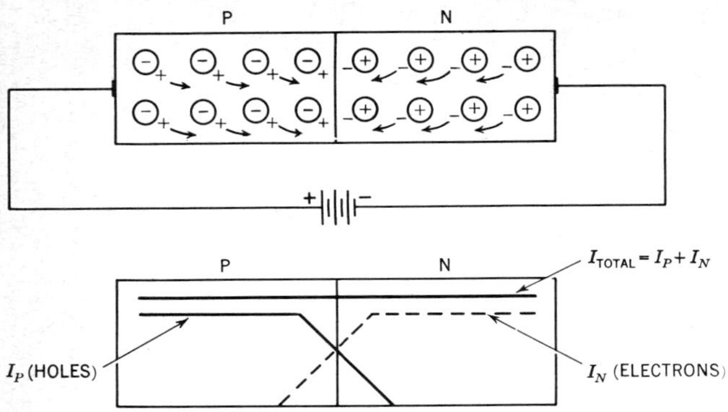

FIGURE 2.13
The current flow in the N region is by electrons; in the P region, the current is carried by holes. In the vicinity of the junction, both types of carriers are present.

As the external voltage is increased, it gradually overcomes the restraining forces present at the junction, and the current rises. Once the restraining forces are completely overcome, the current rises sharply, as shown in Fig. 2.14. If the current flow is permitted to reach too high a value, heat is generated that could permanently damage the junction, and the unit would no longer function in the manner described above.

In the preceding discussion, the diode was biased in its forward, or low-resistance, direction. If we now reverse the polarity of the applied voltage, we find that the battery acts in conjunction with the potential barrier at the junction and practically no current passes at all. The diode is biased in the reverse direction.

The reason for the current decrease is readily understood. With the negative battery terminal connected to the P-silicon section, the excess holes,

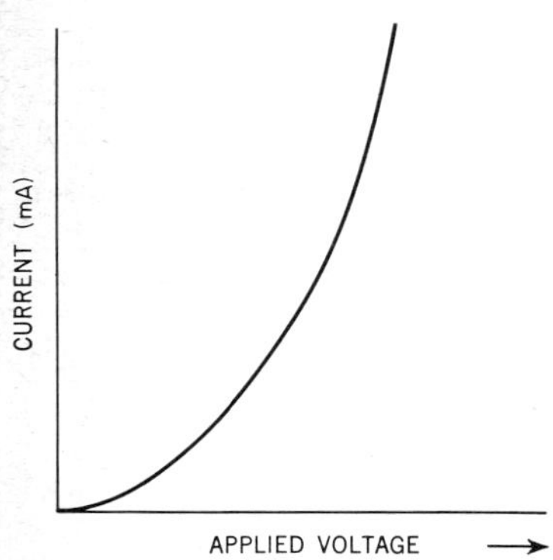

FIGURE 2.14
The behavior of the current flow through a silicon diode as the forward voltage is increased.

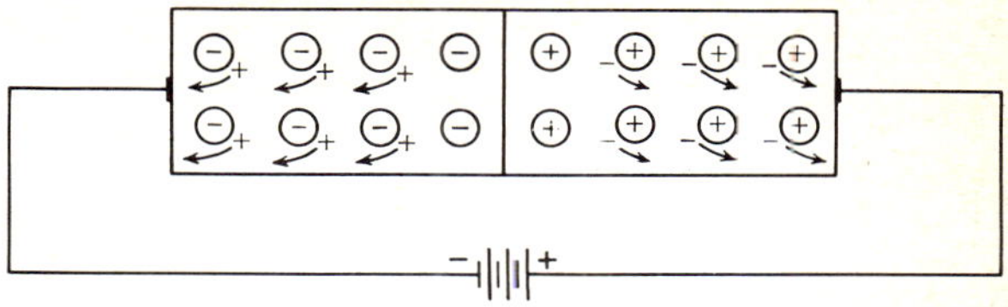

FIGURE 2.15
When the battery connections are reversed, the electrons and holes are drawn away from the PN junction.

with their positive charge, are attracted away from the junction (Fig. 2.15). At the same time, the positive terminal of the battery at the N side attracts the excess electrons away from the junction.

The overall characteristic curve of a silicon diode is shown in Fig. 2.16. The portion of the curve to the right of the zero line represents the forward current. The reason for the upward swing of the curve has already been discussed. The upward swing occurs at relatively low voltage, between 0.6 to 0.8 V for silicon and 0.2 to 0.3 V for germanium. To the left of the zero line we have the diode characteristic under reverse bias. We see that a very small amount of current flows until the reverse voltage is brought to relatively high value, usually well above 20 to 30 V for most diodes.

The source of this reverse current are the minority electrons and holes that receive enough energy from heat and light reaching the crystal to break their covalent bonds. The reverse bias attracts these electrons and holes, and minute current flows through the circuit. The electrons of the P-type material travel to the positive-battery terminal; and as they enter this terminal (or the lead extending from the terminal), an equivalent number of electrons enter the silicon from the negative side of the battery. Similarly, holes of the N-type material move toward the negative terminal, and when they reach

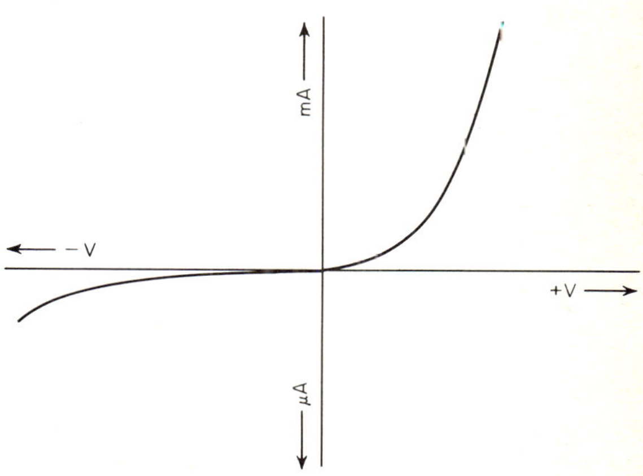

FIGURE 2.16
The overall static characteristic curve of a silicon diode. The portion of the curve to the left of the vertical line is shown dropping faster than it should (by comparison with the curve at the right).

this point, they receive electrons with which they combine. For each such combination, one electron leaves a covalent bond in the crystal near the positive terminal of the battery and enters the battery. This creates a new hole that moves across the crystal under the force of the applied emf. This situation is similar to what it was when the battery voltage was reversed (forward bias) and the majority carriers were involved. Now we are dealing with minority carriers, those electrons and holes that exist on the "wrong" side of PN junction.

The reverse current is small because the number of minority carriers in each section is small. Actually, together with this reverse current is a minute leakage current that flows along the surface of the diode. This is basically an Ohm's law phenomenon governed by $E = IR$. The R is the resistance offered by the diode surface to the applied voltage. Much of the "art" in the manufacturing of semiconductors is involved with controlling this surface resistance. This will be discussed in Chap. 4.

As the value of the applied voltage increases, a point at which there is a sharp increase in current is reached. This steep rise in current is owing to a phenomenon known as avalanche breakdown, in which minority electrons passing across the PN junction gain sufficient energy to knock off valence electrons bound to the crystal lattice and raise them to the conduction band. One minority electron may thus free several valence electrons through collision. In turn, these freed valence electrons may gain enough energy to free two or more additional electrons until a considerable flow of current follows. Again, we must not let too much current flow, because the junction can become permanently damaged from the heat generated. To sum up, then, diodes offer a relatively low resistance when biased in the forward direction and a very high resistance when biased in the reverse direction.

For quite some time avalanche breakdown was called the zener effect after the American physicist Clarence Zener. About 30 years ago, Zener made theoretical investigations of the electrical breakdown of insulators. He came to the conclusion that when the voltage across an insulator rose high enough, electrons could be torn from the valence band and raised to the conduction band fast enough to account for the large breakdown currents. When breakdown was observed in PN junctions, it was thought to occur by the same mechanism. However, as we shall see in Chap. 12, the two breakdown phenomena occur through different actions. Generally speaking, the zener breakdown occurs at voltages below 6 V, whereas avalanche breakdown occurs about 6 V.

JUNCTION TRANSISTORS

An NPN junction transistor is formed by having a narrow strip of P-type silicon between two relatively wide strips of N-type silicon. This is shown in Fig. 2.17. Low-resistance contact is then made to each strip for external circuit

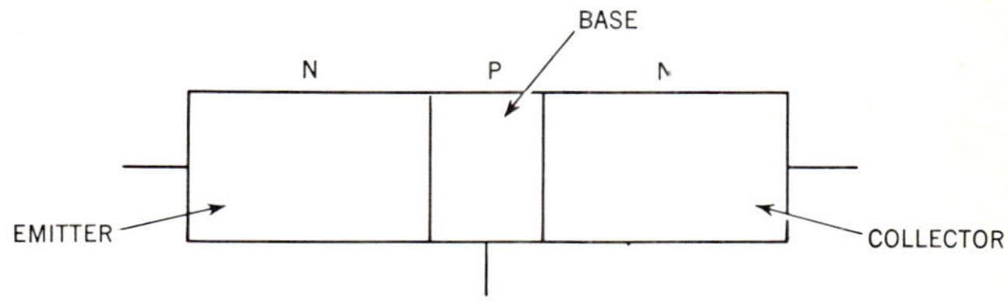

FIGURE 2.17
An NPN transistor and the names of its three sections.

attachment. The N-type crystal at the left is called the *emitter;* the central P-type strip is known as the *base;* and the end N-type silicon is called the *collector.* Although these names have no particular significance as yet, they will tie in with the operation of the transistor. Most modern transistors do not actually take this form, but the sequence of emitter, base, and collector remains unaltered. The form shown in Fig. 2.17 lends itself quite well to the present discussion, which is why it is being used. We will discuss some modern transistor forms in Chap. 4.

As with the PN junction diode, the two end sections of the NPN transistor contain a number of free electrons, while the central P section possesses an excess of holes. At each junction, the action that takes place is the same as that previously described for a diode; i.e., depletion regions develop and potential hills appear. These are then modified by the external voltage that is subsequently applied.

Transistor-Biasing. To employ the NPN transistor as an amplifier, we would bias the emitter and base sections in the forward, or in the low-resistance, direction. This is shown in Fig. 2.18*a*. At the same time, we would bias the base and collector sections in the reverse, or high-resistance, direction (Fig. 2.18*b*). Both bias voltages are shown in Fig. 2.18*c*. Let us now see what happens under these conditions.

Since the emitter and base sections are biased in the forward direction, current will flow across the junction. Every time an electron from the emitter section crosses the junction and combines with a hole of the base section, an electron leaves the negative terminal of the battery and enters the emitter. The battery cannot continue to supply electrons from the negative terminal without receiving an equivalent number at the positive terminal; hence, for each electron leaving the negative terminal, the positive side receives an electron from the base region. This loss of electrons in the base creates holes that then travel to the junction for eventual combination with electrons from the emitter.

Thus far we are on familiar ground—ground that has been previously explored. The main carriers of electricity in P-type silicon are holes. And

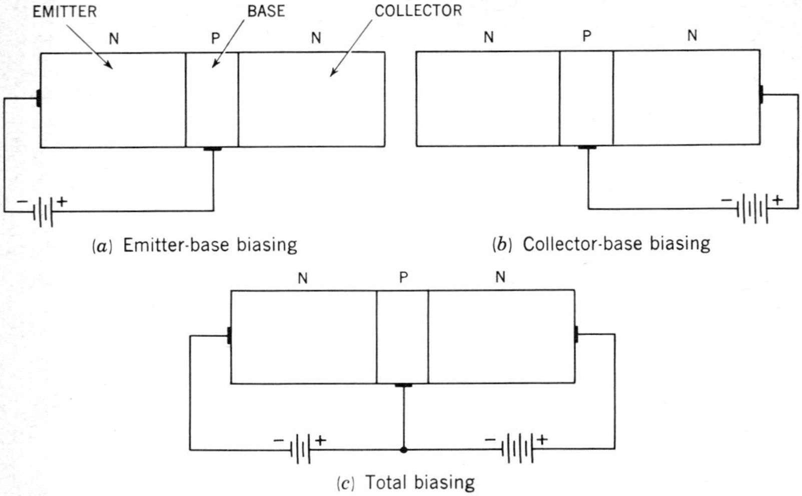

FIGURE 2.18
The proper method of biasing an NPN transistor for use as an amplifier.

this is precisely the situation described above. If the center base region were made quite thick, practically the entire current flow would then occur in the manner described and would be confined entirely between emitter and base. There would be little current between base and collector because of the reverse biasing existing there.

When the base strip is made extremely thin, however, transistor amplifying action is achieved. For with the base thin, electrons leaving the emitter pass right through the base section and into the collector region where they run into a positive force that draws them on. They thus travel through the collector section and around the external circuit back to the emitter again, completing their path of travel.

At this point the reader may wonder why the emitter current flows through the collector when it was specifically stated that the collector was biased in the reverse, or high-resistance, direction. If we disregard the base for a moment and simply consider the path from the collector to the emitter externally, we see that the two bias batteries are connected series aiding. Thus, any emitter electrons that pass through the base region without combining with the holes present there will find the attracting force of the collector battery urging them on through the collector section.

With the base width made very thin, the number of combinations between emitter electrons and base holes will be small, probably no more than 2 percent of the total number of electrons leaving the emitter. The remaining 98 percent of the electrons will reach the collector and travel through it. Thus, while the number of electrons leaving the emitter is a function solely of the emitter-base voltage, the section that receives most of this current is the collector. This emitter-collector current can be varied by changing the emitter-base voltage.

Since the base current is very small, a change in emitter bias will have a far greater effect on the magnitude of the emitter-collector current than it will on base current. This is also desirable, since it is the current flowing through the collector that reaches the output circuit.

Transistor Gain. We achieve a voltage gain in the transistor because the input, or emitter-to-base, resistance is low (as a result of the forward bias), whereas a high-load resistance can be used because the collector-to-base resistance is high (as a result of the reverse bias). A typical value for the emitter-to-base resistance is about 100 ohms (Ω), and a typical value for the load resistor is 10,000 Ω.

As we have noted, the current is 98 percent of the current leaving the emitter. If we now multiply the current gain (0.98) by the resistance gain 10,000/100, we will obtain the voltage gain of the collector circuit over the emitter circuit numerically, that is;

$$\text{Voltage gain} = \text{current gain} \times \text{resistance gain} = 0.98 \times 10{,}000/100 = 98$$

Thus, we see that while the current gain here is actually a loss, this is more than made up for by the extent to which the collector resistance exceeds the emitter resistance. Furthermore, this overwhelming differential in resistance will also provide a power gain. This means that with a small amount of power in the input, or emitter-to-base circuit, we can control a much larger amount of power in the output, or collector-to-base, circuit. Both these characteristics are important; without them the transistor would have only limited application in electronics.

The voltage gain indicated is what would be obtained if the transistor operated into a very high impedance circuit. Actually, one of the problems encountered in cascaded transistor amplifiers is that of matching the relatively high output impedance of a prior stage with the low input impedance of the following stage. This point will be discussed in greater detail in Chap. 6.

Potential Hills. Because a complete understanding of what happens inside a transistor is critical to future circuit applications, another approach, based on potential hills, will be discussed. This method (Fig. 2.19) reveals the effects of emitter, base, and collector voltages and presents a simplified visual picture of junction-transistor operation.

When the emitter is biased in the forward direction and the collector in the reverse direction, electrons leaving the N-type emitter will see only a small potential hill in front of them (at the NP junction), one that many of them can surmount. Once atop the hill, the "ground" levels off and the elec-

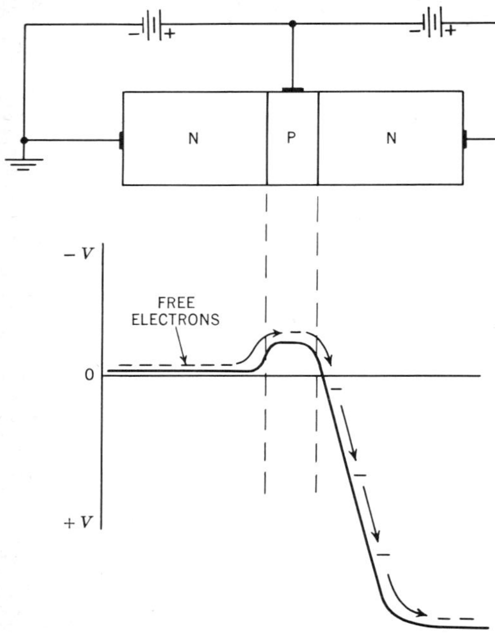

FIGURE 2.19
The potential-hill diagram for an NPN junction transistor. (*After H. K. Milward, Introduction to Transistor Electronics, Wireless World, March, 1955.*)

trons move through the P layer of the base quite readily. When they reach the junction between the base and the N-type collector, the electrons come under the influence of the positive-battery potential and surge forward. In the voltage diagram, this attraction is represented as a downward slope, which electrons (like human beings) find simple to traverse.

If the forward bias on the emitter is reduced, we are, in effect, raising the height of the base potential hill. Electrons leaving the emitter will find the higher hill more difficult to climb, and only those electrons possessing the greatest amount of energy will be able to reach its summit and move from there to the collector. Consequently, collector current will be reduced.

By the same token, increasing the forward bias on the emitter will reduce the height of the hill, enabling more emitter electrons to enter the base region. Thus, the biasing voltages used in a transistor have a very important effect on its operation. Another significant controlling factor is the width of the base. This is demonstrated by the next two illustrations.

Two three-dimensional representations of this potential diagram are given in Figs. 2.20 and 2.21. The difference between the two drawings lies in the width of the base sections. If the base section is wide, the tendency for emitter electrons (represented here by balls) to end up at the base electrode (because of combination with holes) is much greater than it is when the base section is narrow. In a physical model of these illustrations, the potential surfaces through the transistor are formed by a rubber membrane supported

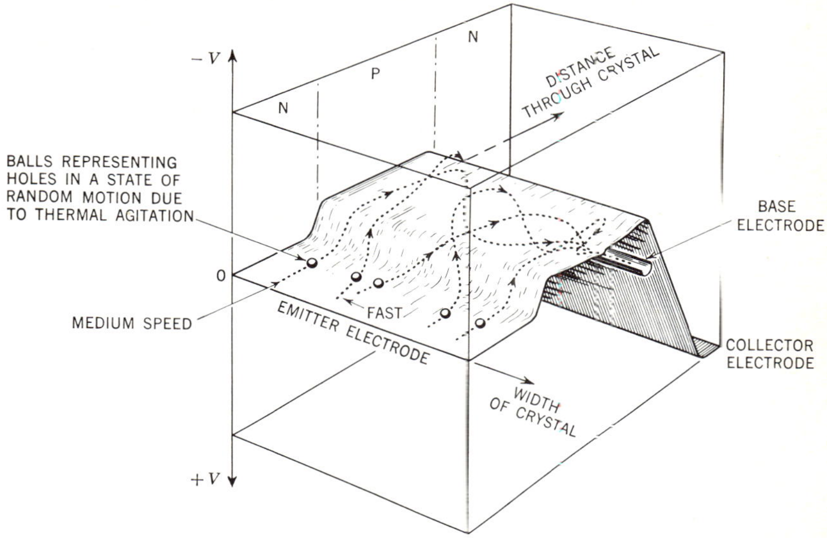

FIGURE 2.20
A three-dimensional representation of the potential levels in an NPN transistor. Suitable bias voltages are assumed to be applied to the various electrodes. In this illustration, the base is made wider than normal to demonstrate its effect. (*Wireless World.*)

FIGURE 2.21
The same illustration as Fig. 2.20 except that the base layer is narrow. (*Wireless World.*)

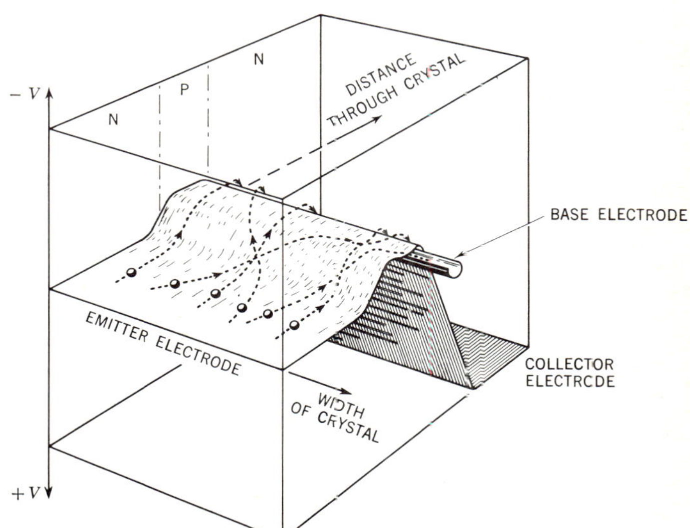

at several points. The holes in the base section are represented by a slight dip, or "valley," at the center of the base section. The wider the base section, the more difficult it is for the balls to roll through the base valley and over the edge of the precipice into the collector region without being trapped by the base dip.

On the other hand, if the base region is made quite narrow, any balls having enough energy to surmount the initial rise of the base hill possess enough energy to reach the far edge of the base and fall down into the collector.

Thus, the base width has a very direct bearing on transistor gain, both voltage and power. For if the percentage of current reaching the collector decreases to very small values, it will reduce the voltage gain in the same proportion. Power, being proportional to the square of the current, will be adversely affected to an even greater extent. Also, as we shall see in the next chapter, base width is extremely important in the frequency response of a transistor.

From the foregoing discussion, we can formulate two rules concerning the normal use of transistors.

1. The emitter is biased in the forward, or low-resistance, direction.
2. The collector is biased in the reverse, or high-resistance, direction.

We can understand why these rules are true if we consider their alternatives. If the emitter is biased in the reverse direction, it will not permit any electrons to reach the base region. And a reverse-biased emitter, with a reverse-biased collector, will produce a transistor in which current never passes. There are occasions when it is desirable to bias the transistor to cutoff. This is achieved by bringing the emitter-to-base bias to zero or even inserting a small amount of reverse-biasing voltage. In the majority of applications, however, statement 1 above is true.

If the emitter and collector are both forward-biased, the general tendency will be for the emitter electrons to flow between emitter and base and for collector electrons to flow between collector and base. In essence, we have two junction diodes possessing a common base. If the collector forward voltage is larger than the emitter voltage, some of the collector electrons will flow back to the collector via the emitter. But in any event, the desired amplification will not be obtained, and the purpose of the transistor will be defeated.

At this point a note of caution regarding the application of reverse voltage to transistors is advisable. As we shall see later, the emitter bias voltage is quite small, on the order of 1 V or less. The collector reverse voltage is generally much higher. If we should mistakenly connect the collector battery in the forward direction, the excessive current flowing through this section could develop enough heat to destroy the junctions, making the transistor useless. Hence, always be certain that the collector voltage polarity is correct before making connections.

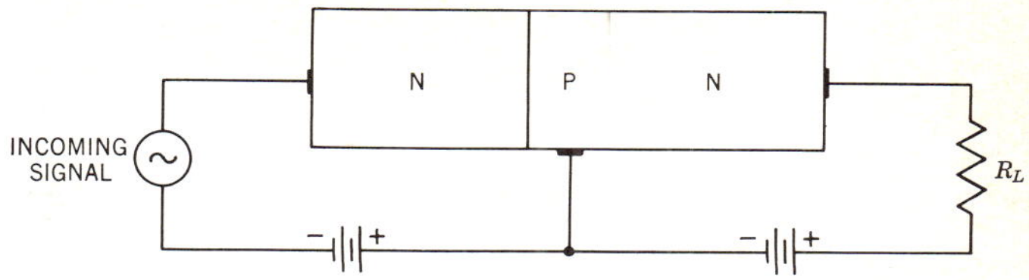

FIGURE 2.22
The NPN transistor connected as an amplifier.

It is interesting to note that a transistor possesses a bidirectional facility. That is, we can forward-bias the collector with a low voltage, reverse-bias the emitter, and then feed the signal in at the collector. The current gain under these conditions will be less than under normal conditions. As a matter of fact, in Chap. 11 there is described a television phase detector in which the emitter and collector sections are structurally identical and each takes turns sending current through the transistor.

Transistor Amplifier. We are now ready to connect the NPN transistor into an actual amplifier circuit with the signal input at one end and the load resistor at the other (Fig. 2.22). The incoming signal is applied in series with the emitter-to-base bias, and the load resistor is inserted in series with the collector battery. When the signal voltage is zero, the number of electrons leaving the emitter and entering base region is determined solely by the forward bias on the emitter. This situation can be represented by the potential-distribution diagram shown in Fig. 2.23*a*. When the signal goes negative, it adds to the forward bias, further reducing the height of the base hill and causing more electrons to flow through the transistor. This is shown in Fig. 2.23*b*. During the next half-cycle, the signal goes positive, reducing the forward bias of the emitter and thereby reducing the number of electrons leaving the emitter and entering the base and, subsequently, the collector regions. This is shown in Fig. 2.23*c*, where the height of the base hill has been increased.

At the other end of the transistor, these current fluctuations produce corresponding voltage variations across R_L, the load resistor. When the input signal is negative and the current increases, the collector end of R_L becomes more negative. By the same reasoning, when the signal goes positive, current decreases and the collector end of R_L becomes relatively more positive.

Thus, amplification is achieved through this transistor without the 180° phase shift that occurs in vacuum tubes. This is not always true of transistors. We shall see as we go on instances when signal phase reversal does occur.

Another point to note regarding this transistor is that an increase in signal polarity (i.e., positive) causes the transistor current to decrease. On

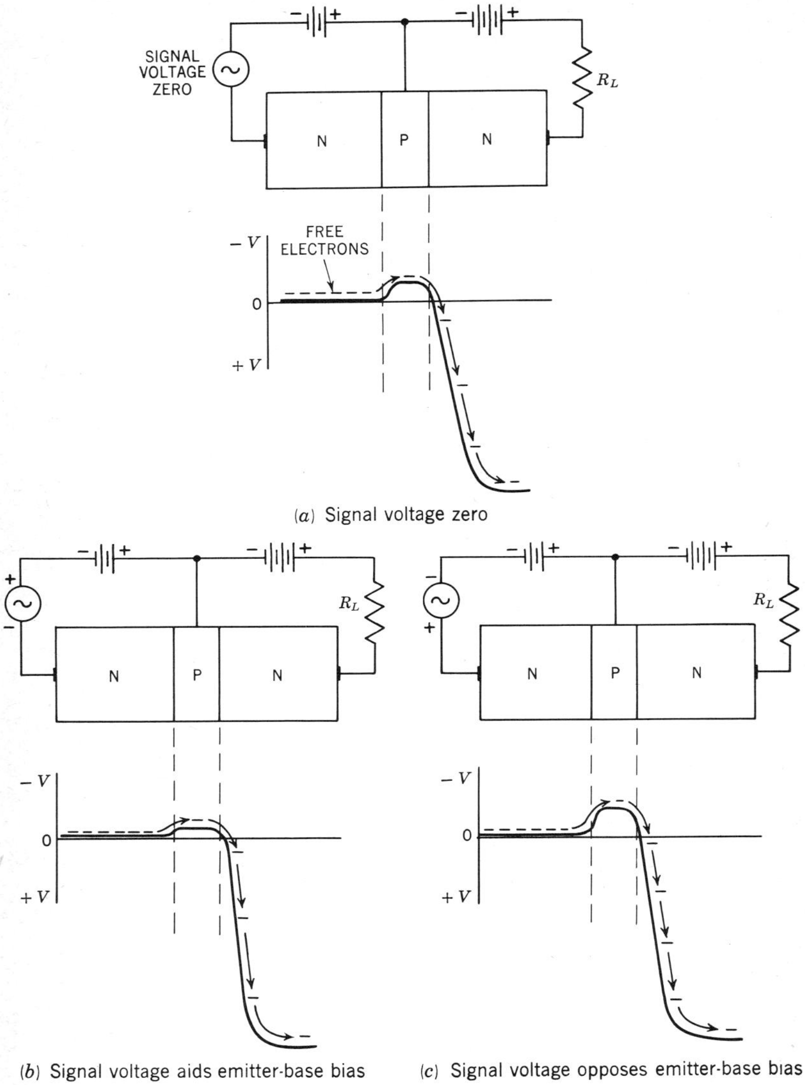

FIGURE 2.23
Transistor amplifier operation demonstrated by potential hills.

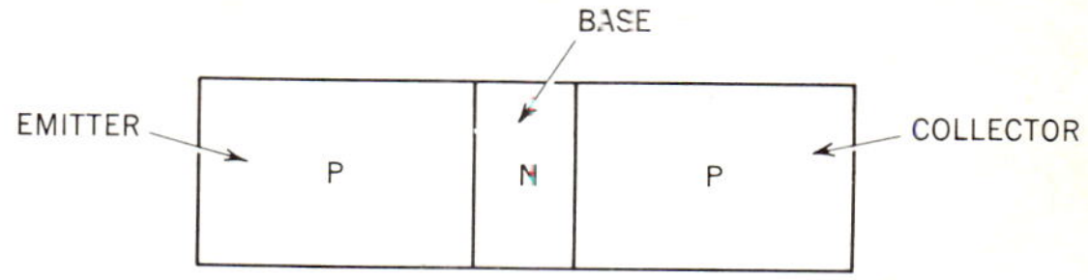

FIGURE 2.24
A PNP transistor.

the other hand, with the PNP transistor to be studied next, a positive increase in signal polarity causes the transistor current to increase.

PNP TRANSISTORS

In the formation of the initial transistor from a PN junction, we added a second N section to evolve an NPN transistor. We can approach the same problem by adding another P section to produce a PNP transistor (Fig. 2.24). The emitter and collector sections are formed now of P-type silicon. Since this is actually the reverse—as far as material structure is concerned—of the NPN transistor, we should expect differences in the mode of operation and in the polarity of the voltages applied to the emitter and collector.

A typical bias setup with PNP transistors is shown in Fig. 2.25. The positive side of the battery connects to the emitter, while the negative terminal of the battery is connected to the base. The collector battery is attached in the reverse manner, with the negative terminal connecting to the collector and the positive terminal to the base. Holes are the current carriers in the emitter and the collector sections; in the N-type base section, electrons are the principal carriers. With the emitter-bias battery connected as shown in Fig. 2.25, the positive field of the battery repels the positive holes toward the base section. At the same time, the negative-battery terminal at the base drives the base electrons toward the emitter. When an emitter hole and a base electron combine, another electron from the emitter section enters the positive-battery terminal. This creates a hole, which then starts traveling toward the base. At the same instant that the first hole and electron combine, another electron leaves the negative-battery terminal and enters the base. In this way, current flows through the base-emitter circuit.

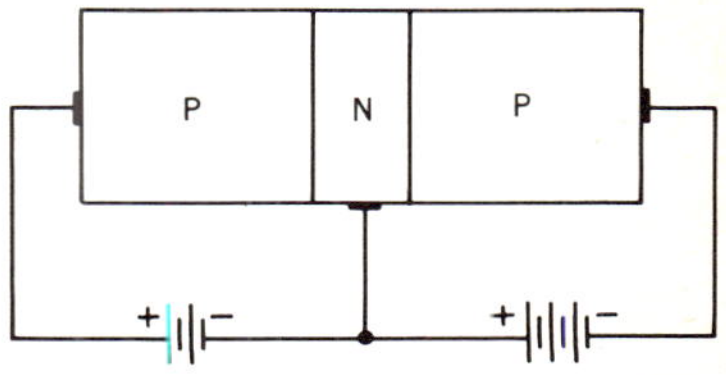

FIGURE 2.25
Method of biasing a PNP transistor.

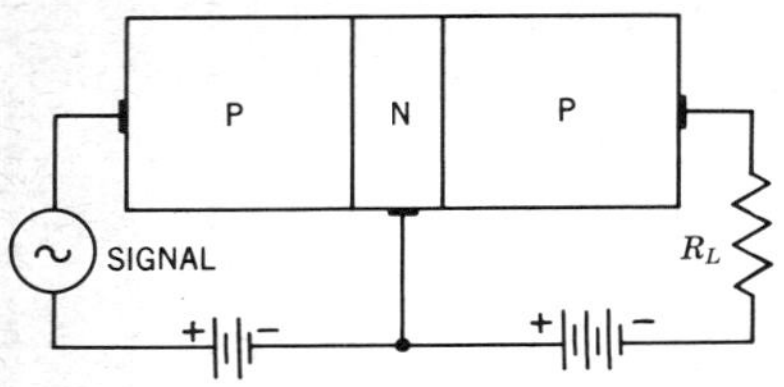

FIGURE 2.26
A PNP transistor amplifier.

In the PNP transistor, the holes are carriers in the emitter section; and when they cross the junction into the base region, a number of them combine with the base electrons. However, well over 90 percent of the holes do not combine with the electrons but rather pass through the base region and continue on to the collector terminal. When this terminal is reached, an electron from the battery combines with a hole and effectively neutralizes it. At the same instant, an electron leaves the emitter region and starts on its way around the outer circuit to the collector battery.

Note, then, that although holes are the carriers in P-type silicon, current conduction through the connecting wires of the external circuit is carried on entirely by electrons. This fits in with the current conduction with which we are familiar.

The incoming signal and the load resistor occupy the same positions in a PNP transistor amplifier that they do in an NPN transistor amplifier (Fig. 2.26). Only the polarity of the biasing voltage is reversed.

QUESTIONS

2.1. Draw a simple diagram showing how the silicon atoms are bound to one another in a crystal.

2.2. Since all the electrons in a silicon crystal are held fairly tightly, explain why silicon is considered a semiconductor rather than an insulator.

2.3. What is a hole in a semiconductor and how is it formed?

2.4. How do holes travel through a semiconductor?

2.5. When an electric field is applied to a semiconductor, what is the effect on holes and free electrons?

2.6. How is N-type silicon formed? List several substances that could be employed in this process.

2.7. How is P-type silicon formed? List several substances that could be employed in this process.

2.8. Are holes ever found in N-type silicon or electrons in P-type silicon? Explain.

2.9. Draw a PN junction showing the donor and acceptor atoms and the free holes and electrons on their respective sides.

2.10. What prevents the wholesale recombination of excess holes and electrons at a PN junction?

2.11. What do we mean by a potential hill?

2.12. How is a battery connected to a PN diode in order to initiate a flow of current? In order to prevent current flow?

2.13. Explain the mechanism of current flow through a crystal diode when the diode is biased in the forward direction.

2.14. Draw the characteristic curve of a silicon diode. Explain the reason for the reverse current.

2.15. What is the avalanche voltage in a crystal diode?

2.16. How is an NPN junction transistor formed? Sketch such a transistor and label each section.

2.17. Describe the rules that must be followed in biasing a transistor.

2.18. Why must the base section be made as thin as possible in a transistor? What happens when it is made too wide?

2.19. Explain how a transistor operates.

2.20. How is gain achieved in a junction transistor?

2.21. Why is there no signal reversal between input and output of the amplifier shown in Fig. 2.22?

2.22. Draw the diagram of an amplifier using a PNP transistor.

2.23. Describe the flow of current through a PNP transistor.

Chapter Three

Transistor Characteristics

Advances in transistor design and manufacturing techniques in the last decade have removed many operating limitations of semiconductor devices. These advances have been responsible for numerous new semiconductor applications and markets.

When transistors were first developed, they could handle only milliwatts of power at fairly low frequencies. Modern microwave power transistors are capable of handling 20 watts (W) of power at 2 gigahertz (GHz), and owing to a highly competitive market, these limits are being extended almost every month.

There is little doubt that the problems limiting present microwave-power-transistor applications will be solved and that the ultimate limitation will be the material from which the device is fabricated. Today this material is epitaxial silicon (see Chap. 4). Tomorrow it could be a new compound semiconductor, pushing the useful frequency and power capabilities of transistors to new horizons.

In this chapter we will examine the more important characteristics, physical as well as electrical, of transistors.

POWER

The power capability of a transistor is limited by three major factors. First, there is a maximum reverse voltage that the collector can stand. When the reverse voltage is raised beyond a certain point, an electrical breakdown is created between the collector and base. This is generally the avalanche effect noted in Chap. 2.

There is also another mechanism of voltage breakdown called *punch-through.* This results from an expansion of the depletion region that exists on either side of the PN junction between the base and collector. When a negative voltage is applied to the collector of a PNP transistor (positive-voltage terminal to the base), the holes in the P-type collector will be attracted to

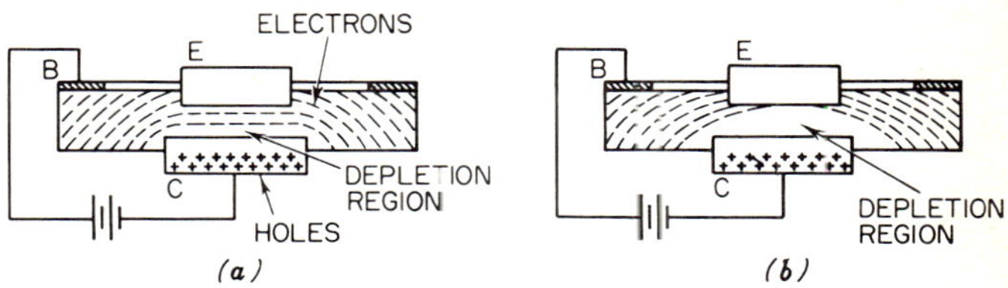

FIGURE 3.1
Mechanism of punch-through voltage.

the negative potential. Similarly, the electrons in the N-type base will be attracted by the positive potential applied to the base terminal. This attraction of the carriers away from the junction creates a depletion region in which there are no free carriers (Fig. 3.1*a*). As the voltage across the junction is increased, the depletion region expands and at some particular voltage will expand through the entire base region and actually come in contact with the emitter junction. This is shown in Fig. 3.1*b*. The voltage at which this effect takes place is called the *punch-through voltage.* When punch-through occurs, and at higher voltages, the normal transistor action ceases and the base can no longer control the current flow. As a practical matter, the resultant dynamic short circuit between collector and emitter permits a flow of current that is limited only by the resistance in the external circuit.

With either of the two conditions above the transistor may be damaged if too much current flows. If the current flow is not excessive, however, the transistor may not be damaged.

A second factor that limits the maximum power capability of a transistor is the decrease in current gain with increased current. This will be examined more fully later because it involves a current factor (beta) that has not yet been discussed. With a decrease in current gain, the power gain similarly decreases and thus the effectiveness of the unit to function as an amplifier is reduced. (Power gain is proportional to the square of the current gain.)

The third factor that establishes a limit to the maximum power output of a transistor is the safe amount of heat that the material or junction can withstand. Another way of stating this is to indicate the maximum power dissipation of the transistor internally, although it is generally the collector dissipation, since the greatest amount of heat is developed in the collector. Since the collector and emitter currents are nearly equal, differing only by the relatively small amount of current diverted by the base, the reader may wonder why more heat is generated at the collector. The answer lies in the fact that power is equal to I^2R, and inasmuch as the collector is reverse-biased, it has a far higher internal resistance than the emitter.

To assist transistors in achieving higher collector-dissipation ratings, special heat sinks (Fig. 3.2) in which these transistors are mounted have been developed. These heat sinks, or heat dissipators, help conduct the heat away

FIGURE 3.2
Typical transistor heat sinks. *(Courtesy Wakefield Engineering.)*

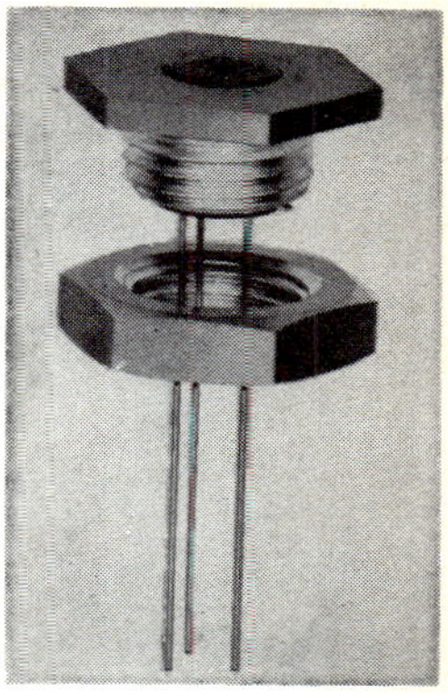

FIGURE 3.3
A power transistor housed in a metallic container. The unit shown will provide a maximum power output of 6 W when operated class A.

from the collector. Also useful for this purpose are metallic housings (cans) for transistors (Fig. 3.3).

It is customary, when the power dissipation of a transistor is given, to specify the temperature (usually 25°C) at which this dissipation can be obtained safely. Low-power transistors are rated in reference to the surrounding, or ambient, air temperature; medium- and high-power units are rated in reference to case or housing temperature. If it is desired to operate the transistor at a temperature higher than 25°C, the power-output rating must be lowered or derated. Either a specific derating factor, such as 1 W/°C, is given or a derating curve is included with the specification sheet.

If a derating factor is given, then for each degree Centigrade rise in temperature, the maximum power dissipation must be lowered by the amount

FIGURE 3.4
A power-transistor derating curve.

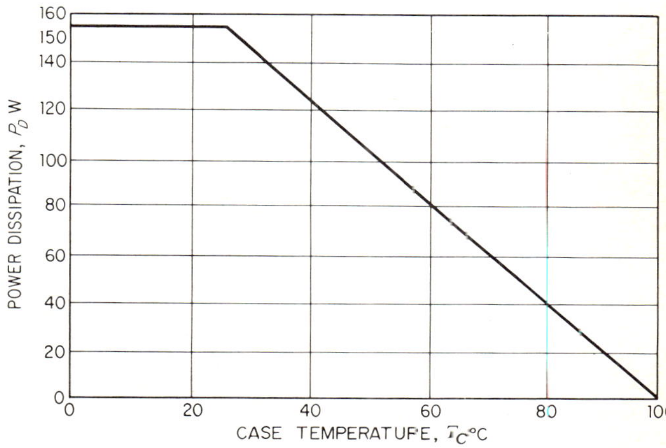

FIGURE 3.5
Typical power transistors.

indicated—here 1 W. If a derating curve is given, this information can be obtained directly. A typical curve is shown in Fig. 3.4. Up to 25°C, the power dissipation is 150 W. Beyond this temperature, the dissipation decreases linearly. Hence, at 60°C case temperature, the power dissipation can be only 80 W, while at 100°C it has been reduced essentially to zero. Failure to heed these derating factors will not only shorten the life of the transistor but will also frequently cause the transistor to burn out in short order. Several types of power transistors are shown in Fig. 3.5.

NOISE

Transistors, like vacuum tubes, develop a certain amount of internal noise. In the early days of transistor development, it was not unusual to find a transistor with a noise figure of 30 decibels (dB) at 1 kHz. Since then, transistor noise figures have been steadily lowered, until at the present the common values are below 5 dB.

Noise figure is a relative term designed to permit comparisor between similar electronic devices or circuits. For example, if the noise figure of a transistor is 0 dB, the transistor will not be generating any internal noise. When a transistor has a noise figure other then 0 dB, however, it contributes noise to the circuit. The higher the noise figure, the greater the ncise being added.

The noise behavior of a typical transistor is shown in Fig. 3.6. The characteristic turns up at either end, but it remains quite flat and generally unchanged in between. At the low-frequency end, the noise figure is inversely proportional to frequency, dropping to the flat base in middle-frequency ranges. The source of the low-frequency noise has been traced to surface conditions. Surface conditions, or surface activity, is related to the number of atoms per centimeter squared (silicon) at the surface, the number of dangling[1] bonds per atom, and the angle the dangling bonds make with the surface. It has recently been discovered that the low-frequency noise figure can be lowered significantly by controlling the growth axis of the silicon.

The central section of the curve is developed by the diffusion of carriers within the transistor and from the thermal agitation of current carriers. The first of these effects is directly related to the shot noise in vacuum tubes, which arises from the physical flow of electrons from cathode to plate. The second, or thermal, noise is analogous to the noise voltages generated in any resistor or other conductor of current. Both noise voltages are completely random in their nature because their effect (i.e., polarity) cannot be predicted at any specific instant of time.

[1] A dangling bond is an incomplete bond between adjacent atoms. Usually such bonds require an additional electron to make them complete.

FIGURE 3.6
Noise behavior of a typical transistor.

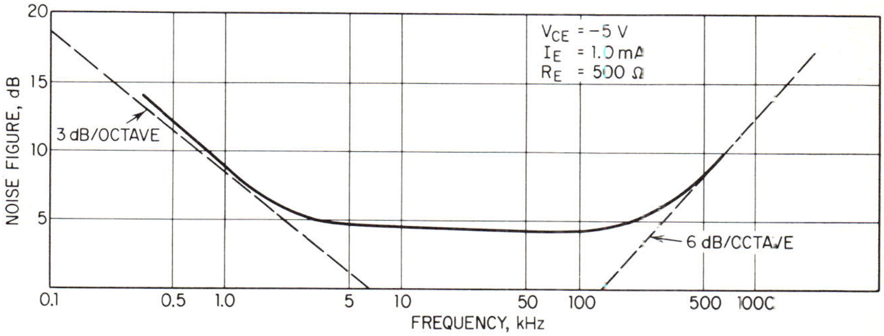

The second upturn of noise, at the right side of the curve of Fig. 3.6, is due to the drop in power gain of a transistor because of the frequency limitation of the device. Since the noise factor by definition depends on the ratio of signal to noise, anything that reduces the signal causes the noise figure to worsen.

Device design engineers take into account the following factors to control noise:

1. *The Growth Axis (Crystal Orientation) of the Silicon.* It affects the surface conditions of the transistor, because different crystal orientations have different densities of dangling bonds at the crystal (silicon) surface. The angle these dangling surface bonds make with the silicon surface, along with their density, are important factors in controlling the characteristics of the oxide-silicon interface of modern planar transistors. The oxide-silicon relationship is important in controlling low-frequency noise.

   An important new type of device depends on this oxide-silicon relationship for its operation. These devices are known as metal oxide semiconductors (MOS), which will be discussed in some detail in Chap. 5.
2. *Device Geometry.* It is known that device size and geometry affect noise. Noise can be reduced by keeping a high emitter area–periphery ratio and by having minimal contact metalization over the device junctions. Also, the smaller the device, the lower the noise figure.

   These considerations are not as significant as are the surface effects just mentioned. However, they are still important factors in device design.
3. *Maximum beta and minimum $R_{b'}$.* Van der Ziel, a researcher in solid-state physics, established the following relationship: F (noise) is minimized by minimizing

$$\frac{1-\alpha}{R_e} R_{b'}$$

where $R_{b'}$ = base resistance
R_e = dc resistance of the emitter
α = dc alpha

This indicates that in order to reduce noise, a designer must take steps to reduce the base resistance and maximize beta. We shall discuss alpha and beta later in this chapter.

FREQUENCY RESPONSE

In any application of a transistor, its frequency limitations must be known not only because of economic reasons (to avoid using a high-cost transistor with an extended high-frequency response where only a low frequency is required) but also because of power gain that falls off rapidly (above a certain point). There is no single factor that by itself completely controls frequency response. The end result is always a compromise of a number of considerations.

An important characteristic in establishing the frequency behavior of a transistor is the time required for a signal to travel from emitter to collector. This, in turn, depends upon the mobility of the carriers within the silicon. We cannot apply signals whose frequency changes so rapidly that the carriers (holes or electrons) are unable to transport the charges from emitter to collector.

The mobility of the holes or electrons in a transistor is the velocity with which they move through the silicon when an electric field is applied. Since electrons move almost twice as fast as holes, we would expect the transistors in which electrons do most of the current carrying to have a higher frequency response than transistors which depend upon conduction by holes. Conduction from emitter to collector in an NPN transistor depends upon the diffusion of electrons from the emitter to the vicinity of the collector. On the other hand, in a PNP transistor, holes are the principal carriers from emitter to collector. Since holes travel slower than electrons, we would expect PNP transistors to have a lower cutoff frequency.

Another factor that limits the high-frequency response is the capacitance between sections of a transistor. The higher the frequency, the lower the impedance of the shunting capacitor and the greater its shunting effect on the applied signal. This is true in both the input and output circuits as well as between input and output.

One of the most important high-frequency transistor parameters is known as the current-gain-bandwidth frenquency, or f_T. This is the frenquency for which the current gain (called beta) is equal to unity. We shall discuss beta later in the chapter. For the moment, let us look at some of the design and processing considerations that affect this important characteristic.

1. *Emitter Area.* The emitter area is kept as small as possible in order to keep the capacitance associated with this junction as low as possible. This capacitance adversely affects the frequency response of the device. This will be discussed in the chapter on applications.
2. *Emitter Periphery.* The ratio of emitter periphery to base area should be kept as large as possible in order to improve the current handling capabilities of the device. It is apparent that processing limitations are imposed by the first two conditions (1 and 2). This has resulted in very long, thin emitters in a confined base area. We shall discuss some modern innovations that have provided large emitter-periphery-to-base-area ratios in Chap. 4.
3. *The Impurity Concentration of the Base at the Emitter-base Junction.* This should be kept as low as possible to minimize the recombination of injected carriers at the emitter-base junction. This recombination increases as the frequency of the applied signal is increased, since a point is reached at which the carriers (emitter) cannot respond to the signal, creating flow

interruption at the emitter-base junction. However, as with the other parameters that control f_T, the base impurity concentration requires trade-offs in device design and processing. Probably the most important compromise is owing to the fact that optimum f_T alone does not ensure maximum power gain at high frequencies. This is because a very significant factor for optimizing power gain is the product of the collector capacitance and base resistance. This product appears in the denominator of a power output (P_o) equation and hence must be minimized. Obviously, this calls for a low base resistance (high concentration). But again, a low base resistance can result in a low punch-through voltage, which has been the cause of many device failures.

In addition, this parameter is particularly frustrating to control, because even after the designer has computed the optimum base impurity concentration, he is confronted with critical process limitations. These limitations are associated with the high-temperature impurity-diffusion process, which we will discuss in the next chapter.

4. *Base Width.* This is the most significant factor in designing for high-frequency performance. The base should be made as small (narrow) as possible to reduce the base transit time, that is, the time it takes the minority carriers to travel through the base to the collector. Besides processing difficulties, the base-width limitation is further aggravated by the fact that with the collector-base junction reverse-biased, the depletion layer could easily spread through a narrow base reaching the emitter-base junction. This is punch-through, which we described earlier in the chapter.
5. *Collector Doping Level.* With the base width extremely narrow, the doping level (resistivity) of the collector predominates in controlling f_T. This is so because a heavily doped material causes the depletion layer in the collector to be narrower and therefore the transit time of carriers through this layer to be shorter. However, a low-resistivity collector will also decrease the breakdown of the collector-base junction, resulting in another trade-off decision. Thus, the value of f_T must be increased as the frequency of operation is increased, and one way this is done is by increasing the collector doping level, which unfortunately decreases the collector-base breakdown voltage. High collector-breakdown ratings are desirable so that the transistor can operate at optimum output-power levels. This will be discussed in Chap. 6.

There are obviously numerous contradictions requiring trade-offs in the design and processing of high-frequency transistors. For example, high-frequency operation requires a small collector-base area for low-collector capacitance and a large collector-base area for current handling and power-dissipation requirements.

Packaging is extremely important in building high-frequency transistors. Low-lead inductance, good thermal conductivity, and isolation of the collector are some of the factors that have to be considered.

All of the parameters (and more) listed for f_T control are part of well-established mathematical equations that the designer of modern high-frequency devices can manipulate with the aid of the computer to optimize device performance. Even the package-design parameters are being optimized with the aid of the computer.

TEMPERATURE EFFECTS

The ability of properly processed silicon to serve as a transistor depends wholly on the electronic bonds and lattice structure existing within the crystal. Conductivity will rise with temperature. An increase in thermal energy will lead to more broken covalent bonds, more free electrons and holes, and a greater current flow in both input and output circuits for the same applied voltages. This, in turn, will reduce control of the collector current by the base, and eventually nullify the transistor action. It is even possible for the thermal action to feed on itself and destroy the transistor. The higher temperature results in more current, which raises the temperature even higher and results in more current, etc., until the unit is permanently damaged. (This effect is frequently referred to as thermal runaway.)

Even if the increase in temperature does not prove detrimental to the transistor, it can be the cause of signal distortion because of a shift in operating point. When a transistor is employed in a circuit, say an amplifier, a suitable operating point is selected. This operating point represents a certain collector current at a certain collector voltage, perhaps 5 milliamps (mA) at 6 V. When the signal enters the stage, it varies the collector current and the collector voltage follows suit. These variations in the collector circuit represent the output signal.

Suppose, now, that because of temperature, the operating collector current increases (with no signal) to 6 mA. This will cause a greater voltage drop across the load resistor and leave less voltage for the collector. Signals now passing through this stage can swing the collector voltage over a smaller range before cutoff occurs. Furthermore, as the collector voltage decreases, the transistor characteristic curves tend to become more nonlinear (Fig. 3.7), and this, too, is a source of distortion. Finally, by raising the collector-current operating-point value, more collector dissipation develops. This leaves less leeway for the signal before the maximum safe-dissipation value is reached.

In transistor characteristic charts, such as we shall study presently, the maximum collector dissipation is specified at a definite temperature (generally 25°C). If the operating temperature exceeds this value, it becomes necessary to lower the collector-dissipation rating.

Sometimes two maximum-temperature ratings will be given: one, the lower one, when in free air; the other, when the transistor is mounted flush against a metallic surface (such as an aluminum chassis) that will conduct the heat away. These heat conductors are heat sinks, and they can make an appreciable

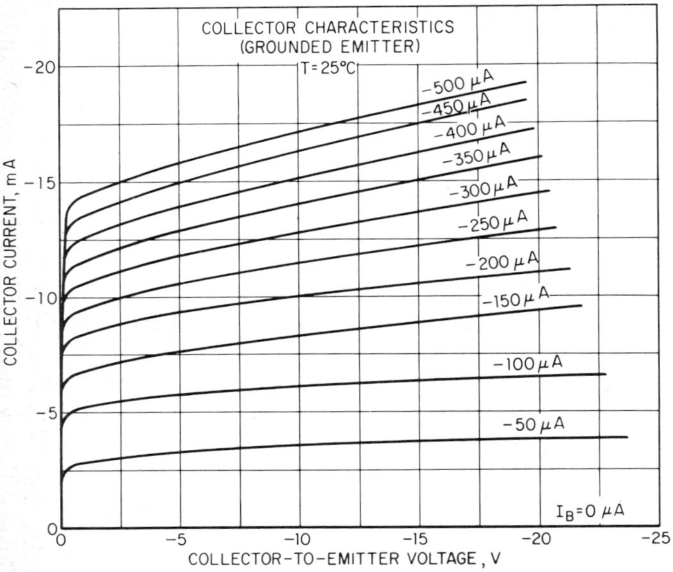

FIGURE 3.7
Typical collector characteristic curves for a transistor.

difference in the maximum-dissipation rating. For example, a transistor with a collector-dissipation rating in free air of 2.5 W will have an approximate rating of 4.0 W when mounted flush against an aluminum chassis. This is a significant point to keep in mind, particularly when the transistor is to be operated near its maximum rating.

In addition to an operating temperature, there is a maximum storage temperature. This is determined by life tests and is limited by the melting point of the materials within the transistor case or parameter stability. Parameter instability (i.e., changes in operating characteristics) results because of the increased chemical action at elevated temperatures primarily at the surface of the transistor. Today's planar transistors are many times more stable than were devices of a decade ago. However, surface changes still occur that affect transistor parameters or characteristics.

For example, an important parameter, I_{CBO}, increases with temperature. I_{CBO} is an important parameter that will be encountered again and again, and some explanation of it is in order. I_{CBO} is variously called the collector leakage current and the collector saturation current. Actually, saturation current is a preferable term, since leakage is only one component of its total value.

I_{CBO} is the current that flows through a transistor when the collector-base junction is reverse-biased and the emitter is open (Fig. 3.8). The notation I_{CBO}

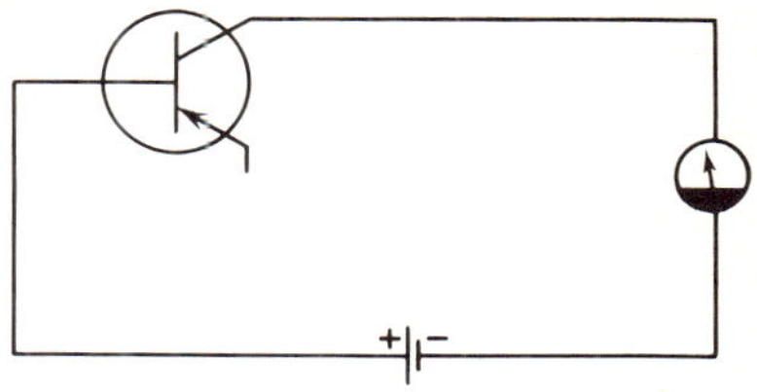

FIGURE 3.8
The manner in which I_{CBO} is measured.

was selected to reflect this fact. Thus, the CB following the capital letter I stands for collector and base. They are the two elements that are reversed-biased. The condition of the third element, here the emitter, is open-circuited, indicated by the letter O.

With the collector-base diode reverse-biased, one would expect zero current. In reality, however, it is impossible to fabricate P-type semiconductors without some free electrons or N-type semiconductors without some excess holes. These minority carriers account for a small current flow even though the collector-base diode is reverse-biased. Current flow takes place in the following manner. The minority carriers (from each section, electrons in the P region and holes in the N region) are forced toward the junction under the influence of the applied battery voltage. Some of the electrons and holes combine and neutralize each other's electrostatic charge. But this loss of electrons from the P section upsets the electrical neutrality that originally existed there, leaving it with a net positive charge. The positive field attracts additional electrons from the negative terminal of the battery. At the same time, the loss of holes (due to the same action above) leaves the N region with a net negative charge. (Remember that both sections were electrically neutral before the battery was connected to them.) To satisfy this charge, additional holes are formed in the N section while the electrons that were combined with these holes move to the positive terminal of the battery.

The foregoing provides a continuous flow of current, with electrons traveling from the negative-battery terminal to the P section, while electrons from the N section enter the positive battery terminal. Since the number of minority carriers originally available was small, the current under reverse-bias conditions is also small.

The small current that flows when the collector-base diode (or any PN junction) is reverse-biased actually consists of two components. One is the current indicated above, i.e., the current produced by the minority carriers in each section. This current is the saturation current and rapidly increases with increasing junction temperature. The heat sensitivity stems from the fact that as the temperature rises, thermal energy is imparted to the atoms forming the diode, enabling some of the previously held valence electrons to break loose from their covalent bonds. And, of course, for every electron that breaks loose, a corresponding hole is created. Thus, the supply of minority carriers increases and with this increase, the reverse current rises.

The second component of I_{CBO} stems from the leakage of current across the outer surfaces of the transistor. This component is voltage-dependent because it basically follows Ohm's law $E = IR$. It is also somewhat temperature-sensitive, probably because the resistance presented by the semiconductor surface itself is temperature sensitive.

WHY SILICON?

The emphasis in the preceding discussion has been on silicon as the semiconductor material from which transistors are made. Up until about 1964, most of the transistors were made from germanium. Of the two substances, silicon is far more abundant in nature than germanium. As a matter of fact, silicon compounds form over 85 percent of the earth's crust. Unfortunately, silicon is never found in the free state, and in order to utilize it for transistor manufacture, extensive separation and refining methods must be employed. The refining techniques have been improved to the extent that the cost of pure silicon is no longer an obstacle.

One of the most important advantages of silicon, which is in large measure responsible for much of its present popularity, is its low collector saturation current I_{CBO}. Table 3.1 presents a comparison between what might be termed typical junction-transistor characteristics for both silicon and germanium, and the

TABLE 3.1
Comparison of Silicon and Germanium Key Transistor Characteristics[a]

Characteristic	*Symbol*	*Silicon (diffused)*	*Germanium (alloy + diffused)*
Volts (max)	V_{CB}	1500 V[b]	300 V[c]
Current (max)	I_C	100 A	60 A
Collector cutoff current	I_{CBO}	0.01 μA	1 μA
Collector capacitance	C_c	7 pF	20 pF
Gain bandwidth product	F_t	100 MHz	1 MHz
Junction and storage temperature	T_{stg}	−65 to 200°C	−65 to 100°C

[a] Silicon transistors are best suited for high-voltage and high-power dissipation and high-frequency applications, while germanium transistors are still used for low-voltage, high-current, and-low cost applications.
[b] Triple diffused mesa.
[c] Alloy diffused epitaxial or post alloy diffused.

great disparity in I_{CBO} values is readily observed. The ratio might be anywhere from 100:1 to 500:1. The rate of increase in I_{CBO} with temperature is about equal for both types of transistor. Since the value of I_{CBO} for silicon is so extremely small at room temperature, however, the unit can be used at much higher temperatures before it becomes troublesome.

Another property in which silicon excels is its collector resistance r_c. This value is higher than the comparable collector resistance of germanium transistors. As we raise the operating temperature, r_c in both types of transistors will decrease; but because we start with a higher value in silicon, it is possible to go to higher temperatures before r_c becomes too small to use. When the behavior of I_{CBO} and r_c is considered, it is seen why silicon transistors possess higher maximum dissipations and why they are useful as high as 150°C.

These characteristics are directly attributed to the larger energy gap that exists between the valence band and the conduction band in silicon atoms. It will be recalled from Chap. 1 that in order for an electron to jump from the valence band to the conduction band, a certain amount of energy is required. More energy is needed to accomplish this jump in silicon than in germanium. This same factor also explains why more bias voltage is needed to produce a certain current in the emitter. It will also be found that the base resistance is higher in silicon transistors. When the silicon unit is employed as a grounded emitter, the higher base resistance requires more driving power from the preceding stage.

The mobility of electrons in silicon is about one-third that of electrons in germanium. This tends to work against the frequency response of silicon transistors. There are, however, other factors that have to be taken into consideration when designing for high frequency. For example, collector capacitance C_c is a very important frequency-determining factor, and in silicon this value is lower than in germanium transistors. Another compensating factor is that higher collector voltages may be employed with silicon transistors; this tends to decrease further the effective value of C_c and thus aid the frequency response.

The development of the silicon epitaxial planar process, with silicon-dioxide passivation, has given more device design flexibility to silicon, which is another major reason why more silicon than germanium transistors are made today.

LIFE EXPECTANCY

An important consideration in the application of any electronic device is its life expectancy. How long will this component last under normal operating conditions? In the case of transistors (and integrated circuits), this is especially pertinent because they form vital links in circuits.

Device failure can be one of three kinds; mechanical failure, bulk failure (arising from some defect in the bulk of the material), or surface failure. We are not concerned here with failure due to misuse, such as a current overload or the application of too high a voltage. Only failures arising from manufacturing difficulties will be considered.

Mechanical Failure. Into this category fall such things as poorly made connections, excessive strains on sections of the transistor, application of too much heat during die mounting or lead bonding, or the differential expansion between adjoining parts. When a transistor is properly made mechanically, it will withstand centrifugal forces with accelerations as high as 31,000 times the force of gravity (i.e., 31,000 g) and impact test with accelerations as high as 1,900 g.

Bulk Failure. Bulk failure arises from changes in the crystalline structure of the transistor body. For example, impurities that exist on the inside of the transistor housing or on the transistor surface can in time diffuse into the body of the transistor and alter the internal structure, particularly in the neighborhood of the junction. Fortunately, at normal operating temperatures such diffusion is quite small and can ordinarily be neglected.

Of far greater importance is the effect of nuclear radiation, which can penetrate into the transistor body and displace atoms from their normal positions and thereby cause them to appear at points where they were not intended to be. This disrupts the crystalline structure and modifies the electrical behavior of the unit. For example, the current gain may be affected, usually by being lowered, because the displacement in the crystal structure in the base prevents as many minority carriers from reaching the collector.

Surface Failure. The greatest number of transistor failures stems from changes at the surface of the transistor. Water vapor, for example, which can condense on a transistor surface, may decrease the collector breakdown and increase the collector current. Neither condition is desirable. A reduction in the collector breakdown voltage means that less collector voltage can be applied to the device, and this limits output power. An increase in collector current means higher internal heat generation at the operating point and less leeway for the signal.

To avoid these surface changes, transistors are hermetically sealed in a dry (inert) atmosphere. Even with these precautions, surface changes may still occur. To minimize these changes, special cleaning procedures are employed during the entire process.

In silicon-planar-transistor manufacture, the completed wafers are sometimes given special high-temperature (300°C) nitrogen bakeouts prior to final sealing for surface stabilization.

Finally, it must be realized that in transistors we are dealing with extremely small dimensions, particularly at the junctions. Even a microscopic particle of dust falling across a junction can completely short-circuit it. Transistors

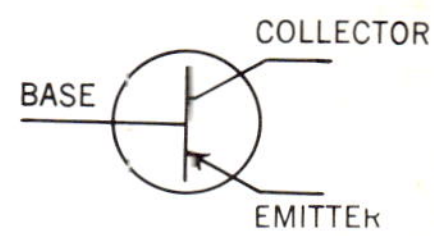

FIGURE 3.9
The schematic symbol for a transistor.

must thus be fabricated under almost sterile conditions, and almost every transistor fabrication plant has one or more "white," or clean, rooms where extensive air-cleansing equipment constantly filters out any dust particles that may be brought in. Some rooms have to be temperature- and humidity-controlled. These precautions extend even to the personnel, who are required in some facilities to pass through special outer rooms where much of the dust from their clothing is removed. They then don white laboratory coats and even special head coverings so that any dust or dirt that they will still carry on their bodies or clothing does not enter the atmosphere of the workroom. Transistor sensitivity to contamination make these precautions necessary.

It has been found that solid-state devices can operate for 20 years or more, a much longer period then we would ordinarily expect for vacuum tubes.

TRANSISTOR SYMBOLS

The basic symbol for a transistor is shown in Fig. 3.9. The emitter element has an arrowhead, the base is a straight line, and the collector element is shaped like the emitter but without an arrowhead.

To distinguish between NPN and PNP transistors, the method shown in Fig. 3.10 is employed. If the emitter is P type, the arrowhead is directed in toward the base. On the other hand, if the emitter is N type, the arrowhead points away from the base.

COMPARISON WITH VACUUM TUBES

Transistors are designed to perform the same functions as vacuum tubes, and it is therefore only natural to want to compare the two electrically to

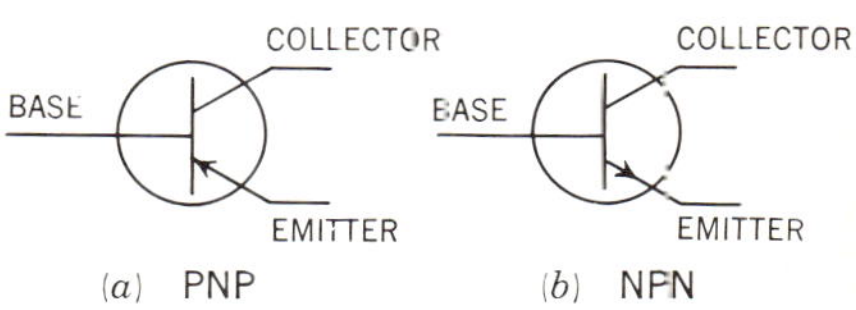

FIGURE 3.10
To distinguish between PNP and NPN transistors, the method shown is frequently employed. This notation will be followed in this book.

FIGURE 3.11
Comparable elements in tubes and transistors; grid and base, cathode and emitter, and plate and collector.

see wherein they differ and wherein they are similar. As a first step, let us consider these two components in the light of their internal operation. In a transistor, current flow through the various sections is initiated by the flow of electrons or holes from the emitter section. In a vacuum tube, this initiation starts at the cathode. We could thus say that the emitter in a transistor is equivalent to the cathode in a vacuum tube. (The word "emitter," of course, is a clue to the function of this element.)

The recipients of this current flow are the collector in the transistor and the plate in the vacuum tube; hence, these two elements can be considered to be equivalent in their actions. This still leaves the grid in the vacuum tube and the base in the transistor, and the equivalence of these elements is seen in the fact that whatever current flows from emitter (or cathode) to collector (or plate) must flow through the base (or grid) structure. Current flow in both devices is governed by the potential difference between emitter or cathode and base or grid. Figure 3.11 illustrates these analogies between transistor and vacuum tube.

The next step is to consider both devices in terms of the dc voltages that are applied to their elements. In a vacuum tube, the grid is almost always biased negatively with respect to the cathode. This makes the grid impedance very high (except at high frequencies, where other effects enter the picture). The plate, on the other hand, is always given a potential that is positive with respect to the cathode. The purpose of the plate is to attract the electrons emitted by the cathode; and since electrons possess a negative charge, a positive potential is needed to attract them.

In the transistor we wish to accomplish the same purpose, although conditions here are somewhat different. To initiate a flow of current, there must first be a flow of current between emitter and base and the bias battery must be connected to produce that current flow. This is what determines the polarity connections of the bias battery. If the emitter is formed by P-type silicon, the base will contain N-type silicon and current flow will occur between these sections when the positive-battery terminal connects to the P-type emitter and the negative-battery terminal connects to the base. We have spoken of this as forward-biasing, and under these conditions the impedance of the emitter circuit is low.

When we employ an N-type emitter and a P-type base, we must reverse the battery connections if we are to obtain the desired current flow through the emitter. Thus, the guiding thought in the emitter circuit is current flow, and we alter the battery conditions to suit the type of silicon being used in order to achieve our objective. From this it can be seen that transistors are current-operated devices, while vacuum tubes are voltage-operated. Alpha (α), for example, is the symbol representing the ratio $\Delta I_c / \Delta I_e$, where ΔI_C is the change in collector current and ΔI_E is the change in emitter current. The counterpart of this symbol in the vacuum tube is μ, the ratio of a voltage change in the plate circuit produced by a voltage change in grid circuit. Again we see the emphasis on voltage in a tube as against current in the transistor.

In the collector circuit the proper battery-biasing is such that the current flow is reduced to a minimum. (Note that it is not zero, although it is only on the order of microamperes.) This is known as reverse-biasing and is always true with collectors. To attain this condition, we must connect the battery in accordance with the type of silicon used in the transistor. If the collector is formed of P-type silicon and the base has N-type silicon (in PNP junction transistor), then the negative terminal of the battery goes to the collector and the positive terminal to the base. Conversely, if the collector has N-type silicon and the base P-type, the reverse is true.

BASIC TRANSISTOR—AMPLIFIER CIRCUITS

All transistor amplifiers can be divided into three classes: grounded emitter, grounded base, and grounded collector. In the sections to follow, the basic differences among these classes will be examined. Further elaboration will then be made in the succeeding chapters dealing with applications.

Grounded-base Amplifier. It is convenient to start a detailed examination with a grounded-base transistor amplifier. This is shown in Fig. 3.12. The input signal is applied to the emitter, and the output signal is obtained at the collector. In a grounded-base transistor amplifier, the input and output signals have the same polarity. To illustrate this, the amplifier of Fig. 3.12 has been drawn using an NPN transistor, and the battery polarities have been chosen accordingly. Assume now that the incoming signal is positive at this instant. This positive voltage will counteract some of the normal negative bias between emitter and base and serve to reduce the current flowing through the transistor. This, in turn, will reduce the voltage drop across R_2, making the collector potential more positive. Thus, a positive-going input signal produces a positive-going output signal.

During the negative half-cycle of the input signal, the emitter will be driven more negative than it normally is with respect to the base. This will increase the flow of electrons (here) from the emitter to collector and cause the negative voltage drop across R_2 to increase. This will drive the collector

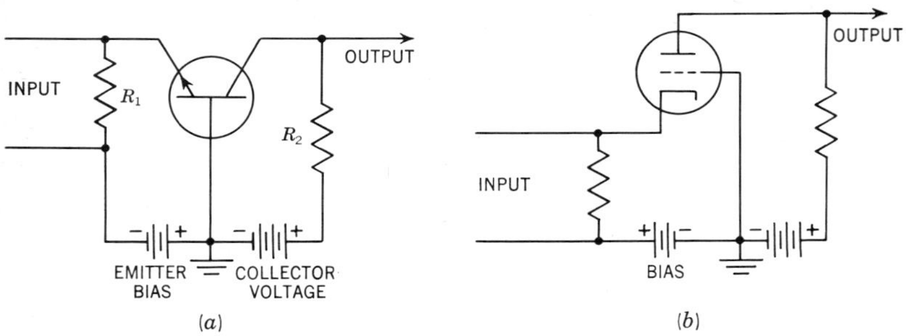

FIGURE 3.12
(*a*) A grounded-base transistor amplifier and (*b*) the analogous grounded-grid vacuum-tube amplifier. The polarity of the voltages used in the transistor circuit will be governed by the doping of the semiconductor used for the various elements. Here an NPN transistor is shown.

less positive. Again we see that the polarity of the output signal is similar to that of the input signal.

For the output, a load resistor of 10,000 ohms (Ω) is a common value. Insofar as current is concerned, there is less at the output (i.e., the collector) than at the emitter. The difference is the 1 or 2 percent that is diverted to the base. The ratio of output current I_C to input current I_E, or I_C/I_E, is the α, or h_{fb}, of the transistor (h_{fb} is the hybrid symbol for α). These hybrid characteristics, or parameters, are employed extensively on transistor data sheets, and it is desirable for the reader to become familiar with them. They are explained at length in Chapter 13. Thus, the current gain of this amplifier arrangement is less than one, which might lead one to believe that the circuit has little utility; but this is not true. A sizable voltage gain may be obtained because the output-load resistance value is so much higher than the input resistance. Thus, if we assume an input resistance of 50 Ω and a load resistance of 10,000 Ω, then the voltage gain (input to output) is

$$\text{Voltage gain} = \frac{E_{\text{out}}}{E_{\text{in}}} = \frac{I_C R_L}{I_E R_{\text{in}}}$$

$$\frac{I_C}{I_E} = 0.98 \qquad \text{for a typical value}$$

Hence

$$\begin{aligned}\text{Voltage gain} &= \frac{E_{\text{out}}}{E_{\text{in}}} = 0.98 \times \frac{10{,}000}{50}\\ &= 0.98 \times 200\\ &= 196\end{aligned}$$

By the same token, a power gain is also possible:

$$\text{Power} = I^2R$$

$$\text{Power gain} = \frac{P_{\text{out}}}{P_{\text{in}}}$$

$$= \frac{I_C{}^2}{I_E{}^2}\frac{R_L}{R_{\text{in}}}$$

$$= 0.98^2 \times 200$$

$$= 0.96 \times 200$$

$$= 192$$

We have indicated above that the input impedance of a transistor in the grounded-base configuration is quite low. The output impedance, if we remove the load resistor and look into the collector, is very high, on the order of 1 to 2 megohms (MΩ). However, when we connect a load resistor of 10,000 Ω, this is the value of the output impedance, since it completely swamps the 1 to 2 MΩ with which it is basically in parallel. It is well for the reader to keep this distinction in mind, because reference is often made in the literature to the high output impedance of the common-base arrangement—and this means, as indicated above, without the load resistor. Once a much smaller load resistance is connected to the collector, its value will essentially determine the output impedance.

Grounded-emitter Amplifier. The grounded-emitter amplifier (Fig. 3.13), is the most popular of the three types. The input signal is applied to the base, and the output signal is obtained at the collector.

It so happens that when the mathematics of grounded-emitter circuitry is worked out, this arrangement possesses a number of advantages over the

FIGURE 3.13
(*a*) A grounded-emitter amplifier and (*b*) its vacuum-tube counterpart. An NPN transistor is used in *a*; for a PNP transistor, the polarities of the batteries would have to be reversed.

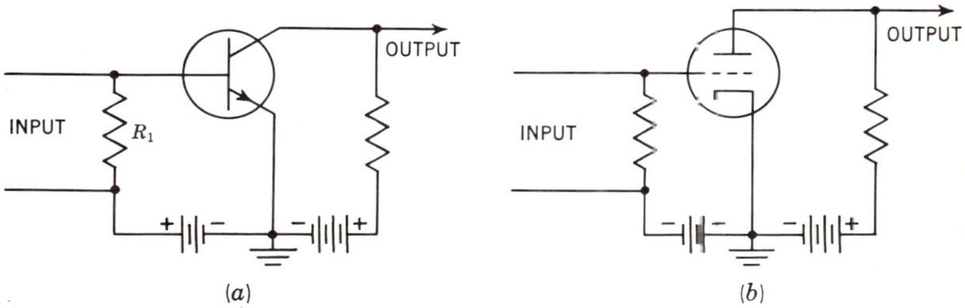

grounded-base approach. For one thing, the input impedance is higher, averaging between 700 and 2,000 Ω. The output impedance, looking into the collector before any load resistor has been connected, is about 500,000 Ω. This is less than the value presented by the common-base amplifier. The same 10,000 Ω of load resistance is usually employed here as well, however.

Since the input signal is applied to the base in the common-emitter arrangement, it is the variations in signal that control the collector current. Hence, current gain here is

$$\text{Current gain} = \frac{I_C}{I_B}$$

$$= \beta \text{ (or } h_{fe})$$

This ratio is called beta (β) or h_{fe} when the hybrid notation is employed. In a typical transistor, if the emitter current is 5 mA, the base will get 0.1 mA and the collector will receive the rest, or 4.9 mA. Substituting these values into the equation, we obtain

$$\beta = \frac{4.9}{0.1}$$

$$= 49$$

Note that a sizable current gain is obtained, in contrast to the small loss occasioned in the preceding amplifier. The gain arises from the fact that very minute variations of the base current produce significant variations in the collector current.

A voltage gain is also obtained because there is not only a current gain but also a resistance gain. If we assume an input resistance of 2,000 Ω and a load resistance of 10,000 Ω, then the voltage gain is

$$\text{Voltage gain} = \text{current gain} \times \text{resistance gain}$$

$$= \frac{I_C}{I_B}\frac{R_L}{R_{\text{in}}}$$

$$= 49 \times \frac{10{,}000}{2{,}000}$$

$$= 245$$

This is somewhat larger than the voltage gain achieved with the common-base circuit. The difference is not very much, however. Power gain is considerably better:

$$\text{Power gain} = \frac{I_C{}^2}{I_B{}^2}\frac{R_L}{R_{\text{in}}}$$

$$= (49)^2 \times 5$$
$$= 12{,}005$$

It is because of higher current and power gains that the common-emitter amplifier is the most popular arrangement employed in transistor circuits.

An interesting feature of the grounded-emitter form of connection is the phase reversal that occurs as the signal passes through the stage. The reason for the reversal can be understood by considering the amplifier shown in Fig. 3.13. The base-emitter circuit is biased in the forward direction, with the negative side of the bias battery to the base. (In this way, the negative battery terminal repels the excess electrons in the N-type emitter toward the PN junction and the positive battery potential drives the holes in the base to the same junction.)

If we now apply a signal to the base, here is what will happen. When the signal is negative, it will tend to reduce the bias potential applied between emitter and base. This means that the electrons in the emitter and the holes in the base will have less compulsion to overcome the inherent separating force at the junction and less current will flow. This, in turn, will reduce the collector current and provide less voltage drop across the load resistor. As a result, potential at the collector will become more positive.

During the positive half-cycle of the signal, the total voltage in the emitter-base circuit will rise. This will increase the flow of current through the emitter, the collector, and R_2. The increased voltage drop across the collector resistor will make the top end of this resistor more negative. Thus, in grounded-emitter amplifiers, the output signal is 180° out of phase with the input signal.

Grounded-collector Amplifiers. The final transistor-amplifier circuit arrangement is the grounded collector. This is shown schematically in Fig. 3.14.

In the grounded-collector arrangement, the input impedance, base to collector (which is the common element), is quite high because of the reverse

FIGURE 3.14
(*a*) A grounded-collector amplifier and (*b*) its vacuum-tube counterpart. The cathode resistor in *b* would have a fairly high value.

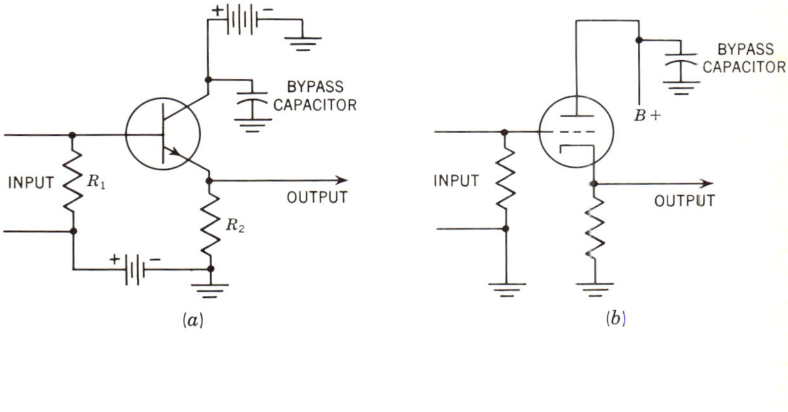

bias that exists between these two elements. (In contrast, the input impedance of the two preceding arrangements was low because the input circuit was between the base and emitter and the diode was forward-biased.) Of course, R_1 in Fig. 3.14 is hung across the input; and if this resistor is low-valued, it will cause the input impedance to be low. If R_1 has a high value, however, we can then obtain input impedances as high as 1 MΩ in this circuit. The output impedance, on the other hand, is low, frequently falling below 100 Ω.

The current gain of a common-collector circuit is slightly higher than β; actually, it is $\beta + 1$. This can be shown quite simply as follows. The input current is the base current I_B. The output current that flows through R_2, Fig. 3.14, is the emitter current I_E and this, we know, is equal to $I_B + I_C$. Hence,

$$\text{Current gain} = \frac{I_E}{I_B} = \frac{I_C + I_B}{I_B}$$

$$= \frac{I_C}{I_B} + 1$$

$$= \beta + 1$$

The voltage gain is always less than 1, although generally it is not much less than 1. This is true here for the same reason that it is true in a cathode follower. The emitter (or cathode) resistor is fairly large and unbypassed. Here, the signal voltage that develops here bucks the input signal at the base, so that only a small portion of the input signal is effective in producing an output voltage. In short, what we have is considerable degeneration.

Power gain is achieved in this stage because of the large current gain, but the gain is less here than it is in the other two configurations. Phase reversal of the signal does not occur in this stage. Any signal applied to the input will appear at the output with the same phase.

TABLE 3.2

Characteristic	*Common emitter*	*Common base*	*Common collector*
Current gain	Large	1, approx	Large
Voltage gain	Large	Large	1, approx
Power gain	Largest	Large	Lowest
Input resistance	Low	Lowest	Highest
Output resistance	High	Highest	Lowest
Signal phase shift between output and input	180°	None	None

Table 3.2 summarizes the general characteristics of the three amplifier configurations.

TRANSISTOR CHARACTERISTIC CURVES

The difference in operation between transistors and vacuum tubes, i.e., that one stresses current while the other voltage, is reflected in the characteristic curves of the two devices. In the characteristic curves for a 6AG5 pentode, Fig. 3.15, the plate-current–voltage curves are given for a series of grid-voltage values. The corresponding set of curves for a transistor is given in Fig. 3.7. Here, collector voltages and collector currents are plotted for various values of base current. Note the great similarity between these two sets of curves. In the case of the 6AG5 pentode, the plate current is relatively independent of the plate voltage above approximately 50 V. The only factor that determines plate current is grid voltage. For the transistor, collector current is likewise independent of collector voltage and is wholly a function of emitter current. This follows from the basic operation of a transistor, wherein the base-emitter potential determines how many minority carriers the emitter injects into the base. Because of its very low value, it is inconvenient to measure base-emitter potential. The base current, which is related to base-emitter potential, is a more easily measured component and is hence employed as the governing

FIGURE 3.15
The E_p-I_p characteristic curves for a 6AG5 pentode vacuum tube.

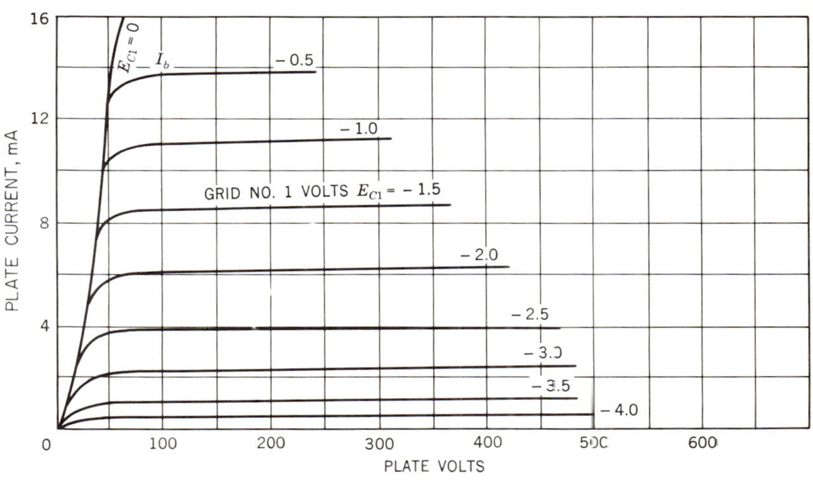

parameter on the family of $I_C - E_C$ curves. Furthermore, the input resistance is not linear. That is, the input resistance will change as the input current changes. If we attempted to use the base voltage as the running parameter, the characteristics would be made unnecessarily nonlinear.

An interesting feature of the transistor-characteristic curves is the fact that when we reduce the collector voltage, the collector current drops to zero. If the chart had extended down, it would have reversed itself and started flowing in the opposite direction (i.e., the forward direction). The latter condition is not desired, since it would quickly lead to excessive current flow with consequent overheating and permanent damage to the transistor.

In the plotting of graphs, it is customary to place the more important (i.e., the independent) variable along the horizontal axis. This should be done with the transistor curves; that is, the collector-current axis should be placed horizontally and the collector-voltage axis vertically. In practice, both types of presentations will be found, with perhaps greater emphasis given to the form shown in Fig. 3.7.

It is not uncommon to find collector-voltage values listed with negative values in these characteristic charts. This is to indicate that the applied collector voltage is negative in polarity, such as we would use with a PNP transistor. For an NPN transistor, the collector voltage would be positive.

Negative signs are also found in front of the current figures, and their presence here is associated with the direction of collector current flow in relation to the conventional indication of current flow through a circuit. Conventionally, electric current is taken to flow from the positive side of a voltage source through the circuit to the negative side. Electron flow, with which most technicians are familiar, travels in the opposite direction. Because of this difference in treating current direction, it is suggested that any negative signs appearing in front of collector-current values be ignored. As long as the proper battery polarity is applied to the collector, the current will take care of itself.

When they give transistor characteristics, most manufacturers include the output characteristic curves for the common-emitter connection. Occasionally, the curves for the common-base connection will be given.

Alpha and Beta. We have seen that the α of a transistor is the ratio of the collector current to the emitter current when the transistor is connected in a common-base arrangement. The collector current should be measured with no load resistor in the collector circuit because by definition α is the short-circuit current gain. The β of a transistor is defined as the ratio of the collector current to the base current when the transistor is connected as a common-emitter amplifier. Here, too, the collector should be measured with no resistance in the collector circuit because by definition β is the short-circuit current gain of the transistor. (By defining α and β in this way, a definite frame of reference is established for all such measurements.)

Since the α and β are characteristics of the same transistor, they are related to each other. We know that the emitter current is equal to the sum of the base and collector currents. That is,

$$I_E = I_B + I_C \tag{3.1}$$

Rearranging, we find

$$I_B = I_E - I_C \tag{3.2}$$

We also know that the collector current I_C is equal to αI_E.

Hence, expression (3.2) can be written

$$I_B = I_E - \alpha I_E$$

$$I_B = I_E(1 - \alpha) \tag{3.3}$$

Now we divide $I_C = \alpha I_E$ by Eq. (3.3)

$$\beta = \frac{I_C}{I_B} = \frac{\alpha I_E}{(1 - \alpha)I_E}$$

$$\beta = \frac{\alpha}{1 - \alpha} \tag{3.4}$$

Conversely, we can express α in terms of β:

$$\alpha = \frac{\beta}{1 + \beta} \tag{3.5}$$

Equation (3.5) is derived by simply rearranging (Eq. 3.4)

TYPICAL TRANSISTOR DATA

An indispensable tool for anyone dealing with the design, operation, or service of electronic equipment is a transistor manual. Here we find the mechanical and electrical specifications for each type of transistor, plus a set of characteristic curves.

Transistor manufacturers' sheets contain the specifications of a particular transistor, including maximum ratings, characteristic curves, and physical outline. Some of the more important items described are;

1. Transistor number
2. Collector-junction voltage rating
3. Emitter-junction voltage rating
4. Current-handling capacities
5. Power rating
6. Temperature limitation
7. Thermal resistance

8. Transistor-case outline
9. Ac h parameters
10. I_{CO}
11. I_{EO}
12. α cutoff frequency
13. Static collector characteristic
14. Temperature variation of transistor parameters
15. Dc β and α
16. Saturation resistance

Each of these parameters may be used in the circuit design and specification of a particular transistor. The location of the parameters on the data sheet need not be the same, and the method of presentation may also vary. Furthermore, not all parameters may be given for any specific transistor; only those that are deemed necessary will be included.

Before we examine a typical specification sheet, some mention should be made of the rules that have been established in reading symbols. To start, it is necessary to note carefully the upper- and lowercase letters used in both the quantities (i.e., major letters) and the subscripts. Following are the standards agreed upon by the Institute of Electrical and Electronic Engineers for semiconductor circuits.

Quantity Symbols

1. Instantaneous values of current, voltage, and power, which vary with time, are represented by the lowercase letters of the proper symbol. Examples: i, v, i_e, v_{eb}.
2. Maximum, average (dc), and rms values are represented by the uppercase letters of the proper symbol. Examples: I, V, I_e, V_{EB}.

Subscripts for Quantity Symbols

1. Dc values and instantaneous total values are indicated by uppercase subscripts. Examples: i_C, I_C, v_{EB}, V_{EB}, p_C, P_C.
2. Varying component values are indicated by lowercase subscripts. Examples: i_c, I_c, v_{eb}, V_{eb}, p_c, P_c.
3. If it is necessary to distinguish between maximum, average, or rms values, maximum or average values may be presented by the addition of a subscript m or av. Examples: i_{cm}, I_{cm}, I_{Cm}, I_{Cav}.

A typical specification sheet (General Electric Company) is shown in Fig. 3.16. The various sections of this listing are numbered 1 to 8, and appropriate explanations of each are given below.

1. The lead paragraph is a general description of the device and usually contains three specific pieces of information: The kind of transistor, in this instance a silicon NPN triode; a few major application areas, here amplifier

and switch; and general features such as leakage current, breakdown voltage, and current gain.

2. The absolute maximum ratings are those ratings which must not be exceeded. To exceed them may cause device failure.
3. The power dissipation of a transistor is generally limited by the junction temperature. Therefore, the higher the temperature of the air surrounding the transistor (ambient temperature), the less power the device dissipates. A factor that indicates how much the transistor must be derated for each degree of increase in ambient temperature in degrees Centigrade is usually given. Note that the 2N2193 (given on this specification sheet) can dissipate 0.8 W at 25°C. By applying the given derating factor of 4.6 mW for each degree increase in ambient temperature, we find that the power dissipation will drop to 0 mW at 200°C. This, then, is the maximum operating temperature of this transistor.
4. All the remaining ratings define what the device is capable of under specified test conditions. These characteristics are needed by the design engineer to develop matching networks and to calculate exact circuit performance.

   There is one important difference between the absolute maximum rating and the design characteristics listed on specification sheets. The absolute maximum rating must not be exceeded under any circumstance; and to exceed it automatically releases the transistor manufacturer from any warranty he may give with the unit. On the other hand, although characteristics may entail some guarantee, they are presented primarily as a guide to the user. Some of the parameters—for example, I_{CO}, h_{fe}, V_{BE}, etc.—have their maximum values guaranteed but not at end of life (i.e., usually after 1,000 h). Other parameters, such as breakdown voltage, are rated on an end-of-life basis. However, none of the typical values listed are guaranteed.
5. Current-transfer ratio is another name for β. In this case we are talking about an ac characteristic, so the symbol is h_{fe}. If the dc beta is meant, the symbol is h_{FE}. β is partially dependent on frequency, so some specifications list it for more than one frequency.

   Base saturation $V_{BE(\mathrm{sat})}$ specifies the base input voltage characteristic under the conditions of both junctions (emitter base and collector base) being forward-biased. The conditions of measurement specify a base current of 15 mA and a collector current of 150 mA. The base-emitter voltage drop is then 1.3 V. This parameter is of particular interest in switch designs.

   Collector saturation voltage $V_{CE(\mathrm{sat})}$ is the electrical characteristic describing the voltage drop from collector to emitter with both base-emitter and collector-base junctions forward-biased. Base and collector currents are stipulated.

   A transistor is saturated when both junctions are forward-biased. The saturation resistance for this condition is equal to the collector-to-emitter

FIGURE 3.16
A transistor specification sheet for General Electric types 2N2193, 2N2193A, 2N2194, 2N2194A, 2N2195, 2N2195A transistors. (*Courtesy General Electric.*)

	NPN
	Silicon Types
(*1*) The General Electric Types 2N2193, 2N2193A, 2N2194, 2N2194A, 2N2195, and 2N2195A are planar expitaxial silicon NPN transistors designed for high-speed switching and high-frequency amplifier circuits. They feature a low leakage current, low saturation voltage, a high breakdown voltage, and a guaranteed current gain from 0.1 to 1,000 mA.	2N2193 2N2193A 2N2194 2N2194A 2N2195 2N2195A

(*2*) *Absolute maximum ratings (25°C unless otherwise specified)*

	2N2193 *2N2193A*	*2N2194* *2N2194A*	*2N2195* *2N2195A*
Voltage			
Collector to base, V_{CBO}	80V	60V	45V
Collector to emitter, V_{CEO}	50V	40V	25V
Emitter to base, V_{EBO}	8V	5V	5V
Current			
Collector, I_C	1.0A	1.0A	1.0A
(*3*) Transistor dissipation, P_T			
Free air 25°C[a]	0.8W	0.8W	
Free air 25°C[b]			0.6W
Case temperature 25°C[c]	2.8W	2.8W	2.8W
Case temperature 100°C[c]	1.6W	1.6W	1.6W
Temperature			
Storage, T_{STG}		−65 to +300°C	
Operating junction, T_J		−65 to +200°C	

(4)	*Electrical characteristics (25°C unless otherwise specified)*					
	2N2193 2N2193A		*2N2194 2N2194A*		*2N2195 2N2195A*	
	Min	*Max*	*Min*	*Max*	*Min*	*Max*
Dc characteristics						
Collector-to-base voltage, V_{CBO} ($I_C = 100\ \mu A$)	80V		60V		45V	
Collector-to-emitter voltage, V_{CEO} ($I_C = 25$ mA)[d]	50V		40V		25V	
Emitter-to-base voltage, V_{EBO} ($I_E = 100\ \mu A$)	8V		5V		5V	
Forward-Current transfer ratio, h_{fe}						
(5) (I_C = 150 mA, V_{CE} = 10 V)[d]	40	120	20	60	20	
(I_C = 10 mA, V_{CE} = 10 V)	30		15			
(I_C = 1000 mA, V_{CE} = 10 V)[d]	15					
(I_C = 0.1 mA, V_{CE} = 10 V)	15					
(I_C = 500 mA, V_{CE} = 10 V)[d]	20		12			
(I_C = 10 mA, V_{CE} = 10 V, T_A = −55°C)	20					
Base saturation voltage, $V_{BE(sat)}$ (I_C = 150 mA, I_B = 15 mA)		1.3V		1.3V		1.3V
Collector saturation voltage, $V_{CE(sat)}$ (I_C = 150 mA, I_B = 15 mA)	0.35V max., 2N2193, 94, 95 only 0.25V max., 2N2193A, 94A, 95A only 0.16V typ., 2N2193A, 94A, 95A only					
(6) Cutoff characteristics						
Collector leakage current I_{CBO}						
(V_{CB} = 30 V)				10 mμA		100 mμA
(V_{CB} = 30 V, T_A = 150°C)				25 μA		50 μA
(V_{CB} = 60 V)		10 mμA				
(V_{CB} = 60 V, T_A = 150°C)		25 μA				
Emitter-base cutoff current (V_{EB} = 5 V)		50 mμA				
Emitter-base leakage current (V_{EB} = 3 V)				50 mμA		100 mμA
(7) High-frequency characteristics						
Current transfer ratio,						
h_{fe}(I_C = 50 mA, V_{CE} = 10 V, f = 20 MHz)	2.5		2.5		2.5	
Collector Capacitance, C_{ob} (I_E = 0, V_{CB} = 10 V, f = 1 MHz)		20 pF		20 pF		20 pF
(8) Switching characteristics, (V_{in} = 15 V, V^q = 15 V)						
Rise time, t_r		70 ns		70 ns		
Storage time, t_s		150 ns		150 ns		
Fall time, t_f		50 ns		50 ns		

[a] Derate 4.6 mW/°C increase in ambient temperature above 25°C.
[b] Derate 3.4 mW/°C increase in ambient temperature above 25°C.
[c] Derate 16.0 mW/°C increase in case temperature above 25°C.
[d] Pulse width $\leq 300 \mu s$, duty cycle $\leq 2\%$.

voltage divided by the collector current. It consists of two components. The first component is the bulk resistance of the material from the collector and emitter terminals to their respective junctions. The second component is due to the transistor action of the device; it decreases as the base current is increased for any given value of collector current. Thus, overdriving the transistor will reduce the saturation resistance.

Collector saturation resistance is generally of considerable importance to engineers who are designing logic circuits in which the transistor itself acts as a switch, going from a very high impedance condition, when it is essentially cut off, to a very low impedance condition, when it is saturated. In such design work it is important to know the saturation resistance, or the resistance of the transistor when it is in the low-impedance condition.

For the 2N2193 transistor, the saturation resistance is equal to the collector saturation voltage (0.35 V) divided by the collector saturation current (150 mA). The result is 2.3 Ω.

The saturation region on the characteristic curves of a transistor is at the extreme left-hand side where the curves appear to come together (Fig. 3.17*a*). Actually, when this portion of the graph is enlarged (Fig. 3.17*b*), it can be seen that each curve is separate and distinct. In this region the curves slope downward in a straight line, and if the ratio of collector voltage to collector current at any point is computed, the saturation resistance value is obtained. (Another way of stating the same thing is to say that the reciprocal of the slope of a curve in this region is the saturation resistance.)

Most manufacturers will list a collector saturation voltage $V_{CE(\mathrm{sat})}$. This voltage is essentially the minimum voltage necessary, at a particular collector current, to sustain normal transistor action, and it occurs when the emitter-base voltage equals the emitter-collector voltage. At lower-collector voltages, the base-collector diode becomes forward-biased and the current-voltage relationship changes abruptly. This is the region where the curve lines slope sharply downward.

It might be noted in passing that there is also a cutoff region on the characteristic-curve plot. This occurs below the curve marked $I_B = 0$. In Fig. 3.17*a* the cutoff region is somewhat below the curve marked $I_B = 0.02$ mA.

6. The collector cutoff or leakage current is the current from collector to base when no emitter current is being applied. This is the I_{CBO} mentioned previously. It varies with temperature changes and must be taken into account whenever any semiconductor device is designed into equipment that is used over a wide range of ambient temperatures. (The notation I_{CBO} is frequently shortened to I_{CO}. Both stand for the same quantity.)
7. The small-signal forward current gain h_{fe} is shown as a minimum of 2.5 at 20 MHz. Included, too, is the collector capacitance, a quantity of considerable importance in high-frequency applications.

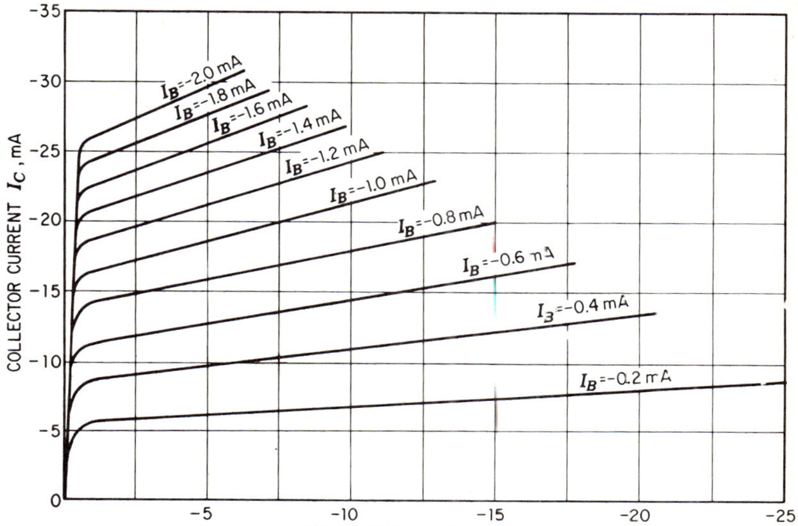

(a)

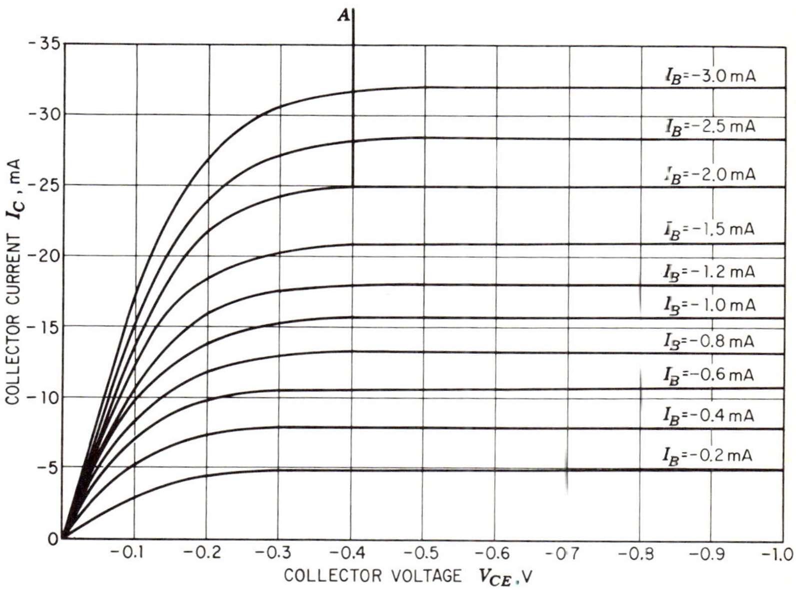

(b)

FIGURE 3.17
***(a)* The saturation region on the characteristic curves of a transistor is at the extreme left-hand side where the curves appear to come together. *(b)* Enlargement of saturation region of *a*. This region is to the left of line *A*.**

8. The switching characteristics show how the device responds to an input pulse under the specified driving conditions. These response times are very dependent on the circuit used. The terms used in this section of the specification sheets are explained in Fig. 3.18.

T_d is the delay time, or the time it takes from the application of the input voltage at point A (Fig. 3.18) until the output voltage has reached 10 percent of its final value.
T_r is the rise time or the time interval required for the output to go from 10 to 90 percent of its saturation value.
T_s is the storage time, or the time it takes from the removal of the input signal for the output to go from its saturation value to 90 percent of that value.
T_f is the fall time, or the time interval required for the output to go from 90 to 10 percent of its saturation value.

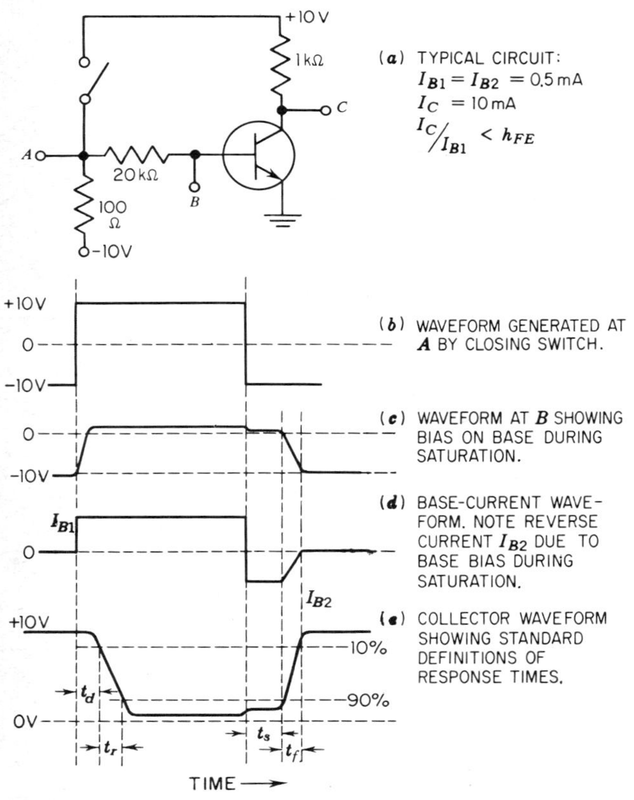

FIGURE 3.18

The delay time is partly due to the time required to discharge the emitter-base capacitance that has been charged to the reverse-space emitter-bias voltage (−10 V in Fig. 3.18) through the base resistance. Secondly, time must be allowed for the emitter current to diffuse through the base region. The rise time refers to the turn-on of the collector current. The storage time is due to the length of time required to sweep out the stored charge carriers in the base region that resulted from the collector-base region being forward-biased during saturation. (During saturation, both the emitter and collector inject carriers into the base region. The emitter normally does this under all circuit conditions; the collector, only when it is forward-biased as it is during saturation.)

The introduction of the epitaxial transistor offered important advantages over the alloy and the diffused uniform-crystal devices. Briefly, the epitaxial process involves growing a very thin, high-resistivity layer of silicon on a very low resistivity silicon substrate. The thin, high-resistivity layer has the same crystal orientation as the substrate, and by very careful process control, uniform layers having any desired thickness and resistivity can be grown. The epitaxial process will be discussed in some detail in Chap. 4.

The high-resistivity epitaxial layer permits the designer to design for high collector-base breakdowns, low capacitance per unit area, and a reduced series resistance resulting in optimum $V_{CE(\mathrm{sat})}$ measurements. This reduced series resistance is possible since the epitaxial layer is made just thick enough to support the depletion layer associated with the collector-base breakdown; and the heavily doped substrate, which is part of the collector, offers very little resistance.

In addition, since the effective minority-carrier diffusion length equals the epitaxial thickness, which is much less than the diffusion length related to the actual lifetime in the epitaxial region, the collector stored charge is significantly reduced. This is true since the recombination rate at the substrate-epitaxial interface is very high due to the high impurity concentration. This results in a reduced storage time t_s.

The development of the epitaxial and planar processes were an important turning point in solid-state technology. Epitaxial planar devices have high current and voltage ratings, excellent dc characteristics, and very high frequency responses.

The final item found on transistor specification sheets is an outline of the transistor housing and an indication of the positioning of the transistor leads. Transistor cases are assigned so-called TO numbers, such as TO-5 and TO-9. The letters TO stand for transistor outline.

Sometimes transistor leads are arranged in a straight line, sometimes they are arranged in a circle, and sometimes they are bunched together at a single point. The illustrations in Fig. 3.19 are representative of the more common arrangements. In all instances, the manufacturer's specification sheets should be checked before any connections are made to transistor leads.

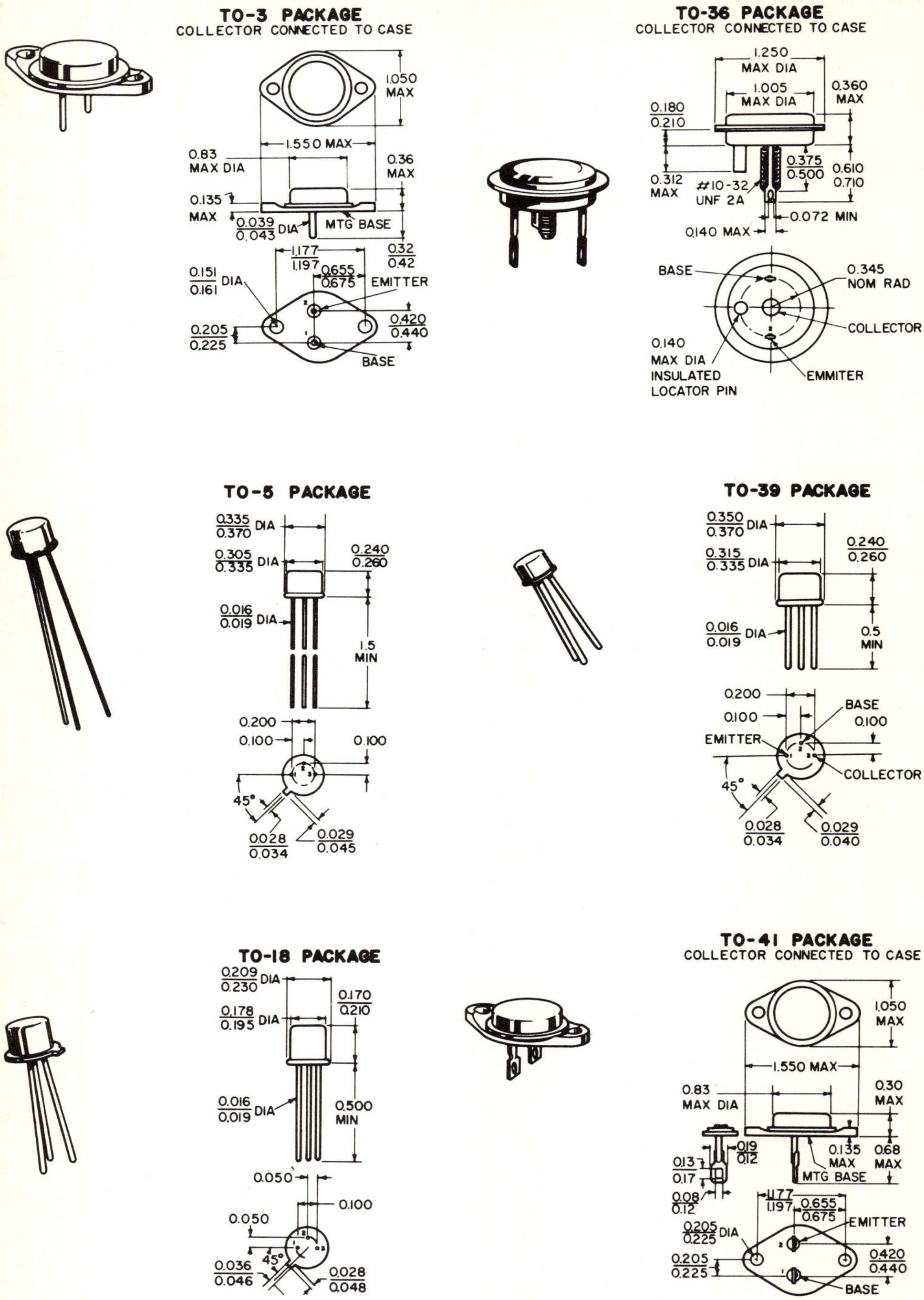

Figure 3.19 (See legend on p. 78)

TO-46 PACKAGE
COLLECTOR ELECTRICALLY CONNECTED TO BASE

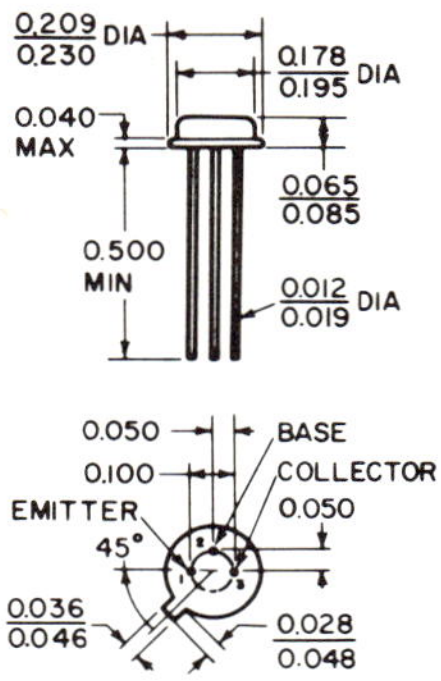

TO-68 PACKAGE
COLLECTOR CONNECTED TO CASE

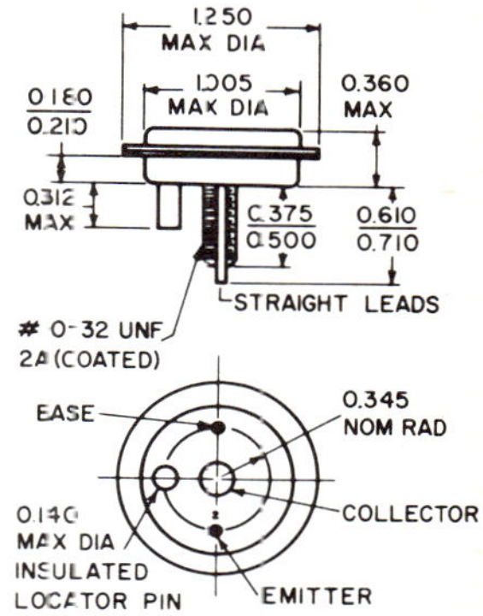

TO-60 PACKAGE

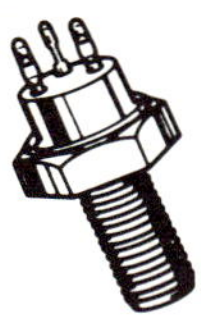

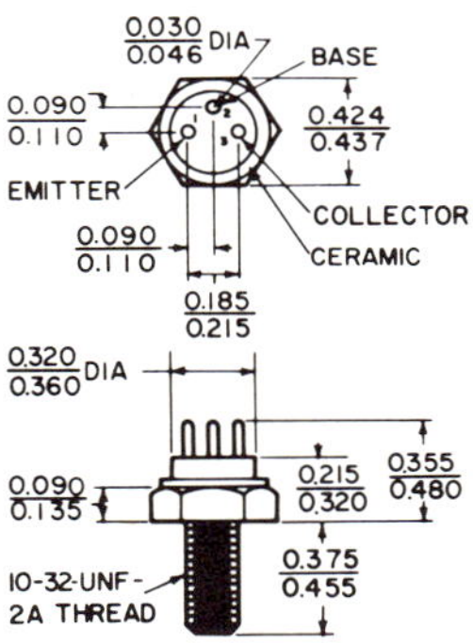

TO-72 PACKAGE

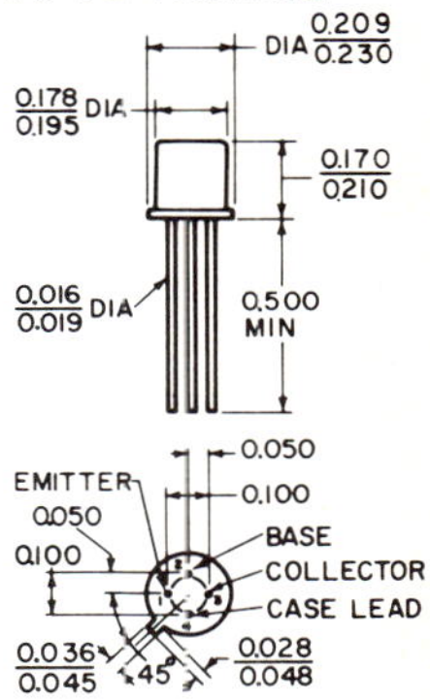

TO-61 PACKAGE

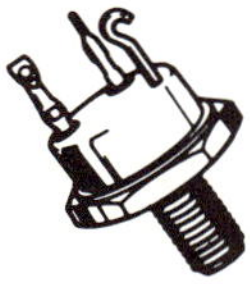

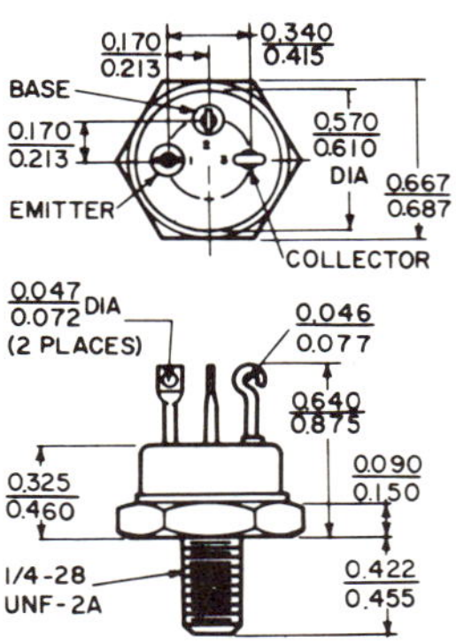

TO-89 PACKAGE
ALL LEADS ISOLATED FROM CASE
LEAD I IDENTIFIED BY IMPRESSION ON UNDERSIDE OF CASE

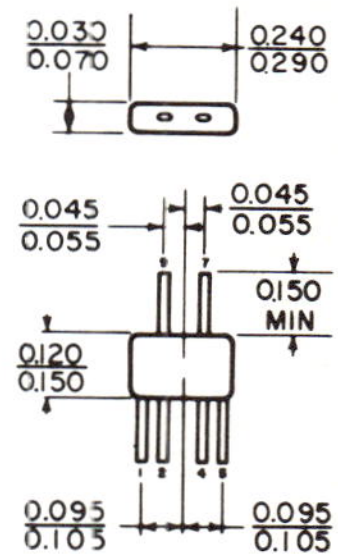

Figure 3.19 (See legend on p. 78)

TO-92 PACKAGE
PLASTIC TRANSISTOR

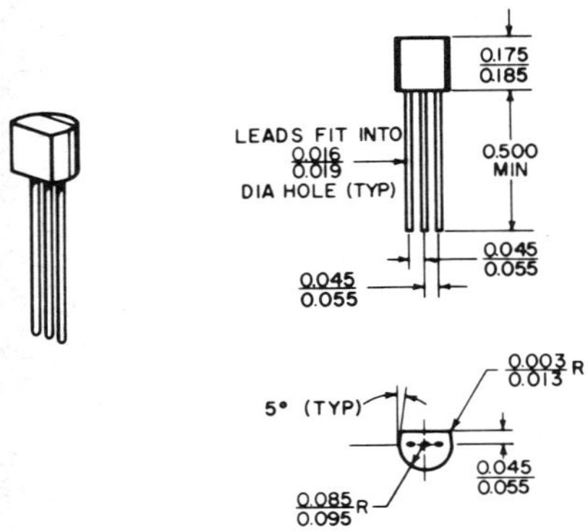

COLLECTOR CONNECTED TO TAB

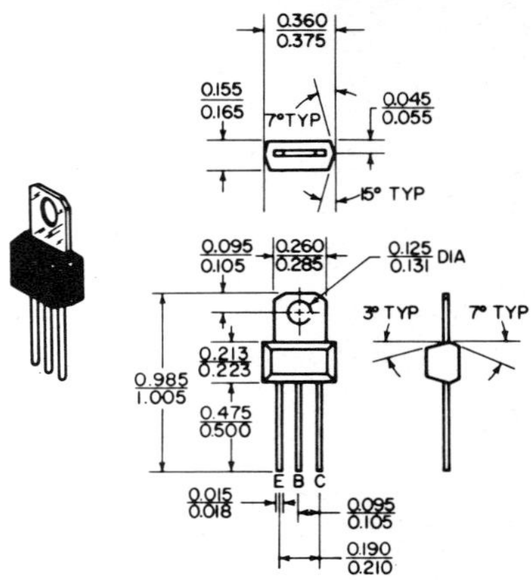

TO-102 PACKAGE
COLLECTOR CONNECTED TO CASE;
STUD ISOLATED FROM CASE

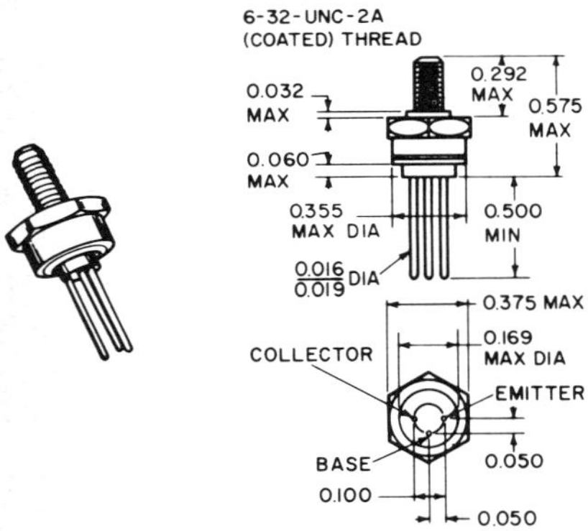

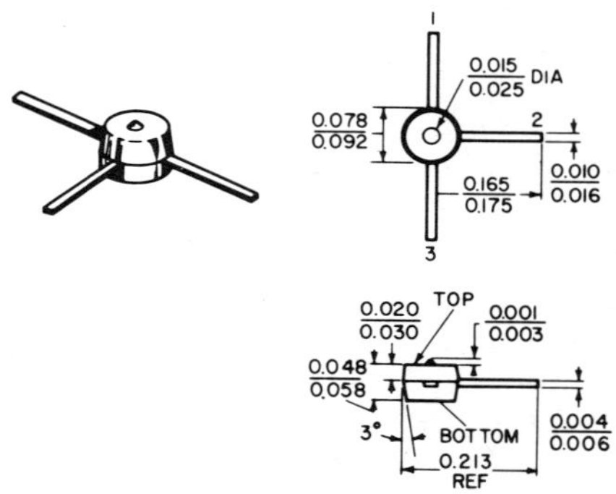

PLASTIC TRANSISTOR

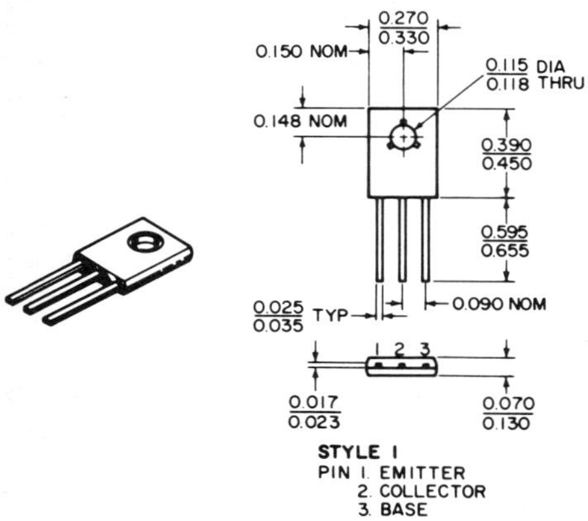

FIGURE 3.19
Above and on the pages preceding are shown typical transistor housings and lead arrangements. In instances when four leads are found, one is usually connected to the case and should be grounded (generally) in the circuit. In the larger power transistors, where only two leads or terminals are found, one is the base and the other is the emitter. The transistor case then serves as the collector connection, being internally connected to the collector element.

TRANSISTOR EQUIVALENT CIRCUITS

As one works with transistors, one finds that there is a considerable degree of dependence between the input and output circuits. This is in direct contrast to vacuum tubes, wherein the input and output circuits are relatively independent of each other. (Feedback effects that may occur can generally be counteracted by suitable means.)

In the common-emitter arrangement, for example, the variation in input resistance with load resistance is as shown in Fig. 3.20. Note how the input resistance decreases with increase in load resistance, eventually leveling off to a value of about 500 Ω when the load resistance becomes inordinately large. This is for a typical junction transistor.

Similar curves showing the effect on the output resistance for different input resistances could be drawn. The reason for these interactions can perhaps be better understood when the equivalent electrical circuit of a transistor is examined. Equivalent circuits are convenient devices that enable an engineer to develop a relatively simple electrical network that will function in the same manner electrically as some complex circuit that he may be investigating. It is interesting to study the engineer's approach to equivalent circuits because it will give the reader a better appreciation of the value and purpose of these circuits. The start is made with a little black box in which the circuit or system to be analyzed is contained. Access to the box is prohibited, and all we have from the box are four terminals, two representing the input and two the output. The procedure then is to take this black box and perform

FIGURE 3.20
Variation of input impedance with load resistance in a junction transistor.

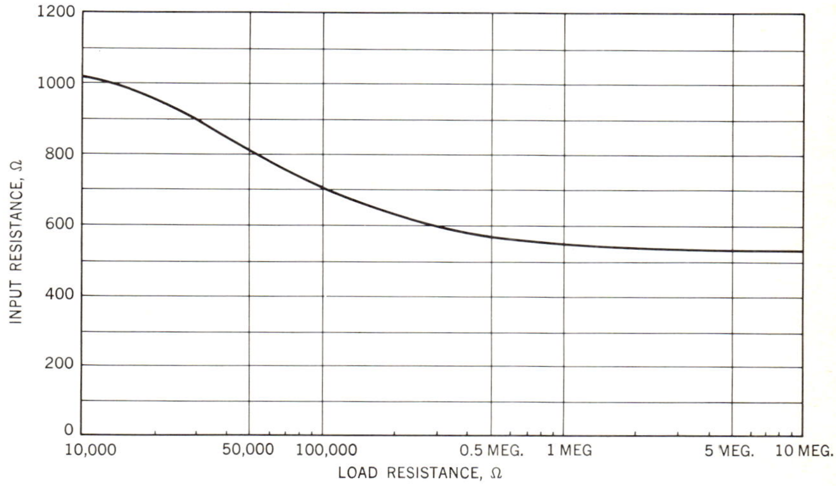

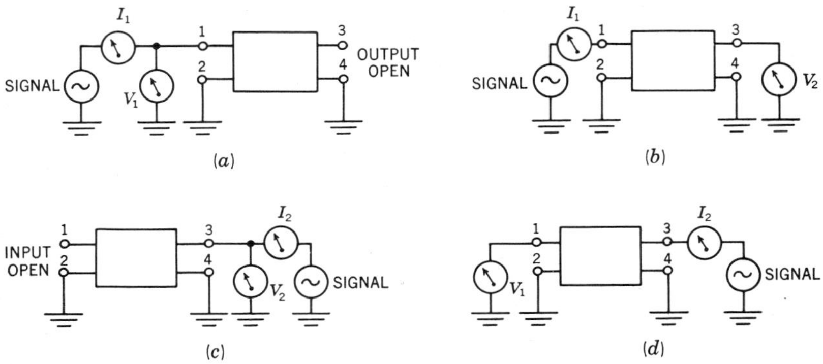

FIGURE 3.21
Voltage and current measurements made on a system in order to determine its equivalent circuit. (*a*) Signal is applied to input; voltage and current measurements are made on input with output terminals open. (*b*) Signal is applied to input; current flowing in input is measured, together with voltage across output. (*c*) Signal is applied to output with input open; voltage and current in output are measured. (*d*) Signal is applied to output; current flowing in output is measured, together with voltage across input.

a series of four measurements on it. One measurement is to apply a signal to the input terminals 1 and 2 and record the voltage that is applied and the current that flows in the input circuit with the output circuit open (Fig. 3.21*a*). This will give us the input resistance R_1 when V_1 is divided by I_1.

A second measurement is made by applying the signal to the input terminals of the black box and recording the current flowing in the input circuit and the voltage developed across the output circuit. This is illustrated in Fig. 3.21*b*. This measurement indicates what effect the input circuit has on the output circuit.

The third test is made with the signal generator connected across the output terminals and the voltage and current meters recording these respective quantities in the output circuit (Fig. 3.21*c*). The input circuit is open.

The final check is made under the conditions indicated in Fig. 3.21*d*. Here we apply the signal to the output circuit and measure the voltage it produces across the input circuit.

The results of these four measurements are then used to draw a simple network that will give exactly the same results when the measurements indicated above are made. If such a network can be found, then we know that it will act under all conditions as the circuit or system in the black box acts, and we can call this latter network the equivalent of the box system and deal with it rather than the generally more complex system it replaces.

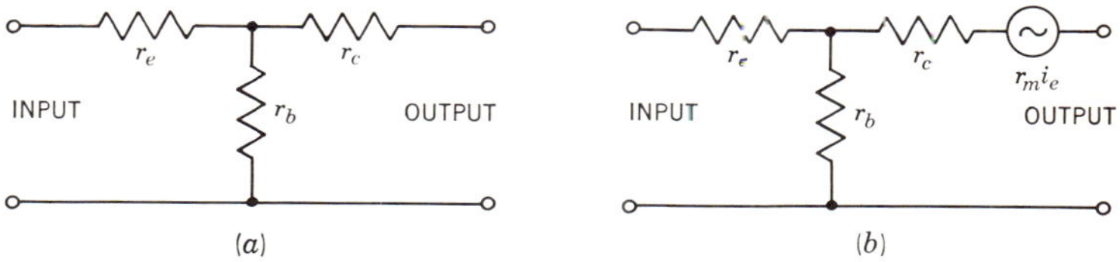

FIGURE 3.22
Steps in the development of an equivalent circuit for a transistor. (*a*) Preliminary equivalent circuit of a transistor. (*b*) A more nearly complete equivalent circuit of a transistor.

Using the foregoing method, one equivalent network obtained for a transistor is as shown in Fig. 3.22, where r_e is the internal resistance of the emitter, r_b is the internal resistance of the base, and r_c is the internal resistance of the collector. Note that the base resistance is common to both the emitter and collector circuits, a fact that we discovered previously when studying the manner in which current is conducted through the transistor.

Now if all we had in our equivalent circuit were these three resistances, then we would have a simple resistive network in which signals (or voltages) could pass from input to output or from output to input with equal ease. This we know is not true of transistors. Furthermore, a simple resistive network could introduce only attenuation, not amplification; and transistors do amplify. Something more is obviously needed, and that something is the small generator placed in series with r_c. For mathematical reasons that are related to the design equations of transistors, this generator is given a value of $r_m i_e$, where i_e is the current flowing through the emitter resistance r_e and r_m is a mutual resistance of the system. For our purpose here, we need simply regard this generator as adding its voltage to that of the input signal to produce a greater (i.e., an amplified) signal at the output. In this way we achieve an equivalent circuit that reveals how a signal applied to a transistor is amplified and just what that amplification will be under various types of load resistances.

For readers who find this added generator strange or confusing, attention is directed to the equivalent circuit for a triode vacuum tube (Fig. 3.23). We note that voltage e, applied between grid and cathode produces the same effect as a voltage in the plate circuit, which is μ times greater. μ, of course, is the amplification factor of the tube.

Returning to Fig. 3.22, we begin to see why the input and output circuits of a transistor are so dependent on each other. Any current flowing in the collector circuit will also flow through r_b, and the voltage developed here will directly influence the current flowing in the input circuit (containing r_e and r_b). And, of course, anything that happens in the input circuit will be immediately felt in the output circuit. In a vacuum tube, where the grid is negative and the frequency is not very high, the equivalent circuit of Fig.

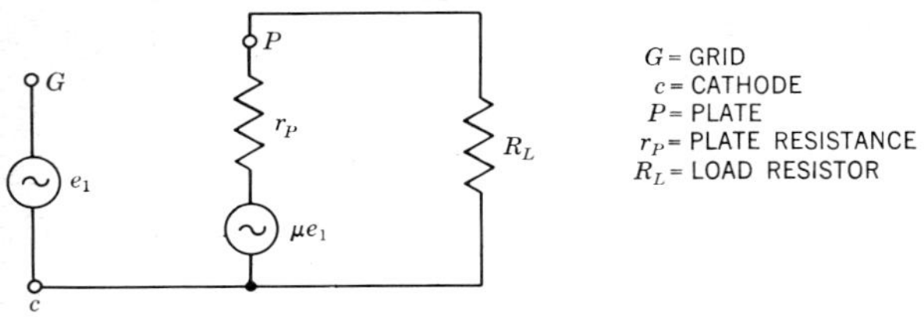

FIGURE 3.23
The equivalent circuit of a vacuum tube. The incoming signal e_1 appears in the plate circuit as a greater voltage μe_1.

3.23 shows quite plainly that the grid and plate circuits are isolated from each other and we do not have the same dependence between the impedances in each circuit that we have in a transistor.

QUESTIONS

3.1. What factors limit the power-handling ability of transistors? Describe each briefly.

3.2. How does the noise figure of transistors vary with frequency?

3.3. What can be done to help transistors achieve higher collector-dissipation ratings?

3.4. Describe briefly how the following features of a transistor affect its high-frequency response.

1. Base width
2. Emitter area
3. Collector doping level

3.5. In what ways can a transistor fail?

3.6. How are the surfaces of planar transistors protected against changes?

3.7. Define I_{CBO}, I_{EO}, V_{CBO}, and BV_{EBO}.

3.8. What differences exist between silicon and germanium in their use in transistors?

3.9. Why is I_{CO} important in transistor operation?

3.10. Why must the collector-dissipation rating of a transistor be reduced when the unit is employed above a certain temperature?

3.11. How are PNP and NPN transistors differentiated schematically? What other conventions are employed in drawing transistor symbols?

3.12. Draw the circuit of a grounded-base transistor amplifier, complete with dc biasing voltages and input and output terminals. Draw the vacuum-tube counterpart of this circuit.

3.13. Follow the same procedure as in Question 3.12 for a grounded-emitter transistor amplifier. Draw the vacuum-tube counterpart of this circuit.

3.14. Answer Question 3.13 for a grounded-collector transistor amplifier.

3.15. Which of the three types of amplifier is best suited for a high-input-impedance—low-output-impedance application? Which arrangement provides the best voltage and power gains? Would the same results be obtained if vacuum-tube amplifiers were employed? Explain.

3.16. Differentiate between the α and β values of a transistor.

3.17. Describe how the β value of a transistor may be determined from its characteristic curves. For your illustration, use Fig. 3.7.

3.18. How can you identify the various element leads of a transistor?

3.19. What characteristics are generally given for a transistor in the manufacturer's listings?

3.20. Why is there greater dependence between the input and output circuits of a transistor than of a vacuum tube?

3.21. Draw the equivalent circuit of a transistor.

3.22. What is the difference between h_{FE} and h_{fe}?

3.23 Define the term current-gain-bandwidth frequency f_T. What features of a transistor affect f_T?

3.24. In what three ways can the life expectancy of a transistor be shortened due to manufacturing difficulties? Briefly explain each.

3.25. Explain the following terms found on transistor specification sheets:

1. Base saturation voltage $V_{BE(\text{sat})}$
2. Forward-current transfer ratio
3. Collector saturation voltage $V_{CE(\text{sat})}$
4. Power-derating factor
5. Collector leakage, or cutoff, current

3.26. Explain the significance of the following terms as they relate to the switching ability of a transistor: Storage time; rise time; delay time.

Chapter Four

Transistor Design and Process Development

Thus far we have studied the structure of the atom and its significance to transistor operation. We discussed how to form a PN junction and then noted transistor characteristics, attempting to explain the electrical and physical relationships of these characteristics that make the transistor such a useful device.

This chapter will examine the development of transistor processing, from the point-contact transistor to the modern silicon epitaxial planar transistor.

The rapid advances made in transistor processing have not only led to the extremely efficient epitaxial planar transistor, but also to the practical production of field-effect transistors (Chap. 5) and the monolithic integrated circuit (Chap. 8). All these achievements have increased the applications of semiconductors by many orders of magnitude.

POINT-CONTACT TRANSISTORS

Point-contact transistors are no longer manufactured. However, from the standpoint of discovery and initial development, the point-contact transistor comes first, and it is of historical interest to examine briefly its mode of operation. As with the junction transistor, it is best to start with a point-contact diode.

Point-contact Diodes. A point-contact diode is shown in Fig. 4.1. One section consists of a small rectangular chip of germanium to which a controlled amount of N-type impurity is added, giving us N-type germanium. The other half of the diode consists of a fine phosphor-bronze catwhisker wire that presses against the center of the germanium chip.

An important step in the fabrication of this diode is the passage of a relatively large current from the catwhisker wire to the germanium chip. The purpose of this "forming" current is to produce a small area of P-type germanium in the regions surrounding the point of contact of the phosphor-bronze tip. The germanium diode now consists of P- and N-type germanium (Fig.

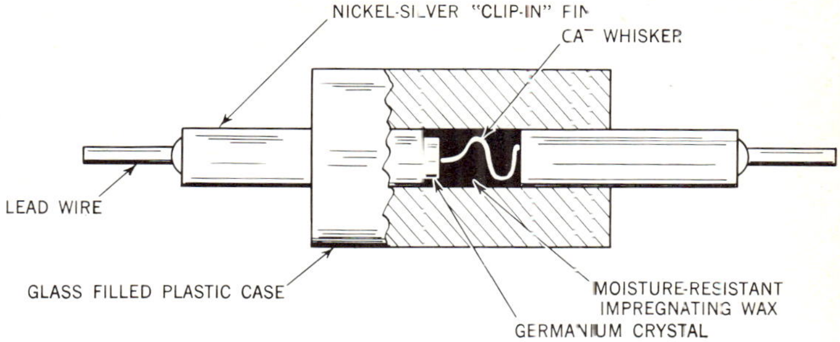

FIGURE 4.1
Cross-sectional view of internal structure of a germanium diode.

4.2), and the explanation of its operation follows exactly along the lines previously discussed for a PN junction diode.

Point-contact Transistor Operation. To form a point-contact transistor, two phosphor-bronze wires are mounted side by side as shown in Fig. 4.3. One wire forms the emitter of the transistor and the other wire, the collector. The third electrode, the base, is formed by depositing metal on the underside of the germanium chip. The final processing step is electrical forming where, as in the diode, relatively large surges of current are passed through wires to the base. This current serves to form small areas, or islands, of P-type germanium under each wire electrode. The area of each P section is extremely small, possibly no more than a few atomic layers thick.

The current flow in a point-contact transistor is similar to the current flow in a junction transistor. There is one interesting variation however; in a point-contact transistor, a change of the 1 milliamp (mA) in the emitter circuit produces a 2-to-3-mA change in the collector circuit. This is a significant departure from what occurs in junction transistors.

The reason for this effect is that on their journey to the collector, the

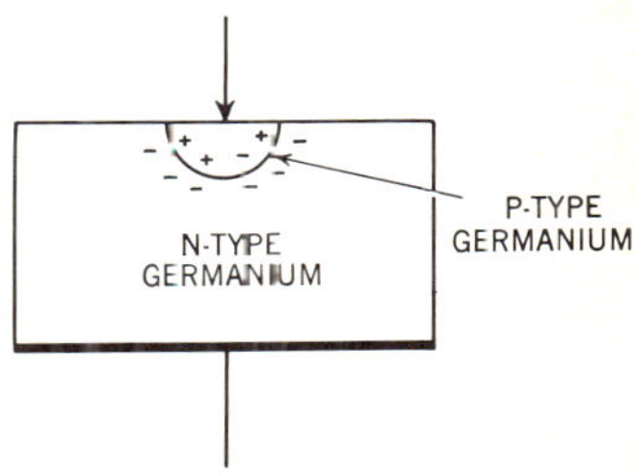

FIGURE 4.2
A point-contact germanium diode showing the approximate distribution of the N- and P-type germanium.

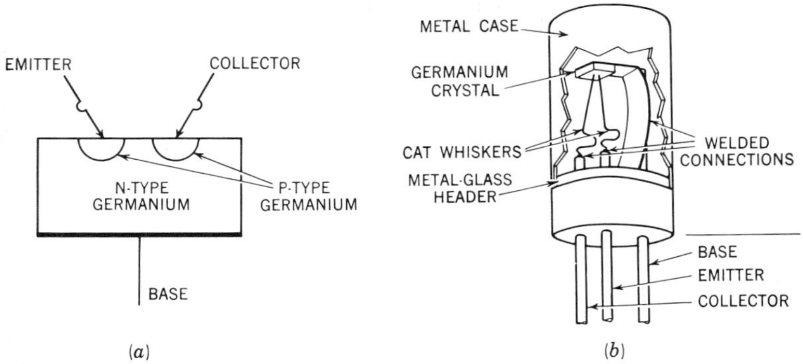

FIGURE 4.3
(*a*) A point-contact transistor. The emitter and collector wires are each held tightly against the germanium block. (*b*) The physical construction of a point-contact transistor.

emitter holes form a positive space charge that attracts electrons from other sections of the germanium crystal and causes them to add to the collector current. These additional electrons are confined to the collector circuit and travel in a path from the negative terminal of the battery through the collector and base sections of the transistor, out through the base lead, and back to the battery again. In effect, what the holes do is reduce the internal resistance of the collector circuit, permitting a greater current to flow for the same applied voltage. This is supported by the fact that whereas the internal resistance of the collector circuit in junction transistors is on the order of 500,000 ohms (Ω) or more, in the point-contact transistor it is typically about 20,000 Ω.

The reason for this difference in behavior of point-contact and junction transistors is primarily because of differences in construction. In the point-contact transistor, there is a large base area from which electrons may be drawn to enhance normal collector current. The holes traveling from emitter to collector serve to attract these excess electrons by their strong positive field. In a comparable PNP junction transistor, the base section is quite narrow and can supply only a limited number of electrons to the collector current. Hence, the same current-gain effect is not observed.

In spite of this, the junction transistor is superior to the point-contact transistor in nearly all respects. For example, a junction transistor can handle larger amounts of power, because each junction has a larger surface area than the corresponding catwhisker wire and its junction. Also, a junction transistor, operated in the common-emitter mode, provides higher current and voltage gains.

Besides being a more efficient device, the junction transistor can be mass-produced more cheaply than the point-contact transistor.

GROWN-JUNCTION TRANSISTORS

Historically, the grown-junction transistor was the first junction type manufactured. Today it is surpassed in performance by other transistor types.

The grown-junction transistor obtained its name from the fact that the junctions were produced by what is basically a growing process. A small crystal of silicon or germanium, called a seed, is dipped into molten silicon (or germanium) contained in a crucible. As the seed is withdrawn slowly, the liquid freezes onto the seed, growing a crystal. The original melt contains a suitable impurity, perhaps arsenic, for N type. After a sufficient amount of crystal has been grown on the seed with N-type impurity, a P-type impurity in minute amounts is added to the melt to produce a P-type silicon. The crystal is grown P-type for about one-thousandth of an inch, and then a second pellet, again N type, is dropped into the melt. An amount sufficient to overcome the P dopant is added and the crystal is returned to N type for the balance of the crystal-growing process (Fig. 4.4).

When the desired crystal has been grown, it is cut into wafers by a diamond saw, lapped to a smooth surface with an abrasive, cut up into little squares,

FIGURE 4.4
Manufacture of grown-junction NPN transistor from single crystal to final unit. (*Western Electric Engineer.*)

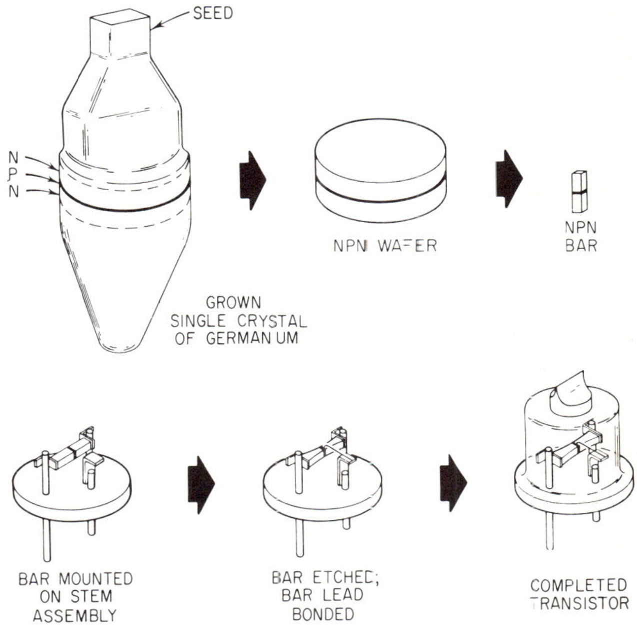

and chemically etched to remove surface damage. The end regions to which emitter and collector leads are attached are N type; the base is P type.

Another grown-junction processing technique, known as rate-growing, depends on the rate at which impurities will leave the molten semiconductor and concentrate in the growing crystal. Most impurities are more soluble in a melted, or liquid, semiconductor than in the solid substance itself. Thus, if a silicon seed is dipped into molten silicon, which has a certain impurity concentration, and then slowly extracted, the concentration of the impurity in the pulled silicon solid crystal will be less than the impurity concentration in the melt (i.e., molten silicon). The ratio of the impurity concentration in the solid to the impurity concentration in the liquid is known as the segregation coefficient. Furthermore, each dopant impurity has its own specific segregation coefficient for silicon and for germanium. Thus, if both donor and acceptor impurities are in the same semiconductor melt, they will segregate independently on solidifying, according to their segregation coefficient.

For example, suppose that a crystal is pulled from a melt containing antimony (N dopant) and gallium (P dopant). At the normal crystal-pulling temperature, both antimony and gallium are segregating from the melt. However, the segregation rate for antimony is higher at this temperature, and the crystal is more N type. Now, if the temperature is suddenly increased, decreasing the growth rate, the antimony segregation rate drops almost to zero, but the gallium is hardly affected by the temperature change. What has happened is that with the increase in temperature, the gallium predominates over the antimony and the crystal becomes P type. When the temperature is lowered to normal, the crystal reverts back to N type. This temperature cycling may be repeated at controlled intervals during crystal growth, forming as many junctions as cycles. This method of producing grown-junction transistors is more economical than the previous technique described.

ALLOY-JUNCTION TRANSISTORS

In the early 1960s, this process accounted for the largest number of junction transistors produced. It can produce a wide variety of transistor amplifiers (both low and moderately high frequency) as well as switching transistors in the low- and medium-speed ranges. The fabrication of alloy-junction transistors has been automated to such an extent that units for the entertainment field are sold for considerably less than a dollar.

The alloy-junction fabrication technique is quite different from the grown-junction technique. Whereas the latter is a batch-processing technique, the present alloy units are essentially individually made (although in large numbers at any one time). The starting point is a germanium chip about 0.080 in^2 and 0.003 to 0.005 in thick. The wafer from which this chip is obtained is grown similarly to that for a grown-junction transistor, except that the crystal

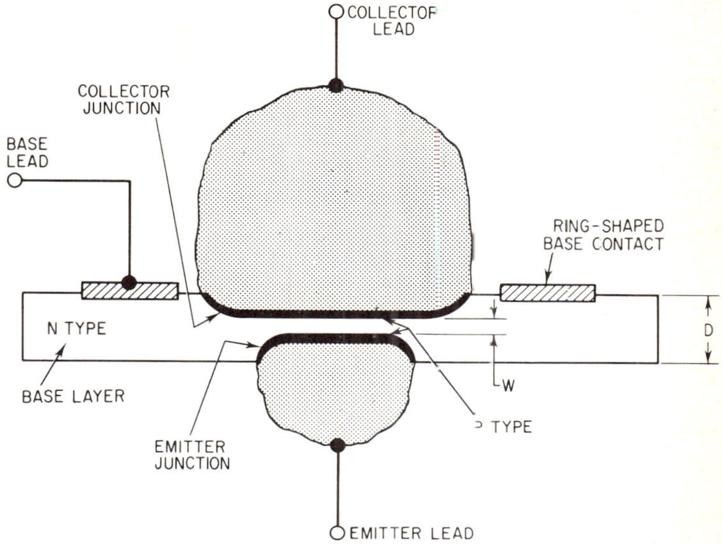

FIGURE 4.5
Cross section of an alloy-junction PNP transistor. The collector and emitter elements are grossly exaggerated in size. (*Western Electric Engineer.*)

is pulled with uniform doping using one type of impurity, either N or P. In this discussion, N-type germanium will be assumed. The crystal is sliced into a number of wafers and then diced into chips of the dimensions given above.

An impurity metal, usually indium, is then placed on opposite faces of the germanium chip and heated in special alloying furnaces. The alloying temperature is approximately 600°C, whereupon the indium alloys into the germanium until a saturated liquid solution of both materials is formed on both sides of the chip. On cooling, the dissolved germanium recrystallizes onto the undissolved germanium. Since it is freezing from a melt containing indium, the recrystallized germanium is highly doped to P type. We have thus a PNP transistor with the emitter and collector P type and the base N type (Fig. 4.5).

Connections to the emitter and collector are made by wires soldered to the alloy. The other ends of these wires are then spot-welded to leads that make contact to the circuit in which the transistor is placed. The base contact is usually made in the form of a ring that completely encircles the emitter. This permits a low-resistance connection to be achieved (Fig. 4.6).

If the chip is cut at a specific crystal orientation, the liquid alloy will

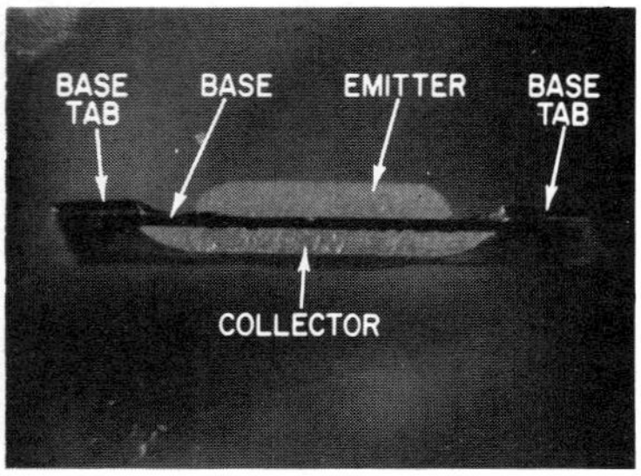

FIGURE 4.6
An alloy-junction transistor.

penetrate the germanium in a flat plane, resulting in plane-parallel junctions. These "flat" junctions influence electrical characteristics of the final device.

With the alloy method of construction, several things have been accomplished. First, the separation between collector and emitter regions is on the order of only 0.0005 in, which permits a significant reduction in transit time between these two elements. Second, the base resistance can be made low by the use of a relatively thick germanium wafer at all points except in the small section between emitter and collector. Also, the emitter and collector diameters, 0.010 and 0.015 in, respectively, are kept small, thereby reducing the various capacitances which these elements introduce.

The alloy method is feasible for both PNP and NPN transistors. For an NPN unit, a P-type germanium wafer would be used and a pentavalent element (such as phosphorus) would be substituted for the indium.

Alloy transistors have excellent low-frequency gain characteristics, switching characteristics, and high current ratings for power applications. They are still being produced in considerable numbers for these applications.

The Diffusion Process. In the sequence of transistor developments, the discovery of how to apply high-temperature-diffusion techniques to semiconductor-device fabrication represents a significant advance. As a matter of fact, the vast majority of modern commercial semiconductors are dependent on the diffusion process.

The diffusion process can be visualized as a mixing process on an atomic scale of two different sets of molecules through the random thermal motion of molecules and atoms. For example, if you place a drop of colored dye in a glass of water, you will find after several hours that the dye will have spread throughout the entire glass of water. The spreading was accomplished by the random motion of the molecules of the dye and of the water.

In a similar manner, gas or liquid will diffuse in solids, although at a much slower rate. For example, if dropped on a cake of ice, the colored dye mentioned above would take many years to completely diffuse through the ice.

All diffusion processes are accelerated by heat. In the formation of semiconductor PN junctions by diffusion, carefully controlled temperatures in the 1000°C range are employed. Let us now briefly discuss PN junction formation by diffusion.

Impurity atoms become mobile in a silicon crystal when it is heated to high temperatures. At these high temperatures, there will be a movement (diffusion) of impurities from the area of high impurity concentration at the crystal surface, to the area of low impurity concentration at the center of the crystal. It is always true that the flow, or diffusion, of impurities is from the region of high concentration to the region of low concentration. In semiconductor processing, the diffusion is halted at precise locations in the crystal by reducing the temperature. This is done by slowly pulling the quartz diffusion "boat," which holds the wafers, from the heat zone of the diffusion furnace by using a quartz pull rod (Fig. 4.7*a*).

The rate of the impurity diffusion is very sensitive to temperature, which has led to the development of modern diffusion furnaces capable of holding the required high diffusion temperatures to within one-fourth of a degree Centigrade (Fig. 4.7*b*). To a much lesser degree, the diffusion rate is a function of the impurity concentration of the crystals (wafers) being diffused and the crystal orientation and perfection. Crystal imperfections normally result in varying diffusion rates, and hence nonuniform junctions.

The sketch in Fig. 4.7*a* shows a cross section of a typical diffusion furnace indicating the source, diffusion, and carrier-gas sections. The impurity is carried to the wafers by an inert carrier gas, usually nitrogen. The impurity concentration of the diffused wafer is controlled by the gas-flow rate [the time the gas flows through the impurity source (deposition time)] and the time and temperature of the diffusion cycle.

FIGURE 4.7*a*
Cross section of a diffusion system.

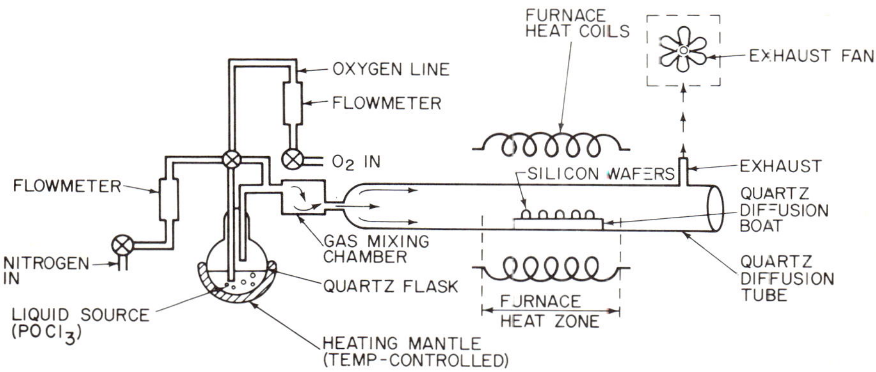

FIGURE 4.7*b*
A modern semiconductor diffusion furnace. (*Courtesy of Thermco.)*

Some of the more common N-type impurities are: phosphorus oxychloride ($POCL_3$), a liquid; phosphine (PH_3), a gas; and phosphorus pentoxide (P_2O_5), a solid. Common P-type diffusion sources are: boron nitride (BN), a solid; diborane (B_6H_6), a gas; and boron tribromide (BBr_3), a liquid.

The diffusion process does require very careful control. For example, in the fabrication of microwave transistors, the diffused junction depths are controlled within tenths of microns. In addition, the surface and junction impurity concentrations must be precisely controlled for optimum device-operating characteristics. The fabrication of monolithic integrated circuits requires a number of different impurity diffusions to form the various circuit components on the same wafer, which also requires precise control. We shall discuss integrated circuit processing in Chap. 8. Today, such extremely precise diffusion cycles are mathematically worked out with the aid of the computer.

For our purposes, however, it is sufficient to know that the following diffusion parameters must be controlled in order to obtain the desired device characteristics:

1. The diffusion times and temperatures
2. The carrier-gas-flow rates and oxygen content
3. The impurity source, which must be uncontaminated; and if the source is a liquid, its temperature must be controlled
4. The wafers being diffused, which must be clean and free of any crystal imperfections
5. The quartz diffusion tubes, pull rods, source containers, and wafer diffusion boats, which must all be kept clean and free of contamination.

THE DRIFT TRANSISTOR

The drift transistor combines diffusion and alloy techniques. From this combination, we achieve a significant improvement in frequency response over transistors formed by the alloy method. The starting point is a wafer of N-type germanium having a fairly high resistivity. Whenever a semiconductor is said to have a high resistivity, i.e., a high resistance per unit area, it is meant that the material contains only a small amount of impurity. In this case, it is N-type impurity. This N-type wafer of germanium is exposed under precisely controlled conditions to arsenic (or some other N-type impurity) at a high temperature. The N-type vapor impurity diffuses into the wafer, leaving the highest concentration at the surface.

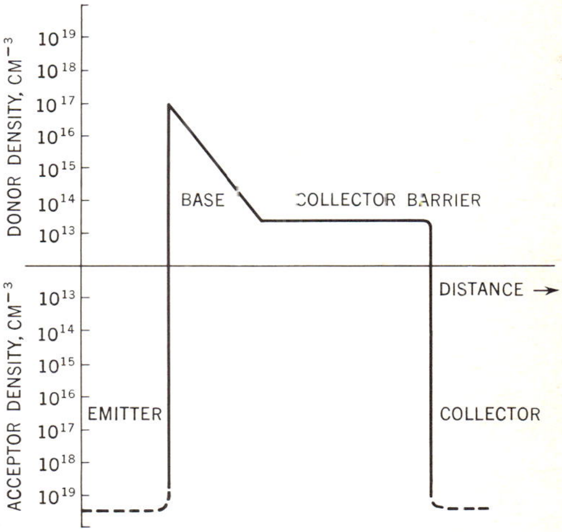

FIGURE 4.8
Impurity distribution for a PNP transistor. The section labeled "collector barrier" has, relatively speaking, very few impurities and can be considered as intrinsic germanium.

The "skin" of graded arsenic is removed from one side of the wafer and a P-type collector junction is alloyed into the germanium. The same P-type impurity is then alloyed into the other side of the germanium chip where the graded arsenic is still present. This forms the emitter junction. The result, as shown in Figs. 4.8 and 4.9, is a transistor in which there is a high density of impurity in the base end nearest the emitter, with a steady decrease until the germanium possesses very little impurity (i.e., reaches the original state of the wafer before the diffusion) somewhat before the center of the base region. From there to the collector, the germanium remains uniform. At the collector, the P-type condition appears.

A major difference, then, between the drift transistor and alloy unit resides in the varying impurity distribution that occurs in the base of the drift transistor. The importance of this can be better understood if we analyze in greater detail the movement of carriers from emitter to collector.

When a signal voltage (or any other voltage) is applied between emitter and base, carriers from the emitter are injected into the base. These carriers must travel across the thin base and arrive at the collector junction. In a PNP transistor, the carriers from the emitter are holes; in a NPN transistor, the carriers are electrons. In either case, the carriers travel across the base by a process of diffusion. This motion, brought on by the thermal energy that the holes or electrons possess, consists of movement in random paths as shown in Fig. 4.10*a*. The electrons (or holes) simply wander about aimlessly, colliding with each other or with the crystal atoms and move in all possible directions. Now, while the individual holes have a random motion, it is possible to obtain a flow of current across the base, because the holes have like charges and tend to move from a region of high concentration to a region of low concentration. This is shown in Fig. 4.10*b*. Furthermore, if the concentration gradient is doubled, thus doubling the number of particles, the flow (or flux) down the gradient is doubled. This basic principle of diffusion is known as Fick's first law. (Note that this law is the same as the one mentioned in our discussion of the diffusion of impurities in solids.)

FIGURE 4.9
Cross-sectional view of the drift-transistor structure. Note that the base tab is a circular ring.

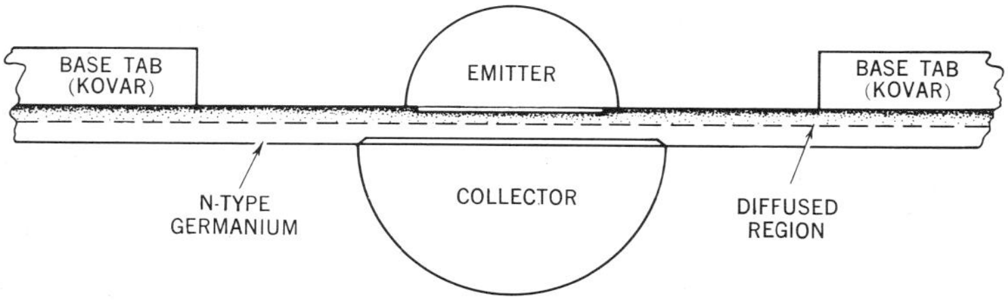

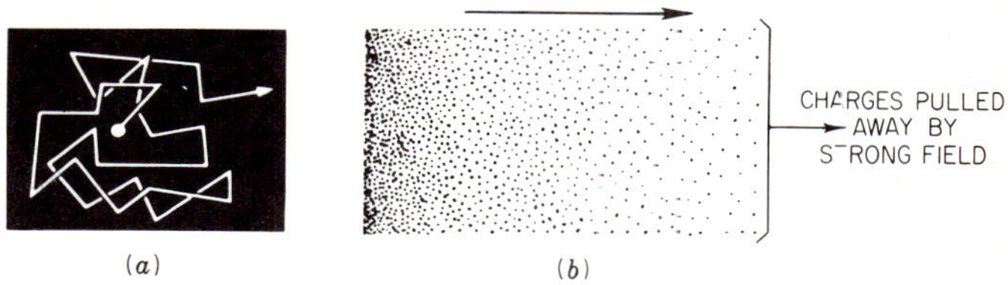

FIGURE 4.10
(*a*) Random motion of holes and electrons implies no direction for diffusion, but (*b*) the holes or electrons will diffuse from a region of high concentration to a region of low concentration. (*Bell Laboratories Record.*)

Now, if we apply a strong attractive electric field at the low-concentration end, the particles are constantly being pulled out of the base, thereby encouraging new particles to take their place. This will produce a continuous flow of particles from the region of high concentration to the region of low concentration.

In a PNP junction transistor, there is a high concentration of holes at the emitter end of the base created by the forward bias between emitter and base. These holes are injected by the emitter into the base. At the collector end of the base, there is a strong negative field due to the negative voltage applied to the collector that pulls in all the holes that reach this point. Thus, there is a steady current flow across the base region due to the diffusive action described above.

A small but measurable time is required for holes injected into the base by the emitter to reach the collector. Note that there are no electric fields within the base region. Whatever bias voltage is applied between emitter and base appears across the junction separating the sections. The same is true at the collector junction, where all the voltage applied between base and collector appears. Once the injected carriers in the base reach the collector (by diffusion), they travel extremely fast because electrical forces are present there.

Now if all the injected carriers required exactly the same travel time, the net effect would be simply to delay the output signal with respect to the input signal. In this random travel, however, not all the carriers take the same path and consequently the carriers (holes or electrons) corresponding to a particular part of the input signal do not all arrive at the collector at the same time. When the signal frequency is low, this minute difference of arrival can be ignored. As we increase the signal frequency, however, some of the late-arriving carriers begin to interfere with the carriers representing the next portion of the signal, with resultant disturbance and cancelation effects. At this point the amplitude of the output signal begins to fall off. The dispersive effect becomes more and more pronounced as the signal frequency rises, and the frequency response continues to decrease.

To minimize this effect, the base section should be made very narrow. As we make the base thinner, however, we steadily decrease the reverse voltage that can be applied between it and the collector section. Also, with exceedingly thin base layers, we not only run into manufacturing difficulties but also encounter irregularities in thickness or in impurity distribution that can result in the collector-to-emitter short-circuit effect of punch-through. (There is also another effect, namely, increased base resistance, which is detrimental to high-frequency operation. This will be considered at a later point.)

In the drift transistor, we achieve the same effect as a thin base region without actually reducing the region to the same extent. By providing a graded impurity distribution in the base region, we establish an electric field there. Holes injected into the base region by the emitter, in a PNP transistor, are accelerated toward the collector. Thus, where previously they traveled aimlessly, they are now more or less directed toward the collector, arriving much sooner than if the base region possessed uniform doping. As the travel time decreases, the maximum operating frequency rises.

The impurity distribution for a PNP drift transistor is shown in Fig. 4.8. The emitter is doped fairly heavily with an acceptor impurity (such as indium). At the junction of emitter and base, the base impurity is at its highest level; thereafter it decreases until the germanium impurity level is quite low. This is somewhere around the center of the base. From there to the collector, the germanium purity remains constant. Since this section of germanium contains very few impurities, it is frequently called an intrinsic-region material; i.e., it is almost pure germanium. The electric field in this intrinsic region is quite strong, and the holes travel through the region quite rapidly. (In an NPN transistor, the base carriers would be electrons.)

THE SURFACE-BARRIER TRANSISTOR

Although the surface-barrier transistor is obsolete, it is interesting to us because it was the first practical semiconductor device to make use of surface conditions. Today, an extremely important device, the field-effect transistor (FET), is also dependent on surface conditions. We shall describe the FET in the next chapter.

Surface-barrier devices are made through a combination of electrolytic etching and plating. A germanium chip is electrolytically etched with jet streams (Fig. 4.11) playing on opposite sides until the chip is etched down to the desired thickness. The jet-stream polarity is then reversed and the same solution acts to electroplate metal ions of a salt solution directly upon the freshly etched surface of the germanium. This forms emitter and collector electrodes. The process can produce very thin base widths with resulting high-frequency response. Unfortunately, the thin base width also limits the useful application of these transistors to low-power applications. Modern junc-

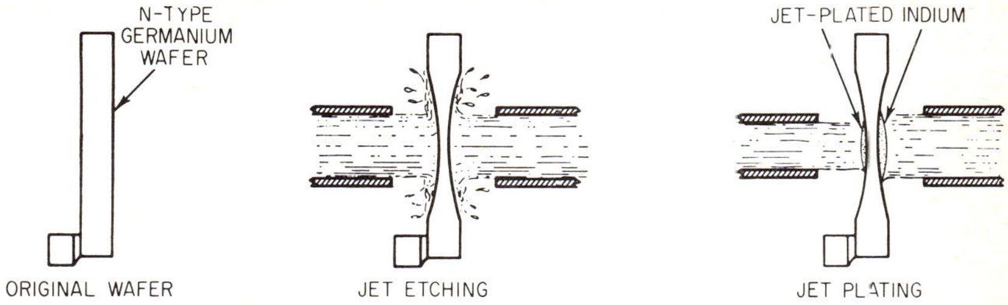

FIGURE 4.11
Jet-etching and plating the surface-barrier transistor.

tion transistors have better frequency performance, with much better power-handling capability.

Briefly, the operation of the surface-barrier transistor depends on special conditions that exist on semiconductor surfaces. It is known that energy levels exist on the surface of a semiconductor that are not found in the interior. These surface-energy levels are due to the semiconductors' dangling (unattached) surface bonds, which we mentioned in our discussion of noise in Chap. 3. The loose bonds interact with available molecules on the surface area and form new structures, with new energy states. In some cases (surface-barrier transistors), these new surface-energy states attract electrons from the crystal interior. A negative field is then created at the surface, which repels other semiconductor electrons toward the crystal interior. It thereby creates a layer right beneath the surface that is less N type than the rest of the crystal. The layer just under the surface can also be considered as more P type. It is known as an inversion layer; and in the device we are discussing it is called a surface barrier.

Now, if a metal electrode is brought in contact with the etched germanium chip (Fig. 4.12) and a negative potential is applied, the electrode will further repel the interior electrons away from the surface and cause the surface barrier to become wider. If the electrode is made positive, the interior electrons will be attracted and the barrier width will be reduced. Hence, the current flow between a surface electrode and the crystal interior can be easily controlled.

To form a transistor with the N germanium, we require an appropriate distribution of holes that will travel from emitter to collector as they do in a comparable PNP junction transistor. In the surface-barrier transistor, it is found that a population of holes exists just under the germanium surface. These holes arise from the valence electrons that are thermally excited enough to leave their atoms and move into some of the surface-energy levels intermediate between the conduction band and the valence band. The electrons come from the atoms located near the surface; and for every such electron departure, a hole is created. This action is confined to the atomic layer just

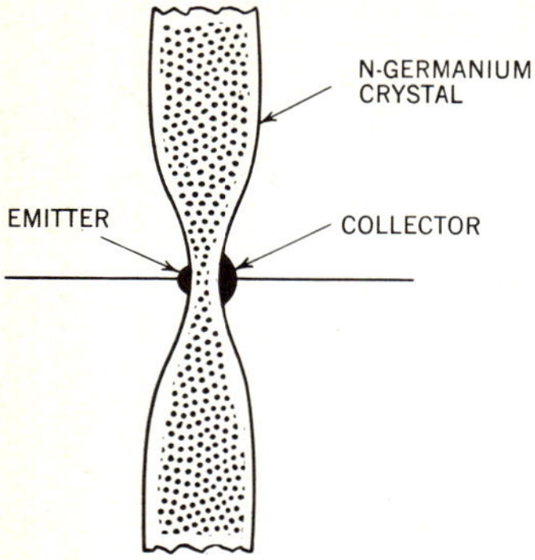

FIGURE 4.12
Schematic cross section of a surface-barrier transistor.

below the surface; the rest of the germanium interior produces relatively few holes. Some metal contacts produce a denser hole population under the surface of the germanium than others. The most useful metals for this purpose are indium and zinc.

In review, then, we see that the surface-barrier transistor owes all its characteristics to the special conditions that exist at the surface of a crystalline structure. The strong electron field at the surface forces free electrons to remain in the interior. Also, because of the presence of intermediate energy levels at the surface, holes are found concentrated just below the surface. When a metal contact to the crystal is made positive, it repels these holes through the barrier. This would be the emitter electrode. The other electrode, the collector, is reverse-biased (i.e., biased negatively), and holes coming within its field after passage through the germanium body will be drawn to the surface.

The surface-barrier transistor thus consists of a germanium crystal forming the base plus two metal electrodes, on opposite faces of the crystal, that serve as the emitter and collector electrodes. A positive emitter will drive the holes toward the collector, but at the same time it will attract the interior electrons. For efficient transistor operation, the electron current should be reduced as much as possible, since only the hole current is desired at the collector. This was achieved by bringing the collector electrode within 0.0002 in of the emitter. The negative charge on the collector drives the germanium free electrons away from the emitter, and at the same time presents a greater attractive force for the holes.

THE MESA TRANSISTOR

The mesa transistor represents still another approach to the fabrication of high-frequency transistors. Its construction is still popular because it lends itself to relatively simple and efficient fabrication techniques.

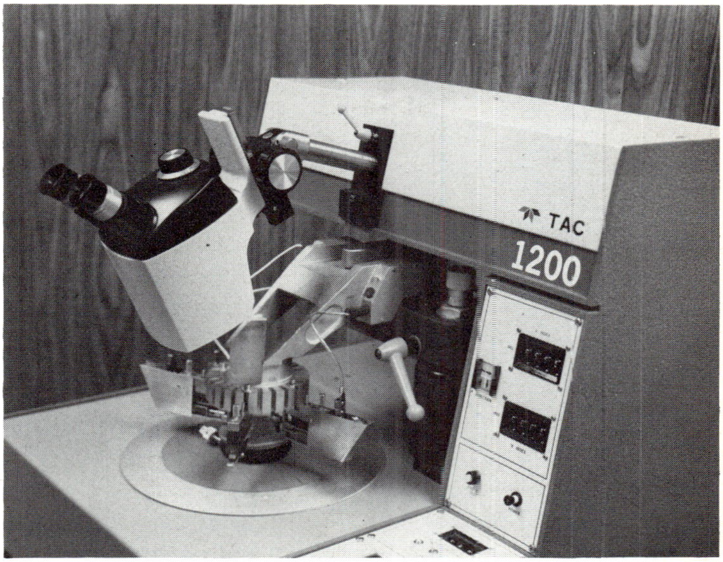

FIGURE 4.13
An automatic wafer prober. Automatic indexing with four-place accuracy is dialed into this wafer prober. Operator controls facilitates rapid orientation of wafer. (*Courtesy Transistor Automation Corp.*)

The mesa transistor was the first to use vacuum evaporation techniques for the formation of the metal contacts to the emitter and base sections. This technique together with new photoresist methods (which we will discuss presently) permitted for the first time the processing of several hundred transistors on the same silicon wafer and in turn led to the development of automatic wafer-probing machines (Fig. 4.13). These machines are capable of rapidly probing each transistor on the silicon wafer for dc characteristics and "inking out" the rejects. After the automatic probe test, the wafers are scribed and "cracked" into individual transistor chips. The inked reject chips are then sorted out and the good chips are assembled into transistor packages. The fact that hundreds of potentially good transistors could be fabricated on a single wafer and automatically tested provided a significant cost saving in device fabrication.

As with any other transistor, silicon or germanium may be employed. Since silicon is now more widely employed, it will be discussed here. Let us now briefly discuss three mesa-transistor fabrication methods: the diffused-base mesa, the diffused-emitter-base mesa, and the alloy-diffused mesa.

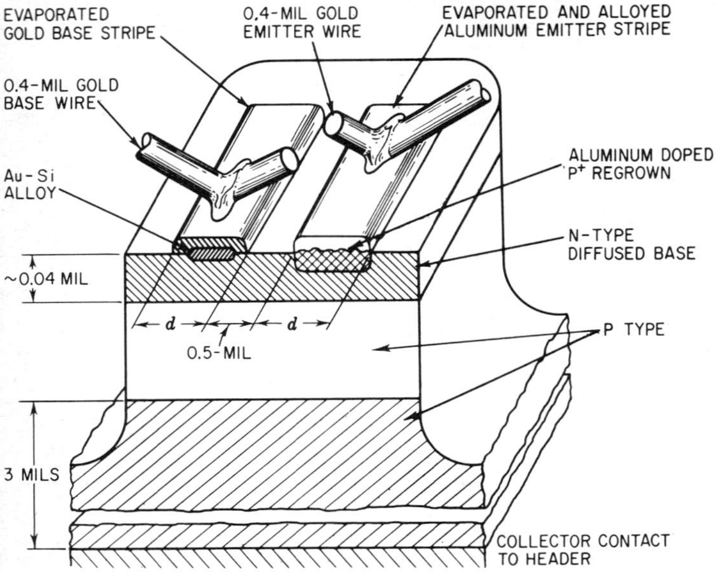

FIGURE 4.14
A mesa transistor.

Diffused-base Mesa. The first junction in the diffused-base mesa is formed by a gaseous diffusion of an impurity opposite in type to that of the substrate wafer. Using a PNP transistor as a model, we would diffuse an N-type base region on one side of a P-type wafer. Modern wafer diameters have reached 3 in but most production wafers have diameters of $1\frac{1}{2}$ to 2 in. The wafer thickness is between 0.008 and 0.010 in. The mesa base depth is in the order of 0.00004 in (see Fig. 4.14). The emitter junction is formed on the base layer by the vacuum evaporation of a P-type impurity, such as aluminum, through a metal precision mask. The emitter stripe evaporation is usually done by the cross-evaporation method (Fig. 4.15), so that the adjacent base contact can be evaporated in the same cycle. The base contact is usually gold with a small amount of N-type impurity. Cross evaporation permits the emitter and base stripe to be very close together, which is necessary for improved high-frequency performance. After evaporation, the wafer is given a short alloying cycle that forms very shallow junctions in the base areas. Finally, through a selective etching and masking process, a mesa is formed, with the collector flaring out at the bottom so that it covers a greater area at the bottom than the area of contact between the collector and base elements. By developing this shape, a smaller collector-base capacitance is formed. This is shown in Fig. 4.14.

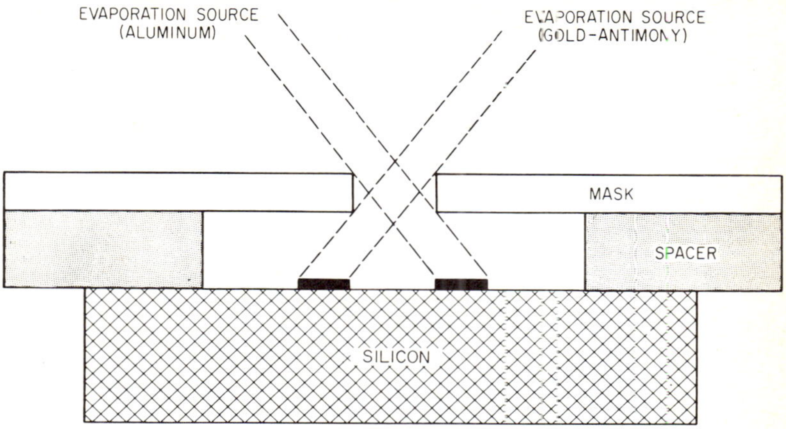

FIGURE 4.15

Diffused-emitter-base Mesa. The diffused-emitter-base process is similar to the one just described, with the exception that the emitter junction is formed by diffusion instead of evaporation and alloying. The base layer is formed by diffusing a P-type impurity into an N-type wafer of silicon. Then a thermal oxide is grown over the entire wafer surface. We will describe the thermal oxidation and photoresist processes when we discuss planar technology later in this chapter. In brief, thermal oxidation provides a uniform layer of silicon dioxide over the wafer surface that cannot be penetrated by the common impurity dopants but can be selectively etched (opened) for impurity introduction onto the underlying silicon in desired locations. In this case, an emitter area is opened in the oxide using photoresist techniques and an N-type emitter is diffused into the silicon through the oxide opening. The mesa is again formed by etching techniques, just as in the diffused-base mesa.

Alloy-diffused Mesa. Although this type of mesa is no longer fabricated, it is interesting from a processing viewpoint, since it utilizes two methods of junction formation during a single high-temperature alloy-diffusion cycle.

The base of a PNP transistor is formed by an impurity that diffuses out from two N-type-doped lead preforms. These two preforms are alloyed into P-type silicon. During the high-temperature cycle, the N-type impurities rapidly diffuse out of the liquid-solid interface and into the silicon, joining together to form the base of the PNP transistor (Fig. 4.16). One of the preforms, the emitter, also contains a high concentration of a P-type dopant (usually aluminum), and as the wafer cools coming out of the furnace, the P-type aluminum segregates from the melt and forms an alloyed emitter junction (Fig. 4.16). This happens because aluminum has a high segregation coeffi-

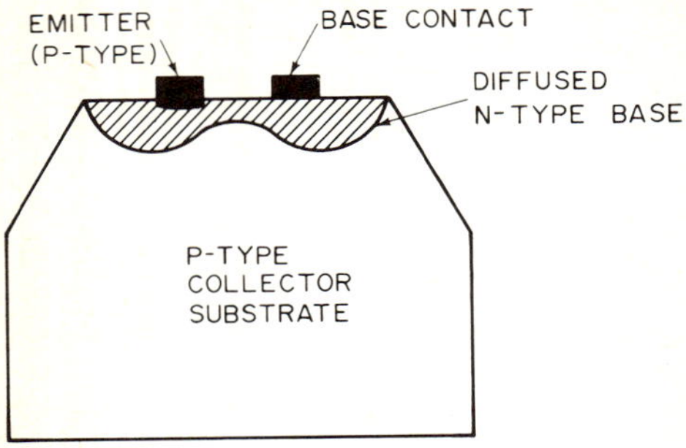

FIGURE 4.16
Alloy-diffused mesa.

cient. Recall that the segregation coefficient is the ratio of the concentration of impurity in the solid to the concentration of that impurity in the liquid. In this case, because of aluminum's high segregation coefficient, there is enough aluminum in the solid to form the P-type emitter.

The reason that this type of mesa is no longer produced is primarily because of the large chip areas required and the development of more efficient processing procedures.

Through improved processing techniques, the modern mesa transistor offers excellent heat dissipation and good high frequency and dc characteristics. It is used as signal and power amplifiers, oscillators, and switching transistors. The processes developed for mesa fabrication have also led to the development of the silicon epitaxial planar transistor. The epitaxial planar transistor comes close to supplying the best in dc characteristics, power dissipation, operating temperature, ultrahigh-frequency applications, and reliability.

We shall now describe the epitaxial transistor and then take a close look at the planar construction that has led to so much in modern semiconductor technology.

EPITAXIAL TRANSISTORS

In 1960, a manufacturing technique was developed for the fabrication of high-frequency transistors, particularly those to be employed for switching operations. Developed initially for the mesa transistor, this process can also be applied to other transistor constructions.

In the epitaxial transistor, the collector consists of two regions instead of one (see Fig. 4.17). The bottom (larger) portion of the collector is heavily doped to provide a very low resistivity. This is shown in Fig. 4.17 as P^{++}, indicating that there is a relatively high concentration of acceptor atoms. Over this is deposited a thin film, 0.1 mil thick, of very lightly doped semiconductor material. The thin film combines homogeneously with the crystalline structure below it (called the substrate), so that there is no discontinuity or break in the crystal structure. From this point, the standard techniques for fabricating

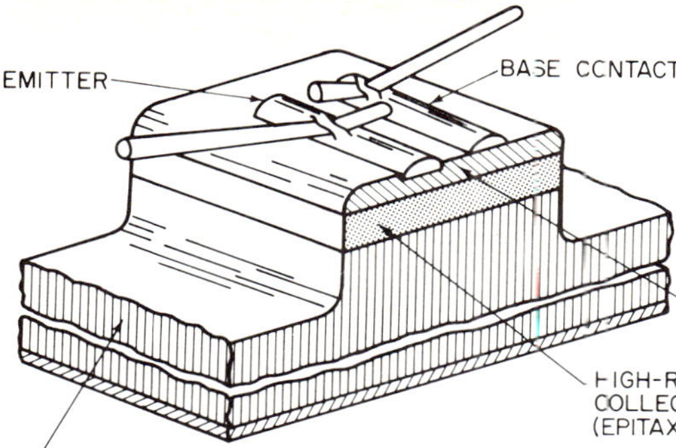

FIGURE 4.17
An epitaxial mesa transistor.

the transistor are used. The main advantage of the epitaxial process is that it optimizes the design of the collector region with resulting improved device characteristics, because a very thin, high-resistivity epitaxial collector eliminates certain design compromises such as those associated with obtaining high collector-base breakdown (BV_{CBO}) and low collector capacitance, with good collector saturation voltages ($V_{CE(\mathrm{sat})}$) and storage times (t_s). We discussed these device characteristics in Chap. 3. The thickness of the epitaxial layer is precisely controlled so that it is just thick enough to support the depletion region of the reverse-biased collector-base junction. If the epitaxial layer is made so thin that the depletion region spreads into the low-resistivity substrate, the collector-base junction will break down prematurely. At the same time, however, the epitaxial layer must be kept as thin as possible so that the total collector series resistance is minimized for obtaining low $V_{CE(\mathrm{sat})}$ values. The resistance of the heavily doped substrate does not add significantly to the collector series resistance. This combination of high collector-base breakdown voltages with low collector voltage saturation values was unobtainable with a uniformly doped silicon wafer.

Also, as mentioned in Chap. 3, the collector stored charge is reduced, since the epitaxial layer thickness is equal to the effective minority-carrier diffusion length, which is much less than the diffusion length related to actual lifetime in the epitaxial region. This is true because of the high recombination rate at the heavily doped substrate-epitaxial interface.

THE PLANAR PROCESS

The planar process was developed in 1960 at the Fairchild Semiconductor Corporation. It has properly been the most important semiconductor processing innovation since the invention of the transistor itself in 1948. In

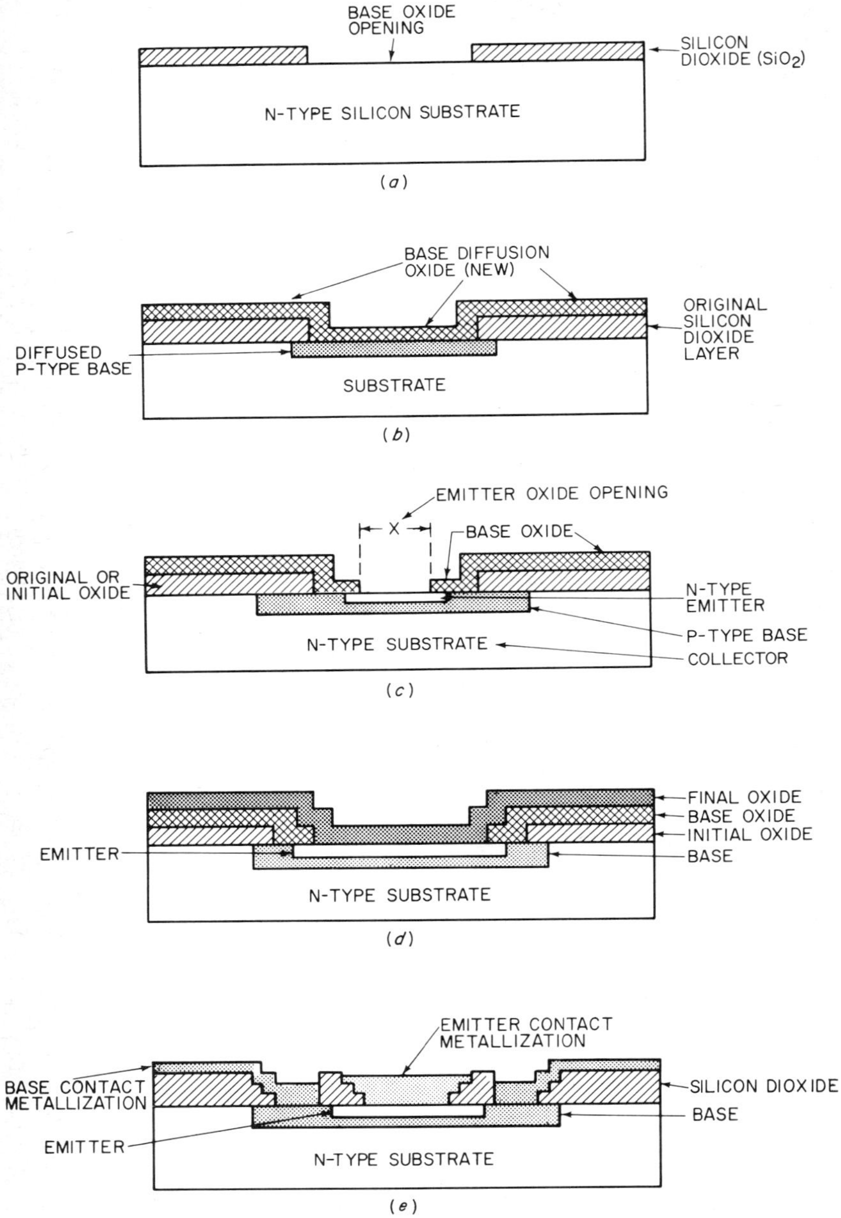

FIGURE 4.18
Planar-transistor formation.

Chap. 5 we shall learn how improved semiconductor surfaces, provided by the planar process, were the key to the development of the first practical field-effect (unipolar) transistors. In addition, the planar process has not only improved the electrical characteristics and reliability of the bipolar transistor, which we have been discussing, but it has also made possible the development of the monolithic integrated circuit. In the planar method of fabricating transistors, the base, emitter, and collector junctions are developed using a patterned oxide diffusion mask.

To start, an N-type silicon wafer, or substrate, is passed through an oven and a silicon dioxide (SiO_2) layer is formed over the entire surface. See Fig. 4.18*a*. This N-type substrate represents the collector. Into this substrate, a P-type base must be formed (to obtain, at the end, an NPN transistor).

To achieve this next step, the entire silicon dioxide coating is covered with a thin, uniform coating of photosensitive material. A mask with a small opening at its center is then positioned above the wafer. Ultraviolet light is now focused onto the mask, with the light reaching the photoresist only through the central opening in the mask.

The exposed section of the resist is then chemically removed (Fig. 4.18*a*) and a P-type dopant, usually boron, is diffused into the N-type substrate. This becomes the base (Fig. 4.18*b*). At the end of the base-diffusion cycle, a new layer of silicon dioxide is deposited, covering the open oxide base pattern. Then an emitter pattern is defined over the newly formed base oxide, again using photoresist techniques. The emitter pattern is etched through the oxide and an N-type dopant, usually phosphorus, is diffused into the P-type base (Fig. 4.18*c*). A final oxide is then grown over the entire wafer, and contact openings are defined and etched through the oxide over the emitter and base areas. A metal (aluminum) is then evaporated over the entire wafer surface and defined and etched using the same photoresist technique but with a different etching solution (Fig. 4.18*d*).

To form an epitaxial planar transistor, the N-type substrate first has a thin, high-resistivity layer of N-type material deposited over it. Then the foregoing steps are followed to produce the complete transistor. See Fig. 4.19.

FIGURE 4.19
Epitaxial planar transistor.

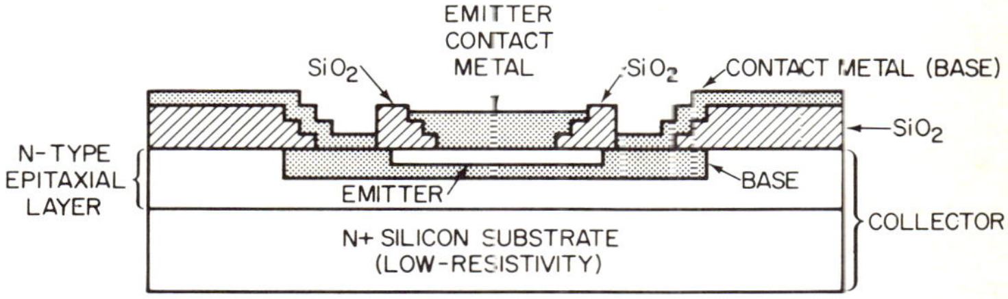

THE ANNULAR TRANSISTOR

There is a phenomenon associated with planar transistor surfaces known as "channeling," which can degrade device operating characteristics or even render the device useless. It is important that we understand this phenomenon and the processing innovation, the annular transistor, developed to correct it.

Channeling occurs when a P-type silicon surface becomes N type or when an N-type surface acquires a P-type charge. When this occurs, a PN junction is formed at the silicon surface.

Channels can be formed by two mechanisms. The first mechanism is initiated during the high-temperature thermal oxidation process (when silicon dioxide is being formed across the transistor surface), and is a result of the redistribution of the impurities near the silicon surface. For example, if an aluminum-doped (P-type) silicon wafer is thermally oxidized, the aluminum tends to migrate, or outdiffuse, and accumulate in the oxide. This leaves an aluminum-depleted region near the oxide-silicon interface, in the silicon, resulting in an N-type layer, or channel, at the surface (Fig. 4.20).

A different effect will occur in a phosphorus-doped (N-type) silicon wafer during thermal oxidation. In this case, the N-type impurity is "rejected" by the oxide and accumulates at the silicon surface. Thus, channels are less likely to form in an N-type-doped silicon wafer, during thermal oxidation than in P-type silicon.

The second channel-forming mechanism is a result of ionic contamination on the surface of the oxidized wafer or, for that matter, on a completed planar transistor. If the ambient conditions are not optimum (i.e., high humidity and/or high temperature), the oxide surface atoms can become mobile and will move under the influence of an electric field (Fig. 4.21). A reverse-biased junction will tend to move the negative ions to the N surface and positive ions to the P surface. The negative ions on the N surface repel electrons toward the center of the silicon wafer, depleting the surface; and if the surface charge is large enough, this will create a P-type channel. The same considerations can be applied to the P side; but now, due to the boron depletion caused in the oxidation cycle, a smaller surface charge is required to form a channel.

FIGURE 4.20
Cross section of PNP planar transistor with N-collector channel.

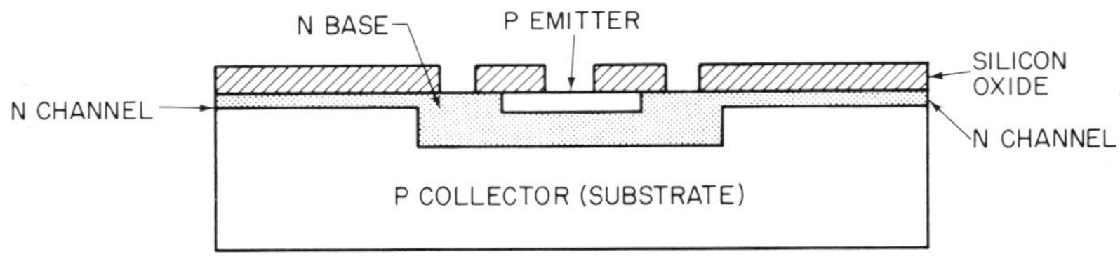

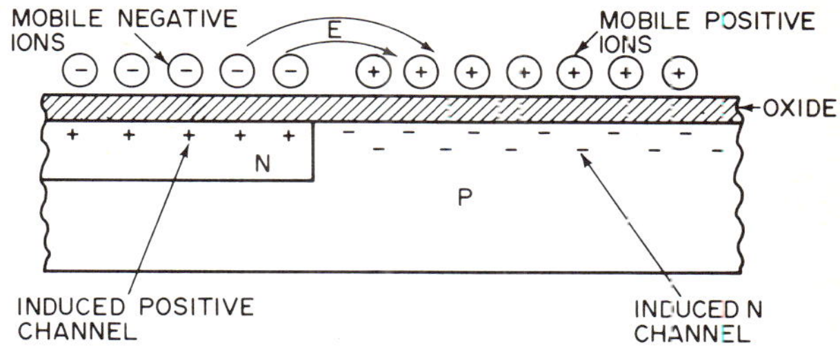

FIGURE 4.21
Channels formed by mobile surface ions under influence of an electric field.

Channels are most prevalent on the collector surfaces of PNP planar transistors (Fig. 4.22). We shall therefore discuss this type of channel. Keep in mind, however, that channels can be formed on the base surface of NPN devices and also on the collector surface if high-resistivity N-type silicon is used. A channel on a collector or base surface can result in a catastrophic increase in leakage current.

The Collector Channel. It can be seen from Fig. 4.22 that the channel increases the area of the base-collector junction. Now, if the collector-base junction is reverse-biased, current will flow through the channel to the base (metal) contact. This results in a lateral voltage along the channel. The voltage at point X (and X') is the full value of the collector-base voltage, but it diminishes to zero as we go along the channel, at points Y and Y' in Fig. 4.22. To the right of point Y (and the left of point Y') there is no voltage across the channel PN junction, and this section of the channel does not contribute any reverse-leakage current. As the collector-base voltage

FIGURE 4.22
Channel on PNP planar transistor (cross section used for channel explanation).

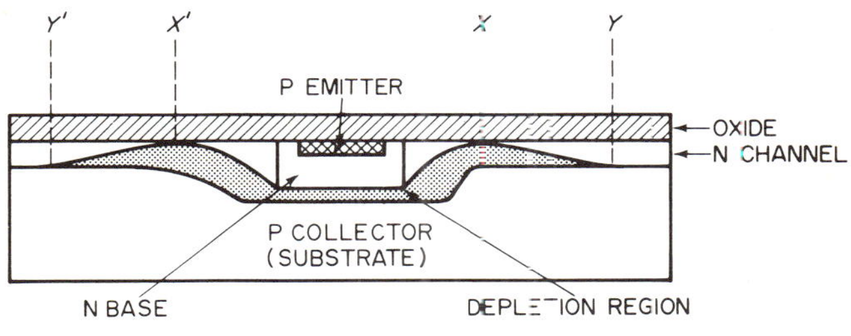

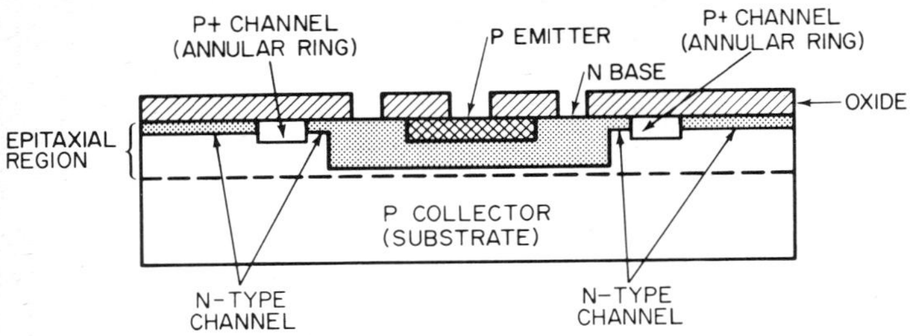

FIGURE 4.23
Annular ring for channel prevention. Cross section of planar annular transistor.

is increased, point Y will extend to the right (and point Y' to the left), increasing the channel current-carrying area.

Channel Control. Suppose that a P-doped wafer has a very high surface concentration, equal to the solid solubility of the P-type dopant. The term "solid solubility" refers to the maximum concentration of that element (boron, in this case) that can be diffused into the silicon at a specified temperature. Now, if such a wafer were thermally oxidized, it is probable that after the oxidation cycle there will be sufficient concentration of P-type dopant left at the surface to inhibit channel formation. This proved to be true, since a channel is formed by a lack of sufficient P-type (in this case) dopant at the silicon surface, and led to the solution of the channel problem.

An annulus, or ring, of the required type of dopant, N or P, is diffused into the transistor base between the emitter and collector, and also around the outside edge of the base area as shown in Fig. 4.23. A collector channel, such as the one we described, can now extend only as far as the annular ring. Since this ring is placed close to the base-collector junction, any channel formed results in a small area increase and a subsequent negligible increase in leakage current. Annular transistors, such as the one shown in Fig. 4.24,

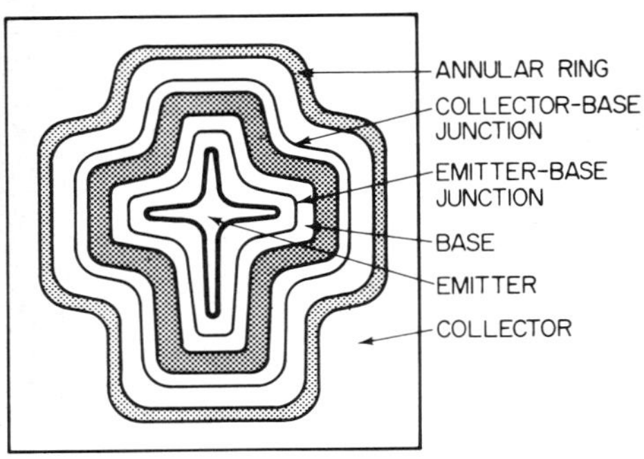

FIGURE 4.24
Top view of annular transistor. Motorola "star" transistor.

have improved power-dissipation ratings. This is true because for a certain bias on the emitter-base junction, as collector voltage is increased the collector current should increase gradually until it reaches avalanche breakdown. Avalanche in itself will not harm a transistor if power dissipation is limited.

THE PLANAR PROCESS

Since the majority of semiconductor devices today are being fabricated using the planar process, it would be instructive to examine in detail the individual steps involved in planar-device fabrication. The discussion to follow is an extended elaboration of the brief description given on page 103.

The planar process can be divided into the following steps:

1. Silicon crystal growth and epitaxial deposition
2. Photoresist
3. Diffusion (page 90)
4. Contact metallization
5. Device packaging
6. Final test

Figure 4.25 is a typical planar-transistor flow chart indicating the sequence of processing, control checks, and appropriate process specifications throughout its production cycle.

Starting Material. In order to obtain the substrate material on which to grow an epitaxial layer, we begin with a specific amount of very pure (uncontaminated) silicon, called a charge. The most popular crystal-forming method involves melting a charge of silicon doped with the desired amount and type of impurity in a high-purity quartz crucible set inside a graphite susceptor. This is confined inside a quartz cylinder containing an inert atmosphere. R-f-induction heating coils surround the quartz cylinder; and when the power is turned on, the charge melts. The temperature is then stabilized just above the melting point of silicon.

A small special piece of silicon known as a seed is then "dipped" into the melt. The seed has a specific crystal orientation, which means that the silicon atoms have a definite arrangement, in specific planes. There are several possible planes (see references). The important point is that the pulled silicon crystal will have the same atomic arrangement (orientation) as the seed. The seed is slowly rotated and pulled from the melt, "growing" as it is pulled (Fig. 4.26).

A typical fully grown crystal is 6 to 10 in long and $1\frac{1}{2}$ to 2 in. in diameter.

The crystal has a flat surface ground on one side, for reference during device-processing. The wafers, or slices, are then cut from the crystal, using

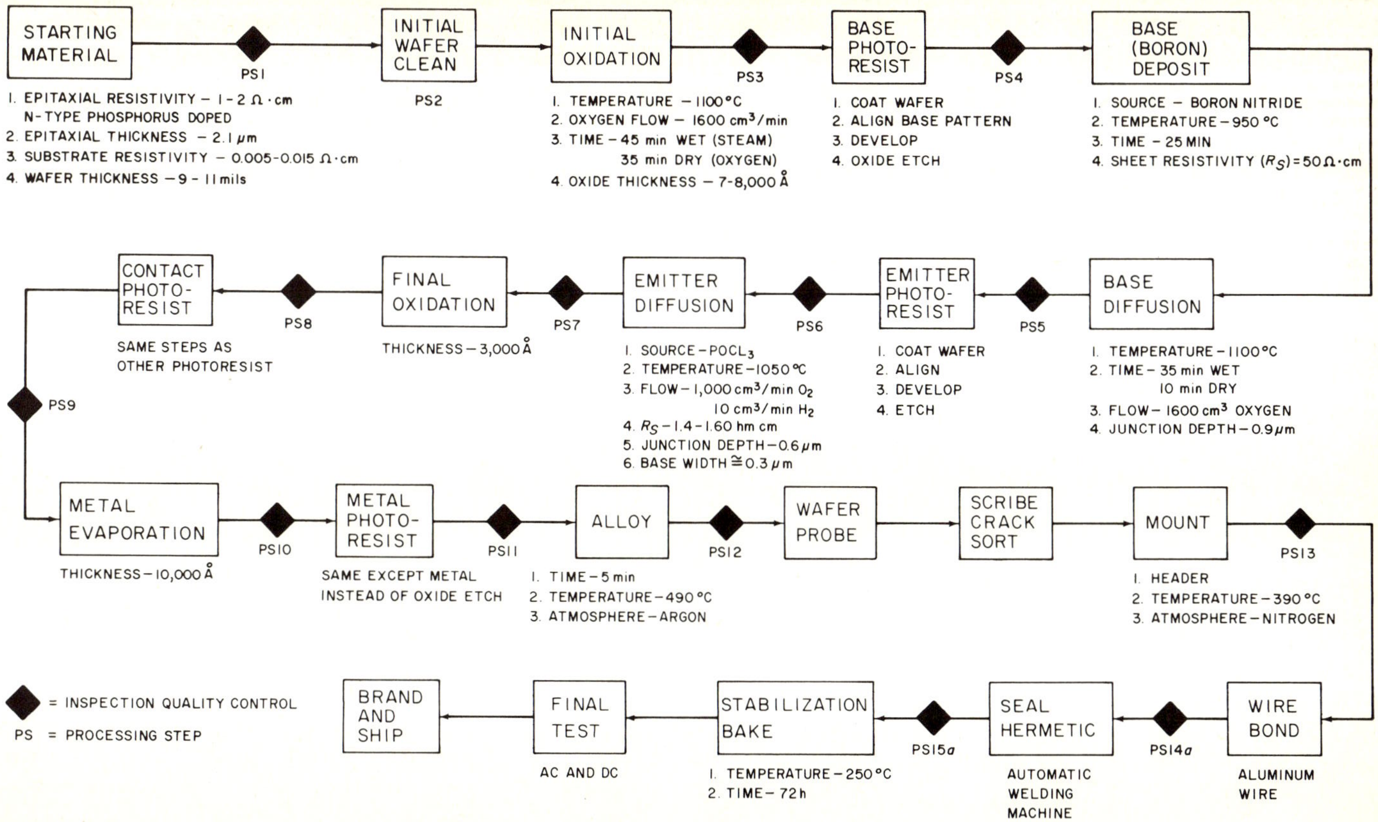

FIGURE 4.25
Planar-process flow chart (microwave transistor).

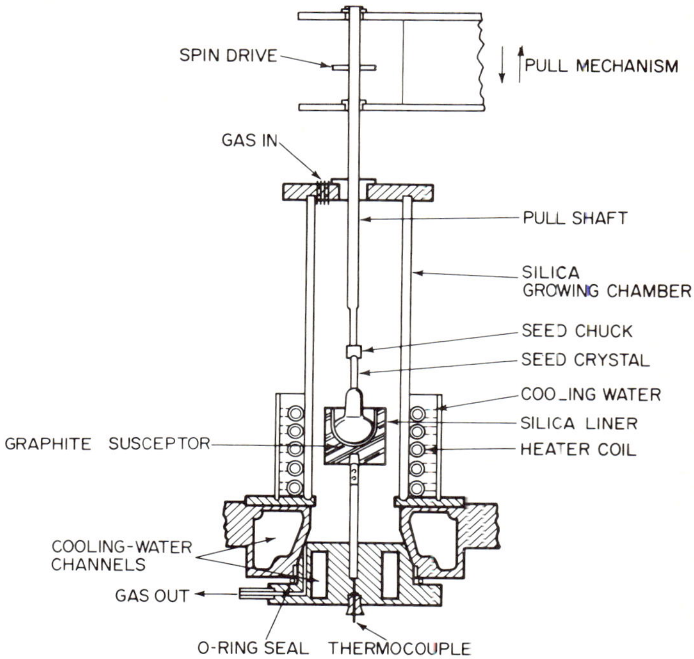

FIGURE 4.26
Apparatus for the growth of a silicon crystal.

a special diamond saw, and are normally 0.010 to 0.015 in thick. The wafers are lapped with a coarse abrasive and then polished with a very fine powder. Both lapping and polishing are done by machine, and many wafers can be lapped, or polished, at one time. The wafers are then chemically etched to remove all subsurface damage that may have occurred during the wafer-slicing operation. The etched wafers are finally ready for epitaxial deposition.

Epitaxial Growth. The most common system for epitaxial growth involves the hydrogen reduction of silicon tetrachloride ($SiCL_4$). The reaction takes place at approximately 1200°C.

In a typical epitaxial deposition system (Fig. 4.27), the etched wafers are placed in a reaction chamber and heated to between 1100 to 1200°C. The chamber is then flushed with nitrogen to remove any residual hydrogen. Hydrogen chloride gas is then introduced into the tube, and this gas etches any oxide or surface damage that may still be on the wafer surfaces.

The epitaxial layer is then deposited, or grown, by using hydrogen as the carrier gas for the silicon tetrachloride.

Silicon tetrachloride is used as the silicon source for a number of reasons: it is inexpensive and nontoxic; and the reaction that takes place forming silicon

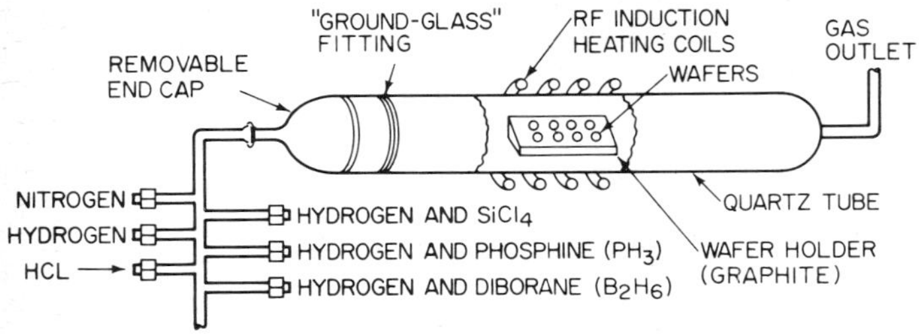

FIGURE 4.27
Epitaxial reaction chamber.

from silicon tetrachloride takes place only on surfaces, not while it is the gaseous state. In addition, conditions in the reaction chamber can be controlled so that the reaction will occur only at the silicon wafer surface and not on the walls of the quarter chamber.

The resistivity of the forming epitaxial film is controlled by introducing precise amounts of either phosphine (N-type) or diborane (P-type) gas into the silicon tetrachloride hydrogen gas stream. Phosphine and diborane have similar decomposition reactions to that of silicon tetrachloride. The result of this is that the phosphorus and boron atoms are uniformly incorporated throughout the growing epitaxial film.

In order to obtain the desired film thickness and resistivity uniformly across the wafers, conditions inside the reaction chamber must be closely controlled. This has been simplified by the development of epitaxial reactor systems, such as the one shown in Fig. 4.28.

Recall that in diffusion the impurity concentration in a silicon wafer is changed by first depositing a high concentration of the desired impurity on the wafer surface. Then, by controlling the temperature and time of the diffusion cycle, we obtain the desired impurity distribution within the wafer. In the process employed here, the impurity to be introduced onto the wafer exists as a gas rather than as a solid, and the impurity atoms wander about the wafer surface of the growing epitaxial film until they find a spot that "fits," after which they become part of that film.

It should be noted that the interface junction of the low-resistivity epitaxial film and the higher resistivity silicon substrate is not a sharp, or step, junction. This is because of the outdiffusion of substrate impurities during the deposition cycle. Although this impurity migration and the resultant slightly nonuniform junction is not ideal, it does not seriously affect the many advantages of epitaxial silicon.

Thermal Oxidation. After epitaxial deposition, the wafers are chemically cleaned and a thermal oxide is grown. In the oxidation process, the clean

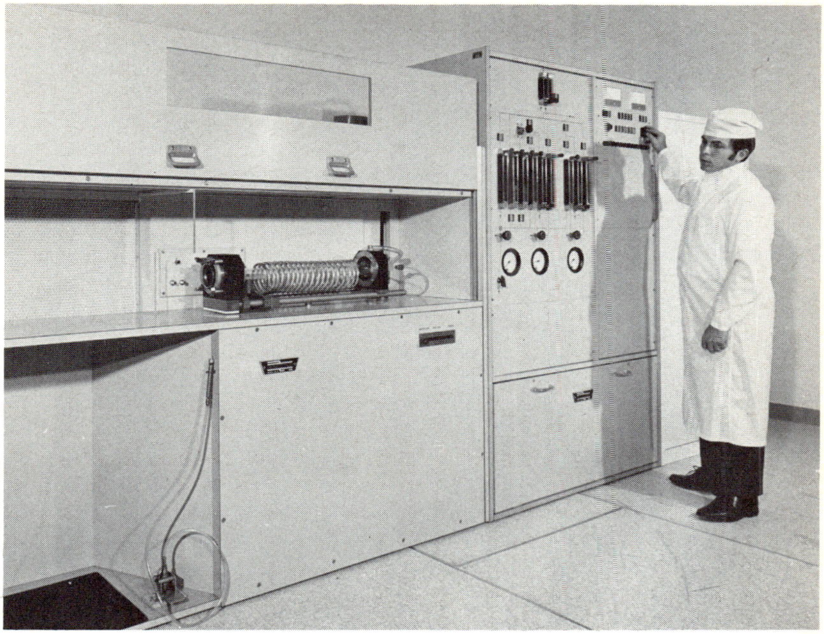

FIGURE 4.28
An epitaxial reactor. (*Courtesy Radyne.*)

wafers are loaded onto a quartz oxidation boat and slowly inserted into an oxidation furnace. Oxidation furnaces are the same as the diffusion furnace shown in Fig. 4.7. The oxidation furnace is not used for impurity diffusion, however, since the quartz tubes can become "contaminated" with the impurity dopant and transfer it to the oxide. As a matter of fact, separate furnaces are used for P-type, N-type, and thermal oxidations for this reason.

Thermal oxides are grown at temperatures between 900 and 1200°C. Oxygen and/or steam is passed over the wafers at carefully controlled flow rates. The percentages of the oxygen and the steam, as well as the time and temperatures, can be varied in order to obtain the desired oxide properties and thickness.

The oxygen combines with the silicon to form a uniform layer of silicon dioxide (SiO_2). As the SiO_2 layer forms, the growth rate declines, since the oxygen and the silicon have to diffuse through the newly formed SiO_2 in order for the reaction to continue.

The importance of silicon dioxide in the planar process is owing to several important properties: it has the ability to "mask out" the most popular N- and P-type impurities during diffusion; it is easily etched (opened) in desired locations on the silicon wafer so that diffusion into the silicon in these areas

FIGURE 4.29
A wafer-alignment machine. *(Courtesy Kasper Instruments, Inc.)*

can take place; and it provides surface passivation, or protection, by "tying up" the dangling surface bonds of the silicon and preventing contaminants from disrupting device performance.

Photoresist. After oxidation, photographic process is used to delineate the areas where the diffusion of N- or P-type impurities through the SiO_2 into the silicon wafers is required. A photographic process is needed in order to obtain the fine detail of modern planar transistors.

The first step in this process is to coat the oxidized wafer with a thin uniform coating of photosensitive material, such as Kodak photoresist (KPR). This uniform coating is obtained by placing a quantity of filtered photoresist on the wafer surface and whirling the wafer on special high-speed whirlers. The acceleration, rpm's, and whirling time are important process parameters, since they control the thickness and uniformity of the photoresist layer. After spinning on the photoresist, the wafer is carefully removed from the whirler chuck and placed in a drying oven to drive off any remaining solvent.

The wafer is then placed in an aligning machine (Fig. 4.29). This machine is equipped with a movable alignment fixture that holds the wafer, a high-power microscope, a movable ultraviolet light source, and a movable mask holder. A mask of ordinary high-resolution photographic plate that contains the pattern to be placed on the wafer is held by vacuum above the wafer. The pattern consists of an array of identical elements, such as the emitter or base patterns, spaced on centers consistent with their size. The masks usually contain aligning key patterns that replace the normal pattern at certain positions and are used as an alignment aid for registration of successive patterns upon the wafer (Fig. 4.30). After the mask is brought into precise alignment, using the high-power microscope and aligning dials, the wafer is brought into physical contact with the mask and exposed to the ultraviolet light for a specified time. Typical exposure times are between 3 and 10 s.

After exposure the wafer is carefully removed from the alignment machine and developed. The developer dissolves the unexposed, or unpolymerized, portions (patterns) of the photoresist from the wafer. If a negative photoresist such as KPR is used, the opaque pattern on the mask would be the part of the photoresist removed by the developer, leaving an exposed oxide pattern to be etched.

With a positive resist, such as Shipley AZ120, the clear portion of the mask would be the part of the photoresist removed in development. Following development, the wafer is baked at a temperature close to 200°C for about 30 min. This drives out additional solvents that may have been absorbed

FIGURE 4.30
A metal mask of 0.002 molybdenum foil used in the production of semiconductor devices. (*Courtesy Industrial Reproductions, Inc.*)

in the film during development and hardens the remaining photoresist to ensure adhesion during the subsequent oxide etching.

After baking, the wafers are immersed in an etching solution containing ammonium fluoride, water, and hydrofluoric acid. This solution removes the the SiO_2 layer where the unpolymerized emulsion had been. It does *not* remove the SiO_2 layer in those places covered by the polymerized emulsion. Silicon dioxide etching takes place at room temperature, the etch time depending on the oxide thickness. Typical oxide etch times are between 2 and 10 min. The remaining SiO_2 now serves to prevent diffusion in the areas that it covers.

The next step is diffusion to form the base element. The diffusant, be it boron for P-type base elements or phosphorus for N-type base elements, will diffuse much more slowly into the SiO_2 than in the silicon. Thus, only where the oxide is *not* present will any useful penetration of the silicon occur.

After the base has been formed, another SiO_2 layer again covers the entire wafer and the entire process is repeated to form the emitter. If a P-type diffusant was used for the base, then an N-type diffusant is employed for the emitter, and vice versa. This sequence of SiO_2, photoresist addition, mask positioning, exposure to ultraviolet light, removal of the unexposed photoresist, selective SiO_2 removal, and diffusion is the repetitive process by which all planar transistors and integrated circuits are produced. It is the basis for present-day semiconductor manufacturing technology.

Contact Metallization. After the emitter diffusion, a final oxide is grown over the wafer surface, and contact openings in the oxide, over the emitter and base, are etched (Fig. 4.18) again, using photoresist techniques.

After the contact "windows" are opened and the photoresist stripped from the wafer, a metal (usually aluminum) is evaporated over the entire wafer surface. The wafer is again coated with photoresist and the metallization contact mask is aligned over the previous patterns on the wafer, which are visible through the aluminum (there are also aligning keys on each mask). After exposure, the metallization pattern is developed out and etched. The metal etch usually consists of varying proportions of nitric acid, phosphoric acid, and water. The etching time is dependent on the etch-solution temperature and the aluminum thickness, which is normally between 8 to 15,000 angstroms (Å) ($1\text{ Å} = 10^{-8}$ cm). The photoresist is removed in a special stripping solution that does not affect the metal.

The wafer is then alloyed, in an inert atmosphere, at 500°C for approximately 5 min. Alloying is necessary to form ohmic contacts between the aluminum and the silicon in the emitter and base contact areas.

Automatic Wafer Probing. The dc parameters, such as B_{VCBO}, B_{VEBO}, B_{VCEO}, I_{CBO}, and β (beta) are automatically probed using a machine like the one shown in Fig. 4.13. Transistors that fail any of the tests are "inked" out. Special counters, "hooked in" to individual test parameters, and different colors of reject inks, are used as an aid in analyzing device failures. For example, if a large percentage of transistors on a wafer are rejected for low beta, the process engineer might want to make the base width smaller by increasing the emitter diffusion time or decrease the base depth by decreasing the base diffusion time. Conversely, if a large percentage of transistors fail because the beta is too high, the emitter diffusion time could be shortened or the base diffusion time lengthened. Thus, device yield per wafer can be improved by appropriate process adjustments indicated by the dc wafer-probe data.

The probed wafer is now ready for "back etching," which is necessary to reduce wafer thickness to facilitate wafer scribing and cracking. The wafer is placed active side down on a quartz plate and held there by a special acid-resistant wax. The quartz plate is then immersed in a silicon etching solution. The final wafer thickness is obtained by controlling the etch time and temperature. Usually, a back-etched wafer is about 0.005 in thick. Wafers are kept thick (0.010 in) through the previous processing to facilitate handling.

Scribing. The wafer is scribed using an automatic scribing machine (Fig. 4.31). This machine is equipped with a diamond-tipped scribing point that

FIGURE 4.31
An automatic scribing machine. *(Courtesy Tempress Research Co.)*

automatically moves in straight lines as the wafer is moved back and forth. The pressure between the diamond tip and the silicon is controlled so that the scribe depth will facilitate the cracking (separation) operation. After cracking, the inked rejected transistors are sorted, and the good devices are mounted and bonded in appropriate packages.

Mounting of the individual transistor die to the header is done at temperatures close to 400°C, in an inert atmosphere. Aluminum or gold wire is used to bond the emitter and base contact areas to the corresponding package terminals.

Many improvements have been made in packaging techniques to reduce cost and at the same time improve device reliability. Some of the well-known new packages include the ceramic "channel" package, the lid, and the microtab packages (Fig. 4.32).

Semiconductor Interconnections. After the basic structure of a transistor has been fabricated, one must provide interconnections between it and the electronic circuitry to which it will be attached.

FIGURE 4.32
Packages currently in use to house semiconductors.
(*a*) Channel package usually filled with epoxy. (*b*) Microtab package. (*c*) LID package, usually filled with epoxy.

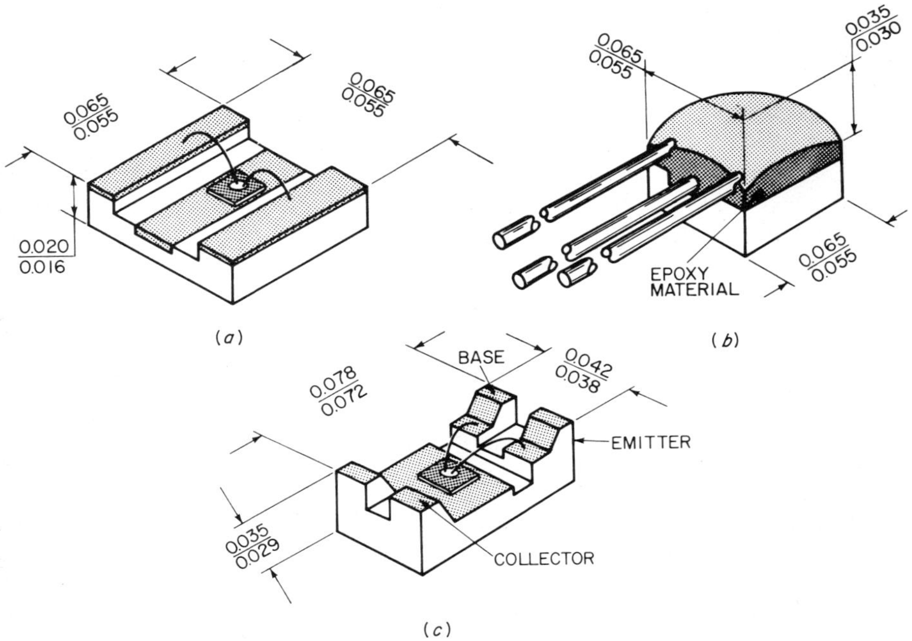

FIGURE 4.33
A hybrid microcircuit containing transistors, diodes, resistors, and capacitors.

A common practice in hybrid microcircuits is to build up an operating circuit, such as an amplifier, by mounting on one substrate the transistors, diodes, resistors, and capacitors needed for that circuit. Here, too, interconnections between the several components are required as well as wires leading to the external circuits of that piece of equipment.

A typical hybrid microcircuit is shown in Fig. 4.33. The various discrete components (here, resistors, capacitors, transistors, and diodes) are each bonded to the common supporting structure or substrate and then interconnected.

Because of the emphasis on miniaturization, not only are the active components extremely small but the passive components (resistors, capacitors) are small as well. To meet the spatial needs of microcircuits, a whole new family of ultrasmall or chip capacitors and resistors have been developed. Some idea of their size can be obtained from Figs. 4.34 and 4.35.

Inductors are more difficult to miniaturize because inductance is directly

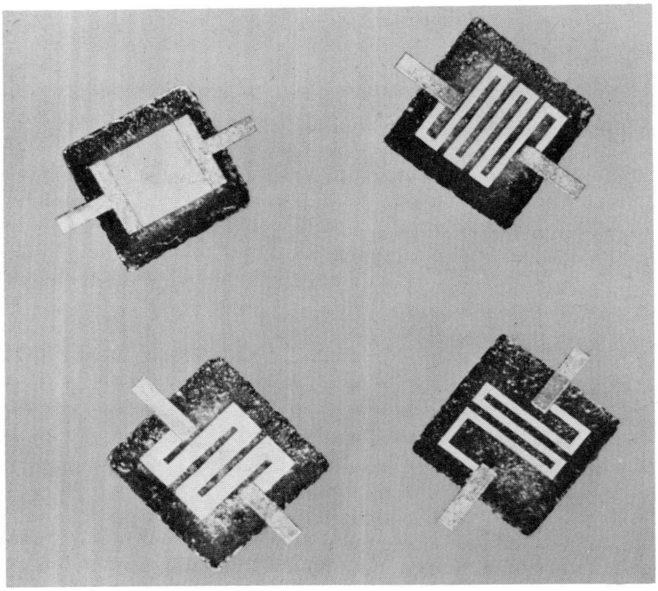

FIGURE 4.34
Chip resistors come in many terminating modes. The units above are 0.050 × 0.050 in. in size. (*Courtesy Motorola.*)

proportional to the length of deposited coil. Micromagnetic discrete components, in the form of toroids and rod cores, are used when higher values of inductance are needed.

Discrete-chip spiral inductors are available in values of from 28 to 230 nH. Designed for use in uhf and microwave hybrid circuits, they are fabricated on a 0.010-in-thick alumina chip and connected to the circuit by fire bonding.

The hybrid microcircuit of Fig. 4.33 employs the "flying-lead" type of interconnection. Aluminum or gold is employed for both the connecting lead and the circuit point where the lead terminates. See Fig. 4.36. The bonding is done by thermocompression bonding, by ultrasonic bonding, or by reflow soldering, among others. Thermocompression bonding depends upon heat and pressure on the surfaces to be mated to achieve the merging of the materials (of the two surfaces) to form a well-bonded structure. Ultrasonic bonding accomplishes the intermetallic bond at room temperature utilizing ultrasonic energy and pressure. Finally, in reflow soldering, the bond is achieved by pretinning the two surfaces to be mated, bringing these surfaces into contact with each other, and then applying heat. Flux is generally used to assist the reflow process.

FIGURE 4.35
Chip capacitors. Note the variety in size and form.
(*Courtesy Vitramon.*)

In Fig. 4.36, the IC chip is bonded to the supporting substrate so that it is held rigidly in position. Then the flying leads are connected.

Face-up Bonding. In face-up bonding, the chip is held face up and then the connections are made. The flying-lead-connection technique is one form of face-up bonding. An alternate connection method is through the use of metal fingers (Fig. 4.37) that are stamped from an aluminum frame and bonded to various points in the microcircuit. In essence, whenever the chip is held with its components (or face) positioned up and suitable leads are attached, we have face-up bonding.

Flip-Chip Bonding. In flip-chip bonding, the chip has a spherical contact (a bump) at each interconnect point on its face, which can range from 2 to 6 mils in diameter and from 0.1 to 3 mils high. Depending on the process, these bumps can be tin-solder material, gold, or aluminum.

The chip is then turned over and the contact bumps are carefully positioned over connection points on a substrate or other circuit pattern to which contact is to be made. Then, through the application of heat and some pres-

FIGURE 4.36
The "flying-lead" type of interconnection.

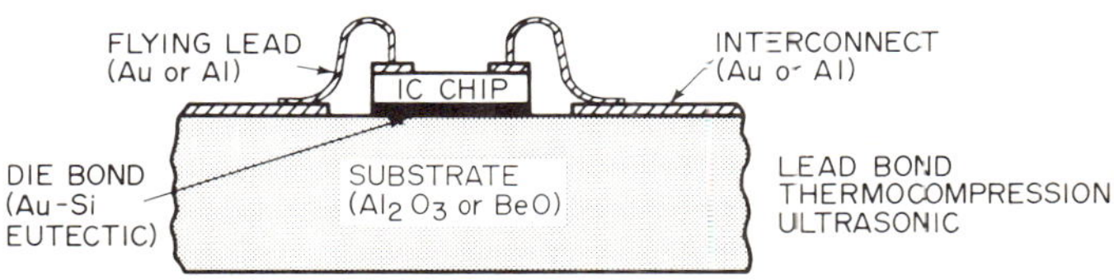

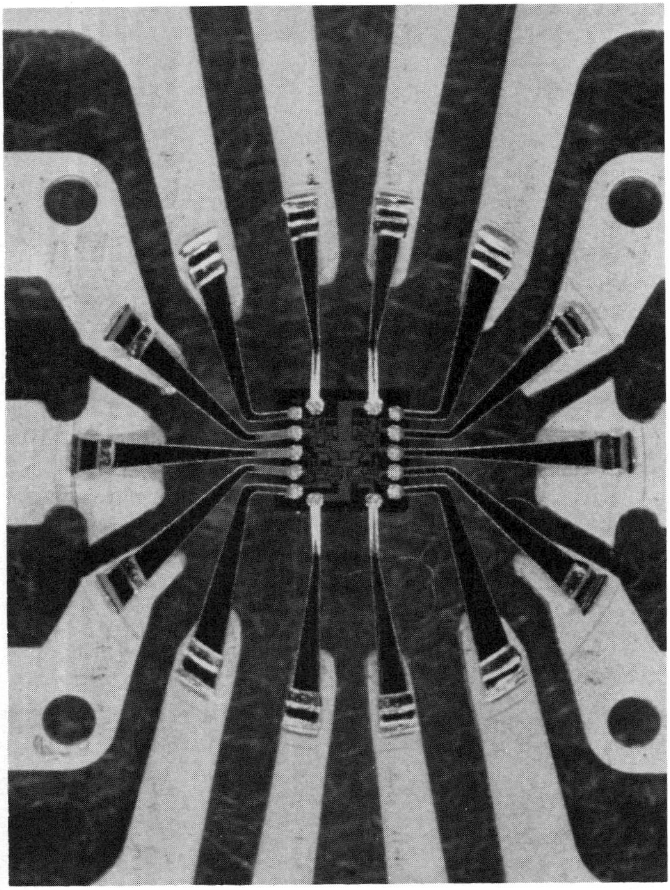

FIGURE 4.37
A face-up bonding technique using metal fingers.
(*Courtesy Motorola.*)

sure, a permanent bond is achieved between the chip bumps and the circuit on which it is mounted. See Fig. 4.38*a*.

Thus, it can be seen that flip-chip mounting results in the chip being positioned upsidedown.

A variation of the technique shown in Fig. 4.38*a* is the arrangement whereby the raised contact bumps are developed on the substrate on which the IC chip is to be placed rather than on the chip itself. In this case, the chip has flat contact points that, when the chip is turned upsidedown, are carefully positioned until they fall directly over the substrate bumps. Then, heat and pressure lead to a permanent bond.

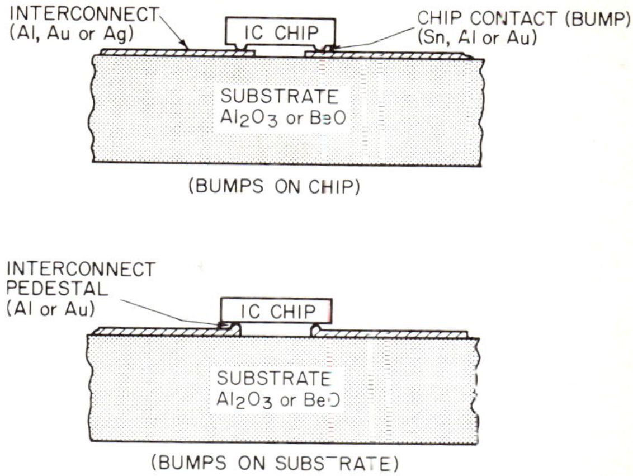

FIGURE 4.38
Flip-chip bonding.

Note that with the flying-lead attachment, two bonds are required for each lead. With the flip-chip method, only one bond is required for each connection between the IC chip and the substrate.

Beam-lead Connections. The term "beam lead" refers to relatively large and heavy leads ($5 \times 2 \times 0.2$ mils) extending over the edge of the chip. The leads are formed during the wafer processing by evaporation of gold in precise patterns on the chip.

The beam lead makes connecting to points on a substrate relatively simple. See Fig. 4.39. It also offers the advantages of using only one bond per interconnect, being of small size, and having the ability to be tested as an unmounted device. It does tend to be somewhat higher in cost to manufacture as well as use up more "real estate" on the wafer than the other methods.

Computer-aided Design (CAD). The successful modern semiconductor-device designer, or process engineer, is almost required to use a computer in his work. Computers are extensively used in device and circuit design, testing, and process control. As a matter of fact, they are used in every engineering aspect of semiconductor manufacturing and design. The computer is the catalyst in the rapid advances into LSI [large-scale integration (Chap. 8)] circuits and new process innovations.

As an illustration of how the computer can be employed in assisting the semiconductor-device engineer, consider some of the problems that this engineer faces. He is frequently called upon to modify an existing semiconductor device so that it will provide perhaps a higher frequency response or present a lower internal resistance under saturation conditions. If he can provide a computer with a set of equations that represent the behavior of this device, then he can alter these equations and note the effect on the device behavior under these new conditions.

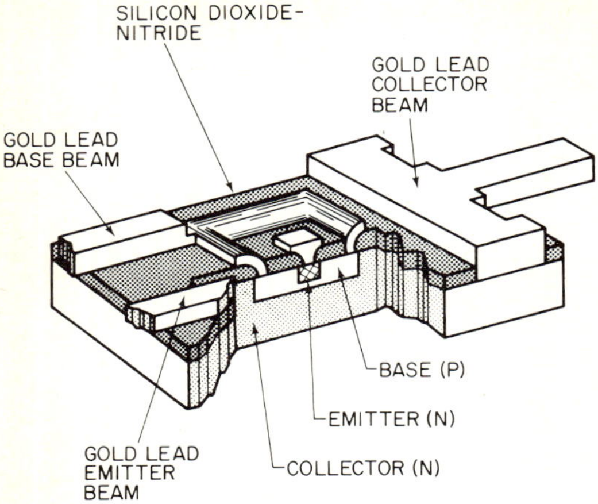

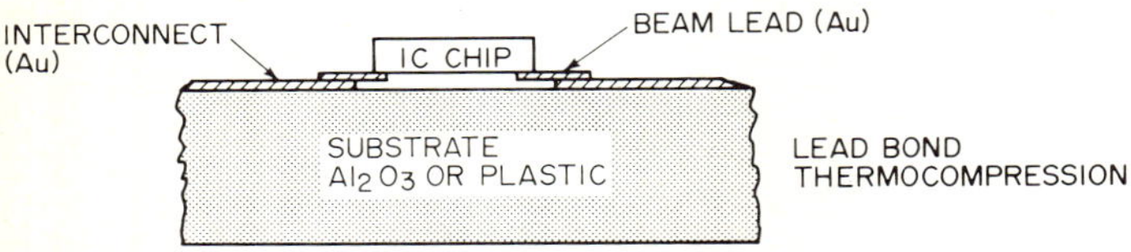

FIGURE 4.39
(*a*) A transistor with a beam-lead construction.
(*b*) Attachment of an IC chip with beam leads to an external circuit.

Thus, rather than go through the long, tedious process of actually constructing different models of the device in question, he can use the computer to at least tell him what will or will not work. When he finds the conditions that give him the results he desires, then he need only build a model to see if it does indeed function as he wishes.

As engineers become increasingly familiar with the semiconductor devices and circuits that they work with, the better able they will be to develop equations that represent their operation. Armed with this knowledge, they will then be able to use computers to help them determine the best design approach and to do this in an incredibly short time. Computers are the most powerful design tool engineers have, and engineers are becoming increasingly adept at employing these new tools.

BIBLIOGRAPHY

1. Phillips, A. B.: "Transistor Engineering," McGraw-Hill Book Company, New York, 1962.

2. Runyon, W. R.: "Silicon Semiconductor Technology," McGraw-Hill Book Cmpany, New York, 1965.
3. General Electric Transistor Manual, 1964.
4. Grove, A. S.: "Physics and Technology of Semiconductor Devices, John Wiley & Sons, Inc., New York, 1967.

QUESTIONS

4.1. What was the first type of junction transistor?

4.2. What is the segregation coefficient as applied to semiconductors? What practical significance does this expression have in semiconductor fabrication?

4.3. In what way does the manufacture of alloy-junction transistors differ from grown-junction transistors?

4.4. In what features is the alloy-junction transistor superior to the grown-junction transistor?

4.5. Describe the diffusion process, including the important processing conditions.

4.6. Name two P-type impurity sources and two N-type impurities.

4.7. What "built-in" advantage does the drift transistor utilize?

4.8. To what special conditions does the surface-barrier transistor owe its operational characteristics?

4.9. Describe the basic structural features of a mesa transistor.

4.10. What is the distinguishing feature of an epitaxial transistor? What electrical advantages does this special epitaxial feature possess?

4.11. Describe briefly the planar process and sketch the base, emitter, and silicon oxide relationships.

4.12. Name two advantages of silicon dioxide (SiO_2) in planar-transistor fabrication.

4.13. What degrading effect was the annual transistor designed to overcome? Explain how the degradation occurred.

4.14. How did the annular transistor construction improve device operation?

4.15. Name the five major steps followed in the planar process.

4.16. Describe two of the processing steps in detail.

4.17. Distinguish between beam-lead and flip-chip methods of connection.

4.18. What is the difference between hybrid and monolithic microcircuits?

Chapter Five

Field-effect Transistors

The second most important electronic device on the market today is the unipolar transistor. (The most important device is the bipolar junction transistor, which we have been discussing.) The unipolar transistor is known as a field-effect transistor (FET), or metal-oxide semiconductor (MOS). The rapidly expanding FET market has led many semiconductor marketing managers to believe that this device will soon become the most important electronic device, primarily because of its integrated-circuit applications.

The field-effect transistor differs distinctly from the conventional junction transistor in that its operation depends on the flow of majority carriers. This is why it is classified as a unipolar transistor. Bipolar transistors are so named because they depend on both majority- and minority-carrier flow.

In this chapter, we will discuss the history, the advantages and disadvantages, some principles of operation, and the characteristics of field-effect transistors. Some applications of field-effect devices will be discussed in Chap. 6.

The invention of the bipolar transistor in 1948 had a profound effect on the development of a unipolar field-effect transistor. At first glance, this may seem strange because research, which had been directed toward the development of a field-effect, solid-state amplifier, was redirected to the improvement of the new bipolar device. Eventually, however, this bipolar research provided the key to the fabrication of practical unipolar devices.

In 1952, William Shockley described a unipolar transistor in which the control electrode was a reverse-biased junction (Fig. 5.1). Shockley's device was later fabricated by I. M. Ross and C. Dacey of Bell labs, who developed a mathematical analysis in 1955 describing the performance limits of such devices. This transistor was similar to the modern junction field-effect transistor, which we shall discuss.

The unipolar devices did not emerge from the laboratory and become commercially practical until the introduction of silicon planar technology in 1960. This technology, with its silicon-silicon dioxide interface, provided the necessary uniform surfaces for the fabrication of useful field-effect transistors.

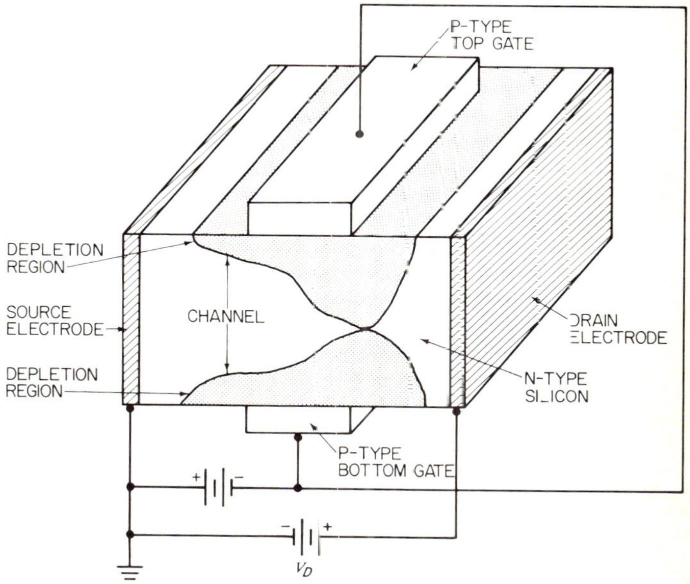

FIGURE 5.1
The basic unipolar transistor developed by Shockley.

Since 1960, the development of the FET has accelerated so rapidly that it now threatens to depose the bipolar transistor as the most widely used semiconductor device.

The need for the FET developed because as modern electronic equipment became increasingly transistorized, it became apparent that there were many functions in which bipolar transistors were unable to replace vacuum tubes. Owing to their extremely high input impedances, FET devices are more like vacuum tubes than are bipolar devices and are hence able to take over many vacuum-tube functions. Thus, because of the FET, electronic equipment is closer today to being completely solid state, and is more reliable, smaller in size, longer lasting, and more efficient overall.

THE JUNCTION FIELD-EFFECT TRANSISTOR (JFET)

There are two basic types of field-effect transistors: the junction field-effect transistor (JFET), which we will discuss now; and the metal-oxide semiconductor FET, or MOSFET, which we will discuss next.

Here is how the JFET works. It is usually fabricated by first diffusing an epitaxial N-type layer of silicon on a heavily doped P-type substrate. Then, by using planar technology (Chap. 4), a P^+-type gate junction is diffused into

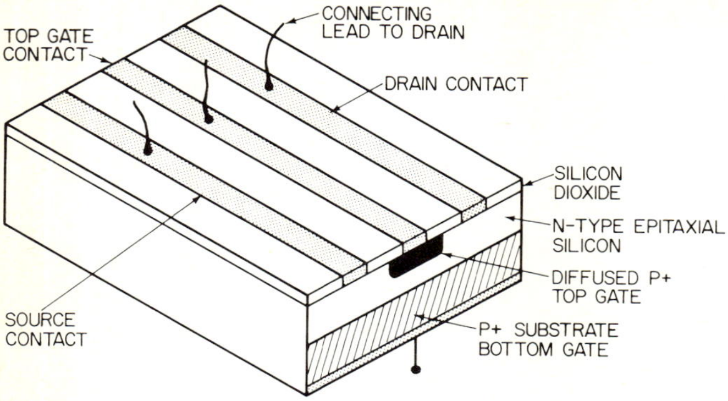

FIGURE 5.2
Planar-epitaxial N-channel-junction field-effect transistor. This is basically the same structure as the unit in Fig. 5.1, but it does reflect more current fabrication technology.

the N-type epitaxial material and contact is made to both top and bottom P^+ gates, and also to the N regions on both sides of the top gate. These last two contact areas are known as *source* and *drain* (Fig. 5.2). The reasons for these strange-sounding terminal names are functional and will become clear as we go on.

To start, let us apply a voltage between the source and drain electrodes with the drain terminal made positive with respect to the source. This is shown in Fig. 5.3. At the same time, let us connect the P-type gate to the

FIGURE 5.3
Voltage applied to the field-effect transistor.

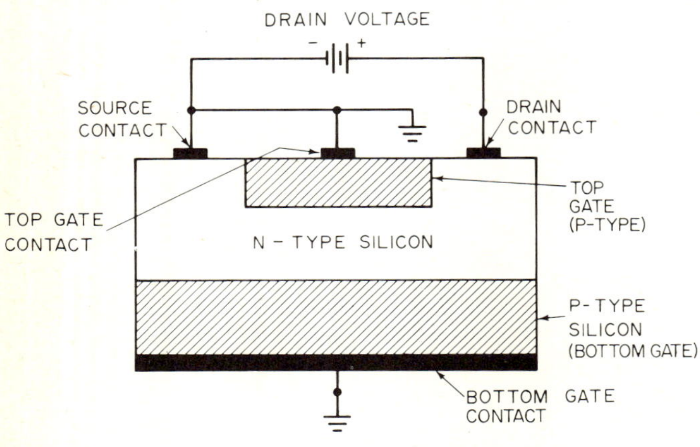

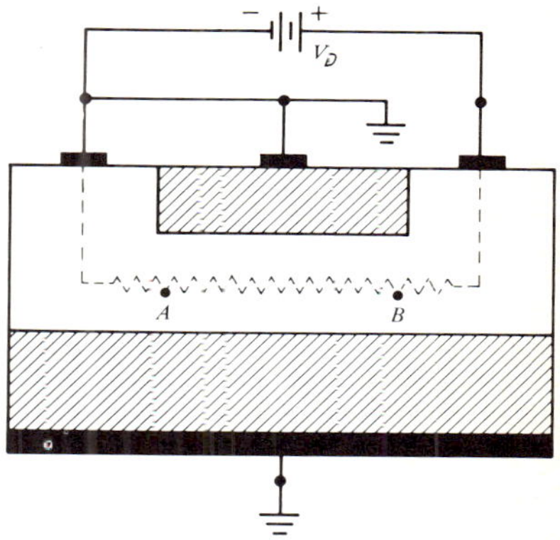

FIGURE 5.4
The internal ohmic resistance of the silicon block is represented by the dotted resistor. Owing to the voltage drop across this resistance, point *B* is more positive than point *A*.

source terminal. Under these conditions, the gate is said to possess zero potential, the source electrode serving as the reference point for the entire unit.

Electrons will travel from the source to the more-positive drain electrode when a voltage is applied to these end terminals. Because a definite potential is being applied across the ends of the N-type channel and because this material possesses a certain amount of resistance, the applied voltage will be distributed equally along the body of the N channel from the source to the drain. In the present arrangement, the potential will become progressively more positive as we travel from the source to the drain. (If the reader has any difficulty visualizing this distributed voltage drop, let him substitute a resistor for the silicon channel. A point on this resistor that is closer to the positive end of the battery will be more positive than any point along the resistor to the left. The same situation holds for the silicon channel.)

Now let us consider the gate. This forms a PN junction with the N-type silicon over the area in which the two are in contact. The end of the gate nearest the source will find the least amount of potential difference between it and the N channel. This is at point A in Fig. 5.4, and the reason is quite simple. The gate itself is at the same potential as the source, since the two are externally directly connected. Inside the N channel, however, at point A, there exists a small positive potential (with respect to the source) because of the voltage drop mentioned above. This positive potential at A repels the holes in the end of the P-type gate just above it and also exerts an attractive force on the electrons in the channel. Thus, a small reverse bias is present here.

As we progress further along the channel, the positive potential (with respect to the gate) increases, making the reverse bias across the PN junction even greater. There is thus no tendency on the part of the electrons in the channel to flow to the gate or for the holes in the gate to cross the junction and move into the channel.

However, holes do exist in the channel. These come from three sources. (1) They may be thermally generated in the body of the semiconductor; (2) they may be developed at the surfaces of the semiconductor; or (3) they may come from the two end contacts, the source and drain metal electrodes. In any event, a certain number of holes exist in the N-type silicon, although the donor electrons exceed these by ratios as high as 10:1 or more.

Under the repelling effect of the positive voltage drop along the channel and the attractive force of the more negative gate, holes in the semiconductor will be drawn up to the gate. The number of holes that are drawn from any section of the channel will be governed by the positive potential present in that area. Thus, more holes will be drawn from the right-hand section of the channel than the left-hand section.

Now consider the P-type gate. It will possess a number of free electrons, possibly for some of the same reasons that the N-type channel possesses holes. And these electrons will be repelled by the negative potential of the gate and attracted by the more positive potentials within the channel. Hence, there will be a movement of electrons out of the gate and into the channel, with the greatest number of electrons leaving the gate at the right-hand side of the channel.

One result of this redistribution of charge is to make the PN junction more reverse-biased than before. Another result is to increase the negative space charge in the channel, with the concentration greatest at the right. The actual space-charge distribution is shown in Fig. 5.5. It rises to a maximum at the right-hand edge of the P gate and then decreases fairly rapidly. This space charge exerts a repelling force on those electrons traveling from the source to the drain electrodes because of the externally applied potential. What it actually does is channel, or direct, the current flow into the regions between the concentrated space charge. These regions are shown in white in Fig. 5.5. The dotted area represents the negative space charge.

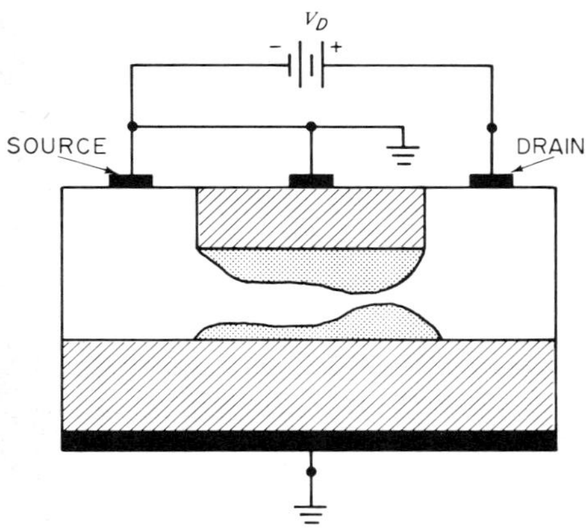

FIGURE 5.5
The shape of the space-charge distribution within the body of a field-effect transistor.

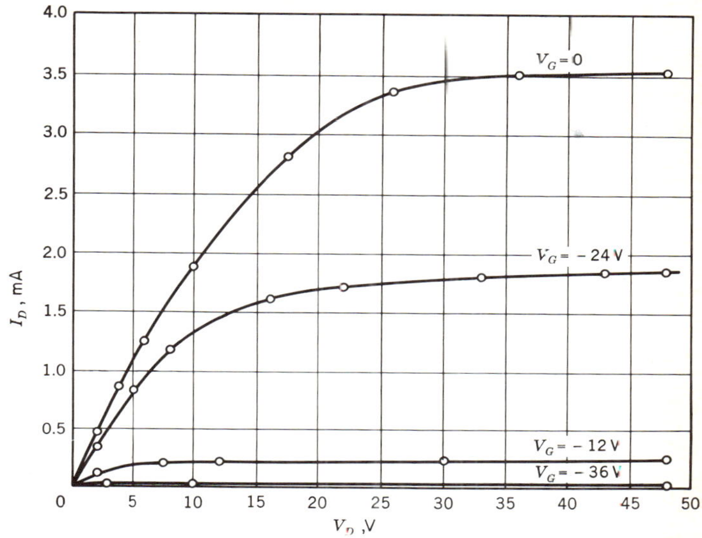

FIGURE 5.6
Typical characteristic curves of a field-effect transistor.

As the drain voltage is increased, the current flow through the semiconductor will rise until the drain voltage reaches a certain critical limiting value, which is referred to as the pinch-off voltage. Beyond this point, no further increases in drain current will result as the drain voltage is made greater. The only effect of higher voltages is to alter the shape of the channel through which the electrons flow.

Thus far, the gate has been held at the same potential as the source. If we now make the gate negative with respect to the source terminal by the insertion of an additional negative voltage, then the amount of external voltage needed between source and drain in order to reach the constant drain-current value will be much less. This is because the negative-gate voltage aids the space charge in the channel to narrow the path for the current to flow. Eventually, if the negative potential of the gate is made high enough, current flow through the device will be halted completely.

The foregoing behavior is depicted by the curves in Fig. 5.6. Maximum saturation current flows when V_G (the gate voltage) is zero with respect to source. For each successively lower current curve, the negative bias on the gate becomes greater. When the gate potential is equal to the pinch-off voltage, the amount of current flowing through the semiconductor is very small.

The circuit in Fig. 5.7 is similar to that of Fig. 5.6 with the addition of a signal in series with the gate bias battery. As the signal varies, it will vary the total voltage applied to the gate. When the signal is negative, the

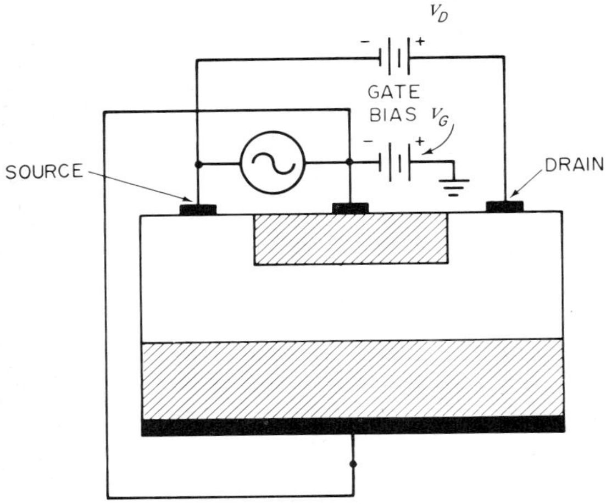

FIGURE 5.7
The field-effect transistor with a negative-gate bias voltage V_G. Also shown is the point where a signal would be injected.

gate will become more negative and serve to reduce the flow of drain current. On the next half-cycle, when the signal goes positive, the overall gate voltage will be made less negative. This will reduce the space charge in the body of the semiconductor, and the drain current will rise. In short, a small voltage variation at the gate will produce a sizable current variation in the channel. In a vacuum tube, this characteristic is labeled "transconductance" and is responsible for the amplification obtained. Obviously the same results should be possible in the field-effect transistor, enabling us to use this device to achieve signal amplification.

The similarity between the field-effect transistor and the triode vacuum tube is quite marked. In the vacuum tube, the grid potential regulates the space charge existing between cathode and plate and by this control determines the extent of plate-current flow. In the field-effect transistor, the polarity of the gate governs the intensity of the negative space charge existing in the N material and, through this, the amplitude of the current flowing from source to drain.

Another interesting feature of this transistor is the fact that essentially only one type of carrier (here electrons) is involved in the process. (A field-effect transistor could be constructed by using a P-type main body and an N-type gate. Battery potentials would have to be reversed, and the current in the source-drain channel would be carried by holes. Similar results would be obtained, however.) This is in sharp distinction to the conventional transistor, wherein both types of carriers play roles. Hence, this device is called a unipolar transistor. It also results in a very significant difference in the way amplification is achieved. In the conventional triode transistor, the carriers must travel from the emitter through the base to the collector. This makes carrier transit time an important factor in determining frequency response. In the field-effect transistor, the signal at the gate serves to modulate the drain current and in this way produces the signal variations in the drain output circuit. Carriers do not have to transport the signal from the gate to the drain; hence, carrier transit time is not involved. We do not even have to worry about transit time from source to drain, because the signal voltage at the gate merely expands or contracts the current stream; it does not alter the rate of travel. This does not mean that the field-effect transistor is without frequency limitation, for there are certain shunting capacitances to be dealt with, as well as variation of transconductance with frequency and other effects. But transit time does not possess the significance here that it does in other transistor structures.

THE METAL-OXIDE SEMICONDUCTOR

Basically, the MOSFET depends on varying the resistance between two electrodes, the source and drain, which is also true of the JFET. However, with the MOSFET the channel resistivity is regulated by controlling the voltage on a gate electrode that is insulated from the body of the field-effect transistor (see Fig. 5.8). For this reason MOSFETs have also been referred to as an insulated-gate FETs (or IGFET).

The MOSFET comes in two varieties, the enhancement type and the depletion type. The following discussion deals first with the more popular enhancement type.

The P-type enhancement-mode MOSFET is fabricated on an N-type silicon substrate into which two heavily P-type impurity regions are diffused. The

FIGURE 5.8
P-type enhancement-mode MOSFET.

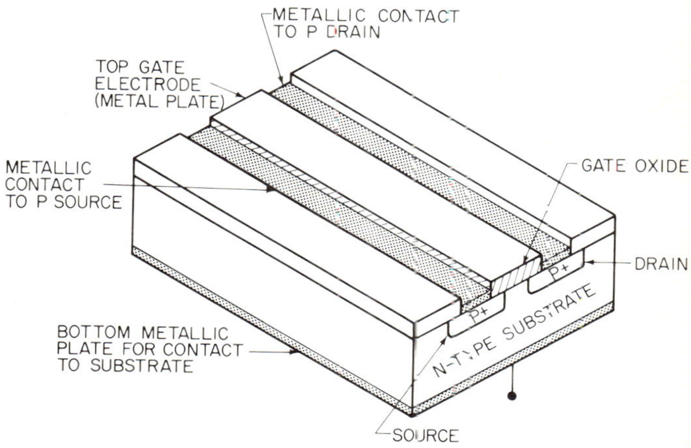

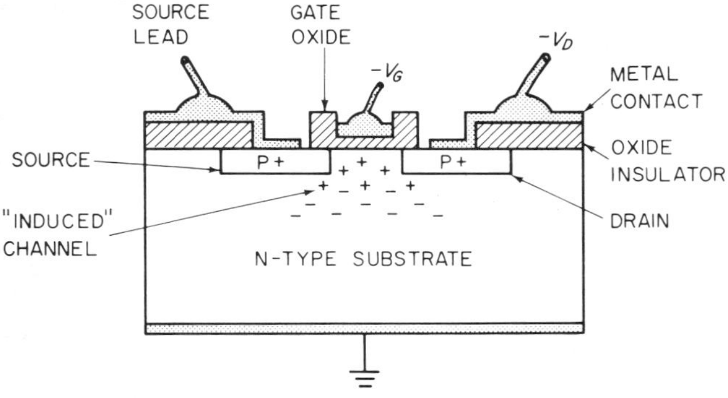

FIGURE 5.9

P-type "islands" are the source and drain. The source and drain can be linked together by inducing a P-type channel between them through the application of the proper voltages to the gate electrode.

The gate covers the channel region between source and drain and is the control element of the device. An insulating oxide layer separates the gate from the channel. This oxide layer is normally made exceedingly thin so that the gate field effect on the channel is optimized.

Let us now ground the gate and substrate and apply a negative voltage to the drain electrode. No current will flow between source and drain because they are isolated from each other by the reverse-biased, drain-to-substrate PN junction. However, if we now apply a negative voltage to the gate, the minority holes in the N-type silicon will be drawn toward the gate (see Fig. 5.9). This is the channel that is close to the surface and initially had an excess of electrons because the substrate material is N-type. The holes, which are drawn by the negative-gate bias, soon neutralize by recombination the electrons in this channel between source and drain. If the negative-gate bias is increased beyond the channel neutralizing point, holes will begin to predominate in the channel area. What has happened is that a thin layer under the insulated gate has "inverted" or become P-type. The channel can now conduct current between the source and drain.

The typical P-channel enhancement-mode transistor will not conduct between source and drain until the gate voltage is at a value of about −5 V. The electrons in the substrate material have to be depleted before any holes can be accumulated in this area. When the channel starts to conduct, the gate voltage is the threshold voltage (V_T). Its value can be controlled by the oxide thickness.

If the gate voltage increases beyond the threshold value, the drain-to-source conductance will increase. In the low-current range, the drain current I_D is directly proportional to the drain-to-source voltage V_{DS}. This is represented in Fig. 5.10 by the region from zero to point A.

As the drain-to-source voltage becomes more negative, one would suppose that the rise in drain current I_D would continue. This does not happen, however. Instead, as seen from Fig. 5.10, the drain current levels off. This section of the curve is known as the pinch-off or saturation region (A-B in Fig. 5.10). Why this occurs is discussed below. Finally, beyond point B the transistor is in the voltage-breakdown region and the drain current increases sharply.

Thus, the drain-current-vs.-drain-source voltage characteristics of all field-effect devices can be divided into three distinct regions. The variable-resistance, or linear, region is the region of the increasing slope near the beginning of low drain-to-source voltages. The curves flatten (constant current flows) for medium drain-to-source voltages, and this is the pinch-off, or saturation, region. As the drain-to-source voltage increases still further, the curves begin to slope upward again at an increasing rate. This is the avalanche breakdown region.

A closer look at what actually happens in the pinch-off, or saturation, region is warranted. The distribution of the *voltage difference* between the gate and the substrate below it is not uniform from source to drain because the source and drain do not possess the same voltage. The source is at (or near to) zero, whereas the drain is quite negative.

This means that the electric field from the gate to each electrode (source and drain) is also not uniform; and since it is this field that determines the

FIGURE 5.10
P-type enhancement-mode MOSFET current—voltage curves.

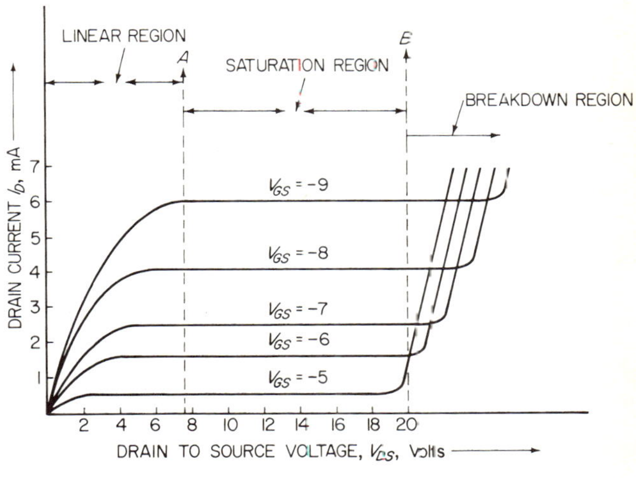

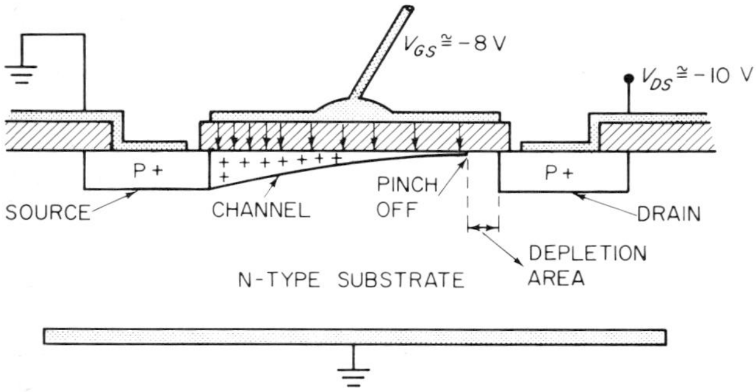

FIGURE 5.11
P-type enhancement-mode MOSFET channel at pinch-off with depletion region.

hole (or positive charge) distribution beneath the gate, this will also not be uniform.

So long as the gate negative voltage is *above* the threshold value (approximately —5 V) and this gate voltage significantly exceeds the source-to-drain voltage, the charges will be distributed beneath the gate along its entire length and the number of holes reaching the drain (from the source) will rise linearly.

However, as the drain voltage becomes more and more negative, a situation develops whereby the end section of the gate insulator near the drain becomes reverse-biased and a depletion layer forms between the drain and the near end of the gate. When this occurs, there are no mobile charges in this depletion layer.

Figure 5.11 shows the mobile-hole-charge distribution under these conditions. Also shown is the depletion area. The channel close to the source still contains a large number of mobile charges, because the source is at ground potential and the potential difference between source and gate is quite large, which means that a much greater electric field exists near the source in the oxide insulator. The number of conducting carriers decreases as we go from source to drain. At pinch-off, the depletion width is quite small; but additional voltage applied from drain to source will be impressed across the depletion area, making it grow wider. The voltage across the remainder of the channel will remain almost constant, which means that for the rest of the channel, from the source to the edge of the depletion region, the current remains constant for drain-to-source voltages above pinch-off. As a matter of fact, the depletion region does not block current flow, and hence has no effect on the saturation current. This is true because of a large electric field in the depletion region that becomes stronger as drain voltage is increased. The current carriers arriving at the depletion region are immediately swept to the other side.

The situation here is analogous, in a sense, to that prevailing in the conventional bipolar transistor. Carriers from the emitter pass through the base and then sweep through the reverse-biased base-collector junction. Here, too, a depletion region exists.

If the drain voltage is increased past the saturation value, the drain current starts to increase. The current increase at first is gradual; but it then rises rapidly. The rapid increase in current in the avalanche region is the result of the avalanche breakdown of this reverse-biased drain junction.

Controlling Threshold Voltage. One of the most important and difficult parameters to control in MOSFET-device fabrication is threshold voltage (V_T). The FET has historically operated at higher voltages than bipolar transistors, which means that level shifters and extra power supplies were required if these two devices were to work together in the same circuit. Recently, an old process has been improved to an extent that permits the FET to operate at competitive voltages with bipolar devices. This permits the use of unipolar and bipolar devices in the same circuit without the added power supplies or level shifters.

The process is silicon nitride passivation and has been used for some time on bipolar chips. It is primarily useful in MOSFET fabrication because of its high dielectric constant.

Recall that current flows from source to drain when sufficient voltage is applied to the gate electrode, over an area between them, creating mobile charges. The voltage V_T required to invert this area, the channel, is directly proportional to the charges present on the surface between the oxide and the substrate, as well as directly proportional to the thickness of the oxide insulator between the gate and the substrate. The threshold voltage is also inversely proportional to the dielectric constant of the gate insulator.

Thus, threshold voltage can be lowered by reducing the surface charge and/or the oxide thickness, or by increasing the insulator dielectric constant. It is not desirable to reduce oxide thickness any more than has been generally practiced, that is approximately 750 angstroms (Å). This is so because pinholes that develop in such thin oxides can have disastrous effects on device yields.

The most interesting and useful processing parameter for controlling threshold voltage is the high dielectric constant of silicon nitride. If silicon nitride is used as insulator between the gate and channel, it can be made physically thick but at the same time electrically thin compared with an oxide insulator. This is true because threshold voltage is inversely proportional to the dielectric constant of the gate insulator.

In addition, this high dielectric constant improves the gain factor of the FET by 50 percent. The gain factor is the amount of current output, relative to the voltage on the device, and it is directly proportional to the dielectric constant of the insulator and inversely proportional to its thickness.

It is not practical to use silicon nitride alone as the gate insulator because

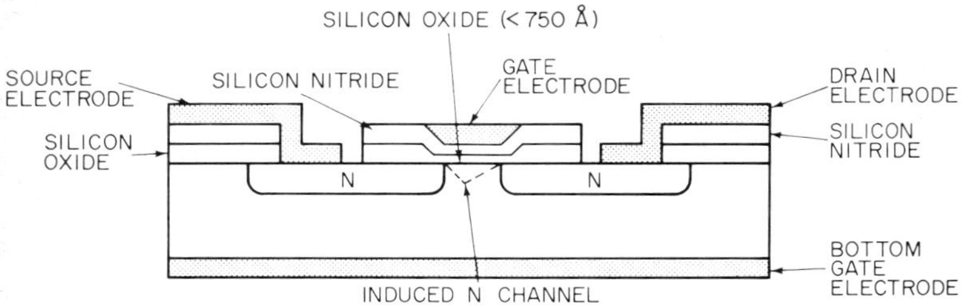

FIGURE 5.12
Use of silicon nitride to decrease threshold voltage (V_T).

it does not interface well with silicon. The uncommitted (dangling) bonds at the nitride interface behave like uncontrollable surface charges. Threshold voltages on devices fabricated with just silicon nitride as the insulator shift substantially in one direction with positive bias and in the other direction with a negative bias. This shift can be controlled by first growing a thin layer of silicon oxide next to the silicon and then depositing a precisely controlled layer of silicon nitride on top of this oxide (Fig. 5.12). This procedure results in improved device yields, since any defects in the oxide will not likely match any defects in the covering nitride layer.

Thus, the nitride process provides not only low-level voltage applications for the FET but also higher gain, more reliability, and greater adaptability to plastic packaging. In addition, the nitride process has provided a great impetus to MOS monolithic integrated-circuit research and production, as we shall see in Chap. 8.

Depletion-mode MOSFET. We described how a MOSFET works by using a P-type enhancement-mode device as an example, because the latter is one of the most popular modern types. There is, however, another type of MOSFET—the depletion type.

A depletion-mode MOSFET is shown in Fig. 5.13. Like the previous enhancement-mode MOSFET, the depletion unit has drain and source regions (here, N type) diffused into a P-type substrate. However, in the depletion-mode MOSFET, the drain and source regions are connected by a very thin channel of N-type crystal. In the enhancement-mode MOSFET, such a connecting channel had to be induced; in Fig. 5.13, it is built in.

The rest of the device is the same as the previous MOSFET. A gate lead is connected to a metal electrode and the latter is separated from the channel by a thin oxide layer.

Because of the built-in channel, charge carriers are present in the channel of the depletion MOSFET. If a reverse-biased gate voltage is applied (i.e., gate negative with respect to the source), the carrier charges are depleted

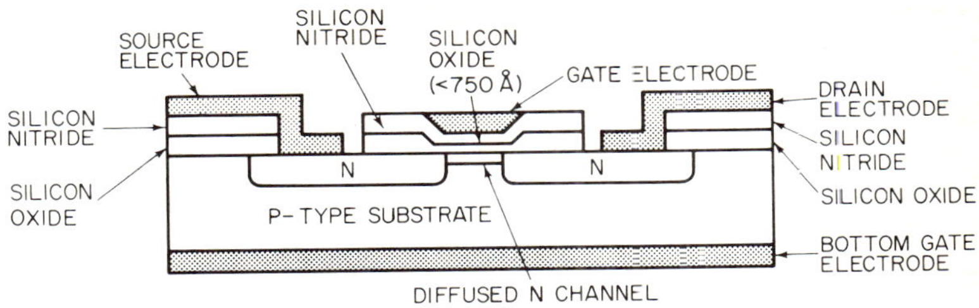

FIGURE 5.13
The internal construction of a depletion-mode MOSFET.

and channel conductivity is reduced. If the gate is forward-biased, more carriers enter the channel and conductivity is increased. Note that the channel region in the depletion-type unit is made from the same type of material as the source and drain. These devices are capable of providing substantial drain current even if no gate voltage is applied.

The current-voltage relationship for a depletion-mode MOSFET device is similar to that of the other field-effect transistors described. At low drain-to-source potential and with gate voltage equal to zero, the channel resistance is essentially constant and current will vary linearly with voltage, which is the familiar linear region. If the drain current is sufficiently increased, the voltage IR drop in the channel produces an increasingly greater voltage difference between the gate and points in the channel, which increases as the drain is approached. As the potential difference between gate and channel increases, the channel constricts and the current will increase much slower with further increases in drain-to-source voltage, which is recognized as the saturation region. Further increases in drain-to-source voltages will, as expected, result in avalanche breakdown.

The enhancement-mode MOSFETs are normally referred to as OFF devices, since the channel does not conduct with zero gate voltage. By contrast, the depletion-type device is normally referred to as an ON device, for reasons that are obvious. There is something common to both types of MOSFETs, however: in both, the conducting channel is changed by surface fields.

Although field-effect-transistor applications will be discussed in the next chapter, it is interesting to note here that they are being used in broadcast receivers, audio amplifiers, FM receivers, limiters, detectors, oscillators, frequency multipliers, phase splitters, choppers, pulse stretchers, current limiters, and voltage-controlled attenuators. With all this, the FET revolution is primarily a result of its advantages in monolithic integrated circuits.

Typical schematic symbols for junction FETs, enhancement-mode FETs, and depletion-mode FETs are shown in Fig. 5.14. We say typical because standardized FET symbols have not yet been established throughout the industry. However, the symbols shown here are widely employed.

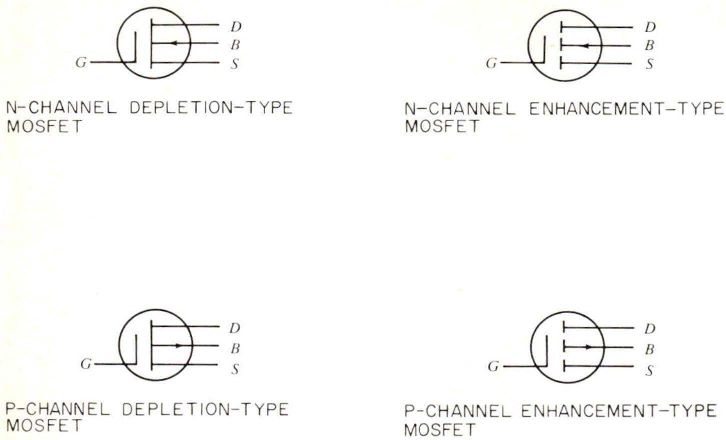

FIGURE 5.14
Schematic symbols for MOSFETs: G, gate; B, Bulk; D, Drain; S, Source.

FET Advantages. Since the FET is a voltage-controlled device rather than a current-controlled device like a bipolar transistor, it has a high-impedancelike vacuum tube. Because of its high impedance, it can be substituted in vacuum-tube circuit designs and driven directly from a high-impedance stage.

The FET is cheaper to produce than a comparable bipolar device, because the FET is smaller and more transistors can be processed on a silicon wafer. Also, the process itself requires only one diffusion rather than several critical and high-temperature diffusions as required in bipolar processing. All of this results in substantially reduced manufacturing costs.

Unipolar devices are more radiation-resistant than bipolar devices, since carrier lifetime effects are relatively unimportant to their operation. In our discussion of radiation effects in Chap. 1, we defined displacement as the knocking of an atom from its original crystal-lattice position to a new position. In Chap. 4, we described how, in bipolar devices, successful operation depends on the introduction of a few parts per million of impurities into the semiconductor crystal. Now if with radiation some of these impurities are knocked out of position, the device will not function properly. Some of the minority carriers traveling from the emitter to the collector through the base will be trapped in the radiation-formed empty lattice sites. Carrier lifetime, therefore, will be affected.

The MOSFET has very high gain at cryogenic temperatures, whereas the gain of bipolar devices decreases with decreasing temperature. This is the case because majority carriers are insensitive to temperature changes. Minority-carrier concentration, however, is proportional to the square of the intrinsic concentration, which is strongly dependent on temperature. Hence, unipolar devices exhibit much more stable operation over a wide temperature range.

Mathematical predication of device operation in the FET is relatively simple. This advantage is significant, since the design parameters can be accurately used (mathematically) to predict the required electrical characteristics. In bipolar-device design, the equations involved are not as straightforward and the predicted operating characteristics are consequently not as accurate.

MOSFET Limitations. Along with the cost-saving advantages of its being small, the MOSFET has acquired some inherent processing problems. The alignment of the gate mask is extremely critical. If the gate is misaligned by only a fraction of a micron in an enhancement-mode device, the inversion layer will not extend completely across the channel. Also, in processing the enhancement-mode devices, the gate metal is sometimes made to overlap the source and drain to compensate for processing (alignment) errors. This overlap increases the intrinsic capacitance associated with the stored changes on the gate and in the channel.

The principal disadvantage of the MOSFET is that it is unsuitable for high-frequency r-f work, mainly as a result of its significant drain-to-gate capacitance. Certain processing steps can be taken to reduce this capacitance, but unfortunately, up to now these have resulted in reducing the gain. The input of the MOSFET is applied to the gate electrode, and the output is taken from the drain electrode. Therefore, the drain-to-gate capacitance is basically a feedback capacitor from the output back to the input. The feedback is negative, since the output is 180° out of phase with the input as a result of the inversion, which tends to reduce gain. In addition, as the frequency increases, the feedback effect increases and gain is further reduced. Improved processing will undoubtedly reduce the interelectrode capacitances and lead to better frequency responses. Bipolar devices today are the logical choice for high-frequency and power applications.

Handling Considerations. The isolated-gate field-effect-device performance depends on the very thin oxide insulating layer between the gate and electrode. If this oxide is damaged by inadvertent application of excess voltage to the external gate connection, the damage is irreversible. Therefore, appropriate precautions must be taken to ensure that the gate voltage ratings are not exceeded.

The greatest threat to the gate insulator is static electricity. Large electrostatic charges can accumulate at the gate electrode in shipping if the device is allowed to slide around in plastic containers. This charge accumulation is usually avoided by use of conductive containers in shipping or by electrically interconnecting the leads in shipping.

Generation of electrostatic charge is also caused by handling. In humidities below 35 percent, people can accumulate an electrostatic potential of 300 V. If a "charged" person causes the gate lead to contact ground before the source and drain leads, the accumulated electrostatic charge will probably "blow" the oxide insulator and thereby destroy the device. This type of damage is usually avoided by using an electrostatic grounding strap during all handling.

Even though the thin oxide layer may be damaged by improper handling of isolated-gate field-effect devices before they are connected into actual circuits, thousands of hours of practical operation have proven that these devices are extremely reliable.

BIBLIOGRAPHY

1. Grove, A. S.: "Physics and Technology of Semiconductor Devices," John Wiley & Sons, Inc., New York, 1967.
2. Engineering Staff, Motorola, Inc.: "Integrated Circuits," McGraw-Hill Book Company, New York, 1965.
3. Richman, P.: "Characteristics and Operation of MOS Field-effect Devices," McGraw-Hill Book Company, New York, 1967.
4. Wallmark, J. F.: The Field-Effect Transistor . . . An Old Device with New Promise, *IEEE Spectrum,* March, 1964.
5. Theory of Insulated-gate Field-effect Transistor near and beyond Pinch-off, *Solid State Electronics,* vol. 9, February, 1966.

QUESTIONS

5.1. Why is the FET known as a unipolar device?

5.2. Describe the main difference between unipolar and bipolar devices.

5.3. What is a majority carrier, a minority carrier?

5.4. What bipolar processing innovation in 1960 accelerated FET development? Why?

5.5. What is inversion?

5.6. Define threshold voltage.

5.7. What is the advantage of silicon nitride in controlling threshold voltage?

5.8. Describe the three regions of the drain-current-vs.-drain-source voltage characteristics of any FET.

5.9. What is the main difference between an enhancement-mode and a depletion-mode MOSFET?

5.10. Sketch a P-type enhancement-mode MOSFET, labeling the gate, source, drain, and channel. Do the same for an N-channel depletion-mode MOSFET.

Chapter Six

Transistor Amplifiers

In the preceding chapters, we noted how a transistor functions internally and how it achieves the desired goal of amplification. Several amplifier circuits were mentioned, but the discussion was incidental to the main discussion of transistor characteristics. In this chapter, we shall concentrate on transistor amplifiers—to help in understanding the forms they take and how they operate. Most of our discussion will concern bipolar transistors, but we will also describe a few unipolar (FET) amplifiers, since a discussion of modern transistor amplifiers would be incomplete if the unipolars were ignored.

This chapter deals entirely with discrete transistor amplifiers. In Chap. 8, we will describe some monolithic integrated circuits that with the fundamentals presented in this chapter will be easier to understand.

Bipolar transistors may be connected into a circuit in one of three ways, or configurations: the common base, the common emitter, or the common collector. The most versatile and widely used configuration is the common emitter. However, the other two configurations have particular advantages and applications; it is essential, therefore, that we become familiar with them.

It is usually desirable to know the voltage-, current-, and power-gain ranges that the circuit can deliver once the configuration has been established. These are tabulated in Table 6.1 together with the input and output impedances, since these are directly related to voltage and power gains.

Normally, an input signal is applied between two terminals of a transistor and extracted from two terminals, one of which is different and one of which is common to both the input and output circuits. This terminal, common to input and output, is connected in practice directly to ground or separated from ground by a fixed voltage source.

In the vast majority of circuit diagrams used in this book, the input signal is developed across an impedance between the input terminal and ground. The output signal is developed across another impedance between the output terminal and ground. All circuits, then, must have a reference voltage against which all other voltages in the circuit, positive or negative,

TABLE 6.1
A Comparison of the Characteristics of the Three Forms of Bipolar Transistor Amplifiers

	Circuit Configuration		
	Common Base	*Common Emitter*	*Common Collector*
Current gain	Less than unity	High	High
Voltage gain	More than 100	Several hundred	Approx. 1
Power gain	Medium	High	Low
Input impedance	Low	Low	High
Output impedance	High	Medium	Low

direct or alternating, can be compared; and the most convenient reference voltage is ground, or zero, volts

COMMON-BASE AMPLIFIER

Practical circuits employing each of the three transistor amplifier arrangements are shown in the illustrations on the pages to follow. In Fig. 6.1, a PNP transistor is used. If we did not know offhand whether this was an NPN or a PNP transistor, we could use the emitter and collector battery polarities as our clue. The emitter must be biased in the forward direction. This means that the positive battery terminal will repel holes.

Therefore, we know that the emitter is P-type silicon. It follows then that the base has N-type silicon and the collector P-type silicon. In short, we have a PNP transistor.

FIGURE 6.1
A common-base transistor amplifier.

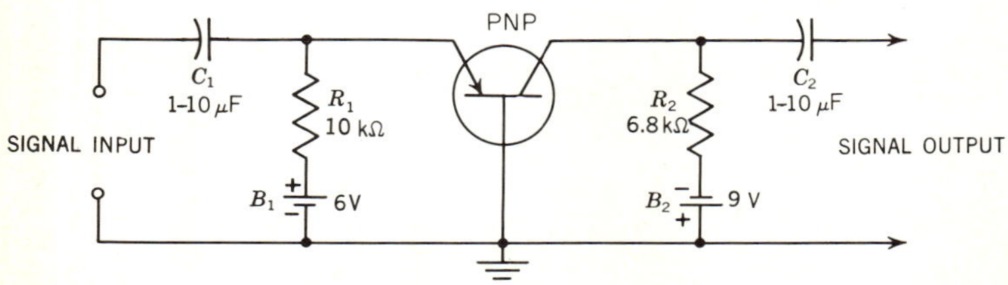

The input signal is RC-coupled to the emitter. The emitter bias is established by battery B_1, 6 volt (V). Current flow through the emitter is governed by R_1, a 10,000-ohm (Ω) resistor. By using Ohm's law, we find that

$$E = IR$$

$$6 = I \times 10{,}000$$

$$I = 0.6 \text{ milliampere (mA)}$$

Actually, by this reasoning we completely neglect the resistance of the emitter-base section. However, the latter is so small with the indicated battery arrangement that it scarcely alters the total current flow. Actually the emitter-base potential is on the order of 0.1 V or less.

The output, or collector, circuit possesses a 9-V battery and a 6,800-Ω load resistor. Signal voltages developed across R_2 are then capacitively coupled to the next stage.

The two coupling capacitors C_1 and C_2 are shown with capacitance ranges from 1 to 10 microfarad (μF). The use of such high values is dictated by the relatively low input impedance of this stage (and for C_2, the relatively low input impedance of the following stage). With an input impedance on the order of 200 to 300 Ω, it is desirable that the impedance offered by C_1 to the lowest operating frequency be no more than 20 Ω. To achieve this, large values of capacitance must be used.

C_1 and C_2 should be as large as possible. When 1-μF values are used, the frequency response is such that the gain at 100 cycles is 16 percent of the gain at 1,000 cycles. When 10-μF capacitors are used, the 100-cycle gain rises to 46 percent of the 1,000-cycle value, which is a marked improvement. The extent of the high-frequency end of the curve is governed by the capacitances shunting the circuit and the manner in which α drops off with frequency.

The use of two batteries is somewhat of a disadvantage, however, which can be remedied by a voltage divider, as shown in Fig. 6.2. Since resistors R_3

FIGURE 6.2
A common-base amplifier using only one bias battery.

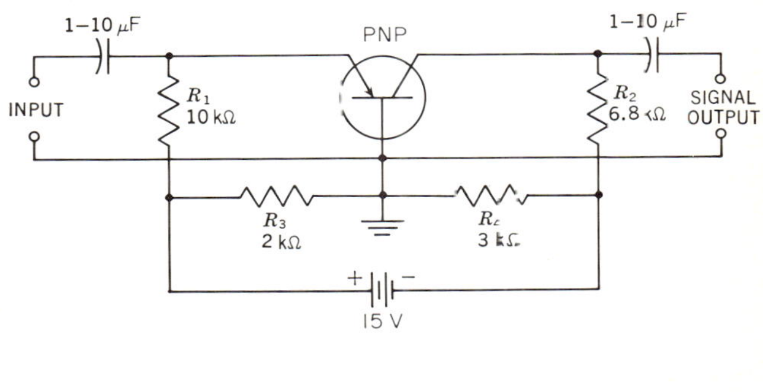

and R_4 have values in the ratio of 2:3, the voltages developed across them will have the same ratio. Thus, 6 V will develop across R_3 and 9 V will develop across R_4. Across R_3 the ungrounded end is positive, whereas it is negative across R_4. This provides the appropriate polarity voltages for the emitter and collector, and the circuit functions in the same manner as that in Fig. 6.1.

In using the voltage-divider arrangement of Fig. 6.2, it is desirable to make R_3 and R_4 as low as possible so that variations in emitter or collector currents will not have any appreciable effect in altering the current through R_3 and R_4 and consequently causing their voltage to drop. On the other hand, if R_3 and R_4 are made too small, the current drain on the battery becomes excessive. The values shown in Fig. 6.2 represent a compromise between these two conflicting goals.

COMMON-EMITTER AMPLIFIER

Much more widespread than the use of common-base transistor amplifiers is the use of common-emitter amplifiers. With the common-emitter arrangement we obtain greater current and power gain. A circuit using a PNP transistor in a common-emitter arrangement is shown in Fig. 6.3. Connection of the input and output resistors and capacitors remains the same as in the preceding amplifier. Note, however, the use of a single battery for both circuits. This is possible because the emitter is common to both input and output circuits and both collector and base require voltages that possess the same relative polarity with respect to the emitter. (In the present discussion, the emphasis is on circuit form rather than on the selection of the most stable circuits. An analytical method of determining the sensitivity of an amplifier circuit to temperature and other factors is given in Chap. 13.)

FIGURE 6.3
A single-stage common-emitter amplifier.

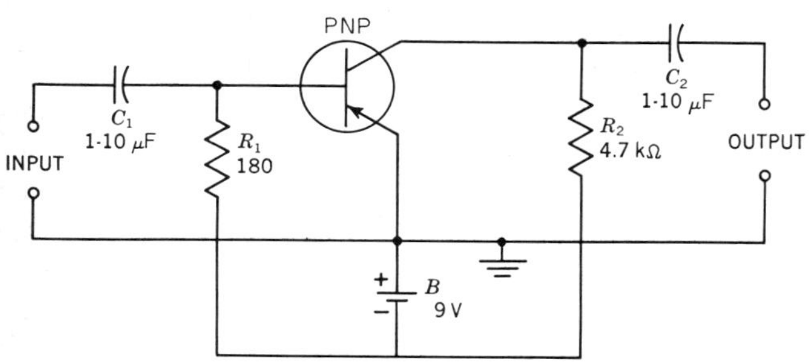

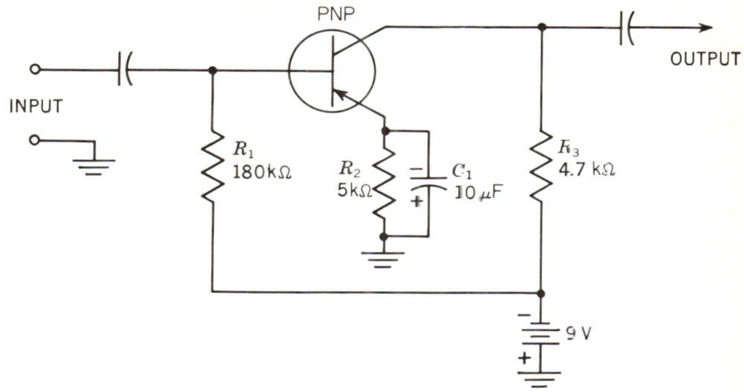

FIGURE 6.4
A common-emitter amplifier that employs a stabilizing resistor R_2.

A form of common-emitter amplifier that is frequently seen is shown in Fig. 6.4. The chief difference between this circuit and that of Fig. 6.3 is the 5,000-Ω resistor R_2 and filter bypass capacitor C_1 inserted in the emitter lead. Resistor R_2 serves to dc stabilize the circuit by compensating for differences between transistors and by reducing the effects caused by temperature drift. Capacitor C_1 is shunted across R_2 to prevent ac degeneration with reduction in gain. In some instances, the added stability provided by ac degeneration may be desired, in which case C_1 would be omitted.

The problem of amplifier stability being affected by temperature changes is more serious in common-emitter and common-collector circuits than it is in common-base circuits because of the presence of a cutoff current I_{CO}, which was mentioned briefly in Chap. 3. I_{CO} is the current that flows through the collector-base diodes when the emitter current is zero. It stems from the presence of minority carriers in the base and collector sections, and it gives rise to a small current when the collector is reverse-biased. I_{CO} is generally below 10 μA, and it is independent of the emitter current. Its value is determined chiefly by the particular transistor being used and by the temperature. It is the latter dependence that is particularly significant.

When a transistor is connected with the base common to both input and output circuits, as in Fig. 6.2., then the total collector current that flows is made up of two components—$I_C = \alpha I_E + I_{CO}$: that is, I_C in a junction transistor is equal to 98 percent (or so) of the emitter current I_E plus the collector cutoff current I_{CO}. Since I_{CO} is in microamperes and I_E is in the milliampere range or higher, changes in I_{CO}, unless drastic, will not seriously increase the heat dissipated at the collector or significantly change the operating point. Hence, we need not take other than the normal precautions with common-base amplifiers.

Consider now the common-emitter circuit (Fig. 6.3). I_{CO} still flows between the base and collector sections, but now the base current determines the amount of collector current flowing. This, too, we noted in Chap. 3; and it was because of this relationship that we developed a second current-gain factor β, which is equal to $\alpha/(1 - \alpha)$, and values of 50 or more are not unusual. The total collector current now flowing is given by

$$I_C = \beta I_B + (1 + \beta)I_{CO}$$

Previously, the factor $(1 + \beta)I_{CO}$ was ignored. When the transistor is subjected to fairly wide ranges in temperature, however, it is possible for this factor to develop values high enough to affect transistor operation seriously. I_{CO} is extremely sensitive to temperature, and any increase in this current will be magnified 50 or more times because of the presence of $(1 + \beta)$. This can have a marked effect not only on the collector current but—what is equally disturbing to the circuit—on the bias or operating point. What it will do is shift this point out of the linear region, giving rise to an increase in distortion.

The insertion of a series resistor in the emitter leg is designed to prevent the foregoing action from occurring. If we use the circuit of Fig. 6.4 as an illustration, all the collector current will flow through the emitter resistor. The voltage drop produced across R_2 serves to make the emitter negative with respect to ground. Note, however, that the base is also negative with respect to ground, and hence the base-emitter voltage will be the difference between these two negative voltages, with the base generally more negative than the emitter. Now let us say that the collector current rises because of a temperature-induced rise in I_{CO}. This will cause the voltage drop across R_2 to increase, making the overall base-to-emitter voltage less negative than it was before. This is actually working against the forward-biasing voltage of the base-emitter circuit, resulting in less emitter current. Hence, we are counteracting the rise in I_C by decreasing I_B and I_E. In this way we achieve stabilization of our amplifier circuit.

A variation of this stabilization circuit is that shown in Fig. 6.5. Here the base is connected to a voltage divider. This arrangement provides greater stabilization than its predecessor, but the additional resistor does absorb more power from the battery and from the signal source. In this sense, then, the circuit is less efficient.

The reader will recognize that every transistor circuit has two kinds of stability, dc and ac (or signal). In dc stability, we desire to maintain the same operating point irrespective of any changes in the transistor or in the values of any of the other components in the circuit. It is generally the changes in the transistor that cause the most trouble, but variations in the values of circuit resistors can also have a marked effect on circuit operation.

Ac stability refers to the ability of the circuit to treat any signals that pass through in the same way. The most common variation occurs in the amplification accorded different-frequency signals, but different-amplitude sig-

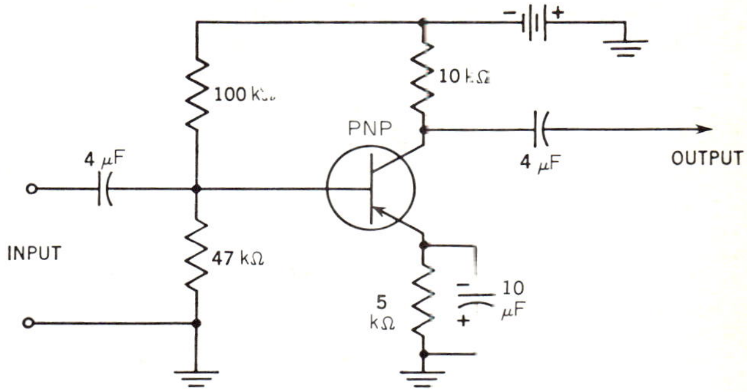

FIGURE 6.5
A variation of the stabilized amplifier for Fig. 6.4.

nals can also be treated differently. For example, a strong signal may receive a greater distortion than a smaller signal. In Fig. 6.5, removal of the 10-μF capacitor shunting the 5,000-Ω emitter resistor will provide more uniform amplification to signals over a wider frequency and amplitude range than if this capacitor were permitted to remain. This is called current feedback. Another method is voltage feedback, which will be illustrated presently.

COMMON-COLLECTOR AMPLIFIER

A single-stage common-collector amplifier is shown in Fig. 6.6. The input signal is applied between base and ground. Since the collector is at ac ground potential because of C_1, however, we can say that the signal is effectively being applied between base and collector. The output is taken from a load resistor between emitter and ground or, what is the same thing, between emitter

FIGURE 6.6
A single-stage common-collector amplifier.

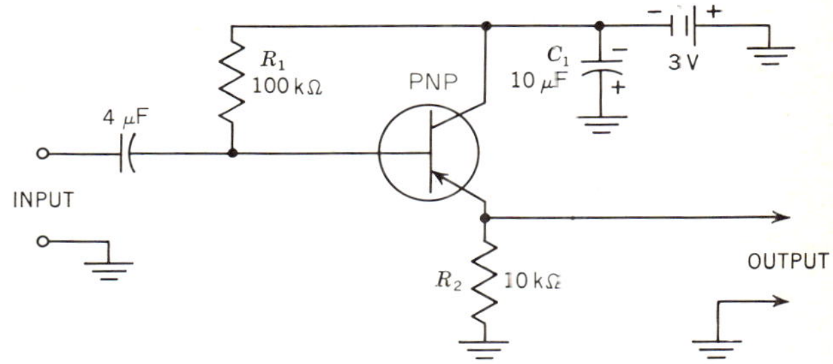

and collector. This circuit has a high input impedance, on the order of 100,000 Ω or more, and an output impedance of 200 Ω. The voltage gain of a common-collector amplifier cannot exceed 1, and in the circuit shown it is about 0.9. Power gain, however, is 15. The voltage gain is relatively independent of frequency, but the current gain falls off with frequency exactly as it does for a common-emitter amplifier. The fall-off of power gain, with frequency, is therefore about midway between that of a common-base amplifier and that of a common-emitter amplifier.

Still another variation of the common-collector amplifier is shown in Fig. 6.7*a*. The load resistor is the 47,000-Ω (R_4) resistor in the emitter leg. In addition, there is feedback (furnished by C_2) between the emitter and base. The latter is designed to decrease the shunting effect of the base voltage divider.

The frequency response of this circuit for two different bias voltages is shown in Fig. 6.7*b*. Note that in neither case does this gain exceed 1. (The change in circuit presentation is purposely being made to help the reader become familiar with the different methods of illustration that he will encounter.

FIGURE 6.7
(*a*) Another common-collector circuit. (*b*) The frequency response of the circuit shown in *a* for two different bias voltages.

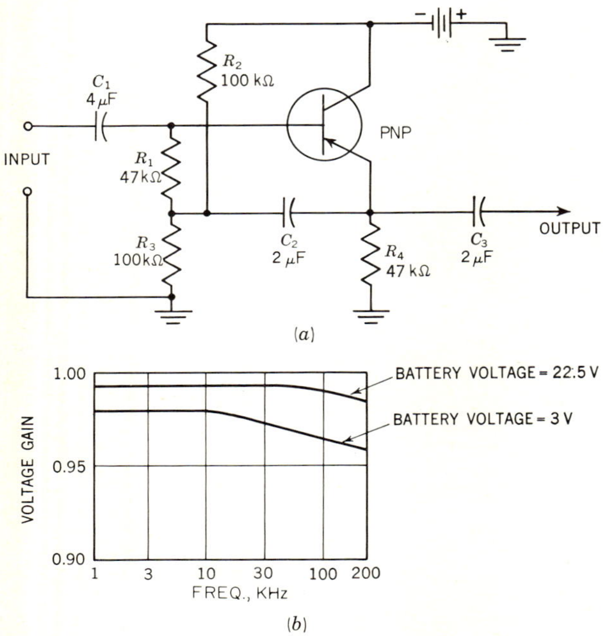

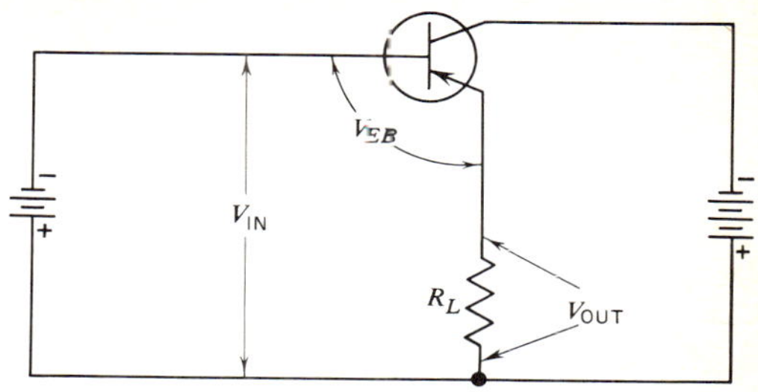

FIGURE 6.8

The important items to look for are the point of application of the input signal and the takeoff point of the output signal.)

Another name for this circuit is bootstrap circuit.

It is relatively simple to derive the current gain of a common-collector amplifier. The input current is I_B, since the incoming signal is fed to the base. The output current is I_E, since the load resistor is in the emitter lead. Hence

$$\text{Current gain} = \frac{I_E}{I_B}$$

$$I_E = I_B + I_C$$

$$\text{Current gain} = \frac{I_B + I_C}{I_B}$$

$$= 1 + \frac{I_C}{I_B}$$

$$= 1 + \beta$$

To all intents and purposes, this is equal to β, because beta is normally greater than 30. Thus, the common-emitter and the common-collector amplifiers provide approximately the same current gain.

The input impedance of a common-collector stage is approximately equal to β times R_L, or βR_L. This can be shown as follows. The input impedance of the stage R_{IN} is equal to the input voltage divided by the input current. From Fig. 6.8, the input voltage is equal to $V_{EB} + V_{OUT}$. These voltages can be added directly because they are always in phase for a resistive load. Thus,

$$R_{IN} = \frac{V_{EB} + V_{OUT}}{I_B}$$

$$= \frac{V_{EB}}{I_B} + \frac{V_{OUT}}{I_B} \qquad (4.1)$$

Now, V_{EB}/I_B is the internal resistance between base and emitter r_{be}; and since this junction is forward-biased, the value is very low. On the other hand,

V_{OUT}/I_B depends on the circuit as well as the transistor. We noted just above that I_E/I_B is equal to $\beta + 1$, or, essentially β. From this we may state that

$$\frac{V_{\text{OUT}}}{I_B} = \frac{\beta V_{\text{OUT}}}{I_E}$$

Since $V_{\text{OUT}}/I_E = R_L$, we have

$$\frac{V_{\text{OUT}}}{I_E} = \beta R_L$$

Substituting this into Eq. (4.1) above, we obtain

$$\begin{aligned} R_{\text{IN}} &= r_{be} + \beta R_L \\ &\cong \beta R_L \end{aligned}$$

This demonstrates that it is desirable to use as high a value of R_L as is practical in order to obtain a high input impedance. In turn, this means that for common-collector amplifiers, the operating bias current should be as low as possible; otherwise, the dc voltage dropped across R_L would be so high that an uneconomically large bias supply voltage would be required.

One word of caution. R_{IN} is governed not only by R_L and β but also by any resistor that connects to the base. Thus, if the base-bias current reaches this element through a low-valued resistor, the input impedance will be lowered accordingly, because the resistor will shunt the input circuit. Also, circuit capacitance across the base will affect the input impedance, particularly as the operating frequency rises. This, of course, is just as true of transistor amplifiers as it is of vacuum-tube amplifiers. For example, at 16 hertz (Hz), a 10 picofarad (pF) capacitor presents an impedance of 1 megohm (MΩ). At 100 Hz, the capacitance impedance has decreased to about 167,000 Ω. The drop is significant. It must be remembered that 10 pF is quite small and readily developed through stray circuit-wiring capacitance alone. Hence, considerable care must be exercised when designing and constructing such stages.

By way of contrast, common-base and common-emitter amplifiers have quite low input impedances; hence, much more shunt capacitance would be required before the input impedance is affected.

Since β can also be represented by h_{FE} or h_{fe}, the above formulas are frequently shown using the latter symbols. h_{FE} is the dc beta and h_{fe} is the ac beta.

COUPLED AMPLIFIERS

Transistor amplifiers are seldom used singly. Rather, it is more common to find them in groups, with two, three, or more stages coupled to each other.

With vacuum-tube amplifiers and, as we shall see, FET amplifiers, it is relatively simple to connect them one after the other because of a much higher input than output impedance. Hence, when the input of one stage is attached to the output of the preceding stage, we ordinarily do not affect the preceding stage.

There are three basic methods used to couple transistor stages: transformer coupling, resistance-capacitance (RC) coupling, and direct coupling.

Let us consider a transistor amplifier—one designed using the common-emitter configuration—with an input impedance on the order of 1,000 Ω. The operating output impedance is more likely to be between 10,000 and 20,000 Ω. Obviously, a direct connection between two stages will result in a significant loss in gain as a result of this mismatch. If we accept the reduced gain, it becomes necessary to use more stages in order to obtain a desired amplification. A more practical solution would be to insert a device that will match the higher output impedance of one stage to the lower input impedance of the following stage. Such a device would be a step-down transformer.

Although this solution has been used efficiently and economically, transformer-coupled amplifiers do not ordinarily possess the same flat frequency response that can be obtained from RC networks. Both methods are used, however, and typical amplifiers of both types will be examined.

A two-stage transformer-coupled grounded-emitter amplifier is shown in Fig. 6.9. The interstage transformers have primary impedances of 20,000 Ω each and secondary impedances of 1,000 Ω each. Capacitors C_1 and C_2 are 10 μF in value, and resistors R_1 and R_2 are 150,000 Ω each. The two resistors are needed to establish the proper base-emitter forward bias, and the two capacitors are inserted to prevent grounding of the base bias through the low dc resistance of the transformer secondary windings. Overall power gain of this particular combination is approximately 50 decibels (dB). The primary advantage of transformer coupling is this transfer of power from one impedance level to another.

If resistance-capacitance coupling is used, some loss in power gain is ex-

FIGURE 6.9
A two-stage transformer-coupled grounded-emitter amplifier.

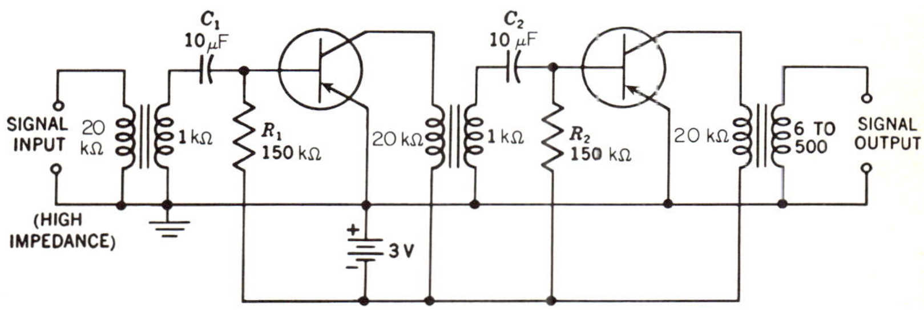

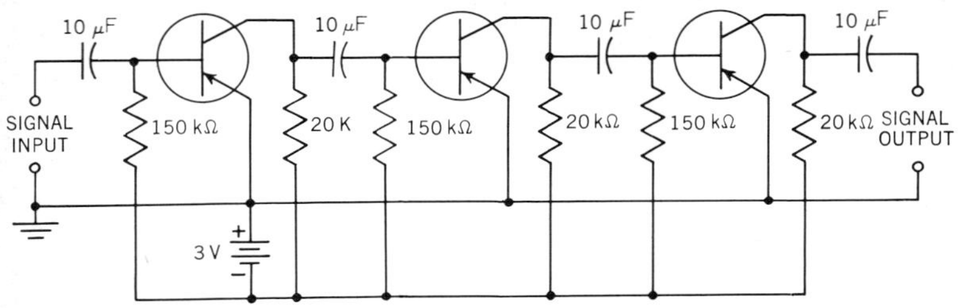

FIGURE 6.10
A resistance-coupled grounded-emitter amplifier that will provide approximately the same amount of overall power gain as the amplifier of Fig. 6.9.

perienced; but gains in circuitry costs and size reduction will at times offset this power loss.

A resistance-capacitance-coupled grounded-emitter amplifier is shown in Fig. 6.10. Note that three stages are required because of the mismatch between the output of one stage and the input of the following stage.

Direct coupling is used when cost is a primary factor. A two-stage amplifier with high input impedance and dc stabilization is shown in Fig. 6.11. The higher input impedance is achieved by the use of a grounded-collector stage, the signal of which is forwarded to a grounded-emitter amplifier. Insertion of 10,000-Ω resistors in the emitter leads of both transistors provides amplifier stabilization against temperature changes. The first 10,000-Ω resistor cannot, of course, be bypassed, since the signal is obtained from this point. In the second stage, however, an 8 μF bypassing capacitor is employed.

FIGURE 6.11
A two-stage amplifier with high input impedance and dc stabilization.

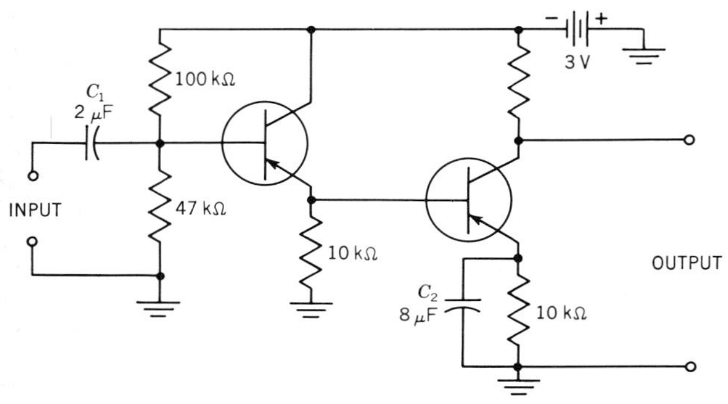

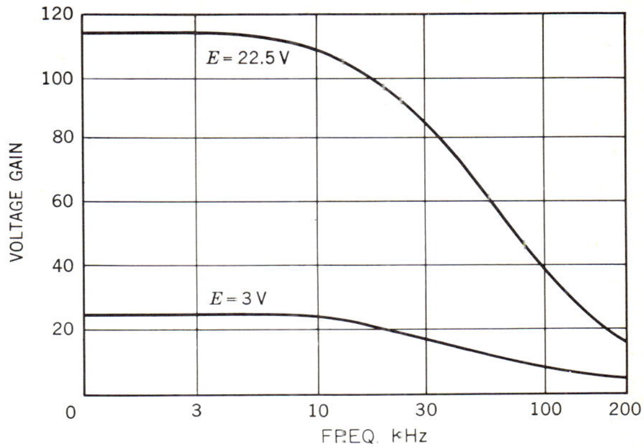

FIGURE 6.12
The frequency-response behavior of the amplifier shown in Fig. 6.11 at two different bias voltages.

The frequency-response behavior of this direct-coupled amplifier, at two different bias voltages, is shown in Fig. 6.12. In transistor circuits, as in vacuum-tube circuits, we obtain more gain for higher voltages. The only precautions to observe are those dictated by the maximum safe-operating currents, the temperature, the breakdown-voltage values, and collector dissipation. Also, the number of stages that can be directly coupled is limited, since any temperature variation of the bias current in one stage may be amplified by all the stages, and severe temperature instability could result.

In amplifier circuits such as we have been discussing, it is frequently desirable to employ a decoupling filter across the battery or power supply. This is shown in Fig. 6.13, where R_f and C_f serve this function. The need for these components stems from the impedance of the power source (be it battery or ac supply) and the necessity of preventing positive feedback from a later stage where the signal level is high to a prior stage where it is low.

FIGURE 6.13
A resistance-coupled amplifier with a decoupling filter C_f and R_f. It might be desirable to add additional filter sections if motorboating is encountered.

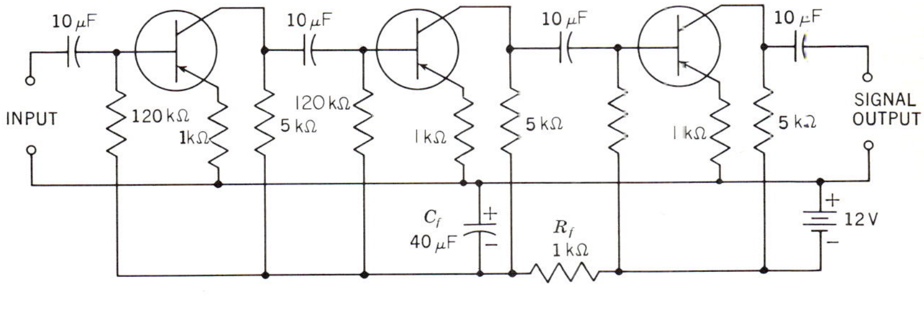

In choosing values for R_f and C_f, it is best to restrict R_f to fairly small values so that the operating voltage to the stages situated prior to R_f is not reduced to a value that will materially affect their gain. Remember, however, that the smaller R_f becomes, the larger C_f must be made in order to obtain effective filtering action. In general, the time constant of R_fC_f should be greater than $1/f$ for the lowest frequency passed by the amplifier. In computing this time constant, C_f is expressed in farads, R_f in ohms, and f in hertz. The values indicated for R_f and C_f are typical.

NEGATIVE FEEDBACK IN TRANSISTOR AMPLIFIERS

Basically a feedback network returns part of an output signal to the input circuit of an amplifier. If the feedback signal is returned in phase with the input signal, it is positive feedback; and if the returning signal is 180° out of phase with the input signal, it is negative feedback. In positive or negative feedback, the returning signal can be made proportional to either the output voltage or the output current and can be applied as either input voltage or input current.

Negative feedback improves amplifier stability, reduces distortion, increases input impedances, and reduces possible variations in gain caused by different transistors. This last advantage has become somewhat insignificant in the past few years in view of the new processing and testing procedures in transistor production that have led to more-uniform transistor types.

FIGURE 6.14
A two-stage amplifier employing negative feedback.

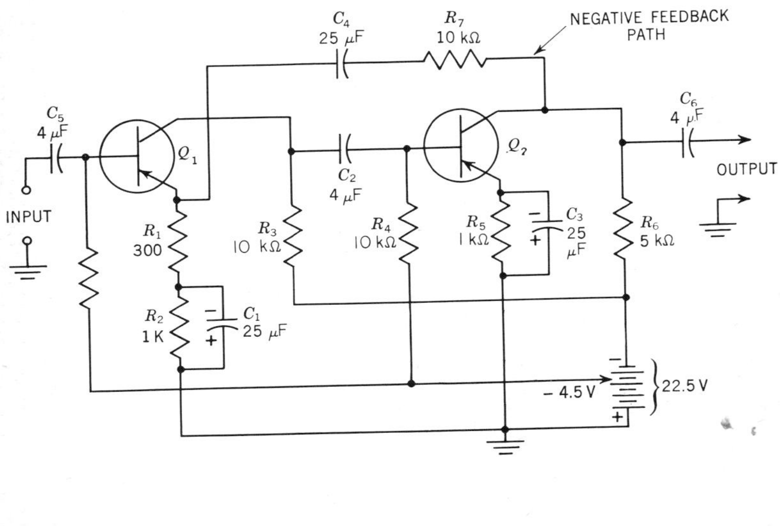

All the advantages of negative feedback are not obtained without some penalty—in this case, a loss in gain. The loss is not a serious one, however, because of the high voltage amplifications available. We seldom lack sufficient gain; we usually have more than we actually require.

A simple form of negative feedback is obtained by leaving unbypassed the dc stabilization resistor in the emitter lead. This is a single-stage approach to be used or not, as desired, by each of the various stages

Another form of negative feedback is shown in the two-stage amplifier of Fig. 6.14. The feedback loop here extends from the output circuit of Q_2 to the emitter circuit of Q_1. Involved in this feedback are two resistors R_1 and R_7 and one capacitor C_4. R_1 is needed to provide a means of inserting the feedback energy into the emitter circuit of Q_1; hence, it is left unbypassed. R_2, in the same emitter circuit, is bypassed by C_1, and no feedback voltage is developed across these two parallel components. R_2, however, does provide dc stabilization for Q_1. R_5 does the same for Q_2.

In the negative feedback of voltage, we know that the phase of the signal feedback must be 180° from the phase of the incoming signal. It may be instructive to check the signal polarities in the circuit of Fig. 6.14 to see if this condition holds true. (The procedure will also help the reader become familiar with the methods of checking signal polarities in transistor circuits.) If we assume that the incoming signal, applied to the base of Q_1, is positive at some instant, then the signal voltage at the collector of the transistor is negative. This stems from the 180° phase reversal that occurs in a common-emitter amplifier.

The negative signal at the collector of Q_1 is also negative at the base of Q_2. This produces a positive signal at the collector of Q_2, and a portion of this signal is fed back to the emitter of Q_1. Thus, we have a positive signal at the base and a smaller positive signal at the emitter. Since these two

FIGURE 6.15
The effect of negative feedback on the frequency response of the amplifier shown in Fig. 6.14.

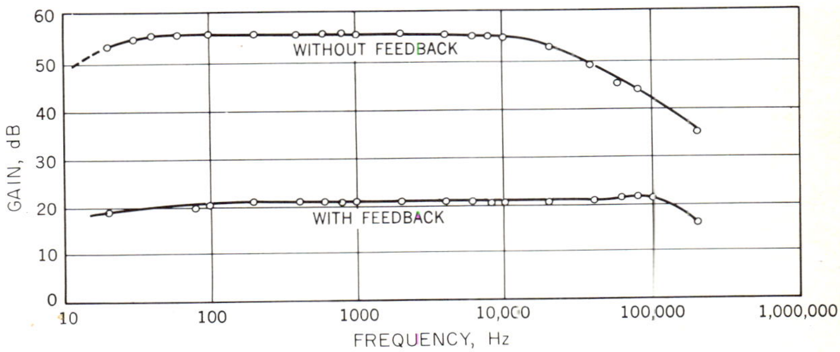

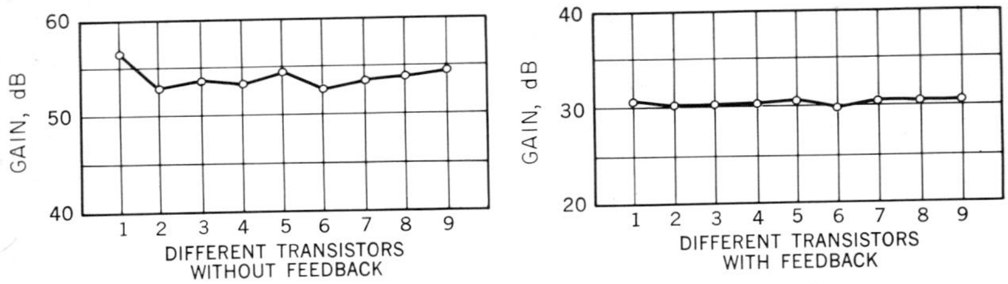

FIGURE 6.16
Negative feedback in an amplifier (such as the one shown in Fig. 6.14) will serve to reduce the variations in overall gain that different transistors will introduce owing to varying characteristics.

voltages will work in opposition to each other in forcing current through the emitter-base circuit, we obtain negative feedback.

The effect of negative feedback on the frequency response is shown in Fig. 6.15. Note how much flatter the curve is with feedback. Also instructive are the two curves in Fig. 6.16. The left-hand curve shows how the overall gain could vary with different transistors when there is no feedback. Note how much better the action becomes when feedback is employed.

The point of feedback return is governed by the phase conditions in the circuit around which the feedback is sent. Consider, for example, the two-stage amplifier shown in Fig. 6.17. The first stage is a grounded collector, which does not introduce any phase reversal in the signal. The second stage is a grounded emitter, which causes a 180° reversal. Under these conditions,

FIGURE 6.17
Another negative-feedback arrangement.

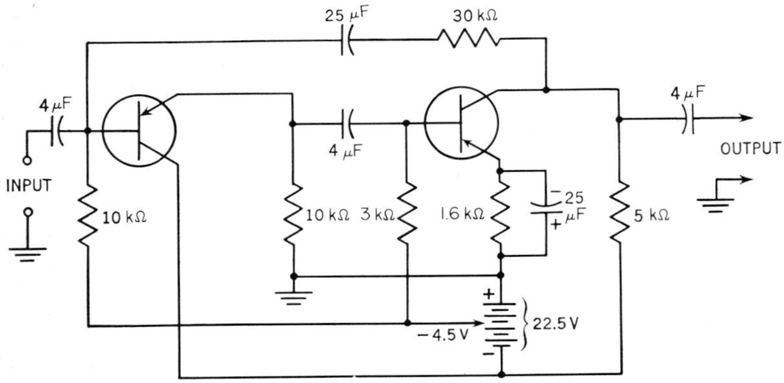

the feedback line from the collector of the second stage can be brought back to any signal point prior to the base of the stage. The point chosen in this particular circuit is the base of the grounded-collector stage, but it could just have easily been the emitter of this stage or the base of the second stage. At all these points, signal polarity is the same. The feedback in Fig. 6.17 is voltage feedback, in contrast to the current feedback obtained when the emitter resistor is left unbypassed.

TRANSISTOR-CIRCUIT CONSIDERATIONS

In dealing with transistors and transistor circuits, a careful distinction must be drawn between the limitations of the transistor as a device and the circuit in which it is placed. Perhaps the most obvious example of this occurs at the low-frequency end of an amplifier-response characteristic. Here it is the external circuit elements—capacitors and resistors—that are responsible for a drop-off in gain. The transistor itself is capable of amplifying down to direct current; and once the signal is brought to the transistor, it will receive as much amplification at 30 Hz as at 300 or 3,000 Hz. It is in bringing the signal to the transistor through the coupling network that fall-off occurs.

At this point, still another factor must be recognized: while the transistor itself has uniform amplification at the low frequencies, it can influence the external circuitry by its input and output impedances. For example, the common-base amplifier has a very low input impedance, which means that any coupling capacitor connected to the transistor input must itself possess a very small impedance so that it does not rob any signal from the transistor. This is shown in simplified form in Fig. 6.18. The current leaving the collector of the preceding stage sees two paths to follow. It can either flow through the collector-load resistor R_L or it can go through C_1 and the input impedance of the following transistor. In both instances, it can return to the emitter of its own transistor, thereby completing the circuit.

Now, in order to bring as much signal to Q_2 as possible, it is desirable to divert as much collector current through C_1 and the input impedance of Q_2 as

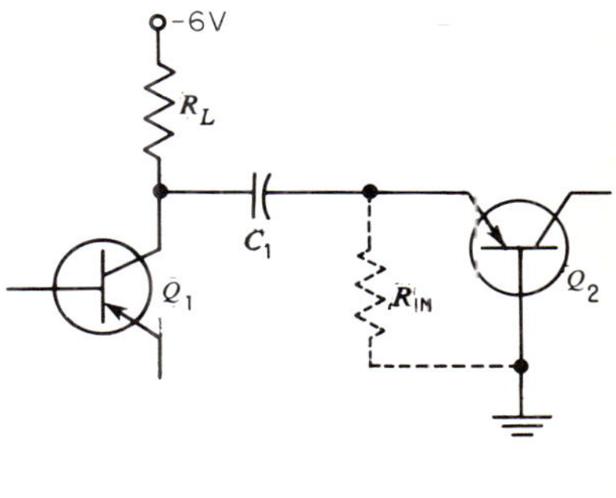

FIGURE 6.18
Simplified diagram of interstate coupling network between two transistor stages.

possible. Since the input impedance of Q_2 is very low, C_1 must be made as large as possible so that its reactance will be small.

Consider, now, a common-emitter circuit at Q_2. Its input impedance will be larger than for a common-base circuit. Consequently, it will not be necessary to employ as large a capacitor C_1 as before because with the input impedance larger, lowering the impedance of C_1 below a certain point will not measurably alter the total series impedance of the input circuit. Finally, with a common-collector arrangement at Q_2, we will see a large input impedance. This means that C_1 will have even less effect on the total series impedance so that a smaller value of C_1 can be employed than in either of the two preceding circuits.

The reader will recognize, of course, that what we are dealing with here is the relative impedance of C_1 and the input to the transistor. If the transistor input is lower, C_1 will need to be larger before its effect can be considered negligible. As the input impedance rises, the best value for C_1 can change accordingly. In this sense, the transistor can affect its connecting circuit. But given a certain transistor, we can then look to the external circuit to establish the low-frequency end of the network.

At the high-frequency end of the amplifier response, it is the transistor and the drop-off in its β that govern circuit behavior. This will be discussed more fully presently. While the transistor is usually the primary cause, poor external circuit design can also affect the high-frequency cutoff. For example, too large a shunting capacitor will lower the upper cutoff frequency. This, too, was noted previously. Also, too small a coupling capacitor can reduce the signal voltage reaching the next stage. But if the circuit has been properly designed, then in most instances it is the transistor that establishes the upper-frequency limit.

Bias Considerations. Another factor to consider when comparing the three transistor-amplifier arrangements is the precaution to observe when biasing these circuits. A very common biasing method, one which helps to maintain amplifier stability, is the circuit shown in Fig 6.19. R_1 and R_2 form a voltage divider that provides the base with the necessary potential to forward-bias

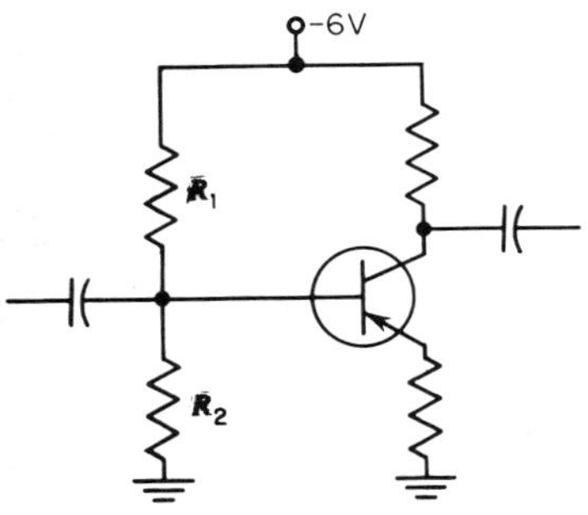

FIGURE 6.19
A widely used biasing network, R_1 and R_2.

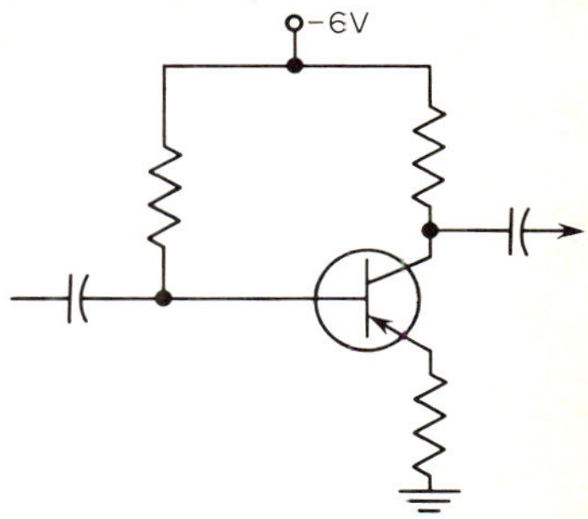

FIGURE 6.20
Another method of biasing a transistor.

the base-emitter diode. With a given battery voltage, R_1 and R_2 can assume a wide range of values to achieve the proper voltage division for the base. The smaller R_1 and R_2 are, however, the more stable the circuit. Now, let us see what complications this brings with the three transistor configurations.

For a common-base amplifier, R_1 and R_2 can have their lowest values, because the input impedance of the common-base arrangement is low. Hence, very little difficulty is encountered here.

In a common-emitter circuit, the input impedance is higher, and we must be careful not to make R_1 and R_2 so low that they bypass the signal current arriving from the preceding stage. Also, if R_1 and R_2 are low, they will load the preceding stage more than the input impedance of the transistor that they are attached to would, and this will lead to a lower over-all circuit gain. In selecting higher values for R_1 and R_2, however, we reduce the temperature stability of the circuit.

Finally, in a common-collector stage, where the input impedance is very high, we would try to avoid using the bias approach of Fig. 6.19 and go to something like the method shown in Fig. 6.20. The temperature stability of the stage with this bias network of Fig. 6.20 is much poorer than it would be by using the bias circuit of Fig. 6.19, but R_1 and R_2 of Fig. 6.19 would act to reduce the input impedance and thus undercut one of the major features of the common collector.

Complementary Symmetry. An amplifier arrangement that is unique to transistors, and impossible with tubes, makes use of the fact that there are two basic kinds of transistors: NPN and PNP units. Each is the symmetrical counterpart of the other, and the polarity of an input signal necessary to increase conduction in a PNP transistor is the opposite of that necessary to increase conduction in an NPN transistor.

A direct-coupled amplifier that makes use of this symmetry is shown in Fig. 6.21. The first transistor is an NPN unit; the second, a PNP type. The first stage is set up so that the collector current flowing through its load resistor R_4 develops just enough voltage here to make the base of the PNP

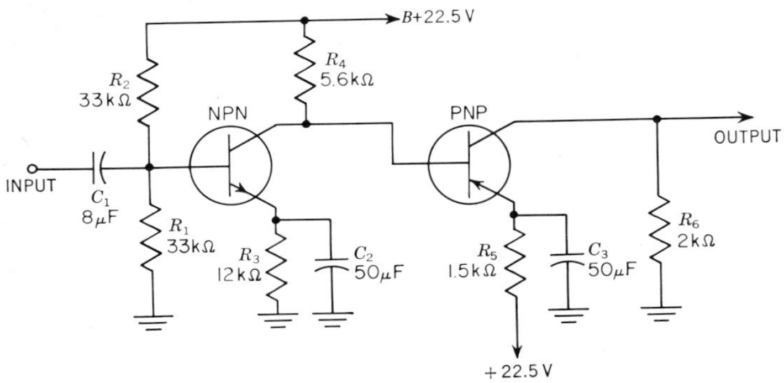

FIGURE 6.21
A direct-coupled amplifier that makes use of the complementary nature of NPN and PNP transistors. Voltage gain of this system is 660; power gain is 53 dB.

transistor negative with respect to its emitter. This establishes the proper conditions in the emitter-base circuit of the PNP unit to bias it in the forward direction. Thus, by the proper choice of resistor values and battery potential, both stages will operate as class A amplifiers. (We shall discuss amplifier classes presently.)

The application of a signal to the base of the NPN stage will result in an amplified signal appearing across R_6. For example, when the signal at the base of the NPN transistor goes positive, an amplified negative voltage will appear across R_4 (collector end negative with respect to battery end). This increasing negative voltage will provide an even greater forward bias for the base-emitter circuit of the PNP transistor and cause an increased flow of current through this unit. Electrons will flow up through R_6, making the top end positive with respect to the bottom end.

Thus, the positive signal applied to the input of this amplifier appears with the same polarity, but in amplified form, at the output. Note the simplicity of this arrangement, requiring as it does no coupling capacitors and only one battery supply. (Both 22½-V potentials shown would come from one source.)

Another two-stage amplifier designed along somewhat similar lines is shown in Fig. 6.22. This system has for its sole purpose the amplification of pulses, and its mode of operation is therefore modified accordingly. For example, if you examine the base-to-emitter circuits of both stages, you will note that no forward bias is employed. The characteristic curves for grounded-emitter operation (such as we have here) reveal that when the base current is zero (that is, $I_b = 0$), the collector current is quite small. In terms of operation, this means that the transistor is biased close to cutoff. This is true in both amplifiers of Fig. 6.22, although the second stage is not so close to cutoff as is the first stage. This is because the small collector

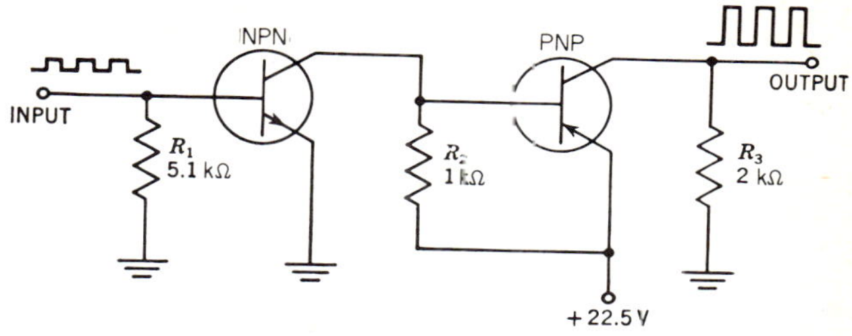

FIGURE 6.22
A direct-coupled pulse amplifier.

current that flows from the first transistor passes through R_2 and the small forward-biasing voltage that is developed shifts the operating point of the second transistor away from cutoff.

In the circuit of Fig. 6.22 a positive pulse of 0.25 V input to the first stage is amplified to a 20-V peak at the output of the second stage. Conduction is required only when the pulses are applied—hence the reason for the cutoff biasing.

Direct-coupled amplifiers must be designed carefully to minimize the stability problems they present. Any change in the dc operating point of one stage immediately alters the dc operating points of all succeeding stages. This in itself might not be so bad if it were not for the fact that most stages are connected common-emitter and a change in input base current produces β times that change in the collector circuit. Since common values of β range from 50 to 100 or more, it will be readily recognized that in a direct-coupled string of stages, a minute change in an early stage will quickly build up, after a few stages, to substantial proportions. To prevent this buildup, considerable dc feedback must be employed, and this will not only add to the expense of the circuit but also generally act to affect the gain adversely.

The principal offender, in most instances, is the change in I_{CO} with temperature. Hence, any arrangement of direct-coupled amplifiers that may be devised must somehow minimize the effect of changes in I_{CO}. This has led to circuit designs such as the one shown in Fig. 6.23. Two silicon transistors are directly coupled, with the signal output from Q_1 proceeding directly through R_4 to the base of Q_2. The two transistors share a common-emitter resistor R_E, and this has the effect of eliminating the degeneration that normally results from an unbypassed emitter resistor. Degeneration is effectively eliminated because the ac signals through R_E from each transistor are 180° out of phase.

The effect of the I_{CO_1} of Q_1 on Q_2 can be seen from the following analysis. Any increase in I_{CO_1} will cause a change in the collector current of Q_1 by a factor of βI_{CO_1}. This increased current flowing through R_3 will produce a greater voltage drop, with the collector end of R_3 becoming more positive.

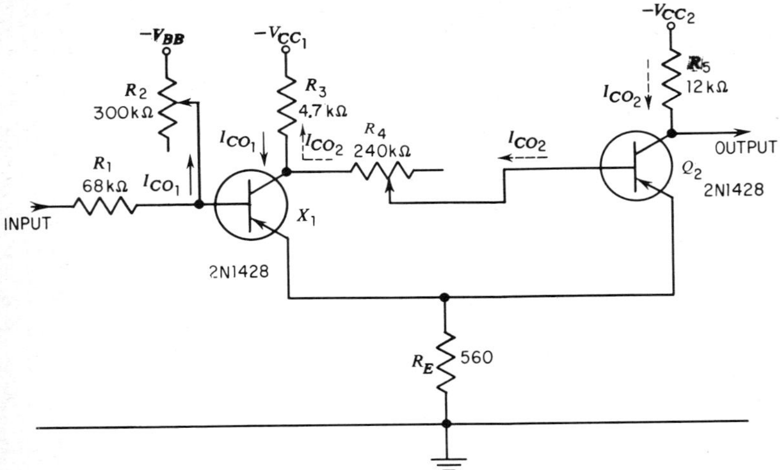

FIGURE 6.23
A low-drift direct-coupled amplifier.

Since Q_2 is a PNP transistor, the increase in positive voltage across R_3 will drive it closer to cutoff; hence, this will alter the operating point of Q_2 and certainly modify its effect on any signals passing through the circuit.

In this arrangement, the I_{CO_2} of Q_2 and the I_{CO_1} of Q_1 are made to offset each other's effects. I_{CO_1} flows in the path indicated by the solid arrows in Fig. 6.23, and I_{CO_2} flows in the path indicated by the dotted arrows. By adjusting the value of R_4, we can regulate the additional current that the voltage across R_3 sends into the base of Q_2. (Consider the additional voltage drop across R_3 as a small battery. Then the current fed to the base of Q_2 will be

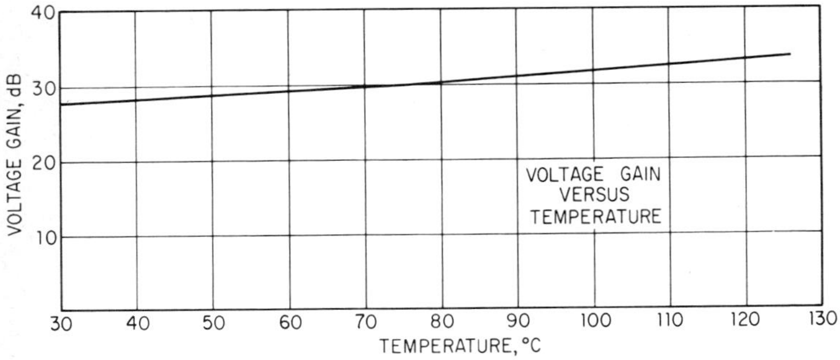

FIGURE 6.24
The variation in voltage gain versus temperature for the direct-coupled amplifier of Fig. 6.23.

proportional to this voltage divided by R_4. This, of course, is Ohm's law.) R_4 is selected to have this current equal I_{CO_2}. Since the two currents flow in opposite directions, changes in I_{CO_1} and I_{CO_2} can be minimized. By using similar transistors for Q_1 and Q_2, we are more likely to have fairly identical variations in I_{CO_1} and I_{CO_2}, which will permit compensation over a wide range of temperatures. Figure 6.24 shows the low drift in voltage gain with temperature for this arrangement.

Two bias supplies are required in the amplifier, because R_3 is small compared with R_5, yet each transistor should have approximately the same operating point. Since the emitter voltage is developed across a common-emitter resistor, the collector supply voltage is necessarily smaller for the transistor with the smaller load resistor.

DARLINGTON AMPLIFIER

An interesting amplifier arrangement that provides very high current gains is the Darlington amplifier shown in Fig. 6.25. Two transistors are employed, with the emitter of one transistor directly connected to the base of the second unit. The input signal is applied to the base of Q_1 and the output signal is obtained from the emitter of Q_2.

The high current gain of this combination occurs in the following manner. The input current i_b produces an amplified current in the collector of Q_1 equal to βi_b. Since the collector current of Q_1 is essentially equal to the emitter current of Q_1, then βi_b flows in the emitter of Q_1. This now becomes the base current for Q_2, and following the same line of reasoning, the collector and emitter currents of Q_2 are $\beta\beta i_b$ or $\beta^2 i_b$, assuming the two transistors are similar.

Thus, if the value of β is 40, the current amplification of this arrangement is 1,600. If the β values are not equal, then the combined amplification is $\beta_1\beta_2$.

FIGURE 6.25
The basic Darlington amplifier.

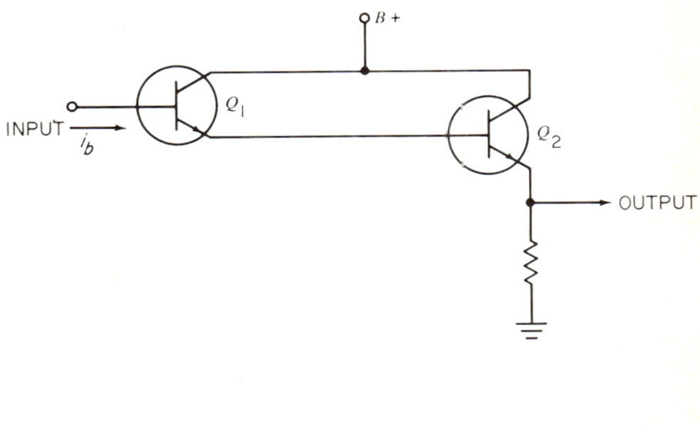

POWER AMPLIFIERS

Since all transistors are current-operated devices, all serve essentially as power amplifiers. In this section, however, we are concerned with high-power units, those which are capable of dissipating powers in excess of 1 watt (W).

In the last few years there have been significant advances in power-transistor capabilities, particularly at the higher frequencies. Semiconductor manufacturers, constantly pushing for more power at high frequencies, have pushed the power-frequency "state of the art" to such levels as 100 W at 150 MHz, 20 W at 2 GHz, and 10 W in the 3-to-5-GHz range. It remains to be seen how far the power-frequency levels can be pushed, but some manufacturing managers are predicting that 200 W at 4 to 6 GHz will be available in the 1970s.

In Chap. 3 we discussed some of the limitations, and trade-offs, involved in the design of high-frequency transistors. One of the trade-offs mentioned was that high-frequency power applications require a small collector-base area for low collector capacitance and a large collector-base area for current-handling and power-dissipation requirements. We also stated in Chap. 3 that a large emitter periphery-to-area ratio is desirable for increased current handling and better emitter efficiency. The result of all this is that long, very thin emitters have been confined in appropriate base areas. Let us look at this a little more closely.

The obvious approach in designing a higher power transistor would be simply to enlarge the dimensions of the emitter, base, and collector elements until the desired amount of current is obtained at the specified heat-dissipation rating.

However, as the transistor dimensions are increased, the emitter efficiency decreases. That is, the emitters' ability to inject carriers into the base decreases. This is so because in a large transistor the base current becomes substantial and in flowing through the base layer produces an ohmic voltage drop. As a result, most of the emitter current is obtained from a narrow strip along the emitter edge, close to the base contact. This has led to the development of long, thin emitters.

All modern planar transistors have a common physical characteristic. That is, the emitter and base contact areas are on the same plane (Fig. 6.26). Note also that the emitter is flanked by base contacts on each side and that the actual emitter-base junction is under the emitter contact area. Thus, there is a relatively significant distance between the base contact and the actual emitter base junction, $x - x'$ in Fig. 6.26. This means that the base drive voltage must approach the emitter-base junction from the side, as indicated in Fig. 6.26. It also means that the base current has to pass between the emitter-base and collector-base junctions or, in other words, laterally along the base width. Recall that the smaller the base width, the higher the dc beta. However, the smaller the base width, the higher the base resistance,

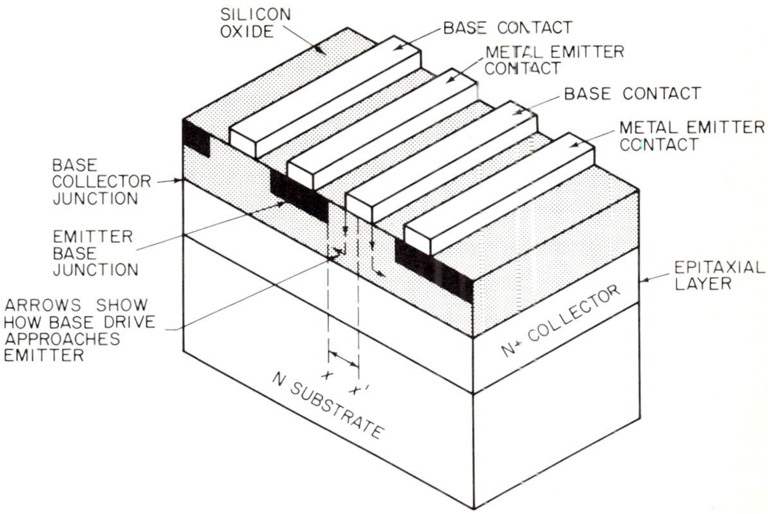

FIGURE 6.26
Cross section of planar (interdigited) transistor showing emitter and base contacts in same plane. Note that all base contacts are connected together (in parallel) and the same is true of all emitter contacts.

which results in a larger voltage drop for a given base drive. Thus, the voltage reaching the emitter (even along the edges) is not as high as the emitter-base voltage V_{BE} applied to the external contacts—which is one of the reasons why the base contacts are placed as close to the emitters as possible.

Hence, we see that the middle section of the emitter will receive less emitter-base voltage than the edges, resulting in a smaller current "turn-on." This is the current pinch-off effect. The current pinch-off effect gets worse as the device is driven harder, because the current-carrying emitter area does not increase as rapidly as it does at lower current levels.

The amount of voltage that reaches the emitter-base junction is also a function of frequency, because the base resistance has a shunt capacitance associated with it—and in fact behaves like an RC filter (Fig. 6.27). This means that less emitter-base voltage will appear under the emitter as frequency increases, which, in effect, decreases the active area of the device. Figure 6.28 illustrates the accompanying current drop under the emitter. Finer geometries provide a higher active-to-physical area ratio for a given frequency and thus a more efficient device at that frequency.

These considerations have led to such base-emitter structures as those shown in Fig. 6.29. We see in three of the four illustrations in Fig. 6.29 the star shape of the emitter, with long fingers radiating from a center section.

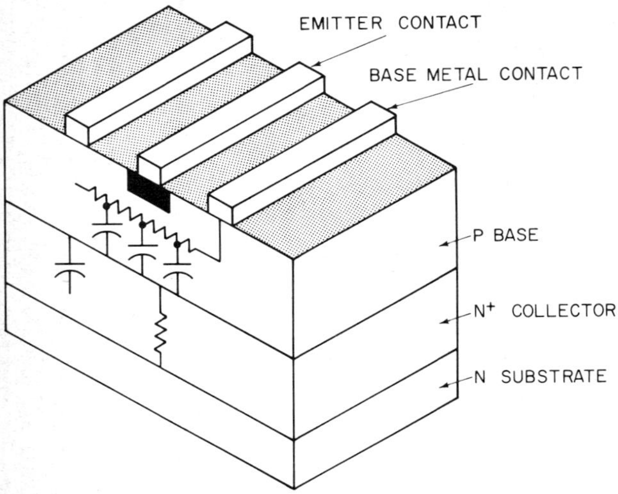

FIGURE 6.27

FIGURE 6.28
A comparison of fine (b) and coarse (a) geometries. Fine geometries provide more active-to-physical area ratio for a given frequency.

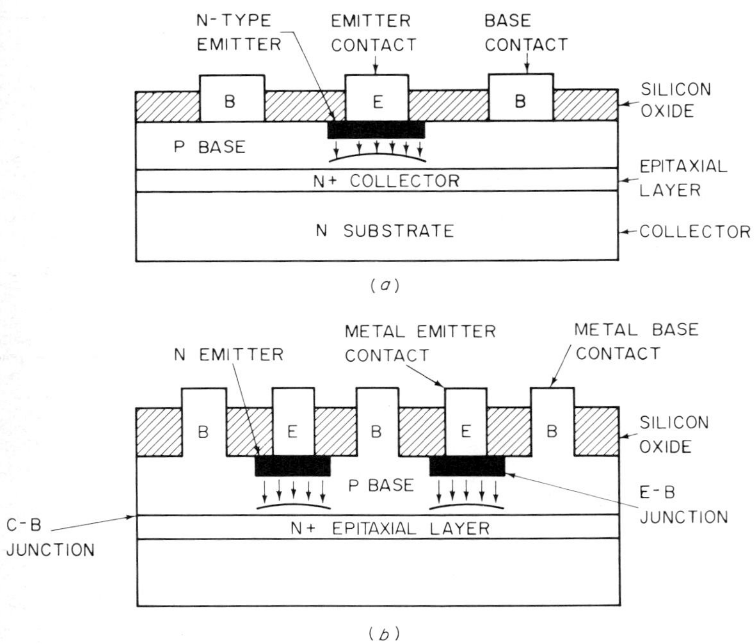

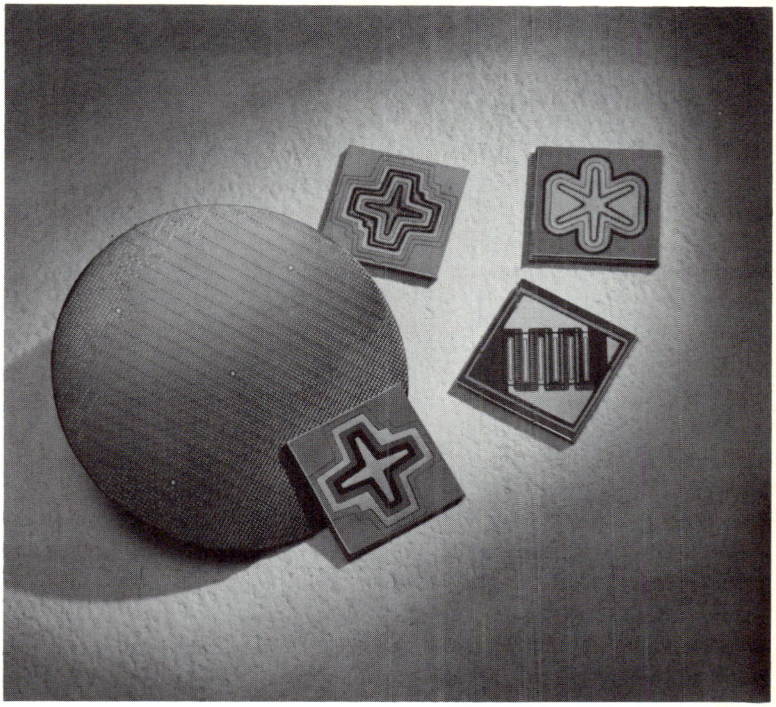

FIGURE 6.29
Internal structure of several power transistors. (*Courtesy Motorola.*)

The fourth illustration in Fig. 6.29 has six parallel emitters lined up in parallel, similar to the configuration shown in Fig. 6.28*b*. All the emitter terminals here are electrically connected together, as are all of the base terminals. An extension of this basic arrangement, shown in Fig. 6.30, is known as an interdigitated structure.

Secondary Breakdown. Despite the utilization of optimum device design parameters and power-temperature ratings, modern high-frequency power transistors are subject to secondary breakdown, a condition that is usually destructive.

If the emitter-base bias is fixed, the collector current will increase slowly with increasing collector voltage, until avalanche breakdown occurs. Avalanche itself is nondestructive if power dissipation is contained within limits. However, if the collector current is allowed to increase to a high value, secondary breakdown can occur.

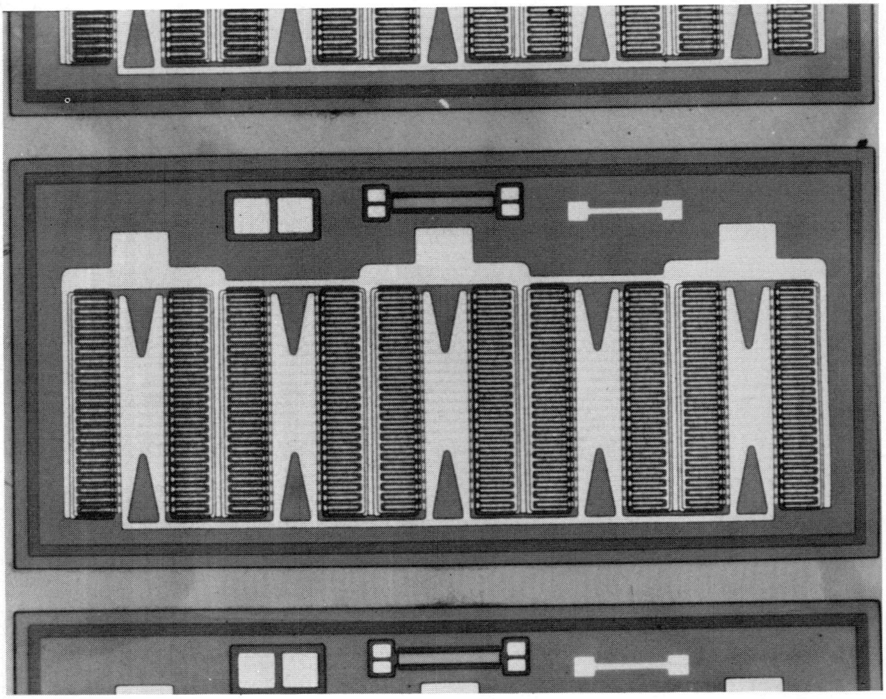

FIGURE 6.30
R-f interdigitated chip. ***(Courtesy Motorola.)***

The exact mechanism of secondary breakdown is still being debated; however, it is associated with a sudden concentration of energy into a small area. The energy is a function of collector current, collector voltage, and time; and when this energy dissipates power in a small area, the temperature can get hot enough to melt the silicon. The result of this is that the two junctions may flow together and produce a collector-to-emitter short.

Safe Operating Areas. Secondary breakdown, as we mentioned, is a function of time, as well as voltage and current; and since transistors can be operated at infinite time intervals and operating points, safe-operating-area (SOA) curves are produced that include time as a parameter.

These curves are generated by the destructive testing of five to ten transistors at specified pulse widths and voltages. Special test circuits are used, usually with the source voltage set at fixed levels in appropriate steps for each class of transistor, and collector current is increased by increasing the base drive until secondary breakdown occurs. The voltage and current failure points are observed on an oscilloscope and plotted. Normally, the safe-area-operating curves are incorporated in the specification sheet of the power device and the published curves are derated 20 percent, in voltage and current, from the lowest observed failures, thus providing an extra margin of safety.

Another cause of high-frequency transistor "burn-out" is due to load mismatches. If the load impedance differs from the transistor's output impedance, some of the power that under matched conditions would be dissipated in the load is reflected back into the transistor. This means that the transistor has to dissipate extra power, which increases its temperature and often results in device burn-out.

Most transistor manufacturers today employ emitter-ballast resistors to help avoid burn-out in high-frequency power transistors. Emitter resistors are either diffused or deposited at each emitter "finger" site (Fig. 6.31). These resistors prevent hot spots by assuring even current distribution to each emitter finger and thus preventing current crowding in a particular section of the emitter area. Another method used to avoid burn-out is to make certain that the device's thermal capacity far exceeds the power-output rating. For example, a transistor with a 50-W thermal capacity (employing emitter-ballast resistors) would be sold as a 24-W device.

Heat Removal. Power transistors develop considerable amounts of current and this current, flowing through the transistor elements themselves, as well as the internal connecting wires will naturally generate a certain amount of heat. As the temperature rises, the efficiency of the transistor decreases. If the temperature rises high enough, the transistor may easily burn out. Hence, it becomes important with power transistors to bring the internally generated heat out of the transistor housing as quickly as possible and then conduct this heat away from the unit as efficiently as possible.

FIGURE 6.31
Top view of emitter-ballast resistors designed to provide uniform current distribution and prevent hot-spot formation.

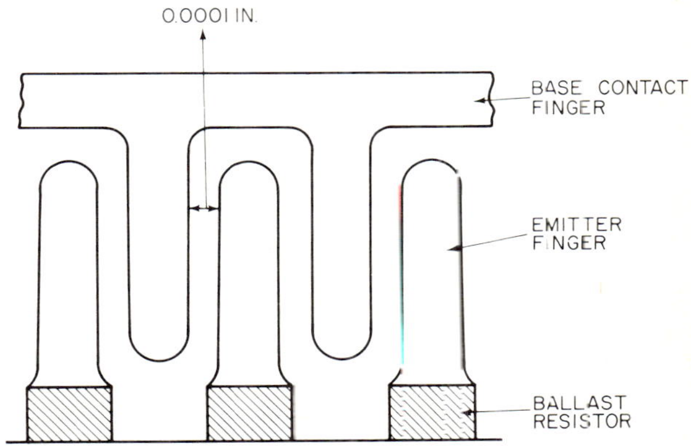

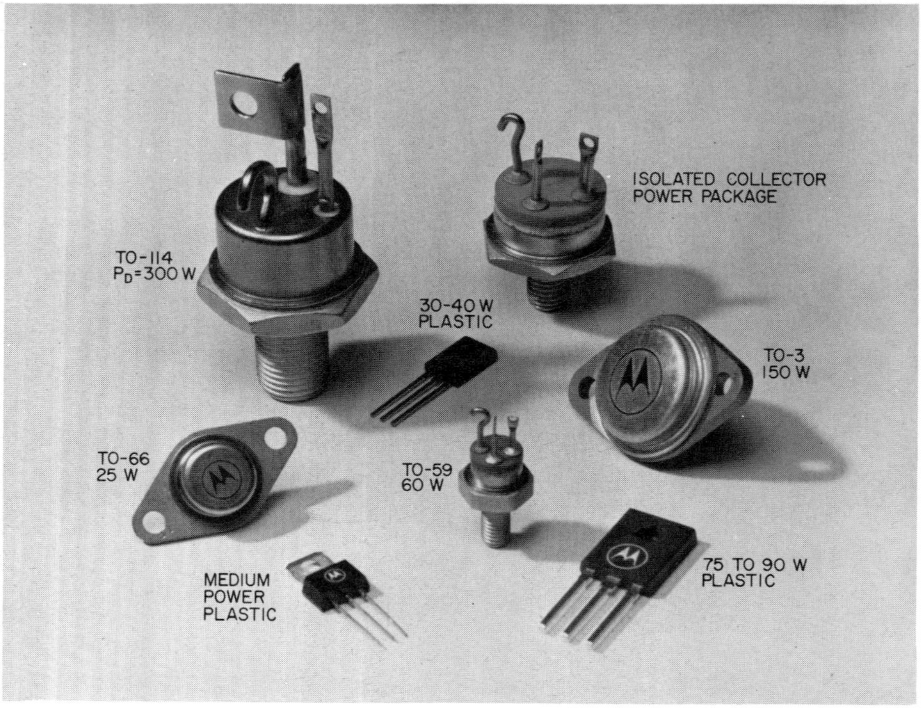

FIGURE 6.32
A variety of physical structures used for power transistors.
(*Courtesy Motorola*.)

Heat can be compared with electric current in the sense that it flows from a point of high heat concentration (i.e., high voltage) to a point of low heat concentration (i.e., low voltage). And, to carry the analogy further, just as some materials have lower resistance to current flow than others, so do some materials have lower thermal resistance than others. For example, copper and gold are good heat conductors (i.e., they have low thermal resistance), whereas glass is a poor thermal conductor.

The objective, then, is to employ materials of low thermal resistance internally in the transistor as well as for its outer case. In addition, the exposable surface of the housing is made large in order to achieve as much cooling as possible, either to the surrounding air or to a chassis.

Figure 6.32 shows a variety of power-transistor housings designed to provide a sufficiently large heat-radiating surface. For low-power applications, attaching a transistor securely to a copper or aluminum chassis will provide sufficient heat removal to maintain normal operation.

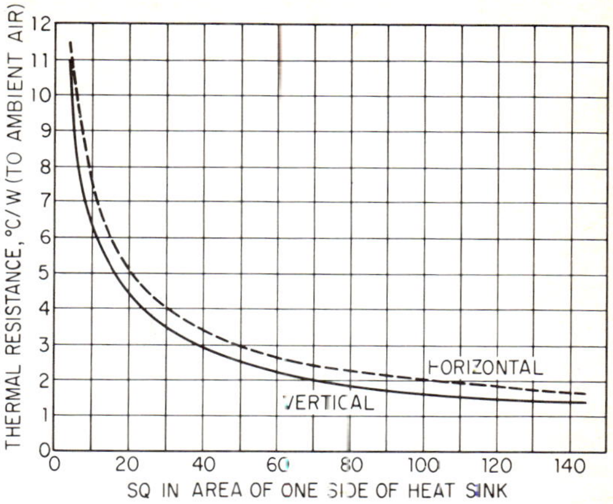

FIGURE 6.33
The variation in heat-transfer characteristics of ⅛-in-thick aluminum as its area and method of mounting are varied.

High-dissipation applications will generate more junction heat than can safely be dissipated by the chassis itself. If unlimited space is available, plain sheets of copper or aluminum may be used as heat sinks. Figure 6.33 indicates the variation that may be expected in the heat-radiating efficiency of ⅛-in aluminum as its area and orientation are varied. Figure 6.34 gives corresponding data for ⅛-in sheet copper.

When space limitations forbid the use of large sheets of metal as heat sinks, it is possible to use finned heat sinks of a design that permits the concentration of large surface areas within a small space. The heat sink shown in Fig. 6.35 compresses 80 in^2 of radiating surface into an overall volume that measures $4\frac{5}{8} \times 1\frac{3}{8} \times 3$ in. Made of extruded aluminum, the heat sink is painted flat black to facilitate the radiation of heat.

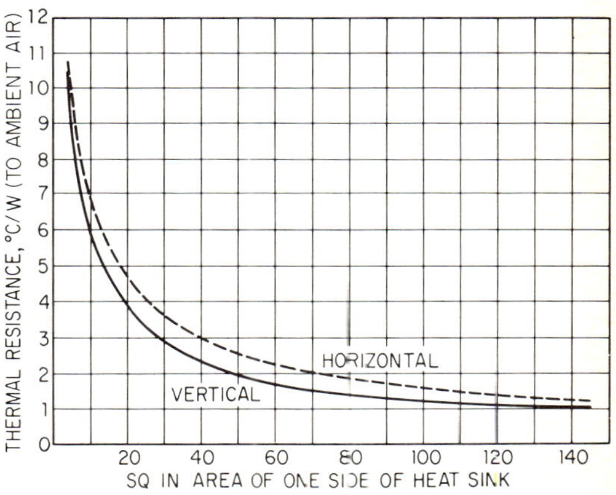

FIGURE 6.34
The variation in heat-radiating efficiency of ⅛-in-thick copper as its area and method of mounting are varied.

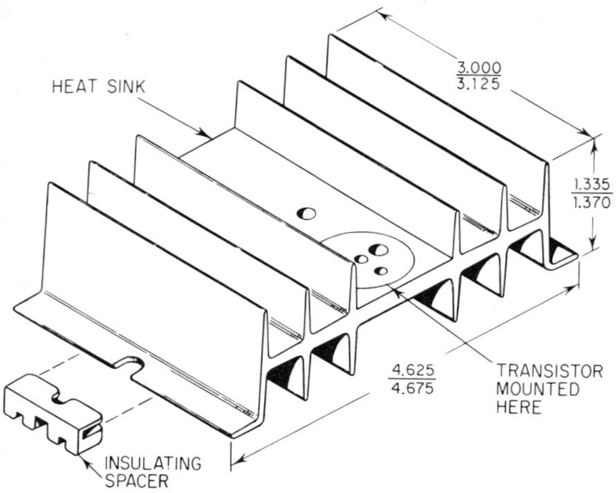

FIGURE 6.35
A heat sink for limited space.

The graph in Fig. 6.36 shows the temperature differential between the collector mounting stud and the ambient air at various collector-dissipation levels, with the heat sink in the horizontal and vertical positions.

Figure 6.35 also shows an insulating spacer that may be used for electrically isolating the heat sink from the chassis. Electric isolation of the transistor from the heat sink can be accomplished by special transistor-mounting kits, shown in Figs. 6.37 and 6.38. The mica insulators, however, will have a thermal resistance that must be added to the thermal resistance of the heat sink. Best heat dissipation is achieved when the transistor is mounted near the lower edge of the vertically mounted heat sink. This heat sink is prepunched to allow mounting either of the illustrated transistors directly to the heat sink. If the transistor is to be insulated from the heat sink, the user may enlarge the mounting holes to allow the use of the appropriate mounting kit.

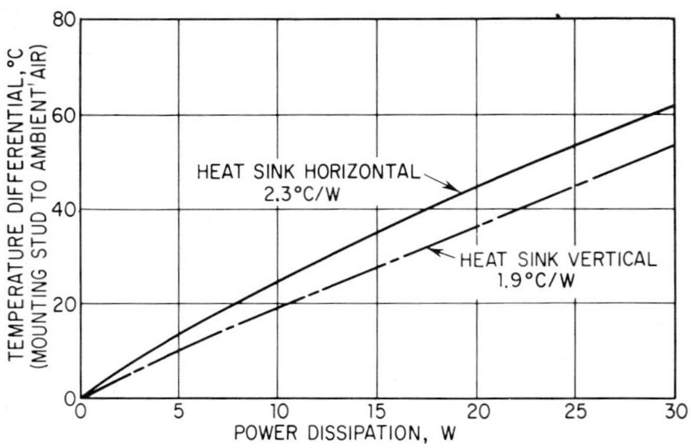

FIGURE 6.36
Thermal characteristics of the heat sink shown in Fig. 6.35.

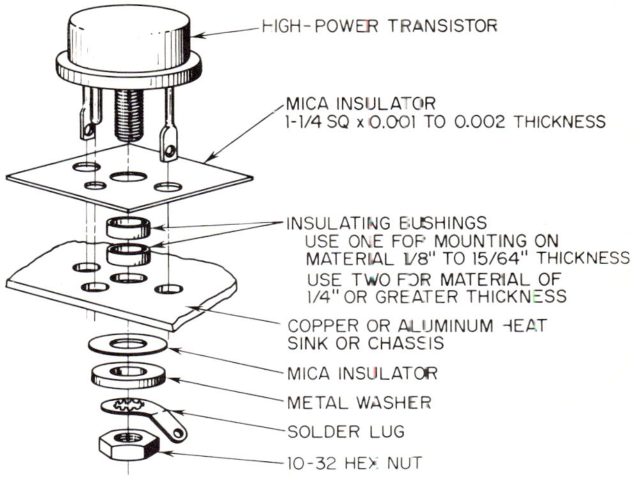

FIGURE 6.37
A method of electrically isolating a transistor heat sink from the chassis.

A variety of heat sinks are available, and some typical units are shown in Fig. 3.2.

As a generalization it may be said that heating is determined by the mode of operaton (switching service, audio amplifier, etc.), the level of bias and signal applied to the transistor, and the waveform of the applied signal.

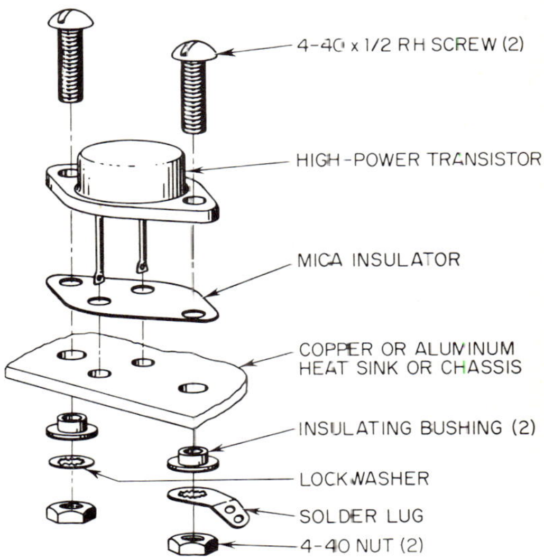

FIGURE 6.38
Another method of electrically isolating a heat sink from the chassis.

Operation of a transistor as a square-wave oscillator or as a class B amplifier of square waves results in less junction heating than does the generation of sine or complex waves or operation as class A amplifier.

In the generation or amplification of square waves, the transistor is alternately cut off and saturated. When cut off, no heating occurs, because there is practically no current flowing through the junction. When saturated, the resistance of the junction is so low that there is practically no voltage developed across the junction that would cause power losses.

Efficient square-wave operation requires the application of relatively high levels of signal to the control element of the transistor in order to assure complete cutoff and saturation. Any rounding of the waveform indicates power losses that are dissipated as heat.

Class A operation of transistors for the amplification or generation of a sine or complex wave results in the dissipation of power in the transistor junction, because the transistor is operating essentially as a variable resistor rather than as a switch. In class A operation, the transistor bias is adjusted to a value that will keep the transistor operating on the linear portion of its characteristic curve. Thus, there is flowing a no-signal collector current that is equal to about one-half the peak value of collector current. The power dissipated in the junction by this no-signal current is in the form of heat.

Class A Power Amplifiers. Transistor amplifying action can be utilized in several ways in electronic circuits, depending on the required function. There are four types, or classes, of amplifier circuits: class A, class AB, class B, and class C circuits. In class A amplifiers, the base bias and alternating signal are such that collector current flows in a transistor continuously throughout the complete electric cycle of the signal. As a matter of fact, current flows even with no signal. Figure 6.39 illustrates two class A power amplifiers designed to drive a loudspeaker. In one instance, two batteries are employed;

FIGURE 6.39
Two class A power amplifiers.

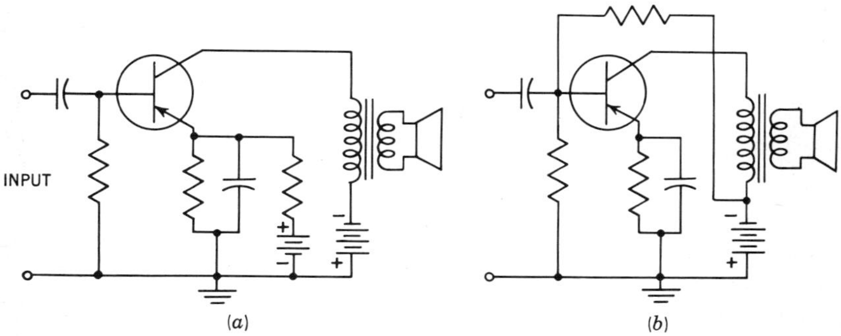

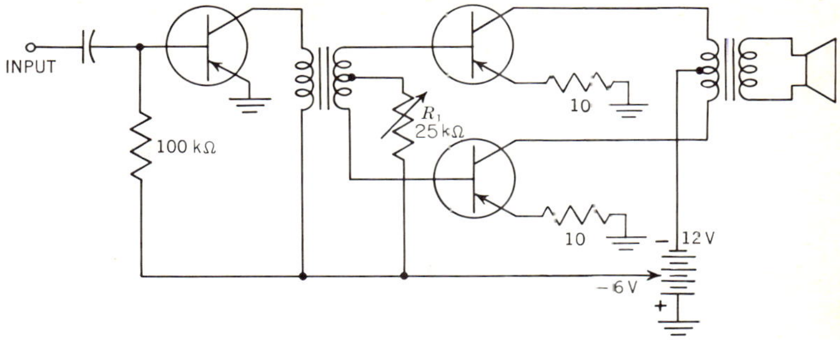

FIGURE 6.40
An amplifier using a driver and two power transistors in push-pull. All these transistors are PNP units. The 10-Ω resistors in the emitter leads are designed to stabilize the transistor against temperature changes.

in the other, a single battery is employed. The output transformer would be designed to match the impedance of the collector on one hand and that of the loudspeaker on the other. The amount of power that may be obtained from this arrangement can be calculated from the transistor characteristics and the voltage supply. A single-ended power amplifier can be operated only class A.

Class A amplifiers can also be operated in push-pull. A typical illustration of an audio amplifier using a single driver stage and a class A push-pull output stage is shown in Fig. 6.40. All transistors are operated with common emitters, and transformer coupling is employed between the driver and output stages and between the output amplifiers and the loudspeaker. The resistance R_1 is variable and is adjusted for a total collector current of 8 mA.

Push-pull amplifier operation results in the cancelation of second and even order harmonics within the stage. For the same amount of distortion, then, a class A push-pull amplifier can be driven harder, thereby providing greater output. This also means that we can obtain more output with push-pull operation than we can get by using two similar transistors as single-ended amplifiers.

Class B Amplifiers. In class B amplifiers, the base is biased to approximately collector-current cutoff, so that practically no collector current flows unless there is an applied signal. This means that there will be collector current for about one-half of each cycle when an alternating signal is used.

The circuit of a class B push-pull amplifier is shown in Fig. 6.41. Three power transistors are employed: the first one serves as a class A driver amplifier and the remaining two serve as a class B output stage. Efficiency of the

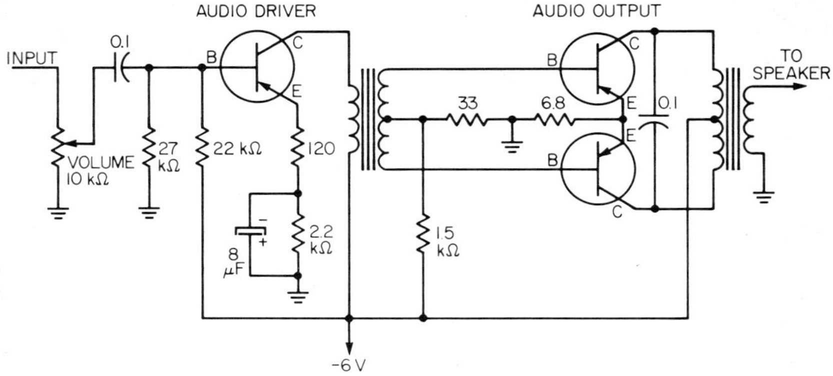

FIGURE 6.41
A two-stage audio amplifier. The output stage is operated class B.

class B stage is close to 75 percent. This is achieved because with no signal, the total class B collector current is extremely low, since the stage is biased near cutoff. In a class A amplifier, the efficiency is perhaps half this amount or less, because a fairly sizable collector current always flows, signal or no.

Input signals are applied to the base of the first audio amplifier. A 2,200-Ω bypassed resistor in the emitter circuit of this driver stage provides only thermal stabilization; it does not introduce signal degeneration. However, just above it is an unbypassed 120-Ω resistor, and this does provide signal degeneration.

The output of the driver stage is transformer-coupled to the class B amplifier not only to allow proper impedance matching but also to provide two signals 180° out of phase with each other (as required by the class B amplifier). A 6.8-Ω resistor in the emitter circuit of this output stage introduces a small amount of signal degeneration to improve the stability of the circuit. The base circuit contains a 33-Ω resistor, across which a small voltage from the negative 6-V supply is developed. The purpose of this voltage is to reduce crossover distortion. (This will be explained presently.) The collector elements of the two output transistors connect to opposite ends of the output transformer, and the dc voltage is brought in at the center tap. The 0.1-μF capacitor across the primary of the output transformer removes the highs from the output signal for a more mellow output tone. The speaker is an extremely small one that because of its dimensions naturally tends to emphasize the higher frequencies. This is counteracted somewhat by the 0.1-μF capacitor.

Class B audio amplifiers are favored in many transistor receivers not only for their greater power and reduced distortion but also for their current drain, which is practically zero when no signal is being received. If two class

A output amplifiers were connected in push-pull, an average current would always flow and impose a constant drain on the battery. Since these are power transistors, their current requirements are fairly large, which means that a significant amount of power would be dissipated.

Now let us examine the reason for the 33-Ω resistor in the base circuit of the output amplifier. Its purpose is not so much to provide base-emitter bias as it is to minimize a condition known as crossover distortion. When transistors are connected back to back in push-pull arrangements and the bias is zero, there is a region near cutoff where their respective characteristic curves tend to become nonlinear (Fig. 6.42*a* and *b*). Note the jog in both curves near the origin. If we now introduce a sinusoidal base current into the input circuit (Fig. 6.42*c*), we will obtain the distorted collector current indicated to the right of the characteristic curve. This distortion becomes more severe as the signal level decreases.

To prevent crossover distortion, a small amount of forward bias is introduced between the base and emitter of each transistor. In Fig. 6.43*a*, the transfer characteristics of the transistors are shown back to back for zero base bias. These curves are not combined. The dashed lines indicate the base-current values when forward bias is applied to provide the overall dynamic operating curve of the amplifier. With this forward bias, the two curves must be shifted until the dashed lines are aligned with each other (Fig. 6.43*b*). Note that now the jog at the center of the curve has disappeared. If we apply a sine-wave signal, the undistorted output shown in Fig. 6.43*c* is obtained. The 33-Ω resistor in Fig. 6.41 provides this forward bias for the output stage. The output of the audio system in Fig. 6.41 is approximately 150 mW to a 3-in speaker.

A second audio system is shown in Fig. 6.44. There is one small circuit variation in Fig. 6.44 that the reader should become familiar with. This occurs in the class B output stage, where two PNP transistors are employed. The voltage from the battery is applied to the emitter and base from the +12-V line. The collectors, however, are returned to ground (negative side of the battery in this circuit). In a vacuum-tube amplifier, this would be equivalent to placing a large negative voltage on the cathode and returning the plate to ground. Since it is the relative potential between the two elements that produces current flow through the device, it makes little difference whether the cathode is made negative or the plate is made positive. The same type of reasoning applies to transistors. Note, however, that because a large positive voltage is applied to the emitter, the same +12-V line must also be directly connected to the base circuit. If the base were similarly grounded and the large positive voltage were applied to the emitter, excessive current would flow through the base-emitter circuit and destroy the transistors.

A similar voltage arrangement is employed in the driver stage preceding the class B output amplifier.

In class AB amplifiers, the base bias and alternating signal are controlled

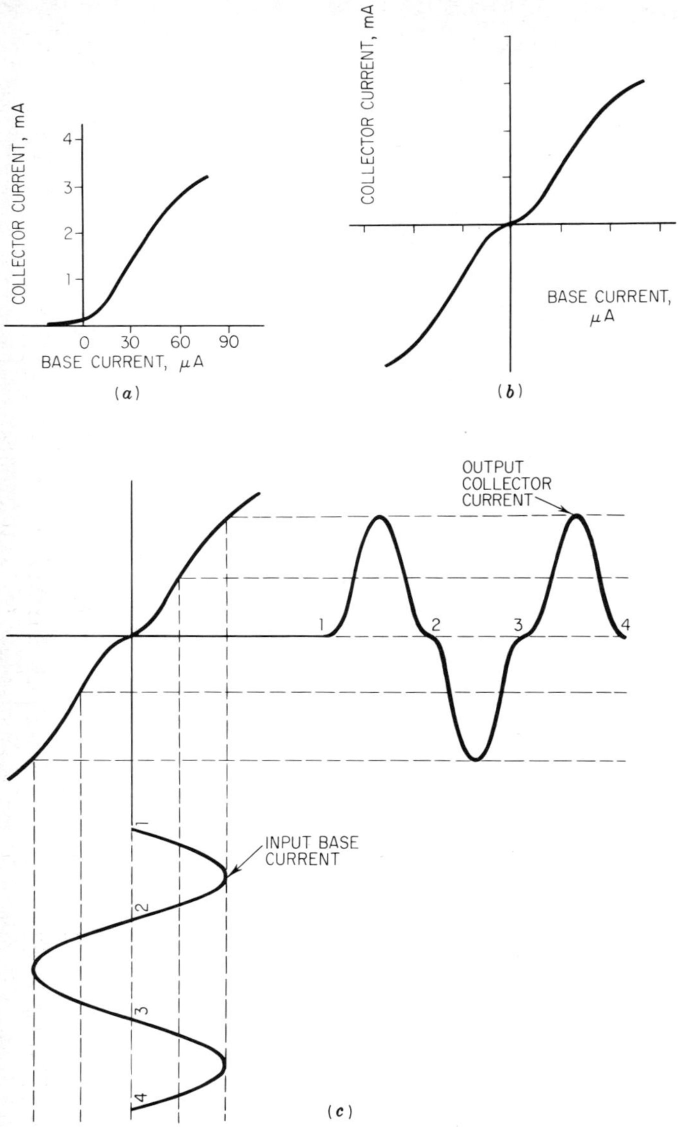

FIGURE 6.42
The mechanism of crossover distortion. (*a*) A single transistor. (*b*) Two transistors connected in push-pull. (*c*) Distorting effects caused by the jog in the transfer characteristic curve.

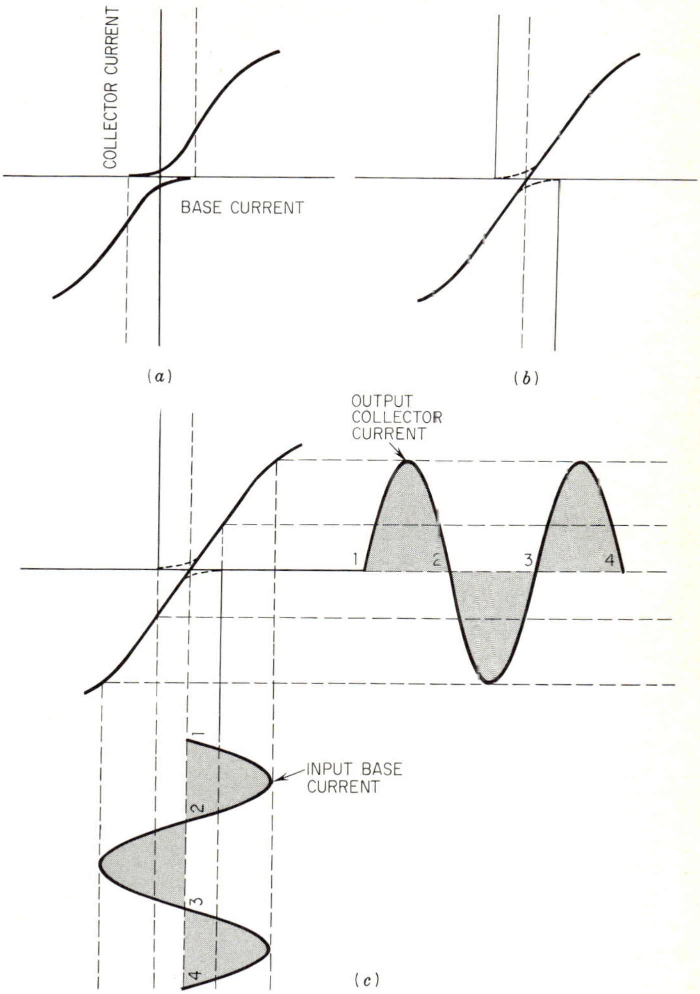

FIGURE 6.43
The introduction of a forward bias in the class B output stage removes the crossover distortion. (*a*) Back to back for zero base bias. (*b*) The two curves shifted. (*c*) Input and output current waveforms.

so that the collector current in a transistor will flow for more than half, but less than the entire, applied signal.

A class C amplifier is an amplifier in which the transistor is biased to conduct less than half of each cycle of an applied signal. Class C amplifiers are used in oscillators or in transmitters where the fact that the stage conducts

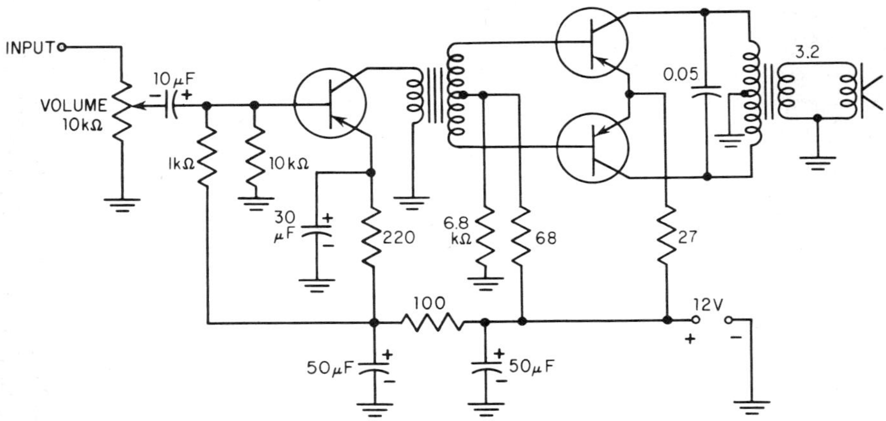

FIGURE 6.44
A second audio amplifier system.

for less than one-half cycle does not introduce distortion. In both oscillators and r-f amplifiers (in transmitters), the resonant circuits that are present maintain the necessary signal form. The advantages of a class C amplifier are its power output and high collector-circuit efficiency. Further, when no input signal is present, the power supply drain and the collector dissipation is low.

COMPLEMENTARY PUSH-PULL AMPLIFIERS

The complementary symmetry of PNP and NPN transistors was employed previously in direct-coupled amplifiers. These same features may also be utilized to obtain push-pull operation without any input or output transformers. This is possible because the collector currents of NPN and PNP transistors react in opposite ways when subject to the same applied signal.

To see this more explicitly, consider the class A push-pull amplifier shown in Fig. 6.45. At the top of the illustration we have a PNP transistor; below, an NPN unit. Both are suitably biased with a 22.5-V battery (one for each transistor). R_1 and R_2 for the PNP transistor and R_3 and R_4 for the NPN transistor serve to establish the base-to-emitter bias suitable for class A operation. The 560-Ω emitter resistors provide dc stabilization. Each of these resistors is suitably bypassed to prevent ac degeneration, which would reduce the gain of the amplifier. The load is a 500-Ω voice coil of a loudspeaker, and it is directly connected to the collectors of the two transistors. A single input line is provided, with the base of each transistor connected to this line.

Assume, now, that a sine wave is being amplified and at the moment in question the positive half of the sine wave is active. This means that both bases will be driven positive simultaneously. In Q_1 this will cause the base-emitter current, and with it the collector current, to decrease. Since the collector current flows up through R_L (i.e., the speaker voice coil), it will serve to make the top, or collector, end of this load impedance less positive, or more negative.

Now let us turn to the NPN transistor. The path for its output current is from the collector to R_L. Hence, the voltage drop across R_L due to this transistor is such as to make the top end of R_L negative.

When the positive half of the applied signal reaches the base of the NPN transistor, it acts to increase the forward bias there, thereby increasing the base-emitter current. This, in turn, increases the collector current, causes more of a voltage drop across R_L, and raises the negative potential present at the top of the load. This serves to work with or strengthen the voltage drop produced by the PNP transistor.

During the next half-cycle, when the negative half of the signal is active, the reverse set of conditions occurs. That is, the current through the PNP transistor increases, producing more of a positive voltage across R_L. At the

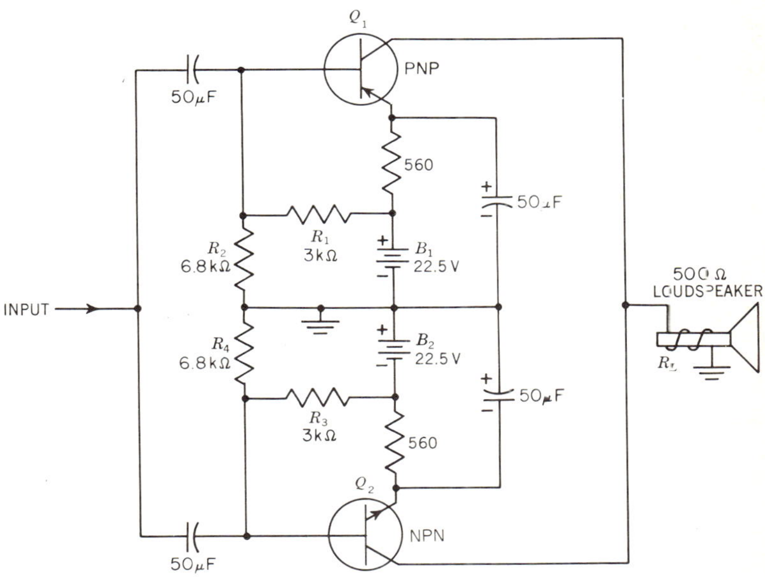

FIGURE 6.45
A push-pull class A amplifier using neither input nor output transformers.

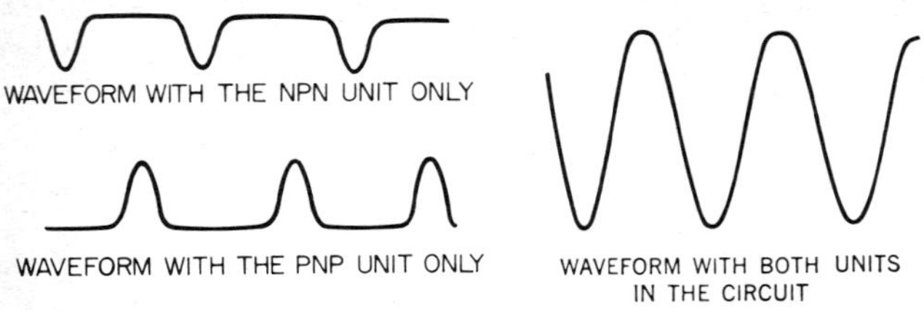

FIGURE 6.46
Waveform in the push-pull amplifier of Fig. 6.45.

same time, the current through the NPN transistor decreases, lowering its negative voltage drop across R_L, which, in essence, is equivalent to a positive increase.

Thus, both sections of this circuit work in unison with each other, producing a larger output than either one could by itself. This is demonstrated in Fig. 6.46, where the individual output waveforms of each transistor are shown, together with the combined waveform. Note the differences in relative sizes.

A class B push-pull transistor amplifier with complementary symmetry that can feed its output directly to the 16-Ω voice coil of a loudspeaker is shown in Fig. 6.47. Across the top section of the diagram we have a PNP transistor directly coupled to an NPN transistor. Across the bottom section we have the reverse situation. Both halves are similar to the direct-coupled amplifier of Fig. 6.21 and operate in the same manner. In addition, the two sections form a push-pull arrangement. Power gains on the order of 30 dB (i.e., 1,000:1 ratio) have been obtained in this manner.

FIGURE 6.47
Another push-pull amplifier using complementary symmetry.

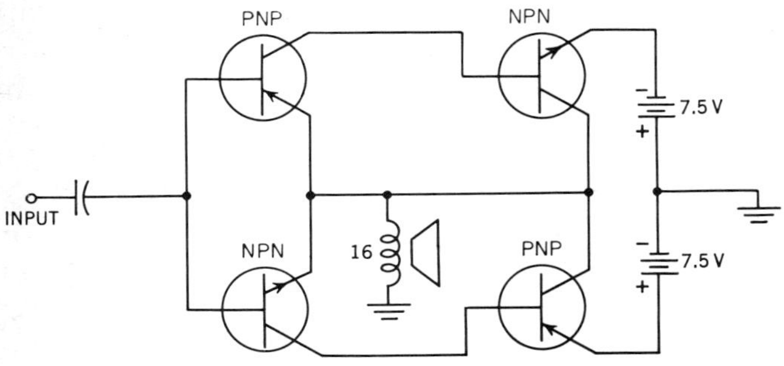

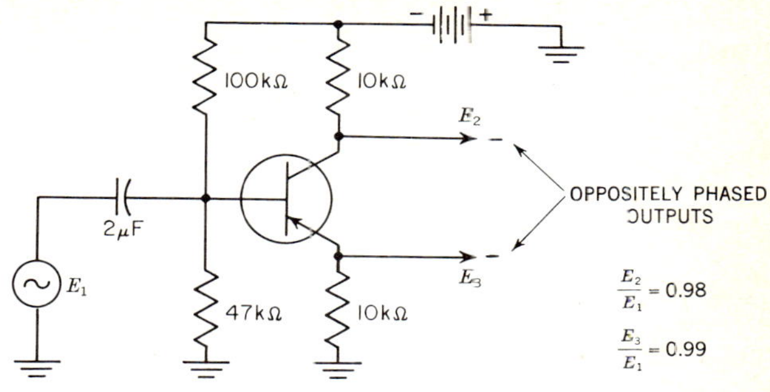

FIGURE 6.48
A transistor phase inverter.

PHASE-INVERTER CIRCUITS

Oppositely phased signals of closely similar amplitudes can be obtained from a single transistor stage as shown in Fig. 6.48. One output voltage is taken from across the emitter resistor and the second output voltage (of opposite phase) is obtained from the collector load resistor. Whereas perfect balance cannot be obtained, because the current gain is not equal to 1, the voltages can be made to approach each other quite closely. Typical values of voltage gain to both outputs are shown in the diagram.

In a vacuum tube, the grid does not ordinarily draw any current. Hence, whatever current passes through the plate circuit also flows entirely through the cathode resistor (Fig. 6.49). By having equal-valued resistors in the plate and cathode legs of the tube, equal output voltages will be obtained. In the transistor circuit (Fig. 6.48), a portion of the emitter current does not reach the collector. Hence, equal-valued collector and emitter resistors will not pro-

FIGURE 6.49
A vacuum-tube phase inverter.

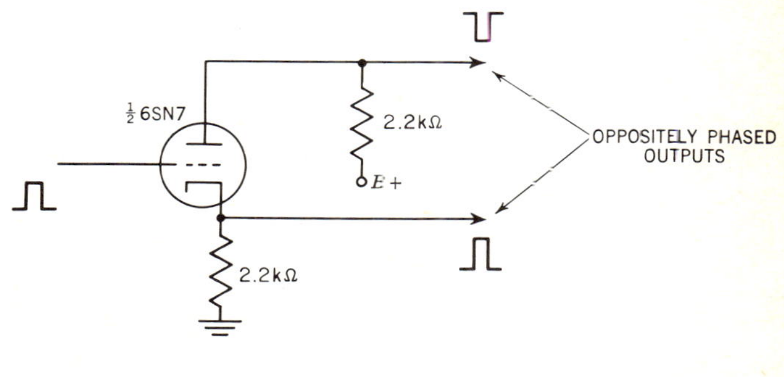

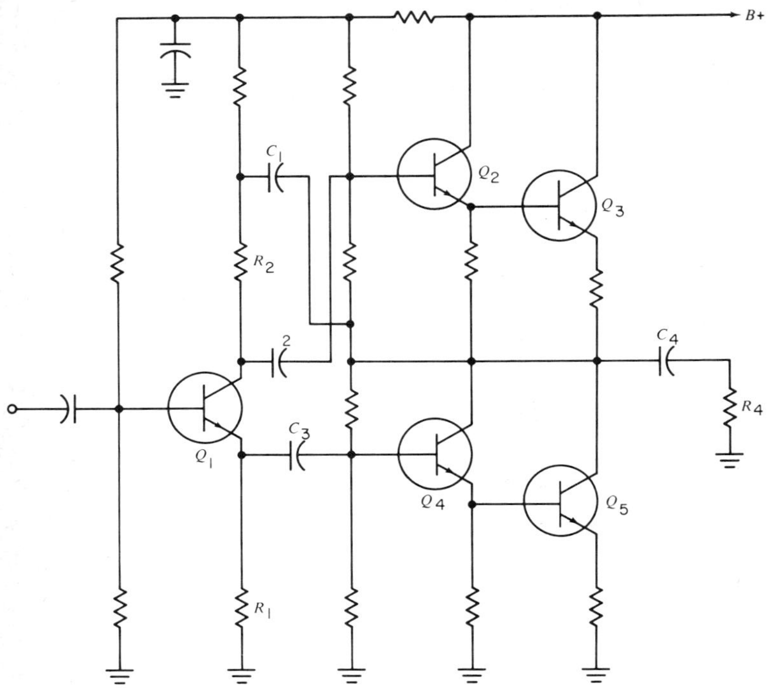

FIGURE 6.50
Push-pull series-output amplifier in which driver and output amplifiers are connected as Darlington pairs. Transistor Q, provides the driving signal for Q_2 and Q_4.

duce equal output voltages. If we alter the resistances to achieve better balance, we change the circuit operating conditions, including the various currents that flow. Thus, while we may come close, we shall not attain a perfect balance.

It is, of course, possible to obtain balanced signals of opposite polarity by using two transistor stages. One stage will serve to provide one polarity signal, while the other stage will take a portion of this signal, amplify it, and invert it, and thereby provide the required second signal.

An interesting audio amplifier system that employs a phase inverter driving two sets of Darlington output amplifiers is shown in Fig. 6.50. Q_1 is the phase inverter that provides equal but opposite polarity signals to the two sets of output transistors, Q_2 and Q_3, and Q_4 and Q_5. Capacitor C_3 feeds one signal to Q_4 (and Q_5), whereas capacitor C_2 feeds the opposite-phased signal to Q_2 (and Q_3). Capacitor C_1 provides the reference line for the emitters of Q_2, Q_3 just as ground provides a reference line for Q_4, Q_5.

Transistor Q_2 and Q_3 form one Darlington combination, whereas Q_4 and Q_5 form the second Darlington set. Both sets of amplifiers operate with equal gain, but one set provides the output signal across R_4 (i.e., representing the speaker voice coil) for one half-cycle and the second Darlington pair provides the signal for the other half-cycle.

DIFFERENTIAL AMPLIFIER

There is one type of amplifier that is employed extensively in solid-state circuit because of its wide versatility. It is the differential amplifier.

To understand the reasons for the popularity of this circuit, we must recognize that whereas solid-state technology as applied to electronics started with the transistor, the developing trend has been toward greater usage of integrated circuits where an entire circuit or system is produced on a single chip. This situation already exists to an extensive degree in digital computers where entire arrays of what are essentially on-off type amplifiers are employed in the memory circuits, the logic circuits, and the processing circuits. In most of these digital computer circuits, transistors and diodes are the principal components, although resistors and capacitors are also employed. These latter components, however, have a fairly restricted range of values simply because they require considerably more space (on the chip) to achieve the range of values useful in the circuit. (This can be seen from the formulas that govern resistance and capacitance. Both equations contain area as one of the governing factors.)

An amplifier that has only two operating states, either on or off, is much simpler in construction than an amplifier that is required, for example, to amplify a signal over a range of amplitudes with a minimum of distortion. It is for this reason that digital (or on-off) circuits have been more extensively developed and employed than so-called linear integrated circuits.

By definition, a linear integrated circuit is a group of components that combine to operate in a linear fashion to perform signal amplification, processing, timing, and control. If integrated-circuit technology is to be fruitfully and economically applied to electronic circuits *other than* digital circuits, then suitable, low-cost linear circuitry must be developed.

It is toward this end that the differential amplifier owes its considerable appeal. Properly designed and built, a differential amplifier exhibits excellent stability, has a high immunity to interfering signals, does not require any capacitors, and has a wide range of applications. For example, the differential amplifier can be employed as an i-f or r-f amplifier, a mixer, a limiter, a modulator, and a detector.

The basic circuit of a differential amplifier is shown in Fig. 6.51. Two transistors are employed, each with its own input circuit and its separate collector or output circuit. The emitters of the two transistors, however, are

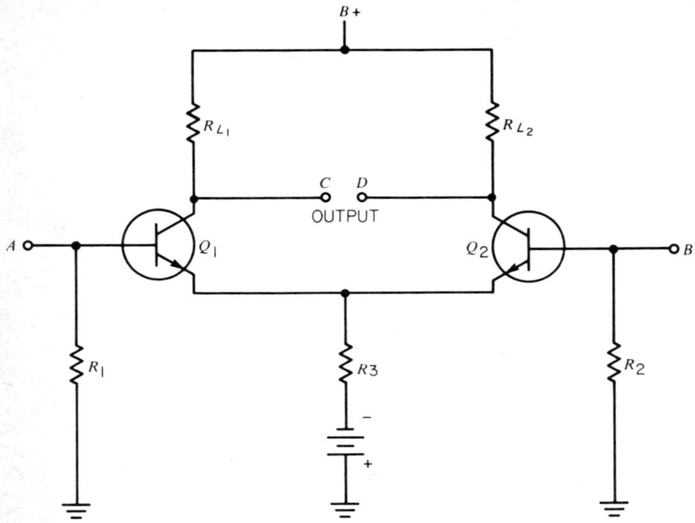

FIGURE 6.51
The basic differential amplifier circuit.

tied together and given sufficient voltage to properly bias the two transistors to the desired operating point. If the two circuits are carefully matched to each other, that is, if $Q_1 = Q_2$, $R_1 = R_2$, $R_{L_1} = R_{L_2}$, then equal signal voltages applied at A and B will produce *zero* output across terminals C and D. That is, any voltage that affects both inputs similarly will have its effect canceled in the output. This is a good characteristic to possess, since interference voltages in a system would tend to have the same effect and the differential amplifier would thereby not respond to such interference.

Signals that affect the input to both transistors equally are called common-mode signals, and the ability of a differential amplifier not to produce any output is called common-mode rejection. If circuit balance is perfect, no output at all is developed for common-mode signals and the rejection ratio is infinite. However, in actual circuits, precise balance is seldom achievable and some common-mode voltage does appear at the output. With proper design, however, this can be kept exceedingly small.

To obtain an output from the differential amplifier, we have to apply different-valued signals to each input. The output between terminals C and D is then the difference between the amplitudes of the two input signals. Or one input terminal, say B, can be grounded, the input signal applied to terminal A, and the output obtained between terminal C and ground or between terminal D and ground. If the circuit is balanced, the output voltage at C or D would be equal. However, the signal polarity at these two terminals is 180° out of phase.

A popular version of the differential amplifier uses a third transistor in place of the emitter resistor (see Fig. 6.52). By using a zener diode to maintain a fixed bias on this third transistor, we obtain a situation where the current through Q_3 is fixed and this, in turn, fixes the total current flowing through Q_1 and Q_2.

Looking at this arrangement another way, we note that Q_3 is a fixed-current generator or a constant-current source. Electrically, it appears as a very high-valued resistor. If, now, a common-mode signal reaches the bases of Q_1 and Q_2, the resistance of these transistors decrease because their total current cannot change. Thus, the voltage across Q_1 and Q_2 decreases, but it rises across Q_3 in like measure. The net result is greater degeneration, reducing the gain of the differential amplifier to common-mode signals.

When signals of opposite polarity are applied to terminals A and B, no net voltage increase appears in the emitter circuit because the tendency for current to rise in one transistor (say Q_1) is offset by an equal tendency to decrease (in Q_2). Hence, the same degenerative action for differential-mode signals does not occur and the stage gain remains high.

Thus, the differential amplifier can be used to amplify signals of opposite polarity across the two input terminals, or one input terminal can have a fixed voltage (as well as zero by being grounded) and the input signal applied to the remaining terminal. Output is then taken from one of the collectors.

FIGURE 6.52
A differential amplifier with a constant current generator in the emitter circuit.

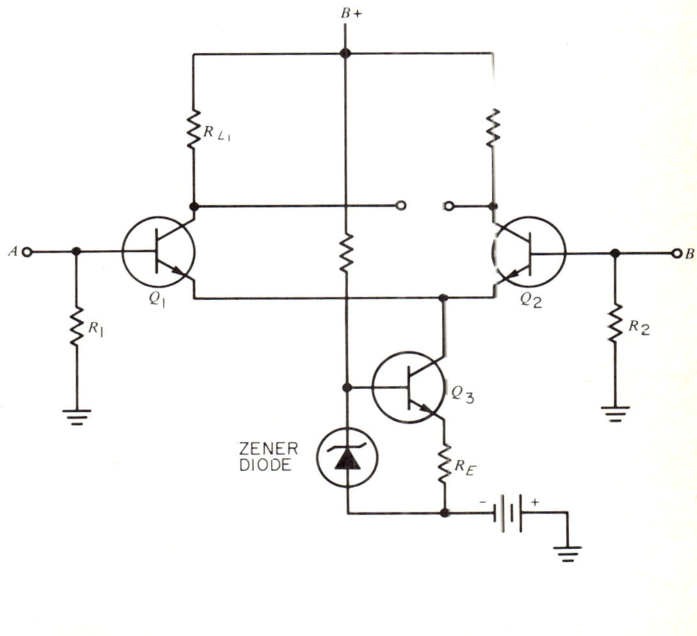

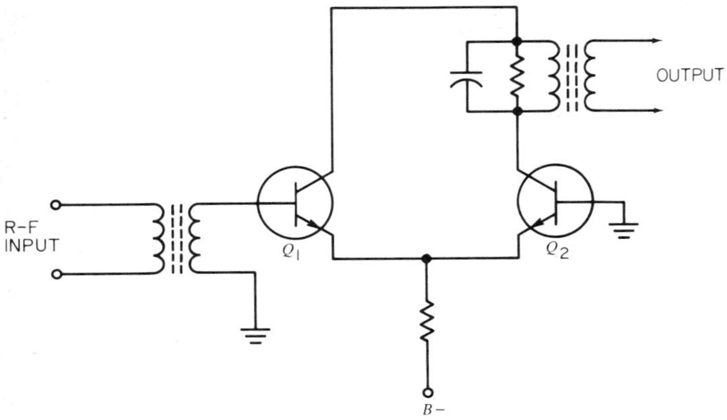

FIGURE 6.53
An r-f differential amplifier.

Figures 6.53 and 6.54 show the differential amplifier used in two different ways. In Fig. 6.53, the circuit is employed as an r-f amplifier, with the input signal applied to Q_1 and the output signal obtained from the collector circuit of Q_2. In this arrangement, the base of Q_2 is put at ground.

In Fig. 6.54, the differential amplifier is employed as a mixer. The incoming r-f signal is applied in equal measure, but opposite polarity, to Q_1 and Q_2. At the same time, an oscillator signal is fed to Q_3. The two signals mix, producing the i-f signal across the output transformer.

Note that because the primary winding of the output transformer T_1 is center-tapped and the collector currents (for the oscillator signal) flow in opposite directions through this winding, no oscillator signal appears at the i-f output terminals. The reader will recognize that in this arrangement, the oscillator signal is in the common mode for the differential amplifier.

RADIO- AND INTERMEDIATE-FREQUENCY AMPLIFIERS

In dealing with transistors in low-frequency circuits, any decrease in stage gain is due not to the transistor but rather to the external circuitry. For example, in an audio amplifier, the gain drop-off as the frequency is decreased stems from the rising impedance of the coupling capacitors, such as the 10-μF units shown in Fig. 6.13. Eventually, so much signal voltage is dropped across these coupling capacitors that very little reaches the transistor. Hence, very little can appear in the output.

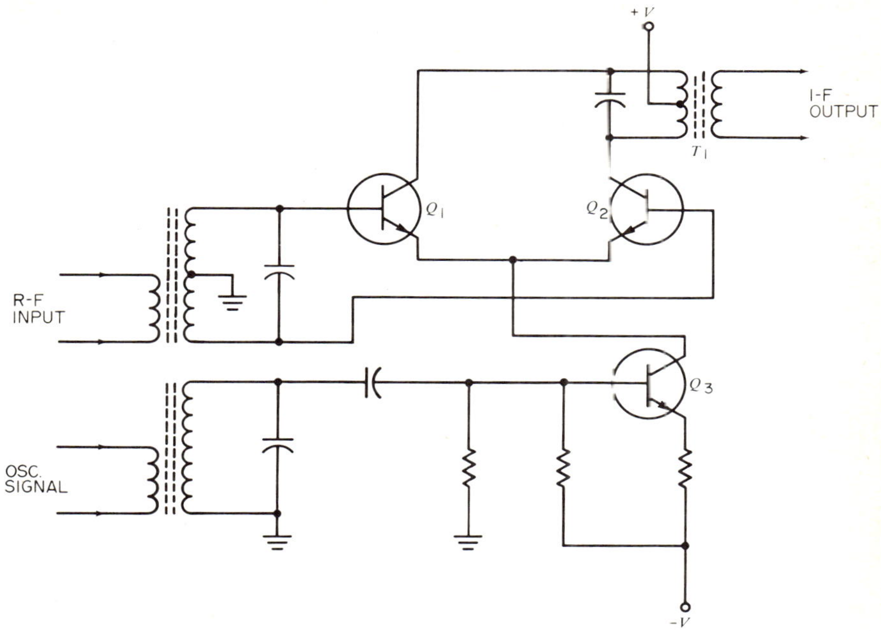

FIGURE 6.54
A differential amplifier employed as a mixer.

At high frequencies, things are quite different because now the current gain β enters the picture. Current gain, or β, can be defined in two ways. The first, dc β, is given as

$$\text{dc } \beta = \frac{I_C}{I_B}$$

where I_C = dc collector current
I_B = dc base current needed to produce this collector current

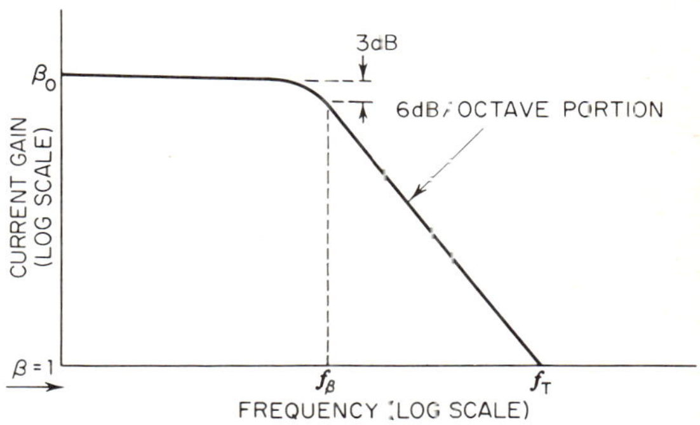

FIGURE 6.55
Typical variation of β with frequency.

The second, small-signal, or ac β, is given as

$$\text{Small-signal } \beta = \frac{i_c}{i_b}$$

where i_c = ac collector current

i_b = ac base current needed to produce this collector current

In addition to the letter β, another symbol is employed extensively for current gain. This symbol, for dc β, is h_{FE}; for ac β, it is h_{fe}.

In order to specify and measure small-signal β (h_{fe}), it is necessary to specify also the frequency at which it is being measured. A typical variation of small-signal β as a function of frequency is shown in Fig. 6.55. Specifically, this graph is for an output that is short-circuited for alternating current (i.e., the ac load impedance is zero). With an actual load resistor, the same shape curve is obtained, but with lower β values.

In Fig. 6.55, β_0 at the left-hand side of the chart represents the value of h_{fe} at some low frequency, perhaps several hundred cycles. This value remains fairly steady over a limited range of frequencies and then starts to decrease. f_β in Fig. 6.55 is the frequency at which the current gain is 3 dB (that is, 0.707 β_0) down from its β_0 value. Beyond f_β, the curve slopes down at a constant rate of 6 dB/octave. This simply means that each time the frequency is doubled, h_{fe} decreases by 6 dB. Eventually, the value of h_{fe} reaches 1, and this frequency is labeled f_T. This latter value is listed frequently in transistor-data sheets.

It should be noted that whereas frequency determines the value of small-signal β, the collector-to-emitter voltage and collector current also affect this characteristic (agc control of stage gain is based on this behavior).

Interstage Coupling Networks. The most widely used amplifier connection is the common emitter. With this arrangement the input impedance is fairly low (on the order of 1,000 Ω or so) and the output impedance is in the neighborhood of 10,000 to 20,000 Ω. It is the purpose of the interstage coupling network not only to provide whatever frequency selectivity is desired but also to match these input and output impedances.

There are a number of interstage coupling networks possible; the more important of these are shown in Fig. 6.56. In the first group, Fig. 6.56*a*, the second amplifier is connected directly into the parallel-resonant circuit, either in series with the inductance or in series with the capacitance.

In the second group, Fig. 6.56*b*, the second stage is connected to the junction of two capacitors C_1 and C_2, which resonate with the inductance. By properly proportioning the values of C_1 and C_2, we can use the network to match the high output impedance of the first transistor to the much lower input impedance of the second transistor.

The third group, Fig. 6.56*c*, employs inductive coupling between stages. In the first illustration of Fig. 6.56*c*, the primary circuit is tuned, offering

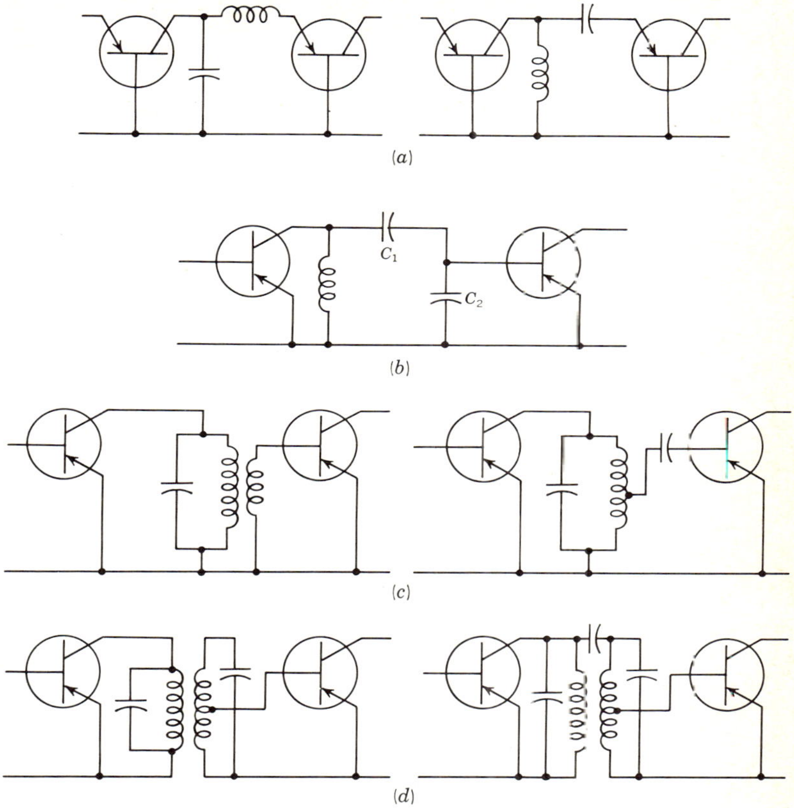

FIGURE 6.56
Various methods of coupling transistor stages together.
(After W. F. Chow, High Frequency Transistor Amplifiers, Electronics, April, 1954.)

a high impedance to the first transistor stage. The signal is then transferred to an untuned secondary containing fewer turns. This step-down action enables the low-impedance input of the second transistor to match the output of the first stage. In the second illustration of Fig. 6.56*c*, we obtain the same electrical action by dispensing with the secondary winding and tapping directly into the primary inductance. In this arrangement, a coupling capacitor is needed to prevent the higher collector bias of the preceding transistor from reaching the base of the second unit.

In the final group of coupling networks, Fig. 6.56*d*, double tuning is employed. Again note how the second stage must be tapped down in order to achieve the proper impedance match.

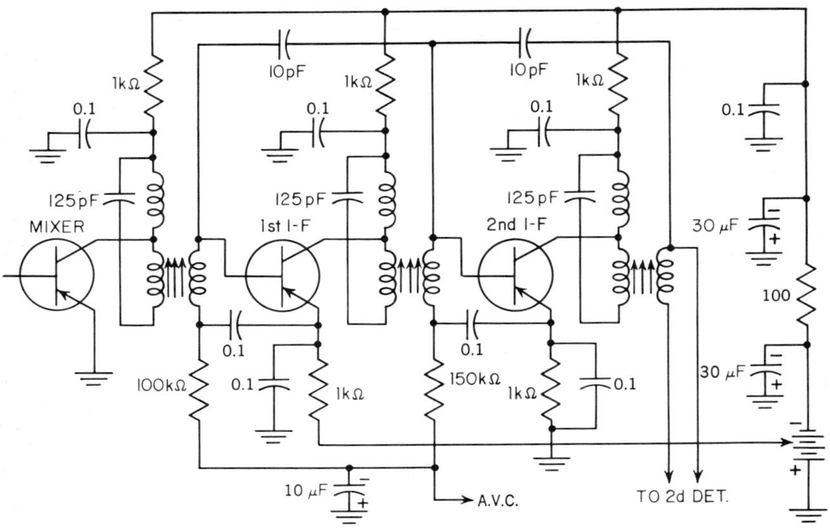

FIGURE 6.57
A two-stage i-f system operating at 465 kHz.

Intermediate-frequency Amplifiers. A two-stage i-f system suitable for a broadcast receiver is shown in Fig. 6.57. The circuit uses two high-frequency PNP transistors in a grounded-emitter configuration. Operating frequency is 465 kHz, and the overall gain is at least 90 dB. The i-f transformers have 155 total turns on the primary, tapped at 55 turns with an 18-turn secondary. The coils are bifilar wound and enclosed in an adjustable ferrite cup. They are tuned by a fixed 125-pF capacitor across the primary.

Each emitter possesses a 1,000-Ω dc stabilizing resistor. Alternating-current or signal degeneration is prevented by the use of 0.1-μF bypass capacitors across these resistors.

It will be noted that each i-f stage is neutralized by connecting a 10-pF capacitor from the base of the following stage to the base of the preceding stage. (These two points are 180° out of phase because of the grounded-emitter arrangement.) Neutralization was deemed necessary because enough internal capacitance existed in the two transistors used to lead to oscillation. There are transistors in which the signal feedback is so small that special neutralizing networks are not required and hence not used.

The bases of both i-f stages connect into an agc line. Operation of agc systems will be explained in Chap. 9.

Another radio-receiver i-f system is shown in Fig. 6.58. The transistors here are also connected with the emitters common to both input and output circuits. Transformers T_1, T_2, and T_3 constitute three bifilar circuits that

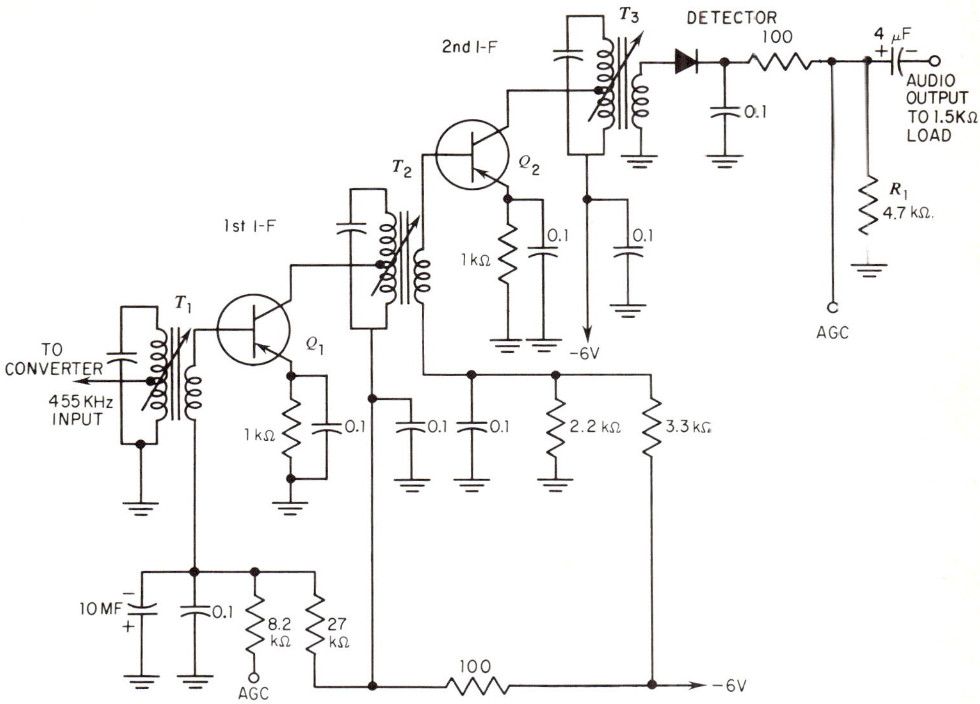

FIGURE 6.58
A second i-f system designed for broadcast receivers.

serve as interstage coupling networks, with essentially unity coupling between primary and secondary windings.

The first i-f transformer T_1 transfers the received signal to the base of the first i-f transistor Q_1. This stage is also provided with agc bias. The control action is accomplished by varying the base current of Q_1 in step with the signal level at the second detector and this, in turn, varies the collector current. Figure 6.59 demonstrates what effect this variation has on the gain and on the input and output impedances of the transistor. Note that transistor gain decreases rapidly as the collector current drops below 0.25 mA. Observe, too, that the input and output impedances rise with emitter-current decrease, causing mismatching in the input and output circuits and further reducing gain.

In the second i-f stage we have essentially the same circuit arrangement, although agc is not applied here. Because of this, the base bias voltage (hence, current) is different. Beyond this stage, the signal goes to a solid-state rectifier where it is demodulated. It is then ready for the audio

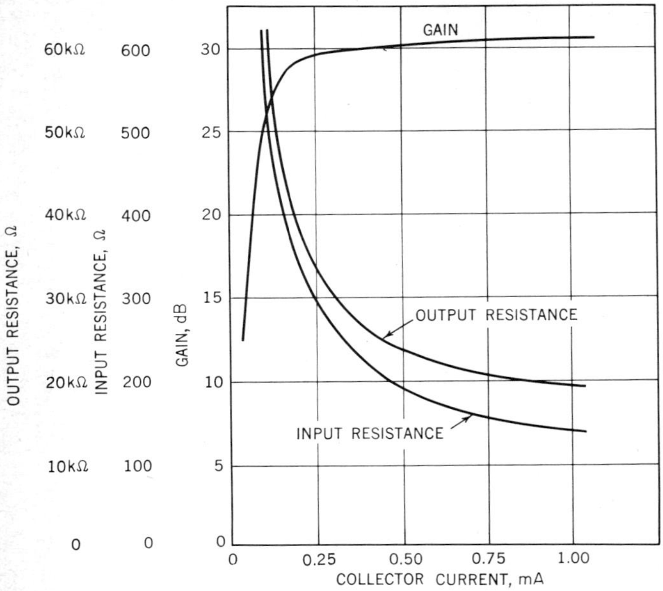

FIGURE 6.59
Variation in input and output resistance of the first i-f stage in Fig. 6-58 with changes in collector current. Variation in stage gain is also indicated.

amplifier stages. The agc voltage is developed across R_1. This resistor is also the load for the detector.

Radio-frequency Amplifiers. The considerations that govern the design of the i-f amplifiers also hold true for r-f amplification. However, because r-f amplifiers operate at higher frequencies, we can expect lower gain.

The r-f stage of a transistor automobile radio is shown in Fig. 6.60. The input transformer T_1 is slug-tuned, with its slug mechanically ganged to the slugs of the converter and the local oscillator (not shown) coils. The Q of T_1 varies from 70 to 50 across the tuning band, 550 to 1,600 kHz. The base of Q_1 is returned to the junction point of R_2 and R_3, where the dc potential is approximately 1.5 V. A dc stabilizing resistor R_4 is placed in the emitter leg of Q_1 to make the stage relatively insensitive to changes in ambient temperature. A small 680-Ω resistor R_5 brings an agc voltage to the emitter. If no agc control is desired, the connection between R_4 and R_5 can be severed. C_4, across R_4, serves to place the emitter at R-F ground. C_3, at the junction of R_2 and R_3, serves also as a low-impedance path to ground for radio frequency.

Transformer T_2 couples the signal from the R-F stage to the mixer. In the midfrequency range of the broadcast band, the output impedance of Q_1 is

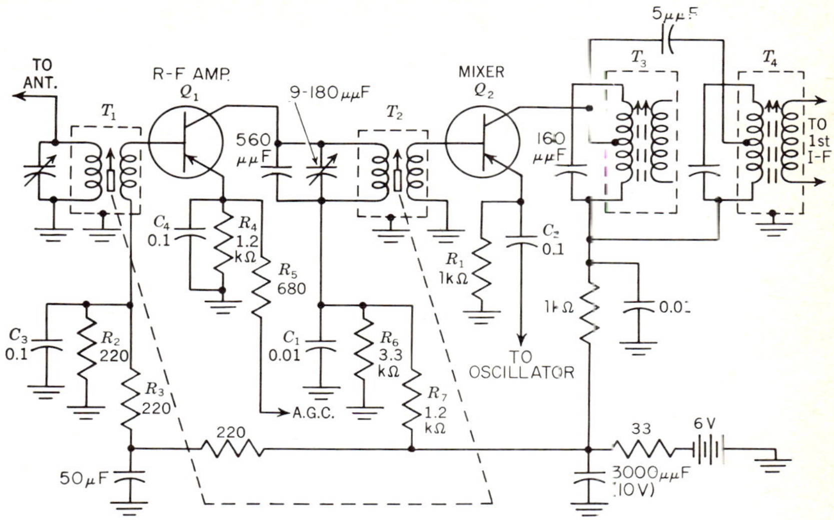

FIGURE 6.60
The r-f and mixer stages of a transistor automobile radio.

10,000 to 15,000 Ω and the mixer input impedance is about 500 Ω. These are the two impedances that must be matched by T_2.

The r-f signal is applied to the base of Q_2, while the locally generated oscillator signal is brought into the circuit by C_2 and developed across R_1. The latter resistor, incidentally, also serves to provide bias stability in the same manner as R_4 in the r-f amplifier stage. Approximately 0.4 V rms of oscillator voltage is injected into the converter stage, this value having been found to provide optimum conversion gain of Q_2. If the oscillator voltage is reduced below this level, the conversion gain drops rapidly, which means that we obtain a smaller i-f signal for a given amount of incoming r-f signal. On the other hand, if the oscillator signal is made larger than this optimum value, conversion gain will again decrease, although this time more slowly.

A fairly elaborate interstage coupling network is employed between the mixer and first i-f amplifier stage. This is designed to achieve the desired signal bandpass, with a fairly rapid fall-off on either side. Output impedance of the converter is in the neighborhood of 50,000 Ω, and it is not affected by signal frequency. Capacitor C_2 and the oscillator circuit that it ties into offer very low impedance to signals of intermediate frequency, so that for i-f signals, R_1 is effectively bypassed and no degeneration results.

Interstage coupling networks other than the network shown in Fig. 6.60 could be employed between R-F stages. These will follow closely the patterns indicated in Fig. 6.56.

FIELD-EFFECT TRANSISTOR (FET) AMPLIFIERS

In Chap. 5 we discussed field-effect transistors and how they function internally. We shall now look at some FET circuit configurations and applications. Before we do this, let us briefly compare the bias conditions of an N-channel field-effect transistor, an NPN bipolar transistor, and a vacuum tube. The similarity between the three biasing conditions is shown in Fig. 6.61.

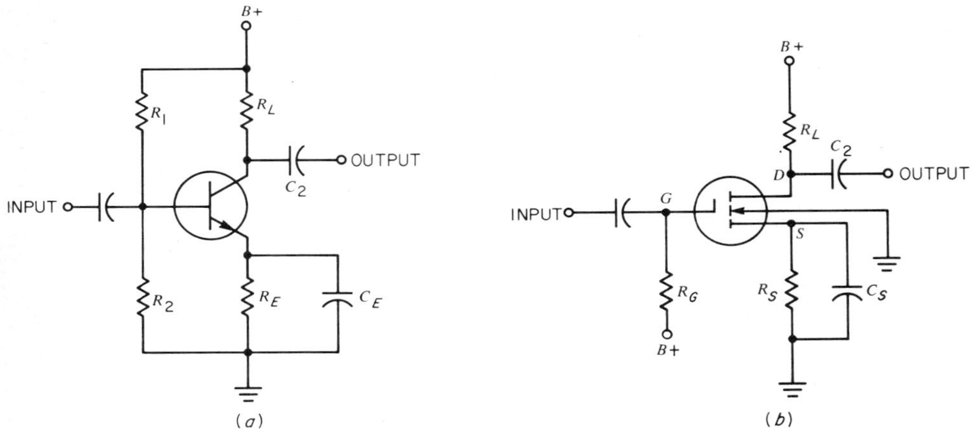

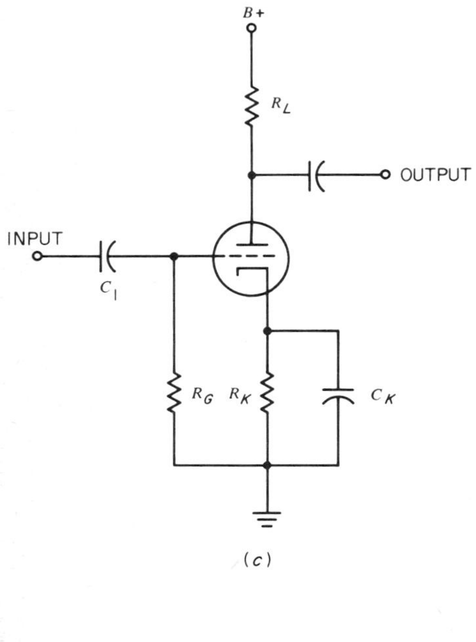

FIGURE 6.61
A comparison of the method of biasing bipolar transistors, MOSFET units, and vacuum tubes. (*a*) NPN Bipolar transistor; (*b*) MOSFET (N-channel type); (c) vacuum tube.

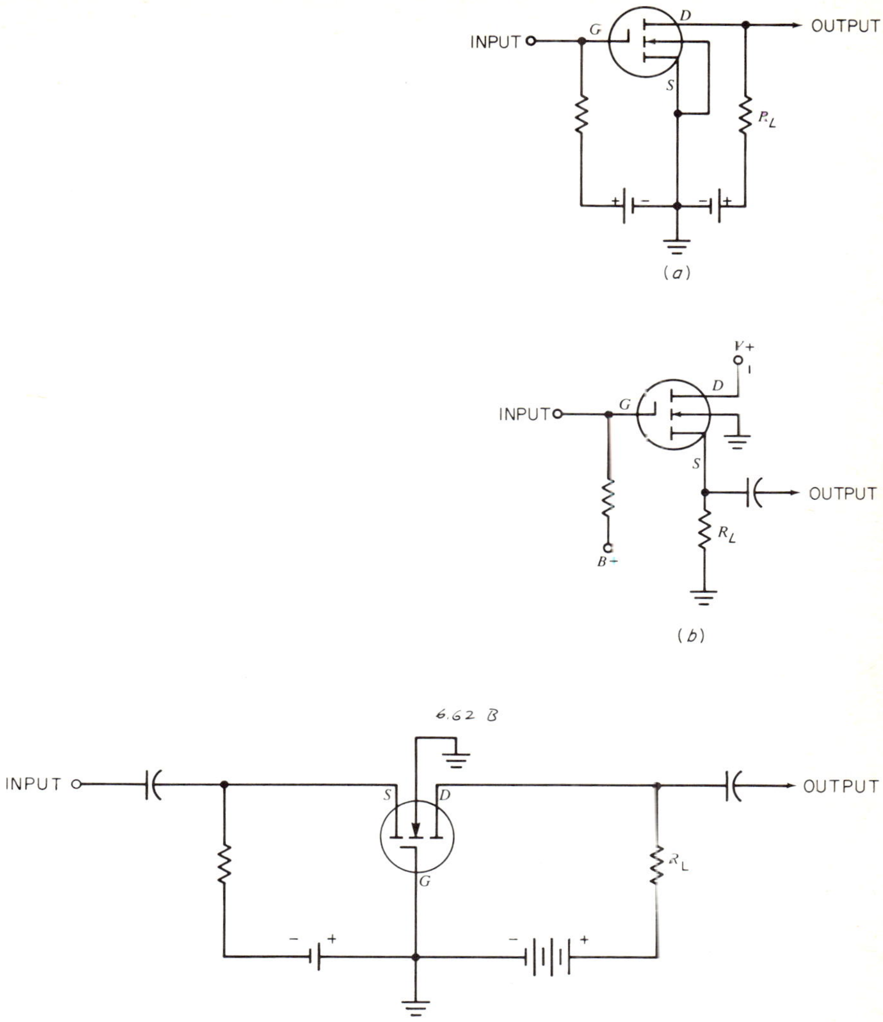

FIGURE 6.62
The three circuit configurations of the MOSFET. (*a*) Common source; (*b*) common drain or source follower; (*c*) common gate.

Notice that the base of the bipolar transistor is biased by the drop across the voltage divider R_1R_2 between the collector supply and ground. There is also some bias voltage due to the emitter-stabilizing resistor R_E.

The FET gate is positively biased, with respect to the source. This positive voltage is required to attract electrons to the area under the gate in order to form a channel between the drain and the source.

In all three devices, a bypass capacitor permits the resistor across which it is bridged to function as a voltage dropper for dc but not ac. That is, C_S is across the source resistor, C_E is across the emitter resistor R_E, and C_K is across the cathode resistor. In all three circuits, the output is taken across a load resistor R_L and a coupling capacitor.

It is important to remember that field-effect transistors are also fabricated with P-type channels. The bipolar counterpart, of course, is the PNP transistor, and the polarities just discussed would be reversed in both cases. That is, in a PNP transistor, the emitter is positive and the collector is negative. In a P-channel FET, the gate is biased negative with respect to the source (the drain voltage is negative and the source is at ground).

Circuit Configurations. There are three basic single-stage amplifier configurations for FETs: common source, common gate, and common drain. These are shown in Fig. 6.62.

The first circuit, Fig. 6.62*a*, is the most common: it is the common source arrangement, providing a high input impedance, good voltage gain, a moderate output impedance, and a 180° phase reversal between input and output signals.

The second circuit, Fig. 6.62*b*, is the common-drain configuration where the input signal is applied between the gate and ground and the output signal is obtained from across R_L in the source leg of the device. This arrangement has a moderate voltage gain, very high input impedance, no voltage gain, and input and output signals that are in phase with each other. As the reader will recognize, this circuit is closely similar to the vacuum-tube cathode follower or the common-collector bipolar transistor amplifier.

The final circuit, Fig. 6.62*c*, is the common-gate configuration. It exhibits moderate voltage gain, low input impedance, moderate output impedance, and

FIGURE 6.63
A MOSFET r-f amplifier.

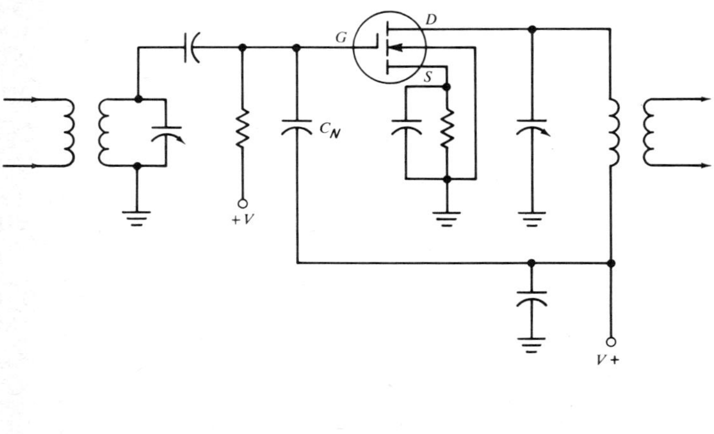

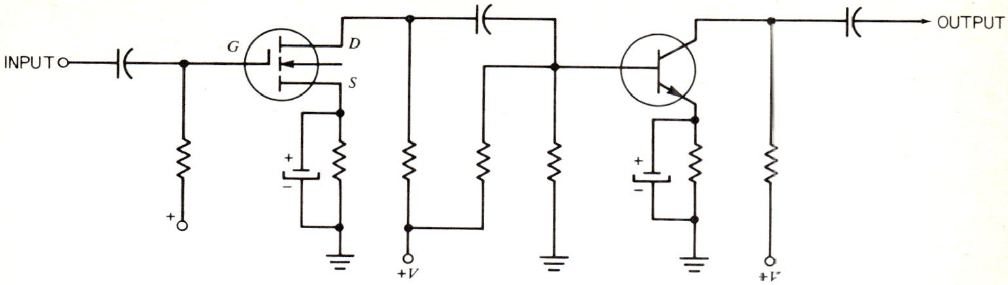

FIGURE 6.64
A common-source MOSFET with a common-emitter bipolar transistor amplifier.

in-phase input and output signals. It is equivalent to the bipolar common-base arrangement.

A FET or MOSFET can be employed in essentially the same manner as bipolar transistors. For low-frequency amplification, the common-source circuit shown in Fig. 6.62*a* is widely employed. Stages can be cascaded in whatever number is required and even designed for push-pull operation if considerable power is needed.

For high-frequency amplification, the circuit of Fig. 6.63 is typical. A tuned-input circuit brings the desired signals into the circuit, and a tuned circuit transfers the amplified signal to the following stage.

Capacitor C_N is a small-valued capacitor that feeds back to the input circuit sufficient *r-f* voltage to neutralize any r-f voltage returning to the input gate through the capacitance that exists between the drain and the gate internally in the MOSFET. This is very similar to the neutralization technique employed for the same purpose in bipolar transistors as well as high-frequency triode tubes.

An interesting combination can be formed with a common-source FET and a common-emitter bipolar transistor amplifier (see Fig. 6.64). The field-effect transistor presents a high input impedance, followed by the lower input impedance of the NPN bipolar transistor. Good voltage and power gain is obtainable from this combination.

The field-effect transistor, whether it be MOSFET or JFET, is a very versatile component, particularly with a high input impedance that permits it to substitute almost directly for tubes and yet with a noise level that is less than either tubes or bipolar transistors. In r-f applications, the FET is inherently more stable than a bipolar device and can handle a wider range of input-signal voltages. Finally, FET parameters tend to be more stable than bipolar units because FET parameters are purely bulk dependent. Thermal runaway in bipolars does not occur in FETs.

QUESTIONS

6.1. How can you determine by looking at a schematic diagram whether a PNP or an NPN transistor is being employed? (Assume that this information is not indicated.)

6.2. Why should coupling capacitors in transistor audio amplifiers possess high values?

6.3. Explain the purpose of C_2 and R_2 in Fig. 6.3.

6.4. Why is I_{CO} more critical in common-emitter amplifiers than in common-base amplifiers?

6.5. How can I_{CO} be controlled in common-emitter amplifiers?

6.6. Name three ways to couple amplifier stages.

6.7. Why can more gain be obtained by using transformer coupling than RC coupling between amplifier stages?

6.8. Draw a diagram of a two-stage transformer-coupled transistor amplifier and explain the purpose of each component.

6.9. Draw the diagram of a three-stage RC-coupled transistor amplifier.

6.10. What considerations govern the choice of values for R_f and C_f in Fig. 6.13?

6.11. Why is negative feedback employed in transistor amplifiers?

6.12. Illustrate a simple method of obtaining negative feedback.

6.13. Explain how negative feedback is obtained in the circuit of Fig. 6.14. Show that feedback voltage is actually 180° out of phase with the voltage existing at the feedback point.

6.14. Would the operation of the circuit in Fig. 6.14 be altered if the feedback line terminated at the base of Q_1 rather than at the emitter? Explain.

6.15. Explain how the input impedance of a transistor amplifier can influence the coupling network bringing the signal from the previous stage.

6.16. Show two ways to bias the base circuit of a transistor amplifier.

6.17. What do we mean by complementary symmetry in transistors?

6.18. Explain how the circuit in Fig. 6.21 operates.

6.19. Draw the circuit of a Darlington amplifier.

6.20. Briefly describe its operation.

6.21. How are the structures of transistors modified to enable them to function as a power amplifier?

6.22. What is emitter efficiency? Why does it decrease as the size of a device is increased?

6.23. Sketch a cross section of a planar transistor and label emitter, emitter-base junction, base, and base width.

6.24. What is current pinch-off? Explain the relationship between current pinch-off and beta.

6.25. Explain why the effective area of the emitter that can be turned on decreases with increasing frequency.

6.26. Why are fine geometries necessary in high-frequency power transistors?

6.27. What is thermal resistance? Contrast it to ohmic resistance.

6.28. What is load mismatch? How does it cause burn-out?

6.29. What is a ballast-emitter resistor and describe its function.

6.30. Draw the diagram of a push-pull class A transistor power amplifier.

6.31. What advantages does class B operation offer over class A operation in the audio range?

6.32. How can complementary symmetry be employed in push-pull amplifiers?

6.33. Draw a transistor phase inverter. Explain its function, and how it operates.

6.34. Draw the circuit of a differential amplifier.

6.35. Explain briefly how a differential amplifier operates.

6.36. Explain the ability of a differential amplifier to reject common-mode signals.

6.37. What is the purpose of transistor Q_3 in Fig. 6.52?

6.38. Illustrate several suitable interstage coupling networks for transistor r-f or i-f amplifiers.

6.39. What is the purpose of the neutralizing circuits sometimes found in transistor i-f or r-f amplifiers?

6.40. How is neutralization achieved?

6.41. What is the difference between ac and dc β? Why is this difference important in high-frequency transistor amplifiers?

6.42. Define f_β, f_T, and β_0.

6.43. What do we mean by crossover distortion in a class B amplifier? How is it minimized?

6.44. What is a heat sink? How is it used? Describe several types of heat sinks.

6.45. Name the three basic single-stage amplifier configurations for FET transistors.

Chapter Seven

Transistor Oscillators

The transistor amplifier circuits we have just discussed are similar to transistor oscillator circuits in many respects. For example, the dc bias-voltage requirements for amplifiers are very much the same for oscillators. The primary difference is that in an oscillator, part of the output power is returned to the input network in phase with the input signal to sustain oscillations. This in-phase return is positive, or regenerative, feedback, as compared with degenerative, or negative, feedback, where the returning signal is 180° out of phase with the input signal. Thus, we can define an oscillator circuit as one that takes power from a dc source and converts it into a fluctuating, or oscillating, output by utilizing regenerative feedback.

The maximum operating frequency of an oscillator circuit is dependent on the frequency capabilities of the transistor being utilized in the circuit. The maximum frequency of oscillation of a transistor is defined as the frequency at which its power gain is unity. The operating frequency of an oscillator circuit must be kept somewhere below the maximum frequency of oscillation of the transistor, since some power gain is used just to overcome losses in the feedback network (Fig. 7.1). If the amplifier has a power gain less than unity, oscillations will diminish with time and eventually stop. Thus, feedback power must be at least equal to the input power plus the losses in the feedback network if oscillation is to be sustained.

LOW-FREQUENCY OSCILLATORS

In addition to the transistor, the elements that determine the frequency of an oscillator can consist of an inductance-capacitance (LC) network, a resistance-capacitance (RC) network, or a crystal (Fig. 7.2).

Figure 7.3 is an audio-frequency oscillator in which the feedback of energy (needed for oscillation) takes place between collector and base of the transistor.

In the design of this oscillator, the windings on the transformer T_1 must

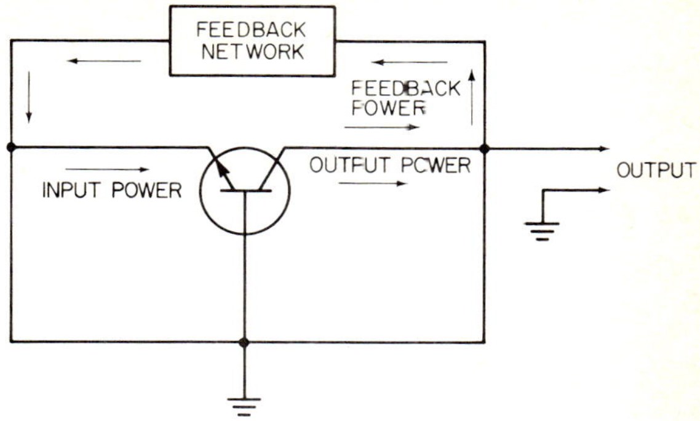

FIGURE 7.1

match the low impedance of the base-emitter circuit on the one hand and the high impedance of the collector circuit on the other. The frequency of oscillation will depend upon the inductance of the windings and their distributed capacitance, and this may be controlled by a shunting capacitor C_2 across the high-impedance collector winding.

One battery supplies the dc operating voltage for the circuit. An output control resistor R_1 with an output coupling capacitor C_1 is one way in which an output signal can be obtained from this circuit. Resistor R_2 aids in maintaining the required base-bias conditions.

The same oscillator, using the transistor in a common-base arrangement, is shown in Fig. 7.4. Two batteries are used, in this case to bias input and output circuits effectively.

In the common-base amplifier, recall that the signal polarity on the emitter (normally the input element) is in phase with the signal on the collector (the output element). Since these two elements are fairly close together, it is relatively simple for some of the collector signal to return to the emitter. This sets up positive feedback between these two elements; and if enough signal voltage is fed back, the stage will oscillate. This feedback is enhanced by connecting a small external capacitor between the collector and emitter, or by transformer coupling, as shown in Fig. 7.4.

The common-base oscillator is a widely used circuit, particularly at the higher frequencies.

Figure 7.5 is a tuned-collector oscillator. The base bias of Q_1 is determined by resistors R_1 and R_3. Capacitors C_1 and C_2 bypass AC around R_1 and R_2. The feedback signal is coupled from the transformer winding 3-4 to the tickler coil 1-2. A third winding of the transformer couples the output signal to the load.

The transistor version of the Colpitts oscillator is shown in Fig. 7.6. Regenerative feedback from the collector to the emitter is achieved by capacitors C_1 and C_2. Resistor R_1 influences the emitter-base bias and hence the extent of current flow through this portion of the circuit during each oscillation cycle. R_1 can be any value between 5,000 and 100,000 ohms (Ω). Resistor R_2 is designed to limit the reverse collector-current flow during that part of the half-cycle when the collector is positive.

If R_2 is 0 Ω, the positive peaks of the voltage waveform will have flat tops, since the collector is driven to overload. The waveform improves as the value of R_2 is increased to approximately 1,000 Ω, and thereafter improvement occurs more slowly as resistance increases. When R_2 is increased to about

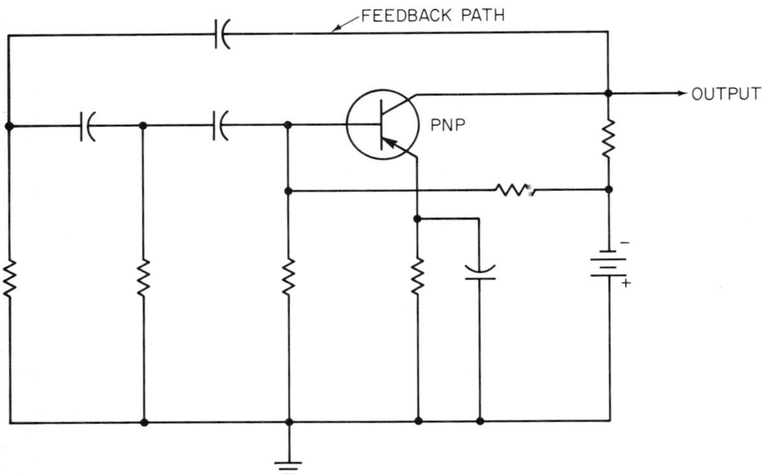

(a)

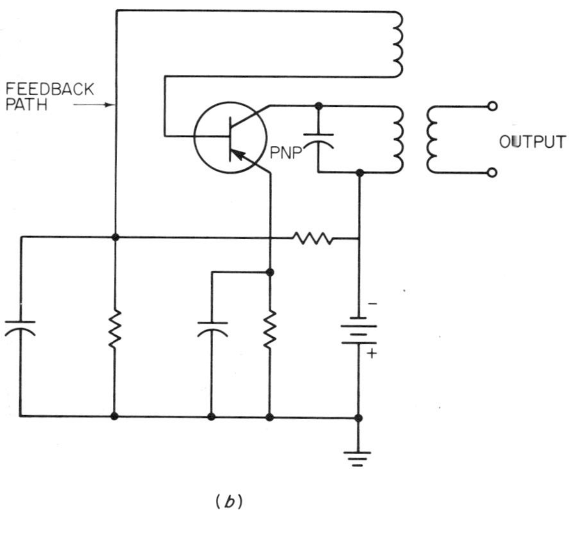

(b)

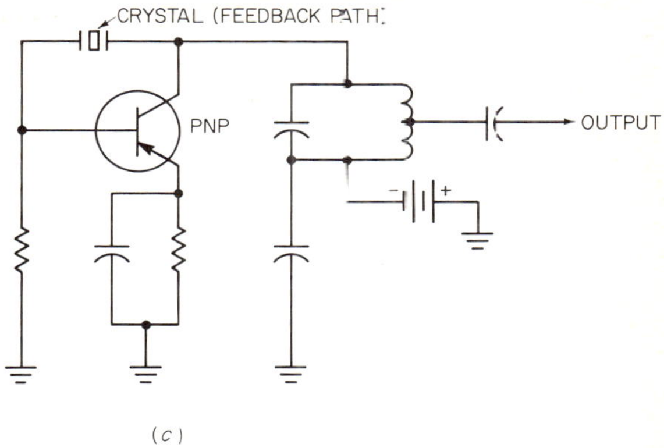

FIGURE 7.2
Three different types of frequency determining elements. (*a*) *RC* oscillator; (*b*) *LC* oscillator; (*c*) crystal oscillator.

40,000 Ω, the collector voltage is reduced to such an extent that oscillations stop.

RC Feedback Oscillators. An oscillator can be designed using resistors and capacitors in place of coils (or transformers) and capacitors. The RC network is positioned between the output and input circuits, which in the common-emitter arrangement means between collector and base. Furthermore, since the output signal is 180° out of phase with the input signal, the RC-coupling network must introduce an additional 180° phase shift in order to provide the base with an in-phase signal to sustain oscillations.

In order to achieve this 180° phase shift, three identical RC sections are generally employed, each shifting the signal by 60° at the frequency of

FIGURE 7.3
A transistor audiofrequency oscillator.

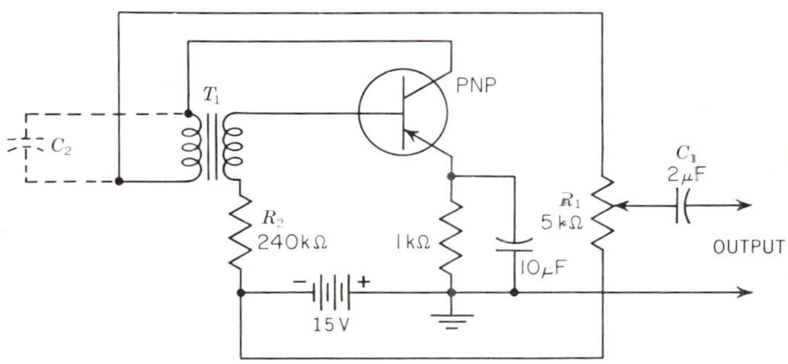

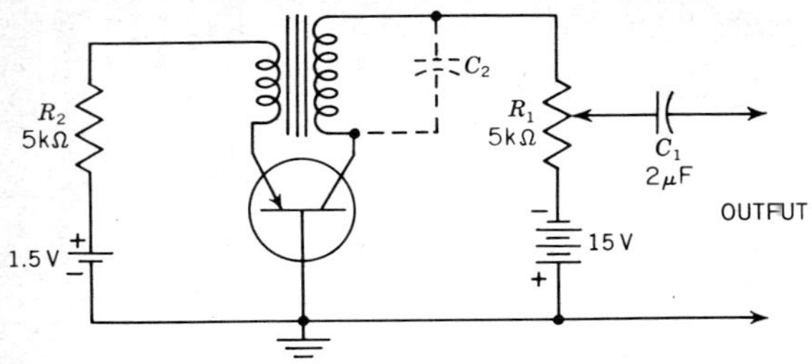

FIGURE 7.4
The same oscillator as in Fig. 7.3 in a common-base arrangement.

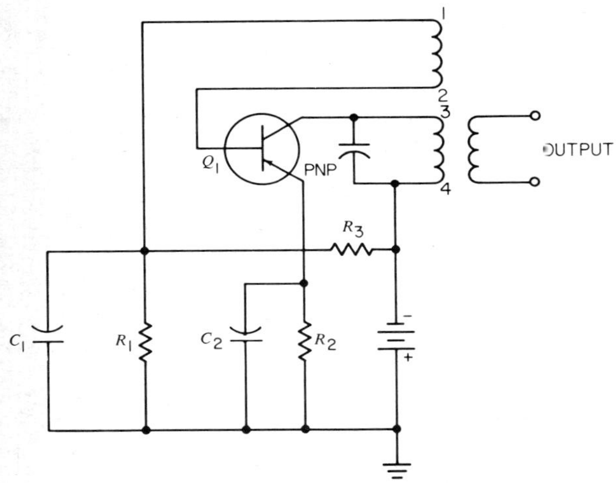

FIGURE 7.5
A tuned-collector (*LC*) oscillator.

FIGURE 7.6
A transistor oscillator using a Colpitts circuit arrangement.

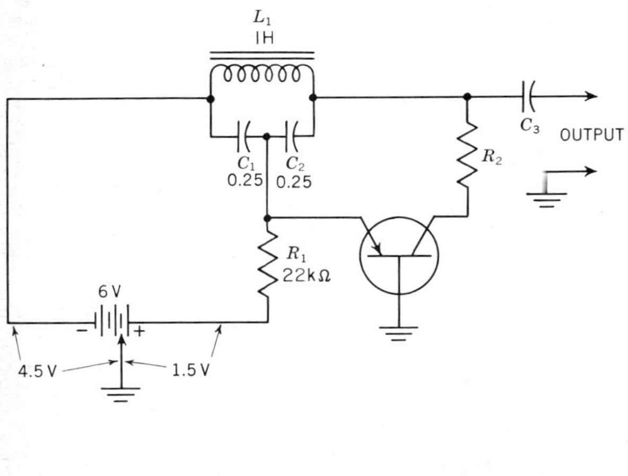

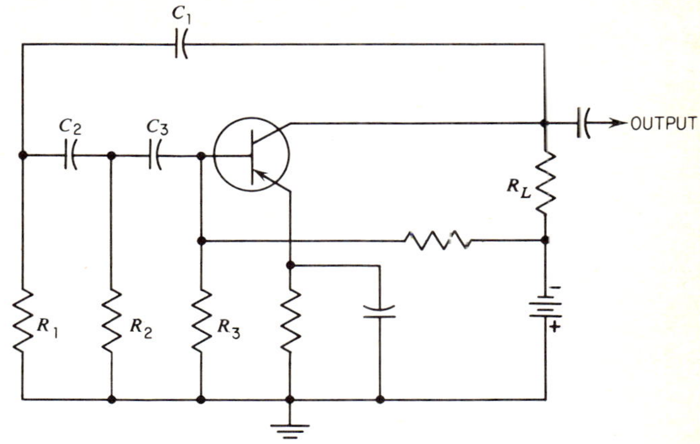

FIGURE 7.7
An *RC* oscillator that generates sine waves.

oscillation. Because the capacitance reactance of the network increases, or decreases, at different frequencies, the 180° phase shift required for the common-emitter oscillator occurs only at one frequency. Thus, the output of this oscillator is fixed. If R or C is made variable, however, a range of frequencies can be obtained.

Each RC section consists of a series-coupling capacitor and a shunt resistor. In Fig. 7.7, these are R_1 and C_1. R_2 and C_2, and R_3 and C_3. The phase shift comes about because R and C in series produce a current that leads the applied voltage by a certain angle. This angle is determined by the numerical relationship of resistance and capacitance. The smaller the capacitance, the more the current will lead the voltage for a given resistance. By properly selecting R and C, a 60° phase shift per section can be obtained.

A similar circuit arrangement employing a field-effect transistor is shown in Fig. 7.8. The same factors that determined the frequency of oscillation in Fig. 7.7 also govern the operating frequency here. In both instances, the approximate equation is

$$f = \frac{1}{5.9\pi RC}$$

Transistor oscillator circuits can be designed simply, as illustrated by the preceding arrangements, or they can be utilized in more complex designs. A low-distortion audio oscillator employing three transistors is shown in Fig. 7.9. It is interesting to note how this circuit was developed.

The first step in the development of the transistor version of the circuit is shown in Fig. 7.10a. The first stage is a grounded-emitter amplifier. The

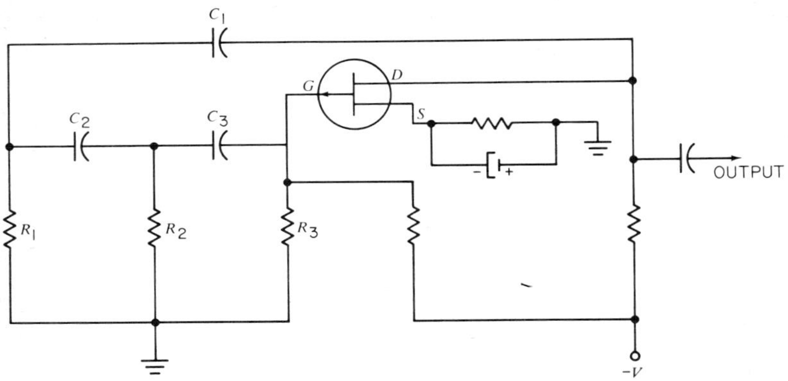

FIGURE 7.8
An FET *RC* oscillator.

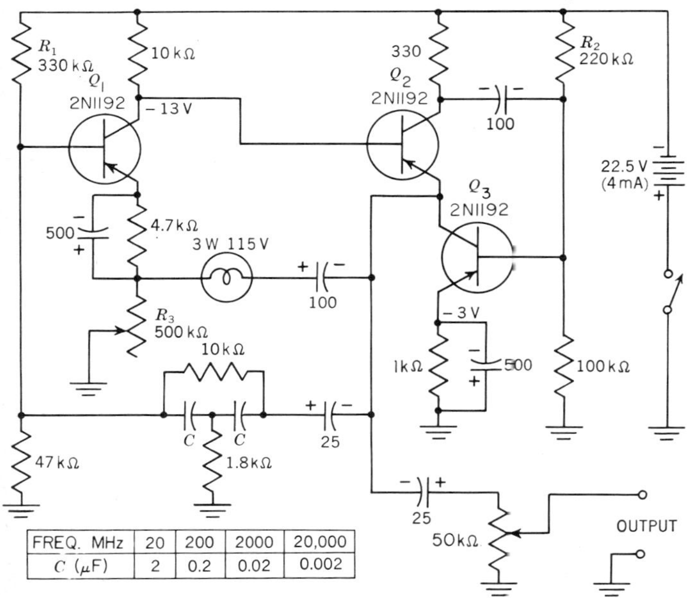

FIGURE 7.9
Completed circuit of low-distortion transistor a-f oscillator. (*Electronics.*)

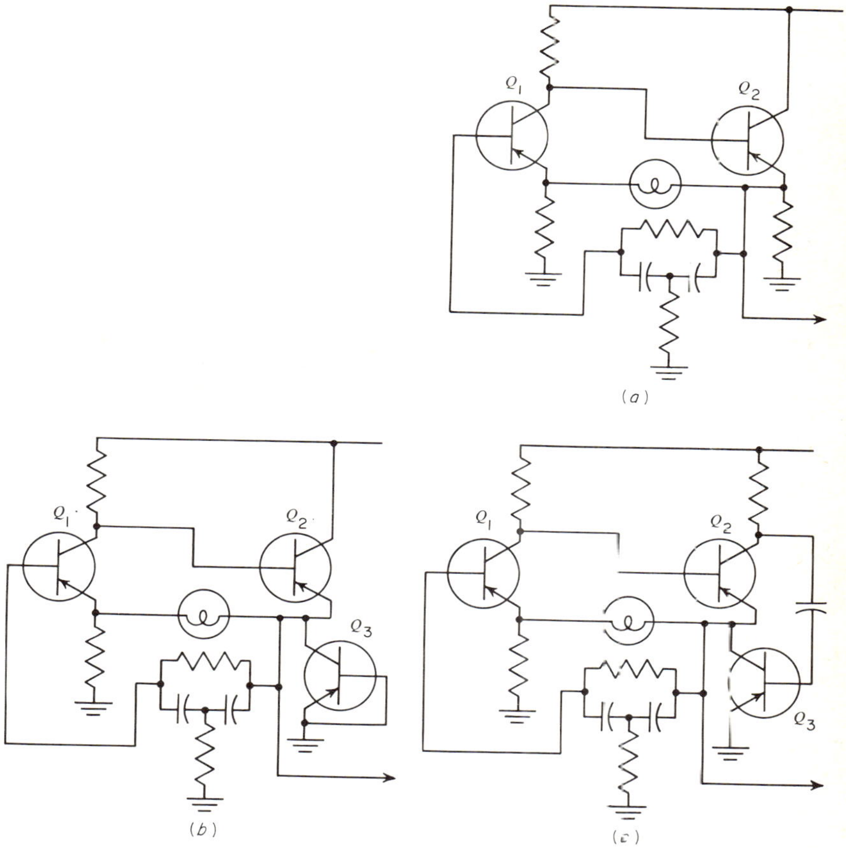

FIGURE 7.10
The various steps in the development of the transistor oscillator shown in Fig. 7.9.

second transistor is operated as a grounded-collector stage. Positive feedback is brought from the emitter of Q_2 to the emitter of Q_1 through a lamp. (A lamp is used to stabilize the amplitude of the oscillations. Lamp filament resistance rises with current, tending to maintain a constant positive-feedback voltage.)

An improved version of the oscillator is shown in Fig. 7.10*b*. The emitter-load resistor of transistor Q_2 is replaced by a third transistor Q_3. The change was made because the high-dynamic-collector resistance of Q_3 permits more efficient operation of Q_2.

The third step in the development of the circuit (Fig. 7.10*c*) was made to reduce the amount of even harmonics present in the oscillator output sig-

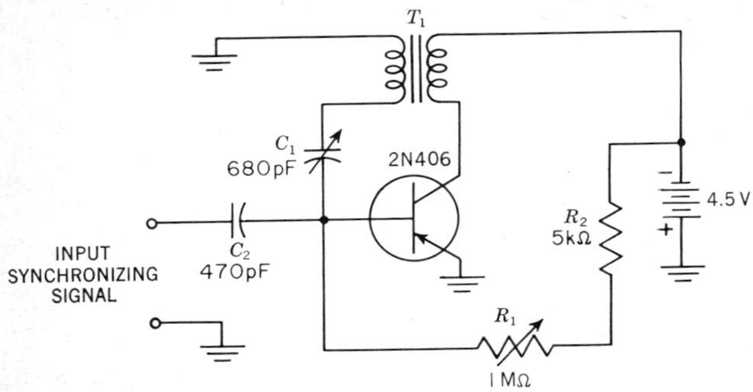

FIGURE 7.11
A blocking oscillator. Capacitor C_1 is made variable to permit the pulse width to be changed.

nal. The base of Q_3 is connected to the load resistor of Q_2, effectively placing the two transistors in push-pull. When the collector current of one transistor increases, the collector current of the other decreases. The signal currents of both transistors flow through the same load resistor, and this provides for a high-output current.

In the final version of this oscillator, Fig. 7.9, dc stabilizing resistors were placed in series with the emitters of Q_1 and Q_3. Additional resistors, such as R_1 and R_2, establish the proper base-operating voltages.

Nonsinusoidal Oscillators. The foregoing are all sine-wave oscillators. Important, too, are relaxation oscillators developing pulses, square waves, sawtooth waves, and other nonsinusoidal forms. One such unit is the blocking oscillator shown in Fig. 7.11. This circuit is closely similar to that shown in Fig. 7.3 except for the addition of a small coupling capacitor C_1 between the base and T_1. The frequency of the circuit is variable between 3 and 60 kHz and is inversely proportional to $R_1R_2C_1$. The blocking oscillator can be synchronized to a pulse or a sine-wave input by coupling the signal to the base through capacitor C_2. Note the extreme simplicity of this circuit, including the manner in which one battery furnishes power to the entire circuit.

Another common nonsinusoidal oscillator is the multivibrator. The basic form of this oscillator, shown in Fig. 7.12, is seen to consist of two resistance-capacitance-coupled amplifiers with the output of the second stage, Q_2, fed back to the input of the first stage, Q_1. Oscillations occur in this system because each transistor reverses the voltage applied to its base by 180°, and two such reversals produce a signal at the collector of Q_2, which is in phase with the voltage at the input of Q_1.

In a multivibrator, one transistor is cut off while the other one is conducting. How long this condition persists is determined largely by the values of resistors R_1 and R_3 and capacitors C_2 and C_4. To see how the shift in conduction is made from transistor to transistor, let us briefly follow one cycle in the operation of a multivibrator.

Assume that the power supply has just been connected across the circuit. Owing perhaps to some slight disturbance in the circuit, the collector current of Q_1 increases. This produces an increase in the voltage across R_2, with the collector end of the resistor becoming less negative or more positive. Capacitor C_2, which is connected to R_2 at this point, likewise attempts to become more positive, and the base Q_2 also assumes the same potential. The net result is a lowering of the current through Q_2 and less of a voltage drop across R_4.

The lower voltage across R_4 means that the collector end of this resistor becomes more negative or less positive to its previous value. Capacitor C_4 transmits this negative increase to the base of Q_1 and, consequently, even more collector current flows through R_2. The process continues in this manner, with the base of Q_1 becoming more and more negative and driving the base of Q_2 increasingly positive by the large positive-charge built up across R_3 and C_2. The collector current of Q_2 is rapidly brought to zero by this sequence of events.

Transistor Q_2 remains inactive until the positive accumulation of charge on C_2 discharges and removes some of the large positive potential at the base of Q_2. The discharge path of C_2 is through the high resistance of R_3, the battery to ground, from ground through R_5 and then through the low resistance of transistor Q_1 (which is conducting strongly). When C_2 has discharged sufficiently, Q_2 starts conducting again, causing the collector end of R_4 to become increasingly positive. This now brings to the base of Q_1 an increasingly positive voltage, reducing collector-current flow in Q_1 and R_2. The reduction in the voltage drop across R_2 causes the collector end of the resistor to increase negatively, and the base of Q_2 (through C_2) receives this negative voltage. The

FIGURE 7.12
A transistor multivibrator.

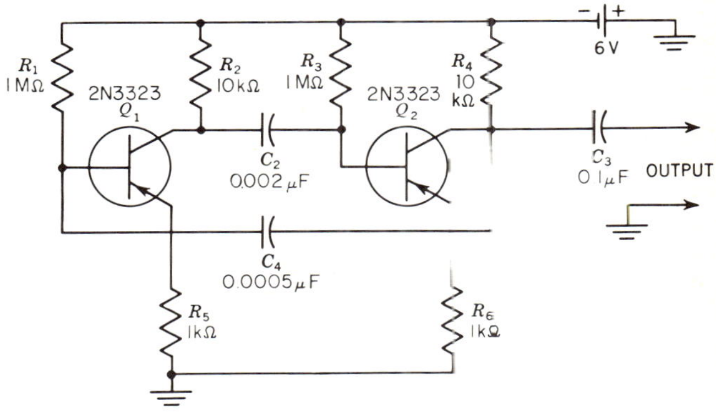

increased current through R_4 quickly raises the positive-base voltage on Q_1 (through C_4) and drives this transistor to cutoff. When the excess charge on C_4 leaks off, the process starts all over again. C_4 loses its accumulative positive charge by discharge through Q_2, R_6, and R_1.

The entire operation may be summed up by stating that first the collector current of one transistor rises rapidly, driving the second transistor to cutoff. This conduction persists until the second transistor is released from its cutoff state and starts to conduct. It is now the first transistor that is cut off. When the first transistor is again permitted to conduct, the second transistor is driven into nonconduction. The switching continues in this manner, with the rapidity of turnover (i.e., frequency) determined by C_2, R_3, and C_4, R_1. If a sawtooth wave instead of a rectangular wave is desired across the output terminals, it can be obtained by connecting a capacitor from the collector of Q_2 to ground. This is satisfactory for low-frequency operation. It will not work too well at high frequencies because of the loading effect of the capacitor.

If it is desired to control the frequency of this multivibrator closely, synchronizing pulses can be applied across R_5 of Fig. 7.12. See Fig. 7.13. To be most effective, the incoming pulses should be positive in polarity, because a positive pulse applied to the emitter of a PNP transistor will initiate a flow of current. Furthermore, the frequency at which the pulses arrive should be fairly close to the natural operating frequency of the multivibrator. If the two frequencies are too far apart, effective synchronization will be difficult.

The incoming sync pulses can also be applied to R_6, with equal effectiveness.

A FET multivibrator, similar to the bipolar circuit, is shown in Fig. 7.15*a*.

FIGURE 7.13
The transistor multivibrator of Fig. 7.12 with an effective method of controlling its frequency.

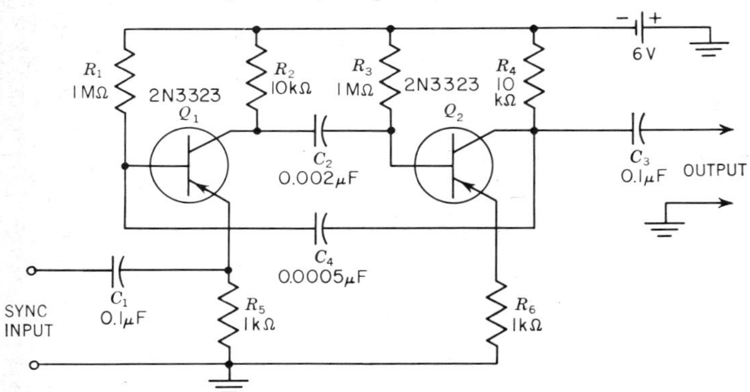

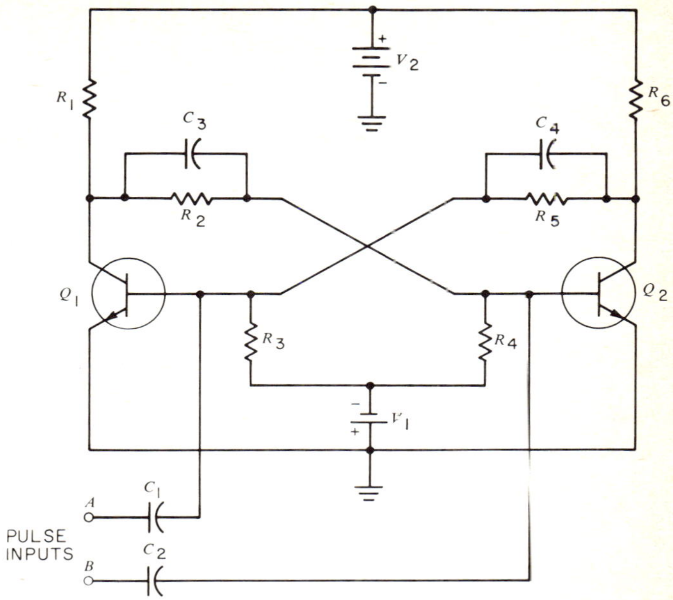

FIGURE 7.14
A bistable multivibrator.

Note that the arrangement of the components is the same and operation of the FET multivibrator is similar in all respects to the description just given.

The multivibrator circuits just discussed fall into the category of astable multivibrators. That is, they are free-running multivibrators in which the cycle of conduction and nonconduction shifts from transistor to transistor. Another type of multivibrator, called a bistable (or flip-flop) multivibrator, is shown in Fig. 7.14. This type of multivibrator is not free-running; it only operates when it is triggered. This type of circuit is extensively employed in computers.

The multivibrator circuit of Fig. 7.14 has two stable states. In one state, Q_1 is conducting and Q_2 is cut off; in the other state Q_2 is conducting and Q_1 is cut off. Either condition can be maintained indefinitely, since there is no action within the circuit forcing the circuit to switch from one state to the next. A change can occur, however, if a triggering pulse of the proper polarity and amplitude is applied to an appropriate point in the circuit.

In operation, assume that Q_1 is conducting. This will cause the polarity at the collector of Q_1 to be close to zero volts. Since the collector of Q_1 is tied into the base of Q_2 through R_2, it means that this zero voltage at the base of Q_2 cannot overcome the negative voltage that is brought here by R_4. Since Q_2 is an NPN transistor, a negative voltage on the base will keep it cut off, hence no conduction will occur through Q_2. Because of this nonconduction,

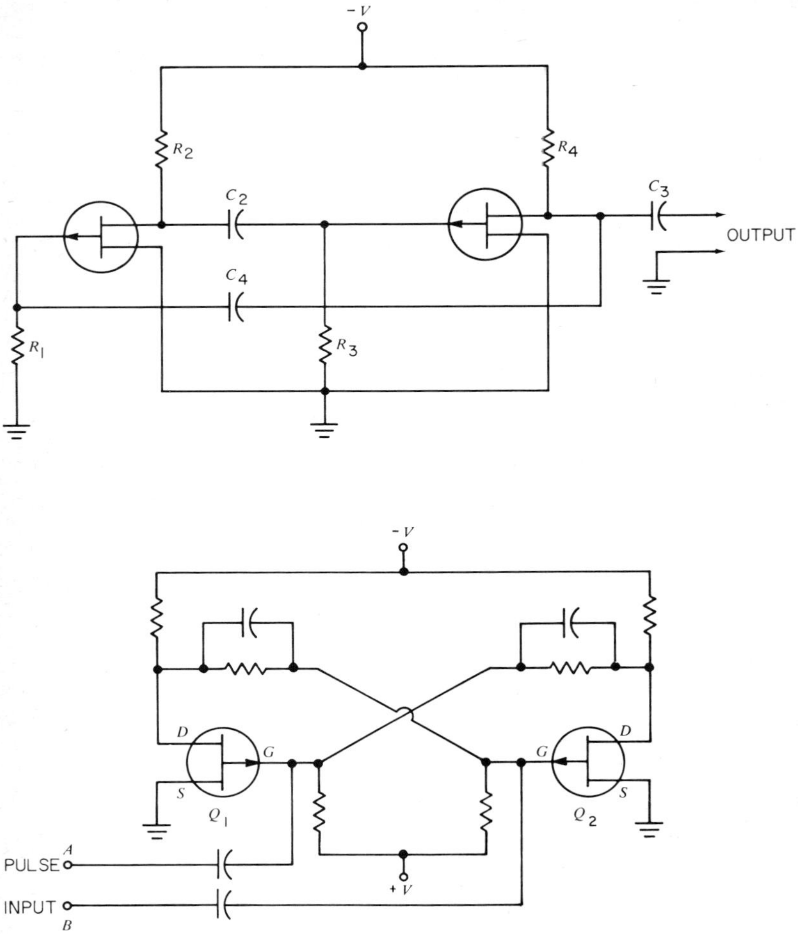

FIGURE 7.15
(*a*) a FET astable multivibrator; (*b*) a FET bistable multivibrator.

the collector of Q_2 will have the full positive voltage of V_2 and this positive voltage will be able to overcome, at the base of Q_1, the negative voltage that is brought to this base by R_3. This will maintain transistor Q_1 in full conduction, and the circuit will remain in this condition unless something is done to upset the voltage relationships just described.

In order to effect a change or switchover in conduction from Q_1 to Q_2, a strong positive pulse can be applied to input terminal B. This positive-pulse input will overcome the negative voltage at the base of Q_2 and cause Q_2 to start conducting. When Q_2 does begin to conduct, the positive voltage formerly at its collector will now drop to essentially zero volts, and this will bring the base of Q_1 under the influence of the negative voltage of battery V_1, causing Q_1 to be cut off. Thus, by the application of a positive pulse to the base of the nonconducting transistor Q_2, we have switched the state of this multivibrator.

Again, the circuit will remain stable, with Q_2 conducting and Q_1 cut off until a positive pulse of sufficient amplitude is applied at input terminal A. This positive pulse will start Q_1 conducting and cause Q_2 to be cut off by the same sequence of events described above. It is because of this action that the circuit is widely known as a flip-flop, since it tends to flip from one transistor to the other. Capacitors C_3 and C_4 in Fig. 7.14 are used to speed up the switching action, since they help to transmit voltage changes at the collectors of the transistors to the bases of the opposite transistor.

A FET bistable multivibrator circuit is shown in Fig. 7.15*b*. Its operation is essentially the same as the multivibrator shown in Fig. 7.14. Even though the circuit uses junction FETs, MOSFETs with the proper applied operating voltages would function equally as well.

Crystal Oscillator Circuits. Transistor oscillator circuits are often crystal-controlled to improve and maintain frequency stability. The frequency stability of an oscillator circuit is dependent on the temperature stability of its components and the sharpness of resonance of its tuned circuit. Variations in inductance and capacitance with temperature will cause frequency variations unless the components can be selected to compensate for these variations.

Certain electromechanical resonators, because of their temperature stability and resonance sensitivity, are widely used for frequency stabilization in oscillator circuits. The most popular of these resonators is the quartz crystal.

Crystalline quartz, among other materials, can exchange energy between electrical and mechanical states. Thus, a mechanical force applied in the right direction upon such a material will create an electrical charge on the surface of the material. Conversely, an applied electrical potential will create a mechanical displacement (vibration) of the material, which is known as the piezoelectric effect.

An alternating voltage of a desired frequency applied across a quartz crystal will cause it to vibrate. This vibration exhibits a resonance at a frequency determined by the crystal dimensions. At this particular resonant frequency, the energy exchange is extremely efficient, which means that very little energy is dissipated in the crystal. The frequency of resonance is very accurately maintained by the crystal structure. In addition, the electrical and physical properties of quartz do not significantly change with temperature fluctuations. Thus, quartz crystals are excellent frequency-stabilizing elements for oscillators.

Figure 7.16 is a crystal-controlled oscillator somewhat similar to the Colpitts circuit arrangement. It is called a Clapp oscillator after its originator J. K. Clapp. The crystal serves as a series-resonant tuning circuit connected

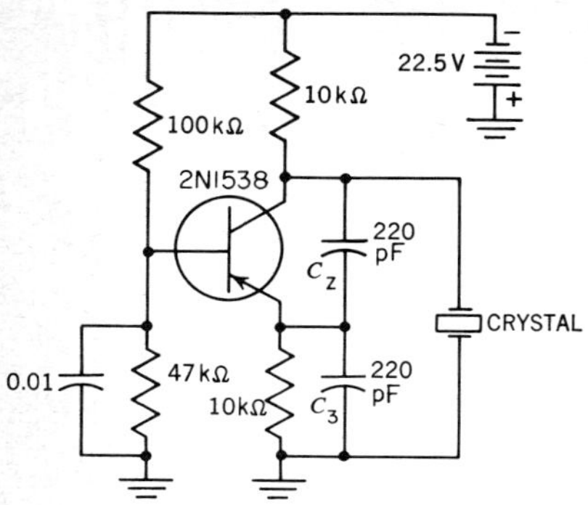

FIGURE 7.16
A crystal-controlled Clapp oscillator.

between collector and ground. Feedback voltage between the collector, emitter, and base is provided by C_2 and C_3.

Figure 7.17 is an emitter-coupled sine-wave oscillator. The frequency stability of the circuit can be improved by connecting a series-resonant crystal between the emitters of the two transistors (see Fig. 7.18). The frequency of the crystal must equal the resonant frequency of the parallel LC network.

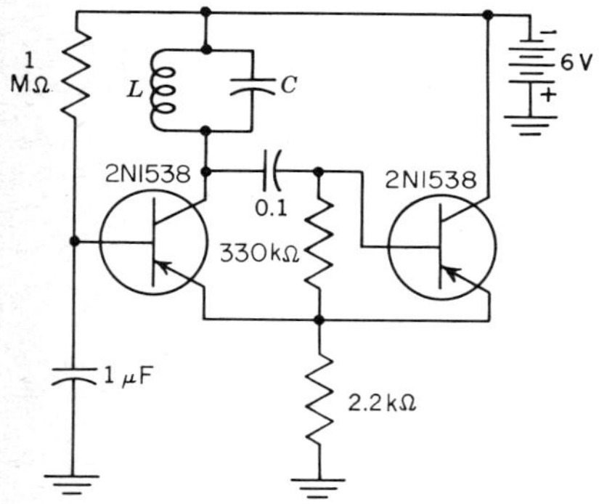

FIGURE 7.17
An emitter-coupled sine-wave oscillator.

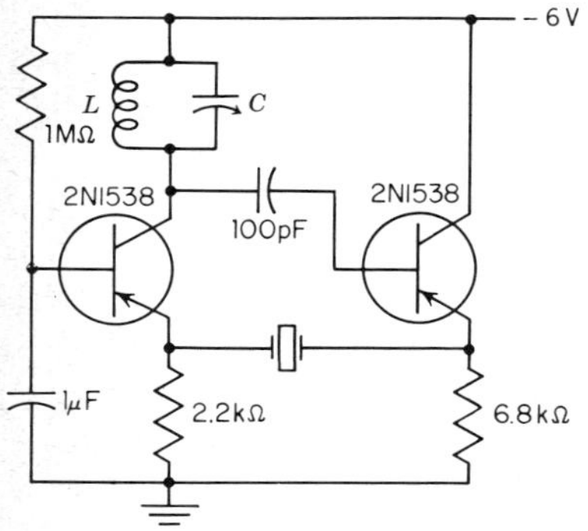

FIGURE 7.18
An emitter-coupled sine-wave oscillator that is crystal-controlled. Resonant frequency of the crystal should be the same as the resonant frequency of the parallel-resonant network *LC*.

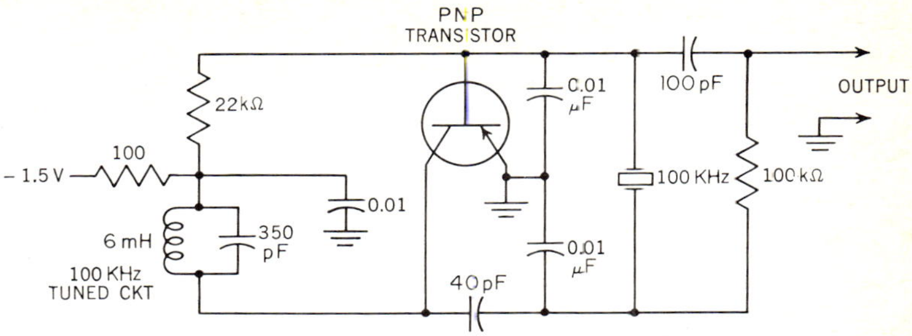

FIGURE 7.19
A highly stable transistor oscillator.

A crystal oscillator circuit with excellent frequency stability is shown in Fig. 7.19. A PNP junction transistor is used here, together with a 100 kHz crystal. Bias-battery voltage is only 1.5 volts (V). Current drain is 100 microamps (μA). The 0.01-microfarad (μF) capacitors connected to ground from each side of the crystal serve to maintain a constant phase shift in the crystal feedback loop. The crystal itself is placed in the path between collector and base. The transistor is operated as a grounded emitter, and it develops 0.8 V across the 100-kHz tuned circuit connected to the collector. Driving current for the crystal is obtained from a capacitive voltage divider consisting of the 40-picofarad (pF) and 0.01-μF capacitors connected in series between the collector and ground.

The stability of this oscillator is excellent. Measurements of frequency with changes in temperature and voltage indicate that the frequency varies approximately 1 part in 100 million per °C and 1 part in 100 million per 0.1 V. Short-time variations are about +3 parts in 10,000 million, and long-interval drift indicates changes of about 3 parts in 1,000 million (i.e., 1 billion) per 24 hours (h).

HIGH-FREQUENCY OSCILLATORS

The use of transistors in high-frequency oscillators is governed by the same considerations as for low-frequency oscillators except that the frequency capabilities of the transistor become the prime limiting factor. At the start of our discussion on oscillators, we defined the term maximum frequency of oscillation $f_{\max}$ as the frequency at which the transistor's power gain is equal to unity. We also said that the operating frequency of an oscillator circuit must be kept somewhere below the maximum frequency of oscillation of the transistor, since some power gain has to be used to overcome losses in the feedback network.

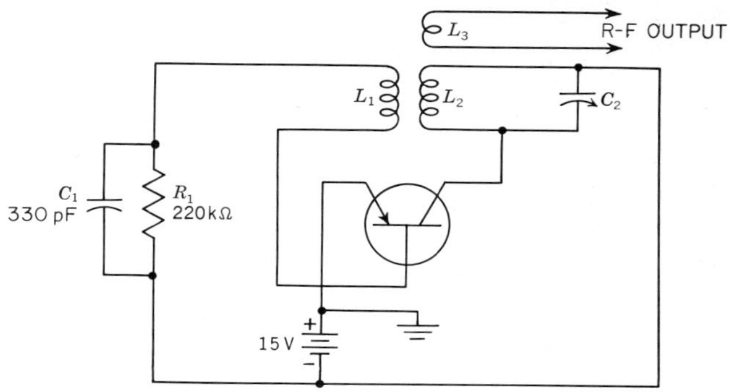

FIGURE 7.20
An r-f transistor oscillator.

Good high-frequency transistors must have high beta, low base resistance, low collector capacitance, and the ability to operate at relatively high collector voltage.

Frequency itself is not the only limitation in designing an oscillator circuit. What is important, and difficult to obtain, is output power plus frequency. The problem is that power output and gain decrease with increasing frequencies (Chap. 6).

A simple r-f oscillator is shown in Fig. 7.20. L_1 and L_2 are two tightly wound coils that provide for the transfer of energy between output and input circuits. L_3 is wound close to L_2, and the energy it absorbs is transferred to whatever external circuit is connected to the oscillator. Capacitor C_2 tunes L_2 and enables the generated frequency to be varied. R_1 serves to limit the emitter current to a safe value; C_1 across R_1 assists in the oscillating action.

The oscillator shown in Fig. 7.21 is similar to the preceding circuit except

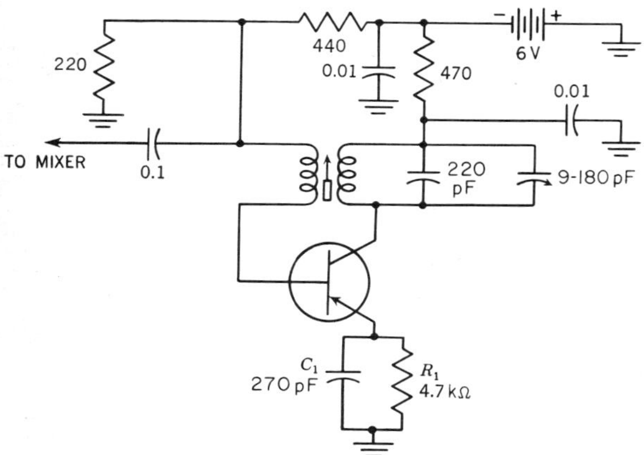

FIGURE 7.21
An oscillator designed for an automobile receiver.

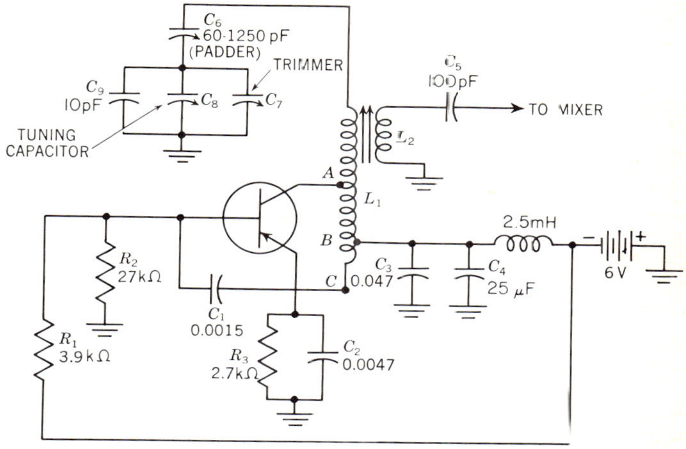

FIGURE 7.22
An r-f oscillator employed in a radio receiver. An r-f PNP transistor would be used.

that a 4,700-Ω (4.7kΩ) resistor and a 270-pF capacitor are inserted in the emitter lead. This type of oscillator is employed in broadcast receivers, and the R_1C_1 network is used to introduce a limited amount of degeneration in the circuit. This offsets some of the positive, regenerative feedback and serves to reduce the loading effect of the transistor input circuit on the oscillator tuned circuit. It was found that when this is done, oscillator tuning becomes relatively independent of the transistor input impedance. A further aid to stability is the use of a relatively high tank capacitance across the collector tuned circuit.

Another oscillator that has been employed in radio-broadcast receivers is shown in Fig. 7.22. This is similar to the Hartley oscillator (Fig. 7.23). The voltage that is developed between points B and C of L_1 represents the energy that is fed back to the base input circuit via coupling capacitor C_1. The

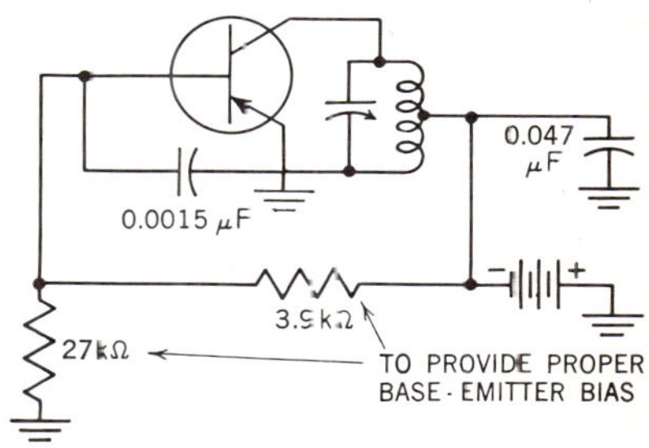

FIGURE 7.23
A Hartley oscillator.

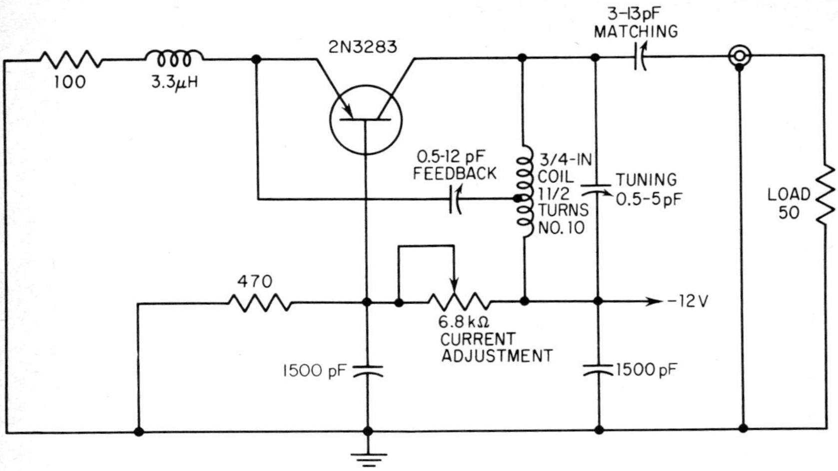

FIGURE 7.24
A common-base oscillator designed to operate at 200 MHz.

collector is tapped down on L_1 to decrease the effect of its (i.e., the collector) capacitance, to provide a better impedance match between the transistor and the tuned circuit, and to improve frequency stability and tracking. (The last feature stems from the application of this oscillator in a radio receiver.) Tracking also explains the reason for the presence of capacitors C_6 (600 to 1,250 pF) and C_7. In this particular design, three-point tracking between oscillator and mixer was obtained by using a slug in the collector coil, a padder capacitor C_6, and a gang capacitor trimmer C_7. The slug takes care of the central portion of the band, the padder provides an adjustment at the low end of the band, and the trimmer is employed to make the high end of the tracking curve fall into line.

Energy from the oscillator is transferred into the mixer circuit (not shown here) by a combination of inductive and capacitive coupling. The initial transfer from L_1 to L_2 is inductive; the second transfer, from L_2 through C_5 to the mixer, is capacitive. The designers of this circuit felt that this arrangement would provide a more nearly constant oscillator injection voltage at the mixer.

The necessary biasing voltage for the transistor collector is brought in through coil L_1. A similar biasing voltage for the base is brought in through R_1 and R_2. The resistor R_3 and capacitor C_2 in the emitter leg serve approximately the same purpose here that they did in the preceding circuit.

A 200-megahertz (MHz) oscillator in a common-base arrangement is shown in Fig. 7.24. Ten milliwatts of output power is provided. The current-adjust resistor is set for a collector current of 2 mA. The matching capacitor serves as an impedance transformer to match the oscillator to the 50-Ω load section.

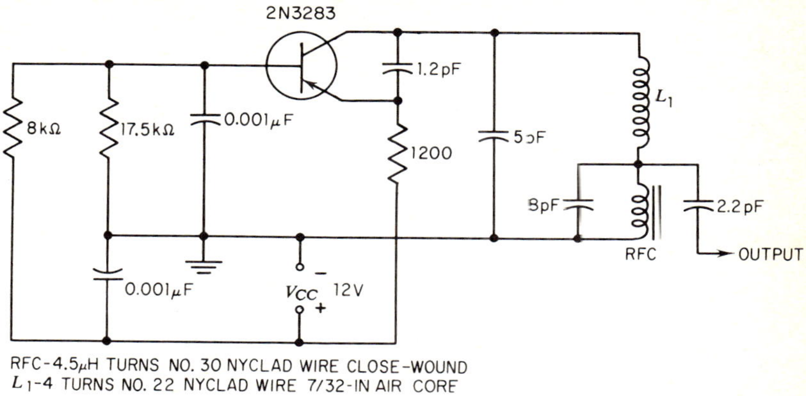

FIGURE 7.25
An oscillator operating at 200 MHz.

Figure 7.25 shows another oscillator arrangement operating at essentially the same frequency.

An r-f Colpitts oscillator is shown in Fig. 7.26. The single tuning circuit L_1 and C_1 is connected in the collector circuit between the collector element and the battery. To provide the proper feedback to sustain oscillations, 22-pF capacitors are connected between collector and emitter and between emitter and ground. R_1 and R_2 provide the appropriate voltage for the base-emitter circuit, while R_3 functions as a dc stabilizing resistor (to minimize the effects of temperature variations).

Modern transistors can be used as oscillators, with considerable power output, well up into the uhf region.

Field-effect transistors can be employed in many of the r-f circuits shown. Typical circuit arrangements for FETs (in these cases MOSFETs) are shown in Fig. 7.27. The first diagram, Fig. 7.27*a*, is a Colpitts oscillator; Fig. 7.27*b*

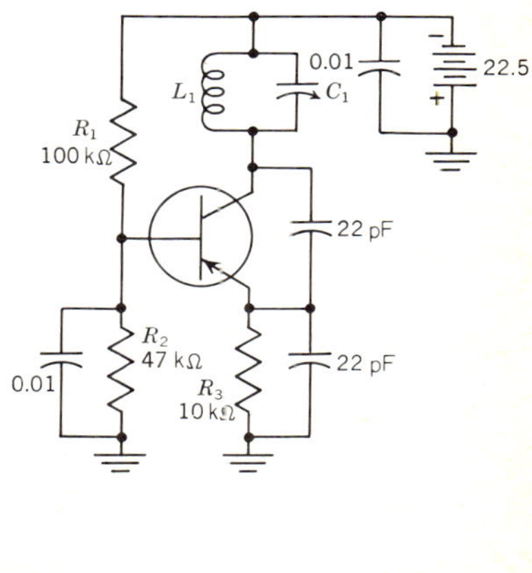

FIGURE 7.26
An r-f Colpitts oscillator.

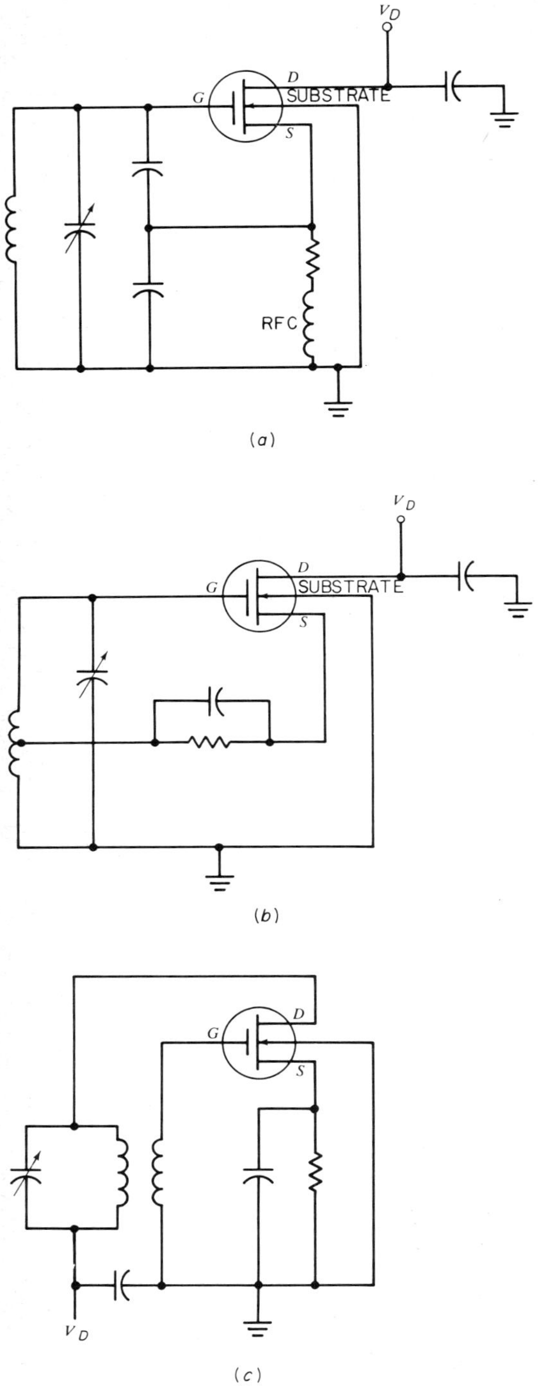

FIGURE 7.27
Three FET oscillator circuits. (*a*) Colpitts oscillator. (*b*) Hartley oscillator. (*c*) Gate-tickler-feedback oscillator.

is a Hartley; and Fig. 7.27*c* is a gate-tickler-feedback oscillator. All three work quite well over the vhf frequencies, and the Colpitts is exceptionally useful even within the uhf range.

QUESTIONS

7.1. What is the primary difference between an amplifier and an oscillator?

7.2. What is regenerative feedback? Degenerative feedback?

7.3. Define maximum frequency of oscillation (f_{max}). Of what use is it?

7.4. What, besides the transistor, determines the frequency of an oscillator?

7.5. Draw the circuit of a Colpitts oscillator using a bipolar transistor. Explain briefly how it operates.

7.6. How does an RC oscillator producing sine waves function? What elements determine its frequency? How do they do this?

7.7. Name four types of nonsinusoidal oscillators.

7.8. Draw a block diagram of a multivibrator.

7.9. Explain briefly how the circuit of Fig. 7.12 functions.

7.10. Differentiate between an astable and a bistable multivibrator.

7.11. What is the advantage of a crystal oscillator? What disadvantage?

7.12. What is the piezoelectric effect?

7.13. Draw the circuit of a Clapp transistor oscillator using a crystal.

Chapter Eight

Integrated Circuits

From Chap. 1, "Modern Electron Theory," through the last chapter, "Oscillator Circuits," we have obtained an appropriate background to discuss the integrated circuit.

The integrated circuit (IC) has dominated semiconductor research and planning for the last decade; and it is safe to say that this fascinating technology has had a greater impact on the electronics industry than have either the vacuum tube or the transistor.

The term "integrated circuit" has been the cause of some confusion, owing perhaps to the various IC fabrication methods. (We shall discuss these methods in this chapter.) However, the integrated circuit can be defined generally as an arrangement of multifunction semiconductor devices in which large numbers of active elements, together with passive elements and some means of interconnection, are produced and combined to perform complex electrical functions. The interconnected circuit elements are all fabricated from a continuous substrate.

Integrated circuits represent a significant development in the field of semiconductor electronics. This chapter will serve as an introduction to their construction.

MONOLITHIC AND HYBRID ICS

There are two major IC categories: monolithic and hybrid. These two categories are the result of different fabrication techniques, which we shall discuss presently.

Monolithic. In the monolithic type of IC, the active elements (transistors and diodes) are fabricated on the same silicon chip, together with whatever passive elements (resistors and capacitors) are required. In the monolithic form of construction, all the elements (or components) are formed from the same semiconductor slab or substrate.

Hybrid. A hybrid integrated circuit can be defined as one in which separately manufactured components, active and passive, are interconnected on a passive substrate, such as ceramic. There are three types, or forms, of hybrid circuits: thin films, thick films, and multichip.

History and Economics. Before we describe IC processing, let us briefly discuss how (and when) the modern IC concept was initiated, and also some of its economic ramifications.

It was way back in 1725 that du Fay proved that the region surrounding a red-hot body conducted electricity. One hundred and fifty-eight years later, Edison showed that electrons could pass from a hot filament to a cold plate. It was not until 1905 that Fleming developed the diode; and it took two more years for De Forest to add a third electrode (a grid), thereby creating the first vacuum tube.

In contrast, consider that the transistor was invented in 1948 and that today we have entire complex solid-state circuits on a single silicon chip, such as shown in Fig. 8.1. In addition, these circuits are being mass-produced! The rapid technological growth in solid-state electronics becomes more astounding when it is realized that if the vacuum-tube timetable just mentioned were applied to solid-state development, the semiconductor industry would not exist until 1980.

FIGURE 8.1
A beam-leaded integrated circuit.

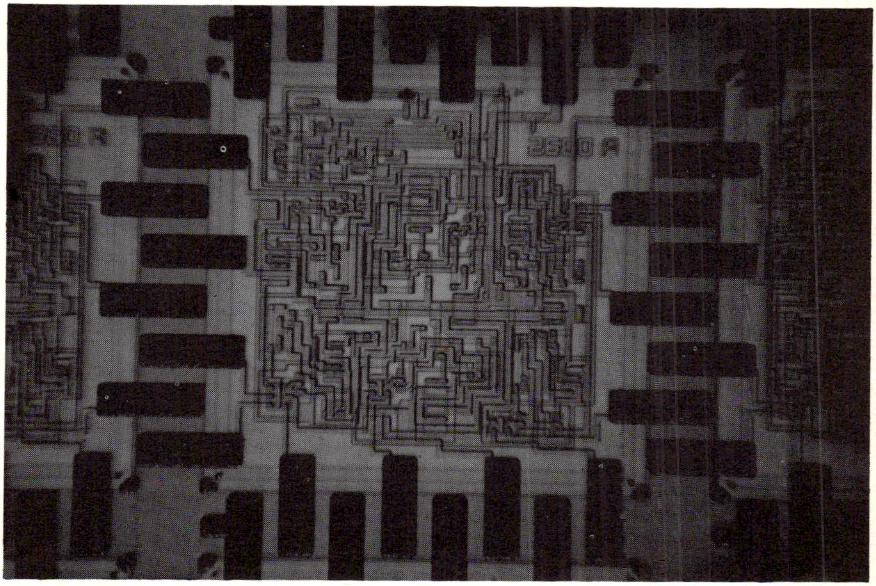

The first monolithic IC was developed by Texas Instruments (TI) in 1959, and three years later TI announced the first off-the-shelf IC. These circuits were operational amplifiers (OP amps), which we shall describe, with a 62-decibel (dB) gain and a maximum frequency response of 60 kilohertz (kHz). In 1965, the Fairchild Semiconductor Corporation introduced its well-known 700 series of OP amps. These circuits designed by Fairchild's R. Widlar have been the basis for many of today's linear IC amplifiers.

It is interesting to note that during the first three months of 1970 there were 79 million monolithic ICs sold, valued at $115 million. During the same period, 32 million hybrid circuits were sold for $25 million. And these figures are increasing at a rate in excess of 15 percent a year.

Why ICs? With the advent of the transistor, electronic designers set in motion a trend with the apparent objective of converting to solid state almost everything that formerly used vacuum tubes. The initial motivation was to produce smaller electronic equipment. In the beginning, these efforts were frustrated because most of the other circuit components (resistors, capacitors, coils) were larger than the new transistors. Soon, however, these other circuit components were miniaturized, thereby pushing ahead the development of smaller electronic equipment.

The armed services invested large sums of money to reduce the bulk of electronic equipment. Early attempts were concentrated in assembly techniques, which led to a modular concept. In this technique, standard elements (resistors and capacitors) were soldered to passive circuit substrates that were stacked and connected together with bars (Fig. 8.2).

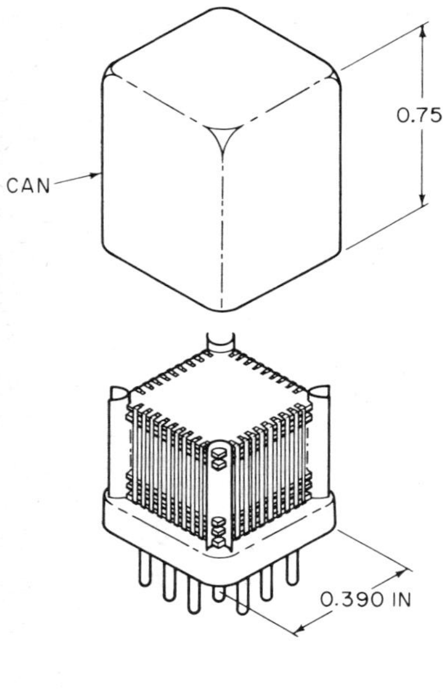

FIGURE 8.2
An early modular package in which resistors, capacitors, and transistors were stacked in layers.

The redesigning of standard circuit components so that they became small enough to be assembled on substrates, or printed circuit boards, and stacked using the modular technique greatly increased the packaging density of circuit components. This permitted equipment producers to specialize in "breadboarding" the circuits and component manufacturers to concentrate on improving their device designs. This is still being done.

The drawback to this approach was that the modules required a considerable number of connections that took up much of the packaging space and increased assembly costs. In addition, it has been shown through life tests that circuit reliability is adversely affected as the number of connections is increased.

A new technique was thus required to improve reliability and further increase packaging density. The solution was microelectronics, or integrated circuits.

Today the prime motivating factor in further IC development is equipment cost. Indeed, the search for more circuit functions at lower prices has resulted in a price war among circuit producers. As a result, to reduce costs, IC manufacturers have developed extremely complex functions on a single silicon chip.

The increased reliability of the IC becomes an obvious economic factor when one considers the cost of unreliability. For example, the government reported that the maintenance cost of some types of electronic equipment far exceeds the purchase price. Also, the Navy has estimated that the improved reliability offered by the integrated circuit can save $2 billion in maintenance costs.

Future electronic upkeep costs will be further reduced, since it is likely that ICs will soon be produced so cheaply that throwaway modules will become available.

The compactness of the IC has not only reduced the size and reliability of electronic equipment, it has also stimulated new industries, such as medical electronics. The IC can be made so small, and reliable, that it can be implanted in the body and coupled to appropriate sensors for diagnosis and treatment.

By far, the largest IC market is the computer industry. Besides the obvious reasons (i.e., size and reliability), ICs offer the computer manufacturers a better cost-per-performance ratio, which has to do with the value of work that the computer can perform. This value can be analyzed in the laboratory, since a computer system is valued in dollars per minute and the time required for various systems (circuits) to perform specific functions can be converted to dollars. The IC is much faster than a comparable discrete circuit, primarily because of the inductances associated with discrete circuit wiring.

Another advantage that the IC offers the computer manufacturer is reduced assembly costs. Discrete components require a considerable amount of hand labor to fabricate a circuit. Finally, the maintenance cost of a computer system is less when using ICs, because of accessibility and the fact that an IC failure is usually catastrophic and therefore much easier to locate.

MONOLITHIC IC PROCESS

One of the most common methods of monolithic IC processing begins with a P-type substrate, onto which an N-type epitaxial layer is grown. The resistivities of substrate and epitaxial layer are determined by the required collector-substrate breakdown voltage B_{VCS}.

Figure 8.3 is a basic-process flow chart for a typical monolithic integrated circuit. The processes used for monolithic IC fabrication are similar to those for a silicon epitaxial planar transistor, i.e., epitaxial deposition, oxidation, diffusion, photoresist, etc. However, there are more diffusion steps in IC processing and because of this and component density, the diffusions and mask alignments are somewhat more critical.

Before we discuss IC processing, let us briefly contrast this flow chart with the one presented in Chap. 4.

The first major difference to be noted between the two processes is that in the IC a high-resistivity P-type silicon substrate is used as the starting material, onto which an NPN transistor is deposited. In the discrete transistor process, the starting material (for NPN transistors) is an N^+ substrate onto which an N-type epitaxial layer is subsequently deposited.

The reason for the P-type IC substrate is that since a number of transistors and other components are fabricated on a single chip (circuit), some form of electrical isolation between each of the circuit components is required. This isolation is achieved by diffusing a P-type impurity through the N-type epitaxial layer, in selected areas, to the P-type substrate (Fig. 8.4*a*). (We shall discuss presently the isolation process and how electrical isolation is achieved.)

The N^+-buried-layer diffusion, prior to epitaxial deposition, is also required only in monolithic IC fabrication. The purpose of the N^+ buried layer is to reduce the saturation voltage of the monolithic IC (NPN) transistor (Fig. 8.4*a*). It can be seen from the illustration that the N^+ layer provides a low-resistance lateral path for the emitter-to-collector voltage. The doping (concentration) level of the diffused buried layer, and the distance between the emitter and the N^+ collector contact, control the total resistance of this lateral path.

Figure 8.4 illustrates the major difference between the monolithic IC transistor and the discrete transistor: the IC transistor must utilize a top-collector contact, whereas the discrete device makes use of the low-resistance N^+ substrate as an ohmic extension of the collector.

Besides requiring an N^+ buried layer, the IC top collector necessitates certain transistor-design adjustments in order to obtain the desired transistor characteristics. A top-collector-transistor configuration limits frequency response and power dissipation (Chap. 4, "Flip-Chip"). However, new designs and processing innovations have extended the frequency response and power

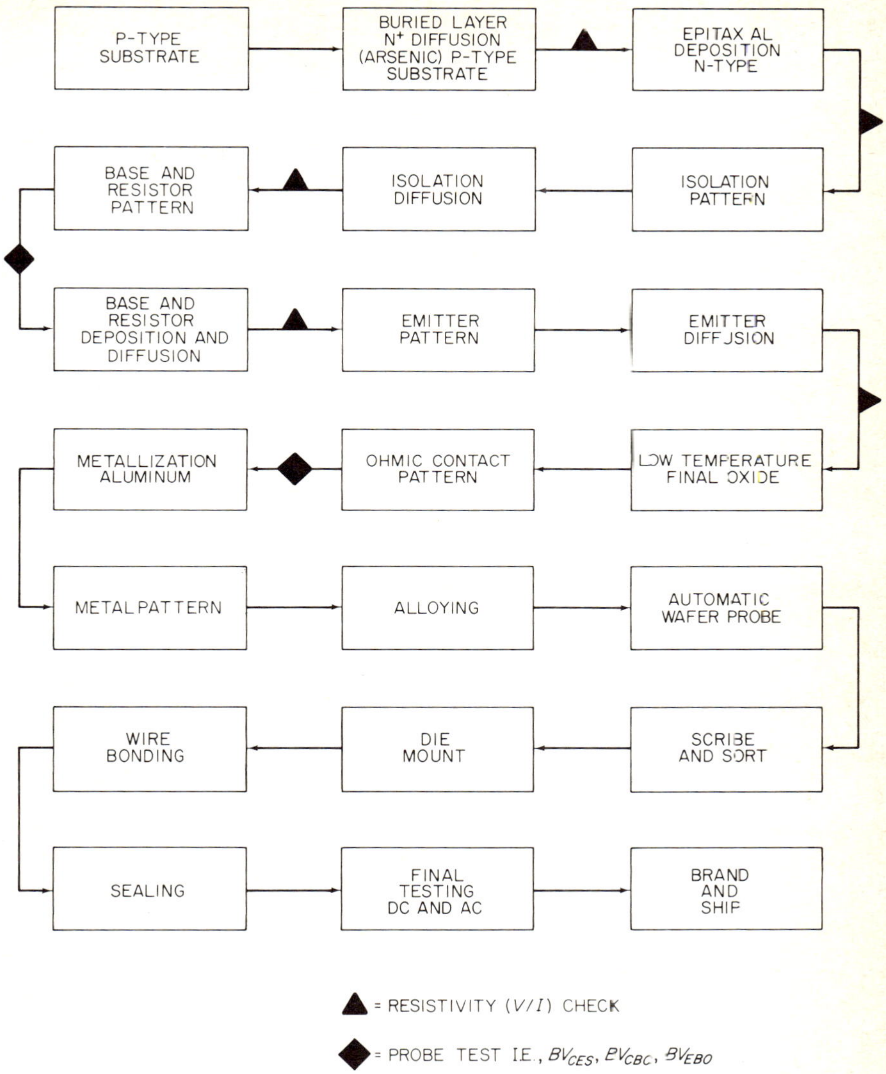

FIGURE 8.3
A chart showing the steps for producing an IC.

dissipation of top-collector devices so that they now cover a wide range of applications.

Thus, the two processing steps peculiar to monolithic ICs are:

1. A P-type substrate with an N-type epitaxial layer, and the associated electrical isolation of circuit components

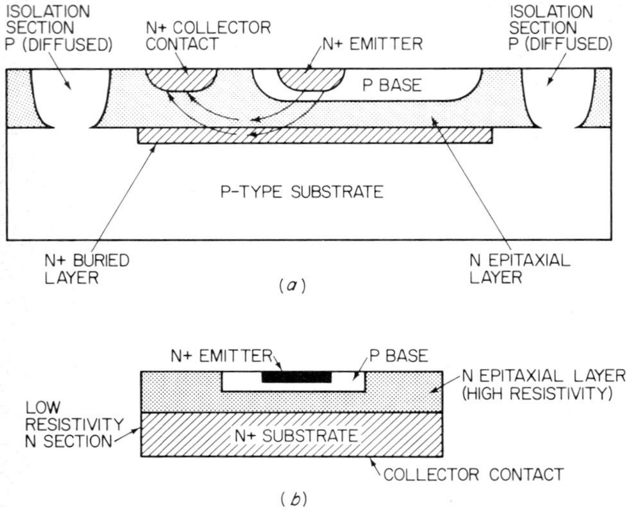

FIGURE 8.4
A comparison of a discrete and an IC NPN transistor. (*a*) Monolithic IC transistor with N^+ buried layers and top collector contact. (*b*) Discrete epitaxial planar transistor.

2. The N^+ buried layer required to minimize the collector series resistance of the top-collector-circuit (NPN) transistors.

There are other IC processing innovations that are not required in discrete planar transistor fabrication. We shall discuss these in some detail, but first let us see how monolithic diodes, resistors, capacitors, and transistors are formed in the same silicon substrate without additional processing steps.

Diodes. Figure 8.5 illustrates, structurally and schematically, five different ways in which the monolithic IC transistor structure can be utilized as a diode. The characteristics, such as breakdown voltages, leakage currents, and forward voltage drops, vary among the five possible diode connections and hence determine the basis for application.

Note that diodes a, b, and d are limited by the electrical characteristics of the emitter-base junction. The average reverse emitter-base breakdown voltage of monolithic IC transistors is between 5 and 7 volts (V). The average forward voltage for the five connections is 1.9 V (at 10 milliamps, mA).

Diodes c and e are limited by the collector-base junction. The average reverse collector-base breakdown voltage is approximately 50 V (unless punch-through conditions exist, Chap. 4).

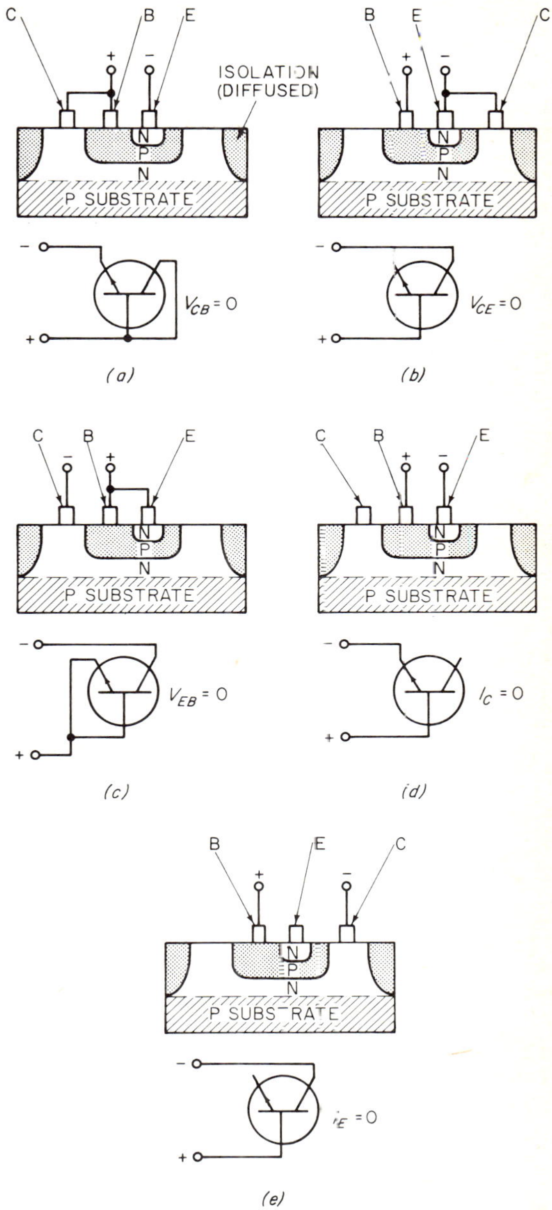

FIGURE 8.5
Five ways in which a monolithic transistor can be connected to function electrically as a diode.

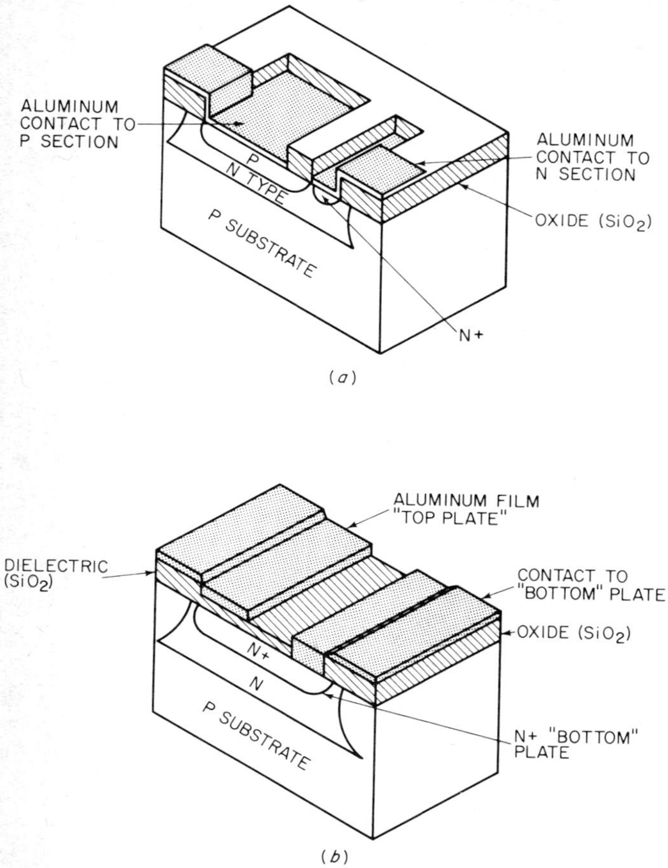

FIGURE 8.6
Two types of monolithic IC capacitors. (*a*) Junction capacitor. (*b*) Thin-film capacitor. (MOS structure.)

Capacitors. There are two forms of monolithic IC capacitors: junction capacitors and thin-film capacitors. Each type can be formed simultaneously with the other circuit elements.

Figure 8.6*a* is a cross section of a junction capacitor. This capacitor is formed by placing a reverse voltage across a PN junction. As we noted previously, when this is done, the carriers present at the junction draw back, leaving a strip on both sides of the junction, called the depletion region. Since this region is devoid of charge carriers, in essence what we have are two conductors (the P and N regions) separated by a nonconductor, the depletion region.

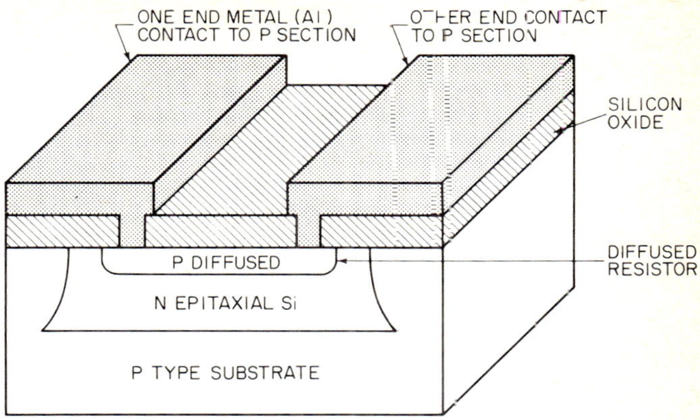

FIGURE 8.7
Cross section of a monolithic IC diffused resistor.

The capacitance value formed will depend on the junction area and the thickness or width of the depletion region. The larger the junction area, the higher the capacitance. On the other hand, the wider the depletion region, the lower the capacitance value. This latter factor is, in turn, a function of the reverse voltage applied across the junction.

It should be particularly noted that the voltage applied to the PN junction must be in the proper polarity for the junction. That is the positive voltage connects to the N-type semiconductor and the negative voltage to the P-type semiconductor. Any reversal of these connections will lead to destruction of the device.

Figure 8.6*b* is a cross section of a monolithic thin-film capacitor. The structure consists of dielectric (oxide) between two conductors: the N^+ (diffused) silicon and aluminum metallization.

Thin-film capacitors have several advantages: they are nonpolar because either of the two plates can be made positive or negative, the capacitance is constant, and the Q is normally higher than for the junction-type capacitors. These advantages result in greater flexibility in circuit design and application.

Typical thin-film capacitor values range from 0.25 to 0.30 pF (picrofarad)/mil^2 using an oxide thickness of from 800 to 1,000 angstroms (Å). The capacitance available with the junction-type capacitor is about 0.2 pF/mil^2.

Resistors. Figure 8.7 is a cross section of a monolithic diffused resistor. A length of P-type semiconductor serves as the resistor, with the final resistive value determined by the photoresist mask dimensions, the formation in the oxide of the desired resistor length-to-width openings by etching procedures, and the diffusion of the proper amount of impurities (generally P-type) through

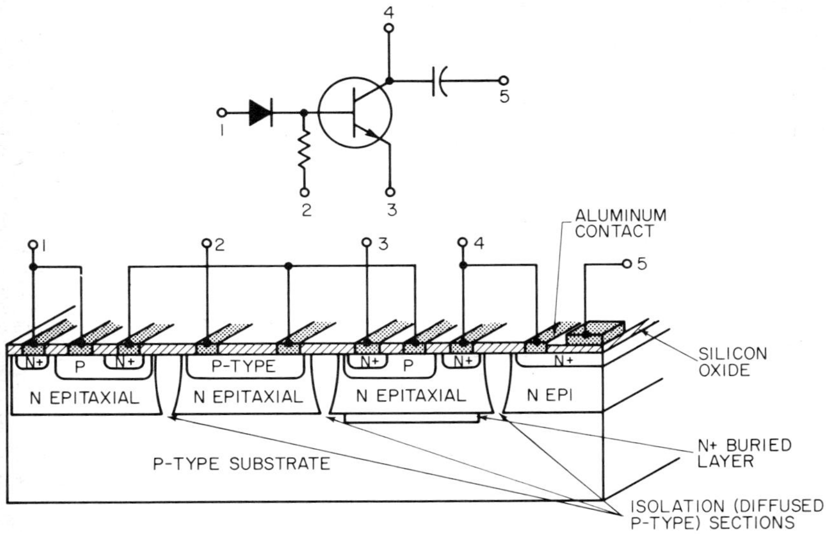

FIGURE 8.8
Monolithic IC showing diode, resistor, NPN transistor, and capacitor components.

the oxide openings to obtain the desired resistivity. Since diffused resistors are used in the vast majority of monolithic ICs, we shall discuss them in more detail when we describe the entire monolithic process in the next section.

Figure 8.8 illustrates the construction of a monolithic IC containing a diode, a resistor, an NPN transistor, and a capacitor.

Parasitics. A parasitic can be defined as a stray, undesired capacitance, inductance, or resistive leakage that could degrade circuit efficiency.

A transition capacitance is associated with each PN junction in a monolithic circuit. For example, in Fig. 8.4*a*, there will be a parasitic capacitance between the N-type collector of the transistor and the P-type substrate. Further, this parasitic capacitance consists of two parts: the side wall and the bottom parts of the junction.

The magnitude of this parasitic capacitance is governed by the impurity concentration and the cross-sectional area of the junctions involved plus the bias (i.e., voltages) present. These unwanted capacitances can serve as side paths for signals, enabling these voltages to move into circuits where they are not desired.

The most predominant monolithic parasitic capacitances are associated with the isolation PN junctions and the collector-substrate junctions of the circuit transistors. Through careful circuit design, the number of isolation areas may be reduced, with a resulting reduction of undesired capacitance. For example, all P-type diffused resistors of a circuit can be placed in a common isolation region.

The isolation junctions are also responsible for another form of monolithic parasitics. A four-layer-diode (NPNP) switch action may occur under certain bias conditions. To see how such a four-layer diode may be formed, consider again Fig. 8.4*a*. The N is provided by the emitter, the P by the base, the second N by the combination of the epitaxial and buried layers, and the final P by the substrate.

The operation of an NPNP structure is described in Chap. 12. In action, this device appears as a PNP and an NPN transistor connected as shown in Fig. 8.9. When the sum of the current gains of the two transistors equals unity, the device is placed in a highly conductive state and the overall resistance drops to a very low value. In essence, it acts as a closed switch.

To avoid this shorting condition, the P substrate is connected to the most negative potential of the circuit, keeping the emitter of the PNP transistor reverse-biased.

Still another possible parasitic circuit can develop when diffused resistors are incorporated into an integrated circuit. Figure 8.10 is a cross section of a P-type diffused resistor and a schematic of its equivalent circuit that takes into consideration structure parasitics. A parasitic PNP transistor is formed, with the resistor as its emitter, the N-type epitaxial region as its base, and the P-type substrate as its collector.

Since the substrate is connected to the most negative potential of the circuit (to prevent four-layer switching), the collector of the parasitic PNP transistor is reverse-biased. This requires that the diffused resistor junction must also be kept reverse-biased in order to keep the parasitic PNP transistor turned off, and at the same time maintain the integrity of the resistor. This

FIGURE 8.9
An NPNP parasitic four-layer diode and its equivalent circuit.

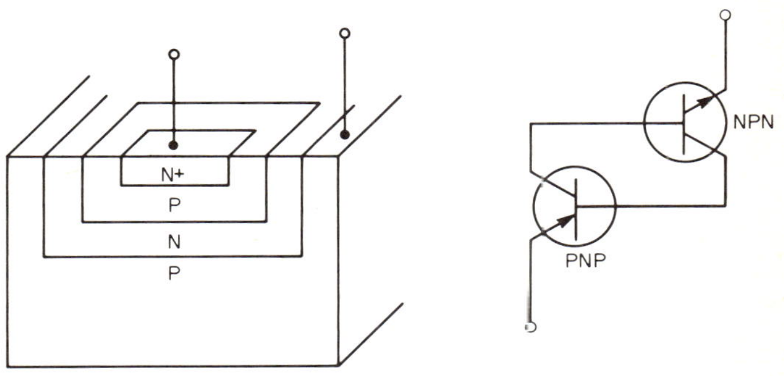

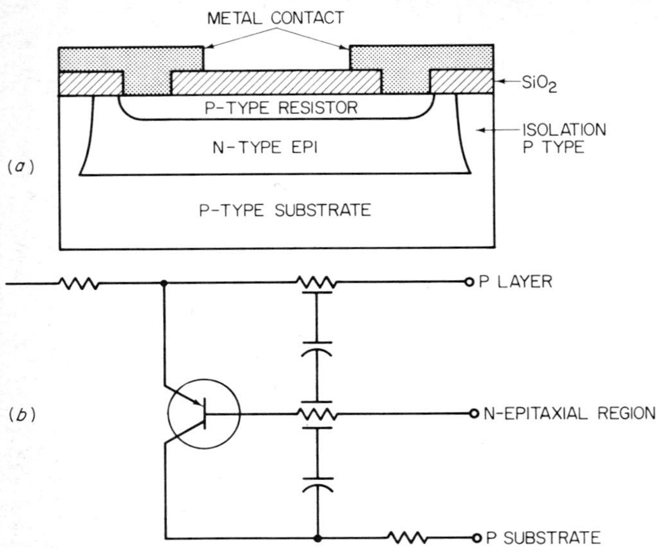

FIGURE 8.10
(*a*) Cross section of a monolithic diffused resistor.
(*b*) Equivalent circuit of *a*, illustrating associated parasitics.

is usually accomplished by connecting the N-type base of the parasitic PNP transistor to the most positive voltage available in the circuit. With all the circuit resistors in a common isolation region, making the N region positive ensures that all resistors are connected to the most positive circuit voltage. Note that the resistors are electrically isolated from each other by the back-to-back junction between resistors.

Now that we have seen how an integrated circuit can combine the various components within its structure, let us take a closer look at IC fabrication.

Buried Layer. The buried layer, Fig. 8.4*a*, is a pocket of heavily doped N-type material located under each of the NPN transistors of the circuit. And, as just mentioned, its purpose is to reduce the collector series resistance of the transistor.

Arsenic is generally used as the N-type source for this diffusion because of its high solid solubility and low diffusion coefficient. This means that a high concentration of arsenic can be diffused into the silicon substrate without damaging the silicon-crystal structure and also that during the subsequent high-temperature processing, only a relatively small amount of arsenic will outdiffuse. Following the buried-layer diffusion, the oxide is chemically removed and the wafer is chemically cleaned prior to epitaxial deposition.

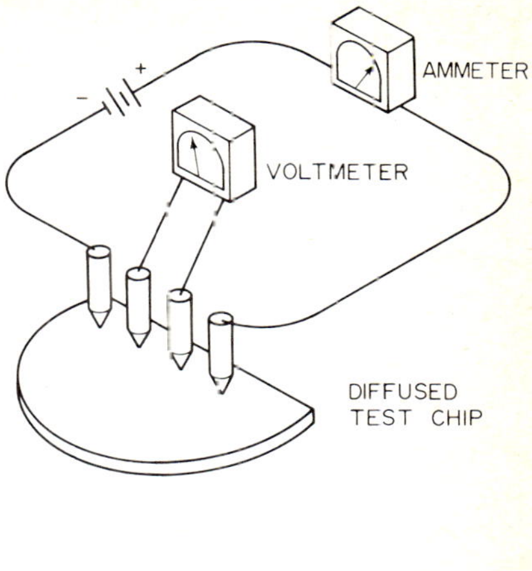

FIGURE 8.11
A four-point probe to measure the resistivity of a wafer.

An important process control measurement is made at this point. The resistivity of a silicon control wafer (usually one-half of a wafer) of the same type (P in this case) and resistivity of the actual circuit wafers, which was diffused in the same diffusion cycle, is measured.

The desired, or target, resistivity is calculated for each diffusion using the appropriate diffusion equations and tables. Normally, resistivity control limits are set for every diffusion cycle, which of course varies with different circuit types. If the measured resistivity value is outside the set limits, processing adjustments (diffusion time, temperature, etc.) will be made. In severe resistivity variations, all the wafers in that diffusion cycle will have to be scraped. This resistivity check actually measures the number of diffused impurity dopants near the surface of the control chip.

A four-point probe, Fig. 8.11, is generally used to measure surface resistivity. A current is passed through the outside probes, and the voltage is measured across the inside probes using a high impedance voltmeter. The probe spacings are kept small and equal.

The N^+ pattern outline can be seen (by microscope) through the epitaxial layer for alignment of the isolation pattern. This is so because the reaction of the N-type impurity, arsenic, with the silicon during diffusion forms a slight depression in the buried-layer pattern (diffusion) area.

Isolation. After the epitaxial layer is deposited, another oxide layer is grown, and the wafer is now ready for photoresist and isolation patterning (Fig. 8.12).

Discrete-device circuit isolation is automatically achieved by the individual packages. In the monolithic IC, isolation is generally accomplished by pairs of reverse-biased PN junctions (Fig. 8.13). The isolation diffusion forms the PN junctions that electrically separate (isolate) the different areas of the circuit that have no common-collector region connections. As Fig. 8.13 indicates, voltages are applied that reverse-bias the PN junction between the isolation section

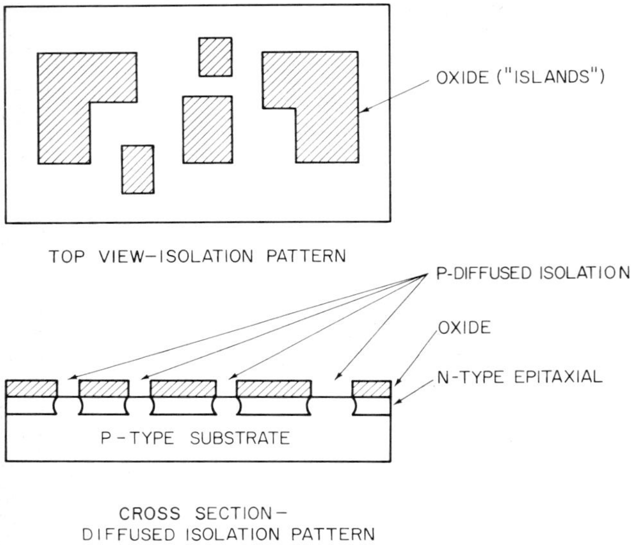

FIGURE 8.12

and the collectors of the two transistors. This provides a high-resistance path between the two transistors and, in effect, isolates them from each other.

The isolation diffusion must penetrate through the epitaxial layer to the P-type substrate, making it the deepest (i.e., longest) diffusion cycle of the entire process. The diffusion conditions (time, temperature, atmosphere) must be carefully controlled because many subsequent high-temperature cycles follow. These cycles could cause outdiffusion and redistribution of the P-type isolation dopant that would degrade circuit performance.

Dielectric Isolation. Another form of electrical isolation of monolithic IC components that has recently become popular is dielectric isolation.

In this process, a thin layer of an insulating or dielectric barrier, such as silicon dioxide (SiO_2), is placed between the circuit components to be isolated (Fig. 8.14). An important advantage of this form of isolation is that it eliminates the parasites of the PN junctions formed in the diffused isolation process.

The starting material, for dielectric isolation, is an oxidized N-type substrate wafer with N^+ buried layers (Fig. 8.15*a*).

First, the isolation areas are defined in the oxide, using normal photoresist and oxide-etching techniques (Fig. 8.15*b*).

Next, isolation grooves, or moats, are etched in the silicon, using the patterned oxide to define the etched areas (Fig. 8.15*c*). A recent innovation has been use of special silicon etchants that etch many times faster in one

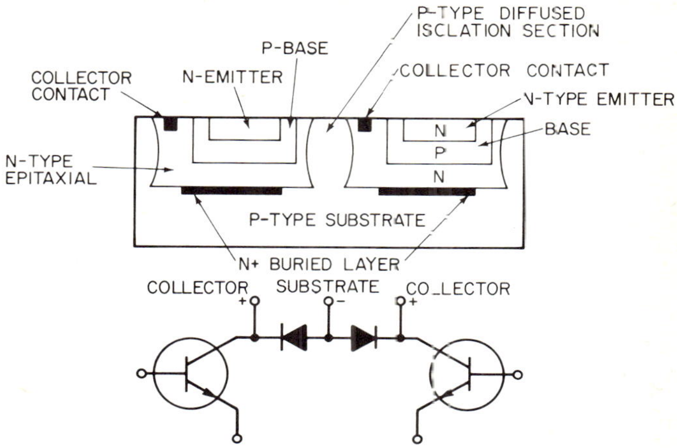

FIGURE 8.13
Cross section of two monolithic NPN transistors with a diffused isolation section between them. Schematic for transistor isolation represented by back-to-back diodes.

crystal plane rather than along other planes. This type of etch is known as an anistropic etch, and its primary advantage in isolation processing is that the grooves can be etched much smaller (narrower) and thus permit more components to be fabricated per unit area of silicon. In addition, the depth of the groove is controlled only by the width of the oxide patterns, as opposed to the critical time and temperature process controls of the normal silicon etches.

FIGURE 8.14
Dielectric isolation.

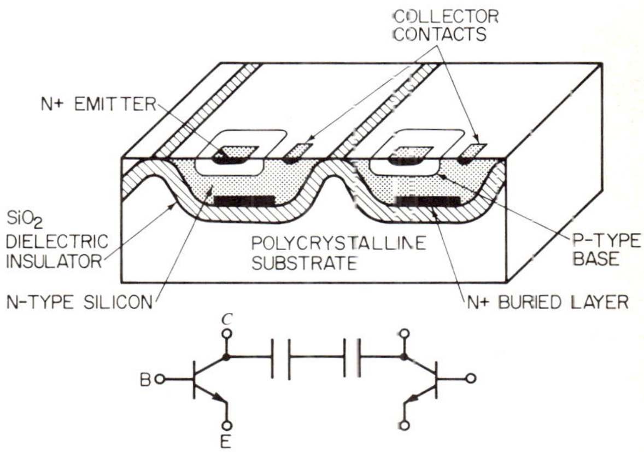

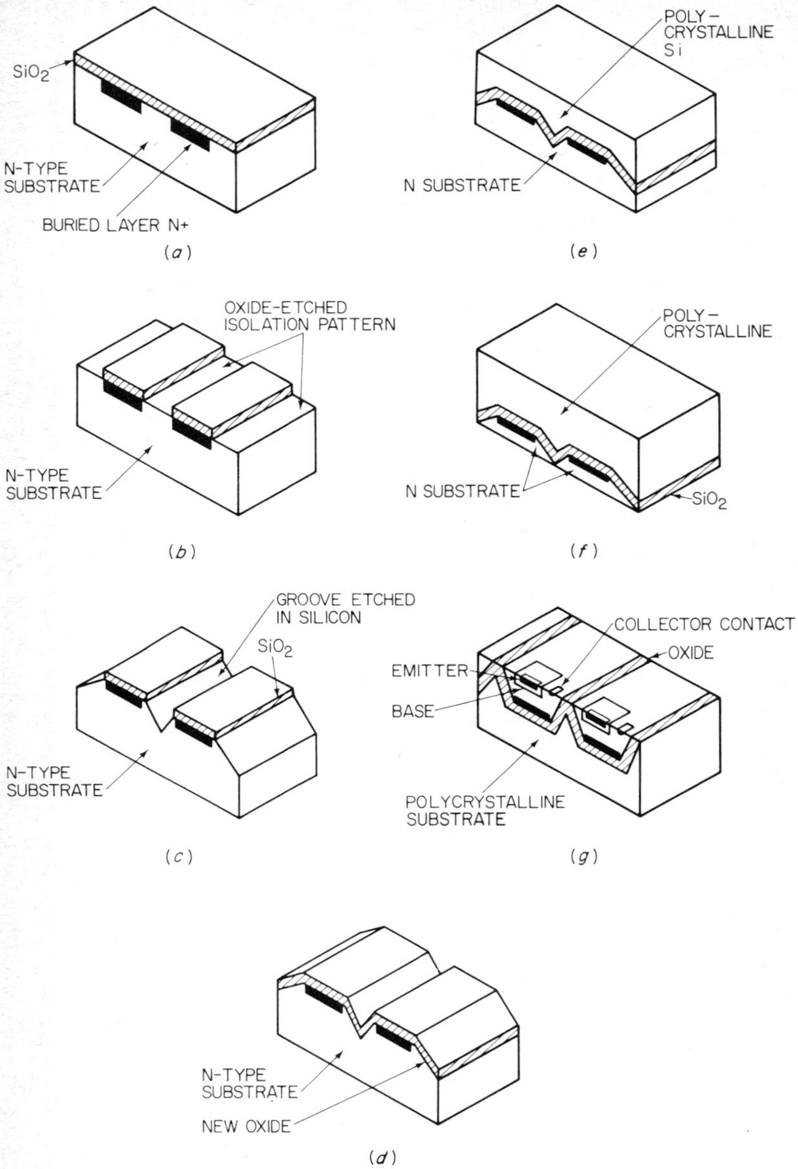

FIGURE 8.15
The processing procedure for producing a monolithic IC device employing dielectric isolation between IC elements.

After the grooves are formed, a continuous layer of SiO_2 is grown over the wafer surface, coating the inside of the grooves (Fig. 8.15*d*).

Several mils of polycrystalline silicon are then deposited, back-filling the grooves (Fig. 8.15*e*). (Polycrystalline silicon has more than one crystal orientation and can be deposited in an epitaxial reactor.) Polycrystalline and monocrystalline silicon have similar properties, most important of which is that they have the same coefficient of thermal expansion. If this were not so, the silicon might crack with the heat generated during circuit operation.

Next, the monocrystalline side of the wafer is lapped, or etched, down to the oxide level (Fig. 8.15*f*). The wafer is then cleaned, and "normal" IC processing is used to complete the circuit (Fig. 8.15*g*).

Dielectric isolation is employed in circuits requiring high collector-base breakdowns (100 V). This is so because high-resistivity substrate material can be used (high-resistivity collector material = high B_{VCEO}). Uniform layers of high-resistivity epitaxial silicon are difficult to deposit.

Circuits utilizing dielectric isolation are also less prone to radiation damage than are diffused isolated circuits, because the diffused impurity atoms can be displaced, creating adverse changes in circuit performance. Radiation hardening was the original purpose of dielectric isolation.

Transistor Base and Resistor (B and R) Formation. After the isolation procedure, the wafers are ready for the base and resistor formation. Note that the diffused isolation system results in the formation of a silicon oxide layer that is used, after photoresist patterning, to mask the base and resistor P-type dopants. If dielectric isolation is utilized, an oxide must be thermally grown, and patterned, for masking the base and resistor dopants.

The B and R diffusion must provide a surface resistivity that does not vary more than ±5 percent from the designed target value. The resistor values are critical to proper circuit function and their values are controlled by:

$$R = R_S \frac{l}{w} \qquad \text{(Fig. 8.16)}$$

where R = resistor value
R_S = surface resistivity
l = resistor length
w = resistor width

The B and R diffusion is generally done in two steps for better control of the surface resistivity. A low-temperature (900°C) P-type (boron) deposition is done first, followed by a high-temperature base diffusion (Fig. 8.3).

Emitter Formation. After the base diffusion, the wafers are processed through photoresist for emitter formation (i.e., alignment, oxide etch, etc.) and are cleaned for diffusion.

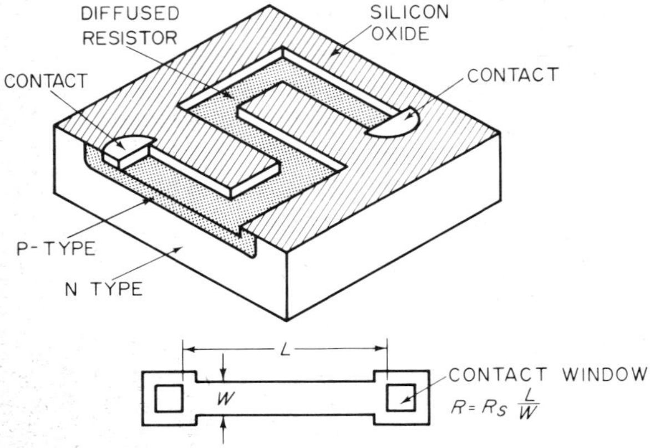

FIGURE 8.16
Diffused resistor.

The N-type emitter (phosphorus) diffusion must be closely controlled, of course, to obtain the desired transistor characteristics, such as beta and frequency response. The diffusion is carried out at temperatures in the 1000°C range for an average time of 15 min, using oxygen and nitrogen atmospheres under similar conditions to those described in Chap. 4 for planar transistors. At the same time as the emitter openings are etched in the oxide, a top collector contact "window" is opened for the formation of an N^+ ohmic contact between the silicon and the aluminum metallization. In addition, the N^+ emitter diffusion can be used to form substrate (diffused) interconnections known as "crossovers." In complex circuits, it is occasionally impossible to use the normal metallization interconnections between circuit components, in which case a low-resistance tunnel (emitter diffusion) under the oxide is then utilized to make the required connections (Fig. 8.17). Crossovers, however,

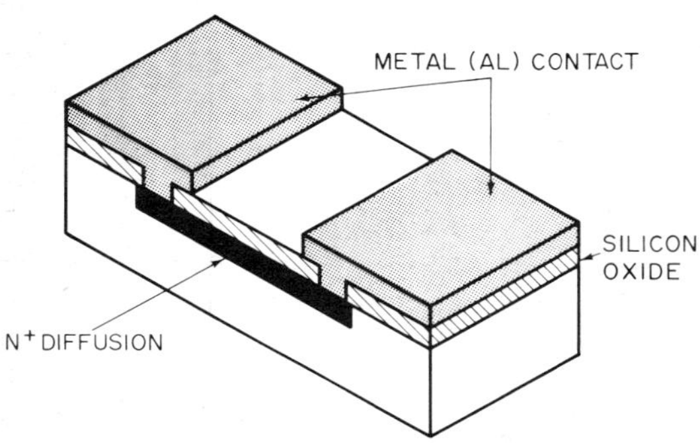

FIGURE 8.17
N^+ crossover, diffused during emitter diffusion.

FIGURE 8.18
Integrated circuits mounted in TO-5 case.

should be utilized only as a last resort, since they require separate isolation areas, which results in a parasitic capacitance, and they take up space on the circuit chip.

After emitter diffusion the phosphorus glass, formed during diffusion, may be removed by special etching techniques. Some manufacturers leave the phosphorus glass on the wafer surface, since it acts as a "getterer," or sink, for any surface or oxide impurities that may be present. At any rate, after the emitter diffusion cycle, a low-temperature (900°C) final oxide is grown or deposited over the entire wafer surface. Low-temperature oxides are used to minimize impurity redistribution (outdiffusion). Contact windows are then etched in the oxide in selected areas for aluminum silicon ohmic (nonrectifying) contacts (Fig. 8.8).

Aluminum is generally used to form the interconnecting patterns and to provide bonding pads for the lead wires that go to the circuit package. The metal evaporation and photoresist procedures are the same as those described in Chap. 4.

As a matter of fact, the remaining processes, i.e., alloying, scribing, sorting, testing, and packaging, are similar to those discussed in Chap. 4. However, the testing procedures are more involved and require expensive equipment, i.e., probers and computers. Also, many more bonds have to be made to ICs, from chip to package (Fig. 8.18).

Monolithic Lateral PNP Transistors. The vast majority of monolithic integrated circuits use only NPN transistors, such as those we have discussed. However,

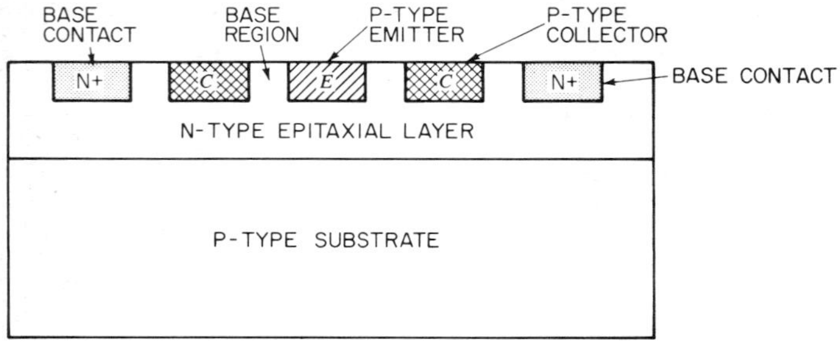

FIGURE 8.19
Cross section of a lateral PNP transistor. The N-type epitaxial layer serves as the base.

there are applications that require both NPN and PNP transistors on the same chip (i.e., in the same circuit).

In the early days of monolithic circuit development, there was no practical method of fabricating both transistor types on the same silicon chip. Now, however, there is a scheme that can produce complementary PNP and NPN transistors with good electrical characteristics, both on the same chip. The key to this technique was the development of the lateral PNP transistor structure.

Figure 8.19 illustrates the basic lateral PNP concept. Note that as with conventional NPN transistor formation, an N-type epitaxial layer is grown on top of a P-type substrate, which means that normal isolation techniques may be utilized—an important consideration.

During the P-type base and resistor diffusion of the NPN transistors, two closely spaced concentric P-type regions are diffused (Fig. 8.20); these will become the emitter and collector of the PNP transistor. (The emitter and collector patterns of the PNP transistors are included on the same mask as the base and resistor patterns of the NPN transistor, thus eliminating additional or special photoresist techniques.)

The lateral type of transistor structure has low gain; but by amplifying the collector current of such a PNP transistor with an NPN transistor, the composite transistor, Fig. 8.21, behaves like a high-gain PNP transistor, and the combined current gain is comparable to that of an NPN transistor of the same circuit.

Electrical Characteristics. The electrical characteristics of the lateral PNP transistor can be derived from conventional junction theory (Chap. 2). The primary difference between the lateral PNP transistor and the "normal" NPN transistor is that in the PNP device, the injected minority carriers flow parallel to the surface—hence the name lateral transistor.

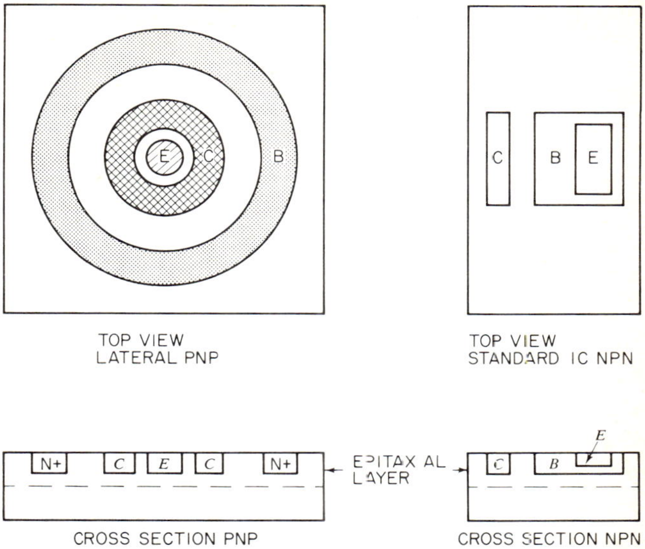

FIGURE 8.20
Top and cross-sectional views of lateral PNP and standard IC NPN transistors.

FIGURE 8.21
PNP and NPN connected in series to compensate for low gain of the lateral PNP transistor.

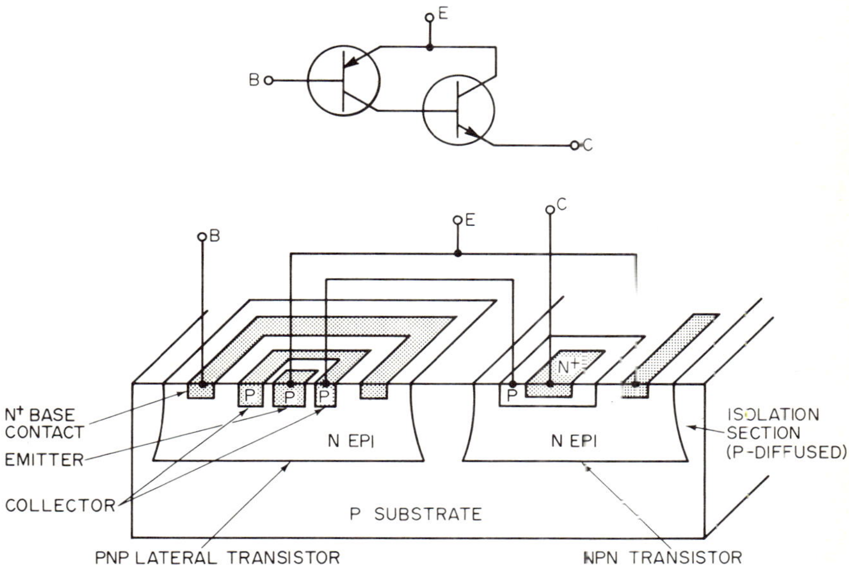

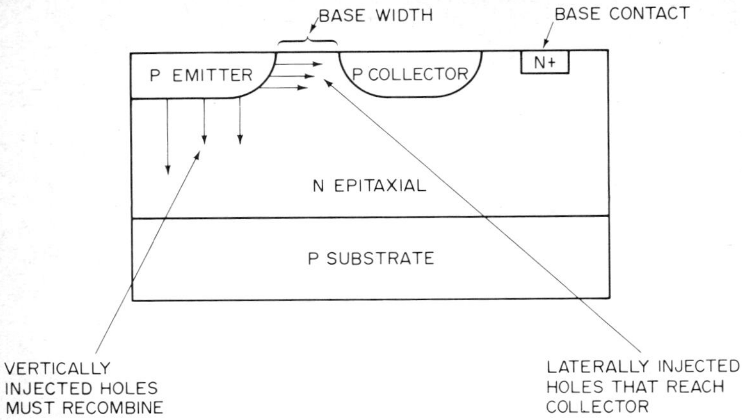

FIGURE 8.22
The flow of carriers in a lateral transistor. The size relationship between the N-epitaxial section and the P-substrate is exaggerated.

Figure 8.22 is a close-up cross section of the lateral PNP transistor. Note that the effective distance between the emitter and collector (i.e., the base width) is dependent on the extent of the lateral diffusion of the P-type impurity plus, of course, the distance between the emitter and collector as determined by the pattern mask. The base width is narrowest at the surface and widest at the bottom.

The injected emitter current consists of three parts; (1) the lateral-hole current injected into the base, (2) the vertical-injected-hole current into the substrate, which is not collected, (3) and the electron current from the N-type base into the emitter. Of the three currents, only the lateral-hole current will be collected by the collector. Therefore, in order to optimize current gain, the emitter resistance and the base width must be kept small. In addition, the distance from the bottom of the emitter-base junction to the edge of P-type substrate must be kept larger than the diffusion length of the vertically injected minority carriers (holes). This is necessary to prevent the PNP transistor that exists between the diffused emitter—the N-type epitaxial layer and the P-type substrate—from turning on.

Recall that for any diffusion, the impurity concentration near the surface is highest. Therefore, more P-type minority carriers will be injected from the emitter into the base near the surface. Another way of saying this is that the injection efficiency of minority carriers increases as the emitter impurity concentration increases. Thus, owing to the higher surface concentration of the emitter and the narrow base width, most of the collected carriers are near the surface.

This method of forming complementary transistors in the same circuit has several drawbacks: higher offset voltage, increased area, and greater phase shift. However, for low-frequency applications, lateral PNP transistor characteristics are quite satisfactory.

MOS MONOLITHIC INTEGRATED CIRCUITS

In the introduction to Chap. 5 ("Field-effect Transistors"), it was mentioned that the MOSFET was rapidly becoming the most important electronic device on the market, primarily because of its potential for integrated-circuit applications. These increasing applications are the result of improved processing and design techniques.

A huge market for MOS ICs is the computer industry. Specifically, MOS IC memories and long shift registers are more efficiently fabricated with MOS circuits (higher circuit density) than with bipolar circuits. Also, memories fabricated using MOS technology are faster than the conventional memory cores (we shall discuss computers in Chap 10).

MOS vs. Bipolar. The basic processes used in MOS IC fabrications are the same as those just discussed for the monolithic bipolar circuits, i.e., oxidation, diffusion, photoresist, etc. However, although the processes, and even the functions, of the bipolar and MOS integrated circuits may be similar, the number of fabrication steps are different. For example, in order to turn on a MOS transistor, a much larger input signal is required, as compared with the bipolar transistor. Consequently, the necessary impedance levels are much higher in MOS circuits, and thus large resistor values are required. Because of this, the use of diffused resistors, such as the ones discussed earlier in this chapter, is not practical for MOS integrated circuits. Fortunately, MOS transistors themselves can be used as load resistors by using the source and drain electrodes as a two-terminal device, with the gate voltage returned to a fixed supply (Fig. 8.23). Also, since the MOS structure itself can be used as a capacitor (Fig. 8.6*b*), a complete circuit involving the functions of transistors, receivers, and capacitors can be fabricated from an array of nothing but MOS structures.

In addition, the MOS IC transistor structure itself is much simpler than the bipolar IC transistor in that the former requires only one diffusion and no isolation for enhancement-mode transistors (the most common form of a MOS IC transistor). (Recall that in a MOS device the current flow is primarily in the plane of the chip, whereas in bipolar units the current flow is primarily normal to the chip.)

The elimination of a long, costly isolation diffusion cycle and the subsequent undesirable effects of the isolation impurity outdiffusion further increases the usable silicon area in MOS integrated circuits. Thus, the reduced size

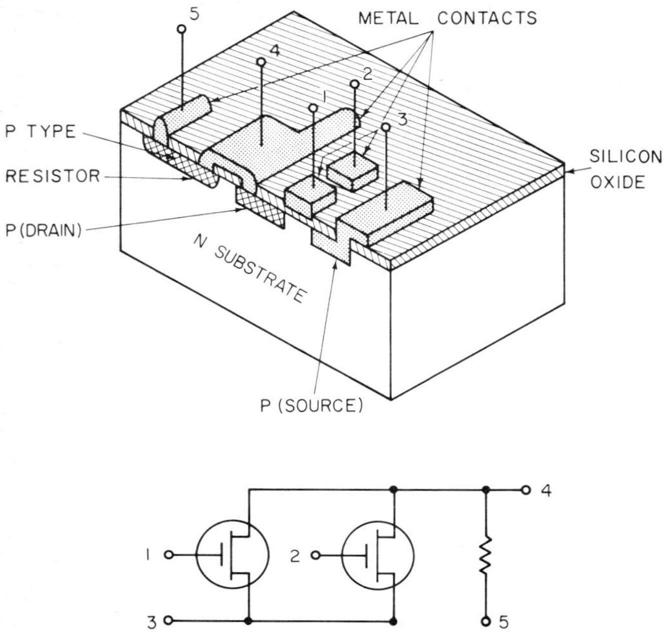

FIGURE 8.23
A MOS transistor and resistor.

(that is, more functions per unit of silicon area) is a major advantage of MOS integrated circuits.

However, because of the high operating impedance levels of MOS transistors, the MOS integrated circuit is about an order of magnitude slower in switching speed than a bipolar integrated circuit. This disadvantage can be overcome somewhat by complementary symmetry, which we shall discuss presently.

The high input impedance of MOS devices has its advantages, i.e., it can act as a voltage-controlled resistor and the input-output isolation is excellent. The only input current that flows in the gate lead charges and discharges the input capacitance.

The reliability of MOS integrated circuits is at least as good as bipolar ICs. MOS circuits are very sensitive to contamination, particularly moisture. This drawback led to extensive research, which has virtually eliminated the contamination type of MOS IC failure modes.

MOS IC Processing. Let us now briefly discuss the processing of a typical MOS (FET) integrated-circuit transistor, i.e., a P-channel enhancement-mode device.

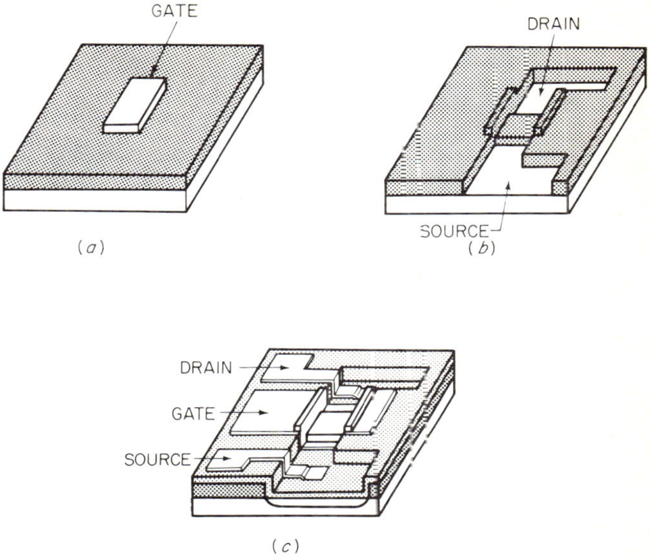

FIGURE 8.24
MOS transistor fabrication. (*a*) Raised-gate oxide. (*b*) Etched source and drain. (*c*) Completed transistor.

The starting material is a lightly doped N-type silicon wafer with a uniform layer of thermal oxide covering one side. Using photoresist techniques, the area surrounding the gate region is defined, and the oxide in this region is then partially etched away, leaving a raised oxide in the gate region (Fig. 8.24*a*).

After this partial oxide etch the photoresist is removed, the wafer is again coated with photoresist, and the source and drain region are defined and etched completely through the oxide (Fig. 8.24*b*). Note that during this oxide etching of the source and drain region, the raised-gate oxide island is also etched. The etching is stopped when the gate-oxide-island thickness is between 800 and 1,000 Å. The gate-oxide thickness is very critical inasmuch as it controls important transistor parameters, such as V_T (threshold voltage).

After the source and drain regions have been opened in the oxide, a P-type impurity dopant is diffused through the openings forming the source and drain. Following the source and drain diffusion, a final oxide is grown over the entire wafer and contact windows are etched through this oxide in the source and drain regions. A layer of aluminum is then evaporated over the entire wafer surface and patterned, using the appropriate metallization (interconnection) mask, photoresist, and etching techniques (Fig. 8.24*c*). The remaining processes, i.e., alloying, scribing, sorting, die mount, etc., are the same as those employed in bipolar monolithic IC processing.

However, this form of MOS transistor can be made to occupy about 2 mils^2, and a typical silicon wafer can hold thousands of such devices. (Bipolar monolithic IC transistors occupy an average of 40 to 50 mils^2.)

MOS resistors are formed at the same time as the transistor just described by the same processing cycle. Thus, with only one diffusion cycle (P-type source and drain), complex circuit arrays can be completed.

Complementary Symmetry. The switching speed of MOS ICs can be improved by fabricating both N- and P-channel devices in the same circuit (chip)—in other words, using complementary symmetry.

The processing involved in complementary symmetry fabrication is difficult to control, since both enhancement-mode N-channel devices and low-threshold P-channel devices must be fabricated on the same chip.

A standard N-type substrate is prepared; using the planar process, a P-type region is formed that shall become the P-type substrate for the N-channel devices (Fig. 8.25*a*). The wafer is then prepared (oxidation and diffusion) to form the source and drain of the P-channel transistor, Fig. 8.25*b*. The wafer is again oxidized and diffused (N type) into the previously prepared P-type region, forming the source and drain of the N-channel device (Fig. 8.25*c*). Next, the oxide from the areas covering both devices is completely removed and a new special oxide (perhaps silicon nitride) that significantly determines device characteristics is grown. Finally, contact windows are etched and a metal pattern is formed using conventional photoresist and etching techniques (Fig. 8.25*d*).

These devices are normally connected in series forming a complementary inverter stage. The gates are driven by the same signal. Thus, with the input zero volts, the P-channel device is on and the N-channel device is off. If the input signal is made positive, the reverse occurs. When either device is on, it supplies a direct current equal to the leakage current of the other device. Also, when the input signal is in transition, the capacitive loads are being charged and discharged through the low impedance of either of the two devices. The time constants associated with these capacitances are short, making this type of circuit faster than noncomplementary MOS ICs. Thus, complementary MOS circuits can provide low-power and high-speed switching.

Ion Implantation. A new impurity-doping technique that has recently received considerable attention, particularly for MOS IC fabrication, is ion implantation.

The initial processing steps are similar to those employed in conventional MOS IC fabrication. The wafers are diverted from the normal process flow, after the source and drain diffusion, for ion implantation in an accelerator (Fig. 8.26*a*). Boron ions are accelerated just fast enough to penetrate the thin gate oxide, but they do not acquire enough energy to penetrate the thicker oxide covering the source and drain regions. The boron penetrating the thin gate oxide forms a thin P-type channel (Fig. 8.26*b*). The advantage of this

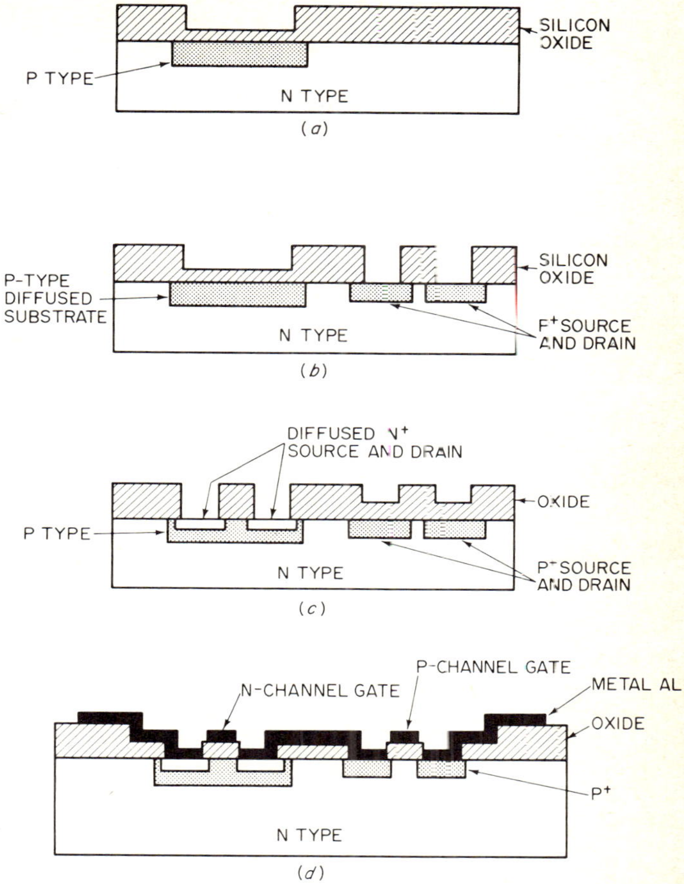

FIGURE 8.25
Formation of N- and P-channel complementary MOSFETs.

type of channel formation is that it lowers threshold voltage (V_T, Chap. 5) from the average 4 to 5 V to 1.5 V. In addition, ion implantations are extremely consistent (normal diffusions cannot be as accurately controlled) in controlling V_T.

This process has made possible the fabrication of new types of circuits, i.e., ones with depletion-mode loads and enhancement-type devices providing logic gates with more speed and power than the conventionally diffused circuits.

Ion implantation also simplifies the fabrication of N-channel MOS ICs, which are normally difficult to fabricate.

It also simplifies the processing of complementary MOS circuits and at the same time improves their performance.

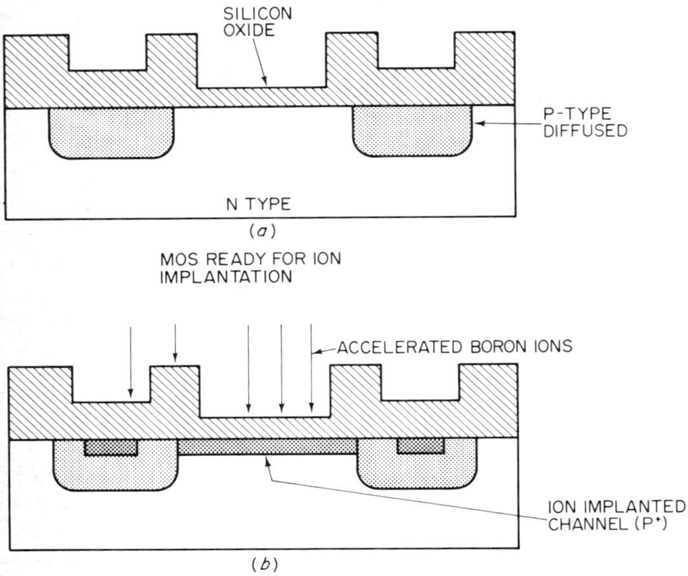

FIGURE 8.26
Ion implantation.

All the advantages are a result of the fact that a large number of dopants using ion implantation can be introduced into the semiconductor material with precise control, over a large range of concentration and at low temperatures.

The drawback to ion implantation is the equipment cost. Sophisticated equipment is required that must be able to generate ions of a large number of materials, accelerate them, and implant them in semiconductor material at

FIGURE 8.27
Ion implantation of source and drain but not gate.

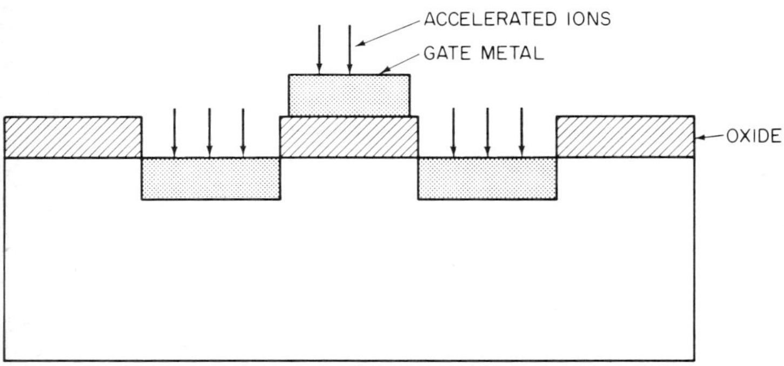

specific concentrations and depths. Such a system can cost as much as $100,000. However, even though ion implantation is not a common production method at present, its many advantages will undoubtedly generate methods for making it an economically feasible production method in the near future.

Before we leave ion implantation, another MOS fabrication method should be mentioned. In this technique, the source and drain regions are doped with ions while the metal electrode of the gate acts as a mask against the ion implantation. The advantage of this method is that precise alignment of the source and drain under the gate is automatic (Fig. 8.27). Gate overlaps, or misalignment, increase parasitic capacitances and frequency instability and also decreases switching speed. The technique just described avoids such deficiencies.

MSI, LSI

As more functions per chip become available, new terms such as medium-scale integration (MSI) and large-scale integration (LSI) have appeared. MSI is generally defined as a circuit having between 30 and 100 gates per chip; and LSI is generally defined as a circuit with over 100 gates per chip. Circuits with well over 100 gates per chip are being produced in large numbers, and the component density of some LSI circuits is equal to $250{,}000/in^2$.

Of course, many process and design innovations had to be developed in order to produce such complex circuits. In addition, LSI design requires the use of a computer in order to achieve optimum circuit partitioning and component configurations. It can be said that without the computer, LSI production would not be possible.

A detailed discussion of LSI is beyond the scope of this book. However, one important innovation that we will discuss is multilayer metallization (Fig. 8.28), which has practically doubled circuit-component density.

Let us say that the cross section of the multilayer circuit shown in Fig. 8.28 is that of a high-density bipolar logic circuit. Its processing sequence can be briefly described as follows:

1. N^+ buried-layer diffusion. The purpose and process would be the same as that discussed for the typical monolithic IC.
2. Epitaxial deposition (N type).
3. The isolation, base (plus resistor), and emitter diffusions would be similar to the normal bipolar IC, following the same sequence and utilizing the regular photoresist procedures.
4. After the first (conventional) metallization pattern is formed, a layer of silicon dioxide is deposited over the entire wafer surface, usually by r-f sputtering techniques.
5. Contact holes, known as "vias," are etched in this new oxide layer above the desired contacts areas (Fig. 8.28) of the first metallization layer.

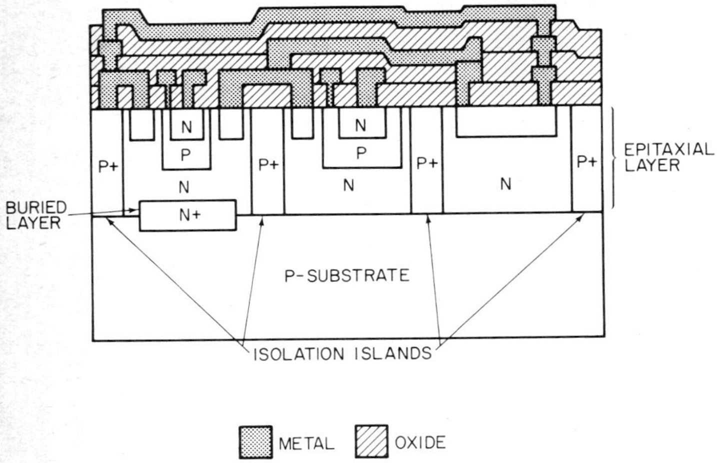

FIGURE 8.28
Multilayer metallization.

6. Metal is again deposited and patterned, making contact to the first layer through the vias.
7. Silicon dioxide is again deposited, vias are etched, and the third metal pattern is formed.

Electrical process control checks are normally made after each metallization pattern is completed to ensure that contact integrity is maintained and that component values remain within the desired values.

Final testing of such circuits is quite complicated and must be done with computers and expensive probers. Scribing and sorting is similar to what we have discussed for conventional devices. Die mounting and lead bonding are also done using familiar techniques. However, the large number of leads has resulted in some unusual packages (Fig. 8.38).

If the reader wonders how the various leads in the flat-pack cases shown in Fig. 8.38 are electrically separated from each other, the answer lies in the fact that the outer supporting, or holding, frame is removed prior to use. The frame is employed initially to prevent the long, thin, and easily damaged leads from breaking off the case during assembly operations. Once the IC chip has been inserted in the center case, connected to the external leads, and sealed, the outer supporting frame can then be discarded.

HYBRID ICS

We defined hybrid integrated circuits at the beginning of this chapter as those circuits in which separately manufactured components, both active and passive, are interconnected on a passive substrate, such as ceramic. We also said that there were three forms of hybrid ICs: thin film, thick film, and silicon multichip. (Thin films are generally defined as films less than a micron thick. Films greater than a micron thick are generally referred to as thick films.)

Thin Films. Conventional thin-film circuits are fabricated by depositing thin-film resistors and capacitors (Fig. 8.29*a* and *b*) on a passive substrate (ceramic) and then adding complete prefabricated active components to the same substrate (Fig. 8.29*c*).

FIGURE 8.29
(*a*) Thin-film resistor. (*b*) Thin-film capacitor. (*c*) Thin-film circuit.

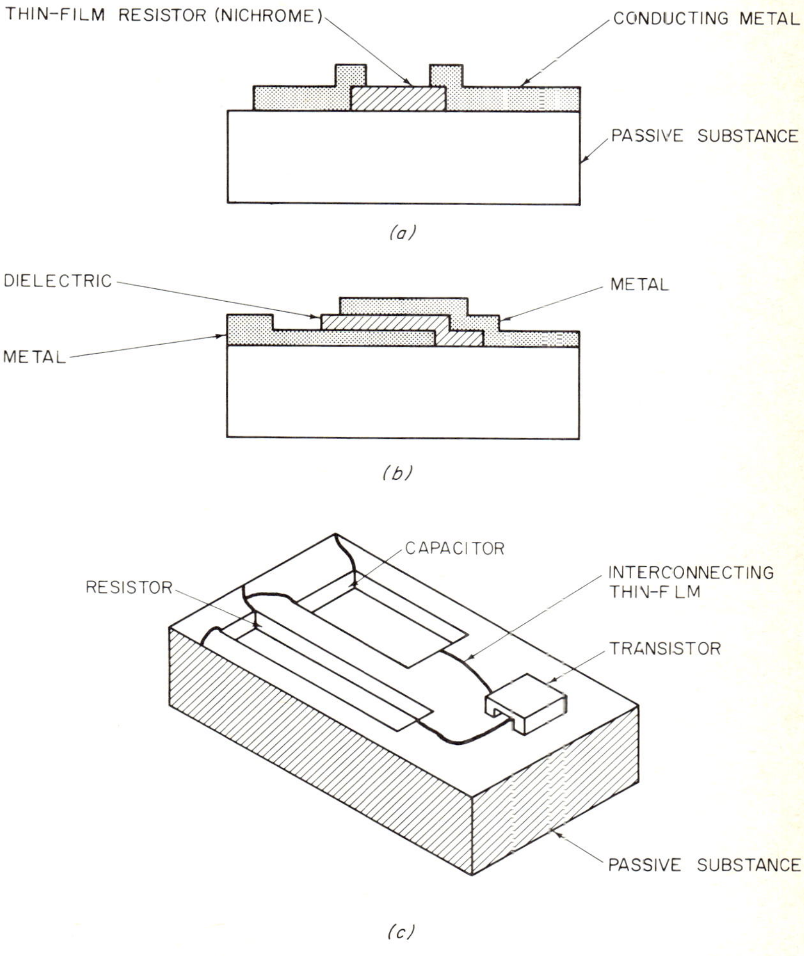

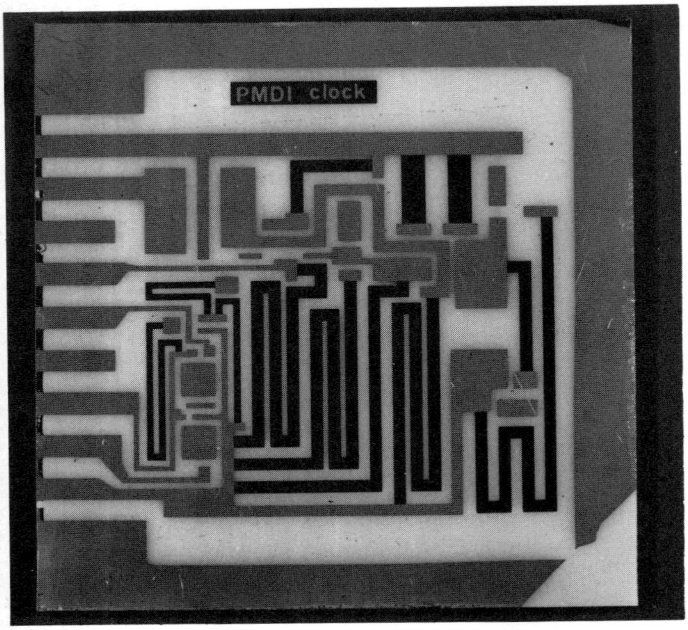

FIGURE 8.30
A thin-film conductor-resistor network.

Interconnection is normally achieved by using thin-film paths of conducting material or wire bonds. Flip chips, both in discrete device and IC form, are often used in thin-film hybrid ICs.

The thin-film elements of a circuit must be compatible with the other circuit components—that is, they must be able to survive various processing conditions (thermal and chemical), sealing and encapsulation, and accelerated life tests that are required for complete circuit fabrication. In addition, the thin-film components must be as reliable as the other circuit components.

The primary advantage of thin-film hybrid circuits is that fairly complex circuits can be quickly and economically fabricated. Figure 8.30 shows an example of this flexibility. In order to make an equivalent monolithic IC economically, large markets have to be available for that particular circuit.

In addition, even though thin-film circuits are larger than monolithic ICs, they are better suited for certain applications because of their low parasitics and good temperature coefficients.

Deposition of Thin Films. Vacuum evaporation is the most common method of thin-film deposition, and both conducting and resistive films can be vacuum-deposited.

Figure 8.31 is a typical vacuum-bell-jar evaporation system. The vacuum pressures used in thin-film depositions have to be kept between 10^{-4} to 10^{-6} millimeters (mm) of mercury. A tungsten filament, containing a "charge" of the thin-film material, is brought to a high temperature by passing an electric current through it. The charge is quickly vaporized; and since the mean free path of the evaporating molecules is many times the bell-jar diameter, they radiate in all directions from the filament.

The target substrate is usually placed directly under the source charge and about 4 in away. The target, which is sometimes heated to improve film adhesion, receives an even layer of the evaporating material. Film thickness is controlled by the size of the source charge, the moveable shutter, and the evaporation time. The film properties can be affected by the rate of evaporation, substrate temperature, source-to-substrate distance, pressure in the jar, and the impact angle of the evaporating molecules.

R-f Sputtering. Sputtering is the term used to describe the process of disintegrating a solid surface by bombarding it with tiny particles, which are normally ions, accelerated toward the surface by a high voltage (Fig. 8.32).

The energy of the imparting ions is transferred to the surface atoms of the source material, ejecting these surface atoms with resulting high velocities of their own. These displaced atoms are then deposited (electrically attracted) on the substrate target, thus forming extremely uniform films.

FIGURE 8.31
Simplified sketch of a vacuum evaporation system.

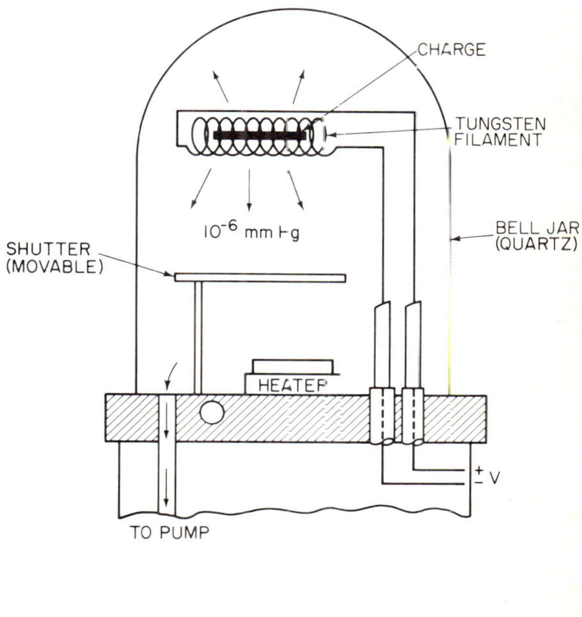

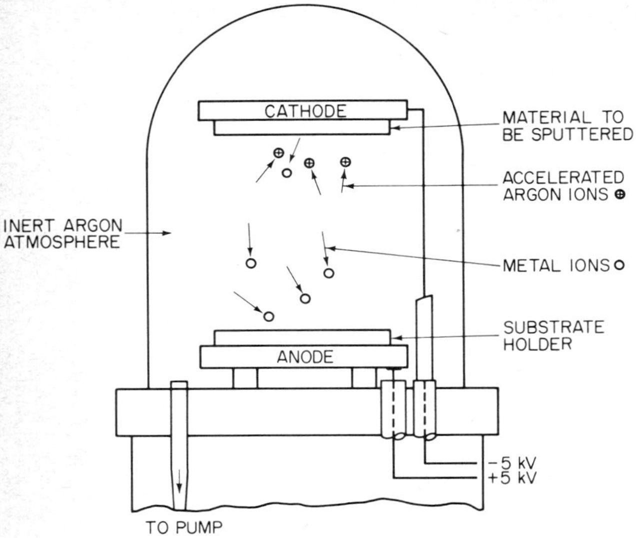

FIGURE 8.32
R-f sputtering system.

This technique has become popular for semiconductor processing in general, since it is a nonevaporative process that permits the deposition of high-temperature melting materials such as tungsten, tantalum, and even ceramics. In addition, sputtering can be used for noncoating purposes, such as surface cleaning and selective etching of appropriately masked materials.

Electroplating. The process is used to coat the conducting (interconnecting) metals on the passive substrates of hybrid circuits.

An ionic metal salt such as copper sulfate disassociates when dissolved in water into the Cu^{+2} and SO_4^{2-} ions. If an electric current is passed through such a solution, using a copper anode and the "target" (substrate) as the cathode, a uniform layer of metal (copper, in this case) can be deposited at the cathode.

Films of precise thickness, uniformity, and good adherence can be deposited if the following processing parameters are controlled:

1. Solution temperature
2. Current
3. Agitation of the solution
4. Concentration of the metal salt
5. pH of the solution
6. Substrate cleanliness

Electroless Plating. In this plating process, a metal ion in solution is reduced (gains electrons) to the free metal and deposited as a thin film on the desired target, without the use of an electric current. This form of thin-film plating is also used primarily for depositing the conducting films for hybrid integrated circuits.

Thin-film Resistors. Nichrome (80 parts nickel, 20 parts chromium) is probably the most popular thin-film resistive material. Nichrome is usually vacuum-evaporated, with the substrate maintained at a temperature between 250° to 350°C for film stabilization.

Nichrome films offer a wide range of sheet resistance (40 to 400 ohms Ω per square). The films in the higher resistivity range (200 Ω per square) must use special passivation techniques to improve stability.

In order to obtain the desired film resistivity and stability, the rate of film deposition, the source composition, the substrate temperature, and the vacuum pressure must be carefully controlled. Nichrome is patterned generally using photoresist techniques.

Tantalum films are also used to form thin-film resistors. This type of film must be r-f sputtered because tantalum has a very high melting point. An advantage of using tantalum resistors is that the film can be anodized in an oxygen or nitrogen atmosphere to form a thin stabilizing passivating layer over the film. Tantalum films with sheet-resistance values between 100 to 500 Ω per square are generally used for thin-film resistors.

Cermet films, which are thin films of combinations of ceramic or dielectric materials with metals, have become quite popular because of their ability to produce stable high-resistance values.

Cermet films are generally deposited by vacuum evaporation or r-f sputtering. Multiple sources are used and must be closely controlled.

A typical cermet film is composed of silicon monoxide and chromium. The sheet resistance of such films can be varied by varying the film thickness and the concentration of the chromium in the silicon monoxide.

Thick Film. Thick-film circuits, like thin films, owe their name to the thickness of the circuit film resistors and conducting film paths. Thick-film resistors are generally formed with cermet materials. Thick-film resistors and the interconnecting conductive patterns are normally deposited on a passive substrate by silk-screening.

Silk-screening is an adaptation of a process used by artists and printers. All areas of the screen except the desired patterns are blocked out. A very thin stainless-steel screen is placed in close contact with the substrate and a layer of conducting or resistor material is squeezed through the openings of the screen onto the substrate. The substrate is then fired at temperatures in the 800°C range. Note that resistors and the conducting patterns require separate firing cycles. This is a relatively simple process, not meant to produce

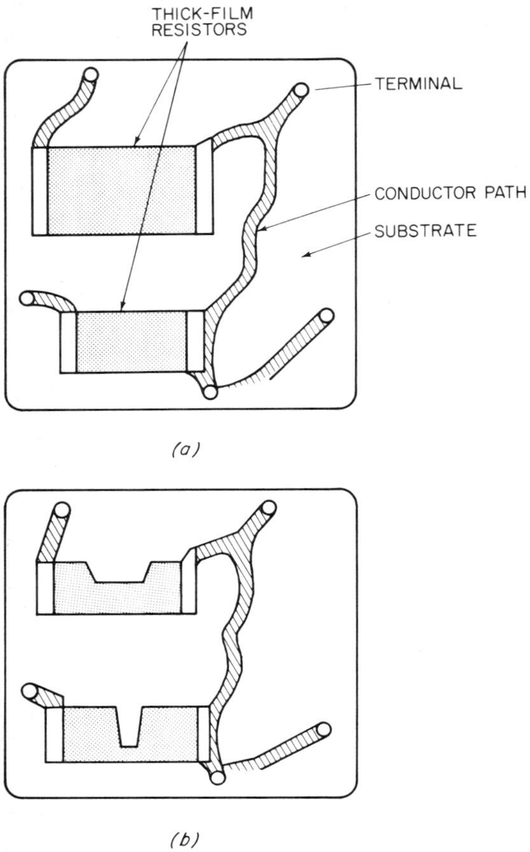

FIGURE 8.33
Thick-film resistors can be trimmed to value by removing sections of the resistors by sandblasting. (*a*) Untrimmed thick-film resistors. (*b*) Trimmed thick-film resistors.

precise resistor values. Final resistor values are obtained by using special trimming methods such as sandblasting, which can be controlled electronically with the sandblasting (trimming) halted when the desired resistor values are obtained.

The trimming of thick-film resistors is an advantage, since the resistors can be deposited oversized, usually as much as 20 percent, and then economically trimmed to ± 1 percent of the target value (Fig. 8.33).

After trimming, the active elements are mounted into place on the circuit, and wire bonding can be used to connect these added components to the rest of the circuit.

The packaging, or encapsulation, of thin and thick films is extremely important because it influences, to a large extent, circuit stability. Epoxy encapsulation is often used for hybrid circuits, although it has unfortunately not proved as reliable as hermetically sealed units. Often epoxy and hermetic packaging are combined to ensure against electrical changes caused, usually, by moisture.

Silicon Multichip. This is the simplest method of hybrid IC fabrication, involving the mounting of individually fabricated semiconductor devices on a passive substrate that normally has a conducting pattern already evaporated on it. Interconnection, plus the normal wire bonding to the top contacts of the circuit elements, is provided by these patterns. See Fig. 8.34.

The technique is best suited for small-volume custom circuits, or as an interim design stage for monolithic ICs.

Ceramic Multilayer Structures. Another approach to increased circuit density is the use of multilayer ceramic substrates that carry the interconnection lines, or patterns, leaving the top and bottom surface layers with more free "land," or area on which to mount various IC chips, plus other components (such as resistors and capacitors).

FIGURE 8.34
Sixteen ICs mounted on a ceramic substrate and interconnected with thick-film conductors.

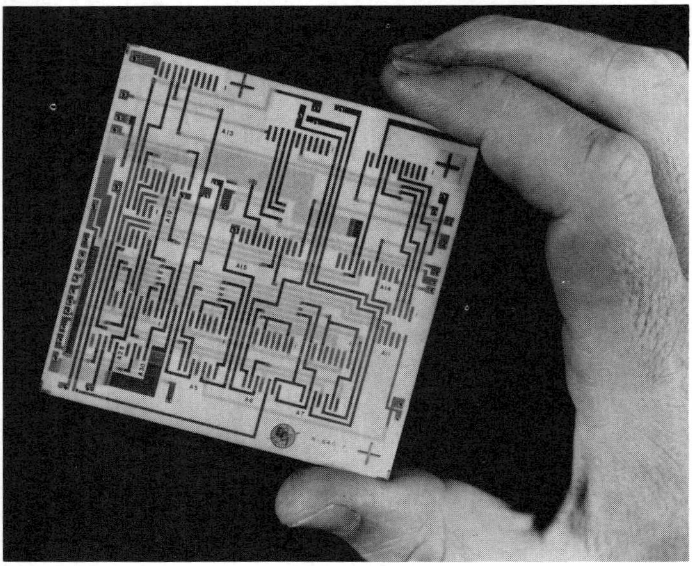

FIGURE 8.35
A four-layer ceramic board with conductive through-holes connecting selected points on each side of the board.

One type of ceramic multilayer structure is shown in Fig. 8.35. This is a four-layer board with conductive through-holes connecting selected points on each side of the board. By properly designing the conducting paths on each layer with whatever components and IC assemblies are mounted on the top and bottom surfaces, fairly complex circuit arrangements can be formed.

Figure 8.36 illustrates the steps in the formation of a ceramic multilayer structure.[1] Sheets of green (unfired) alumina are cut to shape and drilled where through-holes between layers are needed. Metallization is applied in the required pattern and the sheets are stacked on top of each other. The assembled green board is fired at high temperatures, resulting in the alumina (i.e., ceramic) and metallization sintering together into a "monolithic" multilayer structure. Additional thick- or thin-film wiring may then be added to the top (and bottom) and the IC, hybrid, and microdiscrete components mounted at appropriate positions. The result is a very compact, high-density microelectronic assembly.

Beam Leads in ICs. The beam-lead concepts (Chap. 4) has influenced IC design, particularly hybrid LSI circuits.

[1] J. J. Cox and L. C. Hoffman, New Materials For Level I and II Packaging, *Electronic Packaging and Production,* March, 1969.

Hybrid LSI circuits using beam leads for certain functions have become a better design approach than a monolithic LSI. This is particularly true for proven requirements where heat dissipation from a single chip becomes a problem. The hybrid-LSI-array concept employs less-complex monolithic circuits mounted and interconnected on a single passive substrate to perform complex functions. This circuit can be fabricated using beam leads and is thus extremely reliable. Using the beam-lead procedure, the metallization does not terminate at the bonding pad of the chip but is extended beyond the chip as a lead. The beam leads are made thick enough to support the chip and can be used to bond the chip directly onto the hybrid substrate. The circuit does not have to be packaged prior to mounting on the substrate, and wire bonding has been eliminated, both of which factors improve circuit reliability.

Beam leads are also successfully used on MSI individual circuits, but they are not commonly used, primarily because of the difficult processing and high costs associated with them.

FIGURE 8.36
Structure of all-ceramic multilayer circuit board. Conductive wiring is burned in the aluminum oxide by printing on successive layers (*a* and *b*) with hermetic risers between layers. The buried wiring and the ceramic board are made as integral unit (*c*), followed by firing the entire unit. Leads follow, then cover (*d*) for hermetic seal.

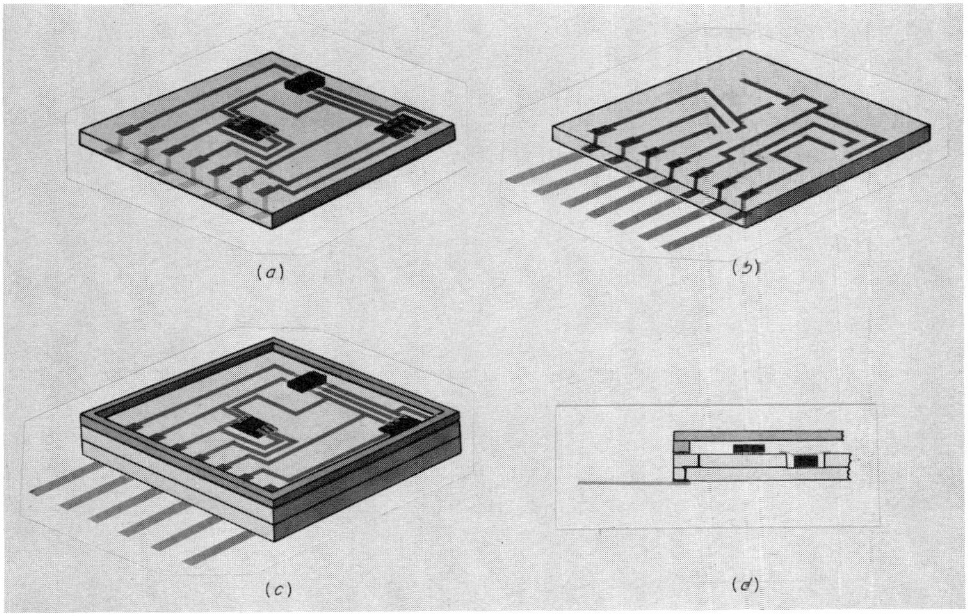

FIGURE 8.37
Radial-lead- or TO-type packages. Left to right, TO-18; coined TO-5; 8-lead TO-5; 10-lead TO-5.

INTEGRATED-CIRCUIT PACKAGING

Once the integrated circuit has been processed and produced, it still requires a housing that can protect it from the deteriorating effects of moisture and dirt, as well as any other deleterious elements found in the surrounding atmosphere. A suitably strong housing is also required for handling on the assembly line, as well as for connection into the system where this device is to function.

A number of packages have been developed. Actually, the variety of housing keeps changing as newer techniques evolve that enable package manufacturers to provide units that possess economic and/or structural advantages. Undoubtedly this change will continue. At the present time three types of packages or housings are in common use: TO cans, flat packs, and dual-in-line packages.[1]

TO OR RADIAL LEAD TYPE

The most common radial-lead packages are the TO-5 and the TO-18 (Fig. 8.37). Both are predominantly made using a matched-seal technology (the

[1] G. Fehr, A Survey of Today's Microcircuit Packaging, *Proceedings of National Electronic Packaging and Production Conference,* 1970.

thermal expansions of the glass and the metal are very closely matched). The matched-seal TO-5s and TO-18s have proven to be of outstanding reliability. The TO-5 package family is made with leads numbering from 2 to 12 in a 0.200- or 0.230-in-diameter pin circle. The leads range from 0.016 to 0.026 in diameter, with lengths between 0.5 to 1.5 in. The TO-18 is simply a smaller version (0.100-in-diameter pin circle) of the TO-5. The round shape provides an even distribution of the stresses in the seal.

The leads are brought to the upper surface of the TO-5 and TO-18 through small beads of glass; it is important that a hermetic seal be achieved between the metal of the leads and the can openings and the glass. The IC chip is then positioned between these protruding leads and appropriately connected to them. A metal cover is then welded over the top to encase hermetically the IC assembly.

With respect to cost, the TO types are quite economical, primarily because of the vast experience that the industry has had with these units, which stem from the earliest days of the transistor. However, the number of leads which can be employed with TO cans is probably no more than 12, and the leads are easily bent and require the use of special carriers for automatic handling—both decided disadvantages.

FLAT PACKS

The flat packs are produced in both round and rectangular shapes (Fig. 8.38). Common sizes range from $\frac{1}{4} \times \frac{1}{8}$ in to 1×2 in. The $\frac{1}{2} \times \frac{1}{8}$ in and

FIGURE 8.38
Flat packs in both round and rectangular shapes.

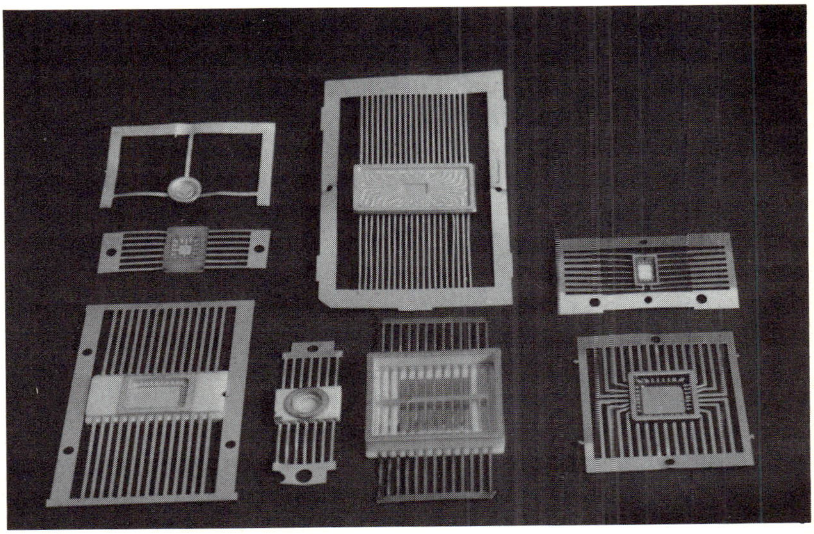

$\frac{1}{4} \times \frac{1}{4}$ in are the larger volume units at this time. Most of the packages available utilize a matched-seal technology, with either a metal or ceramic base and ring. The leads are kept thin (0.004 in) and flexible to reduce weight and to prevent stresses from being transmitted to the seal. The main advantages are:

1. Light weight and small size
2. Relatively large die area per package size

The package type lends itself to being expanded to large sizes to contain ceramic-based hybrid circuits containing several chips. Depending on its construction, it can be either welded, brazed, or solder-glass closed. Its major disadvantages are a relatively high cost [about 25 cents each for the small package in large volume (100,000) up to $5 to $6 each for small quantities such as 1,000 of the large packages]. It is not a particularly easy package to handle and, again, requires a carrier for automatic handling. The smaller packages with the short seal patch (0.030 in) have had a general history of hermeticity problems when mishandled or when the packages were not made under very controlled conditions.

DUAL-IN-LINE PACKAGES (DIPS)

There are many types of construction of DIPs; each has its advantages and disadvantages.

The multilayer ceramic-to-metal seal is shown in Fig. 8.39. Since the seal is formed between pieces of identical ceramics, this package will withstand temperature shocks better than any other construction. Also, it can utilize reliably a very short seal patch (25 to 35 mils), resulting in a large cavity per package size. Packages with 0.170-in-wide cavities can be bent in 0.300-in-lead centers. To gain the width, the leads are brazed on the package bottom, causing some handling problems. The top brazed package results in smaller cavity width per package width, which is particularly important in the

FIGURE 8.39
A multilayer ceramic, hermetically-sealed package. The cavity contains the microcircuit assembly.

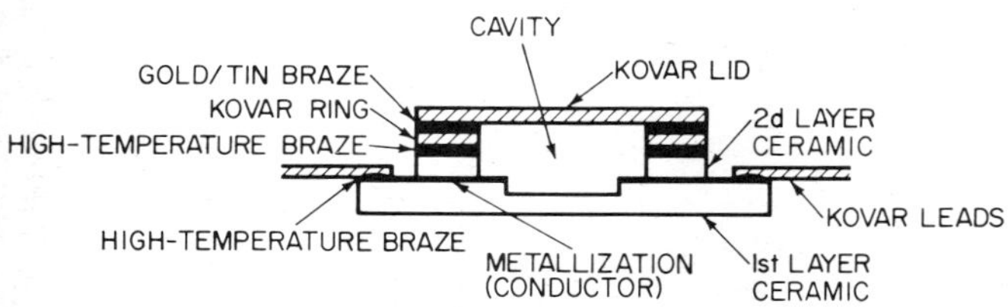

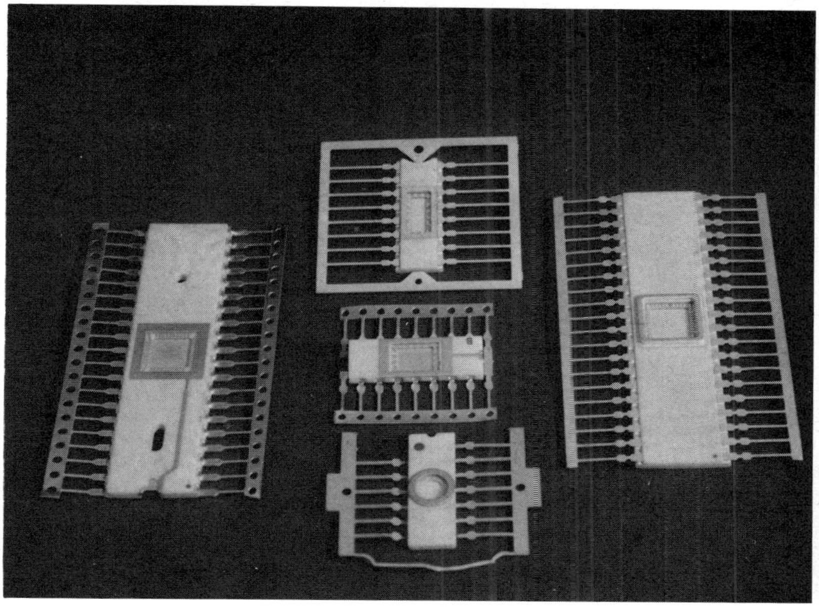

FIGURE 8.40
Ceramic multilayer DIPs.

16-lead package bent in 0.300-in centers. One drawback in any of these packages presently available (Fig. 8.40) is their relatively low lead strength—the leads bend very easily. Another critical point is to ensure, particularly in the wide-cavity-bottom-braze packages, that there is a braze fillet between the top of the lead and the side metallization connection. Without the fillet, the lead cracks away from the side relatively easily. The package has a tendency to bow and warp during manufacture, causing some sealing problems. This package is probably the most expensive DIP in common usage. A 16-pin package costs from 75 cents to $1 each in quantities of 10 to 25 thousand. It is too thin to be handled by automatic insertion equipment.

The side-braze package shown in Fig. 8.41 is slightly lower priced (65 to 70 cents each in like quantities) than the ceramic-to-metal. It utilizes a glass between the two ceramic pieces which closely matches the expansion of the ceramics. Its temperature-cycle capabilities although not as good as the ceramic-to-metal are adequate. Its cavity width is limited to 0.150 to 0.160 in for a 0.300-in center-to-center lead package, which is adequate for most MSI products. It has a relatively clean bottom (some metallization close to the leads) and fairly rigid leads. Its disadvantage is its cost, plus the fact that it is different from most other packages; however, it can be used in the automatic insertion equipment.

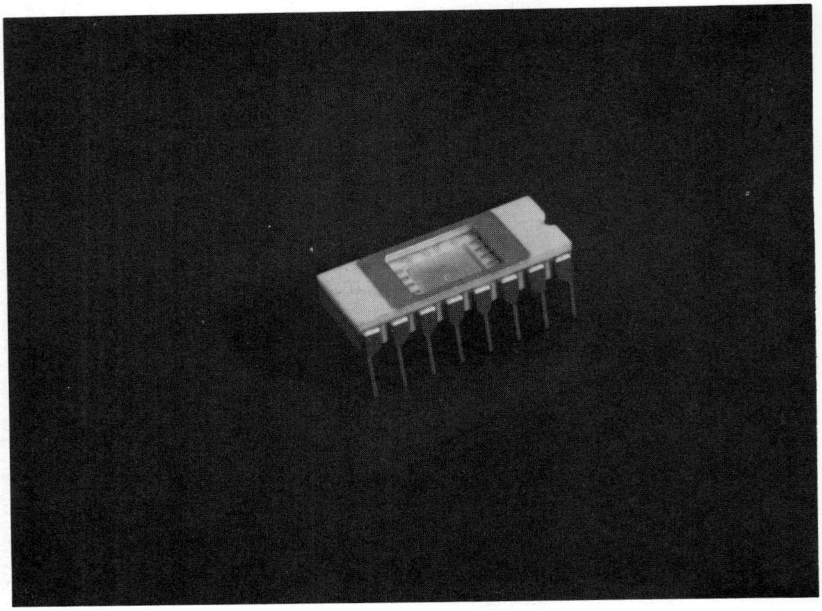

FIGURE 8.41
Side-brazed DIP.

The glass-to-metal sealed DIP (Fig. 8.42) is priced from 50 to 75 cents for equivalent volumes. In large volumes, the price can drop to from 25 to 35 cents. However, this package has had a general lowering of capabilities. Its seal path should be at least 0.050 to 0.060 in. It is necessary to have 0.015 to 0.020 in from the edge of the package to the lead bend, so that it can be clamped sufficiently not to crack the glass when bending the leads. This, plus the seal length required, limits the cavity width for a 0.300-in lead center-to-center package to approximately 0.100 to 0.120 in. This is a serious limitation for MSI manufacturers. The thermal shock capability is lower than the other two packages. The main selling point for this package is cost.

The ceramic DIP (or C-DIP) package shown in Fig. 8.43 is assembled on the component assembly line. The header-oriented costs are in the 10 to 15 cent range. The technology required for successful sealing is critical and needs to be well controlled. Problems can result if the glass is either over- or underdevitrified. This package has about the same limitations as the glass-to-metal DIP previously discussed. While this package has a lower cost than the earlier package, its quality is more sensitive to the manufacturing conditions and requires a well controlled line.

The plastic-injection molded DIP is the lowest-cost DIP (Fig. 8.44). The package is solid and hence has little problem with mechanical shocks. Dis-

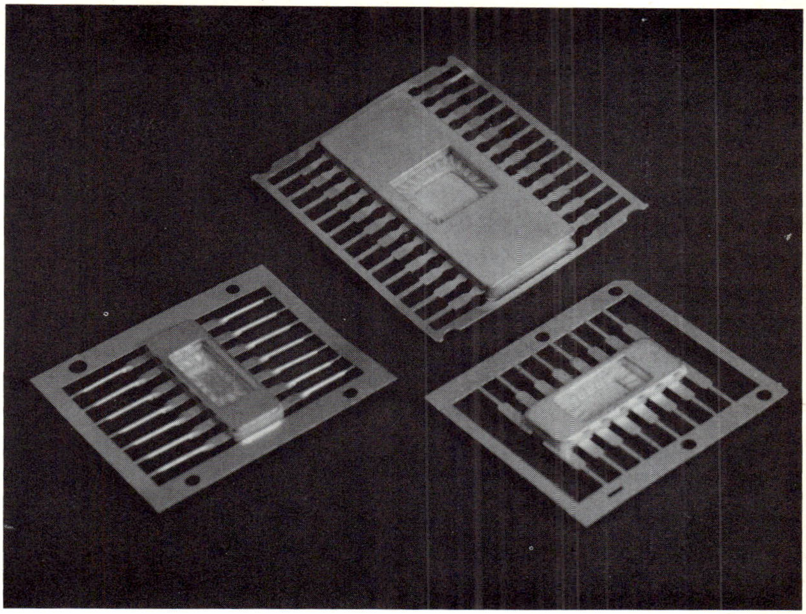

FIGURE 8.42
Glass-to-metal DIPs.

advantages are its nonhermeticity, lower resistance to moisture, high thermal resistance, temperature-cycle failures, and temperature limitations. However, when used in a controlled environment, it offers adequate protection to the die at a greatly reduced cost.

Plastic packages are made by assembling and interconnecting the chip on a substrate or lead frame and molding the entire structure in plastic with the exception of the leads to form the body of the component. Epoxy, phenolic, and silicone are the most commonly used resins for this purpose.

Plastic packages have gained their popularity not only because of their significantly lower manufacturing costs (part of which is the elimination of the necessity for hermetic testing) but mainly because of the molding process, which can be adapted to high production rates. The most formidable problems to be solved for this type of package is to improve its ability to resist moisture contamination and thermal cycling.

MISCELLANEOUS PACKAGES

This category includes the packages of unique design for particular usage or because an individual company likes them. They are used in a much lower

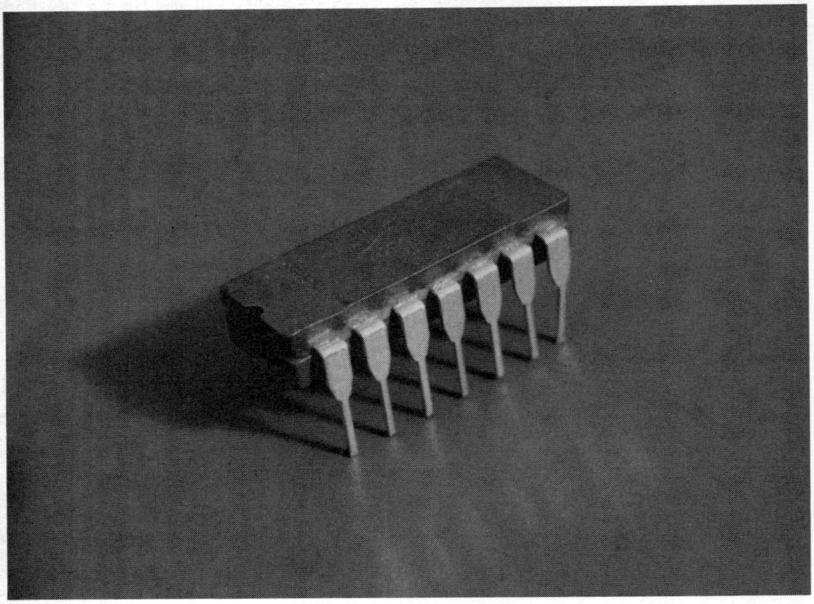

(a)

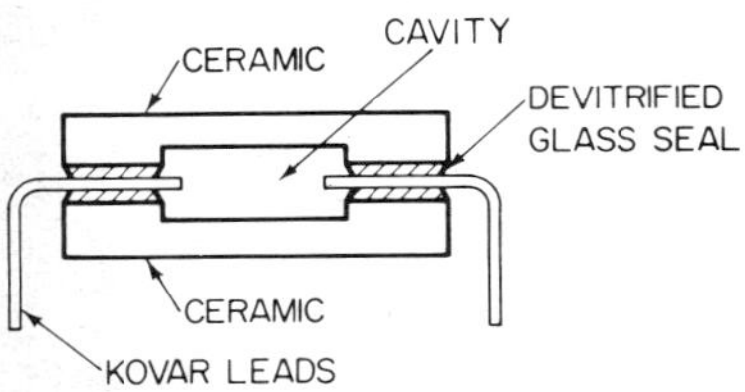

(b)

FIGURE 8.43
(*a*) The ceramic DIP (or C-DIP). (*b*) Construction of this package.

volume than either the dual-in-line or the flat-pack packages. Examples are high-power packages and the IBM outline (Fig. 8.45).

SOME MONOLITHIC IC APPLICATION CONSIDERATIONS

Since one IC can perform functions that were previously executed by numerous discrete transistors and other associated components, the design (IC) has been aimed at subsystem level applications. (LSI can be designed at the system level.)

Most ICs designed for computer applications are designed toward specific subsystem functions, i.e., logic applications or memory functions.

Digital ICs. In general, digital circuits operate as on-off switches and are used in computers to count or compute in response to the presence or absence of a signal.

When circuit designers first began building IC logic devices, they tried to convert from proven discrete component logic schemes. This proved to be impractical, however, since discrete resistors and capacitors are not as easy to reproduce in monolithic form as diodes or transistors. Fortunately, the cost of additional transistors or diodes in a monolithic IC is insignificant. This has resulted in digital ICs using direct-coupled active components, eliminating capacitors and resistors where possible. Thus, active components have been substituted to perform passive functions such as constant-voltage and constant-current sources. By the proper arrangement of active components in the monolithic circuit, impedance and DC levels can be shifted. Coupling arrangements can also be replaced by complementary transistors.

The result of this has been that many familiar discrete circuits have become difficult to recognize even though their functions have not changed. We shall see examples of this in the following chapters.

FIGURE 8.44
The plastic DIP.

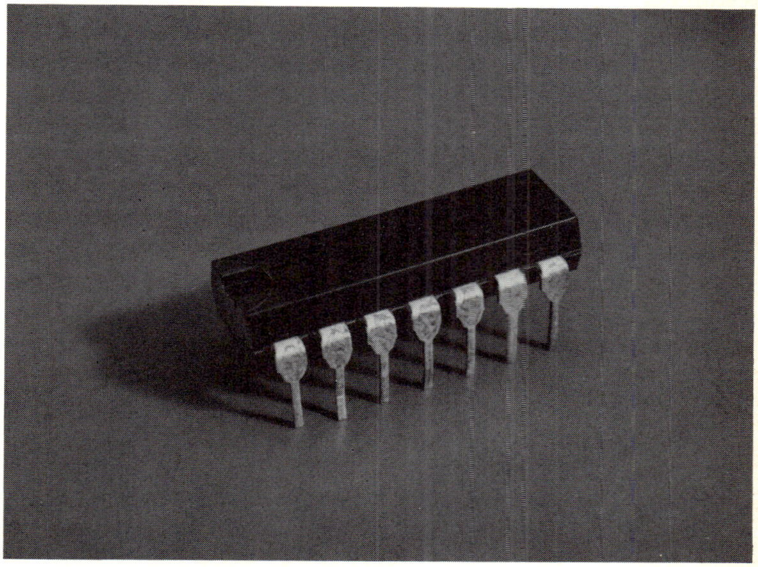

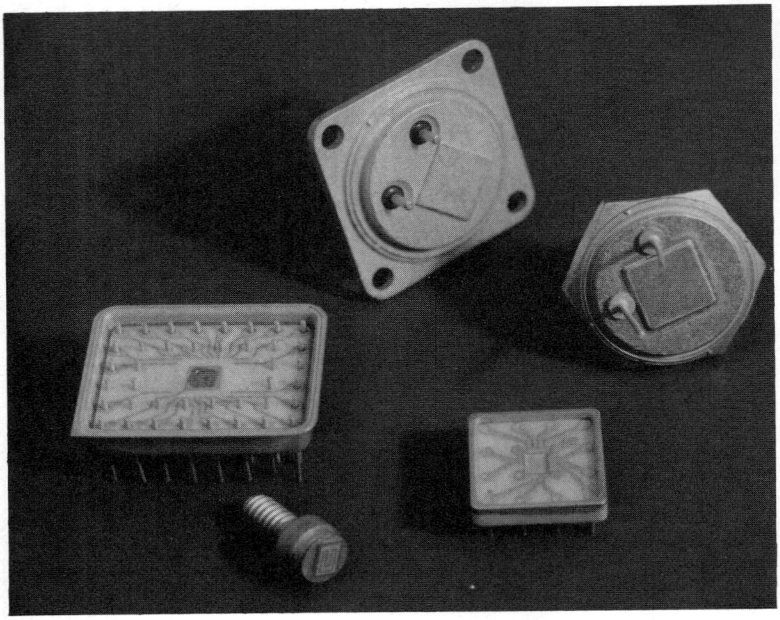

FIGURE 8.45
Miscellaneous packages.

Linear ICs. Linear circuits operate on an electrical signal to change its shape and amplitude or to modify it in some manner to perform a specific function.

In the early days of monolithic linear ICs there were more problems in converting from discrete component circuits than were involved in digital IC conversion.

Linear circuits have a continuous output that must be directly proportional to the input, which presents certain monolithic conversion problems. For example, early linear ICs lacked feasible inductance and acceptable capacitance values. Initially, resistor values could not be fabricated accurately enough to meet the rigid linear requirements. In addition, alternating current influenced linear-circuit design and the necessary filtering and wave-shaping requirements could not be duplicated in monolithic form.

The answers to these problems was to give up the idea of copying the discrete-component linear circuit and to adapt a new simulation program. For example, capacitors were moved out (off the chip) and were replaced by functionally equivalent active devices. Again, direct coupling is used wherever possible to eliminate coupling capacitors. Biasing arrangements and constant-current sources are employed to eliminate impractical passive components. The power limitations of linear ICs have been improved using special techniques such as diodes and coupled transistor cascades that are able to accommodate the large signals of power amplifiers and regulators.

Operational Amplifiers (OP amps). The most common linear IC is known as an operational amplifier. This circuit type acquired its name from computer language that describes a circuit capable of performing an operation on the input signal and resulting in mathematical type of function, such as addition, subtraction, integration, etc. We shall discuss this more fully in Chap. 9. OP amps are also employed to compare, regulate, control, and detect linear and nonlinear signals. Because of this, the OP amp has many applications.

Heat Transfer and ICs. Although individual integrated circuits do not dissipate much heat, when they are stacked close together, the heat generated per cubic foot becomes a serious problem.

Several methods are used to transfer heat, a few of which are cold plates, forced air, cooling fans, etc.; but these methods are not enough to significantly increase the circuit density.

Experiments have been run with liquid cooling (Freon) of the bare chip. It was found that an IC that would become overheated in air at a power level of about 300 milliwatts (mW) would be safe in a liquid coolant if that level were raised to over 10 W. This improvement would have a significant effect on both circuit and system designs. Faster systems would become available with denser packing (less wiring); and the size and reliability of equipment would be optimized.

QUESTIONS

8.1. Define the following:
a. Integrated circuit (IC) **b.** Monolithic IC **c.** Hybrid IC
d. Thin film **e.** Thick film **f.** LST **g.** MSI

8.2. Why have integrated circuits become so important?

8.3. Why are integrated circuits so attractive to the computer industry?

8.4. Contrast the construction and lead connections between a monolithic IC transistor and a discrete transistor.

8.5. Why are isolation sections required in integrated circuits? Name two types of monolithic IC isolation.

8.6. Name two advantages of dielectric isolation.

8.7. What is the purpose of the N^+ buried-layer diffusion?

8.8. Sketch an NPN monolithic IC transistor, showing the N^+ buried layer and the isolation section.

8.9. How are diodes formed in monolithic ICs? Sketch three different methods.

8.10. Name two forms of monolithic IC capacitors? Use sketches.

8.11. Sketch a monolithic diffused resistor and describe how resistor values are obtained.

8.12. Define parasitics and describe one way in which semiconductor-device designers control parasitics.

8.13. What is a four-point probe, and what important process control check does it perform?

8.14. What structural differences exist between a lateral transistor and a standard monolithic IC transistor?

8.15. Contrast the flow of carriers in lateral and standard IC transistors.

8.16. Why can larger circuit-component densities be obtained with MOS integrated circuits than with bipolar circuits?

8.17. What is MOS complementary symmetry, and what advantage does it offer?

8.18. What is the primary advantage of thin-film hybrid circuits?

8.19. What is ion implantation? Describe its advantages.

8.20. Name two methods of depositing thin films.

8.21. What advantage do thick-film resistors offer?

8.22. What advantages do beam leads offer?

8.23. What is a digital circuit? A linear circuit?

8.24. What limits the packing density of IC circuits? Describe a possible remedy.

8.25. Describe briefly two types of cases employed for integrated circuits.

Chapter Nine

Semiconductors in Radio

Among the first commercial uses to which transistors were put was in small, portable radio-broadcast receivers. The transition from vacuum tubes to transistors was relatively easy, since it was readily shown that superior performance at lower costs were obtainable with transistors. Lighter weight and smaller size were additional inducements for change. However, the additional advantages offered by ICs, such as reduced power consumption and further decreases in weight and size, were not as quickly accepted by the radio manufacturers. The reason for this was primarily one of economics: when they first became commercially available, ICs tended to be more expensive than the discrete components they replaced. That is, the cost/performance ratio did not justify their use.

At the present time, however, IC designs and yields have improved, resulting in reduced costs and better performance. These factors combined with further reductions in weight and size have made integrated circuits ideal for radio receiver applications.

In this chapter we shall discuss the development of radio receivers from vacuum tubes to transistors to integrated circuits.

TRANSISTOR AM RECEIVER

A fairly typical portable radio receiver is shown in Figs. 9.1 and 9.2; it takes advantage of every space-saving feature afforded by transistors and their associated miniature components. Overall dimensions of the unit are $5 \times 3 \times 1\frac{1}{4}$ inches (in), so that the entire receiver can fit easily into the pocket of a man's jacket. Weight of the set, with the battery, is only 12 ounces (oz).

A schematic diagram is shown in Fig. 9.3. There are four transistors and five stages. The extra stage is the second detector, and its function is performed by a diode. The transistors are of the NPN variety.

The first stage, containing transistor Q_1, is essentially a self-oscillating

FIGURE 9.1
A transistor radio small enough to fit into the pocket of a man's jacket.

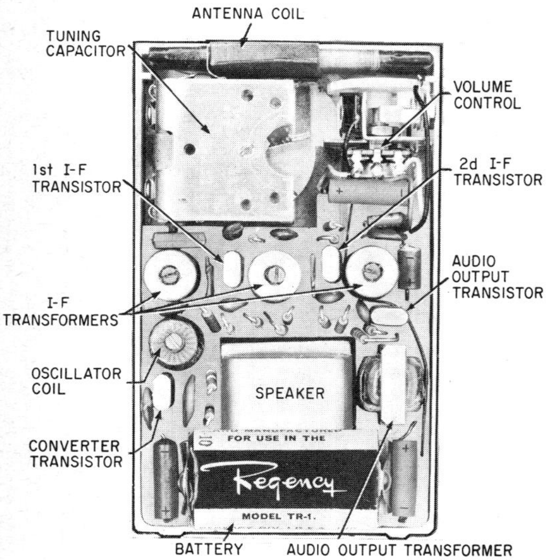

FIGURE 9.2
Inside view of transistor receiver showing layout of components.

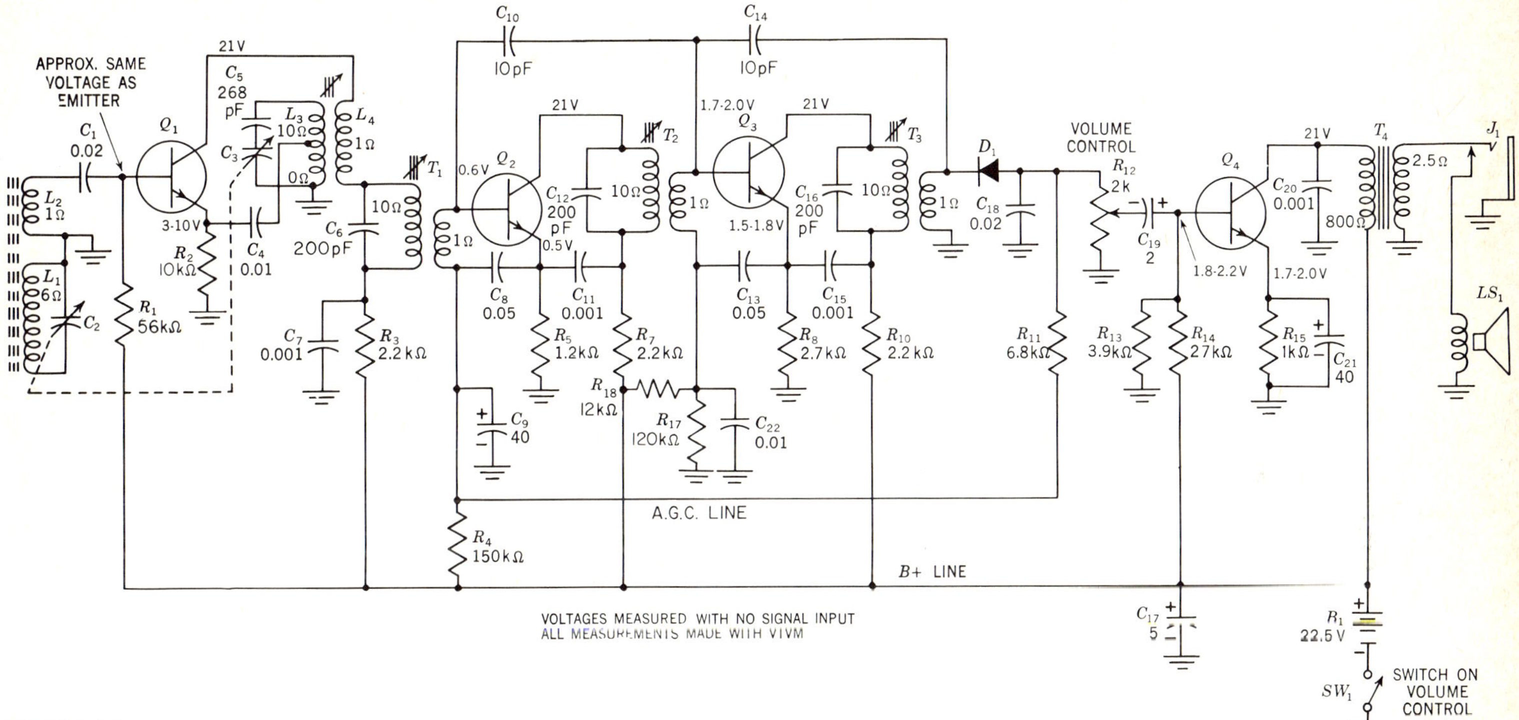

FIGURE 9.3
Schematic diagram of the receiver shown in Fig. 9.1.

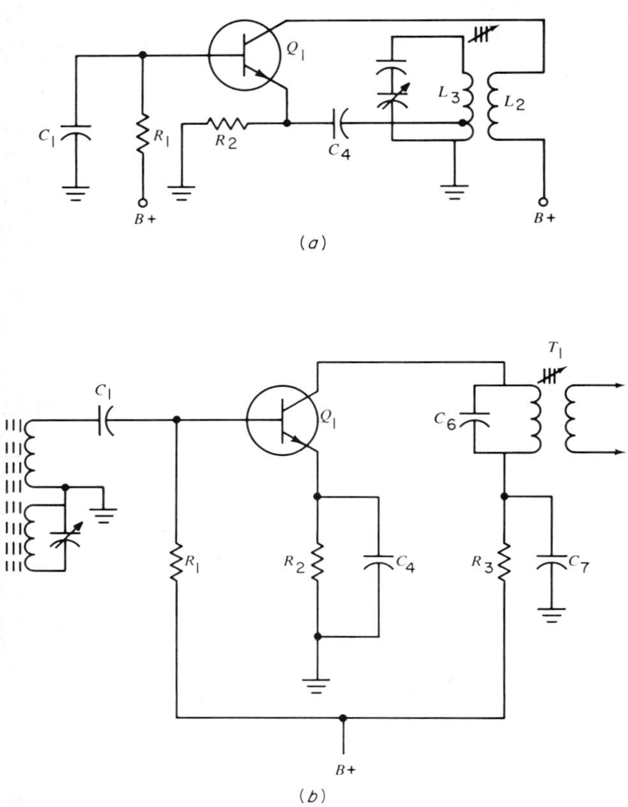

FIGURE 9.4
Equivalent circuits of (*a*) oscillator and (*b*) mixer sections of the autodyne-converter stage Q_1 of Fig. 9.3.

converter, called an "autodyne converter." The input signal is picked up by a tuned ferrite-core coil that possesses a high Q. A low-impedance winding on the ferrite-core coil couples the signal to the base of Q_1.

Local oscillations are generated by a parallel-resonant circuit in the emitter network, which is inductively coupled to a coil in the collector circuit. The low-impedance emitter is tapped down on the tuning coil in order to provide the proper impedance match without lowering the Q of the circuit.

In essence, the autodyne converter is two circuits in one. The oscillator portion of this stage is shown in Fig. 9.4*a*. Oscillations are sustained by voltage feedback between emitter and collector elements of Q_1. The mixer section is shown in Fig. 9.4*b*. Here, the incoming signal and the oscillator voltage are combined, thereby developing an appropriate i-f signal that is then applied to transformer T_1 and the i-f stages beyond.

A 10,000-ohm (Ω) resistor is placed in the emitter circuit to provide dc stabilization against temperature changes and variations among different replacement transistors. The positive voltage that the emitter current develops across R_2 is counterbalanced by a positive voltage fed to the base from the battery. The actual voltage difference between these two elements is only on the order of 0.1 volt (V).

The proper biasing voltage for the collector of Q_1 is obtained from a 2,200-Ω resistor that is tied to the 22½-V B+ line. A 0.001-microfarad (μF) bypass capacitor C_7 keeps the signal currents out of the dc distribution system.

There are two stages in the i-f system, both of which operate at 262 kHz. This frequency is considerably below the 455 kHz common in vacuum-tube radio receivers, and it possesses the disadvantage of making the receiver more susceptible to image-frequency pickup. However, the lowered frequency of operation is advantageous in that it provides greater gain and more stability.

The primary of each i-f transformer is tuned with a fixed capacitor, whereas the secondary is untuned. This is done to match the high collector impedance of the preceding stage to the low input impedance of the following stage. Peaking of each i-f coil is achieved by varying the position of an iron-core slug.

Each i-f stage is neutralized by feeding back a voltage from the base of the following stage to the base of the preceding stage. The feedback occurs through a 10-picofarad (pF) series capacitor.

Automatic gain control is applied to the first i-f stage only. A negative voltage is obtained from the second detector and applied to the base of Q_2. Its purpose is to regulate the emitter and collector currents and, by this means, the stage gain. When the incoming signal becomes stronger, the negative agc voltage rises, reducing the collector current of Q_2 and, with it, the gain. The opposite condition prevails when the signal level decreases. This method is quite effective and provides a wide range of control. (A detailed discussion of automatic gain control in transistor receivers will be given shortly.)

The bias for the second i-f stage is obtained from the voltage-divider network of R_{17} and R_{18}. This bias voltage is bypassed by C_{22} and then further bypassed by C_{13}, a 0.05-μF capacitor.

Both emitters have dc stabilizing resistors. (If it were not for the presence of C_8, C_{11}, C_{13}, and C_{15}, signal degeneration would also occur. As it is, only the direct portion of the current passes through R_5 and R_8.) Note, however, that the emitter resistor of the first i-f stage is 1,200 Ω in value, whereas the emitter resistor of the second stage is 2,700 Ω. The reason for this difference stems from the compromise that must be reached in the first i-f stage between good agc action and the dc stability of the amplifier. A value of R_5 greater than 1,200 Ω is desirable for stability purposes, but the degeneration that produces the stability would result in reduced gain-control action.

Each of the collectors of Q_2 and Q_3 receives its operating voltage through a 2,200-Ω dropping resistor. C_{11} and C_{15}, at the collector end of the resistors, serve as decoupling and bypass capacitors.

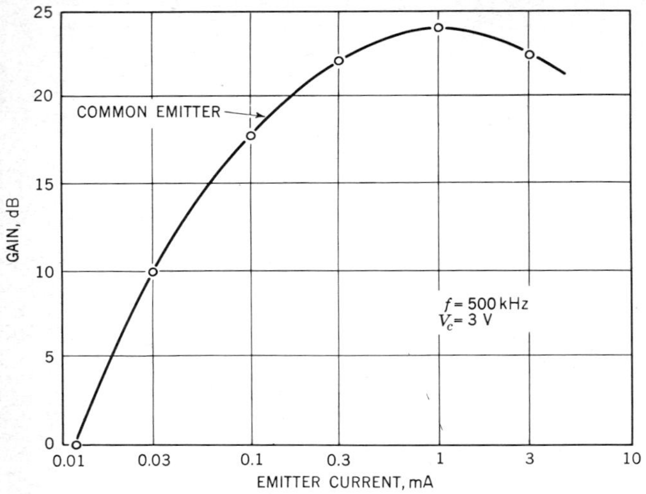

FIGURE 9.5
Transistor-amplifier gain as a function of the emitter current.

Beyond the second i-f stage is the second detector D_1, the function of which is performed by a germanium diode. The load resistor for the detector is the volume control. Note that the impedance of the control is 2,000 Ω; this low value is needed to match the input impedance of the audio output stage Q_4.

The final amplifier is operated with the emitter grounded through a 1,000-Ω resistor. Base bias is obtained from the voltage-divider network formed by R_{13} and R_{14}. The output transformer matches the 10,000-Ω collector impedance of Q_4 to the low voice-coil impedance of the miniature speaker. Diameter of the speaker is only $2\frac{3}{4}$ in. There is also provision for a small earphone plug that can be inserted into a small jack on the side of the receiver. When the earphone is in use, the speaker is disconnected.

The total power for the receiver is furnished by a hearing-aid-type $22\frac{1}{2}$-V battery. Total current drain is on the order of 4 milliamperes (mA).

The compactness of this receiver can be seen by an inspection of Fig. 9.2. All components, including the two-gang tuning capacitor and the speaker, are miniaturized. Operating voltage on electrolytic capacitors C_9, C_{21}, and C_{19} is 3 V; on C_{17}, it is 25 V.

AUTOMATIC GAIN CONTROL OF TRANSISTOR AMPLIFIERS

The automatic control of the gain of a transistor r-f or i-f amplifier in radio receivers is achieved almost universally by varying the emitter or collector

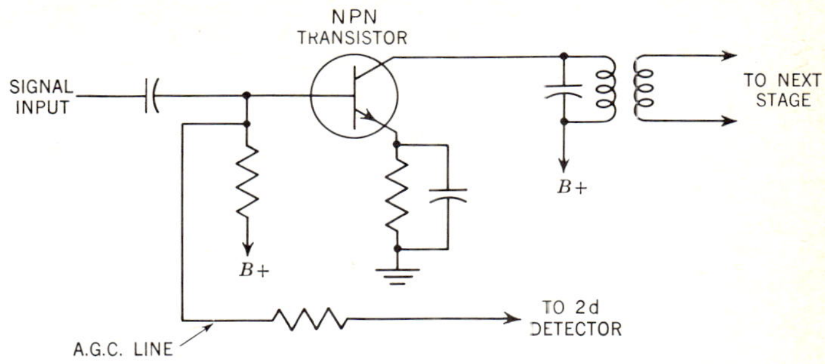

FIGURE 9.6
An example of I_e control of an i-f amplifier. The agc voltage is applied to base.

current in accordance with the amplitude of the incoming signal. Figure 9.5 illustrates the variation in gain that can be achieved. (No mention is made in the above discussion of the changes in transistor input and output impedances accompanying changes in emitter-current flow. These do occur, and they also contribute to the changes in gain.) Now, to provide this control of gain, a certain amount of dc power is necessary, because current from the control source is required. In vacuum-tube amplifiers, little or no power is required because the control voltage is fed to the grid of a tube and this element, being negative with respect to the cathode, draws no current. A transistor, on the other hand, is a current-operated device; and to alter its current, we must have the control stage supply a suitable amount of its own current. This, in turn, means that power must be expended.

The control voltage is obtained almost invariably from the second detector; hence, this stage must supply the control power. To assist the detector in this task, the controlled transistor i-f amplifier is usually made to function as a dc amplifier for the control signal. For example, in the stage shown in Fig. 9.6 the emitter current is agc-controlled. Instead of varying the emitter current directly, however, the control voltage is applied to the base of the transistor, and the resulting changes in direct base current are amplified and appear as larger changes of emitter current. When the incoming signal is strong, a voltage is then fed back that serves to reduce the emitter current and the stage gain is reduced. Conversely, when the incoming signal is weak, the voltage fed back is reduced in value, permitting more emitter current to flow and raising the gain.

By reducing the emitter current, we also reduce the collector current and, hence, the strength of the signal developed across the output tuned circuit.

If sufficient power is available in the detector circuit, an attempt can be made to control the emitter current directly by introducing the control voltage

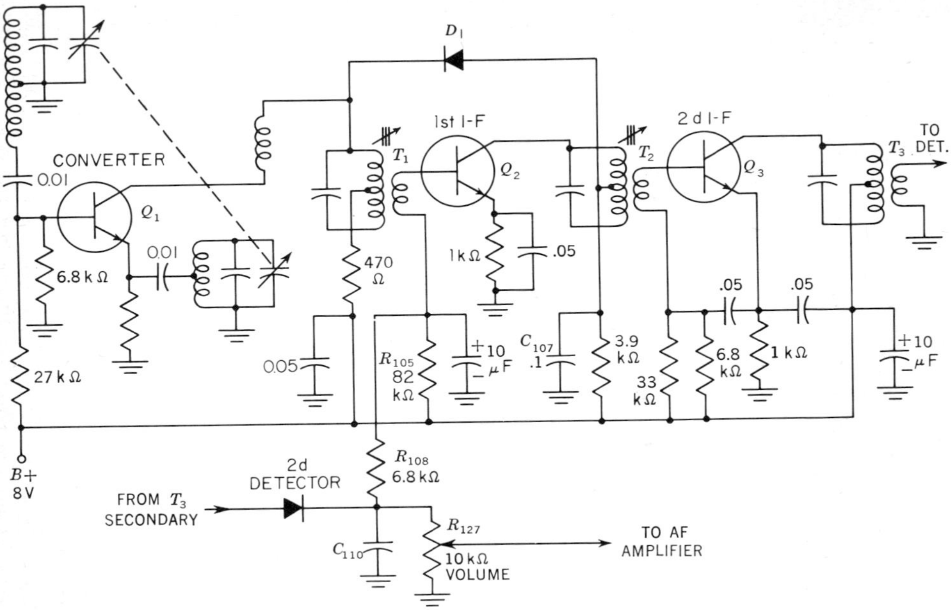

FIGURE 9.7
An agc system that is supplemented by an auxiliary diode D_1.

in the emitter circuit of the i-f stage. However, in the absence of this power, the control voltage can be applied to the base of the i-f stage, as shown.

The advantage of the foregoing system of agc is its simplicity and economy. There is the serious disadvantage, however, that the transistor is unable to cut off the signal effectively, thereby giving rather incomplete agc action. To overcome this deficiency and provide a greater range of control, a delayed agc circuit is frequently added to supplement the primary agc. This addition consists of an auxiliary diode, as shown in Fig. 9.7. Here, the primary agc network extends from the detector through R_{108} to the base of the first i-f stage, an NPN transistor. The control voltage developed across C_{110} and R_{127} is negative, and it works to reduce the B+ voltage present at the top of R_{105}.

The auxiliary germanium diode is connected between the center tap of T_2 and the top end of the primary of T_1. On strong signals, the amount of agc voltage applied to Q_2 is quite high and conduction through the transistor is reduced, thereby lowering the effect of the stronger signal. This causes the positive voltage at the collector to rise—in this instance toward 8 V. Actually, 8 V would occur only if the transistor ceased conducting completely; but the voltage value does get slightly above 7.5 V. When it does, D_1 will conduct, because the anode will be more positive than the cathode. The potential on the latter element is normally about 7.5 V, obtained from the connection at T_1.

When D_1 conducts, it acts as a low-valued resistor in series with the 0.1-μF capacitor C_{107}. This combination loads the primary of T_1 and reduces the signal fed to the first i-f stage and all subsequent stages. On moderate and weak signals, the current passed by Q_2 is enough to keep its collector potential below 7.5 V. In such instances, D_1 cannot conduct and T_1 is permitted to pass a stronger signal to the first i-f amplifier.

A SECOND TRANSISTOR RECEIVER

Another radio receiver, designed along somewhat similar lines as the prior set but containing a greater number of stages, is shown in Fig. 9.8. The lineup of stages includes a converter, three i-f amplifiers, a germanium-diode detector, and three audio amplifier stages. The maximum power output is in excess of 200 mW, which is more than ample for this particular purpose. The B+ voltage is 6 V, and the current drain, at low levels, is 10 mA. At high peak levels it may go as high as 20 mA. A set of four 1½-V batteries could be expected to have a life in excess of 250 h if used at the rate of 2 h day.

The converter stage is closely similar to the same stage in the receiver previously discussed. The antenna is a ferrite rod on which is wound a tuned primary coil for station selection and a secondary coil to apply the signal to the base of Q_1. Tuning range extends from 532 to 1,620 kHz.

The oscillator transformer is somewhat more extensive: it has one winding connected into the emitter circuit, one winding connected to the collector, and one winding serving to couple the other two windings to each other. This intermediate winding, in conjunction with the oscillator tuning capacitor, establishes the oscillator frequency. A small trimmer capacitor provides an alignment adjustment at the high end of the band, whereas a small adjustable slug permits alignment at the low end. The intermediate frequency produced in Q_1 is 455 kHz.

Only part of the primary winding of T_1 is utilized by the collector of the converter transistor in order to achieve an impedance match. As is customary in transistor i-f transformers, the primary is tuned to develop the necessary high impedance for Q_1, whereas the secondary is untuned to match better the low input impedance of the following stage. Tuning of the transformer is accomplished by a movable slug in T_1.

A diode D_1 is connected between the primary of transformer T_1 and the center tap of transformer T_2. This is a delayed agc diode that supplements the normal agc network. Operation of this circuit has already been described.

Before we leave the converter stage, we should note the special switch S_1 in the base-bias network. The operating bias for the transistor is provided by R_2 and R_3. S_1 is normally open and does not take part in the operation of the stage. If it is desired to measure the battery drain for servicing purposes, however, the receiver is then turned off, a milliammeter is connected across

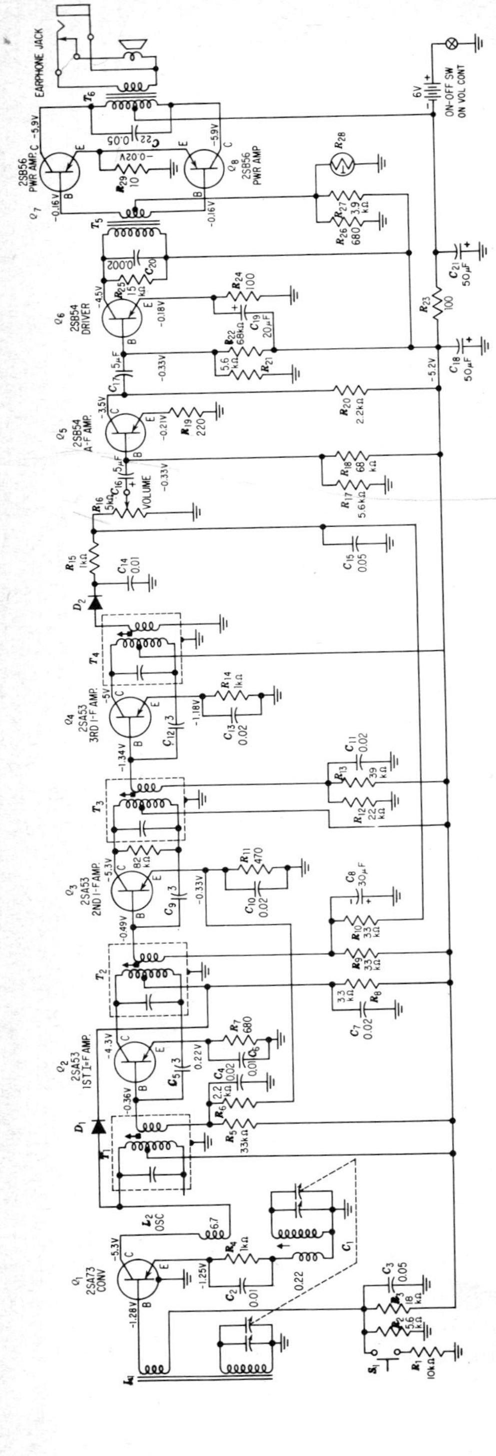

FIGURE 9.8
A second, more extensive transistor receiver. *(Motorola, Inc.)*

the on-off switch, and S_1 is depressed. The 10,000-Ω resistor that S_1 brings into the battery circuit establishes the same current drain as a normally operating receiver will. In this case, the milliammeter should read 9.5 to 11.5 mA. This, then, is a simple circuit arrangement to check current drain quickly.

There are three stages of i-f amplification, each providing more than 25 decibels (dB) gain. Each transistor is connected with the emitter common to input and output circuits. Input windings are untuned to present a low impedance to the base of the following stage, whereas the output windings are tuned to develop a high impedance.

In each instance, only part of the collector winding is utilized as an output load for the transistor to which it is connected. Part of the winding also provides a suitable neutralizing voltage that is fed back to the base (of that transistor) through a 3-pF capacitor. Transformers T_1, T_2, and T_3 are basically the same insofar as primary and secondary windings are concerned. Transformer T_4 has a different set of impedances to match the loading imposed by the diode detector.

Automatic gain control is applied to the base of the second i-f amplifier. A positive voltage is obtained from the diode detector and fed back to the base of the second i-f transistor. Note that this agc voltage is applied in series with a negative dc voltage (from the battery), and it is the latter that establishes the bias on the second i-f transistor base when no signal is being received. When a signal is received, the agc voltage reduces the negative biasing voltage by an amount dependent on signal intensity.

The same i-f stage also has a small stabilizing resistor in the emitter lead. Its value is lower than the corresponding emitter resistors of the other i-f stages, the reason for which was discussed previously. Base bias for the other i-f transistor amplifiers is provided in each instance by a voltage divider.

The detector is a germanium diode with a 5,000-Ω volume control as its dc load. A 0.01-μF capacitor C_{14} acts as an i-f bypass. Capacitor C_{15} is a filter capacitor for the agc voltage.

Three stages of audio amplification follow the second detector. The first two stages employ similar transistors and are similarly biased. In each instance there is a voltage-divider network in the base circuit and an emitter resistor for dc stabilization. In Q_5 the emitter resistor is unbypassed, so that both signal and dc degeneration or stabilization are provided. In Q_6 the emitter resistor is bypassed by C_{19} and C_{18}; hence, only dc degeneration takes place.

The output stage is a common-emitter push-pull class B amplifier. A thermistor is provided in the base-biasing circuit to maintain a steady base current with temperature changes. This is an effective stabilization technique. Some additional dc stabilization is also provided by the 10-Ω resistor in the common-emitter return path. No signal degeneration is introduced by leaving R_{29} unbypassed, because the signal currents of the transistors

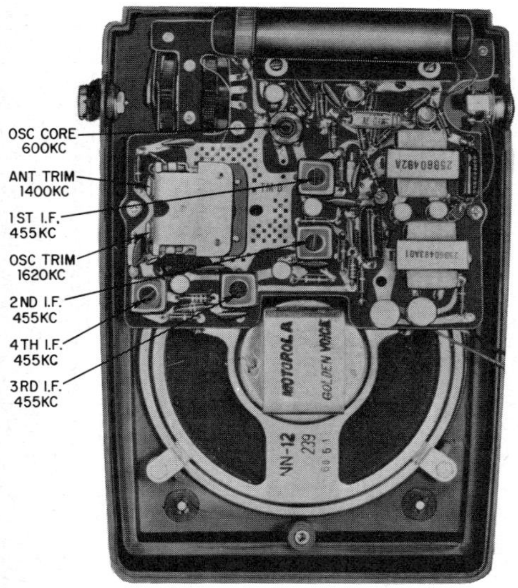

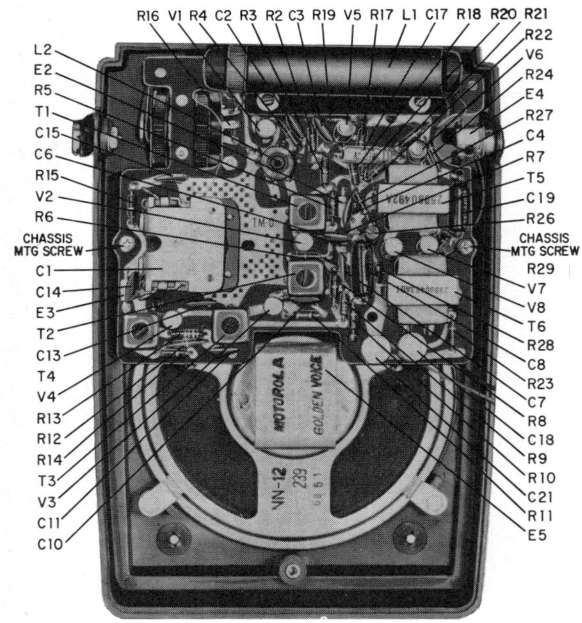

FIGURE 9.9
(*a*) Alignment-point locations for the receiver of Fig. 9.8. (*b*) Parts location in the receiver of Fig. 9.8. (*Motorola, Inc.*)

are 180° out of phase with each other. The a-f response is limited by C_{22}, the 0.05-μF capacitor across the output transformer.

A 15-Ω earphone can be connected to the output for personal reception if desired. When this earphone is in use, no sound is heard from the loudspeaker.

Two internal views of this receiver are shown in Fig. 9.9. The photograph in *a* shows the location of the alignment adjustments; the other identifies the various components.

RECEIVER WITH A TRANSISTOR DETECTOR

The schematic diagram of another transistor portable radio is shown in Fig. 9.10. This circuit contains a separate mixer and oscillator, two i-f stages, a transistor second detector, an audio amplifier, and a class B push-pull output. Direct-current power is supplied by five 1½-V flashlight batteries, and the audio-power output is in excess of 250 milliwatts (mW).

The converter-oscillator combination (with minor modifications) was discussed previously. The two i-f stages employ common-emitter arrangements, and both stages are agc-controlled. The agc voltage is obtained from a class B power detector and applied to each base. Since these are PNP transistors, the base should be negative with respect to the emitter. A negative voltage is supplied to each base from the dc battery line. To vary the gain of each stage, the output voltage from the detector becomes less negative as the incoming signal level increases. This has the effect of reducing the bias between the base and emitter and, in essence, lowering the emitter and collector currents through the transistor. As the current decreases, the gain of the stage drops.

In the first i-f stage, the bottom end of the emitter resistor R_7 connects to the —3-V terminal on the battery. An additional —0.75 V develops across the 1,000-Ω emitter resistor, so that the total emitter voltage with respect to ground is —3.75 V. The voltage of the base is —3.90 V, which is obtained through the agc line and the connection of this line to R_{14}. The latter resistor ties in, at its opposite end, to the battery. The net bias, then, between base and emitter is 0.15 V, with the base more negative than the emitter. When the agc bias is active, upon the arrival of a signal, it will act to reduce the bias difference between base and emitter.

The reason for returning the emitter of Q_3 to a negative tap on the battery is that it permits the gain of this stage to be reduced sufficiently to prevent overload of the second i-f stage or the detector. In the second i-f stage, the emitter resistor is returned to ground; as a result of this, the agc voltage cannot reduce the transistor current to as small a value as it can in the first stage.

The immediate frequency is 455 kHz. C_7 (10 pF) and C_{12} (10 pF) are neutralizing capacitors.

The stage following the second i-f stage is the second detector; a PNP transistor is employed here in a class B power-detector arrangement. This

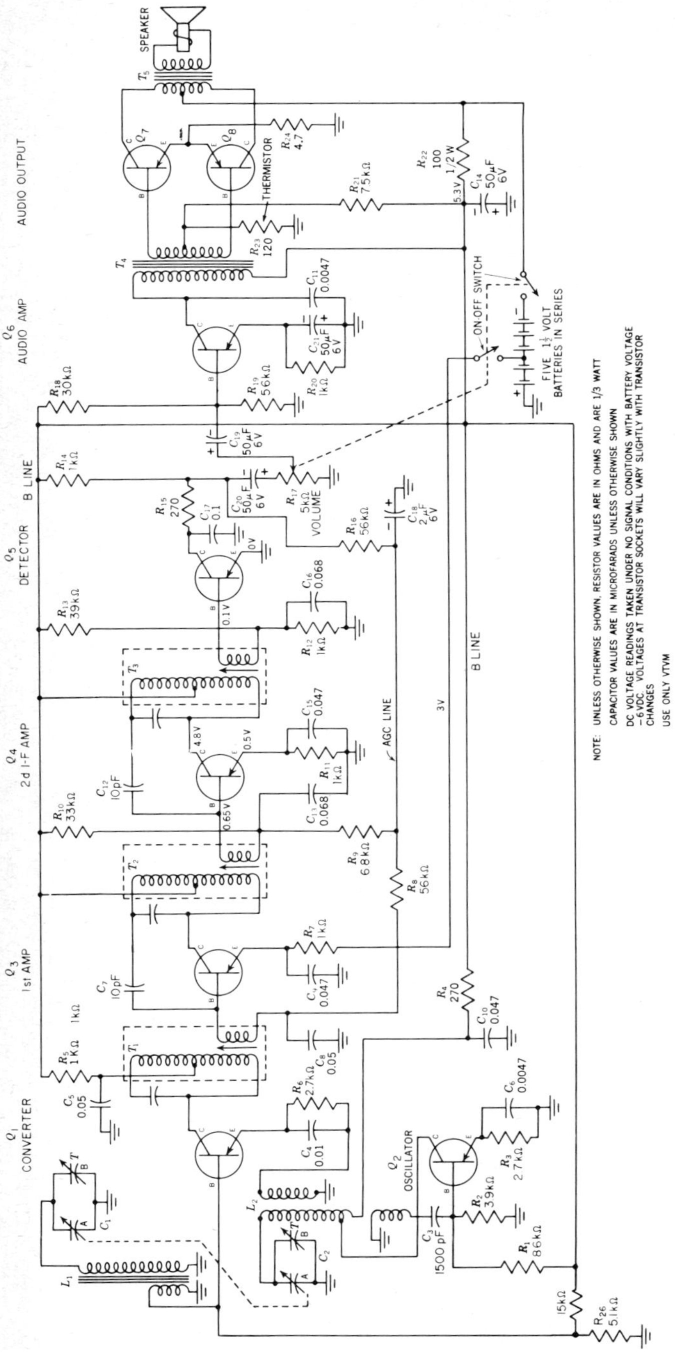

FIGURE 9.10
Schematic diagram of a receiver employing a transistor as a detector.

type of detector was once fairly popular in vacuum-tube circuits, a good deal of the popularity stemming from the fact that it will amplify the signal—an advantage that accounts for its use here. In its present application, 10 dB of gain is obtained. It would be simpler and cheaper to use a germanium diode, as in the two preceding sets, but a diode introduces a loss, and gain is desired here. Also, a transistor detector can provide more agc power than a diode detector.

The operation of a power detector depends on the transistor's being biased very nearly to cutoff. Thus, when the incoming signal is applied to the base of the detector transistor, only the negative half of this input signal will produce appreciable collector-current flow (because this is a PNP transistor and to operate the base must be more negative than the emitter). The positive portion of the signal will produce little or no current flow because the emitter-base circuit will be driven into cutoff. Essentially, we have rectified (i.e., detected) the signal, and if we remove the i-f component, we shall obtain the desired audio intelligence.

In the transistor detector of Fig. 9.10, the emitter is connected directly to ground. The potential of the base is established by the divider network of R_{12} and R_{13}, and this voltage is so low that the stage is close to cutoff. Under no-signal conditions, the collector voltage is very close to the full B voltage. The agc line connects also to the collector of Q_5, and it is through this connection that the base elements of Q_3 and Q_4 receive their operating voltages.

When a signal is received, the collector current of the detector increases; and since this is a PNP transistor, electron flow will be from the battery to the collector, which will produce a voltage drop across R_{14} such that the collector will become *less* negative or *more* positive. This change will be transmitted to the bases of the controlled i-f amplifiers and result in a current decrease through these transistors. In this way, the gain of the two i-f stages is controlled. R_{16}, R_9, R_8, C_{18}, C_{13}, and C_8 serve to filter out any audio components of the agc voltage and to establish the time constant of this network.

The remainder of this receiver circuit is fairly straightforward and will present little difficulty to the reader. The signal from the detector appears across the volume control R_{17} and is taken from there and applied to the base of Q_6, an audio amplifier. R_{18} and R_{19} form a voltage-divider network to provide the desired bias for the stage. A stabilizing resistor is present in the emitter circuit, and it is suitably bypassed by a 50-μF capacitor. The 0.0047-μF capacitor from collector to ground serves to remove any stray i-f voltage that may have reached this point.

The final stage is a class B push-pull output amplifier. The full 7.5 V is applied to the collector elements to obtain the desired power output. A small base-to-emitter bias is used to minimize crossover distortion and to make it easier to substitute other output transistors should replacement become necessary.

REFLEX AMPLIFIERS

In an effort to economize and save space, a number of designers have resorted to the use of reflex amplifiers in portable transistor receivers. The reflex action usually occurs in the final i-f amplifier that functions not only as a normal i-f amplifier but as the first audio amplifier as well. The two functions are carried on simultaneously without interference with each other. This can be done because the i-f and audio frequencies are so widely separated.

A typical reflex circuit is shown in Fig. 9.11. As a normal i-f amplifier, Q_1 receives the i-f signal from the preceding stage through T_1, amplifies the signal, and then transfers it by way of T_2 to the diode second detector that follows. At the detector, the signal is demodulated, with the audio and dc components appearing across the volume control R_1. From this point, the dc voltage, properly filtered, is fed back to the base of the first i-f amplifier for control of the gain of that stage.

To this point, circuit operation as described is normal. However, the audio signal at the volume control is now coupled back to the base of Q_1, the second i-f amplifier, by C_1 and R_2. The audio signal sees the secondary of T_1 as just another length of wire, with negligible impedance. Likewise, after it has passed through Q_1, the audio signal sees T_3 and not T_2 as its load. Transformer T_3 is an a-f transformer, with its primary in the collector circuit of Q_1 and its secondary connected to the bases of the output class B amplifiers.

FIGURE 9.11
A reflex-amplifier circuit.

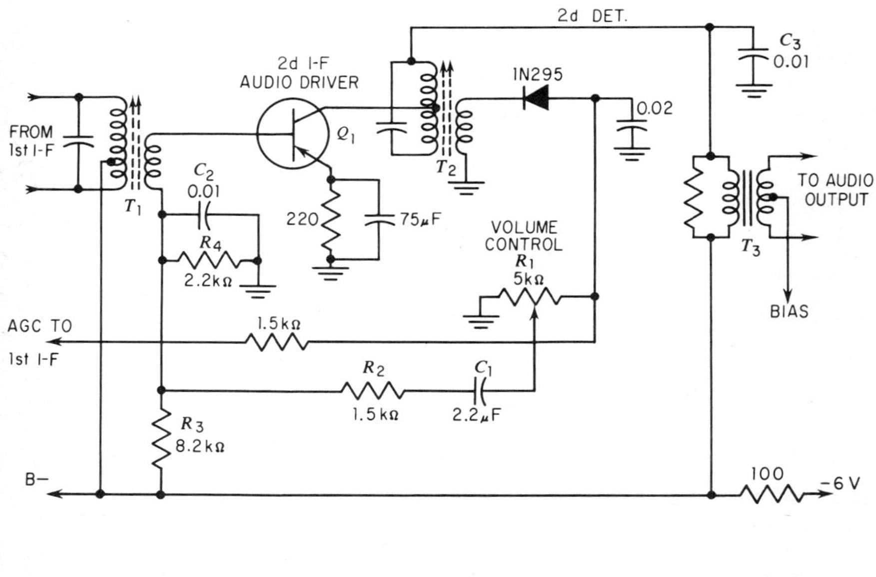

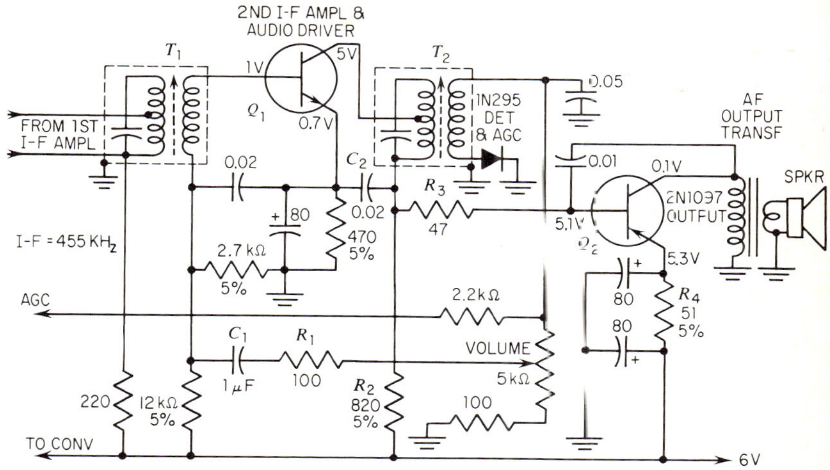

FIGURE 9.12
A second reflex circuit.

Just as the i-f transformers do not impede the audio signal, so must the audio circuits be suitably bypassed for the i-f signals. In the input circuit of Q_1, capacitor C_2 keeps the i-f signal away from the volume-control network. In the output of Q_1, C_3 permits the i-f signal to travel around T_3 on its way back to the emitter. Thus, the two signals are kept in their respective paths while Q_1 amplifies them both (although generally not equally).

Capacitors C_2 and C_3 have some effect on the audio signal, but it is small in comparison with the signal that does not pass through them. Resistors R_3 and R_4 establish the dc base bias for Q_1.

The reflex circuit in Fig. 9.12 is similar to the preceding circuit with several variations. The audio signal is fed from the volume control through R_1 and C_1 to the base of Q_1. After the audio signal has passed through the transistor, it is developed across the 820-Ω resistor R_2, which is connected in series with the primary of T_2. From R_2, the audio signal is transferred directly to the base of the audio-output transistor Q_2 through R_3. Intermediate-frequency signals are kept away from R_2 by capacitor C_2.

In the audio-output stage, the collector dc potential is close to zero because its output transformer winding is grounded at its opposite end. Note, however, that the emitter resistor R_4 is returned to +6 V, so that the potential difference between emitter and collector is 5.3 V. The base element is also provided with a fairly high dc voltage, here 5.1 V, to bring its dc level up to that of the emitter. The actual potential difference between base and emitter is on the order of 0.2 V.

TRANSISTORIZED FM CIRCUITS

One of the difficulties that had to be surmounted in order to produce commercially a transistorized FM receiver was a low-cost high-frequency transistor. The FM r-f range extends from 88 to 108 megahertz (MHz), far beyond the radio-broadcasting frequencies. Even in the i-f section, FM receivers use a center frequency of 10.7 MHz, a value still considerably above the radio-broadcast frequencies. Fortunately, it is now possible to produce high-frequency transistors in quantity, with fairly uniform characteristics and at relatively low cost.

FM broadcast receivers are usually designed to incorporate AM as well, since the additional expense involved is quite small. This is because one set of transistors can handle both signals, and only a few additional i-f transformers and a detector diode constitute the major portion of the added cost. In the discussion to follow, a combination AM/FM receiver will be considered.

Block Diagram. A block diagram of an AM/FM receiver is shown in Fig. 9.13. The complete front end of the set, from the r-f amplifier to the second i-f stage, is utilized in common by both AM and FM signals. It is only necessary to change the tuning circuits in order to switch from AM to FM, or vice versa. As a matter of fact, the i-f tuned transformers remain series-connected without

FIGURE 9.13
Block diagram of a transistorized AM/FM receiver.

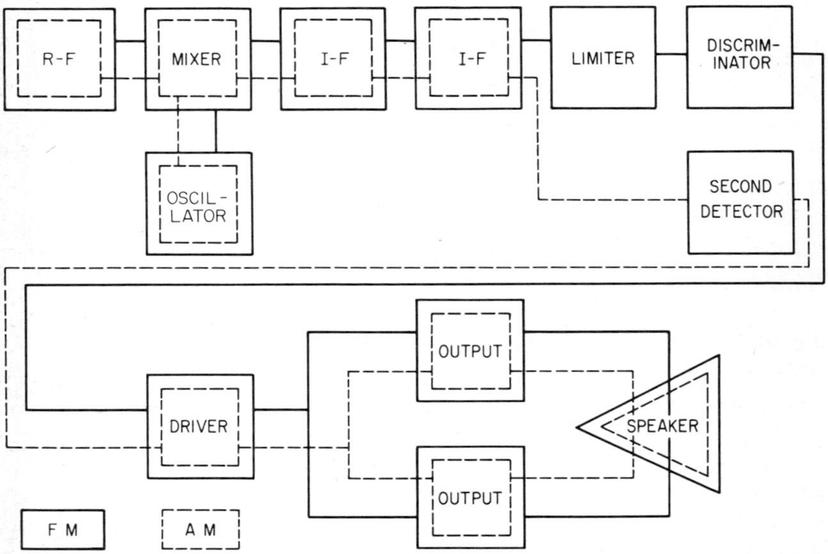

any change. Switching in the front-end section occurs principally in the oscillator, r-f amplifier, and mixer.

Beyond the second i-f stage, the signals separate, with the AM signal going to a diode detector where it is demodulated. At the same time, a dc control voltage for agc is developed and fed back to one or both i-f transistors. Both actions are completely conventional. Thereafter, the audio signal is amplified first by an a-f driver and then by a class B power-output amplifier. At this point, it is sufficiently powerful to drive one or more loudspeakers.

For the FM signal, a limiter stage follows the second i-f amplifier. The signal then goes to a discriminator, where the sound intelligence is abstracted. The signal, now in its audio form, is fed to the driver, where sufficient power is developed to drive the class B output stage adequately.

In the entire receiver, then, there are only three stages that the AM and FM signals do not use in common. This makes for a highly efficient arrangement, not only from the standpoint of design and cost but also from that of servicing.

CIRCUIT DESCRIPTION

The Tuner. The front-end section, the tuner, of this AM/FM receiver consists of an r-f amplifier, a local oscillator, and a mixer (Fig. 9.14). The r-f stage is a common-emitter neutralized tuned amplifier utilizing a h-f transistor. The AM and FM input circuits are switched at the base of Q_1. The FM antenna is a 75-Ω whip type that simply extends straight up in the air. The AM antenna is a 6-in ferrite loop. Both input circuits receive agc voltages from separate agc systems. When Q_1 is AM-operated, reverse agc is employed just as it is in the broadcast receivers previously described. When Q_1 is FM-operated, forward agc is employed. This differs markedly in operation from reverse agc. (It will be described presently.) However, both methods serve the same purpose—to control the signal level reaching the FM and AM detectors with varying input-signal levels.

In the collector circuit of Q_1, L_1 is the tuning coil for FM signals. The coil is resonated by C_6, a small variable capacitor, which is part of a six-gang capacitor that tunes all of the tuning circuits in the front-end section. The other members of this assembly are C_1, C_3, C_8, C_9, and C_{18}. Each of these capacitors is shunted by small trimmer capacitors that permit tracking of each set of capacitors over its respective band, AM or FM. A small tap at the bottom of L_1 provides a neutralizing voltage that is fed back to the base of Q_1 by capacitor C_N. AM operation, occurring at much lower frequencies, requires no neutralization, and hence none is provided.

During FM reception, coils L_2 and L_3 are inactivated by the short circuit placed across L_3 by the AM/FM selection switch. Resistor R_1 is the collector-

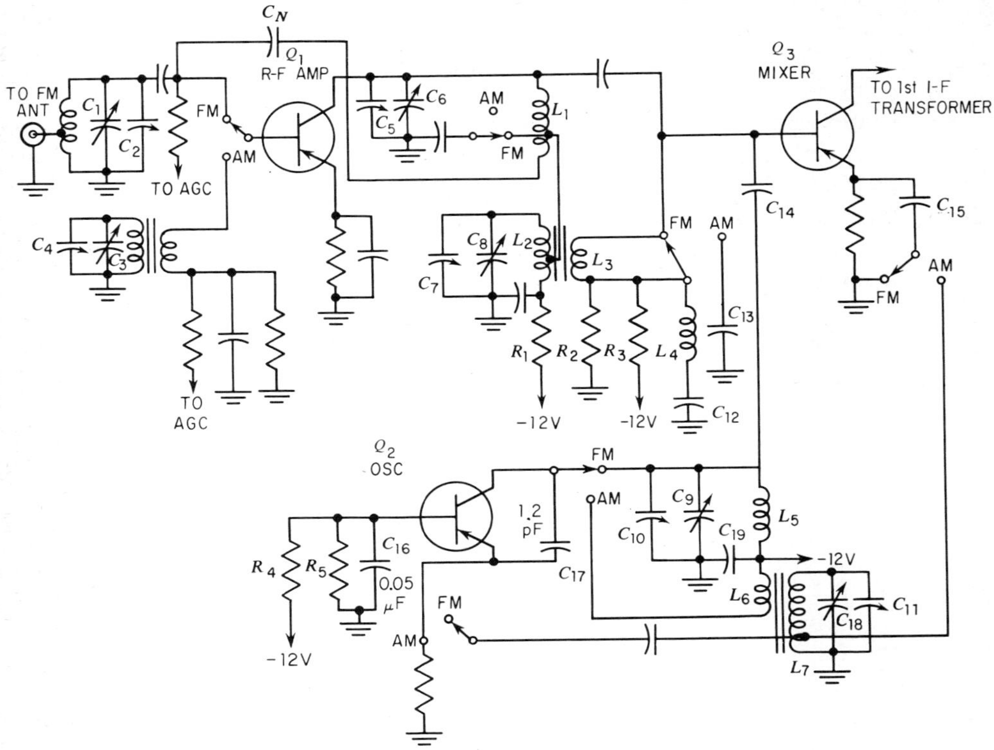

FIGURE 9.14
The front-end section of the AM/FM receiver.

dropping resistor for Q_1, whereas R_2, and R_3 form a voltage-divider network for the base of the mixer Q_3. L_4 and C_{12} form a 10.7-MHz trap to prevent any 10.7-MHz voltage from developing in the mixer-base circuit and leading to possible oscillation in the mixer. In essence, these two components serve to neutralize the stage. The i-f trap also prevents spurious signals from the r-f stage at the i-f frequency from penetrating further into the receiver. When AM signals are received, the trap is shunted by C_{13}, a relatively high-value capacitor that effectively shunts the network out of the circuit.

The oscillator transistor Q_2 appears at first glance to be connected in the common-emitter configuration. A closer look, however, reveals that the base is placed at ground potential by capacitor C_{16}. Hence, this is a grounded-base oscillator. Capacitor C_{17} is externally added to supplement the internal emitter-collector capacitance of Q_2 to encourage oscillations. L_5, C_9, and C_{10} form the tuning circuit for FM, developing the required oscillator signal that is injected at the base of Q_3 by capacitor C_{14}. The AM tuning network, composed of L_6, L_7, C_{18}, and C_{11}, is effectively kept out of the active circuit by C_{19}. R_4 and R_5 provide the proper biasing voltage for the base of Q_2.

On AM, L_5, C_9, and C_{10} are switched out of the oscillator circuit, whereas L_6, L_7, C_{18}, and C_{11} are brought in. Injection of the oscillator signal at the mixer is now made at the emitter of Q_3 with C_{15}. (During FM operation, C_{15} serves as an emitter bypass capacitor.)

The mixer is completely conventional in design, employing Q_3 as in the common-emitter configuration. In the collector circuit of Q_3, the FM and AM transformers are series-connected (Fig. 9.15). This arrangement is possible because the impedance of the FM transformer windings is low at AM frequencies whereas the capacitor shunting the AM transformer winding provides a low-impedance path for FM signals when these are active.

I-f Amplifiers and Limiter. The i-f section consists of three stages of FM i-f amplification and two stages of AM (Fig. 9.15). All stages employ the common-emitter arrangement for both FM and AM signals. If we examine these stages from the standpoint of the AM signals, we find that interstage coupling is achieved by T_1, T_2, and T_3. An agc voltage is brought in at the base of Q_4, where it combines with the normal dc bias. This agc voltage is a reverse bias in that it tends to reduce the current flowing through Q_4 when the signal is too strong and increase the collector current when the signal is too weak. An amplified agc voltage is also obtained in the emitter circuit of Q_4, and it is used to control the gain of the r-f stage. Since the emitter voltage follows the base voltage, the r-f amplifier is also subjected to a reverse agc action.

The second i-f stage operates without any agc voltage. The AM signal is further amplified here and then transformer-coupled to a diode detector. Small 220-Ω resistors are inserted in the collector circuits of each i-f transistor because it has been found that under strong-signal conditions, the collector-base diode becomes forward-biased, allowing feedback of the proper amplitude and phase to produce oscillation on the signal peaks. These series collector resistors limit the signal fed back and maintain good stability.

For FM operation, transformers T_4, T_5, T_6, and T_7 couple the signal from stage to stage. The first two stages are neutralized, but the third stage, a limiter, is not. The neutralization capacitors on FM add to the internal capacity of the transistors on AM. To prevent oscillation on AM, a certain amount of mismatching is purposely incorporated into the AM transformers. This will reduce the AM gain, but so much is still available that no particular difficulty ensues.

Even though the third i-f stage is labeled as the limiter, in practice, the first and second i-f stages also provide some limiting. The major portion of the limiting, however, occurs in the final stage. Additional limiting is required because when the third stage is strongly overdriven, it becomes inoperable.

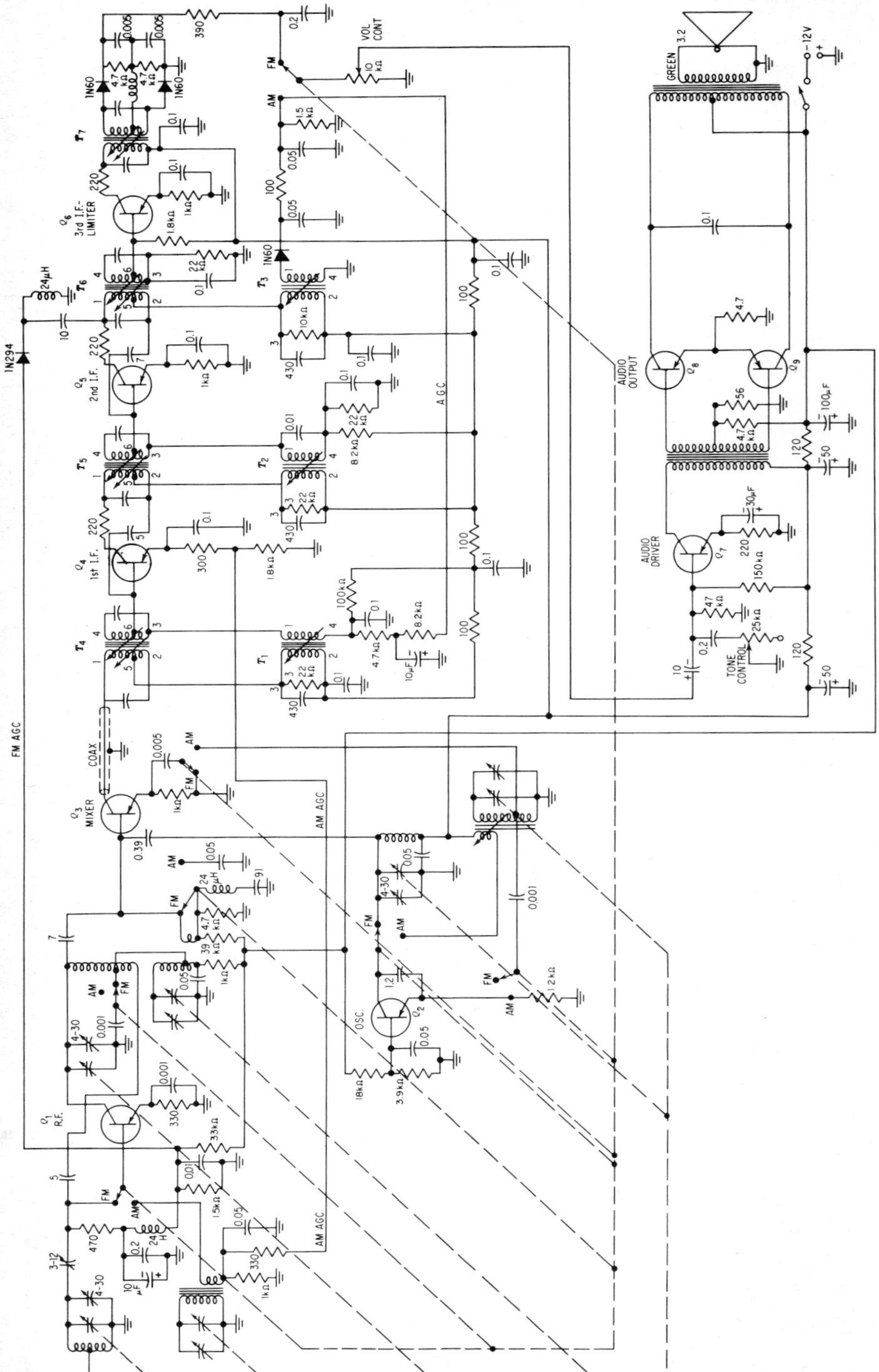

FIGURE 9.15
Schematic diagram of entire AM/FM transistor receiver.

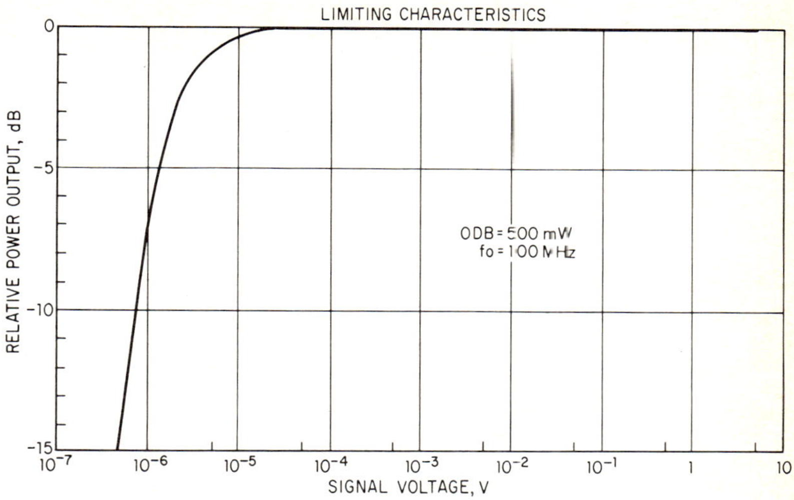

FIGURE 9.16
The limiting characteristics of the i-f system.

Figure 9.16 shows the limiting characteristics of this circuit. Notice that full limiting occurs with an input signal of about 13 μV.

Detector and Audio Circuit. The AM detector is a conventional circuit providing a demodulated audio signal for the audio amplifiers and a reverse agc voltage for the first i-f amplifier (Fig. 9.17). The 10,000-Ω volume control was selected primarily because of the low input impedance of the following audio-amplifier stage.

The FM detector is a Foster-Seeley discriminator utilizing two diodes. The FM-detector output is applied to the same 10,000-Ω volume control; the latter is switched to the system in operation. (For additional information concerning the operation of this and other FM detectors, see Milton S. Kiver, "F-M Simplified," 3d ed., Van Nostrand Reinhold Company, New York, N.Y., 1960.)

The audio circuit consists of a PNP driver, transformer-coupled to a pair of PNP power transistors operated in class B push-pull. The output stage is transformer-coupled to a 3.2-Ω speaker providing a maximum undistorted output in excess of 500 mW.

FM agc Circuit. The FM agc circuit obtains its drive from the second i-f amplifier through a 10-pF capacitor (Fig. 9.15). The 10.7-MHz signal is rectified and filtered to provide a negative voltage that is then applied to the base of the r-f amplifier. When the signal level increases, the negative agc voltage increases. Since the r-f transistor is a PNP type, a greater negative

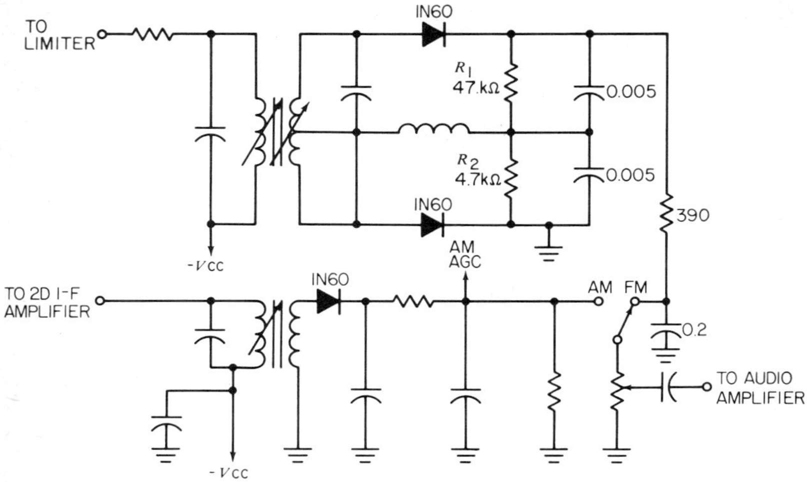

FIGURE 9.17
FM discriminator and AM detector circuit.

base voltage means more collector current. Thus, what we have here is a forward agc voltage in distinction to the reverse agc voltage provided during AM reception.

Forward agc operates not by current variations directly affecting transistor gain but by changing the collector-to-emitter voltage to alter gain. A resistor is inserted in the collector lead. When the transistor current is increased, more voltage is dropped across this transistor. Less voltage is left between the collector and emitter, as a result of which, the gain drops. The stage gain is increased by reducing the collector current, which, in turn, reduces the voltage drop across the resistor and thus permits the dc collector voltage to rise.

As a broad distinction, then, forward agc achieves its control through variation of the collector current. To employ forward agc effectively, the transistor must be internally fabricated to provide a wide, linear reduction in gain at reasonable collector-to-emitter voltages. Without such a characteristic, useful control by this method is not feasible.

Any of the preceding receivers could be readily adapted for use in automobiles. The power would be furnished directly from the 6 or 12-V batteries employed in automobiles, and the physical structure of the sets themselves would be sturdier because of the jolts and other stresses and strains that electronic equipment is necessarily subjected to in a moving vehicle. Either AM or combination AM/FM receivers can and are used, generally coupled with a number of push buttons to make tuning easier for an otherwise occupied driver.

FIELD-EFFECT TRANSISTORS IN RADIO RECEIVERS

In the front-end or r-f stages of an FM receiver, increasing use is being made of field-effect transistors because their overall performance is much better than bipolar transistors. As FM r-f amplifiers and mixers, FETs (and MOSFETs) provide very low noise figures, excellent large-signal handling capability, and circuit simplicity (fewer additional components). For power gain, FETs are comparable to bipolar transistors.

Noise figure relates to the amount of noise voltage (or current) that an electronic device itself generates. It is an important consideration in stages (such as the receiver front end) where the signal level is exceedingly low. If the noise level is of the same magnitude as the signal, then obviously the noise will be subsequently amplified with the signal, thus producing at the output as much noise as signal. Hence, the system requires at its input a signal strong enough to override the noise. Therefore, if we can lower the noise voltages or currents generated up front, then weaker signals that will produce a useful output can be received. Hence, the emphasis on low-noise figure.

An FM front-end section using an FET r-f amplifier, an FET mixer, and a bipolar oscillator is shown in Fig. 9.18. The r-f amplifier is an N-channel silicon FET, with a single-tuned input and a single-tuned output. An agc voltage is applied to the gate of the FET, although here it is somewhat delayed until fairly strong signals are received. The bandwidth of the r-f stage is sufficiently wide so that almost no detuning of the front end is observed.

The mixer is another N-channel FET. The mixer is biased at a low drain current for best conversion gain. Input signal is fed into the gate while 440 millivolts (mV) of oscillator voltage is injected into the source element of the FET. Injecting oscillator voltage into a terminal different from the signal results in almost no tuning interaction between the r-f and the oscillator circuits.

The oscillator is a low-cost silicon bipolar transistor in a conventional common-base Colpitts arrangement. A three-section tuning capacitor provides for frequency tuning of the r-f amplifier, mixer, and oscillator.

Dual-gate MOSFET. A recent development holds considerable promise of making field-effect transistors even more useful in communications receivers (such as radio, television, etc.). This development is the dual-gate MOSFET in which two gates are employed instead of just one. The structure of such a unit is shown in Fig. 9.19. The transistor includes three diffused regions connected by two channels, each of which is controlled by its own independent gate.

Source 1 and drain 2 are equivalent to the source and drain of a single-gate MOSFET. The center diffused region, here labeled drain 1 and source 2, has no external connection. In essence, the MOSFET shown in Fig. 9.19 can be regarded as two units connected in series.

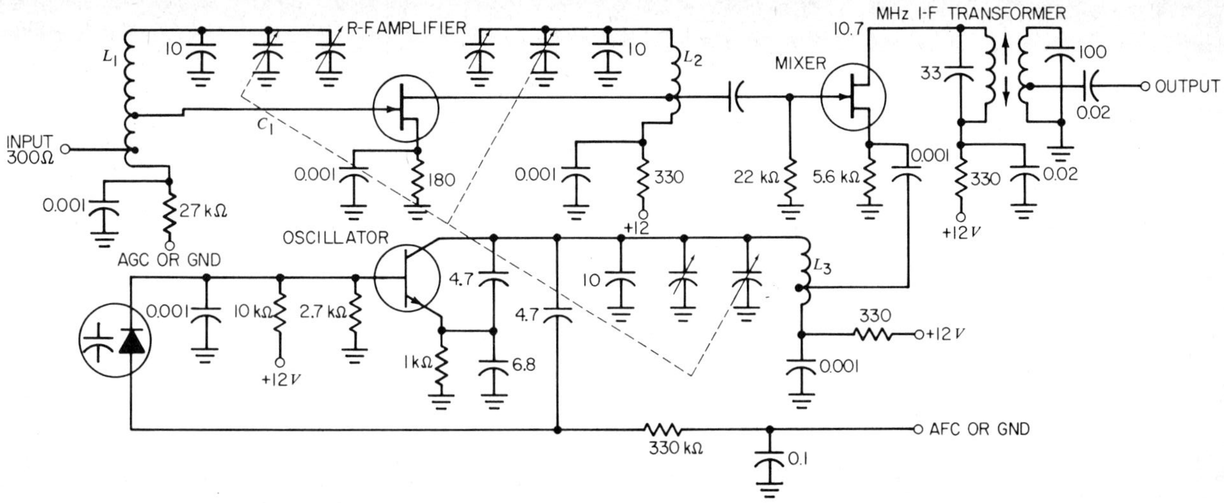

FIGURE 9.18
The front-end section of an FM receiver using an FET r-f amplifier, an FET mixer, and a bipolar transistor oscillator.

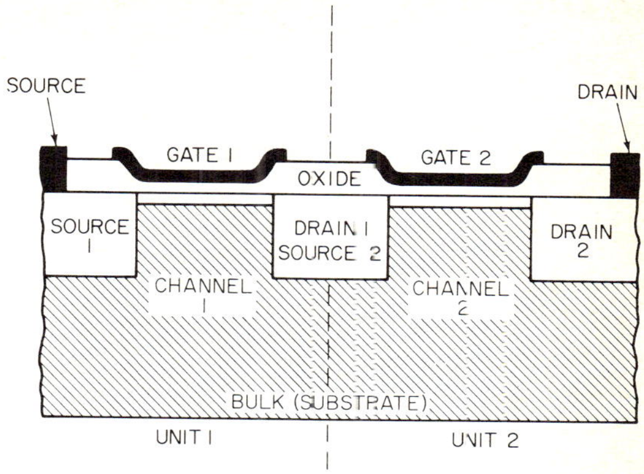

FIGURE 9.19
Internal construction of a dual-gate MOSFET.

A dual-gate MOSFET can be cut off if either gate is made sufficiently negative with respect to the source (i.e., source 1). This, of course, is another way of saying that both gates exert a strong influence on the operation of the current flowing through the device. Dual-gate MOSFETs have been used as automatic-gain-controlled amplifiers, frequency converters or mixers, detectors, and color demodulators. They exhibit the features of the single-gate MOSFET, i.e., high input impedance, low noise, and the ability to handle a wide variation in signal strength. In addition, they display a better ability to handle agc control of an amplifier and they do possess a significantly lower feedback capacitance.

As an r-f amplifier, the dual gate is used as a grounded-source amplifier driving an r-f grounded-gate amplifier (see Fig. 9.20). Since the second gate is at r-f ground potential (but not at dc ground), it forms a shield between the output drain and the input signal gate, resulting in a low feedback capacitance of 0.01 to 0.02 pF. This frequently obviates the need for special neutralizing circuits and aids the stability of the amplifier.

Gate 2 can be used for the application of an agc voltage. As this voltage varies, it alters the current flow through the MOSFET, in essence altering its internal impedance. It can be regarded as placing a variable resistance (representing unit 2) in series with an amplifier (unit 1). As the resistance of unit 2 increases, less voltage reaches the drain of unit 1, thus lowering its gain.

The dual-gate MOSFET can be effectively used as a mixer with the local oscillator connected to gate 2 and the incoming signal applied to gate 1. This provides good conversion gain, low spurious signal response, and low oscillator signal feedback from gate 2 to gate 1 and from here to the antenna where it could be radiated to other nearby receivers and produce interference.

The spurious signals referred to above are the signals generated when

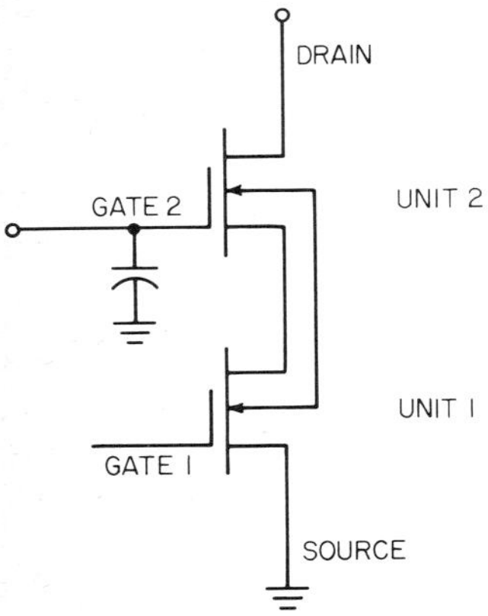

FIGURE 9.20

harmonics of the local oscillator signal beats with harmonics of the incoming signal to form a signal at intermediate frequencies.

The circuit of Fig. 9.21 shows how dual-gate MOSFETs would be employed as an r-f amplifier and mixer in the front end of an FM receiver. The oscillator uses a bipolar transistor.

ICs IN RADIO RECEIVERS

The fact that imported small radio receivers, utilizing discrete components, can be purchased for approximately $5, has exerted pressure on IC manufacturers to produce economically competitive receiver circuits.

One result of this has been the development of ICs that have great application flexibility and that can be mass-produced at low costs. The operational amplifier, which we will describe, is an example of such a multipurpose integrated circuit. The operational amplifier can also perform many routine functions.

The use of ICs in radio receivers offers the following advantages.

1. Improved cost/performance ratio
2. Improved reliability
3. Reduced space and weight
4. Good thermal tracking, which is a result of the close spacing of components on the same chip.

All the functions in AM and FM receivers can essentially be fabricated in integrated (monolithic) circuit form. However, before we discuss some specific receiver IC applications, let us become familiar with the operational amplifier.

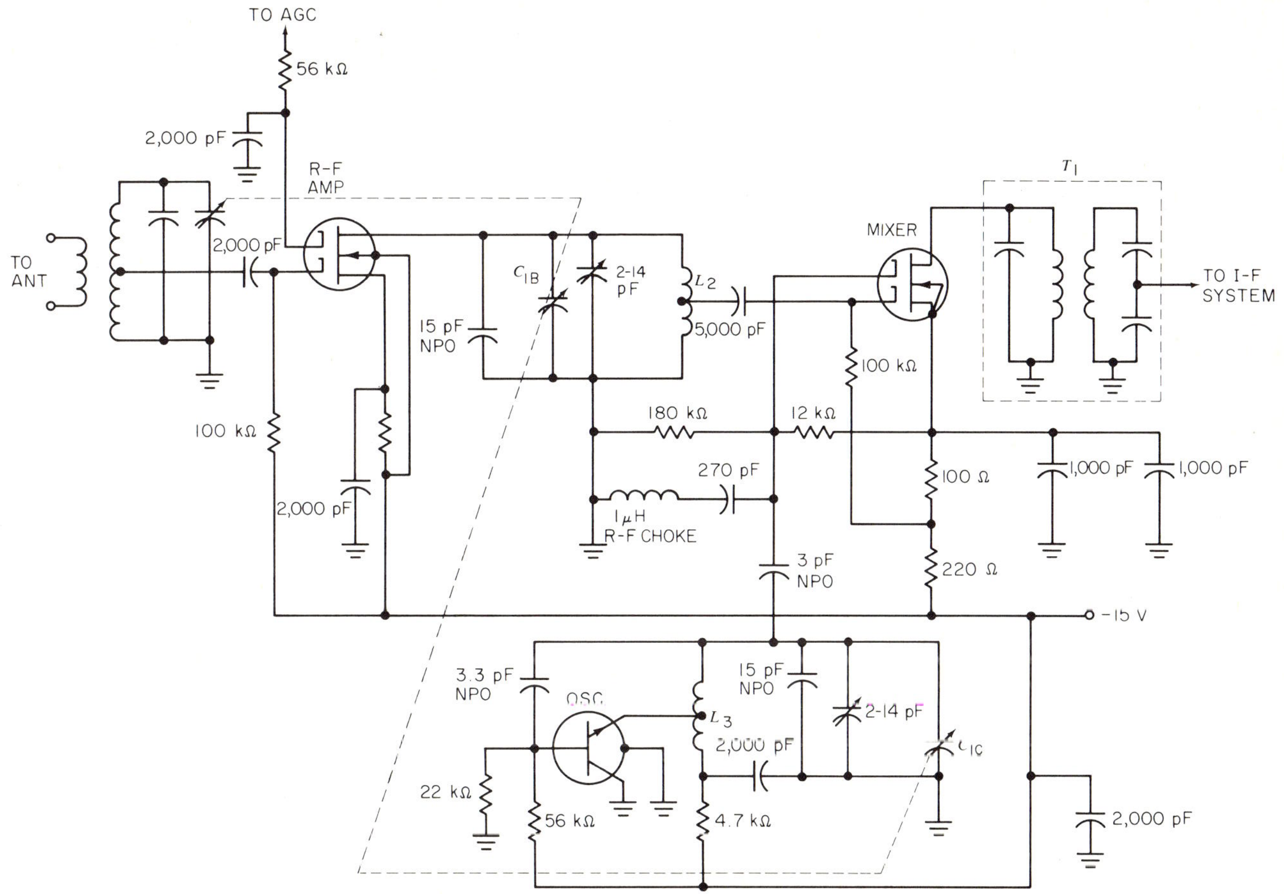

FIGURE 9.21
Schematic diagram of an FM tuner using dual-gate MOSFETs in r-f amplifier and mixer.

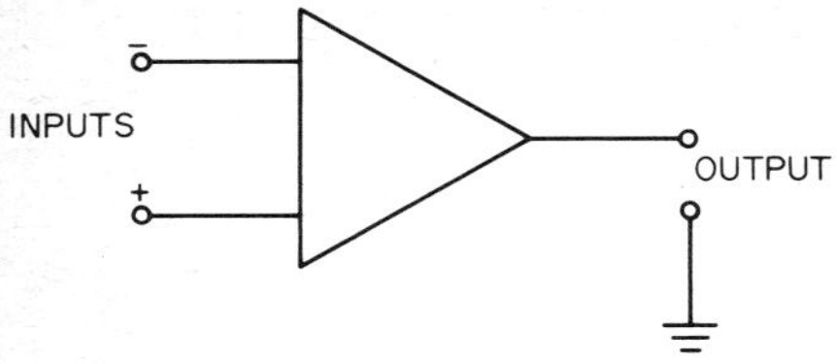

FIGURE 9.22
The symbol employed for an operational amplifier.

OPERATIONAL AMPLIFIERS

The operational amplifier has become one of the most versatile and widely employed devices in electronics, particularly so when IC linear circuits are required—in receivers, audio amplifiers, communications equipment, servo systems, and analog computers, among others.

Assisting this growing usage is the ability of semiconductor device manufacturers to produce operational amplifiers (commonly called "op amps") at very low prices. As a matter of fact, IC op amps are now so inexpensive that they can be used in hundreds of applications whereas before in discrete form they would not have been considered on the basis of cost alone.

The prime features that make op amps so desirable include a circuit performance that is very predictable, a high input impedance, a wide range of obtainable gain values, and the ability to amplify dc as well as ac signals. All this is brought about by the use of a substantial amount of negative feedback. As a matter of fact, the feedback is so large that it essentially determines the characteristics of the amplifier. By alteration of the feedback circuit, the characteristics of the op amp can be suitably varied to provide whatever operating features are desired.

The symbol employed for an op amp is shown in Fig. 9.22.[1] There are two outputs, one marked minus (−) and the other plus (+). Any signal applied to the minus terminal will appear at the output with its phase shifted 180°. That is, a positive signal at the minus terminal will appear negative at the output, and vice versa. On the other hand, any signal applied to the positive terminal will not have its phase altered at the output.

The gain of an op amp, without any feedback, can range from *several thousand* to *several million*. However, when negative feedback is added, as shown in Fig. 9.23, the overall gain decreases. The input impedance will depend on how the input terminals are connected. For example, if the positive terminal is grounded, as shown in Fig. 9.23, and the signal is applied to the minus or inverting terminal, then the input impedance is approximately equal to R_1. Amplification of this arrangement is equal to R_F/R_1.

[1] Actually, the triangular symbol represents all amplifiers, whether or not of the operational variety. However, by showing a positive and a negative input terminal, the notation designates an operational amplifier.

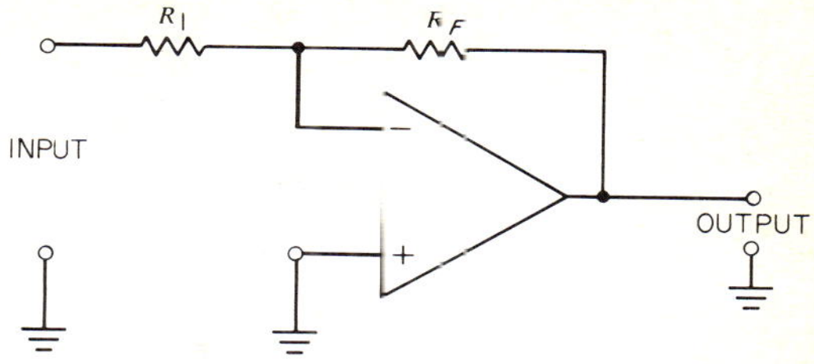

FIGURE 9.23
An operational amplifier showing the feedback path.

If the input circuit is modified to the form shown in Fig. 9.24*a*, then no phase inversion occurs between input and output signals. The input impedance of this arrangement is extremely high, often in the megohm range. Overall amplification is equal to

$$1 + \frac{R_F}{R_1}$$

which makes it approximately the same as the previous setup. This circuit is very useful as a buffer because it does not invert the signal.

A variation of Fig. 9.24*a* is the circuit of Fig. 9.24*b*. Here, resistor R_1 is removed, making it infinite in value, with an even higher input impedance. The gain of the stage, under these conditions, is unity. Output impedance is low.

Special terminals are provided on all operational amplifiers to facilitate adding the feedback networks as well as making any other adjustments that

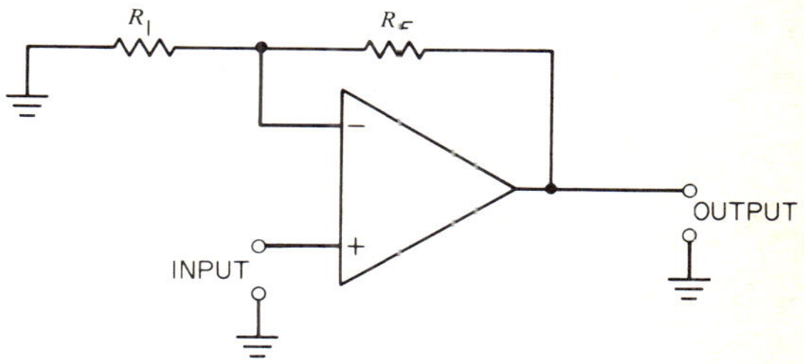

FIGURE 9.24a
To obtain no phase inversion between input and output of the op amp, the input is applied as shown above.

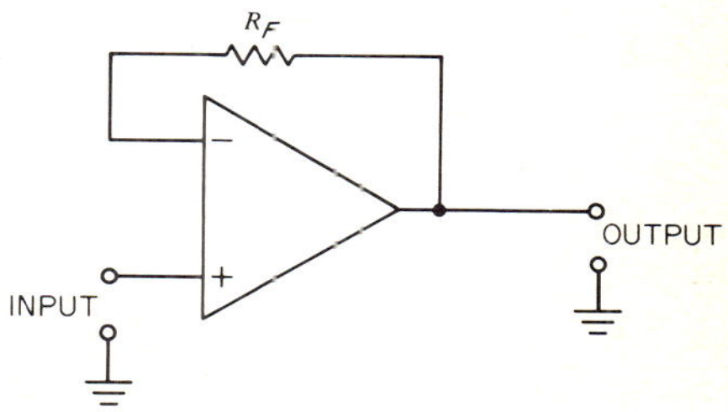

FIGURE 9.24b
An arrangement that provides an exceedingly high input impedance, with unity gain and low output impedance.

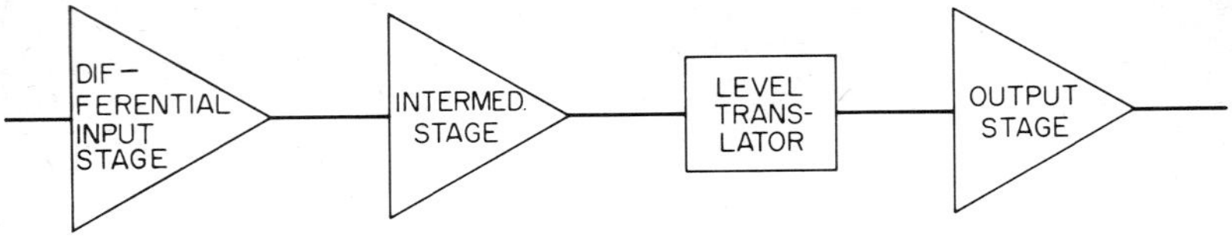

FIGURE 9.25
Block diagram of an IC operational amplifier.

may be required. Several of these will be discussed after we see what an op amp consists of internally.

It is undoubtedly apparent from the preceding discussion that an op amp, internally, is not a simple *one-stage circuit.* A block diagram[1] of a typical operational amplifier is shown in Fig. 9.25. It is seen to consist of four distinct sections. The input section employs a differential amplifier providing a significant amount of gain. In the present instance (Fig. 9.26), a Darlington amplifier connected for differential operation is employed. This arrangement is formed by Q_1, Q_2, Q_3, and Q_4. Transistor Q_6 is the current source for Q_2 and Q_3; Q_1 and Q_4 each employ separate current sources (Q_5 and Q_7). The result of this circuit arrangement is to produce a stabilized Darlington amplifier that provides high gain and rejects common-mode signals.

Pins 3 and 4 are external terminals available for any frequency compensation that may be desirable to introduce into the op amp. If no such compensation is required, the pins are left unconnected.

The second section of the op amp (Q_8 and Q_9) serves primarily as a buffer between the input and output stages, although some small amount of gain is provided by this differential amplifier. Note that a special constant-current network is not employed in the emitter circuit of this stage because common-mode rejection is achieved primarily in the first stage and little need be done here to enhance it.

The "level translator" stage is inserted between the intermediate and output stages because the quiescent (or zero signal) level is above or below ground. It is necessary to move (i.e., translate) this level so that with no signal, the dc output voltage will also be zero.

Transistor Q_{12} is the driver for the complementary symmetry output stage. This stage is biased (by D_1 and D_2) to class AB.

A split power supply, that is one having equal positive and negative voltages with respect to ground, is employed here. This is a common practice with operational amplifiers because it not only makes the system easier to balance, but also tends to keep direct current out of the load. Any slight mismatch of components in those places where balance is necessary (as in the differential amplifier circuits) will cause a small dc voltage to appear at the output even if no signal is being applied to the input. By applying a small dc voltage

[1] Brent Welling, Monolithic Operational Amplifiers, *Electronics World,* July, 1968.

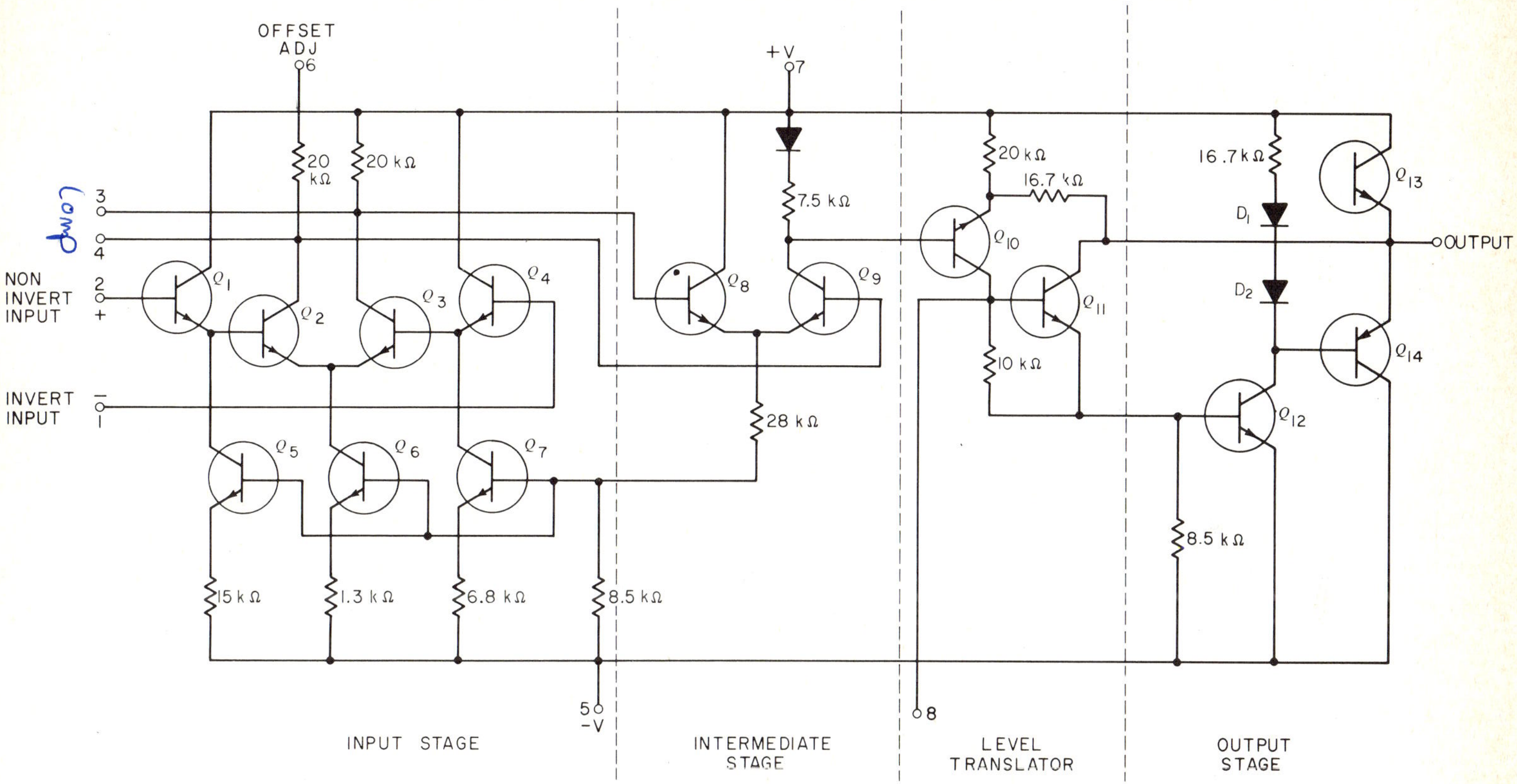

FIGURE 9.26
Schematic diagram of a high-gain IC op amp. ***(Courtesy Electronics World.)***

of proper polarity to terminal 6 (labeled "Offset ADJ"), we can overcome this imbalance.

Operational amplifiers are available from many manufacturers. With the continuing drop in costs that is being achieved with integrated circuits, it can reasonably be expected that usage of op amps will keep increasing.

INTERNAL VOLTAGE REGULATORS FOR INTEGRATED CIRCUITS

Since they contain active elements in the form of diodes and transistors, integrated circuit networks require suitable biasing arrangements so that these devices will operate at the desired operating points.

A simple resistive voltage-divider network could readily provide the necessary bias voltages, but an IC resistor—or worse still, a resistor network—would occupy so much surface of the semiconductor chip that it would not be a feasible solution. Furthermore, it is difficult to produce resistor values closer than 20 percent and this would, of course, affect the voltage drops developed across the resistor accordingly.

A more effective solution is to use the voltage drop developed across a forward-biased diode to provide an operating bias voltage for a transistor. A silicon diode, when forward-biased, will have a voltage drop of 0.7 V. If we

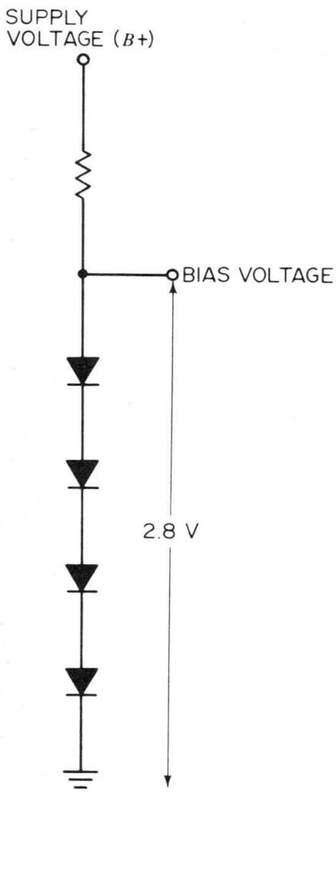

FIGURE 9.27
A diode string that can provide 0.7, 1.4, 2.1, and 2.8 V for use in biasing transistors.

connect one or more such diodes in series and forward-bias the entire string, then each diode will provide 0.7 V. Thus, if four diodes are strung in series, a total bias of 2.8 V can be obtained. See Fig. 9.27.

The advantage of using diodes is that they are readily fabricated, require very little "land," or area, on the surface of an IC chip, and are as sensitive to temperature changes as a transistor fabricated on the same chip.

For example, as the temperature of the chip increases, the diode bias voltage decreases. The same temperature increase should cause the collector current of a transistor to increase. However, if the bias applied to that transistor decreases (as it will, if supplied by a diode), then the collector current rise will not occur.

Another advantage offered by the arrangement shown in Fig. 9.27 is that changes in supply voltage have a much smaller effect on bias voltage than a purely resistive network. To illustrate, if the supply voltage rises, more voltage will be applied to the diodes; this will increase the current flowing through the diodes, increase the voltage drop across the series resistor, and in essence change the resultant bias voltage very little. Using this approach, supply-voltage stabilization will reduce current variations in the controlled circuits to about one-fifth that obtained when resistor or thermistor bias is used for a germanium transistor and one-fifteenth that obtained for a silicon transistor.

The diode string may be composed solely of diodes, or the base-emitter diode of transistors may be used. In the latter case, the collector is connected to the base, effectively shorting these two elements together. See Fig. 9.28.

The output of a diode voltage regulator need not be restricted to multiples of 0.7. By using the arrangement shown in Fig. 9.29, one can obtain any value between 0.7 V and the supply value. In essence, what happens here is that the ratio of R_1 and R_2 determines the bias applied to Q_1 and this governs the current through the transistor and the load resistor R_3. More current produces less voltage at point A, whereas less current produces a higher voltage at point A.

Also, if a momentary increase in supply voltage occurs, more positive voltage appears at point A. This causes an increase in current through R_1 and R_2 and produces an increase in current through Q_1, which acts to lower the voltage at point A. Thus, the momentary rise in supply voltage is counteracted. The same reaction occurs with a decrease in supply voltage.

MONOLITHIC i-f STRIP FOR AM/agc

New methods of i-f strip partitioning have been made possible by the monolithic integrated circuit. Previously, the best way to make use of available power gain was by matched, tuned interstage networks, such as those we have discussed. Owing to the limited agc range available by varying dc emitter current,

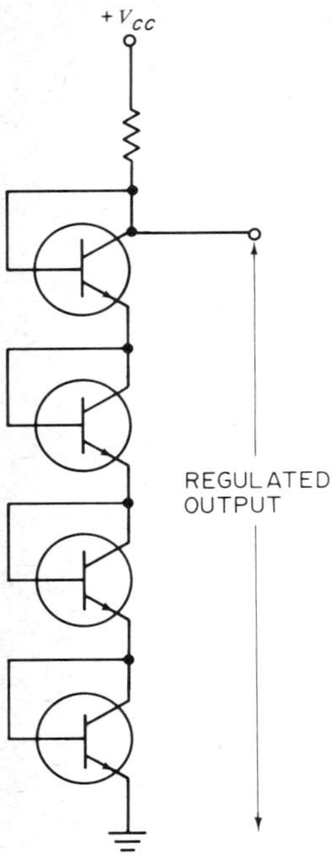

FIGURE 9.28
Four transistors, connected as diodes, provide 2.8 V of regulated output voltage.

in a conventional common emitter i-f stage, several stages receive agc voltage from the detector at once. This frequently results in cumbersome dc biasing in order to obtain the correct agc characteristic. It also makes input and output transformers necessary for each stage to decouple the dc operating point shifts due to agc operation.

Conventional detector schemes are kept simple, primarily for economic reasons, and they usually consist of a simple diode biased from a tuned trans-

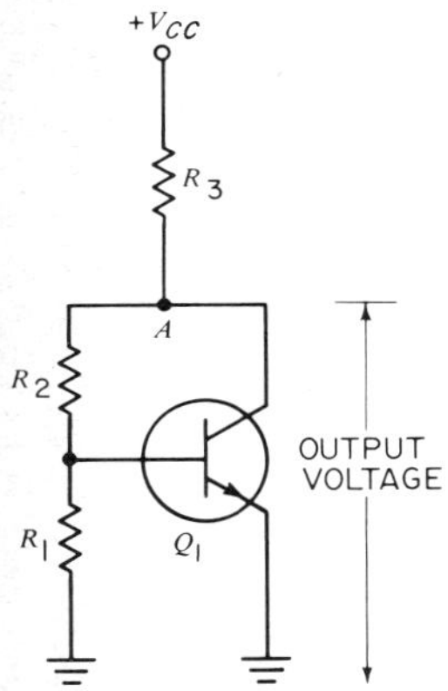

FIGURE 9.29
A transistor voltage regulator.

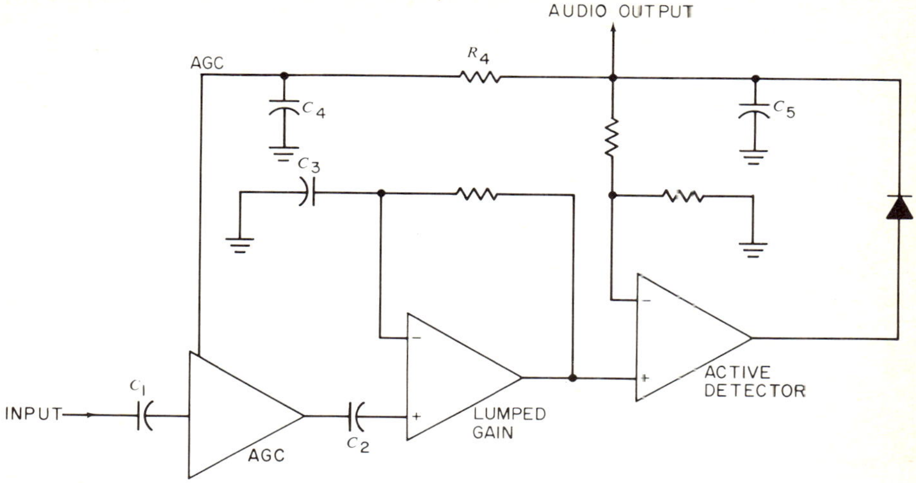

FIGURE 9.30
Block diagram of a monolithic i-f strip for AM receivers.

former secondary. The agc voltage normally is taken directly from the detector. In addition, because of large tuned gain and the instability in conventional common emitter stages, power-supply decoupling for each stage is required.

Now, however, as a result of the availability of almost unlimited monolithic complexity, and internal biasing, these requirements may be substantially altered and a more efficient method employed. For example, all the power gain could be provided by a single stage, whereby the design would be maximized toward this end without concern for gain variation (i.e., agc) or bandpass limiting. These other two functions could be obtained from other stages, each optimally designed for its purpose.

An AM i-f system that typifies this approach is shown in block-diagram form in Fig. 9.30. The unit is the National Semiconductor's AM i-f monolithic strip LM172. The first block, labeled agc, is a variable attenuator stage positioned between the input bandpass filter (not represented in Fig. 9.30) and the lumped-gain stage. This is followed by the section that is essentially responsible for all the gain of the strip. Finally, there is the AM detector where not only is the audio signal recovered, but a suitable control voltage for the agc stage is developed.

Figure 9.31 is a schematic for National Semiconductor's AM i-f monolithic strip LM172. The capacitors are all external to the 8-pin TO-5 package, and they provide the minimal amount of decoupling and time constants necessary to operate a complex high-gain bandwidth monolithic circuit such as the LM172.

Figure 9.32 is the agc section of the LM172, in which an emitter-coupled transistor pair is used as a series-shunt variable attenuator. The base of Q_2

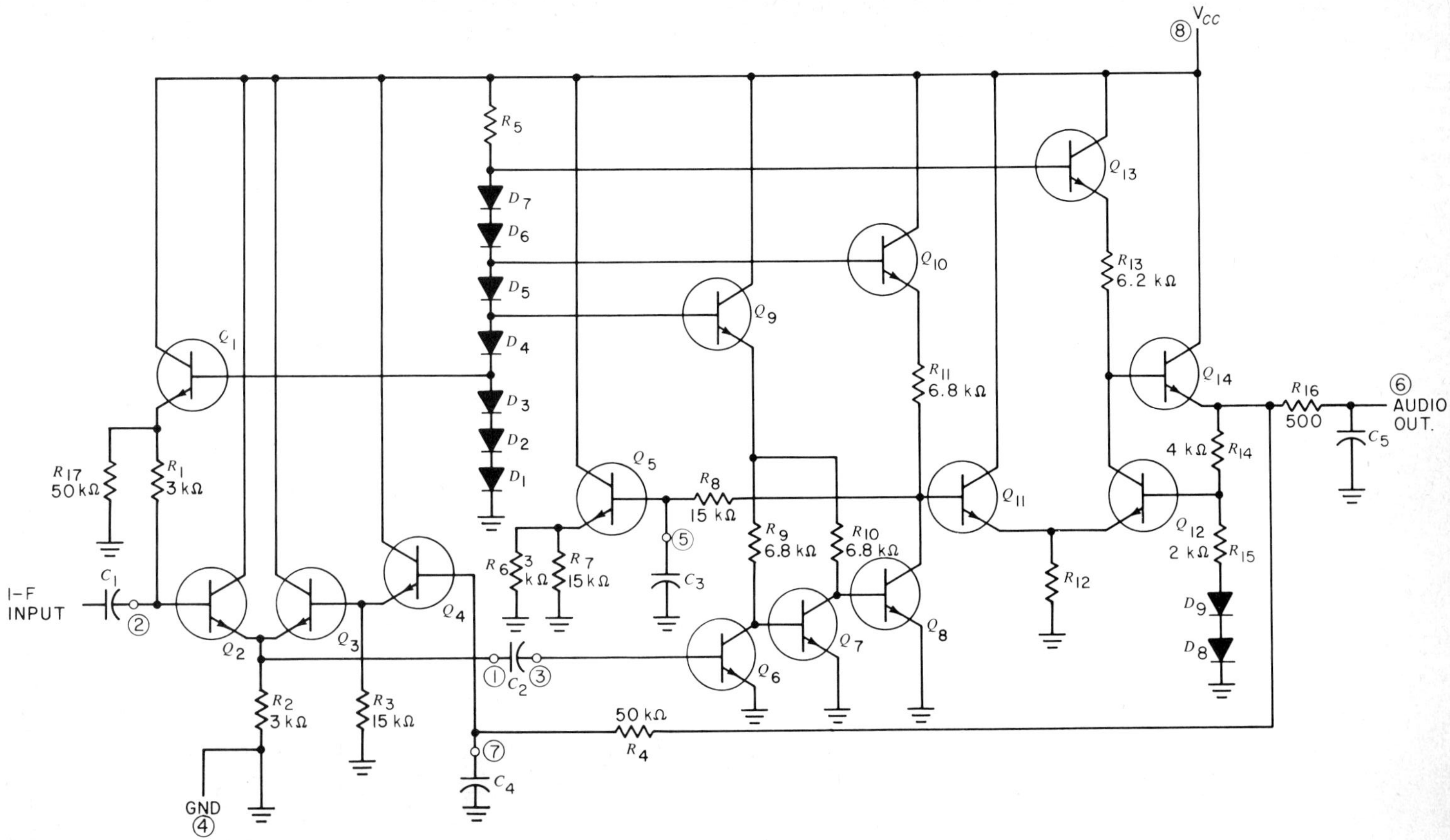

FIGURE 9.31
Schematic diagram of the National Semiconductor LM172 AM i-f strip.

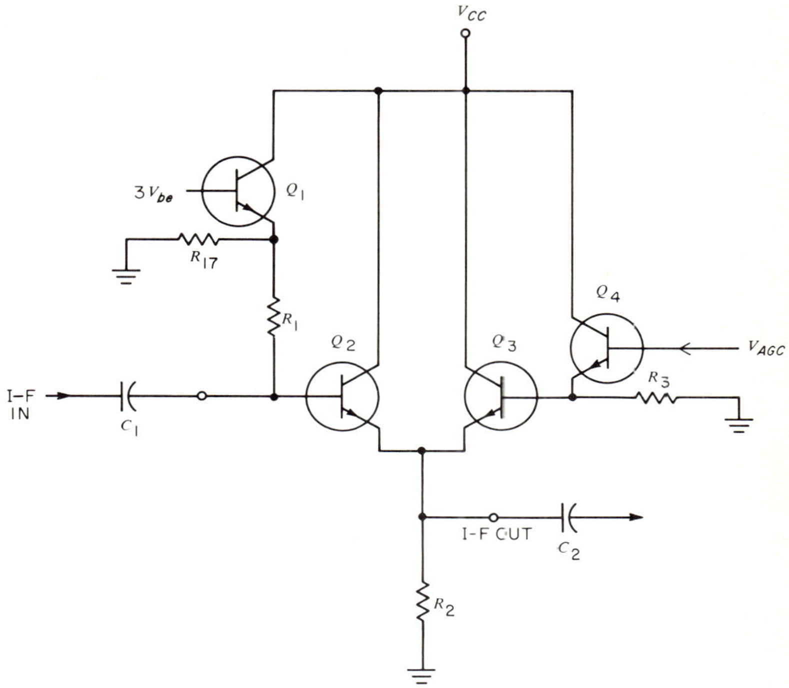

FIGURE 9.32
The agc system of the LM172.

is held at a dc voltage of two forward diode drops, or $2V_{be}$ by Q_1, an emitter follower, and R_1, with only ac signals coupled through an input capacitor C_1. Transistor Q_3 will be turned off if the agc voltage is held below $3V_{be}$, and Q_2 then will act as an emitter follower, with R_2 as the load.

If the agc voltage is equal to $3V_{be}$, Q_2 and Q_3 become a balanced differential pair and conduct equal emitter currents from R_2. As the agc voltage is increased, Q_3 is turned further on, and Q_2 turns off. At the same time, the emitter resistance of Q_2 increases, and the emitter resistance of Q_3 decreases, shunting across the signal. Thus, Q_2 and Q_3 form a series-shunt attenuator.

The base of Q_2 will remain at fixed bias, while the bias of Q_3 increases with agc, and the dc output voltage at the common emitter point will rise slightly with the increasing gain. This requires a decoupling capacitor (C_2) between the agc stage and the lumped gain stage, so that the gain is not affected (biased) by agc fluctuations.

Figure 9.33, the lumped-gain stage, is a cascade of direct-coupled, common-emitter amplifiers. In the conventional discrete-device version of the stage, complex dc biasing would be necessary. Note that no emitter resistors

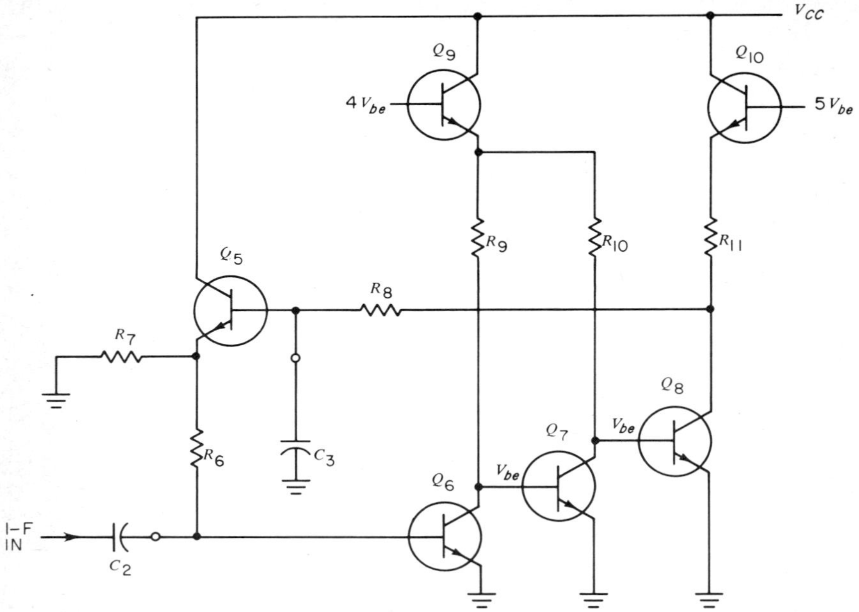

FIGURE 9.33
The diagram of the lumped-gain stage of the LM172.

are used, as a result of which there is maximum voltage gain per stage. The overall dc feedback loop R_8, C_3, Q_5, and R_6 automatically sets the dc output voltage of each transistor to the correct level to bias the next transistor accurately. Owing to the R_8C_3 rolloff with frequency, the feedback is effective only for dc, as a result of which there is maximum ac gain with dc stability.

The collectors of Q_6 and Q_7 are operated at V_{be} to accommodate biasing of the following stages. Thus, Q_6 and Q_7 operate with very low collector-to-base voltage; but because of their very small geometries and low saturation voltages, they have excellent current gain and bandwidth product. The collector load resistors R_9, R_{10}, and R_{11} are biased from the emitter-follower voltage regulator Q_9 and Q_{10}. This eliminates any supply decoupling problems and permits the active part of the circuit to function with constant bias conditions, regardless of the power-supply voltage. The supply current increases linearly with supply voltage, as in conventional designs, but remains relatively constant, since each stage of the circuit is regulated in the same manner. The circuit, then, remains extremely efficient at low supply voltages with insignificant drain at higher voltages.

For detection, a conventional diode must be slightly forward-biased by additional circuitry in order to respond to small input signals due to the voltage necessary to overcome forward V_{be}. In addition, diode detectors are not

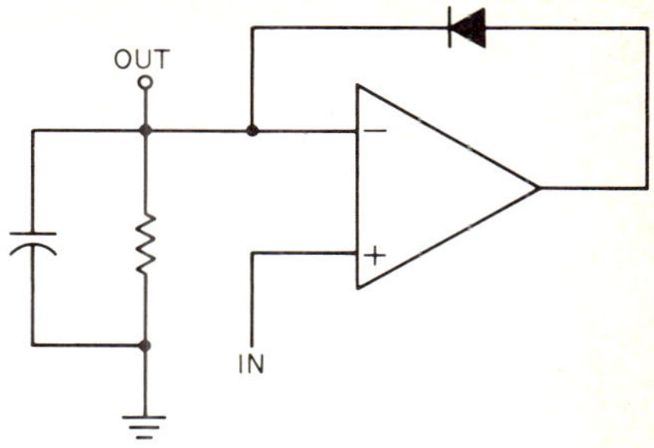

FIGURE 9.34
Unity gain detector of LM172.

efficient, since they generally give less audio output than is available from the modulated carrier. Thus, the conventional diode detector can be improved.

Figure 9.34 illustrates a more efficient detector, which is found in most modern operational amplifiers. With this scheme, audio output tracks the modulation envelope if the gain of the op amp is high enough. Since the diode is within a feedback loop, it is automatically biased by the operational amplifier to respond to small signals. The dc output voltage is zero when no carrier is present, rising to one-half the peak-to-peak r-f level with an unmodulated carrier. If audio modulation is superimposed on the carrier, it will not affect the dc output voltage but will cause the RC network to "follow" the modulation envelope on the positive side of the carrier.

With the addition of a resistive divider, this detector can perform some of the audio preamplification necessary in the receiver (Fig. 9.35). This is the case because the detector now has an audio and dc voltage gain equal to

$$\frac{R_1 + R_2}{R_2}$$

The detector employed in the LM172 is actually a differential amplifier (Fig. 9.36). The emitter follower performs the function of the feedback diode.

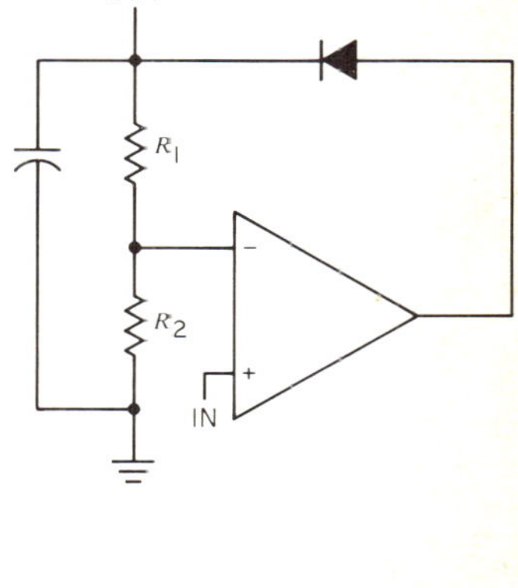

FIGURE 9.35
Detector with resistive divider.

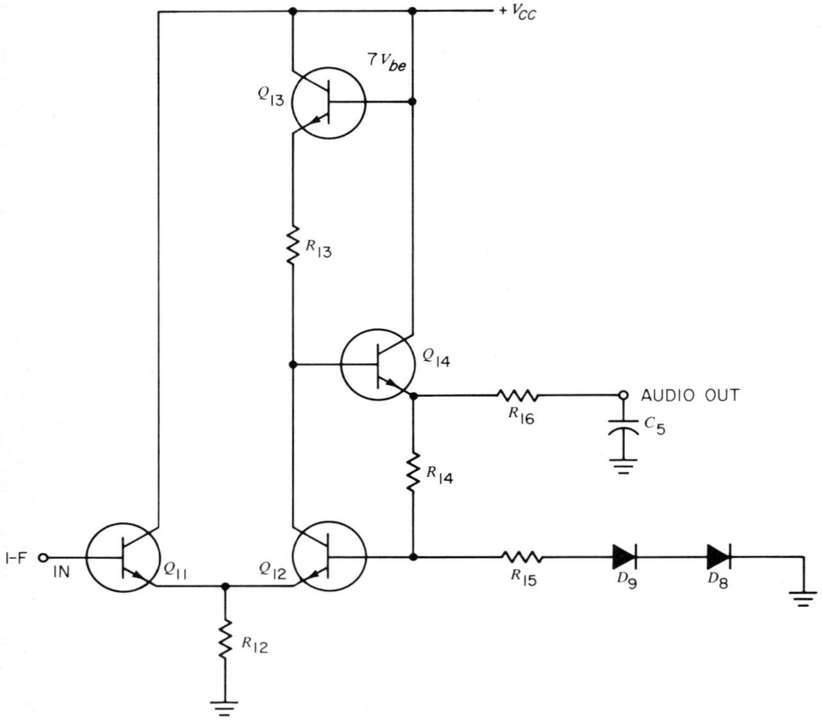

FIGURE 9.36
The actual LM172 detector.

The circuit has a voltage gain of about 40 dB, which is quite sufficient for good detection. Since the emitter follower has been substituted for a diode, output impedance is low. As a matter of fact, if it were not for the addition of R_{16}, the output impedance would be too low for carrier ripple filtering by C_5.

The resistive ladders R_{14} and R_{15} provide the detector with an audio voltage gain of 3. Diodes D_8 and D_9 compensate for the dc voltage, $2V_{be}$, which is superimposed on the i-f input voltage by the lumped-gain stage. Transistor Q_{13} acts as a supply regulator.

Figure 9.37 is a block diagram of a typical superheterodyne receiver system utilizing a LM172 monolithic IC. Conventional circuitry may be used ahead of the IC, and LC filtering may be used instead of the 455-kHz ceramic filter shown.

The circuit works efficiently at frequencies between 50 kHz and 2 MHz, which is dependent on input band-pass components. The receiver can operate with a supply voltage between +6 and +15 V, with no obvious performance variations. This is because of the built-in supply regulator.

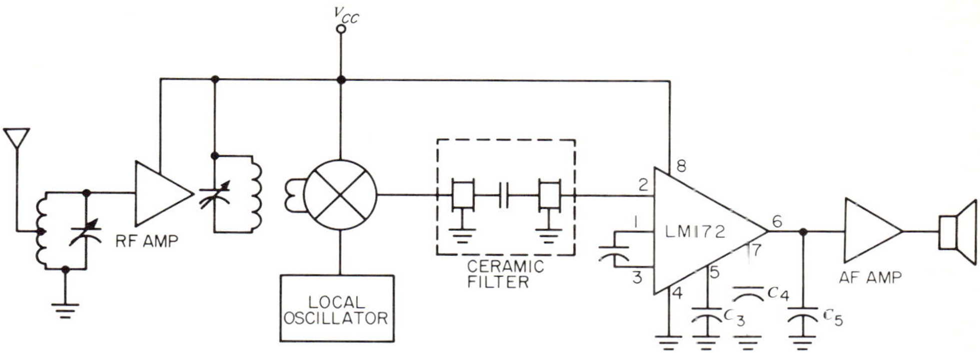

FIGURE 9.37
The use of the LM172 in an AM superheterodyne.

FIGURE 9.38
Closeup view of the LM172 chip. (*Courtesy National Semiconductor.*)

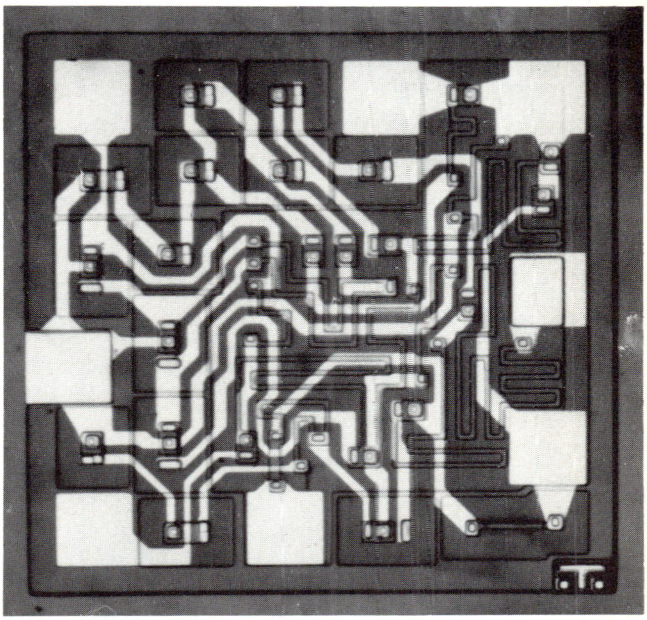

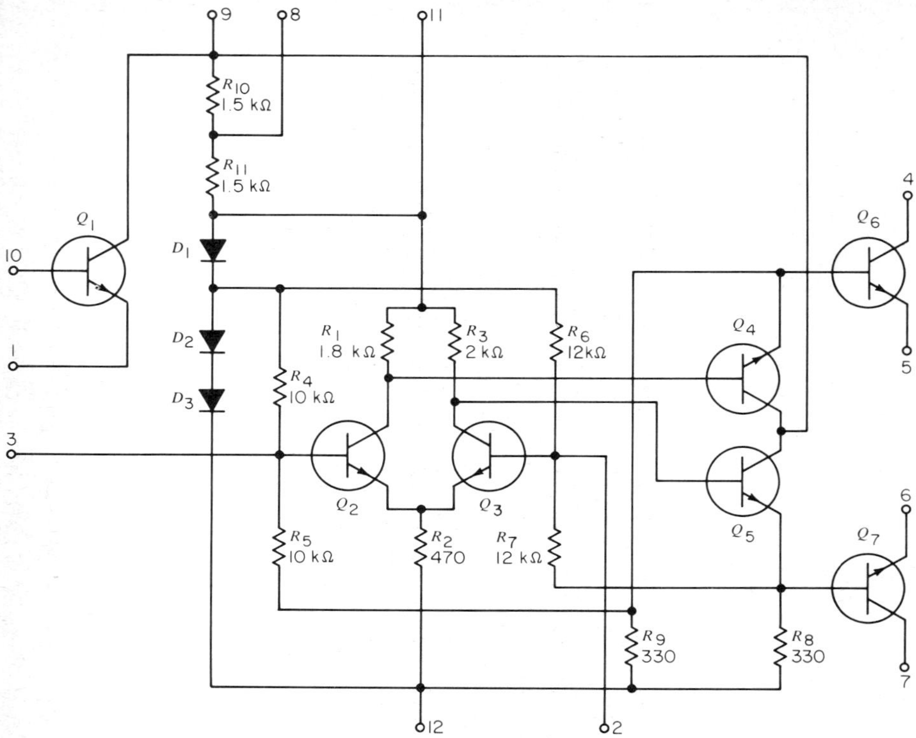

FIGURE 9.39a
Schematic diagram of the CA3020 integrated-circuit multipurpose wide-band amplifier.

In spite of the extensiveness of the circuit of the LM172, all the components fit onto the small, square chip shown in Fig. 9.38. This is placed in a TO-5 can and sealed. Eight external leads are all that is required to bring signals and dc voltages into this network.

IC AUDIO AMPLIFIER

The previous IC performed the functions of i-f amplification, agc, and AM detection. Now, a second integrated circuit can take the detected audio signal and amplify it sufficiently to drive a loudspeaker. Many such circuits are available, with the example selected being fairly typical.

Figure 9.39*a* is a schematic diagram of the Radio Corporation of America CA3020, a universal linear integrated circuit. Figure 9.39*b* is a photograph of the actual circuit. Figure 9.39*c* is a block diagram of the CA3020.

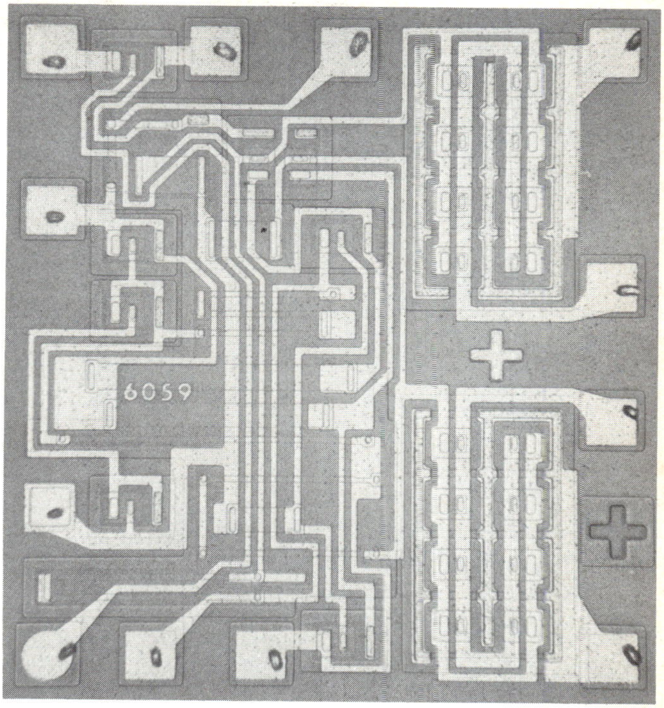

FIGURE 9.39b
Photo of the CA3020 chip. ***(Courtesy RCA)***

The CA3020 is a stabilized, multipurpose, wideband, direct-coupled amplifier that can perform preamplifier, phase-inverter, driver, and power-output functions without the use of transformers. Supply voltages can vary between +3 and +9 V.

With a +3-V supply the circuit is capable of delivering a power output of 65 mW; and with a +9-v supply the circuit can deliver 550 mW.

The voltage regulator consisting of diodes D_1, D_2 and D_3, with resistors R_{10} and R_{11}, provides stable operation between −55° and +125°C. The diodes are actually formed by biasing the emitter-base junctions of transistors. The regulator supplies two voltages to the differential amplifier: a base supply voltage of 1.4 V, and a collector supply voltage of approximately 2.1 V.

Transistors Q_1 and Q_3 and collector resistors R_1 and R_3, with biasing resistors R_4, R_5, R_6, and R_7, form the differential amplifier and the splitter circuits. The signal voltage may be applied to terminal 10 (buffer amplifier Q_1) and then coupled into terminal 3, or it may be capacitor-coupled directly into

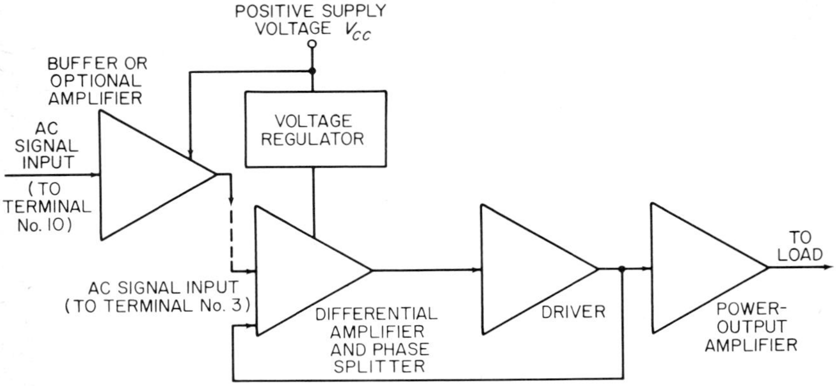

FIGURE 9.39c
Functional block diagram of the CA3020.

terminal 3. The required 180° phase shift, for push-pull operation, is obtained between the collectors of Q_2 and Q_3.

The ac signal is then applied to the driver stage, the emitter-follower transistors Q_4 and Q_5. Resistors R_5 and R_7 provide stabilization for the differential amplifier. Without these resistors, the dc voltage between the collectors of Q_2 and Q_3 might fluctuate with changes in beta or V_{be}. Thus, the application of the emitter-follower voltages from Q_4 and Q_5 to the bases of Q_2 and Q_3 through resistors R_5 and R_7 compensate for any imbalance.

FIGURE 9.40
Audio amplifier using the CA3020.

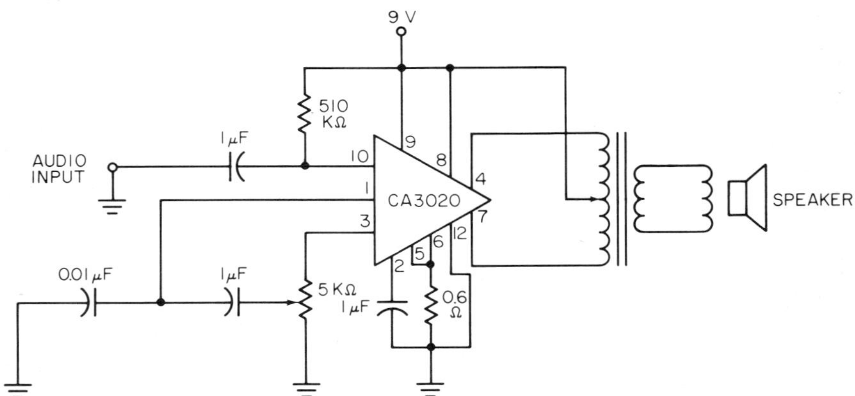

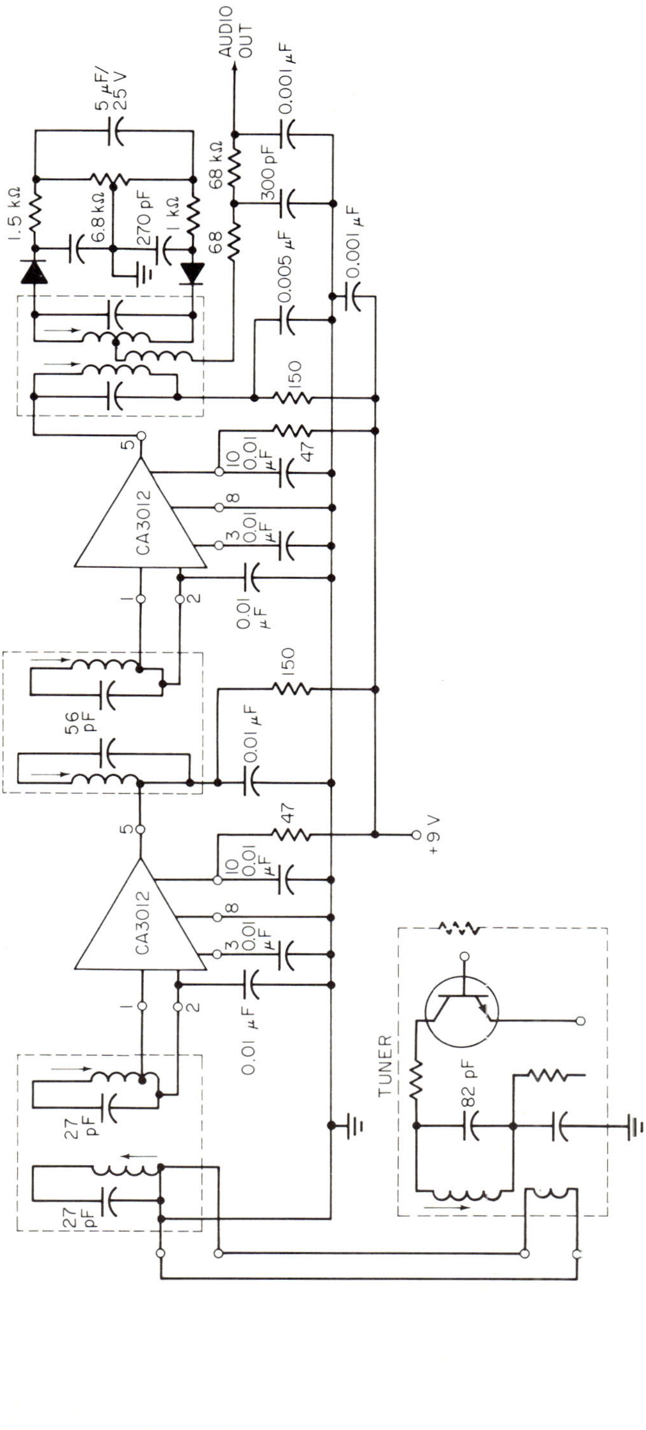

FIGURE 9.41
Complete 10.7 MHz i-f-amplifier strip using two CA3012 integrated circuits.

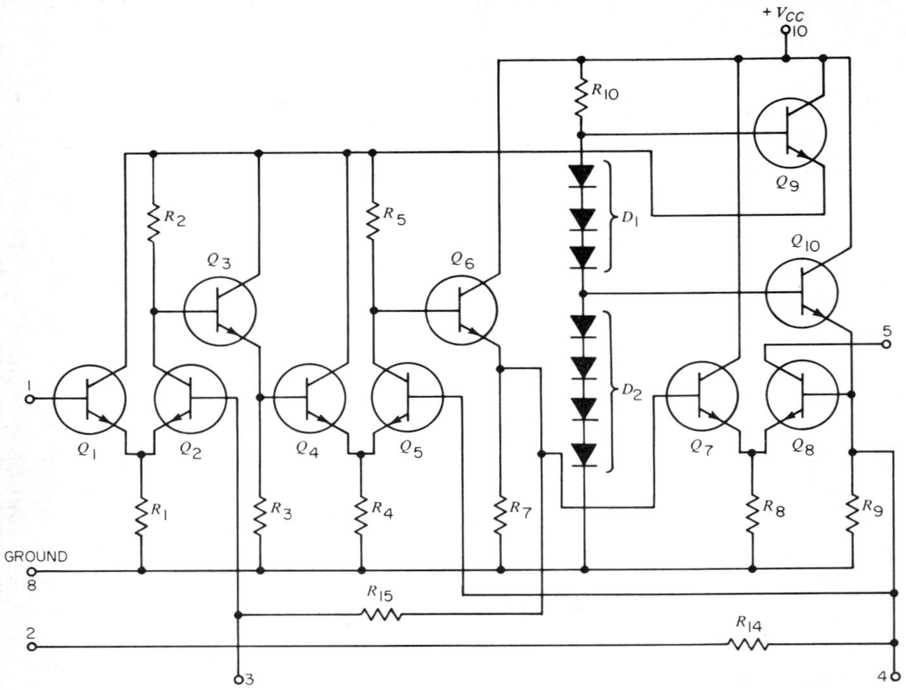

FIGURE 9.42
Schematic diagram of the CA3012 integrated circuit.

Transistors Q_6 and Q_7 are power transistors (six "normal-size" transistors in series-parallel) and receive the ac signal from Q_4 and Q_5 delivering power to the load in a class B push-pull mode of operation.

Figure 9.40 shows the CA3020 used as a complete audio amplifier. The triangle is used as a linear IC amplifier symbol and the numbered terminals are shown emanating from the triangle. (Use Fig. 9.39 to see how these terminals are connected to the schematic and to the actual circuit chip.)

This amplifier can be used for portable receivers or as a driver for higher-powered amplifiers. It has a ½-W power output, with a 45-mV input signal. The input impedance is 50,000 Ω and distortion is 3 percent or less.

IC FM INTERMEDIATE-FREQUENCY AMPLIFIER

An IC intermediate-frequency amplifier (10.7 MHz) for an FM receiver is shown in Fig. 9.41. It contains two IC amplifier networks interconnected by appropriate tuning circuits. At the output of the second IC network, the signal is fed to a ratio detector where the audio intelligence of the broadcast is obtained. The detected signal is then directed to an appropriate audio amplifier.

The internal circuitry of each CA3012 (RCA) IC device is shown in Fig. 9.42. The amplifier consists of three direct-coupled cascaded differential-amplifier stages and a built-in regulated power supply. Q_1 and Q_2 form the first differential amplifier, with the signal output from the collector of Q_2 passing on to the second differential amplifier (Q_4 and Q_5) through an emitter follower (Q_3). The same arrangement is employed at the second stage, with the signal from Q_5 going to Q_7, Q_8 through emitter follower Q_6. Output from the i-f strip is obtained at the collector of Q_8. This terminal connects to the external tuning circuit that drives the ratio detector.

Regulated dc biasing voltages are provided by D_1, D_2, Q_9, and Q_{10}. Amplification stabilization is also provided by negative feedback from the emitter of Q_6 to the base of Q_2. Incidentally, it is possible to tie these two points together by a dc path (i.e., through R_{15}) because the operating conditions in this network have been selected so that the dc voltage at the output of each stage is identical to that at the input to the stage.

The entire circuit fits inside a TO-5 can, with 10 external leads.

The foregoing are representative samples of IC linear networks applied (here) to AM and FM receivers. The trend toward the use of ICs in electronic equipment can only increase, not only because of their increasingly low cost but also because of their offering designers the opportunity to obtain a wide range of operating conditions with excellent stability and repeatability.

QUESTIONS

Answer the first four questions using the transistor receiver circuit of Fig. 9.3.

9.1. **a.** How many transistors does the receiver employ?
b. How does the first stage (Q_1) function? Draw its equivalent vacuum-tube circuit.

9.2. **a.** What purpose does R_2 serve? R_3? C_7?
b. Why is the emitter of Q_1 tapped down on L_3?

9.3. **a.** Why is the secondary of transformer T_1 untuned?
b. What is the purpose of C_{10} and C_{14}?
c. Why is R_8 much larger in value than R_5?

9.4. How does the agc system operate in the receiver of Fig. 9.3?

9.5. What points of similarity exist between the agc systems of vacuum-tube and transistor receivers? What are the differences?

9.6. Answer the following questions about the audio stages of the receiver in Fig. 9.8.

a. Identify the agc circuit. Include all components.

b. How is the crossover distortion minimized in the output stage?

c. How do the output transistors receive their biasing voltages?

Answer the following four questions about the transistor receiver circuit of Fig. 9.10.

9.7. **a.** How does the oscillator signal reach the mixer stage?

b. Where is the incoming signal applied to the converter?

9.8. Explain how the agc system operates.

9.9. Explain how the detector stage functions?

9.10. What advantages does this form of detection offer? What disadvantages in comparison with diode detectors?

9.11. What is a reflex amplifier? Where is it most likely to be used in a transistor receiver? What are its advantages?

9.12. Explain how the reflex amplifier of Fig 9.12 operates. Trace the paths of both the i-f and audio signals.

9.13. Which stages in the circuit of Fig. 9.15 are in operation during reception of an FM signal? During reception of an AM signal?

9.14. What is the difference between reverse agc and forward agc? Explain how each operates to control the level of the output signal. When is each used in the circuit of Fig. 9.15?

9.15. Trace the FM agc and AM agc circuits in the receiver of Fig. 9.15.

9.16. How is it possible to employ the same i-f transistors for AM and FM signals? How are these two signals kept in their respective paths?

9.17. In what respects are field-effect transistors better for FM receivers than for bipolar transistors?

9.18. What is meant by the term noise figure, and how is it important to receiver design?

9.19. Describe briefly an operational amplifier.

9.20. Draw the block diagram of a typical operational amplifier. Explain the function of each block.

9.21. How can we pass a signal through an operational amplifier without causing its phase to reverse? Illustrate by a block diagram.

9.22. Why is it impractical to use resistive networks to obtain the necessary bias voltages for IC transistors? What is a more desirable solution? Why?

9.23. What advantages does an arrangement for an i-f system, such as that shown in Fig. 9.30, offer over the more conventional i-f system (such as that employed in the receivers of Figs. 9.3 and 9.10)?

9.24. Explain briefly how the agc system in Fig. 9.31 functions.

9.25. What bias voltages could be obtained from the voltage regulator of Fig. 9.31?

9.26. Describe briefly the circuit of Fig. 9.39*a*.

9.27. Draw a block diagram of the circuit shown in Fig. 9.42.

Chapter Ten

Semiconductors in Computers

In 1946 the ENIAC, the first electronic computer using vacuum tubes, did an addition in one five-thousandths of a second. Today's computers, using solid-state circuits, can perform over one billion operations a second.

There is little doubt that the computer is having a profound impact on our civilization. For example, the mathematical problems posed by the landing of men on the moon would have been impossible to solve less than 20 years ago. The moon program called for millions of additions and subtractions to calculate effects of the ever-changing gravitational pulls of the earth, the moon, and the sun on a rocket in space. These calculations were made by computers at ground control centers and on the spaceships themselves, computing the forces acting on the ship and tracking its progress through the heavens while at the same time making continuous speed and direction adjustments.

Of far greater significance in our everyday life, computers are being used to diagnose disease, automate industrial production, address magazines, record insurance premiums and statistics, and to handle countless other applications that touch upon every facet of American life. It is little wonder, then, that the computer industry is today among the industrial giants of our economy.

Electronic computers can be divided into groups, or generations, dependent on the types of circuitry employed to perform the functions. First-generation computers used vacuum tubes, or relays, as the active elements. These were replaced by transistors in second-generation computers. And in recent third-generation computers, both transistors and integrated circuits are utilized as the active elements.

In this chapter we shall discuss some of the basic circuits of computers. Furthermore, although there are two basic types of computers, analog and digital, we shall confine our discussion to the digital type, since this is by far the more popular.

THE DIGITAL COMPUTER

A digital computer counts quantities by using discrete numbers, whereas the analog computer measures quantities on a continuous (i.e., nondiscrete) basis.

Since numbers are employed in digital computers, very precise calculations can be made by simply adding an appropriate number of digits. The computer's logic system, whether it consists of diodes, transistors, vacuum tubes or integrated circuits, responds to on or off states, and can hence represent a 0 or a 1. (The significance of this will become apparent in our discussion of binary numbers.)

Figure 10.1 is a block diagram of a typical digital computer. The basic units are: (1) the logic section, where the arithmetical operations are performed; (2) the memory, where the instructions, problems, and solutions reside; (3) the control unit, which directs the computer operations; (4) the input mechanism, which converts all the input information into a form usable by the computer; and finally (5) the output section, which interprets the computer outputs and translates them into understandable form.

The logic section accumulates information and organizes it. The information can be stored in the logic section or recalled from the memory section. To illustrate, a number can be taken from the memory section and moved to the logic section by an "order." It can then be added to a number already present in the logic section, thereby producing the sum of the two numbers. The sum can then be either retained in the logic system or transferred to the memory section.

If the number happens to be negative, it is recognized by the logic section as such and is subtracted. Multiplications are performed by a series of additions, and divisions are performed by a series of subtractions. The series, or sequence of operations, is a program, the individual steps of which are instructions. We shall discuss arithmetic operations in more detail presently.

The memory section has a vast number of individual locations, each of which is capable of storing data. Each location, or address, is identified so that its contents may be recalled. A group of addresses is called a register.

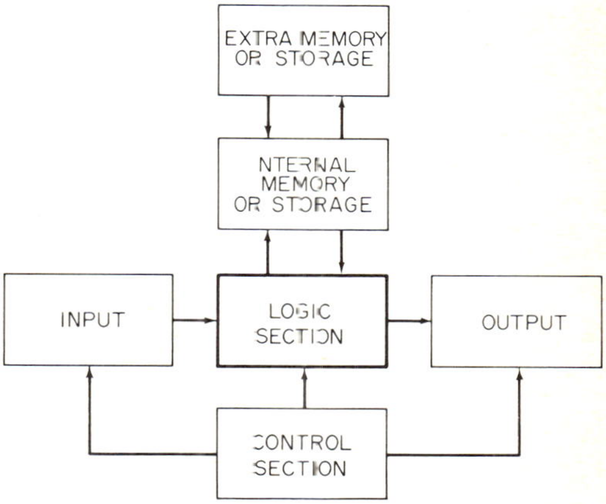

FIGURE 10.1
Diagram of a digital computer.

Information may be recalled as often as desired, the process of which usually includes all the steps, or program, necessary to solve a problem. Once the solution has been computed in the logic section, the result may be stored in the memory until required, at which point it is delivered through the output device.

Many of the latest computers make use of LSI (large-scale integrated) circuits, both bipolar and MOS (metal oxide semiconductor), in the memory section. Heretofore, the basic internal memory was a group of magnetic cores.

Magnetic tapes are often used as secondary storage units. Data can be placed on, or obtained from, a tape, as a supplement to the core or LSI primary memories.

The conditions set forth in the program are fulfilled by the control section, which directs the operation of the computer. The control center "reads" the instructions and prepares for the appropriate execution by following certain operational principles. For example, this section might translate a multiple order into the necessary series of additions.

The input and output sections are alike in operation, but they perform opposite jobs. That is, the input receives information from punched cards, keyboards, disks, or magnetic tapes, and then codes the information for the computer to manipulate. The output section converts the computed results into a practical, readable form, such as typed sheets.

LOGIC

Complex electronic systems, such as the computer, cannot be studied in the same manner as less-complex systems, such as radio receivers. In order to understand how a radio receiver operates, one learns how each stage functions and the order in which these stages are "hooked up."

As systems become more complex, it becomes impractical, if not impossible, to learn the details of each circuit's operation. Fortunately, it is essential to know only how the circuits are interconnected and what particular function, such as amplify, shape, oscillate, switch, etc., each performs—not how it performs this function.

The basic operations of a computer are divided into fundamentally simple categories. In solving complex mathematical problems, the computer performs these operations, over and over, at fantastic speeds. This factor has led to the development of a number of standardized circuits for computer design, which are known in the industry as "building blocks." These blocks, used in sufficient number, form the entire complex system.

A typical diagram of a portion of a computer system will contain these building blocks ("black boxes") labeled AND, OR, FLIP-FLOP, etc. These unusual-sounding circuit descriptions are types of logic circuit functions, which we shall discuss in this chapter.

The arrangement of these circuits is governed by symbolic logic, making it necessary to know a little about logic in order to understand computer operation.

Symbolic logic, or Boolean algebra, was developed by an English genius, George Boole, in 1854, as a method of establishing the validity of thought, or reason. His concept was that if the simple propositions of basic logic could be represented by precise symbols, these propositions could be read as precisely as an algebraic equation.

Boolean algebra thus deals with variables that are permitted to assume only two different "values," like the on and off states of an elementary switch. Boolean algebra is consequently a convenient mathematical tool, particularly suited to the analysis and design of switching circuits. As a matter of fact, Boolean algebra is sometimes referred to as switching algebra.

The type of problem determines whether a Boolean variable might have the values of on or off, true or false, yes or no, open or closed, etc.

In order to take a mathematical approach to a problem, we assign an 0 and 1 as the two possible values of the variable and let the 0 and the 1 represent the two possibilities of the particular problem.

For example, the 0 might represent false and the 1 true in an investigation of uncertain statements. Or, the 0 might indicate an open switch and the 1 a closed switch.

A type of switch that can be opened or closed is known as a gate. A gate has a number of possible inputs, with an output signal obtained only if the input signal meets specific conditions. There are two gate types—the AND and OR.

AND and OR. Formal logic begins with the concept that a statement is either true or false. A convenient analogy is that a switch must be closed (true, or 1) or open (false, or 0).

Two series switches are shown in Fig. 10.2. In order for current from source E to reach the light L, both S_1 and S_2 must be closed. This is the AND function, which stated logically says that only if each of two statements is true,

TABLE 10.1
Truth Table for AND Logic

S_1	S_2	*Lamp*
1	1	1
0	1	0
1	0	0
0	0	0

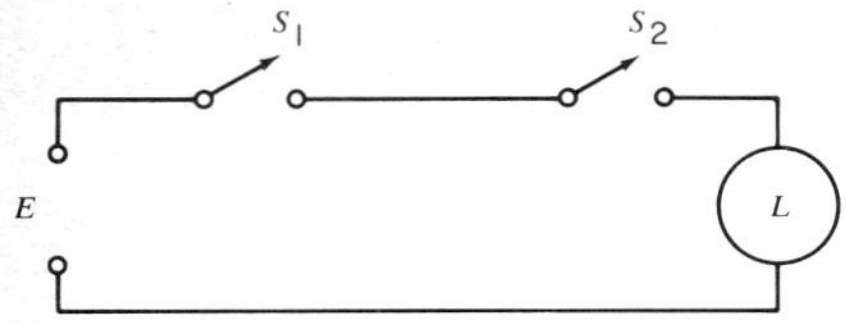

FIGURE 10.2
Series switches, representing an AND logic gate.

will the circuit function—or in electronic terms, will an output be produced. In the present case, switch S_1 and switch S_2 must be closed for the light to go on. In symbolic language, we say: Statement C will be true (i.e., the light will go on) only if statement A (switch S_1) and statement B (switch S_2) is true (i.e., closed).

Now, if we arrange the two switches in parallel, such as is shown in Fig. 10.3, current will reach L if either S_1 or S_2 are closed or if both are closed. Stated logically, this says: Statement C will be true (i.e., the light will go on) if either statement A (switch S_1) or statement B (switch S_2) is true (or both). This is the OR function.

With the AND and OR functions in mind, it is possible to set up an algebra (symbolic logic) to describe the two switching arrangements.

From the circuits of Figs. 10.2 and 10.3, it is quite simple to visualize the various conditions shown in Tables 10.1 and 10.2.

For the series circuit, Fig. 10.2, S_1 can have a value of either 1 or 0 (closed or open). So may S_2—hence, the four possible combinations shown in Table 10.1. However, there is only one combination that results in current flow (or truth), and that is when both S_1 and S_2 are closed. (The tabulation of all the conditions that can exist in a circuit in a chart is called the truth table for that circuit.)

In equation form, this table can be represented by

$$L = S_1S_2 = S_1 \cdot S_2$$

where L is the light, S_1 is switch 1, and S_2 is switch 2. The light will go on (i.e., assume a value of 1) only if S_1 and S_2 are both 1 (i.e., closed). The notation S_1S_2 or $S_1 \cdot S_2$ has been adopted to represent the AND function.

There is also a special symbol that has been adopted for the AND function; it is shown in Fig. 10.4.

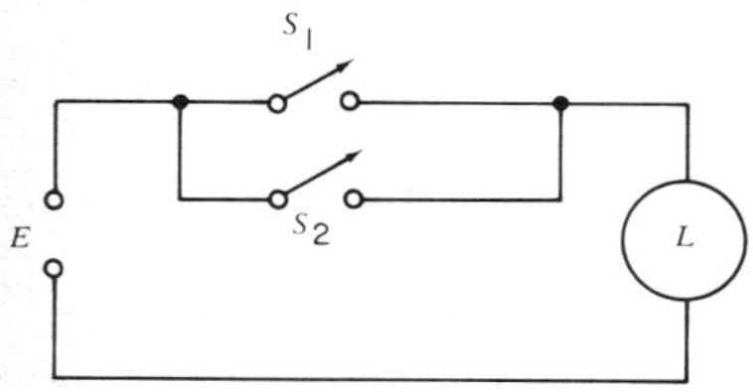

FIGURE 10.3
Parallel switches, representing an OR logic gate.

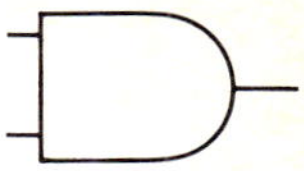

FIGURE 10.4
The AND logic symbol.

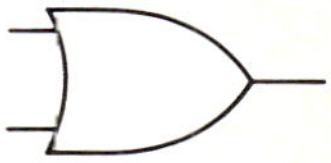

FIGURE 10.5
The OR logic symbol.

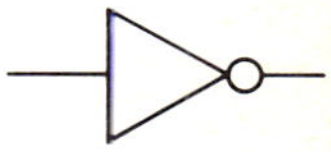

FIGURE 10.6
The logic symbol for an inverter.

For the parallel circuit shown in Fig. 10.3, we can have the same four combinations. However, if either S_1 or S_2 or both is closed, current will flow. This results in the possible combinations shown in Table 10.2.

The OR function is mathematically represented by

$$L = S_1 + S_2$$

where the + sign means or and is read as OR. Symbolically, OR functions are represented by the diagram shown in Fig. 10.5.

Negation, or the denial of a statement, is essential to logic. Since 0 and 1 are the only two values we have, it follows that not 0 (written $\overline{0}$) must be 1, and not 1 (written $\overline{1}$) must be 0. This form of negation can be referred to as invert or simply not. If the normal position for switch S_1 is open, then the notation $\overline{S}_1$ means that the switch is closed. Conversely, if the normal position for switch S_1 is closed, then the notation $\overline{S}_1$ means the switch is open.

As the reader may readily guess, passing an 0 signal through an amplifier will produce a 1 at the output. This represents a 180° phase reversal or an inversion. Symbolically, the inverter, or NOT, function is shown in Fig. 10.6.

Before we see how actual electronic circuits provide logic functions, let us pause and look at the binary (or two-valued numbering) system employed

TABLE 10.2
Truth Table for OR Logic

S_1	S_2	*Lamp*
1	1	1
0	1	1
1	0	1
0	0	0

in digital computers. Like Boolean algebra, the binary number system works with only two numbers or values, 0 and 1.

BINARY NUMBERING SYSTEM

The system, in most widespread use, the decimal system, requires 10 symbols—0 and the digits 1 through 9—to write any number, no matter how small or large. The numbers of this system are built from blocks valued at 1, 10, and powers of 10. For example, 1 is 10^0, 10 is 10^1, 100 is 10^2, or 10×10, 1,000 is 10^3, or $10 \times 10 \times 10$, etc.

From Fig. 10.7 it can be seen that the number 3 requires three 1s whereas number 16 requires one 10 and six 1s. The number 59 requires five 10s and nine 1s. The number 130 needs one 100, three 10s, and three 1s. Reading Fig. 10.7 from left to right, we see that every decimal number is a summary of the orders of 10s and 1s.

Other numbering systems are possible, but for electronic systems, particularly the digital computer, which utilizes two-state (on-off) devices, the binary, or two-number, system is the most efficient.

The binary system utilizes 0 and 1. The number 1 is represented by the number 1 or, what is the same thing, 01. If we wish to represent the number 2, we add a 1 to 01, or

$$\begin{array}{r} 01 \\ +\ 1 \\ \hline \end{array}$$

Since there is nothing higher than 1 in the binary system, we move the 1 to the next position left and place a 0 in the right-hand position. This then gives us

$$\begin{array}{r} 01 \\ +\ 1 \\ \hline 10 \end{array}$$

If this seems strange, it is really equivalent to adding 9 (the highest number in the decimal system) and 1. The result, as we all know, is 10. In short,

100 OR 10^2	10 OR 10^1	1 OR 10^0	
		3	3
	1	6	16
	5	9	59
1	3	0	130

FIGURE 10.7
Decimal counting by powers of 10.

$64 (2^6)$	$32 (2^5)$	$16 (2^4)$	$8 (2^3)$	$4 (2^2)$	$2 (2^1)$	$1 (2^0)$	
					1	1	3
			1	1	0	1	13
	1	1	0	0	0	1	49
1	1	1	1	1	0	1	125

FIGURE 10.8
Binary counting by powers of 2.

no matter whether it is the decimal system or the binary system, when you try to go beyond the highest single digit, you must then start your numbering with the next position left.

$$9 + 1 = 10$$
$$1 + 1 = 10$$

Table 10.3 shows binary and decimal equivalents up to 16. In each instance, whenever a 1 is to be added to 1 in the binary system, the result is 10.

Another way of representing the positions in a binary number is in terms of powers of 2. This is clearly shown in Fig. 10.8, for four numbers. Thus,

TABLE 10.3
Decimal Numbers to 16 and their Binary Equivalents

Decimal value	*Binary equivalent*
0	0000
1	0001
2	0010
3	0011
4	0100
5	0101
6	0110
7	0111
8	1000
9	1001
10	1010
11	1011
12	1100
13	1101
14	1110
15	1111
16	10000

in order to obtain the number 3, we need one 1 and one 2, and this is written 11. If we want to obtain 13, we need one 8, one 4, no 2s, and one 1, and this is written 1101. The number 49 requires one 32 (which is 2^5), one 16 (which is 2^4) no 8s, no 4s, no 2s, and one 1. It is written in binary form as 110001. In the same manner, 125 may be written 1111101.

In the yes-no language of the digital computer, writing the number 100 using binary notation is 1100100. This translates into "computer talk" as: one 64 yes, one 32 yes, a 16 no, an 8 no, a 4 yes, a 2 no, and a 1 no.

Arithmetic Operations. Now that we have seen how the familiar decimal numbers can be represented using binary notation, let us take a look at binary-arithmetic operations.

Addition. In general, addition, subtraction, multiplication, and division using binary numbers are carried out in the same fashion as in ordinary decimal arithmetic.

First, let us add two easy numbers, such as 3 and 4. Referring to the binary table of Table 10.3, we find that this becomes 0011 plus 0100, and adding in the familiar manner, we get

$$\begin{array}{rr} 3 & 0011 \\ +4 & 0100 \\ \hline 7 & 0111 \end{array}$$

Checking the binary table again, we find that 0111 is 7.

Another easy example—"carrying"—will serve to illustrate an important point. If we add 2 plus 6, we have

$$\begin{array}{rrl} 2 & 0010 & \\ +6 & 0110 & \\ & 11 & \text{carry} \\ \hline 8 & 1000 & \end{array}$$

It is apparent that we cannot add 1 plus 1 in the binary system and obtain 2, since the binary system utilizes only 0s and 1s. Thus, 1 plus 1 yields 10, one and zero, so we write down 0 and carry 1, as in the example.

Larger numbers are added in a similar manner, thus:

$$\begin{array}{rrl} 46 & 101110 & \\ +22 & 10110 & \\ & 1111 & \text{carry} \\ \hline 68 & 1000100 & \end{array}$$

In binary addition, then, we see that all situations can be covered by

$$\begin{aligned}0+0&=0\\0+1&=1\\1+0&=1\\1+1&=0 \qquad \text{plus a carry over of 1}\end{aligned}$$

Note that carryovers are handled in the same way as in the decimal system, except that with binaries we have only 0 and 1, whereas in the decimal system we go from 0 to 9.

To illustrate with a somewhat more complex example, consider the addition of

$$\begin{array}{r}1\\1\\1\\\underline{1}\end{array}$$

To work this problem, first add the bottom two 1s, for a total of 10. Now, the problem has become

$$\begin{array}{r}1\\1\\\underline{10}\end{array}$$

The top two 1s are added to provide another 10; and when the two 10s are added, we obtain

$$\begin{array}{r}10\\\underline{10}\\100\end{array}$$

which is equivalent to the decimal number 4.

Subtraction. Subtraction of binary numbers can be done in two ways. The first method is straightforward:

$$\begin{array}{r}6\\\underline{-4}\\2\end{array}\qquad\begin{array}{r}0110\\\underline{-0100}\\0010\end{array}$$

This is direct subtraction and is governed by the following rules:

$$\begin{aligned}0-0&=0\\1-0&=1\\1-1&=0\\0-1&=1 \qquad \text{with a borrow of 1}\end{aligned}$$

In achieving subtraction, the computer finds it easier to add "complements." The complement of a number in the binary system means the opposite of that number. Thus, the complement of 1 is 0, and similarly the complement of 0 is 1. The complement of 0100 (4) then is 1011. That is, wherever there is a 0 in the original number, we place a 1 in the same position in the complement. Now, let us work the above example of subtracting 4 from 6 by the "complements" method. For this, we will add the complement of 4 (which is 1011) to 6. Doing this, we obtain:

$$
\begin{array}{l}
0110 \\
1011 \\
11 \\
\hline
10001
\end{array}
$$

This result, it seems, is not what we need, that is, 2.

In adding a binary complement there will always be an extra bit (binary digit) or, in this case, a 1. The computer "gets rid" of this by taking the extra 1 up front and adding it to the rear (or right-hand side), which results in

$$
\begin{array}{r}
1001 \\
1 \\
\hline
0010
\end{array}
$$

The answer (2) is now correct, and moving the extra bit is again referred to as carrying.

Multiplication. Multiplication can be accomplished by a series of additions. For example, if we want to multiply 20 by 15 in the decimal system, we obtain the same result if we add the number 20 fifteen times. The same thing is true in binary multiplication.

We can also multiply two binary numbers in the same fashion as we learned in grammar school to multiply. For example,

$$
\begin{array}{rr}
10111 & 23 \\
101 & \underline{5} \\
\hline
10111 & \\
00000 & \\
10111 & \\
\hline
1110011 & 115
\end{array}
$$

Note that the binary multiplication carries the same implicit rules as decimal multiplication. That is

$$
\begin{aligned}
0 \times 0 &= 0 \\
0 \times 1 &= 0 \\
1 \times 0 &= 0 \\
1 \times 1 &= 1
\end{aligned}
$$

Here multiplication is a series of shifts and additions; with 101 as the multiplier, each 1 requires a shift left and an addition, while each 0 results only in a shift left.

Division. Dividing binary numbers is done in the usual way, by long division. That is, the divisor is subtracted from the dividend and a 1 is placed in the quotient. If the divisor cannot be subtracted, a zero appears in the quotient. This continues until all numbers in the dividend have been brought down and the subtractions tallied in the quotient.

$$\begin{array}{r@{}l} & \underline{\ \ 111\ \ \ \ } \\ 1101/ & 1011011 \\ & \underline{1101} \\ & 10011 \\ & \underline{\ 1101} \\ & \ \ 1101 \\ & \ \ \underline{1101} \end{array}$$

From the foregoing discussion of binary numbers and their manipulation, we can see that the OR function $S_1 + S_2$ is, in essence, binary addition. If S_1 is open (Fig. 10.3) and S_2 is closed, or if S_1 is closed and S_2 is open, we obtain the same result. This was also indicated under binary addition, where it was shown that

$$0 + 1 = 1$$
$$1 + 0 = 1$$

By the same reasoning, the AND function $S_1 \cdot S_2$ is binary multiplication. If S_1 differs from S_2, then one of them is 0 and the other is 1 and, by multiplication.

$$0 \times 1 = 0$$

In other words, no output. Only when both S_1 and S_2 are both 1 (or closed) will there be output.

Thus, the binary number system, Boolean algebra and computer logic circuits are all tied together.

BASIC LOGICAL CIRCUITS

Diodes and Logic. Diodes have played a significant role in computer development. This is not surprising, since a diode can act as a switch having a closed (on) state when conducting and an open (off), or nonconducting, state

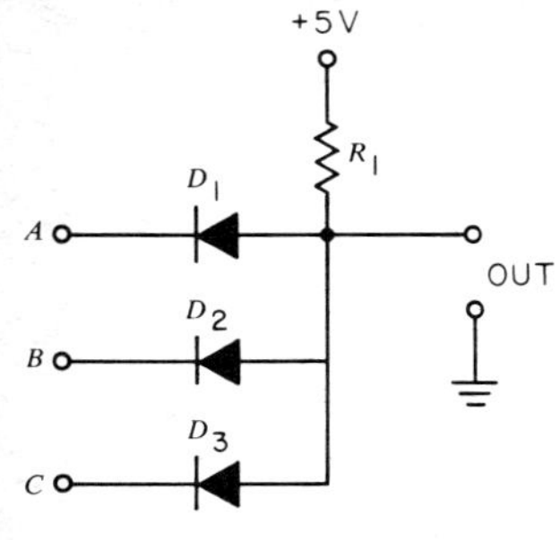

(a)

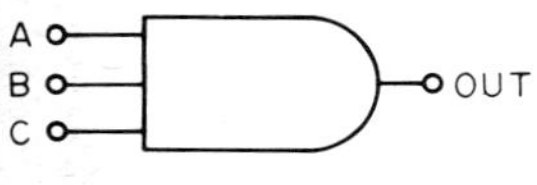

(b)

A	B	C	OUT
0	0	0	0
0	0	+5	0
0	+5	0	0
+5	0	0	0
0	+5	+5	0
+5	0	+5	0
+5	+5	0	0
+5	+5	+5	+5

(c)

FIGURE 10.9
(*a*) A diode AND logic circuit, (*b*) logic symbol, and (*c*) truth table.

when reverse-biased. Thus, a diode can be made to perform the logical functions of AND and OR that we have just discussed.

1. *Diode* AND *Circuit.* In Fig. 10.9, the anodes of three diodes are connected in parallel through a common resistor R_1 to a positive voltage source. The circuit has three inputs, and the value of the output voltage will depend on the voltage (or lack of it) applied to terminals A, B, and C.

If point A is grounded, diode D_1 will conduct. Since a conducting diode possesses very little impedance, all of the $+5$ volts will be dropped across R_1, leaving approximately† 0 V for the output. Hence, grounding A

† We say approximately zero, because actually about 0.7 V does develop across a conducting silicon diode. Thus, the voltage at the output will be $+0.7$ V, but for the moment this can be ignored.

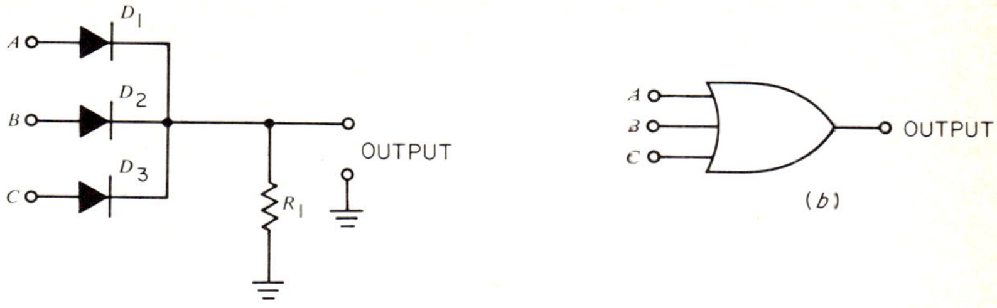

(a) (b)

A	B	C	OUT
0	0	0	0
0	0	+5	+5
0	+5	0	+5
+5	0	0	+5
+5	+5	0	+5
0	+5	+5	+5
+5	0	+5	+5
+5	+5	+5	+5

(c)

FIGURE 10.10
(a) A diode OR logic circuit, (b) logic symbol, and (c) truth table.

will result in a zero voltage output. The same condition will occur if points B or C are grounded. Thus, when any or all of the input terminals are grounded, zero output results.

Let us now see what happens when we apply a positive voltage of 5 V to terminal A, leaving B and C grounded. The 5 V on the cathode of D_1 will reverse-bias this diode and cause it to cease conducting. However, diodes D_2 and D_3, being grounded, are still conducting, causing the output to be zero volts, as discussed above.

Let us now assume that +5 V is applied to each of the input terminals. This will now reverse-bias D_1, D_2, and D_3 and cause them to cease conducting. Under this condition, the +5 V of the supply will appear at the output terminals.

A truth table in Fig. 10.9 lists the various combinations of input and output voltages just discussed. Note that only when A, B, and C are all positive will a positive voltage be obtained at the output. This, then, is a diode AND circuit.

From this action, we can formulate a definition of an AND circuit: a logic 1 will be obtained at the output only when there is a logic 1 signal at each input terminal.

2. *Diode* OR *Circuit.* A diode OR circuit is shown in Fig. 10.10. The truth table in the illustration shows the eight possible combinations of input voltages and the resulting voltage at the output. Note that if a +5 V is applied to any input while the other two have zero volts, a positive output will occur. Thus, a positive voltage at A or B or C will produce a positive output.

   Therefore, in an OR circuit, a logic 1 will be obtained at the output when a logic 1 signal appears at *any* input terminal.

Transistors and Logic. Transistors can perform the same logical functions as diodes and, from the standpoint of integrated circuits, just as economically. That is, it costs no more to fabricate a diode than a transistor in an IC device. Furthermore, transistors can provide amplification, which is an important asset in any circuit of even moderate complexity. Hence, transistor logic circuits are in widespread usage.

1. *Transistor* OR *Gate.* A three-transistor OR gate is shown in Fig. 10.11. Since the output is obtained from the common-emitter circuit, this is an emitter-coupled-amplifier arrangement.

   If zero voltage (i.e., ground) is applied to the base of each transistor, no base-emitter current flows because these two elements are at the same potential. Hence, the output voltage is zero. However, if a positive voltage

FIGURE 10.11
A transistor OR circuit.

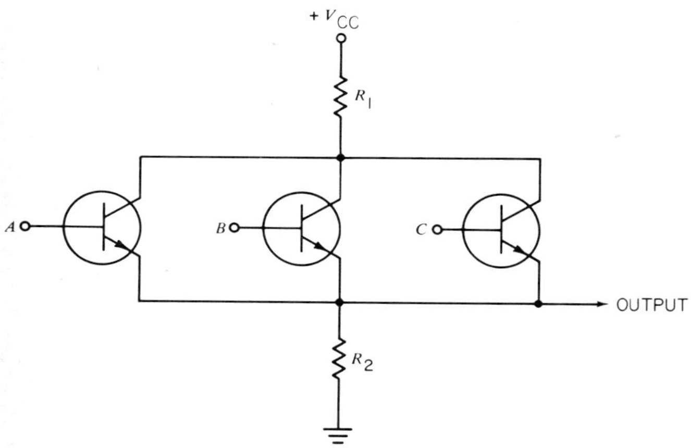

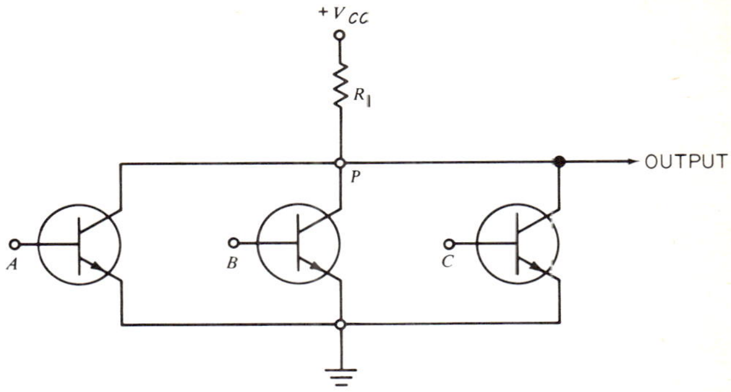

FIGURE 10.12
A transistor NOR circuit.

is applied to any base, that transistor will conduct and a positive voltage will appear across R_2. From this behavior, it is apparent that the circuit in Fig. 10.11 represents a transistor OR circuit.

If we shift the output terminal of the circuit of Fig. 10.11 to the collectors of the three transistors, as shown in Fig. 10.12, two things happen. First, any signal applied to any input receives amplification as it passes through the transistor. A stronger signal is thus obtained at the output.

Second, since we are now dealing with the common-emitter configuration, a phase shift of 180° takes place, which means that if we now apply a positive pulse at any input, we will obtain a zero output voltage because the positive input pulse will drive the transistor to which it is applied into conduction, thereby reducing sharply the internal impedance of the transistor. This will bring point P essentially to zero volts.[1]

However, if a zero voltage is applied to each of the three input terminals, none of the transistors will conduct and the voltage at point P will be equal to $+V_{CC}$.

Note that the foregoing behavior is just the reverse of the previous OR gate. A moment's reflection will reveal that what we have here is a not-OR circuit or, combining the two words, a NOR circuit. If we call a positive pulse at the base of the first transistor A then what we obtain at the output is $\bar{A}$ or the reverse of A. On the other hand, if we have an $\bar{A}$ pulse (here, zero) at all inputs, we obtain A (a positive pulse) at P.

The symbol for a NOR circuit can be that shown in either Fig. 10.13a or Fig. 10.13b. In the first illustration, we combine an OR circuit with an

[1] The actual voltage at point P will be equal to the voltage drop across a transistor that is in full conduction, i.e., in the saturated state. This voltage drop is generally on the order of a few tenths of a volt.

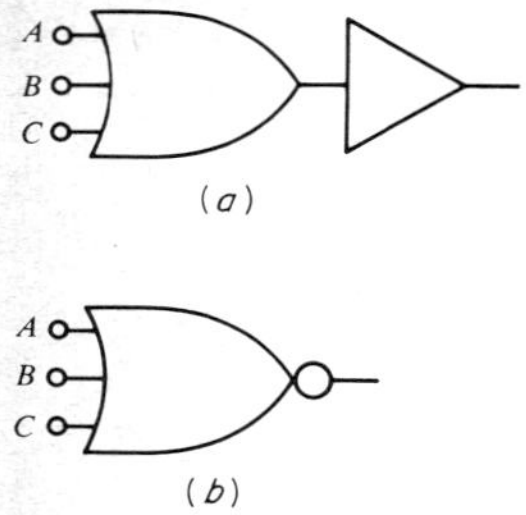

FIGURE 10.13
Two symbols for a NOR gate.

inverter. In the second illustration, Fig. 10.13*b*, a small circle is placed at the tip of the OR symbol to represent the NOT or inverter action.

2. *Transistor* AND *Gate.* To have transistors perform the AND function, we can use the circuit shown in Fig. 10.14. The two transistors are connected in series, and unless both are placed in the conducting state, no current flows through the circuit. Thus, if we apply $-V$ volts to either terminal A or B or both, neither one of the transistors will conduct and the output voltage will be $-V$. In essence, this represents the logic 0 state.

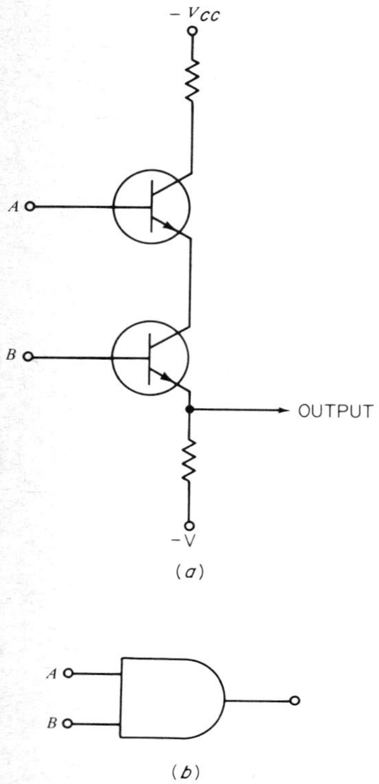

FIGURE 10.14
(*a*) Transistor AND circuit. (*b*) Logic symbol for AND circuits.

If we apply a $+V$ voltage to one of the input terminals, and $-V$ volts to the other, the transistor receiving the $-V$ volts at its base will not conduct and hence no current will flow through the series transistor network. The output voltage will thus be $-V$ volts (representing the logic 0). Only if $+V$ volts are applied to both input terminals will the output voltage go to $+V$, representing the logic 1.

Figure 10.14*b* shows the symbol for an AND logic gate.

If we obtain the output from the collector circuit rather than the emitter circuit, as shown in Fig. 10.15*a*, then we will find that whereas the circuit still functions in the same fashion, the output voltages for the various input combinations will be exactly reversed.

Thus, when a $-V$ voltage is applied to either input or both, the output voltage will be $+V_{CC}$. On the other hand, if $+V$ is applied to both input terminals, the output voltage will be $-V$ volts. In essence, what we have done is combine phase inversion with an AND circuit, to obtain a not-AND, or NAND, combination. Figure 10.15*b* shows the symbol used for a NAND arrangement.

We might pause here and note that the actual voltage values (either at the input or output of a logic gate) used to represent logic 1 or logic 0 will be dependent upon the voltage requirements of the stages employed in that

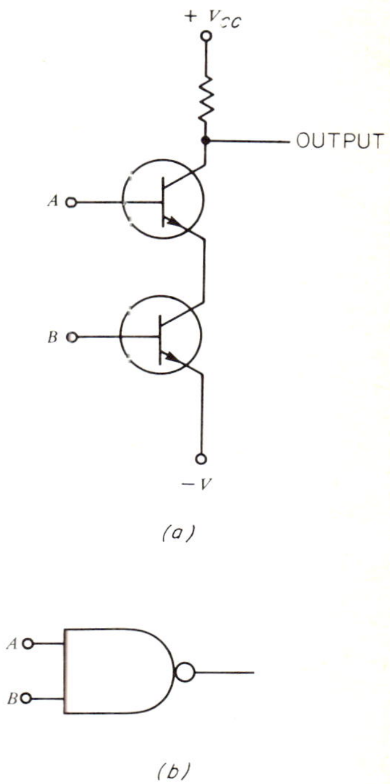

FIGURE 10.15
(*a*) A transistor NAND circuit. (*b*) Corresponding logic symbol.

system. In Fig. 10.12, for example, logic 0 can indeed be represented by a zero input voltage and logic 1 by a positive input pulse of +1 V or more. However, in Fig. 10.14, logic 0 at the input is represented by $-V$ volts and logic 1 by $+V$ volts.

Thus, it is not the actual voltage values themselves but their values relative to each other that determine whether they are logic 1 or logic 0. This brings us to the concept of positive and negative logic.

If the more positive of two voltages is defined as logic 1, and the more negative of the two voltages as logic 0, this is called positive logic—the logic we have been using thus far in our discussion of the various logic gates. It is perhaps a simpler system for the beginner to grasp, but it does not always result in the simpler electronic system (from a schematic or component point of view).

Conversely, negative logic is that system in which logic 1 is represented by the more negative of two voltages and logic 0 by the more positive voltage of the two.

To illustrate, in a positive logic system, +5 V would represent logic 1 and −2 V a logic 0. In a negative logic system, +5 V would represent logic 0 and −2 V a logic 1. The actual voltage values are unimportant, being actually governed by operating levels required by the circuitry. What is important, and the determining factor, is the *relative* levels of the two voltages with respect to each other.

All this may seem somewhat bewildering to the reader, particularly on first exposure. However, by carefully studying and rereading the foregoing, much of the "strangeness" will disappear in time.

Thus far we have discussed diode logic (or DL) circuits and resistor-transistor logic (or RTL) circuits. These were the first two types of logic circuits that designers developed. The prime advantage is their low cost, since they contain a minimum of components and their production processes have been long established. In short, production has been "debugged," and good yields are readily and consistently achievable.

There are some limitations in these circuits, however. Diodes, of course, provide no amplification, which means that we must either employ sufficiently strong signals (i.e., pulses) or else insert amplifiers at appropriate points within the system. In the case of the RTL, amplification is inherent in the circuit, and this problem is not encountered. However, RTL circuits have a somewhat low immunity to noise, and their ability to drive a number of other logic circuits is more limited than other circuits that have been developed. The ability of a logic circuit to drive one or more other logic circuits adequately is called its "fan out." This is an important operating feature of a logic circuit and is included with the other characteristics provided by the manufacturer.

Every digital circuit has an input load factor (fan in), assigned to each input signal terminal. Similarly, an output drive factor (fan out) is assigned to each output terminal. These factors (numbers) are a measure of the input-current requirement, and output-current capability, of the circuit. The fan out of a given device must be at least equal to the *sum* of the input load factors of the input terminals of the other devices connected to its output. If it is not, then circuit operation will be erratic.

A number of other logic circuits other than DL and RTL have been developed, each possessing some feature that makes it attractive for use in a digital circuit. We will now examine some logic circuits that have gained wide usage.

Diode-transistor Logic. A simple diode-transistor logic (or DTL) circuit is shown in Fig. 10.16. There are three input diodes (D_1, D_2, D_3) with their cathodes tied together and connected to the base of an NPN transistor. Input pulses are applied to the individual anodes of the diodes, and the output from this circuit is obtained from the collector of Q_1.

If points A, B, and C are grounded, the three diodes will conduct, producing a negative voltage at point P of 0.7 V. (We are assuming here that the diodes are formed of silicon, 0.7 V being the voltage drop across a forward-biased silicon PN junction.) This −0.7 V will not be able to bring Q_1 into conduction (because Q_1 is an NPN transistor), so that an output voltage of +5 V will appear at the output.

If we apply +2 V to any of the input terminals in place of zero volts, point P will develop a voltage of +1.3 V, because 0.7 of the applied voltage will be dropped across the diode. The +1.3 V is sufficient to drive the NPN transistor into conduction, and the voltage at the collector will decrease to close to zero volts.

Thus, we can see that an input zero voltage at A, B, or C will produce a positive output, whereas +2 V at any input terminal will cause the output voltage to decrease to zero.

Note that in this circuit the logic function is performed by the diodes, with the transistor serving as an amplifier. Because of its action, this circuit would be classified as a NOR gate.

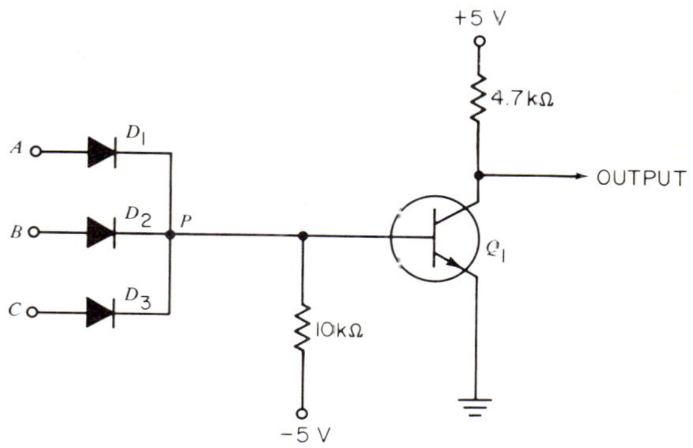

FIGURE 10.16
A diode-transistor logic, or DTL, circuit.

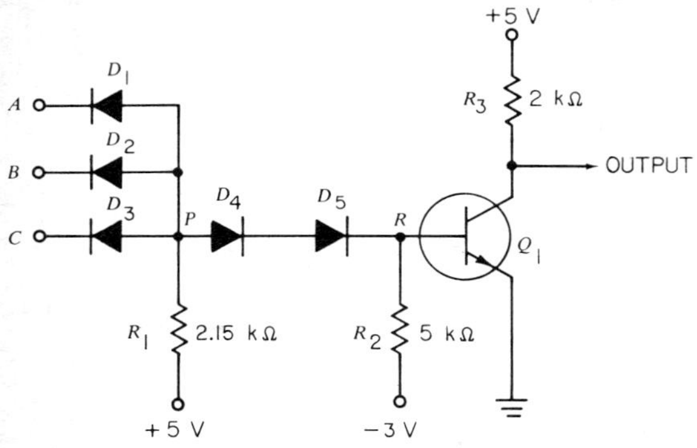

FIGURE 10.17
A second type of DTL gate.

A second type of DTL arrangement is shown in Fig. 10.17. The three input diodes, D_1, D_2, and D_3, receive the incoming pulses and the logic action occurs here. Diodes D_4 and D_5 serve basically as biasing diodes, each providing 0.7 V when forward-biased. Transistor Q_1 functions as an amplifier.

In operation, assume that the three input terminals are each at zero volts. The three diodes conduct, and if we assume that they are silicon diodes, each will have a 0.7-V drop across it. Point P, then, will have a positive potential of 0.7 V.

Current will also flow through D_4 and D_5, because the voltages applied to them through R_1 and R_2 are such that the diodes have a positive potential at each anode and a negative potential at each cathode. Each diode will have a voltage drop of +0.7 V so that in terms of diode voltages, we have the situation shown in Fig. 10.18. The voltage at point P opposes the two voltages of D_4 and D_5, with the result that the voltage at R, the base of Q_1, is −0.7 V. Since Q_1 is an NPN transistor, −0.7 V on its base will *not* turn it on, so that Q_1 remains nonconducting. Under these conditions, the output is +5 V, representing a logic 1.

Let us now see what happens when a strong positive pulse is applied to each of the three input terminals, A, B, and C. The three diodes will become reverse-biased and will not conduct. Diodes D_4 and D_5 will still be conducting, however, but the negative 1.4 V that they develop (together) will be more than offset by the positive voltage that appears at point P by the +5-V supply at the other end of R_1. As a result, the base of Q_1 will have a positive voltage applied to it and the transistor will conduct heavily. The collector voltage of Q_1

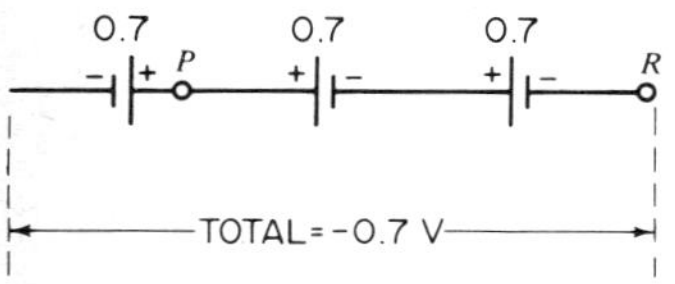

FIGURE 10.18

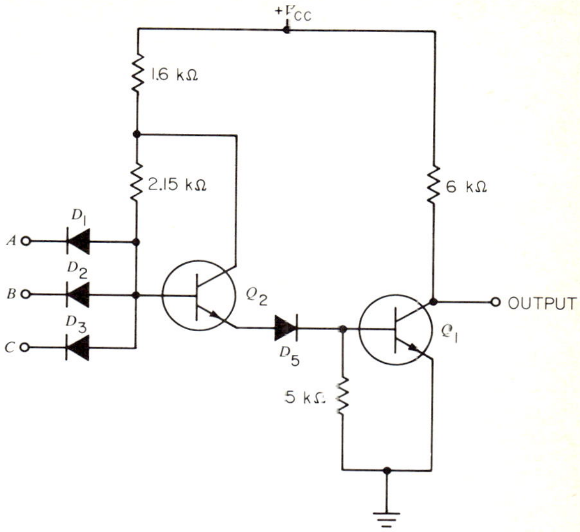

FIGURE 10.19
A variation of the circuit in Fig. 10.17.

will be brought very close to zero, and this will be the value of the output voltage.

From the foregoing description, we see that we have here a NAND logic circuit in which it takes a logic 1 (i.e., a strong positive pulse) at each of the three input terminals to produce a zero output. If any one of the inputs has a zero voltage, the output will be positive (i.e., a logic 1) even if the other two input terminals do have positive voltages on them. This is in accordance with the behavior of a NAND circuit.

A variation of Fig. 10.17 is shown in Fig. 10.19. Here, diode D_4 of Fig. 10.17 has been replaced by a transistor. Circuit operation is still the same, since the base-emitter diode of the transistor functions exactly as diode D_4, even to the point of providing the same voltage drop (0.7 V), assuming that the transistor is also fabricated from silicon.

The new arrangement provides greater driving current for the output transistor Q_1, enabling the circuit to drive more logic circuits. Also, the power-supply requirement is eased because only a single voltage is needed.

There is a modification of the circuit of Fig. 10.19 that is specifically designed to provide immunity under high-noise conditions, such as might exist in an industrial environment where there are usually many strong electrical noise pulses present.

The high-noise-immunity circuit is shown in Fig. 10.20. If this illustration and the previous one are compared, it will be seen that D_5 of Fig. 10.19 is replaced by a zener diode. Thus, where D_5 requires a voltage potential of 0.7 V to conduct, zener diodes requiring much greater driving voltages are available.

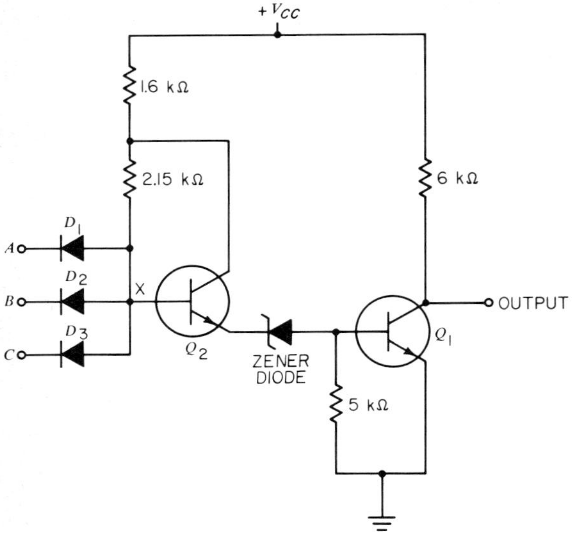

FIGURE 10.20
A high-noise-immunity DTL logic circuit.

Thus, if a 5-V zener diode is used in place of D_5, then a driving voltage at point X of 6.4 V positive will be needed to turn on the base-emitter of Q_2, the zener diode, and the base-emitter of Q_1. Contrast this with the 2.1 V only needed for Q_2, D_5, and Q_1 of Fig. 10.19.

Transistor-transistor Logic. A circuit which is essentially an outgrowth of diode-transistor logic and which has been gaining in popularity is the transistor-transistor logic (or TTL or T^2L) circuit. The basic arrangement is shown in Fig. 10.21. It contains an input transistor with multiple emitters, each of which takes the place of one of the input diodes in DTL circuits. What we have here essentially is a transistor with a collector, a base, and multiple emitters, each of which has its own connection. Such multiple-emitter transis-

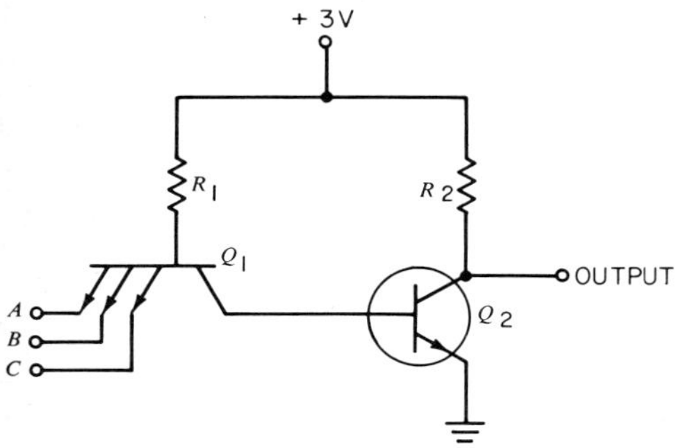

FIGURE 10.21
The basic TTL circuit.

tors can be fabricated in less area than corresponding diodes, resulting in lower capacitance and hence higher operating (i.e., switching) speed.

In operation, assume that +3 V are applied to input terminals A, B, and C. Since the base also has +3 V applied to it, the emitter-base junction does *not* conduct. (Remember that a silicon diode requires a potential difference of 0.7 V of the proper polarity to conduct.) However, the base-collector diode does have sufficient applied potential to conduct, and since Q_2 is in series with the collector of Q_1, it is turned on and conducts heavily. The output voltage thus drops to a very low value, generally about 0.3 to 0.4 V, this being the voltage appearing between collector and emitter $V_{CC(\text{sat})}$ of a transistor driven into saturation.

If one of the input emitters of Q_1 now has zero voltage applied to it (i.e., it is grounded), that emitter becomes forward-biased and current flows in the emitter-base circuit. This decreases the base potential of Q_1 sufficiently from its previous value, so that the base-collector diode of Q_1 cuts off and, with this, Q_2 also, since the latter is no longer receiving any current from Q_1. The voltage at the output terminal now rises to +3 V.

Thus, what we have here is a NAND logic circuit, since a zero input voltage at any input terminal produces a high output voltage but requires +3 V on all input terminals to produce a zero output voltage.

A TTL circuit that is in widespread use because it provides better performance than the circuit of Fig. 10.21 is shown in Fig. 10.22. Q_1 is the multiple-emitter transistor that provides the logic for the circuit, Q_2 is the phase splitter (and amplifier), Q_4 is the output transistor, and Q_3 serves to keep the output impedance low when Q_4 is nonconductive. The advantage of this will be seen from the ensuing discussion.

As before, with Fig. 10.21, if a positive voltage equal to $+V$ is applied to each of the three input terminals, the base-emitter section of Q_1 will be nonconductive. However, the base-collector diode will conduct, and there will be sufficient positive voltage on the base to turn on Q_2 and Q_4. Under these conditions, the output voltage will be equal to the saturated voltage between collector and emitter of Q_4, approximately 0.4 V.

Q_3 is held cut off when Q_2 and Q_4 are conducting, because the collector voltage of Q_2 is too low to turn Q_3 on.

If any one of the input terminals is grounded, that emitter-base portion of Q_1 conducts, lowering the base voltage to such an extent that insufficient voltage is present here to keep Q_2 and Q_4 on. Hence, both transistors turn off. For Q_2, this causes the collector voltage to rise sharply; and since the base of Q_3 connects to this collector, Q_3 is turned on. Finally, with Q_4 off, the voltage at the output terminal increases to $+V$.

From the foregoing behavior of this circuit, it is apparent that what we have here is a TTL NAND logic circuit.

Note that when Q_4 is on, Q_3 is off, and vice versa. To see how this contributes to better circuit performance, let us first consider the circuit of Fig. 10.21

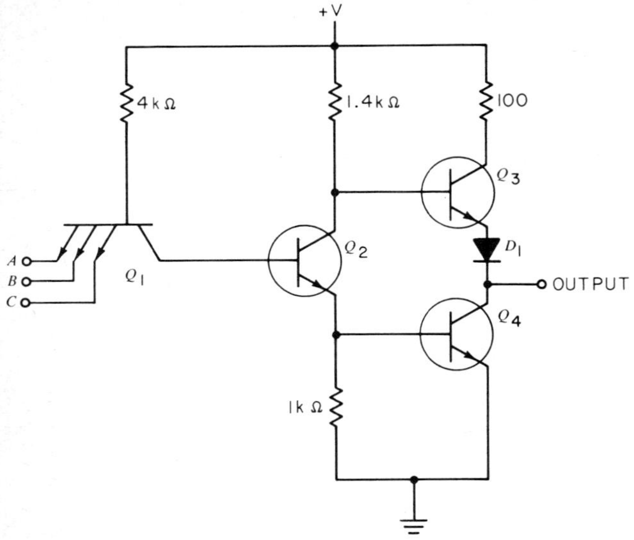

FIGURE 10.22
A TTL circuit in widespread use that provides better performance than the TTL circuit of Fig. 10.21.

(where Q_3 and Q_4 are not employed). It must be recognized that every logic circuit, be it Fig. 10.21 or 10.22, has a certain amount of capacitance across the output as a result of the other circuits that are connected to the output terminals. When, in Fig. 10.21, a zero voltage is applied to one of the input terminals, Q_2 cuts off and the output voltage rises and charges any capacitance present across the output terminals. This charging process will take place through R_2, and this resistor may have a value as high as 6,000 or 7,000 Ω.

On the other hand, when Q_2 is driven into conduction, the capacitance across the output discharges very quickly through the low resistance offered by a saturated Q_2. Thus, the charge and discharge times of output capacitance are quite unequal, and the longer charge time acts to decrease the switching time of this logic network.

Now, let us see how the Q_3, Q_4 combination alters this behavior. When Q_4 in Fig. 10.22 is turned on, any charged capacitance across the output terminals discharges very quickly through Q_4. When Q_4 is turned off and the output voltage jumps in value, Q_3 (as well as D_1) is turned on, and any capacitance across the output terminals will now charge very quickly through D_1, Q_3, and the 100-Ω resistor in the collector circuit of Q_3. There is thus very little difference in capacitor charge and discharge times, and the operating speed of the circuit has increased.

The TTL gate can be employed in arrangements other than a NAND circuit:

it can also be designed to operate as a NOR circuit or an AND circuit. All of these three configurations are commercially available from various semiconductor manufacturers.

Figure 10.23 shows the circuit for a TTL NOR gate. Each input consists of two stages in series with each other. Terminal A has Q_1 and Q_4 following it, and terminal B has a similar arrangement, using Q_2 and Q_3. Both Q_3 and Q_4 have a common-emitter resistor and a common-collector resistor. The output stage contains the same type of arrangement (Q_5, Q_6, and D_1) that appeared previously in Fig. 10.22.

In operation, a positive voltage applied to terminal A will reverse-bias the base-emitter circuit of Q_1. However, the collector base circuit will conduct and drive Q_4 into conduction. As a consequence, R_3 will develop a positive voltage that will turn Q_6 on and produce a very low level output voltage. At the same time, the voltage appearing at the collector of Q_4 will be low enough so that Q_5 will be kept nonconducting.

If instead of applying a positive pulse to terminal A we apply a positive pulse to terminal B, the same sequence of events will occur, except that we are now dealing with transistors Q_2 and Q_3 in place of the previous Q_1 and Q_4. The output will again have a low voltage value.

If we ground terminal A, the base-emitter circuit of Q_1 will conduct but the voltage at the base will drop low enough so that Q_4 will not be driven into

FIGURE 10.23
A TTL NOR gate.

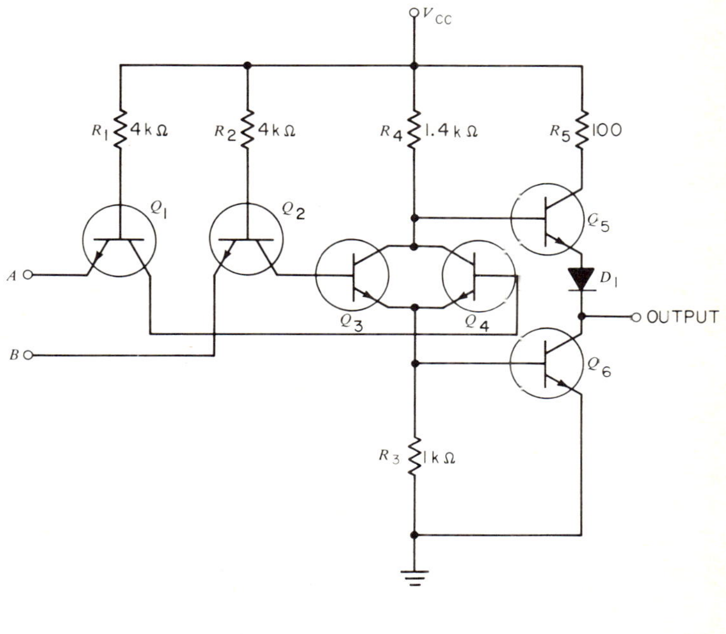

conduction. A zero voltage at terminal B will have the same effect on Q_2 and Q_3.

To summarize, then, a positive voltage at either A or B will produce a very low output voltage, whereas zero voltages at terminals A and B will produce a high output voltage with Q_5 conducting and Q_6 cut off. This is the action of a NOR logic gate circuit.

The TTL gate possesses a higher operating speed than either the DL or DTL circuits. It possesses good immunity to noise that may be present in the system and it can be designed to perform AND, OR, NAND, and NOR functions.

EMITTER-COUPLED LOGIC (ECL) OR CURRENT-MODE LOGIC

The logic circuits that have been discussed up to this point fall into the general category of saturated logic circuits, since the transistors in these circuits operate under saturation conditions when they are conducting. As we noted in an earlier chapter, both the emitter and collector inject carriers into the base region during saturation. The emitter normally does this under all circuit conditions; the collector only when it is forward-biased, as it is during saturation. When such a transistor is switched from the saturated condition to cutoff, it cannot fully return to an unsaturated state (in which there are no stored charges in the base region) until the charge carriers that have been stored in the base region during saturation have had a chance to be removed. This takes a short but finite time, and it is this action that results in slower operation for saturated transistors than we would obtain if the transistors never became saturated during conduction. Emitter-coupled logic circuits are designed so that at no time do they become saturated. As a result, such gates possess a very high switching speed, a feature that can be very attractive for many computer circuits.

A typical ECL circuit is shown in Fig. 10.24. The input stage is a differential amplifier in which Q_2 is given a fixed base bias of -1.15 V and the input signal (i.e., logic 1 or 0) is applied to the base of Q_1.

If we disregard Q_1 for a moment, we see that Q_2 is conducting, because even though a negative 1.15 V is applied to the P-type base, a rather large negative voltage (-5.2 V) is applied to the N-type emitter.

If we check the voltages in the base-emitter circuit of Q_2, we have -1.15 V at the base, which adds to a negative 0.75 V across the base-emitter junction. Opposing this is a -5.2 V applied at the bottom of R_3. The difference between the 5.2 V and the sum of the other two voltages (1.90) represents the voltage that is developed across R_3. This difference is 3.3 V, and to produce this voltage across a 1,240-ohm (Ω) resistor requires a current flow of 2.66 milliamperes (mA).

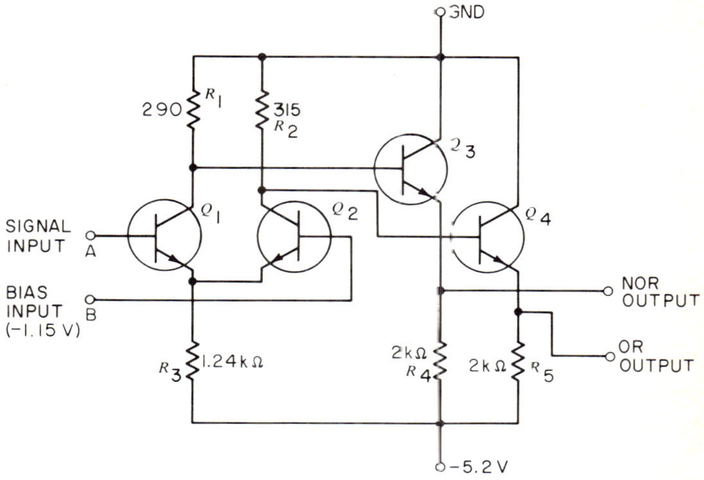

FIGURE 10.24
An emitter-coupled logic circuit manufactured by Motorola.

If we assume that 0.95 of the emitter current reaches the collector ($\alpha = 0.95$), then 2.53 mA flows through R_2, developing 0.8 V here (and placing the collector of Q_2 at −0.8 V with respect to ground). Since the base of Q_2 has −1.15 V on it, the collector-base junction is still reverse-biased and the transistor is not being driven into its saturated state. In other words, it is unsaturated.

Another way to tell that Q_2 is not saturated is to determine the voltage between collector and emitter. The collector of Q_2 is −0.8 V and the emitter −3.3 V, or a difference of 2.5 V. Were Q_2 saturated, its emitter-collector voltage would be on the order of 0.3 V.

Let us now apply a logic 0 pulse to terminal A. For this circuit, using the way the power-supply voltages are applied, a logic 0 pulse would be −1.55 V. When this is applied to terminal A, Q_1 does *not* conduct, because insufficient voltage appears across the emitter-base diode of Q_1 to turn that transistor on. This is true because the emitter potential (as we noted above) has a value of −1.90, and when −1.55 is applied to the base of Q_1, a difference of only 0.35 V is available to turn on the emitter-base diode section. Since at least 0.70 to 0.75 V is needed to do this, Q_1 remains cut off

With Q_1 cut off, the potential at its collector is zero volts. The connection from this point to the base of Q_3 places the same zero voltage here. Since Q_3 has a sufficiently large negative voltage on its emitter, this transistor is conducting, developing at its emitter a potential of −0.75 V, which represents the potential drop across the base-emitter diode of this transistor. The −0.75 V is equivalent to a logic 1 because it represents the most positive potential that will be obtained at the emitter of Q_3. Thus, with a logic zero applied to terminal A, the output from Q_3 represents a NOR output.

By the same token, Q_2 is conducting and its collector potential is −0.8 V. This value is applied to the base of Q_4, which when added to the −0.75 V dropped across its base-emitter circuit, provides an output at its emitter of −1.55 V. Since this value represents the low value at the output terminals, the output of Q_4 represents the OR output. Remember that Q_4 is conducting because of the high negative potential of the −5.2 V voltage source.

If we now apply a logic 1 signal to terminal A, the manufacturer recommends a potential of −0.75 V. If we subtract this −0.75 V from the −1.90 V present at the emitter of Q_1, we have a difference of −1.15 V, which is sufficient to turn Q_1 on. As Q_1 begins to conduct current, the tendency for the circuit is to have the voltage drop across R_3 start to rise. However, as soon as this begins, it reduces the potential that is placed across the base-emitter circuit of Q_2 so that this transistor begins to reduce its current flow. In essence, what happens is that Q_1 takes over the conduction of the current and Q_2 is essentially made nonconducting. The conditions are such that the voltage present across R_3 tends to remain fairly constant whether Q_1 is conducting and Q_2 is cut off or Q_2 is conducting and Q_1 is cut off. Also, whether Q_1 is conducting and Q_2 is cut off, or vice versa, neither transistor ever goes into the saturation state.

When Q_1 is conducting, the voltage that appears at the NOR output has a value of −1.55 V. By the same token, the output from the OR output terminal is −0.75 V. This gate can thus provide an output voltage that has either the same polarity as the input pulse or one that is 180° out of phase with it. The switchover time from Q_1 to Q_2 is extremely short because of the nonsaturated condition of the transistors.

MOSFET LOGIC CIRCUITS

Since field-effect transistors can perform the same functions as bipolar transistors, it is reasonable to assume that MOSFETs can be employed in logic circuits in essentially the same way as the bipolar arrangements just shown. A typical circuit of a NOR logic gate consisting of two MOSFET transistors and a single load resistor is shown in Fig. 10.25. There are two input terminals to which pulses can be applied. If a pulse has a voltage that is less than the threshold voltage for the MOSFET, no conduction will take place through the transistor. If low voltages are applied to both terminals A and B, there will be no current conduction in the circuit and the output at terminal C will be high.

By the same token, if voltages exceeding the threshold voltage are applied to both input terminals, then both transistors will conduct and the output voltage will be low.

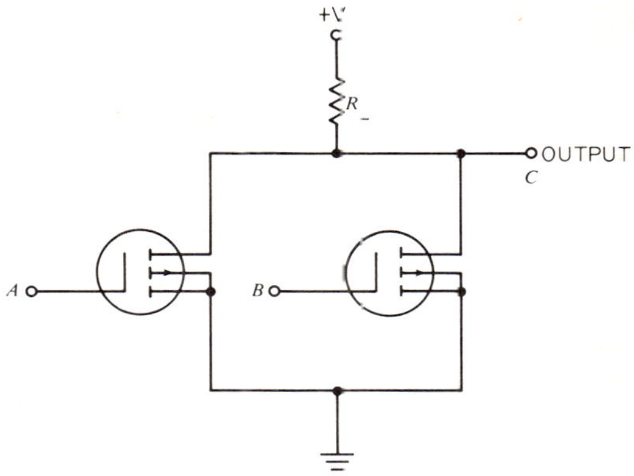

FIGURE 10.25
A simple NOR logic gate using MOSFET transistors.

A high voltage at either terminal produces a low output voltage irrespective of the condition at the other input terminal. On the other hand, if both input terminals have low voltages, the output is high. This is the operation of a NOR logic circuit.

The N- and P-channel MOS devices are related in the same way as PNP and NPN bipolar devices. That is, if the channel conductivity type (N or P) in a MOS device is reversed, the applied voltages must be reversed, thereby changing the current carriers. It is possible to employ this complementary behavior to form different types of logic circuits.

A complementary NAND gate is shown in Fig. 10.26. The P-channel devices are in parallel and the N-channel devices are in series. Unless all the inputs are high, the output of this NAND gate is high (i.e., a logic 1).

If only one or two of the inputs are high and the remaining input is low, the N-channel MOSFET receiving the low voltage will remain off and the output voltage will be high. This is because the P-channel unit receiving this low voltage will be on and will provide a low-resistance path from the power supply ($+V$) to the output terminal.

However, if all inputs are high, the three series-connected N-channel units will conduct, whereas the P-channel devices will be cut off. The output voltage will now be zero, for a logic 0 state.

Figure 10.27 is a complementary NOR diagram using MOSFETs. Note that the order in this circuit is reversed. That is, the P-channel devices are in series and the N-channel devices are in parallel.

The result is that if any of the inputs is high (positive), one of the parallel N-channel devices is on and the output is in the logical 0 state. Logical 1 can only be obtained if all inputs are low, turning both P-channel devices on.

MOSFET logic circuits offer several significant advantages that make them quite attractive. Their circuitry is simple, in many cases requiring only a resistor or two. Their input impedance is very high, which means that many

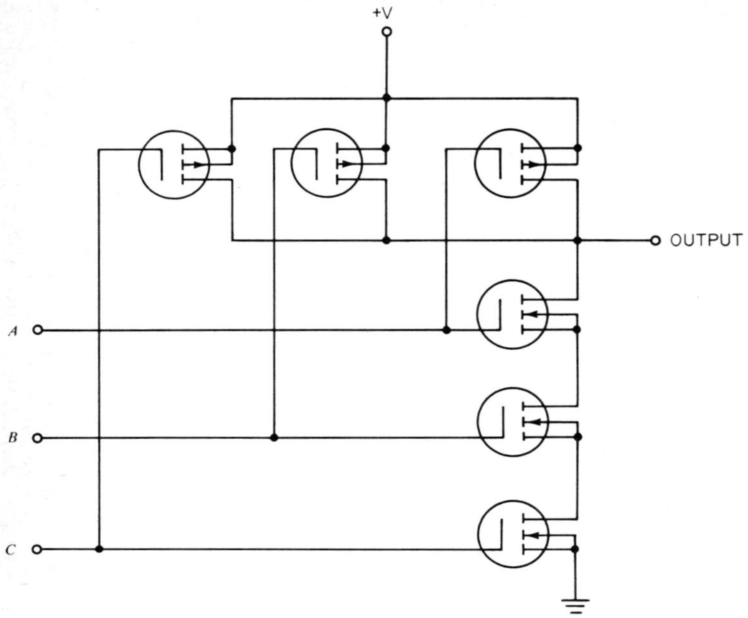

FIGURE 10.26
A complementary NAND gate using P- and N-type MOSFETS.

FIGURE 10.27
Complementary NOR gate using P- and N- type MOSFETS.

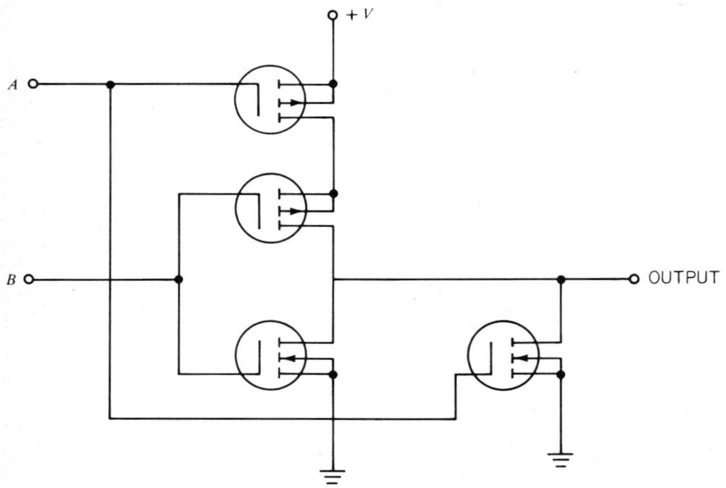

logic gates can be attached to the output of a prior logic gate without producing an undue loading effect. Current drawn by MOSFETs when they are not responding to high input pulses is very low, reducing the power needed from power supplies but also causing less heat development within the device itself. In today's closely packed integrated circuit structure, this is a highly desirable feature.

MOSFETs will also be found to produce high-output-signal swings or variations. On the disadvantage side, a MOSFET logic circuit will introduce a longer delay in signal passage through it than a bipolar and this will limit applications in high-speed computer circuits.

HALF-ADDERS AND FULL ADDERS

The basic structure of the various logic gates (OR, AND, NOR, NAND) and their mode of operation are fairly straightforward and relatively easy to understand. From these basic building blocks come the multitude of operations that computers are capable of performing.

Half-adder. As a simple illustration, consider the manner in which two binary numbers can be added. Figure 10.28 shows a logic circuit that can add 0 and 1 and provide an output not only when there is a single digit but also when 1 plus 1 produces 10.

Let us examine the operation of Fig. 10.28 for the four possible combinations of inputs to A and B. These combinations are listed in the truth table shown in the illustration.

When $A = 0$ and $B = 0$, AND gate 1 will have the zero from A applied to it and the 1 from line B because of the inverter I_1 that is interposed between terminal B and AN 1. Under these conditions, no output (i.e., 0) is obtained at the output of AN 1.

The same situation occurs at AN 2 only here an inverter I_2 is in the A line leading to AN 2. Thus, OR gate 1 receives two zero inputs and produces a zero output.

AN 3, a third AND gate, likewise has two zero voltages applied to it and a zero output is developed. All this is indicated by the first line of the truth table shown in Fig. 10.28.

When $A = 0$ and $B = 1$, AN 1 produces a zero output because AN 1 has two zero signals applied to it. AN 2, however, receives two 1 signals because the zero signal at A is inverted to a 1 by I_2. Thus, AN 2 produces a 1 output signal. OR 1 likewise produces a 1 output signal because its input is a zero signal from AN 1 and a 1 from AN 2. The third AND gate, AN 3, develops a zero output because of the dissimilarity of its input signals.

For the third condition, when $A = 1$ and $B = 0$, the same action occurs and again there is a 1 output at the S terminal and a zero at the C terminal.

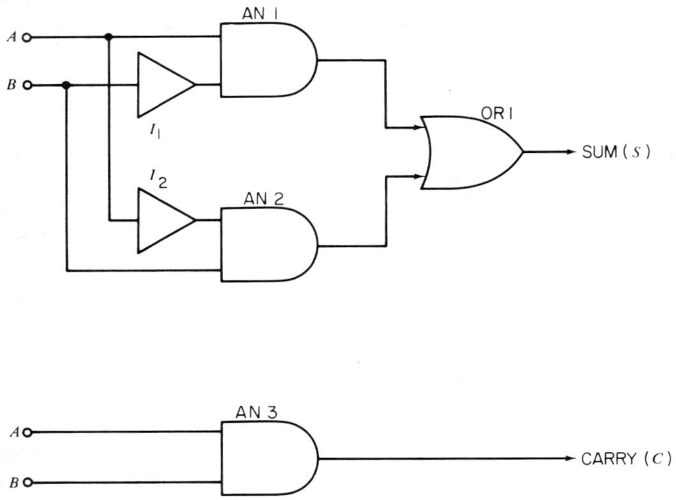

(a)

A	B	S	C
0	0	0	0
0	1	1	0
1	0	1	0
1	1	0	1

(b)

FIGURE 10.28
(*a*) Half-adder and (*b*) its truth table.

The final combination in the truth table, when $A = 1$ and $B = 1$ does serve to produce an output at the carry terminal. With A and B both 1, neither AN 1 or AN 2 produce a 1 output because of the presence of I_1 and I_2. Hence, at terminal S, a zero is developed.

However, at AN 3, a 1 output does appear, which would be carried over to the next column of binary figures to be added.

We might pause here for a moment and note that if we consider merely the logic-gate combination that produces the sum output in Fig. 10.28, we have a circuit known as an exclusive-OR circuit. This arrangement will produce a 1 output when either A or B, but not both, is 1.

Full Adder. Whereas a half-adder will add two binary numbers together, it does not have any capability for handling carrys from any previous additions.

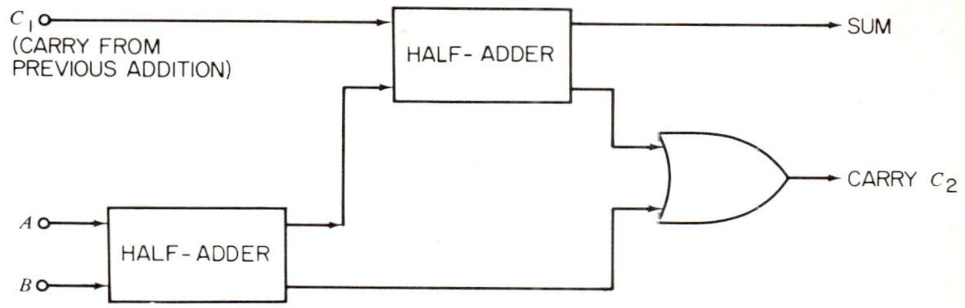

FIGURE 10.29
Block diagram of a full adder.

To deal with this situation, a full adder is required. A typical full adder is shown in Fig. 10.29, with a further elaboration of the AND and OR gates indicated in Fig. 10.30. The truth table in Fig. 10.30 shows the various combinations of input signals possible, including a carry from a previous addition. It should be a fairly easy procedure for the reader to verify that the indicated outputs are indeed obtained for the eight given inputs.

COMPARATOR

Another useful application of logic gates is the two-bit comparator circuit shown in Fig. 10.31. This circuit compares the two inputs A and B. If they are equal, a 1 is developed at the output; if they are unequal, a 0 is produced. These conditions, for various combinations of input, are listed in the truth table, also shown in Fig. 10.31.

The comparator consists of two AND gates, two inverters, and an OR gate. When $A = B = 0$, AND gate 2 produces a 1 output because of the two inverters that precede it. When $A = B = 1$, AND gate 1 produces a 1. For the remaining two conditions, neither AND gate produces a 1 and the OR gate output is zero, denoting the fact that A and B are not equal.

FLIP-FLOPS

Up to this point we have investigated the mode of operation of a variety of logic gates and how they achieve their decision-making functions. However, it must be recognized that for many of the purposes of computation, it is necessary that some form of memory system be incorporated into the circuitry to perform such functions as counters, accumulators (accumulating counts as basic addition), ON and OFF switches, temporary-storage registers, plus any other process that depends upon information that had previously been incorporated into a system.

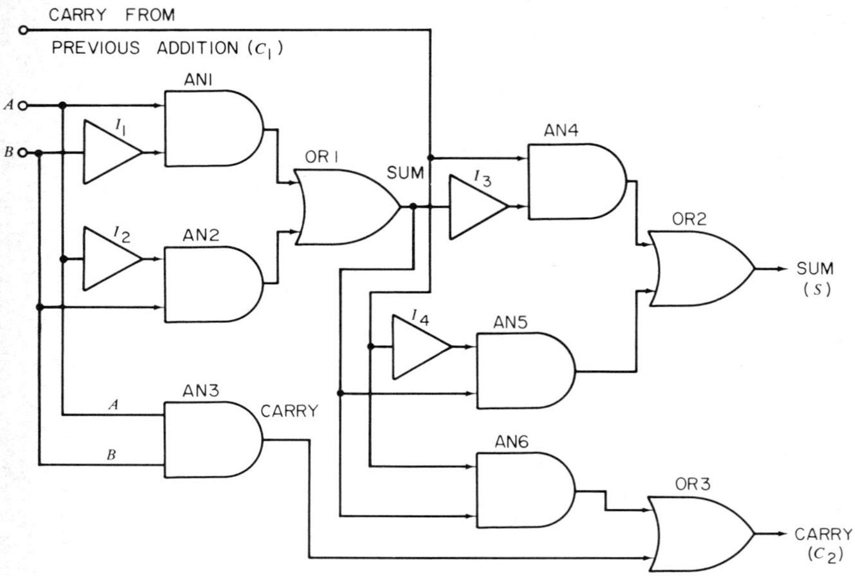

INPUT			OUTPUT	
A	B	C_1	S	C_2
0	0	0	0	0
1	0	0	1	0
0	1	0	1	0
0	0	1	1	0
1	1	0	0	1
0	1	1	0	1
1	0	1	0	1
1	1	1	1	1

FIGURE 10.30
(a) The individual gates forming a full adder.
(*b*) The truth table of a full adder.

The basic electronic digital memory or storage circuit is the bistable multivibrator, or flip-flop. We discussed multivibrators previously in Chap. 7 and touched briefly on both the astable multivibrator and the bistable multivibrator. It was noted that the astable multivibrator oscillates continuously, with the current alternately flowing first through one transistor and then through the other. This switching back and forth represents the stable condition for this circuit, so that in essence it can be said to possess one stable condition.

On the other hand, a bistable multivibrator, as its name suggests, has two stable states. That is, one transistor will conduct while the other transistor is cut off, and the circuit will remain in this condition indefinitely until an incoming pulse causes the conducting transistor to cut off, forcing the transistor

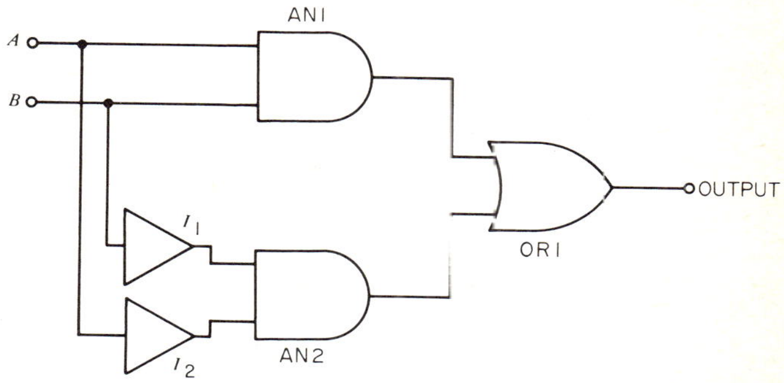

(a)

A	B	OUTPUT
0	0	1
1	0	0
0	1	0
1	1	1

FIGURE 10.31
A two-bit comparator and its truth table.

(b)

previously nonconducting to become conducting. Again, the bistable multivibrator will remain in this condition until it is pulsed back to the alternate condition.

It should be noted that the astable multivibrator cannot serve as a memory, since its condition is entirely dependent upon the values of the components existing within its own oscillating circuit. A bistable multivibrator, on the other hand, does not in itself possess a special frequency but shifts from one conducting state to the other, dependent upon the frequency of the pulses applied to it. Furthermore, the bistable multivibrator will remain in one state indefinitely until pulsed into the opposite state. Since it will produce one value of voltage at a given output terminal when one transistor is conducting and the other is not, and another value of voltage when a switchover occurs, we can readily determine what the previous state of this flip-flop was and use this as a basis for initiating further action by the circuit.

The bistable multivibrator shown previously in Fig. 7.14 is reproduced in Fig. 10.32. When Q_1 is conducting, its collector voltage is very close to zero volts. At the same time, Q_2 is not conducting and its collector voltage is essentially equal to the battery voltage applied to the collector. Under these condi-

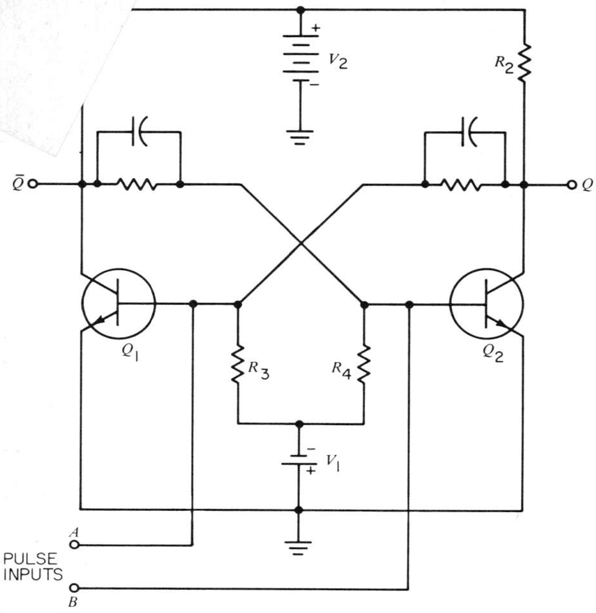

FIGURE 10.32
A bistable multivibrator.

tions, we might say that point Q is in a logic 1 condition and point $\bar{Q}$ is in a logic 0 condition.

If we now switch conducting states, where Q_2 is conducting heavily and Q_1 is cut off, then Q has a logic 0 value and $\bar{Q}$ has a logic 1 value. This information can be fed to other circuits in order to count the number of times that a changeover has occurred in this circuit; this, in turn, will be dependent upon the number of pulses that have come into this circuit. Thus, we see how a bistable multivibrator can be employed for counting.

RS **Flip-flop.** It is interesting to approach the construction of a flip-flop using two NOR logic gates. A typical two-input NOR logic gate is shown in Fig. 10.33. Briefly, the output voltage at the collectors of the two transistors will be essentially equal to V_{CC} when both A and B are at zero volts. Under these conditions neither Q_1 nor Q_2 will conduct. However, if a positive voltage is applied to either terminal A or B, the transistor receiving that positive base voltage will conduct strongly and the output voltage will drop very near to zero volts. Thus, this is a NOR logic circuit for all logic 1 input signals. Also shown in Fig. 10.33 is the logic symbol for this gate as well as the truth table that indicates the manner in which it operates.

Incidentally, the gate in Fig. 10.33 is a NOR gate for positive logic. Let us see, however, what would happen if we applied negative logic signals to the

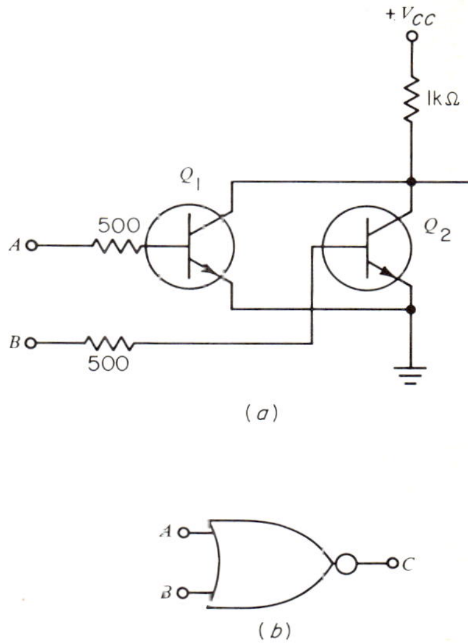

A	B	C
0	0	1
1	0	0
0	1	0
1	1	0

(c)

FIGURE 10.33
(*a*) Schematic diagram, (*b*) logic symbol, and (*c*) truth table for a two-input NOR gate.

same circuit. Assume that a pulse of +2 V represents a logic 0 and zero volts represents a logic 1. We will find that for a logic 1 signal, the output is $+V_{CC}$ and for a logic 0 input voltage (i.e., +2 V) the output is close to zero volts. Thus, logic 1 signals at terminals A and B will produce a high output, which in a negative logic system represents a logic 0. On the other hand, a logic 0 voltage at A or B will produce a logic 1 at point C.

The truth table for these applied signals is shown in Table 10.4. It is seen to be the opposite of the truth table for the same circuit when positive

TABLE 10.4
Truth Table for a NAND Gate

A	B	C
0	0	1
0	1	1
1	0	1
1	1	0

logic is employed. If we now examine the definition for a NAND gate, we see that it fits this truth table. That is, a NAND gate is a gate whose output is a logical 1 unless all inputs are logical 1—which is why gate manufacturers will frequently label a gate such as that shown in Fig. 10.33 as a two-input NAND/NOR gate.

If we take two of the NOR logic gates of Fig. 10.33 and connect them back to back, we obtain the circuit shown in Fig. 10.34a, which is another form of

FIGURE 10.34
An *RS* flip-flop, with its logic symbol and truth table.
(*a*) schematic diagram; (*b*) logic symbol; (*c*) truth table.

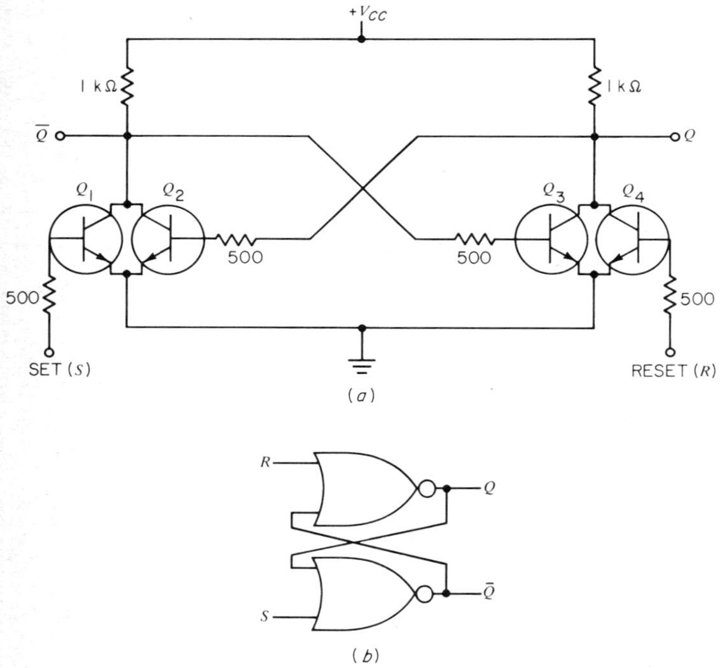

R	S	Q^{n+1}
0	0	Q^n
0	1	1
1	0	0
1	1	0

(c)

FIGURE 10.35
A widely employed representation for *RS* flip-flops.

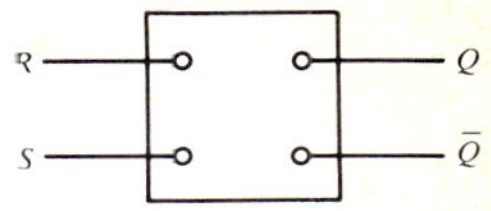

the bistable multivibrator where input pulses to change the state of the multivibrator, formed by Q_2 and Q_3, can be applied to the bases of Q_1 and Q_4. Thus, where we used only capacitors to bring input pulses into the circuit in Fig. 7.13, here we are doing this through two transistors, Q_1 and Q_4, which are connected in parallel with the other two transistors that form the bistable multivibrator circuit.

To produce a logic 1 at the output terminal Q and a logic 0 at the output terminal of $\bar{Q}$ (assuming positive logic), it is necessary that Q_2 be conducting strongly and that Q_3 be cut off. To achieve this state, we can apply a logic 1, or positive, voltage to the terminal marked S, which will cause Q_1 to go strongly into conduction and thereby bring the voltage at terminal $\bar{Q}$ essentially to zero. With a zero voltage at $\bar{Q}$, the base of Q_3 will not be receiving any current, so that Q_3 will be cut off and its collector voltage will be essentially equal to $+V_{CC}$. This high voltage at Q will drive Q_2 into conduction, so that its state is the same as the transistor paralleling it, Q_1.

To achieve a switchover, where $\bar{Q}$ has a logic 1 at its terminal and Q has a logic zero at its terminal, we apply a positive voltage to terminal R. This will cause Q_4 and Q_3 to conduct strongly while Q_1 and Q_2 are cut off.

Note that once the circuit has been placed in a certain state, removal of any input pulses at terminals R and S will not cause that state to change. Only when an input pulse is brought into the circuit that causes a conducting transistor to cut off or a nonconducting transistor to turn on will a switchover occur.

The two outputs are always at opposite levels except when R and S both receive positive pulses. When R and S are both 1, Q_1 and Q_3 are both on and Q and $\bar{Q}$ are both zero.

The logic symbol and truth table for the RS flip-flop are shown in Fig. 10.34. Mention might be made here of the fact that there are actually two logic symbols for this circuit: one is shown in Fig. 10.34, where its origin from a NOR gate is clearly evident; and the other one is shown in Fig. 10.35, where it is simply represented by a rectangular box with four labeled terminals. Both methods of representation of an RS flip-flop will be encountered in the literature.

If we examine the truth table, then conditions 2 and 3 represent the two states that were just discussed. When R is zero and S is 1, the output at terminal Q is 1. The notation Q^{n+1} stands for the state of the flip-flop or the output level of Q that exists after the particular R and S input signals have been applied. The term Q^n represents the state of the flip-flop that

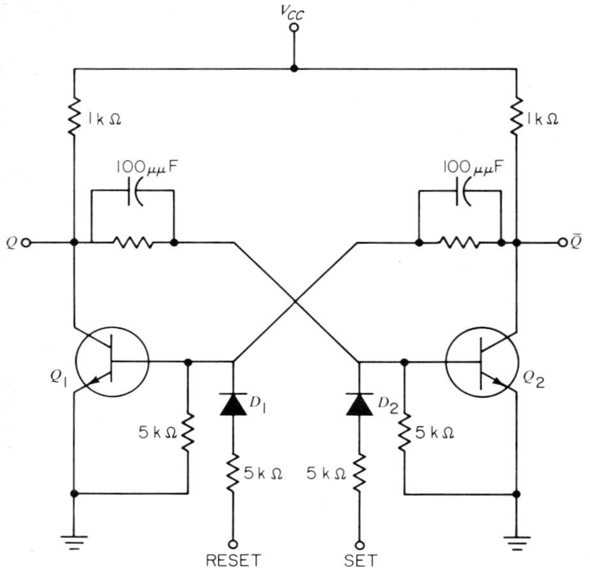

FIGURE 10.36
Another means of bringing triggering pulses into a bistable multivibrator.

existed prior to the application of the input signals to terminals R and S. The term Q^n is shown in the first line of the truth table when zero voltages are applied simultaneously to terminals R and S. If prior to the application of these zero voltages Q had a logic 1 voltage at its terminal and $\bar{Q}$ had a logic 0 voltage at its terminal, then this situation would tend to remain if the two input voltages were applied simultaneously. However, since there may be and generally will be some slight difference in the application of input voltages, the voltage that arrives last will tend to be the controlling or prevailing voltage and it will establish the condition for the flip-flop. At best, this is an indeterminate situation and one that designers avoid.

Likewise, if simultaneous logic 1 voltages are applied to terminals R and S, the two outputs will be zero. Here, again, if one pulse should arrive slightly later than the other one, then it is possible that it will act to alter the condition of the circuit. Thus, this fourth situation tends to become indeterminate and is also avoided by circuit designers.

The significance of the words "set" and "reset" stems from the fact that when a positive pulse is applied to the RESET terminal, the voltage at point Q goes to zero. This can be looked upon as clearing the flip-flop, that is, putting the output terminal, which is generally taken from Q, to a zero state. At the same time, $\bar{Q}$ goes to 1. In place of reset, the term "clear" is sometimes employed. This again relates to the action of clearing the multivibrator just discussed.

To produce a logic 1 at the Q terminal, we would apply a positive pulse to the SET input. This can be looked upon as setting the flip-flop by applying the positive pulse to the SET terminal.

Another way of bringing triggering pulses into the flip-flop is shown in the circuit of Fig. 10.36. Both input lines have a 5,000-Ω resistor and a diode; when a positive pulse is applied to one of these terminals (for example, SET), this positive pulse will tend to turn transistor Q_2 on. If Q_2 happens to be on at the time of arrival of this positive pulse, then its condition will be unchanged. However, if Q_2 is off at this point, then the positive pulse will turn it on and, through the action of the circuit, cause Q_1 to turn off.

RS/T **Flip-flop.** In most applications of flip-flops in computer circuits, it is desired for control of the action of the flip-flop to stem from a specific controlling oscillator elsewhere in the system, called a "clock," so that the control for the operation of the entire system rests at a central point. The flip-flop of Fig. 10.37 provides not only for SET and RESET terminals but also for reception of such clock input pulses from a central point.

FIGURE 10.37
An *RS* flip-flop whose state can be controlled by incoming clock pulses.

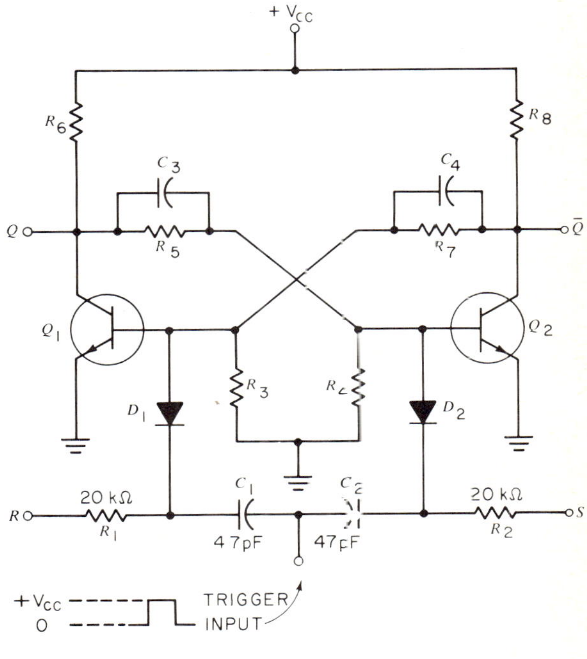

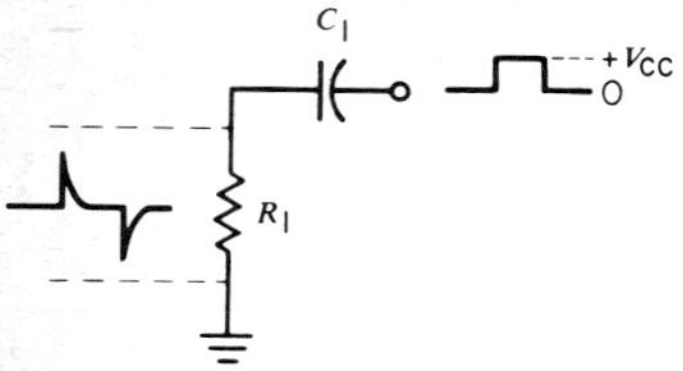

FIGURE 10.38

The specific effect that a trigger, or clock, input pulse will have on Q_1 and Q_2 will depend in this circuit upon the voltages present at R and S at the time that the trigger pulse arrives.

To illustrate, assume that R has a zero voltage applied to it while the voltage at point S is positive by an amount equal to the supply voltage V_{CC}. Under these conditions, R could be said to represent a logic 0 condition and S to represent a logic 1 condition. Now, let us see what happens when a positive trigger pulse having an overall amplitude equal to $+V_{CC}$ arrives. Since R has a zero potential, in essence this point is grounded. Therefore, if we concentrate simply on the circuit formed by C_1 and R_1 and redraw it as shown in Fig. 10.38, we will find that because of the values selected for C_1 and R_1 the square trigger pulse produces a two-peaked waveform across R_1. The first peak, the positive peak, occurs when the voltage at the trigger input terminal goes from zero to the value $+V_{CC}$. This change in circuit conditions across C_1 and R_1 produces a sharp peak of current, after which capacitor C_1 is charged and no further current flows in that circuit. The positive peak developed across R_1 by the leading edge of the trigger pulse cannot get through diode D_1 because the positive voltage is being applied to the cathode and the diode remains nonconducting.

However, when the lagging edge of the trigger pulse occurs, a sharp negative pulse of current flows through the circuit and develops a negative pulse across R_1. This negative pulse is strong enough to drive D_1 into conduction, enabling the negative pulse to reach the base of Q_1 and turn this transistor off. By this action, Q_1 is brought into cutoff and Q_2 is turned on.

We might also see what happens in the C_2R_2 circuit for the same incoming trigger pulse. Since point S is returned to a positive voltage $+V_{CC}$, no current flows in the C_2R_2 circuit because the positive voltage at point S counterbalances the positive leading edge of the trigger pulse. For the lagging edge of the trigger pulse, when the voltage drops from $+V_{CC}$ to zero, the negative pulse that is developed in the C_2R_2 circuit simply brings the voltage at the junction of R_2C_2 close to zero and this does not produce sufficient voltage to turn D_2 on.

To summarize: because of the voltages existing at point R and point S, only diode D_1 is brought into conduction and transistor Q_1 is cut off. Diode D_2 is not brought into conduction for the same trigger input pulse and therefore Q_2 is not affected by the incoming trigger pulse.

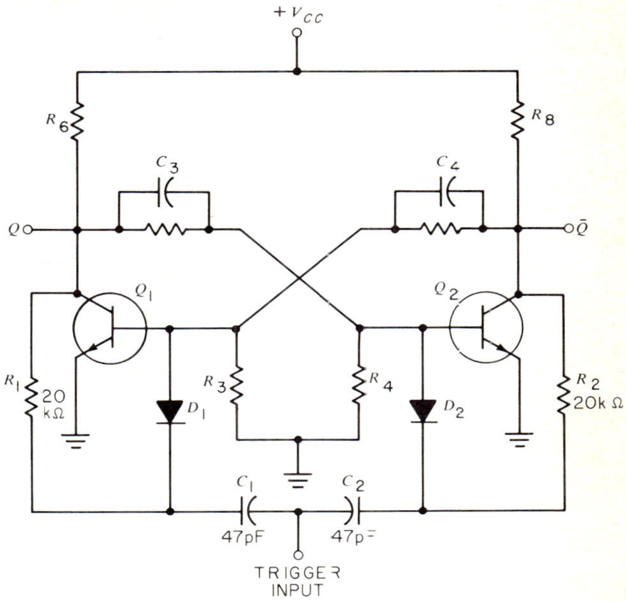

FIGURE 10.39

If we were now to reverse the voltages at terminals R and S and then apply the same trigger pulse, we would find that Q_2 would now be turned off and Q_1 would be brought into conduction because of the forced turnoff of Q_2. It will be noted that the voltages that should be applied to terminals R and S should be opposite in value so that the incoming trigger pulses will automatically be directed to the transistor that is conducting in order to turn it off. In other words, if there is a logic 0 voltage at terminal R, there should be a logic 1 voltage at terminal S. Such voltages are available within the circuit itself at the output terminals Q and $\bar{Q}$. Hence, if we were to connect terminal S to $\bar{Q}$ and terminal R to Q, the voltages at the R and S terminals would then automatically be correct and all incoming trigger pulses from the clock generator would be directed to the proper transistor. For these conditions, the circuit would appear as shown in Fig. 10.39. Now, the condition of the output voltages at Q, $\bar{Q}$ is directly determined by the incoming trigger pulses, and every time such a pulse is applied, there is a change in voltage condition at these output terminals. This provides the type of complete control that is essential in most computer circuits.

The symbol for the flip-flop shown in Fig. 10.37 is indicated in Fig. 10.40a. Note that only the input and output terminals are indicated. If we internally connect the S and R terminals to the Q and $\bar{Q}$ terminals, then we need only show the two output terminals and the input triggering terminal. The

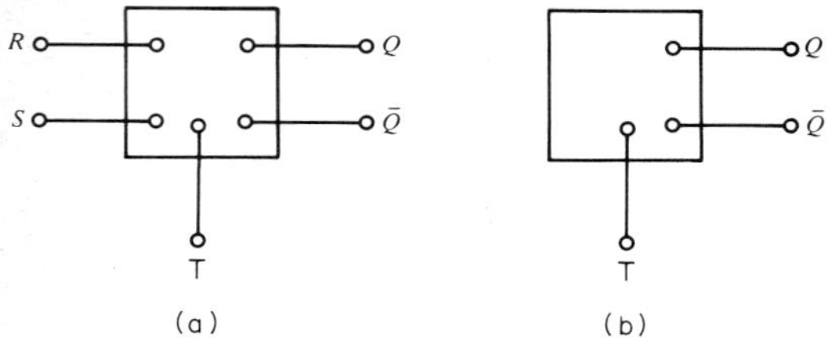

FIGURE 10.40
Two methods of representing *RS/T* flip-flops.

block diagram could now be drawn to the form shown in Fig. 10.40*b*, which is a very simple way of depicting flip-flops and greatly facilitates drawing a system in which a number of these units are employed.

Counters. Flip-flops can be connected together to develop a counter. For example, suppose that it is desired to count the number of pulses being generated by a clock. Let us further suppose that we wish to count up to 32. By arranging four flip-flops in the manner shown in Fig. 10.41, we can count to 16. Initially, all of the flip-flops would be cleared so that a zero voltage would appear at each Q terminal. Then, on the application of an input pulse, flip-flop FF1 would respond to this pulse when the incoming pulse went from level 1 to level zero and produce at Q_1 a value of logic 1 level. Flip-flop FF2 would not be effected by this action, even if Q_1 were connected to the trigger input terminal of FF2, because the voltage at Q_1 would be going from zero to a positive level and the flip-flop shown here will only be triggered when the input level goes from a positive voltage to zero.

When the second input pulse at FF1 arrives, Q_1 will be brought to zero on the downward swing of the lagging edge of this second pulse, and this negative swing at Q_1 will trigger FF2 and cause its Q output to go to a positive level representing logic 1. Thus, at this point, FF1 shows a zero at its output, FF2 shows a 1 at its output, and flip-flops 3 and 4 both show zeros. The result is 0010, which is the binary number for 2.

By following through this line of reasoning and using the series of waveforms shown in Fig. 10.42, we can see that the combination of these 4 flip-flops will pass through the 16 states shown in Table 10.5. When the sixteenth

FIGURE 10.41
Four flip-flops, connected as shown, will count up to 16.

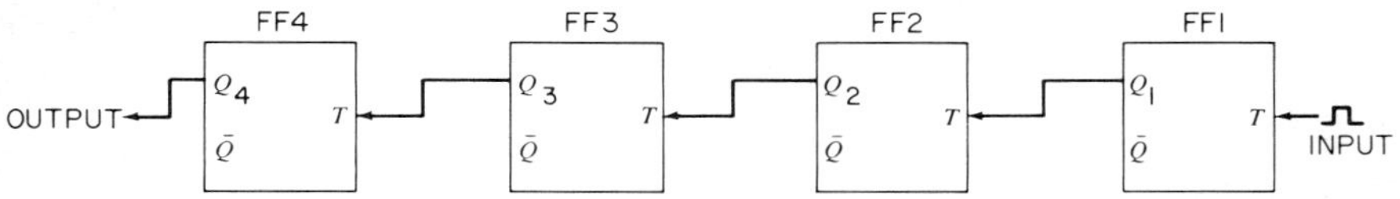

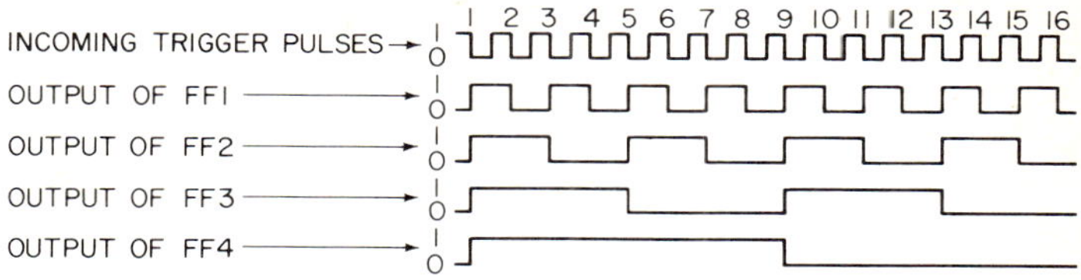

FIGURE 10.42
Relationship of the input clock pulses to the circuit of Fig. 10.41 and the pulses appearing at the output of the four flip-flops that constitute this system.

pulse has been applied, then all the flip-flops will be in the logic 1 state and the next flip-flop to be applied to this chain will serve to clear all of the flip-flops and produce a logic 0 at their outputs. The counter is now ready to go through this same series of events until we once again reach number 16.

To count to a higher number than 16, we simply add more flip-flops to the chain. In this fashion we can count as high as we desire and a readout

TABLE 10.5
All Possible States of the Four Flip-flops Shown in Fig. 10.41

	Four Flip-flops			
State	*FF1*	*FF2*	*FF3*	*FF4*
1	0	0	0	0
2	1	0	0	0
3	0	1	0	0
4	1	1	0	0
5	0	0	1	0
6	1	0	1	0
7	0	1	1	0
8	1	1	1	0
9	0	0	0	1
10	1	0	0	1
11	0	1	0	1
12	1	1	0	1
13	0	0	1	1
14	1	0	1	1
15	0	1	1	1
16	1	1	1	1

device can be attached to the output of the chain in order that the actual number of pulses received to a specific point in time will be indicated visually on this readout. Obviously, if we are to show the readout in conventional decimal fashion, we will need a circuit to convert the binary pulses to the equivalent decimal figure.

We have, of course, barely touched on the subject of counters and their associated circuits. However, the foregoing will show the reader how these circuits are developed and what their relationship is to logic gates.

QUESTIONS

10.1. What are the major sections of a digital computer? Explain briefly what each section does.

10.2. What do we mean by symbolic logic? How is it used?

10.3. What differences exist between the binary system of numbers and the decimal system?

10.4 Distinguish between an AND function and an OR function. Which is represented by a parallel circuit? Which is represented by a series electrical circuit?

10.5. How do we represent the inverse of an action in logic notation?

10.6. Write the following numbers in binary notation: 672, 39, 271, 402, and 1,200.

10.7. Subtract 10010100 and 11100110.

10.8. Divide 1011101000 by 11000.

10.9. Draw the circuit of a diode AND circuit having four input terminals. Explain briefly how it operates.

10.10. Differentiate between an OR and a NOR circuit. Draw the truth table for both functions, assuming three inputs and a single output.

10.11. Draw the symbols for the following circuits: OR, NAND, NOR, and inverter.

10.12. Does a logic 1 signal have to have a voltage of 1 V? Explain.

10.13. What is the difference between positive and negative logic?

10.14. Define the following: DL, DTL, RTL, T^2L, and ECL.

10.15. Draw the diagram of the basic DTL circuit and explain its operation.

10.16. What would happen if negative logic pulses instead of positive logic pulses were employed in Fig. 10.17?

10.17. How does the DTL circuit of Fig. 10.20 achieve its greater immunity to noise?

10.18. How does the output circuit of Fig. 10.22 help this circuit achieve faster operating speeds?

10.19. How does the circuit of Fig. 10.23 achieve the NOR function?

10.20. Why are saturated transistor circuits slower operating than the same circuit when the transistors are not driven into saturation?

10.21. How can you tell whether or not a transistor in a given circuit is operating under saturated conditions?

10.22. Draw the circuit of a MOSFET NOR logic circuit. Explain briefly how it operates.

10.23. Show how a half-adder can add two binary numbers.

10.24. How does the circuit of Fig. 10.31 compare two binary numbers?

10.25. What is the difference between astable and bistable multivibrators? Which is capable of serving as a memory?

10.26. Why can the gate circuit shown in Fig. 10.33 be labeled as a two-input NAND/NOR gate? Explain.

10.27. Explain the operation of the circuit of Fig. 10.34*a* when Q is at a logic 0 and a positive pulse is applied to the SET terminal.

10.28. How do the voltages at the R and S terminals of the circuit of Fig. 10.37 determine to which transistor the input trigger pulses go?

10.29. Draw the circuit of a counter capable of counting up to 64.

Chapter Eleven

Transistors in Television Receivers

The task of producing an all-transistor television receiver is considerably more difficult than that of bringing forth a commercial all-transistor radio receiver for two reasons. First, many of the circuits in television receivers operate at fairly high frequencies; second, the power requirements of a number of stages, particularly those in the vertical- and horizontal-deflection systems, are quite high. Both these requirements must be met by transistors that are economical, stable, and (within each type) uniform in characteristics from unit to unit.

Transistorized television receivers are being produced in considerable quantity and, in a few years, will be the only type of receiver available. This is true for black-and-white receivers as well as color-television receivers.

In this chapter, the various sections of a television receiver will be analyzed and typical circuits will be discussed.

Considerable research is being done by the industry in this direction, and transistors that meet the foregoing criteria are gradually being developed. Already, several fully transistorized television receivers have been designed and marketed. The discussion to follow will analyze the various sections of a television receiver and indicate typical circuits that can be employed in them.

The block diagram of a black-and-white television receiver is shown in Fig. 11.1. The number of stages contained in each box is not shown and, to a certain extent, is not important. What is important is that the function represented by that box be achieved.

THE r-f STAGES

The front-end section of the receiver contains the tuner and the r-f amplifier, mixer, and local oscillator. In the vhf band, signal frequencies extend from 54 to 88 megahertz (MHz) and from 174 to 216 MHz. For uhf reception, a frequency coverage from 470 to 890 MHz is required. The local oscillator

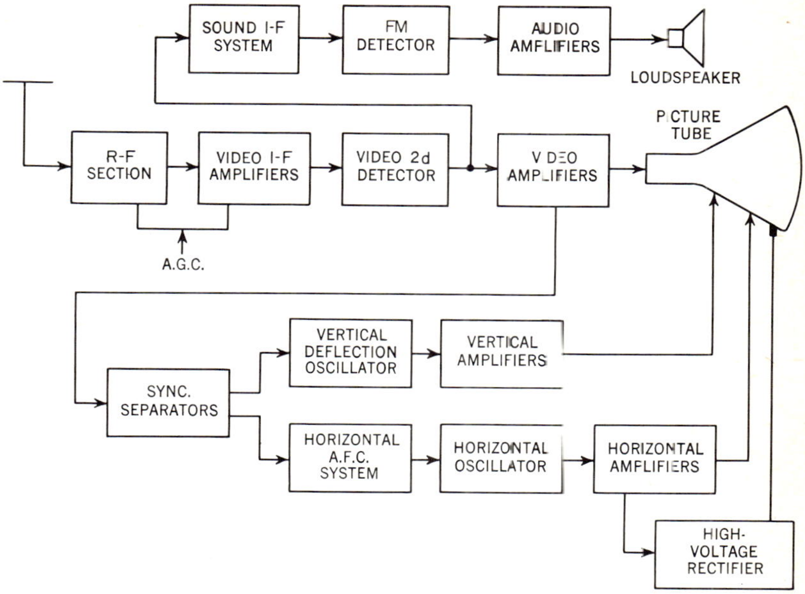

FIGURE 11.1
A simplified block diagram of a black-and-white television receiver.

often generates frequencies that are 25 to 45 MHz above the incoming signals. For a transistor to be usable in these stages, it not only must be capable of operating at these frequencies but must also do so with a fair amount of gain. Also, the noise factor of a transistor is important here because of the very low level of the incoming signal.

Of the three front-end stages, the requirements of the local oscillator are probably the easiest to satisfy, because noise is not a significant operating characteristic and transistors will almost always oscillate at considerably higher frequencies than their cutoff value. On the other hand, for purposes of amplification, it is desirable to keep well below the transistor cutoff frequency, and this will restrict transistor application.

The r-f Amplifier. The r-f amplifier is perhaps one of the most important stages in the television receiver. The signal that it receives from the antenna is exceedingly weak, perhaps no more than 30 or 40 microvolts (μV). It is the function of the r-f amplifier to amplify this signal to whatever extent it can, at the same time keeping the amount of noise voltage that it (the amplifier) introduces at the lowest possible value. Thus, selection of a suitable r-f transistor is primarily based on two factors: (1) an ability to amplify at the television frequencies and (2) as low a noise figure as possible.

A widely employed configuration for the r-f amplifier is the common-emitter, (Fig. 11.2). The input signal from the antenna is brought to the base of the transistor through a filter network which, in Fig. 11.2, consists of a balun, an FM trap, and a 45-MHz trap. The balun serves primarily to match the balanced 300 ohms (Ω) impedance of the lead-in line to the unbalanced input impedance of the transistor. With a collector current of 2.5 milliamperes (mA) and a collector-to-emitter voltage of 10 volts (V), the input impedance will generally be under 200 Ω over the range of frequencies to be received. The purpose of the FM trap is to prevent any 88- to 108-MHz FM signals from entering the tuner in strength. And the 45-MHz series trap is designed to keep signals in this range from also affecting the receiver. Such signals can be particularly destructive to the final picture, because the video i-f section operates in the 45-MHz range and the selectivity of the tuner circuits is not great enough to kill a strong 45-MHz incoming signal effectively.

The input tuning coil for Q_1 is L_1. This, in conjunction with C_1 and C_2, provide a certain amount of selectivity. L_1, as well as L_2, would be changed with each channel. C_1 is an adjustment capacitor that is adjusted on channel 10 and then left alone for the remaining channels.

Gain of the r-f amplifier is controlled by the agc voltage which is brought

FIGURE 11.2
The r-f amplifier of a television tuner.

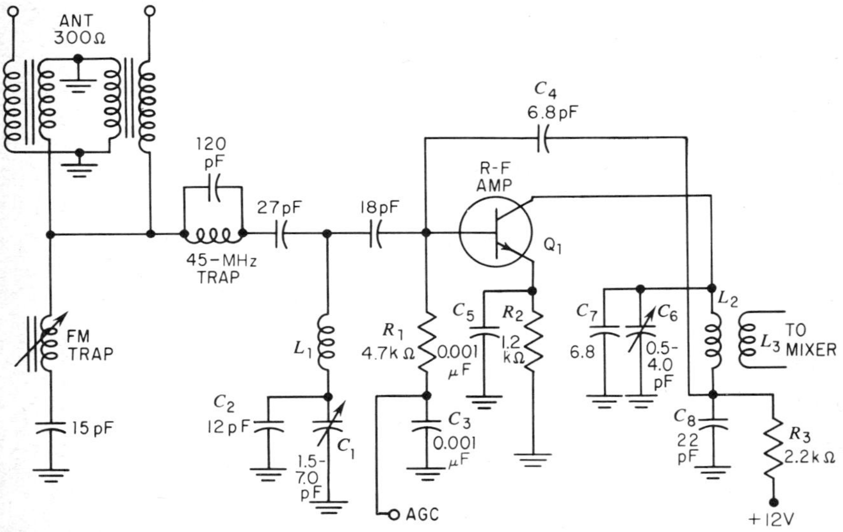

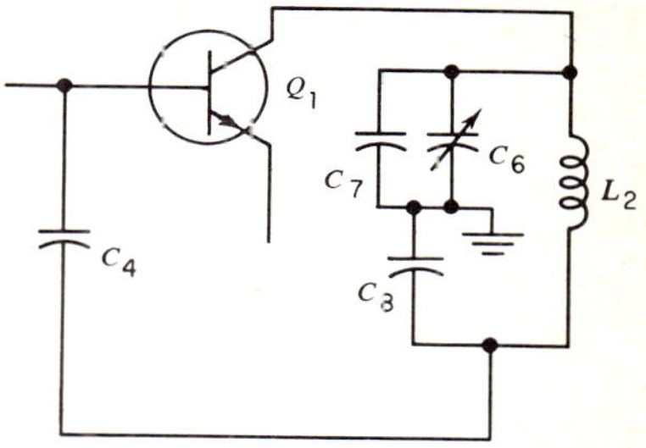

FIGURE 11.3

to the base. This agc voltage is positive in polarity because Q_1 is an NPN transistor.

The output circuit consists of L_2, C_6, C_7, and C_8. To see this circuit more clearly, it is redrawn in Fig. 11.3. Note that the connection point of C_6 and C_7 with C_8 is grounded, and this has the same effect as placing a ground connection on L_2. As a result, the r-f voltage at the bottom end of L_2 is 180° out of phase with the r-f voltage at the top (or collector) end. Capacitor C_4, by tying into the bottom end of L_2, is able to bring to the base of Q_1 a neutralizing voltage to counteract any energy feedback that may take place between the collector and base internally. Capacitor C_4 is thus the neutralizing capacitor.

The 2,200-Ω resistor R_3 is inserted in the collector circuit because forward agc is applied to the base. As indicated in a preceding chapter, with forward agc an increase in signal produces an increase in collector current through the agc voltage change. This increase in collector current produces an increased voltage drop across R_3, thereby reducing the collector potential (with respect to the emitter); this, in turn, reduces the stage gain. A signal reduction produces the reverse action, leading to a rise in gain.

The r-f Mixer. Coils L_2, L_3, and L_4, (Fig. 11.4) are wound on the same form and all are inductively coupled together. These coils would be changed, with L_1, for each different television channel. The signal from L_2 is brought into the base of Q_2 by L_3 at the same time the oscillator signal is received by L_3 from L_4. The two signals mix in Q_2, producing a difference-frequency signal that is then transferred to L_7. This difference signal is the i-f signal and has a frequency range from 41.25 to 47.00 MHz. The sound carrier is at 41.25 MHz, while the video carrier is at 45.75 MHz.

Note that the first i-f amplifier taps down on L_7. This is done because the input impedance of the next stage (in the i-f system) is less than the output impedance of the mixer. The same effect could have been achieved by using a transformer in place of L_7 in which the secondary, connecting to the i-f amplifier, had fewer turns than the primary.

Capacitors C_9 and C_{10} parallel L_3 and help establish the resonant frequency range for this tuned circuit. C_{10}, being variable, is an alignment adjustment.

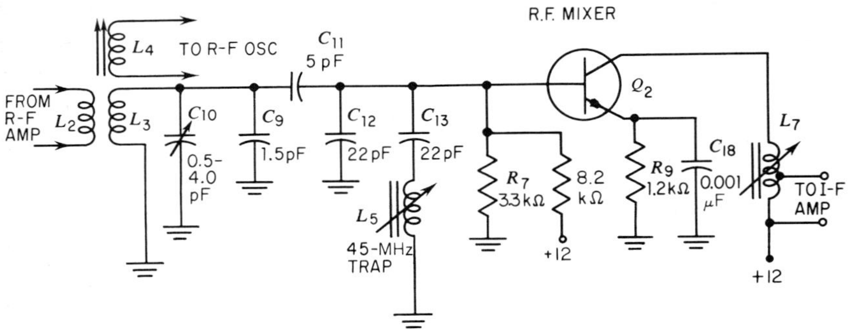

FIGURE 11.4
The r-f mixer of a television tuner.

Through its manipulation on a certain channel, generally 10, the proper response-curve shape is achieved. This then generally holds for all the other channels as new coils are brought in for L_2, L_3, and L_4.

Capacitors C_{11} and C_{12} serve to match impedances of the tuned circuit and the input of Q_2. C_{13} and L_5 form a low-impedance path for i-f signals (in the 41- to 47-MHz range). By placing such a low-impedance network across the input, any i-f signal fed from the collector to the base is, in essence, grounded and regeneration of the mixer at 45 MHz cannot occur. In this way, the mixer is neutralized.

Resistors R_7 and R_8 provide the proper bias current for Q_2. R_9 is employed for thermal stability. At the same time, it is bypassed by C_{18} to prevent signal degeneration. The operating voltages are selected to bring the noise level of the mixer to as low a value as possible. Also instrumental in this action is the amount of oscillator signal voltage injected into the mixer base circuit. This is shown in Fig. 11.5. Note how the gain of the stage rises and the noise generated drops as the injected oscillator voltage rises. Beyond a certain point, however, the gain starts to drop and the noise begins to rise. From these curves, it is apparent that too little oscillator voltage at the mixer base is just as bad as too much.

The r-f Oscillator. The final stage in the front-end tuner is the r-f oscillator. This is most frequently a common-base arrangement, as shown in Fig. 11.6. The base is brought to ac ground potential by capacitor C_{14}. At the same time, to encourage oscillation, a small capacitor C_{15} is shunted between emitter and collector. In the common-base arrangement, the ac voltages at emitter and collector are in phase, and through the internal capacitance in Q_3 between emitter and collector, plus the assisting action of C_{15}, the stage oscillates readily over a wide range of frequencies.

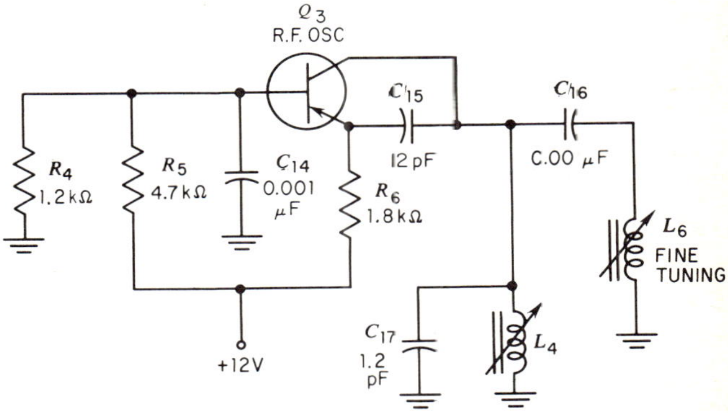

FIGURE 11.5
Mixer gain and the noise-figure variation with oscillator injection level.

L_4 and C_{17} help establish the operating frequency for the oscillator. L_6, an adjustable inductance, shunts L_4, and hence variation of the inductance of L_6 will alter the operating frequency. In a television receiver, the adjustment of L_6 can be made from the front panel to tune in a station properly. L_4 also has an adjustable core to permit major adjustments in the event that rotation of the core of L_6 does not bring in the station properly. Thus, the adjustment in L_4 can be considered as a coarse adjustment and the adjustment of L_6 a fine adjustment.

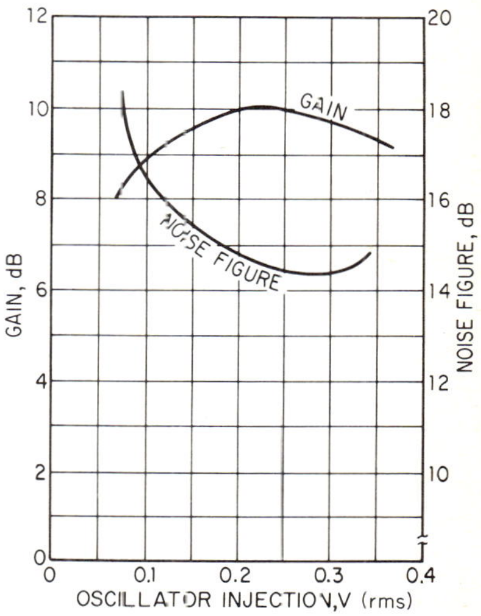

FIGURE 11.6
The r-f oscillator of a television tuner.

Resistors R_4 and R_5 establish the proper operating point for Q_3. R_6 also ties into the +12-V line to bring the emitter potential close to that of the base. The dc path for the collector is completed to ground through L_4.

The circuit diagram of a complex tuner developed with the three stages just discussed is shown in Fig. 11.7. The overall gain will range from about 25 decibels (dB) on the high channels (7 to 13) to about 40 dB on the low channels (2 to 6). The noise figure will be fairly comparable to that of a tube tuner.

The preceding discussion used bipolar transistors because these are the types most widely employed at the present time. However, television receivers are being designed using FET and MOSFET units and these will begin to appear in TV sets in increasing numbers. Insofar as the circuitry of FET and MOSFET tuners are concerned, this will assume the same general configuration shown for these circuits in Chap. 9.

VIDEO I-F SYSTEM

The video i-f system of a television receiver supplies most of the signal gain as well as most of the selectivity. To give some indication of the quantities involved, assume that a 30-μV signal is developed across a 300-Ω antenna feedline. This must become 50 V of composite video signal at the cathode of the picture tube. To achieve this, approximately 110 dB overall gain is required. Of this total value, 20 dB may be assigned to the tuner and 30 dB to the video amplifiers following the video second detector. This then leaves 60 dB of gain to be developed in the i-f amplifiers at the picture carrier frequency.

Both common-base and common-emitter configurations have been employed for the video i-f system. The first arrangement permits using a transistor with a lower cutoff frequency, while the second provides much higher gain per stage. (The reason for this behavior will be explained mathematically in Chap. 13.) Because of the higher gain and because high-frequency transistors are readily available, the common-emitter arrangement is most frequently employed. For either approach, essentially similar circuits are used; they involve either single-tuned or bifilar coupling coils and trap circuits for the sound and adjacent channel carriers. These combinations possess the same form they do in vacuum-tube video i-f stages, and any knowledge of these stages in vacuum-tube systems will apply here in large measure.

A typical video i-f amplifier capable of operating in the 45-MHz range is shown in Fig. 11.8. The i-f signal is brought to the transistor by the bifilar coil arrangement L_1 and L_2. The signal is applied to the base of the transistor, and it appears in the collector circuit in amplified form. Here it is coupled by L_3 to L_5 and on to the next stage. In the arrangement shown, 15 to 20 dB

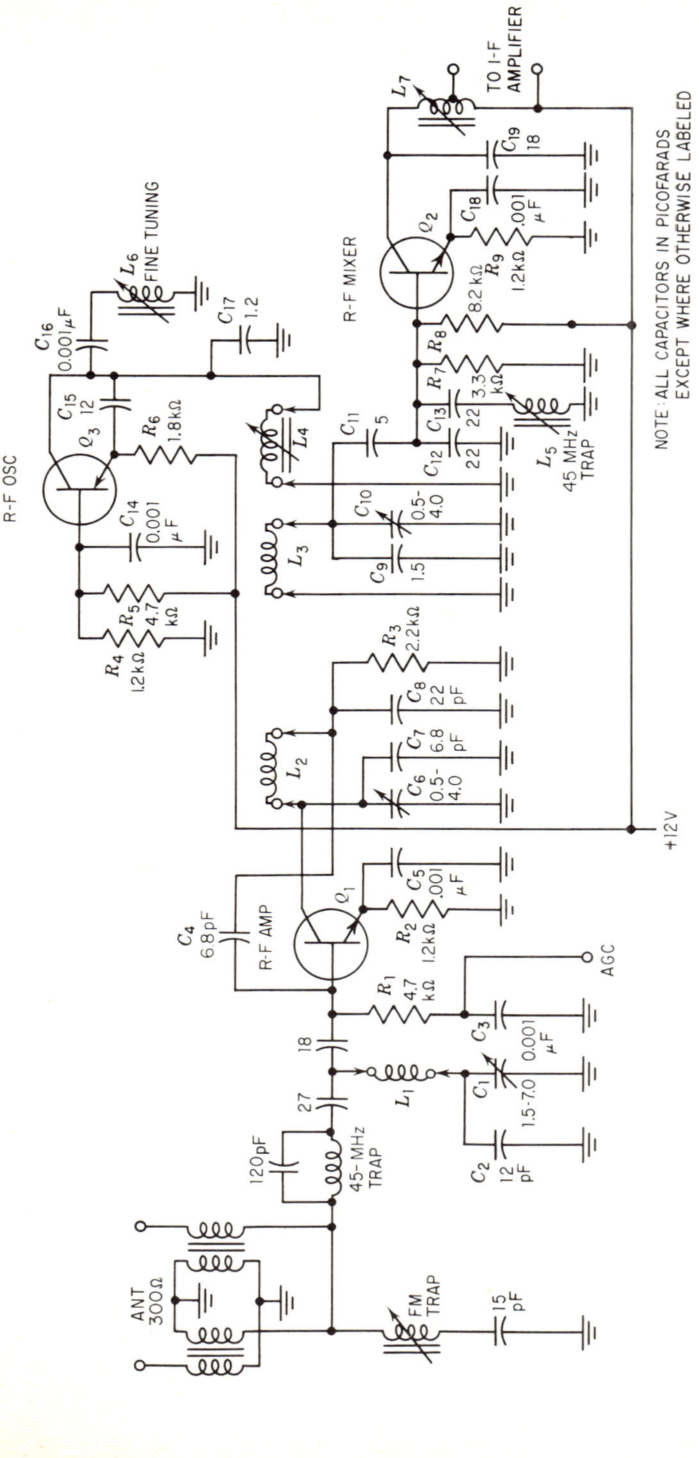

FIGURE 11.7
A complete transistor tuner for a television receiver.

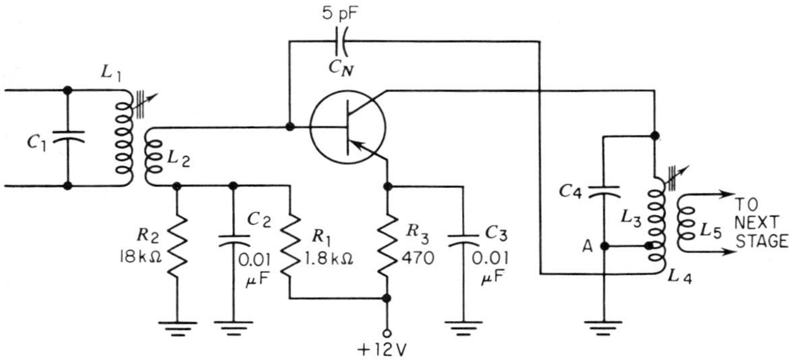

FIGURE 11.8
A typical video i-f amplifier.

of gain is obtainable. Thus, for a total of 60 dB overall i-f gain, three to four i-f stages would be required.

A 12-V supply is required to power the stage. R_1 and R_2 provide the base with the proper current (or voltage) to establish the operating point. C_2 places the bottom end of L_2 at signal ground in order that the full incoming video signal will be applied between the base (connected to the top end of L_2) and the emitter (connected by C_3 and C_2 to the bottom end of L_2). R_3 provides dc stabilization for the transistor, while C_3 prevents signal degeneration across R_3.

In the output circuit, point A is at ground so that L_3 and L_4 have reverse voltages appearing across them. By connecting the end of L_4 to the base through a neutralizing capacitor C_N, any feedback voltage fed internally from collector to base can be neutralized. The value of C_N will depend on the internal feedback capacity of the transistor and on the number of turns in L_4.

Both the input and output sets of bifilar coils have movable tuning slugs to permit frequency adjustment for alignment purposes. As in vacuum-tube video i-f circuits, stagger-tuning of the interstage coupling circuits is done, although in transistor circuits the amount of stagger-tuning is much less than in vacuum-tube circuits. The reason for this difference is the much lower impedance presented by transistors. With this lower impedance shunting the tuning circuits, a wider bandspread is obtained. Hence, it is not necessary to offset the resonant frequencies of the various tuning coils as much to obtain the desired bandspread required by the video signal.

Neutralization may also be achieved by using an inductance in place of C_N of Fig. 11.8, as in Fig. 11.9. If this method is employed, care must be exercised in the placing of parts to minimize any mutual coupling between the tuned circuits and the neutralizing inductance.

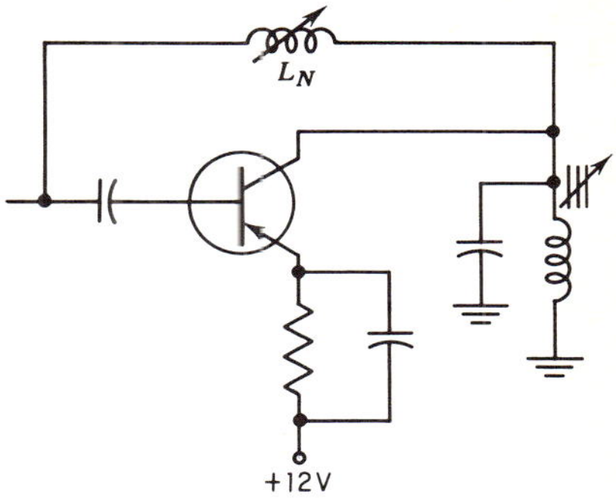

FIGURE 11.9
Neutralization of an i-f amplifier by using a neutralizing inductance L_N.

A three-stage video i-f amplifier system operating at 45 MHz is shown in Fig. 11.10. The input to the system from the preceding r-f mixer occurs through a short length of coaxial cable that links transformer T_1 to transformer T_2. A 41.25-MHz shunt trap is inserted in this path to reduce the amplitude of the sound portion of the received signal to approximately 5 to 10 percent of the peak value of the video signal (Fig. 11.11). The 47.25-MHz trap reduces interference that might arise from the sound carrier of the next-lower adjacent channel. At the same time, this trap also provides the proper slope for the high-frequency end of the video i-f response curve (Fig. 11.11). (By the same token, the 41.25-MHz trap helps to fashion the low-frequency end of the same video i-f response characteristic.) Overall bandwidth is on the order of 3.5 to 4.0 MHz.

The stages are coupled together by bifilar transformers T_2, T_3, T_4, and T_5. These transformers are peaked at slightly different frequencies to provide the necessary bandpass. The amount of detuning, however, is quite small. The primary of each unit has a high impedance (10,000 Ω) to match the relatively high collector impedance of the transistor feeding this transformer. The secondary of each transformer has a low impedance to match the low impedance of the following base input circuit.

Each secondary winding has an extension containing a few turns to provide the necessary out-of-phase neutralizing voltage. An 8.2-pF capacitor returns this voltage to the base of the stage.

Each of the transistors possesses a bypassed emitter resistor to provide temperature stability. In the first stage, the emitter resistor is 330 Ω; in the second stage, it is 680 Ω; and in the third stage, it is 470 Ω. In each instance a 0.005-μF capacitor provides the bypass action. In the second stage, the additional 68-Ω resistor is designed to decouple this stage from the third stage. This might occur through the common 12-V power supply.

Forward agc is applied to the base of Q_2. With an increase in signal, an increasing negative voltage is brought by the agc network to the base of Q_2. This increases the current through the transistor, resulting in a lowered collector

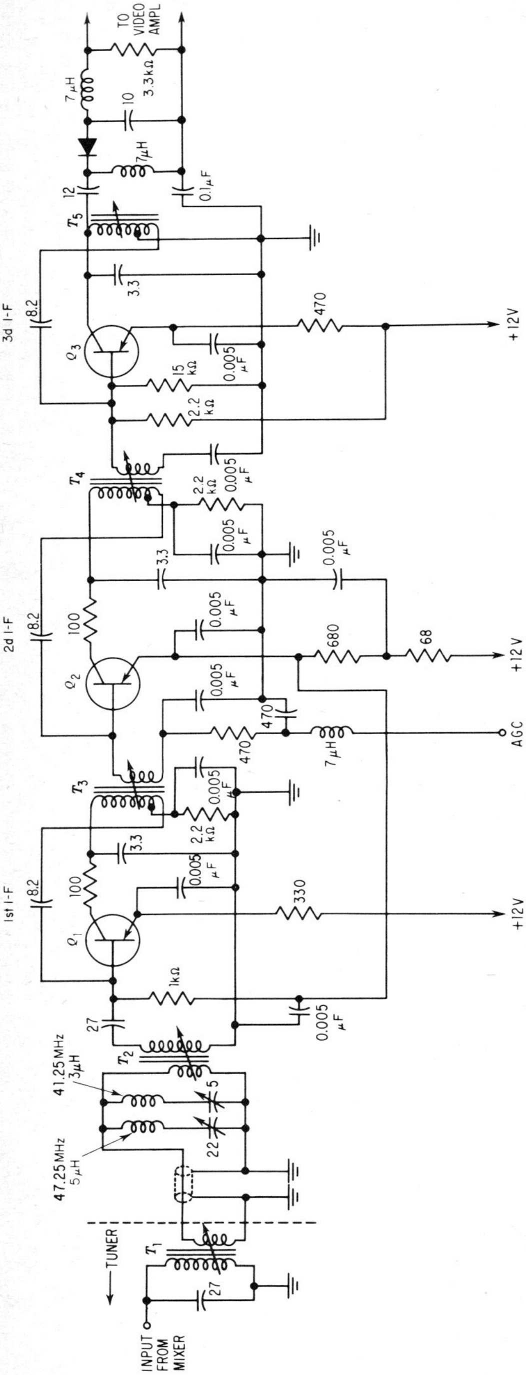

FIGURE 11.10
A three-stage video i-f amplifier system operating at 45 MHz.

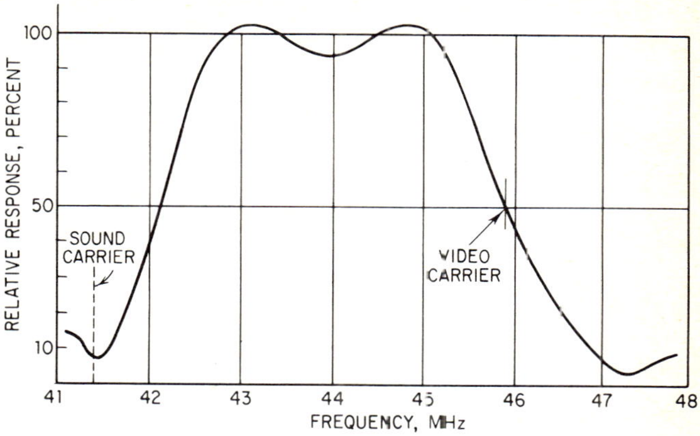

FIGURE 11.11
Overall response curve of the i-f system of Fig. 11.10.

voltage because of the increased voltage drop across the 2,200-Ω resistor in the collector circuit. This decrease in collector voltage results in less stage gain, thereby tending to counteract the signal rise.

The first stage is also controlled by the agc voltage because the base of Q_1 connects to the emitter of Q_2 through a 1,000-Ω resistor. An increase in current through Q_2 produces a rising negative voltage at the emitter, and this, fed to the base of Q_1, likewise causes the current Q_1 to rise. (Both Q_1 and Q_2 are PNP transistors.) Since the collector circuit of Q_1 also has a 2,200-Ω resistor, it will also suffer a reduction in collector voltage and a subsequent drop in gain. Thus, both stages are directly controlled by the agc line. Stage 3 has fixed bias only and is not involved in the agc action.

The collector circuits of the first two stages also have 100-Ω series resistors in them. These are inserted to prevent the appearance of spurious oscillations that are sometimes found to develop under strong-signal conditions.

The output of the final amplifier Q_3 is coupled to the diode detector without any special impedance-matching network because the two impedances are approximately equal. The video detector, in turn, is dc-coupled to the video amplifier that follows it. Because of this direct connection, it is not possible to ground the bottom ends of the detector network components directly. An ac ground is provided by the 0.1-microfarad (μF) capacitor. The diode load is formed by the 3,300-Ω resistor and the 10-picofarad (pF) capacitor. The 7-microhenry (μH) series choke resonates at 45 MHz and serves to keep the i-f signal out of the video-amplifier system.

Overall gain of this three-stage system is about 60 to 70 dB. With the agc circuit, the gain can be reduced all the way down to 0 dB.

One final point: the emitter of each stage is tied into the 12-V power line and the base must also be given a voltage that is somewhat comparable, since the two element voltages must be very close together. This is done in the first two stages by injecting 10.5 V dc onto the agc line (not shown in Fig. 11.10). This voltage is then made to vary by the agc network from 10.5 to 9.0 V, at which point the overall system gain is 0 dB. In the third stage, the base return line connects to the 12-V supply.

Since the emitters connect to the positive 12-V line, each of the collectors can have its circuit returned to ground. This makes each collector properly negative with respect to its emitter (and base), which is the desired condition for a PNP transistor.

The foregoing is a typical video i-f system. Other variations may be developed, but the basic overall operation will remain the same.

VIDEO DETECTOR AND AMPLIFIERS

The video second detector in most vacuum-tube television receivers is currently formed by a germanium diode; hence, transistorization is not required here. A typical circuit arrangement is shown in the i-f system of Fig. 11.10, and this is seen to be similar to video detectors in vacuum-tube television receivers.

The video-frequency amplifiers following the second detector must be capable of amplifying a band of frequencies extending from about 30 cycles to 4 MHz. Actually, with present monochrome receivers, the gain begins to fall off at about 3.2 MHz. A suitable two-stage video amplifier using high-frequency transistors is shown in Fig. 11.12.

The circuit uses a common-collector to common-emitter combination with two supply voltages, +12 and +150 V. A common collector was selected for the first stage so that it would present a high enough input impedance to permit a fairly high video-detector load resistor. This assures a high detector efficiency and a high detector output voltage. A common-emitter or a common-base amplifier, in the same position, would present a low input impedance, and a 6,800-Ω load resistor, such as that being used in Fig. 11.12, could not be employed effectively. What the video detector would see would be the lower input impedance of the common-emitter or common-base stage and not the 6,800-Ω load resistor.

The bias for the first amplifier stage is provided by the 220- and 1,000-Ω network. These two resistors are connected in such a way that the biasing voltage *does not* bias the diode detector. This precaution is required here because of the dc coupling between the diode and the first video amplifier.

A 4.5-MHz sound takeoff coil is inserted in the collector circuit of Q_1. Whatever signal develops here is transferred to the sound i-f system. Also present in the first stage is the takeoff for the agc circuit. This is a good place to make the tap because the full video signal, including the dc

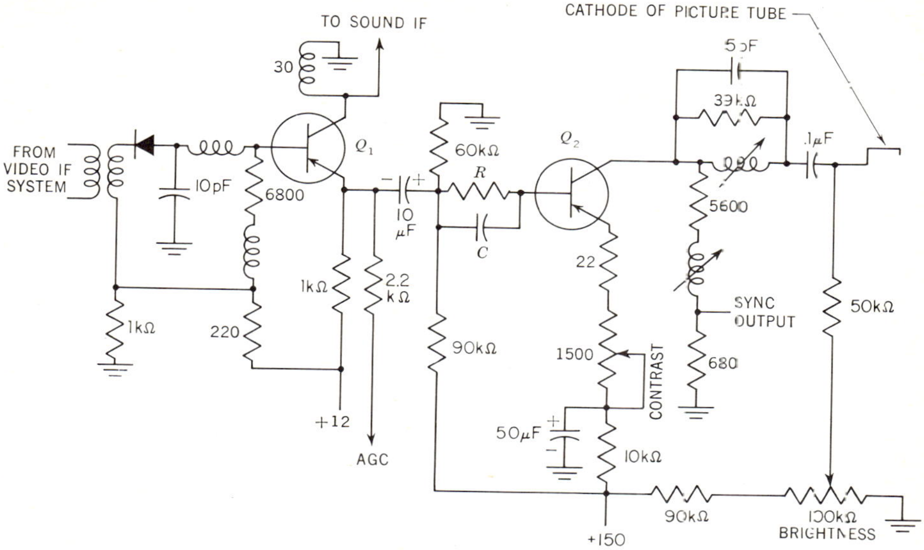

FIGURE 11.12
A two-transistor video-amplifier system.

component, is available. The latter is needed for some agc systems, such as keyed automatic gain control.

The second video amplifier, Q_2, operates from a +150-V supply, and for this reason a 10-μF capacitor is inserted between the emitter of Q_1 and the base of Q_2. Without this blocking capacitor, the bias conditions in the first stage would be completely upset. The 90,000- and the 60,000-Ω resistors establish the bias for Q_2. Resistor R and capacitor C form a peaking circuit to improve the frequency response of the system. The same purpose is served also by the peaking coils in the output of the video detector and in the collector circuit of Q_2.

Because a high bias voltage is applied to Q_2, it is possible to employ a 10,000-Ω resistor in the emitter circuit. This high a resistor, in this section of the transistor circuit, provides the stage with a high degree of stability and makes it relatively insensitive to temperature variations. The resistor is adequately bypassed by a 50-μF capacitor, so that no signal degeneration is introduced. The contrast control, on the other hand, which is also in the emitter leg of Q_2, does operate by varying the amount of signal degeneration it introduces. The video output will be lowest when the contrast control is fully in the circuit.

Sync-pulse output is obtained from a 680-Ω resistor in series with the

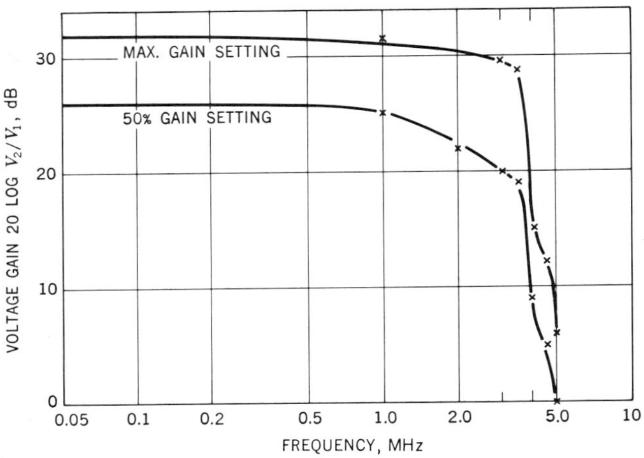

FIGURE 11.13
Frequency response of the two-transistor video amplifier of Fig. 11.12.

load circuit for Q_2. Also present between the collector of Q_2 and the cathode of the picture tube is a 4.5-MHz trap circuit.

A peak-to-peak signal of the order of 50 V is supplied to the cathode of the picture tube. The cathode is chosen because it requires 15 to 20 percent less drive than the control grid of the picture tube. When the cathode is driven, the video signal is in the direction to cancel the screen voltage and thus increase the effective output, since the drive required is proportional to the screen voltage.

Curves showing the frequency response of this two-stage amplifier are given in Fig. 11.13. Overall voltage gain is 30 dB for a bandwidth of 3.5 MHz. Note that the bandwidth changes slightly with contrast control setting, although the change is negligible so far as the actual viewing image is concerned.

Voltage-doubler Video System. Still another approach to a video system that can produce a large output swing is shown in Fig. 11.14. The incoming video signal is fed to the germanium diode where it is rectified and then transferred to the base of the first video amplifier, Q_1. Since a dc path is employed between the detector and Q_1, the video stage receives the full video signal, including the dc component. This is useful in that Q_1 can also act as an agc amplifier. The variations in this dc component, due to fluctuations in signal strength, will vary the voltage across the 4,700-Ω resistor in the collector circuit of Q_1 between 2 and 3 V, and this is sufficient to control the gain of the video

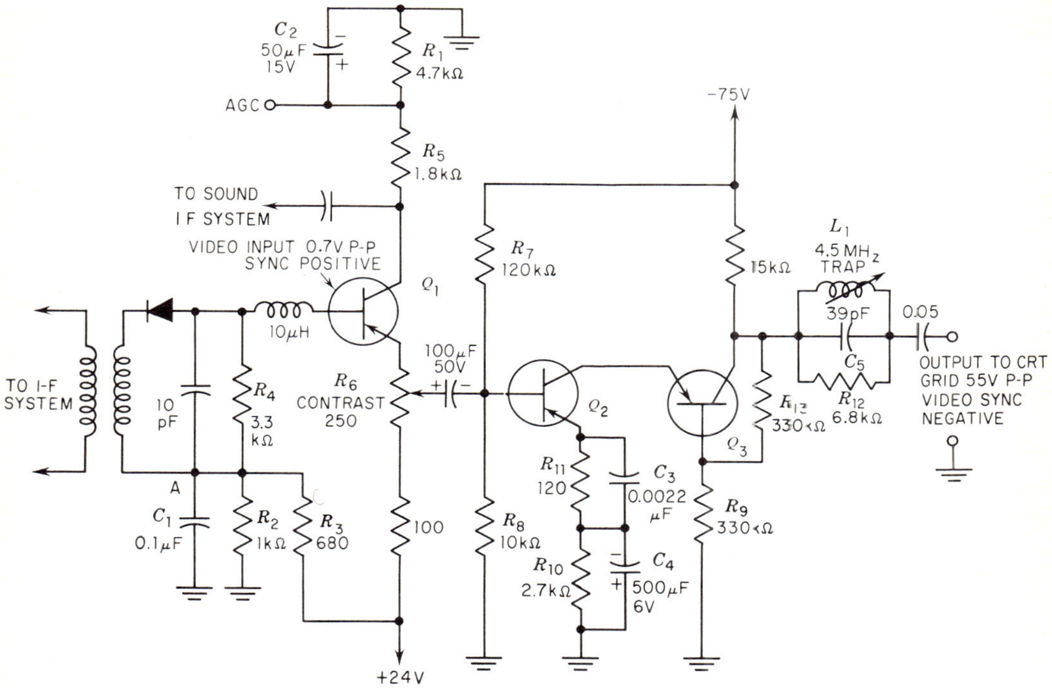

FIGURE 11.14
A voltage-doubler video output circuit and driver.

i-f system adequately. Q_1 is operated from a +24-V supply, producing a +10 V at the agc takeoff point at R_1. This is the reference voltage required by the i-f system of Fig. 11.10. Voltage variations across R_1 due to changes in the dc component of the demodulated video signal then produce the necessary agc action.

Note that R_1 is shunted by a 50-μF capacitor that prevents video signal variations from affecting the voltage developed here. This is necessary, of course, since the agc voltage should not be affected by the video content of the signal.

Q_1 is connected as an emitter follower in order not to load down the video detector. Base bias for Q_1 is established principally by R_2 and R_3, although video load resistor R_4 is also in this bias path. C_1 is needed to provide an ac ground for the detector circuit. A dc ground is not feasible at point A because of the direct connection between the detector and Q_1.

A signal for the sound i-f system is taken from R_5. The signal here is the full video signal as well as the sound signal. However, a subsequent

4.5-MHz resonant circuit, not shown, removes all but the sound intermediate frequency and this is then passed on to the sound i-f amplifiers.

Video-signal drive voltage for Q_2 is obtained from the contrast control R_6. Q_2 is operated as a conventional common-emitter amplifier, except that its collector is directly attached to the emitter of Q_3, a common-base amplifier. This latter stage functions as a voltage doubler, multiplying by 2 whatever voltage it receives from Q_2. Using a common-base amplifier at this point provides several advantages. First, it provides a wide bandwidth, reducing the need for peaking circuits. Second, the high output impedance of the common-base amplifier enables it to better provide a sizable voltage swing to the high input impedance of the cathode-ray picture tube. And finally, this arrangement can better withstand a greater output-voltage swing than can a common-emitter amplifier.

Proper bias for Q_2 is provided by R_7 and R_8. The bias for Q_3 is set by Q_2 and R_{13} and R_9. Resistor R_{10}, heavily bypassed, provides thermal stability. Resistor R_{11}, however, has a very low valued capacitor C_3 across it, and this arrangement provides peaking to help maintain the frequency response of the system. At the low frequencies, C_3 offers a high resistance, increasing the effective signal impedance in the emitter circuit and thereby lowering the current through Q_2 and its gain. As the frequency rises, the effect of C_3 is to lower the emitter impedance, permitting the current through Q_2, and with it the gain of the stage, to rise. The net result is an improvement of the high-frequency gain of the system.

L_1, C_5, and R_{12} form a 4.5-MHz trap to prevent any sound signal from reaching the picture tube and producing a visible sound beat. The overall frequency response of the amplifier in Fig. 11.14, from the second detector to the output, is shown in Fig. 11.15.

THE SOUND SECTION

The sound system of a television receiver functions initially at 4.5 MHz and, beyond the FM detector, at ordinary audio frequencies. Neither the i-f nor the audio stage offers any particularly difficult problems other than, in the case of the i-f system, that of obtaining transistors with a suitably high β cutoff frequency. Typical stages for both sections were discussed in preceding chapters.

Either the FM detector can be transistorized, as shown in Fig. 11.16, or a pair of matched germanium diodes can be employed in one of the arrangements shown in Fig. 11.17. The latter two circuits are quite familiar by now, being direct germanium-diode equivalents of vacuum-tube Foster-Seeley and ratio detectors. One of the important features here is the use of closely matched diodes; the greater their differences, the less effective the circuit in minimizing distortion and combating amplitude modulation.

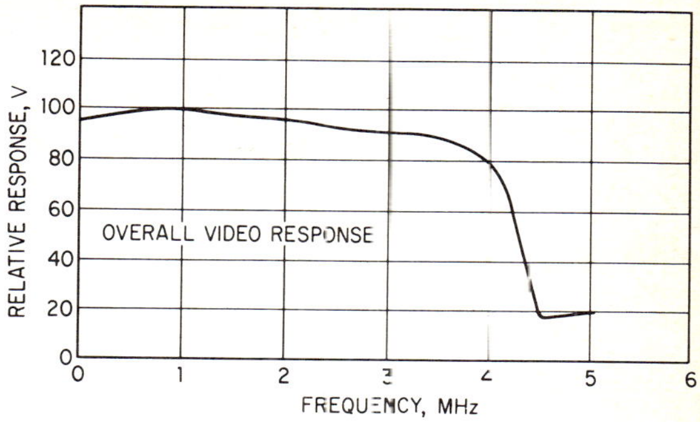

FIGURE 11.15
Overall response of the amplifier in Fig. 11.14.

A particular point of interest is the use of low-valued load resistors in both circuits in place of the fairly high values employed in the vacuum-tube versions. This change is necessary in order for the detector output impedance to match the low input impedance of the audio amplifier that follows.

It is possible to design a transistor FM detector using a symmetrical transistor. Briefly, this is a unit in which the emitter and collector sections are made identical so that with the proper biasing voltage either section could operate as the emitter or collector. (We shall refer to this type of transistor again in connection with a horizontal phase detector.)

The circuit of this FM detector is shown in Fig. 11.16. The FM signal appearing across L_1 is applied to the base, and the voltage developed across L_2 is applied to the emitter. During the positive portion of the signal applied to the base, the emitter-collector path is open and there is no current flow through the load resistor R_1. During each negative swing, current does flow. We are discussing here a PNP transistor; for an NPN unit, the periods of conduction would be reversed.

FIGURE 11.16
A transistor FM detector.

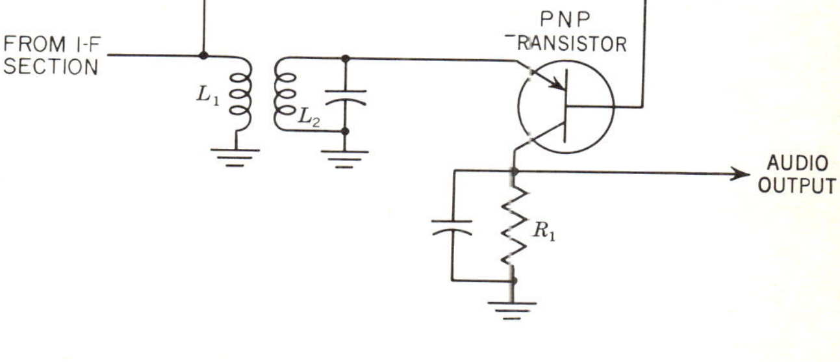

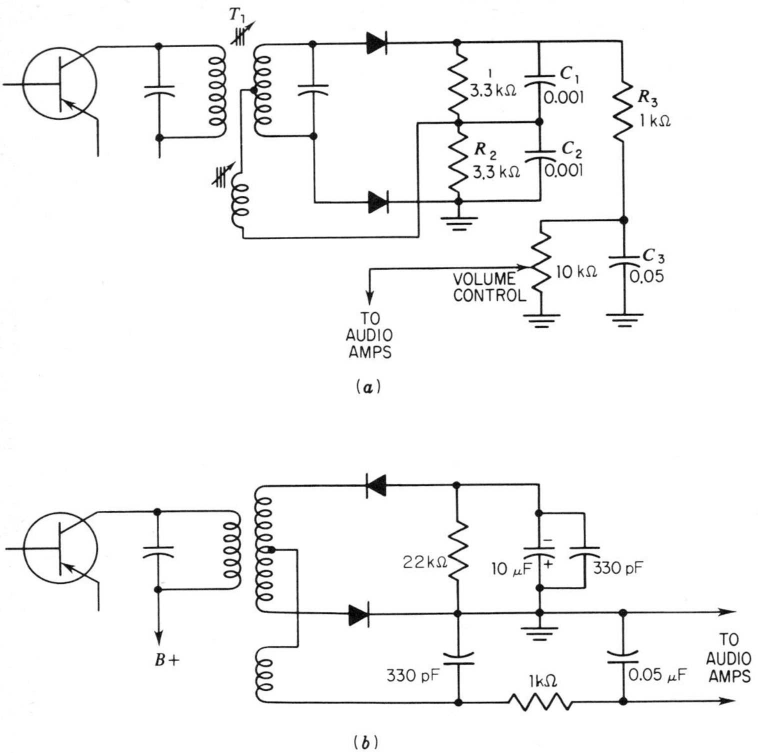

FIGURE 11.17
Two types of FM detector circuits: (*a*) Discriminator; (*b*) ratio detector.

Now, the amplitude and direction of the current flow depend upon the phase relationship of the signal developed across the secondary with respect to the primary signal. At the resonant frequency of L_2, the voltage it develops is 90° out of phase with the voltage across the primary. During this condition, the average voltage drop across R_1 will be zero. As the applied frequency is changed, the secondary voltage lags the primary voltage by an angle less than 90° if the frequency rises, or it will lag by more than 90° if the frequency drops below the resonant (or mid) frequency of L_2. As the phase relationship changes, the voltage developed across the load resistor will vary in step with the frequency modulation. (A full discussion relating phase changes to FM detection will be found in Milton S. Kiver, "F-M Simplified," 3d ed., Van Nostrand Reinhold Company, New York, N.Y., 1960.)

Changes in amplitude of the incoming signal will not affect the output as long as the signal amplitude is strong enough to operate the transistor beyond the knee of its characteristic curve (Fig. 11.18). Once past this region,

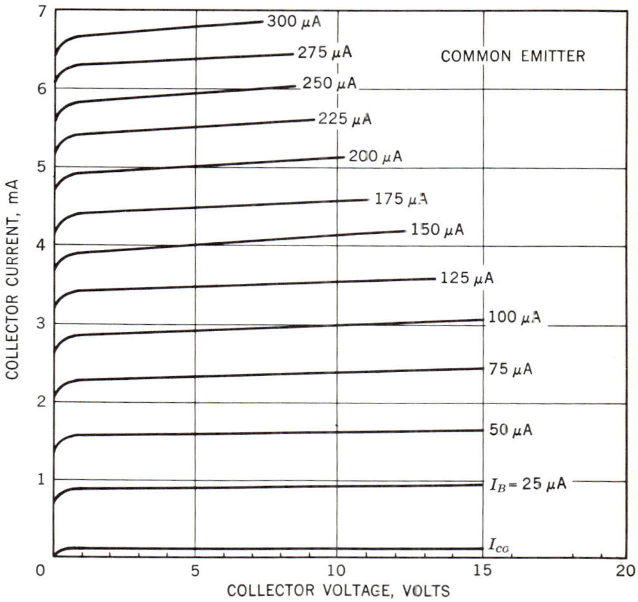

FIGURE 11.18
Characteristic curves of a typical transistor. Note that collector current remains fairly constant with changes in collector voltage beyond the knee of each curve. This behavior is utilized in sync separators and limiters.

the collector current remains fairly constant with changes in collector voltage. Thus, because of its characteristics, the transistor will function as a limiter, too.

IC Sound i-f System. With the decreasing cost of integrated circuits, increasing usage is (and can be expected to continue) being made of such components for various sections of a television receiver. One such section that lends itself readily to such usage is the sound system.

Figure 11.19 shows a 4.5 MHz i-f system and ratio detector in which a General Electric PA-189 IC module is employed in the sound i-f system of a television receiver. The ratio detector, with its built-in self-limiting-at-low-signal levels, improves the amplitude-modulation rejection, while the PA-189 with its 70-dB gain achieves limitation with as small an input signal as 200 mV.

The PA-189 has high gain because it has two differential amplifiers with high-impedance emitter coupling in cascade followed by an emitter follower (Q_5). See Fig. 11.20. Following this is another differential amplifier (Q_7 and Q_8). The differential amplifier may be thought of as a grounded collector stage

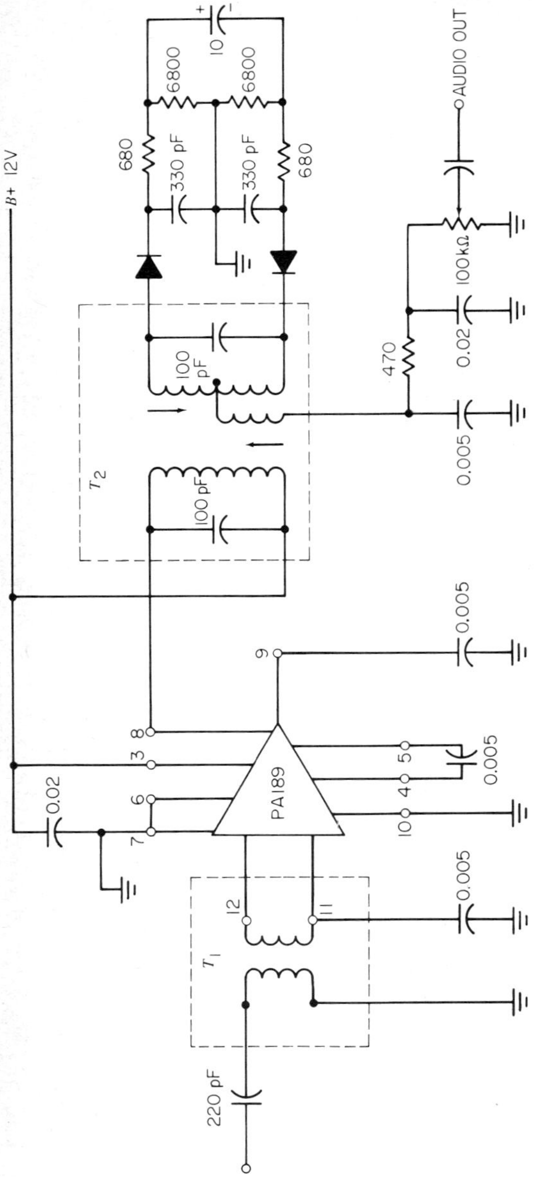

FIGURE 11.19
The audio i-f system in a television receiver using an integrated circuit.

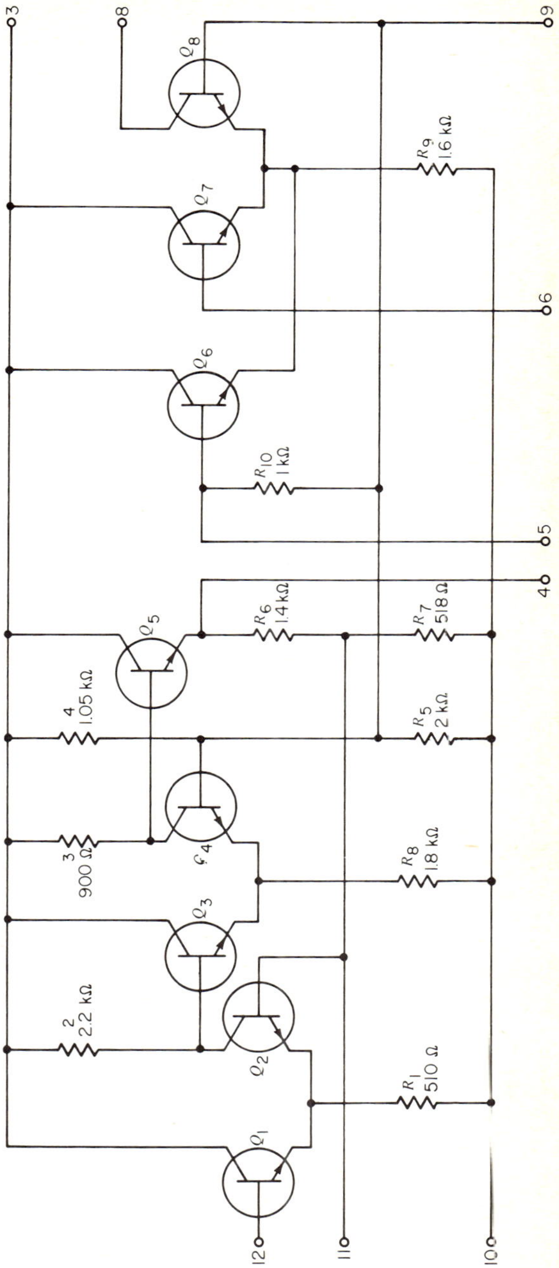

FIGURE 11.20
Schematic diagram of the IC module in Fig. 11.19.

driving a grounded base stage. This configuration gives rise to low feedback, good stability, and a high output impedance. Any output shunt capacitance present is tuned out by the primary of the ratio transformer primary.

In place of the conventional tuned transformer of Fig. 11.19, a ceramic filter tuned to 4.5 MHz can be employed. The use of crystal and ceramic filters with ICs seems to be the trend of the future. By simplifying alignment procedures, they can reduce a manufacturer's costs considerably, a factor that is particularly important in the highly competitive field of consumer electronics.

SYNC SEPARATORS

Returning to the video system, we note that a portion of the signal is taken from one of the video amplifiers and applied to the sync section. Here the vertical and horizontal sync pulses must be separated from the rest of the video signal. The latter voltages are then suppressed, while the sync pulses are passed on to their respective deflection systems. It is also desirable during this separation process to suppress or at least reduce the effect of any noise pulses that may be present.

In adapting transistors for sync separation, advantage can be taken of the fact (just noted) that beyond the knee of the characteristic curves, the collector current changes very little with change in collector voltage. Thus, if we drive a sync separator amplifier from cutoff to saturation, a double-clipped output voltage can be obtained possessing an amplitude that is only a few tenths of a volt less than the collector supply voltage. This is useful not only in securing a flat-topped output pulse but also in clipping any noise spikes that may be present at the sync-pulse level.

Standard video transmission is specified to have no more than 75 percent video information. The remaining 25 percent of the composite signal is reserved for sync. Therefore, in a standard video signal about 25 percent of the total peak-to-peak voltage is made up of the sync. It is up to the sync separator to extract the sync signal without taking any of the video information. To accomplish this, some type of clipping network is usually used.

The most obvious method of clipping the sync signal is with a circuit similar to the one used in Fig. 11.21. In this circuit the transistor is so biased that the sync tips drive it into saturation. (Saturation in this application can be defined as that condition of a transistor in which the collector-to-emitter voltage is essentially zero. It can also be defined as that condition in which the collector current does not increase materially, no matter how hard the base is driven in the forward, or conducting, direction.) The bias in the circuit of Fig. 11.21 must also be selected so that all video information is above the cutoff level of the transistor.

The operation of the circuit in Fig. 11.21 can be better understood by looking at Fig. 11.22. If the transistor bias is adjusted so that the sync tips

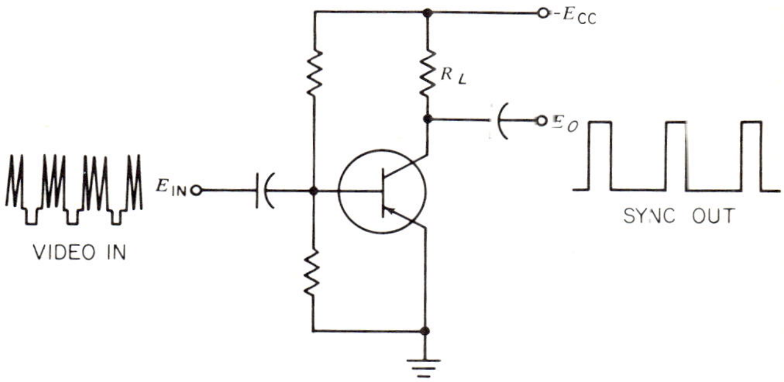

FIGURE 11.21
Simple sync clipper or separator.

extend to $V_{BE(\text{sat})}$ (Fig. 11.22*a*), each tip will result in a collector current that will saturate the transistor and produce at the collector the output pulse shown in Fig. 11.22*b*. As long as the total sync tips are somewhat greater than $V_{BE(\text{sat})}$, all video information will be above the zero voltage (cutoff) and will not appear in the output signal. (We are dealing here with a PNP transistor; hence, a negative input signal will produce conduction, whereas a positive signal will not.)

To obtain clean sync from the circuit in Fig. 11.21, the total composite video input signal need be only slightly over four times $V_{BE(\text{sat})}$. The reason is that 25 percent of the signal is sync. The base-to-emitter voltage required to saturate many transistors in sync-separator circuits is less than 0.8 V. Therefore, transistor sync-separator circuits may be designed to operate with about 2-V peak-to-peak of composite video input.

The circuit in Fig. 11.21 has one disadvantage—the video signal is

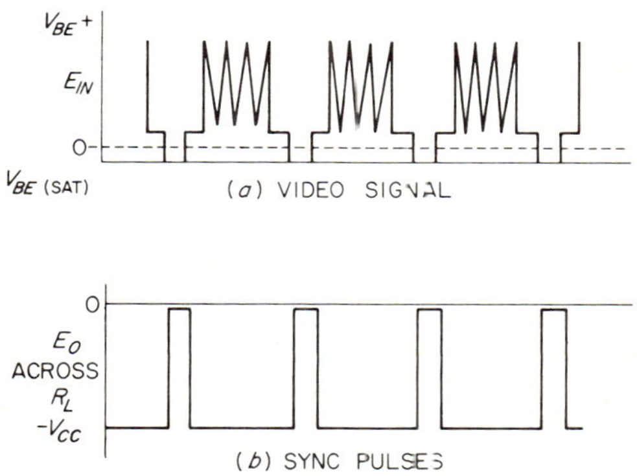

FIGURE 11.22
Illustrating how the circuit in Fig. 11.21 works.

brought to the transistor by ac coupling. During program changes and at other times during a broadcast, it is quite common for the picture content to disappear completely, leaving only the sync signal. Because of the ac coupling, the sync will distribute itself about the bias point, and unless there was a very large signal to start with, the transistor will not be driven into saturation. Under these conditions, the sync output signal will be somewhat less than can be obtained under normal saturated conditions.

A circuit sometimes employed to solve the problem is shown in Fig. 11.23. No fixed bias is used, and with no input signal the transistor is cut off because of the lack of forward bias on the base-to-emitter diode. When a negative sync pulse does arrive, current will flow in the base circuit and charge C_1 to some positive voltage. (In a PNP transistor, electrons flow from base to emitter.) The charging-time constant of C_1 is fairly short because of the low input resistance of the transistor when saturated. Therefore, during the time the horizontal-sync pulse is active, capacitor C_1 will charge. The magnitude of the positive voltage across C_1 will depend on the amount of base current and the duration of the sync pulse. At the end of the sync pulse, the transistor is turned off as the signal level rises from the sync tips to the blanking level and then to the video level. The blanking and video information will always be more positive than the cutoff voltage of the transistor.

While the transistor is turned off, C_1 will discharge slowly through R_1 because the transistor has a high input impedance in the off condition. As C_1 discharges, the base voltage becomes less positive. When the next sync pulse arrives, the voltage across the capacitor will have decreased to the point where the transistor is easily driven into saturation. By the proper choice of R_1 and C_1, it is possible to control automatically the base-bias voltage over a wide range of input signals and still permit sync pulses to drive the transistor into saturation.

If the input signal suddenly decreases, the signal will still average around the voltage established by the capacitor. If the level has changed so much that the next sync pulse does not drive the transistor into saturation, less base

FIGURE 11.23
A sync separator with better operating characteristics than the circuit in Fig. 11.21.

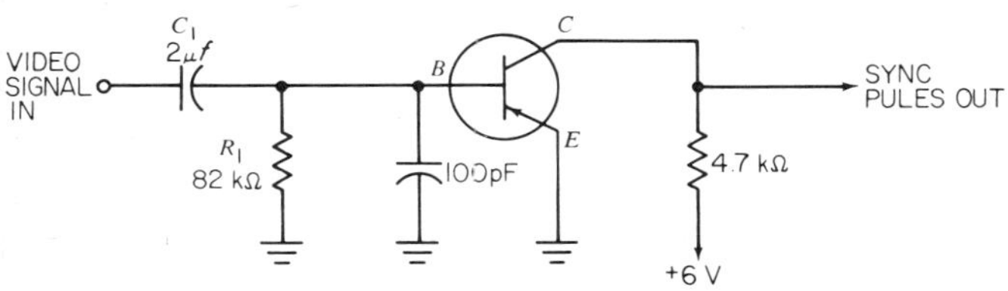

current will flow into C_1 during the charging cycle. Therefore, the capacitor will charge to a smaller positive value, and the base voltage will be closer to zero when the next sync pulse arrives. Thus, in the event of a very great signal change, only a very few horizontal-sync pulses will be lost at most.

The similarity of this action to that in a grid-leak biased vacuum-tube amplifier is quite marked. In both instances, the bias will vary and produce a constant-amplitude output signal for a wide range of input signals.

The circuit in Fig. 11.23 also provides excellent protection with very strong signals. Strong signals will drive the transistor very hard into saturation and result in more base current. This will charge C_1 to a higher positive value and thus tend to reverse-bias the transistor even more. The circuit in Fig. 11.23 usually requires a somewhat larger input video voltage (several volts) to ensure sufficient signal to charge the capacitor. The circuit in Fig. 11.21 can be made to operate with less input voltage, provided the proper transistor type is chosen.

Note that the circuits in Figs. 11.21 and 11.23 both operate with the transistor saturated. Under these conditions, the peak-to-peak sync output should be nearly equal to the supply voltage, less any drop in the emitter resistor.

Several practical sync separators that have been used in television receivers will now be discussed, to illustrate some of the techniques in sync-separator design. The first circuit, shown in Fig. 11.24, combines the basic concepts of Figs. 11.21 and 11.23. The video signal is ac-coupled to the base of the transistor through the 2-μF capacitor. The 0.05-μF capacitor and 4,700-Ω resistor combination provides distortion of the vertical-sync signal. The transistor is slightly forward-biased by the 3.3-MΩ and 82,000-Ω resistors and the 2-μF capacitor charges as described earlier. The 82,000-Ω bias

FIGURE 11.24
A sync separator employed in a television receiver.

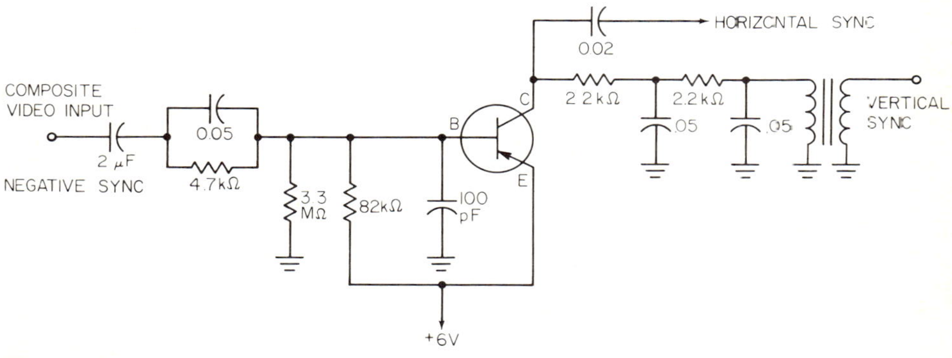

resistor provides the discharge path when the transistor is off. The two-stage vertical-integrator circuit in the collector provides the load for the vertical-sync takeoff. A transformer is used to invert the integrated vertical-sync signal.

The circuit in Fig. 11.24 has been designed to operate with typical transistors on as little as 0.4 V peak-to-peak of composite video input signal. With the poorest combination of transistor parameters, it still delivers saturated sync output with 1 V peak-to-peak or less of composite video input.

An interesting one-stage sync separator is shown in Fig. 11.25. This circuit not only produces flat-topped sync pulses but also provides a certain degree of immunity against blocking caused by strong noise pulses. It achieves this action by the combination of two time-constant networks and a germanium diode. Here is how the circuit functions:

Whenever a sync pulse appears, the emitter current of the transistor will bias D_1 in the forward direction and thereby bring the long-time-constant network of R_3, R_4, and C_3 into the circuit. This will maintain the emitter bias for the duration of the vertical-sync pulse and permit their effective separation from the video signal.

Now, if a strong noise pulse should come along, it will also cause D_1 to conduct and charge capacitor C_3 to the peak value of this pulse. After the noise pulse has passed, the weaker, normal sync pulses will be unable to overcome the voltage developed across C_3. As a result, D_1 will become reverse-biased and hence nonconductive. Now the shorter-time-constant network of R_1, R_2, and C_2 will become effective in the circuit (because it is no longer being overshadowed by the far longer time-constant network of R_3, R_4, and C_3). Horizontal-sync-pulse separation will be carried out, using this short-time-constant network, until the excess charge on C_3 has been drained off (by R_3 and

FIGURE 11.25
A one-stage sync separator with a dual-time-constant network.

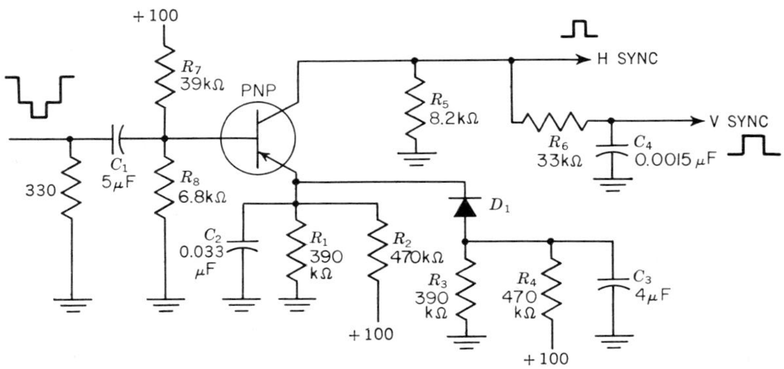

R_4) and the long-time constant network can reestablish itself in the circuit. This reactivation occurs before the next vertical-sync pulse arrives.

It is permissible to use a long-time-constant network to assist in the separation of the horizontal- and vertical-sync pulses, as long as no strong noise pulses are present. If a noise pulse does appear, it can readily block, or inactivate, the sync separator for many horizontal-sync pulses and thus throw the picture out of horizontal synchronization. This is prevented here by the presence of D_1 and the alternate short-time-constant circuit.

A sync separator that also incorporates a noise gate is shown in Fig. 11.26. The purpose of the noise gate is to prevent strong noise pulses from disrupting the operation of the vertical and horizontal oscillators that follow the sync separator.

In Fig. 11.26, the video signal is taken from a split collector load consisting of R_{38} and R_{37} in the video-output circuit. The split load reduces the video voltage applied to the base of Q_1. The video signal is ac-coupled to Q_1 through C_{15}. C_3 and R_7 form a shunt-biasing network, similar to the circuit in Fig. 11.23. The RC combination (R_{10} and C_8) distorts the vertical-sync pulse.

FIGURE 11.26
A combination noise-killer and sync-separator stage.

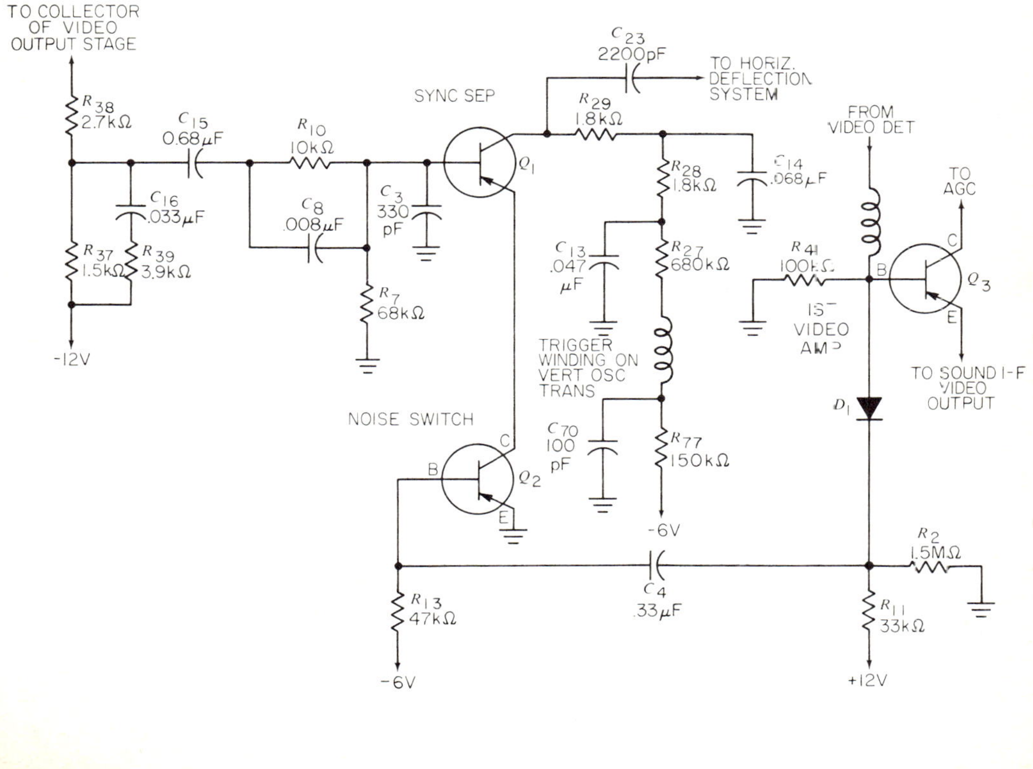

This is required to prevent C_3 from charging to a high positive voltage during vertical sync. Otherwise, there could be a loss of horizontal sync for a few lines after the vertical-sync pulse. C_{16} and R_{39} also improve vertical-sync stability. The combination of these two RC networks reduces the amplitude of the vertical-sync signal and predistorts it so that there is no loss of horizontal sync; at the same time, the vertical operation is not affected.

The emitter of Q_1 is connected directly to the collector of noise switch Q_2. Under normal conditions the noise switch is biased to saturation. The collector is then essentially at ground potential, and so is the emitter of the sync separator. This represents normal circuit operation. However, a large positive noise pulse, arriving at the second detector, is coupled through diode D_1 and capacitor C_4 to the base of Q_2. The pulse turns off Q_2, opening the emitter circuit of Q_1 and disabling the sync separator. Thus, the noise switch opens the sync-separator circuit to prevent large noise pulses from being coupled through the sync separator to the oscillators. The combination of R_2 and R_{11} provides a reverse bias on D_1. The noise pulse must therefore have sufficient amplitude to overcome this bias before being coupled to the noise switch.

The collector load for Q_1 consists of the series combination of R_{29}, R_{28}, R_{27}, and R_{77}, through the winding of the vertical blocking-oscillator transformer. The peak-to-peak sync output at the collector of Q_1 is about equal to the supply voltage (6 V). The horizontal-sync signal is coupled from the collector of Q_1 to a phase-comparator network through C_{23}. The vertical integrator is a two-stage network consisting of R_{29}-C_{14} and R_{28}-C_{13}. The auxiliary winding on the vertical-oscillator transformer inverts the integrated vertical-sync signal. The opposite polarity is required for triggering the vertical oscillator.

As noted earlier, NPN transistors are also used in sync-separation circuits. The operation is the same as for PNP, except for the polarities of the voltage and the incoming composite signal. In some applications, a sync amplifier may also be employed beyond the sync-clipper stage. Amplifiers are more common in circuits where a high-level horizontal oscillator requires a fairly strong control voltage from the AFC network. Sync amplifiers, if used, are similar in form to the separator circuits. They receive only the tips of the sync signal from the preceding sync separator.

VERTICAL-DEFLECTION SYSTEM

Vertical Oscillator. Beyond the sync separators, the pulses are fed to the vertical- and horizontal-deflection sections. Because transistorized vertical-sweep circuitry is simpler and has progressed farther, let us examine it first, starting with the oscillator. In vacuum-tube circuits, either multivibrators or blocking oscillators have been utilized. In transistorized receivers, the first vertical deflection systems to appear utilized blocking oscillators. However, now, both blocking oscillators and multivibrators are being employed. Let us consider typical examples of each, starting with the blocking oscillator.

In many ways the blocking oscillator in Fig. 11.27 is similar to its vacuum-tube counterpart. For example, the blocking transformer T_1 serves to provide feedback from the collector output circuit to the base input circuit. Also, sync pulses are fed to the base while a sawtooth deflection wave is developed across the 8-μF capacitor from collector to ground. Furthermore, charge capacitor C_1 receives its voltage buildup while the transistor is cut off and then discharges when the transistor is pulsed into conduction. There are, however, some significant differences between the two circuits, and these stem from the dissimilarity in operational characteristics of tubes and transistors. The differences will be evident from the following analysis of circuit operation.

The transistor is biased beyond cutoff by the negative voltage present across R_1. The value of this cutoff voltage is determined by the resistance setting of R_2; hence, R_2 is equivalent to the conventional hold control. At the same time, capacitor C_1 charges through R_5 until the voltage across R_3 becomes more negative than the voltage at the emitter. When this occurs, current will flow through the transistor, and the conductive part of the cycle will begin. The current flowing through the collector circuit causes a voltage to be induced in the base side of T_1, increasing the forward bias and the transistor current. This, in turn, couples a greater voltage into the base winding and increases the collector current even more. As a result, the transistor very rapidly becomes a virtual short circuit, permitting capacitor C_1 to discharge quickly through T_1 and R_1. The discharge is practically complete, and the voltage across C_1 essentially drops to zero.

At this point the voltage across R_3 and R_4 is also zero, and the transistor is driven into cutoff by the difference between the emitter voltage at the arm of R_2 and the voltage induced at the base by the collapsing field around T_1. The sequence of events now repeats itself at a frequency determined by the various resistance and capacitance values in the circuit.

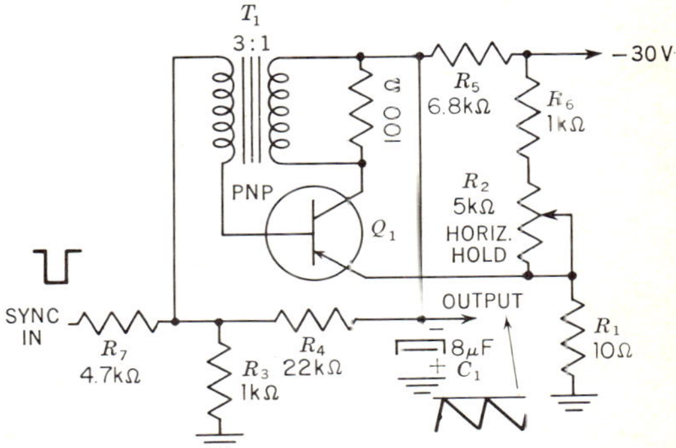

FIGURE 11.27
A transistor blocking oscillator.

The major frequency-determining components are C_1, R_5, and R_2; C_1 and R_5 also develop the output wave. Here is an interesting feature of this particular transistor oscillator. One set of components serves both to determine the oscillator operating frequency and to form the deflection wave which is then sent to the following deflection amplifier.

The sawtooth output waveform produced by the circuit of Fig. 11.27 is negative; if a positive-going signal is desired, an NPN transistor is employed in conjunction with a positive voltage source. Obtaining reversal in this way is another unique feature of transistors and cannot be duplicated by tubes.

Another sawtooth blocking oscillator, in which the output wave is developed in the emitter leg, is shown in Fig. 11.28. The circuit is similar to the preceding arrangement in that one RC combination (here, R_3 and C_2) provides both the timing and the output deflection waveform. It differs, however, in the way in which the sawtooth wave is produced. It has become customary in vacuum-tube circuits to have the charge time of the deflection capacitor correspond with beam trace and the discharge period with beam retrace. In the arrangement of Fig. 11.28, C_2 charges through the transistor during the retrace period and discharges through R_3 during trace time. How this is accomplished can be seen from the following discussion.

The base-emitter circuit is biased in the forward direction by R_1 and R_2. This produces a current flow through the transistor, which starts charging capacitor C_2. The collector current flowing through the primary of T_1 induces a voltage in the base winding that acts to increase the base current. This serves to raise the collector current and the induced base voltage further until the transistor is conducting to its fullest extent. This current buildup through the transistor is exceedingly rapid; and during this interval, capacitor C_2 is charging through the resistance of the transformer primary and the transistor.

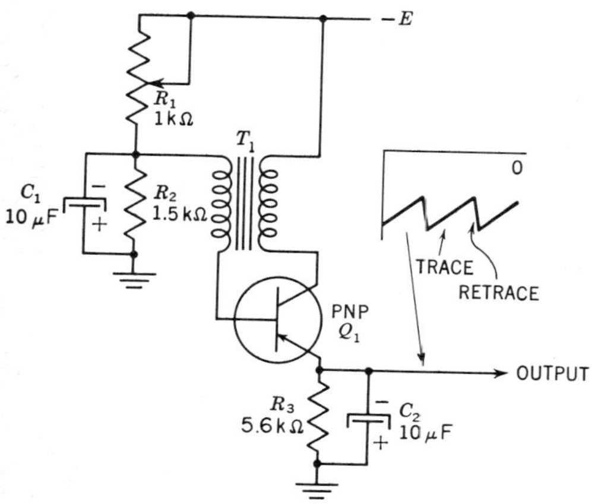

FIGURE 11.28
Development of the sawtooth wave in this circuit differs from that in Fig. 11.27.

The ultimate value of the charge is determined by the fixed base bias and the induced voltage in the transformer secondary winding.

When this point is reached, the base current ceases and the transistor is driven into cutoff because the voltage across C_2 is sufficient to bias the base-emitter circuit in the reverse direction. While the transistor is cut off, C_2 discharges slowly through R_3, thereby developing the trace portion of the deflection wave. Note that this wave gradually rises toward zero from its initial high negative value.

When C_2 has discharged to the point where its voltage is equal to the fixed bias voltage, current again starts flowing in the base-emitter circuit, and the same sequence of events recurs. Since R_1 determines the value of fixed base bias, it determines the frequency of operation (together with the time constant of R_3 and C_2) and hence would function as a hold control.

A negative sync pulse could be introduced to the base circuit through either the use of capacitive coupling or a third winding on the blocking transformer. Both methods have been employed and both are satisfactory.

A positive-going sawtooth wave is produced by the above circuit. If a negative-going sawtooth is desired, an NPN transistor can be used with a positive power supply.

All the circuits discussed above are practical designs that have been employed for the purposes indicated, and each has performed in an entirely satisfactory manner.

Complete Vertical-deflection Systems. In a vacuum-tube deflection circuit, the vertical oscillator is followed by the output amplifier. In order to develop the necessary current swing, this stage requires a certain amount of driving voltage. In transistor circuitry, a similar two-stage arrangement requires that a power transistor be used in the oscillator stage because of the power needed by the output amplifier in delivering sufficient sweep current to the vertical-deflection yoke.

A two-stage vertical-deflection system is shown in Fig. 11.29. The blocking oscillator employs feedback between the collector and base to sustain oscillations; the emitter element is grounded and serves as the common element between these two circuits. C_3 and C_4 have the sawtooth wave developed across them, and it is then applied to the base of Q_2.

Q_1 is kept cut off until C_1 discharges sufficiently to permit the transistor to conduct again. The discharge path for C_1 is through R_1 and R_2. By making R_1 variable, the discharge time can be varied. R_1 is then the hold control. Oscillator synchronization is achieved by coupling a negative trigger pulse into the base, through a tertiary winding on the pulse transformer.

C_2, C_3, and C_4 charge through R_3 while Q_1 is cut off and discharge quickly, for retrace, when the transistor is triggered on.

C_4 and R_7 form a linearity-correction network. The sawtooth current flowing through emitter resistor R_9 is integrated in a parabola by C_4 and fed to

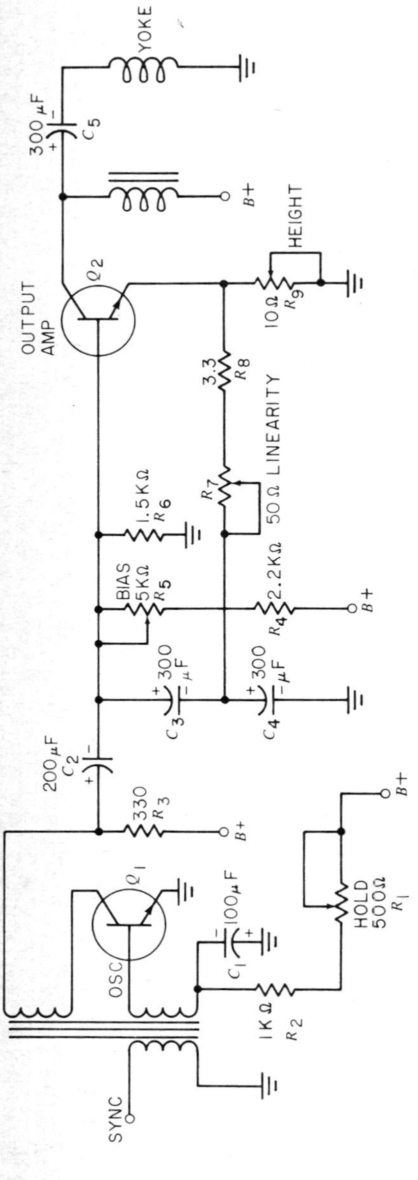

FIGURE 11.29
Two-stage vertical-deflection system.

the base of Q_2 through C_3. This special correction is needed to compensate for the nonlinear characteristics of transistor Q_2. Adjustment of linearity control R_7 affects the amplitude of the parabola and hence the degree of linearity compensation. The linearity control also affects the amplitude of the yoke current; hence, readjustment of height control R_9 is necessary after each linearity adjustment. Finally, bias control R_5 permits adjustment of the bias applied to Q_2 for replacement transistors. This is needed because the beta values may vary considerably among transistors of the same type.

The yoke itself is in the collector circuit of Q_2. However, because of the presence of C_5, no dc current flows through the yoke and hence no vertical decentering of the image occurs. Only the vertical-deflection currents are present here, swinging the electron beam of the picture tube vertically across the face of the screen.

A three-stage vertical-deflection system is shown in Fig. 11.30. The blocking oscillator again possesses essentially the same form as the circuit in Fig. 11.27 except for minor variations. For example, the 8,200-Ω resistor between collector and base not only provides the dc biasing potential for the base but also introduces a small amount of ac degeneration that helps the circuit stability. The oscillator frequency is established by the resistors in the base circuit, together with the 6-μF capacitor C_1. In the collector circuit of Q_1, the sawtooth deflection wave is developed by C_2; C_3 and R_3 then apply this wave to the base of Q_2.

The driver stage employs its transistor (Q_2) as an emitter follower in order to minimize loading on the oscillator. Note that the collector is grounded and, in effect, is returned to the −12-V terminal of the battery. On the other hand, the emitter returns to the +12-V terminal, bringing the full battery voltage across the transistor. Output for the next stage, Q_3, is obtained from a 5,000-Ω potentiometer R_4. This control, the height adjustment, operates like a volume control—the setting of the center arm determines the amplitude of the signal forwarded to the following stage.

The signal from R_4 is brought to output power transistor Q_3 by a 500-μF capacitor. The need for such a high value of capacitance stems from the low signal frequency (60 Hz) and the low input impedance of Q_3 (no more than 300 to 400 Ω).

R_5, R_6, and R_7 provide bias for the output transistor. R_6 is variable to permit adjustments as circuit components change and when Q_3 is replaced. In the collector circuit, Q_3 receives its dc through 60-hertz (Hz) choke L_1. A 9-mH yoke has its two windings parallel-connected and capacitively coupled to the choke. In this way, dc is kept out of the yoke so that the beam will not be decentered.

The vertical-linearity network consists of C_4, C_5, R_3, R_9, and R_{10}. C_4 feeds the waveform, at the collector of Q_3, to this network. Here, it is differentiated and fed back to the base of the driver stage. The two linearity controls and the bias adjustment make it possible, using a number of output

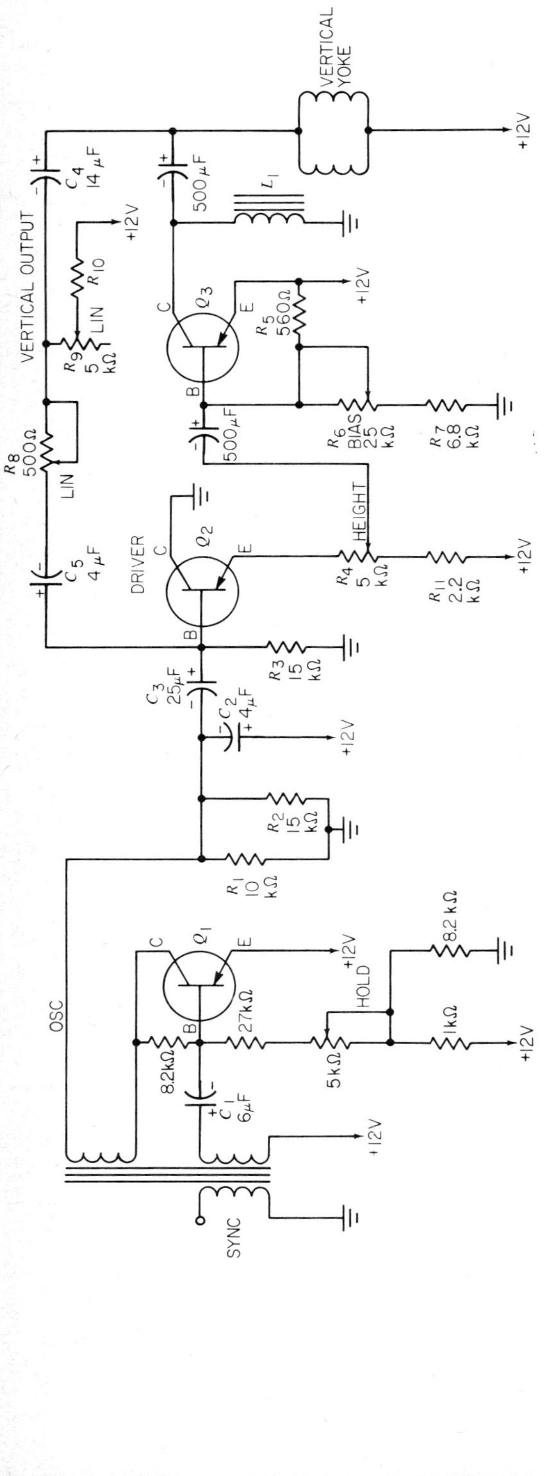

FIGURE 11.30
A three-stage vertical-deflection system.

transistors, to deliver a 500 to 600 mA peak-to-peak linear sawtooth signal to the yoke.

Multivibrator Vertical-sweep System. The foregoing illustrations of oscillators dealt only with those of the blocking-oscillator variety. In widespread use is also the multivibrator (the operation of which has been previously discussed).

The multivibrator circuit shown in Fig. 11.31 is fairly typical of the type used in transistorized television receivers. Incoming sync pulses from the sync separator first pass through a fairly conventional integrator circuit in which the relatively long vertical synchronizing pulses are able to pass through but the much shorter horizontal synchronizing pulses are suppressed. In Fig. 11.31, the integrator circuit consists of R_1, C_1, R_2, and C_2. These incoming pulses trigger Q_1 and in this way control the frequency of operation of Q_1 and Q_2.

In order to sustain oscillations, feedback pulses travel from the collector of the output stage Q_2 to the base of Q_1 through C_3, R_3, and R_4. In this fashion, the oscillation of this circuit is maintained. A vertical hold control, R_5, is shunted across this feedback path, and by changing the total resistance in this network, it is able to cause the frequency of the oscillator to vary within a fairly narrow range. This vertical hold control would be mounted on the front panel of the television receiver where it is readily accessible to the user.

Two additional controls are employed in this multivibrator. One is the

FIGURE 11.31
A multivibrator sweep oscillator.

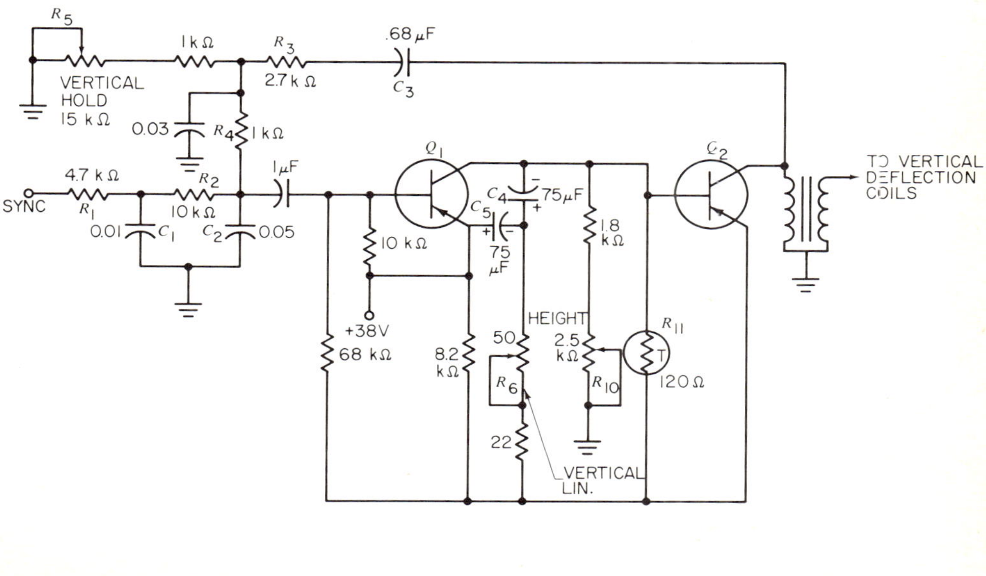

height control, which determines the amplitude of the voltage transferred from Q_1 to Q_2. In turn, this controls and governs the vertical size of the image on the face of the picture tube. The second control is the vertical linearity control which is in series with capacitor C_4, across which the vertical-deflection voltage is developed. By varying R_6 we can vary the shape of the voltage that reaches the base of Q_2 and hence the linearity of the image that is traced out vertically on the picture-tube screen. In essence, this linearity control controls the feedback between the emitter and the base of the output stage through capacitors C_4 and C_5.

An interesting component in the base circuit of Q_2 is a thermistor, R_{11}. A thermistor is a special resistor that possesses the characteristic that as more current flows through it and it becomes warmer, its resistance drops. The purpose of a thermistor is to act in a manner opposite to that of the transistor so as to maintain circuit conditions at their proper operation point. All transistors are subject to current variation as they heat up. In the case of Q_2, where a significant amount of current is passing through, the tendency for the collector of Q_2 is to become hotter with continued operation. This, in turn, causes an increase in the average current flowing through the transistor, even though the various bias voltages remain the same. This particular point, the reader will remember, has been discussed before. Through the insertion of a temperature-compensating element such as the thermistor, the voltage being applied to the base of Q_2 is maintained at a much more level amplitude, because as the current tends to increase, the resistance of R_{11} will decrease, offsetting this increase in current and maintaining a fairly constant voltage across the thermistor element.

HORIZONTAL-DEFLECTION SYSTEM

A second sync-pulse output from the sync separator stage is directed to the horizontal-deflection system. Let us now turn our attention to this section of the receiver.

Horizontal Phase Detectors. The susceptibility of the horizontal-sweep oscillator to noise pulses and other forms of interference has led to the universal use of automatic-frequency-control networks ahead of the horizontal oscillator. Whatever the form of the control system, its method of achieving control is by comparing the frequency of the generated sweep voltage with the frequency of the arriving horizontal-sync pulses. If a frequency difference exists, there is developed a corrective voltage that, when fed back to the horizontal oscillator either directly or indirectly, causes the generated frequency to change until it is equal to that of the incoming pulses.

A widely employed afc circuit in vacuum-tube television receivers is shown in Fig. 11.32. D_1 and D_2 are shown as solid-state diodes, although they can

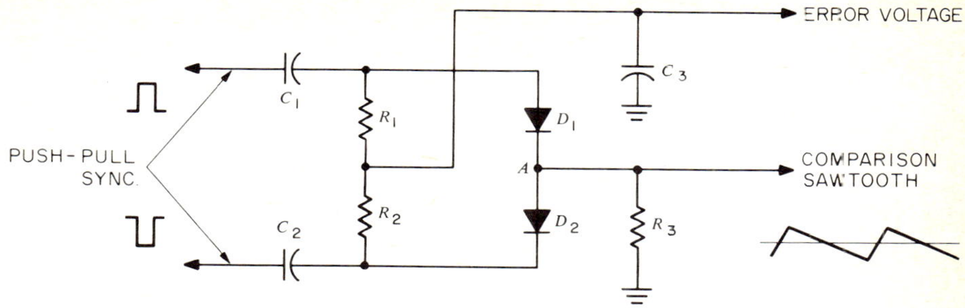

FIGURE 11.32
A double-diode phase detector.

be vacuum-tube diodes as well. In a transistor television receiver, they would be solid-state diodes, and that is the way the circuit will be shown here.

The two diodes are seen to be connected in series with each other at point A, and from this point, a resistor connects to ground. Coming into this network are two horizontal-sync pulses of opposite polarity (representing the received signal) and a sawtooth wave (representing the generated deflection voltage). The latter signal is obtained from a point beyond the horizontal-sweep oscillator so it will reflect the frequency being generated by that stage.

As a first step, let us disregard the sawtooth voltage and observe the effect of the two sync pulses. The positive sync pulse is applied to D_1, and if we were concerned only with this sync pulse and the circuit of D_1, then the simplified circuit would appear as shown in Fig. 11.33. Application of the positive pulse causes current to flow from D_1 to the right-hand plate of C_1 and from the left-hand plate of C_1 through the signal source (i.e., a prior stage) to ground and then up through R_3 to D_1. The time constant of this circuit is low enough that C_1 charges to the peak value of the applied pulse. During the

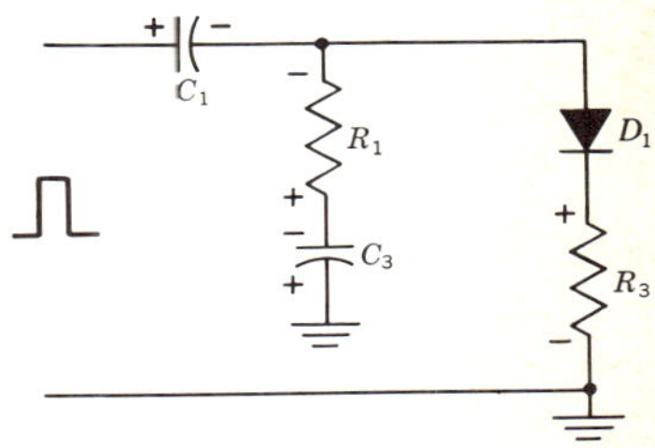

FIGURE 11.33
The circuit of D_1 in Fig. 11.32.

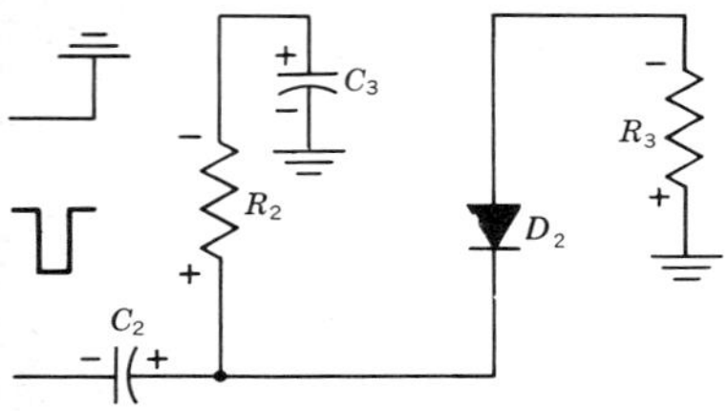

FIGURE 11.34
The circuit of D_2 in Fig. 11.32.

interval between pulses, C_1 discharges through R_1 and C_3, developing voltage drops across these two components with the polarity as indicated in Fig. 11.33. When the next pulse arrives, C_1 is recharged to the full peak value. The current flow through D_1 is thus in spurts that are generally shorter than the applied pulses themselves.

At the same time that this is happening, negative sync pulses are being applied to D_2 and causing current to flow through this diode. A simplified arrangement of this portion of the network is shown in Fig. 11.34. The current travels from D_2 down through R_3 to ground and from there to the signal source and C_2 and then to D_2. The polarity of the voltage drop across R_3 and C_2 caused by this current is indicated in Fig. 11.34. Note that the voltage drop across R_3 produced by the current from D_2 is opposite in polarity to the voltage drop developed across this same resistor by D_1. If, as is usual, both incoming sync pulses possess the same amplitude and both diodes conduct equally well, then the net resultant voltage across R_3 is zero.

During the interval between pulses, capacitor C_2 discharges through R_2 and C_3, developing voltage drops across these two components with the polarity indicated in Fig. 11.34. The net resultant voltage across C_3, due to the two discharge currents that flow through it, is zero. It is the voltage present across C_3 that represents the corrective or error voltage to the horizontal oscillator. With the sync pulses acting by themselves, no net voltage is produced.

By the same reasoning, if we ignore the sync pulses and concern ourselves solely with the sawtooth wave applied to R_3, then we see that since D_1 and D_2 will be driven alternately into conduction for equal periods of time and with equal-amplitude voltages, the net output voltage across C_3 will again be zero.

With both types of voltages applied to this circuit simultaneously, comparison of the two signals will take place only at the instant that the sync pulses arrive, for it is only at this moment that D_1 and D_2 conduct and are therefore in a position to respond to the sawtooth voltage applied across R_3. Three situations are possible.

First, if the sync pulses arrive at a time when the sawtooth wave is passing through zero, then we have a situation similar to that discussed above when the sawtooth voltage was ignored. The net voltage developed across C_3 is zero, which indicates that the frequencies of the sweep oscillator and the sync pulses are in step with each other .

The second situation occurs when the sync pulses arrive and the sawtooth voltage is positive at this instant. Under this condition, D_2 will conduct more strongly than it will if the sawtooth wave is zero and C_3 will charge to a higher peak value (because now two series-aiding voltages are driving current through D_2). At the same time, the positive sawtooth voltage is also being applied to D_1, and for this diode it is working against the applied sync pulse. Hence, the total current through D_1 will decrease and produce a smaller voltage drop across C_3. The net voltage across C_3 will be governed by the current from D_2 and will be positive with respect to ground. The horizontal-sweep oscillator will thus receive a corrective voltage that, if the circuit is designed properly, will serve to alter its frequency so that the sawtooth voltage at R_3 will be passing through zero when the sync pulses arrive.

The third situation occurs when the sawtooth voltage is negative when the pulses arrive. Now D_1 conducts more strongly than D_2, and a net negative voltage will develop across C_3. This opposite-polarity voltage will have an opposite effect on the frequency of the horizontal-sweep oscillator.

A two-diode circuit in which only one set of sync pulses is required is shown in Fig. 11.35. The two cathodes of the diodes (here, germanium

FIGURE 11.35
A phase detector that requires only one set of input sync pulses.

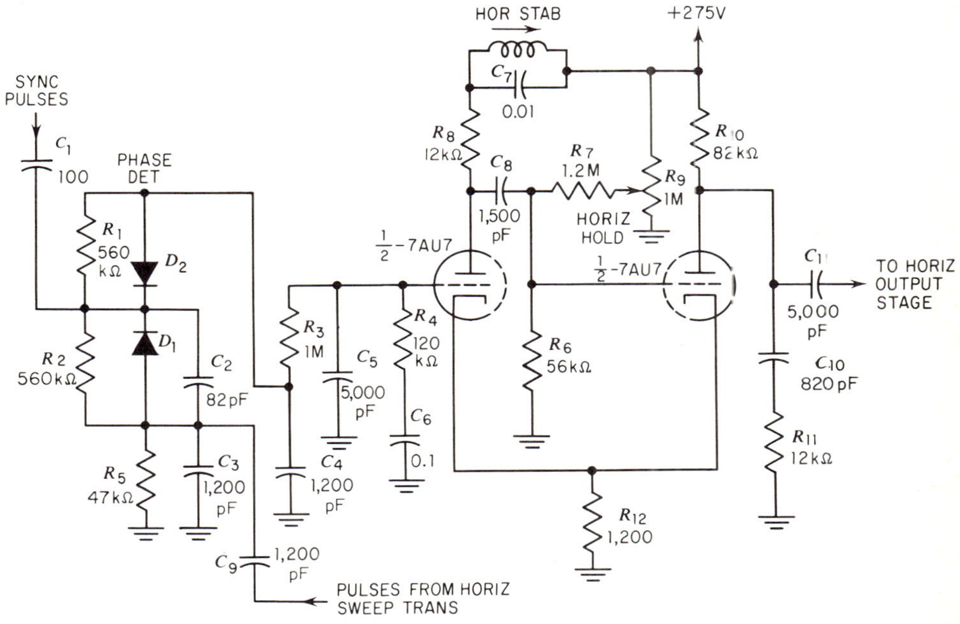

diodes) are connected together, and a negative-going sync pulse is applied to their junction. This applies the sync pulse equally across D_1 and D_2, because C_3 and C_4 are so much greater than C_1 that D_1 and D_2 are effectively connected in parallel. This being the case, current will flow in each diode, causing equal currents to flow in the load resistors R_1 and R_2. The currents, of course, flow in opposite directions, and the voltage drops across R_1 and R_2 will have opposing polarities and, therefore, will cancel out, producing zero volts output.

The sawtooth voltage, formed from pulses from the horizontal output stage, is a sample of the horizontal-oscillator frequency. This voltage is applied across D_1 and D_2, effectively bringing one-half of the original sawtooth wave across each diode. It can be shown that the sawtooth wave across D_1 will be going positive when the voltage across D_2 is going negative, and vice versa (Fig. 11.36). The currents of the two diodes will be equal but opposite in polarity, so equal and opposite voltages across R_1 and R_2 will produce zero volts output.

From this it can be seen that the incoming sync pulses alone will not cause the phase detector to produce any voltage output. In like manner, the sawtooth wave alone will not cause the phase detector to produce any voltage output.

The sync pulse, being of much greater amplitude than the sawtooth wave, keeps the diodes biased so that they operate only when the sync pulse is applied to them. Therefore, only that portion of the sawtooth wave that occurs at the instant of the sync pulse has any effect on the output of the phase detector.

FIGURE 11.36
A simplified diagram of the phase detector of Fig. 11.35 showing the waveforms in the circuit. (*General Electric Co.*)

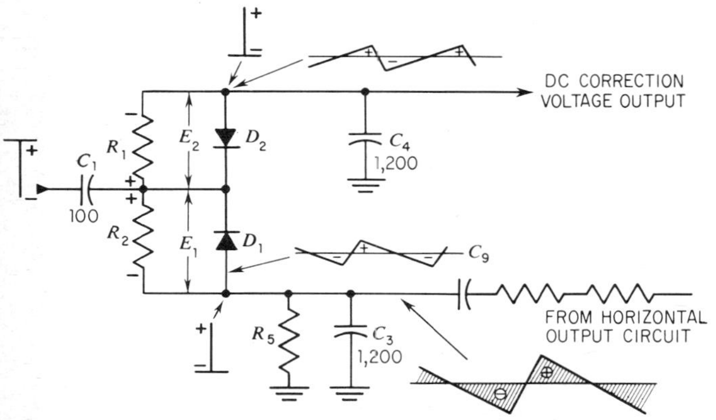

Now, if the sync pulse occurs in the exact center of the sawtooth retrace (i.e., retrace passing through its ac axis), Fig. 11.37*a*, equal but opposite currents will flow and no output voltage will be developed.

If the oscillator is slow, the sync pulse will occur before the sawtooth retrace passes through its ac axis (Fig. 11.37*b*). On D_2 therefore, some of the sawtooth voltage will be added to the sync-pulse voltage because the sawtooth voltage is on the positive half of its cycle when the sync occurs. Some of the sawtooth voltage on D_1 will be subtracted from the sync-pulse voltage because the sawtooth retrace there is still in the negative half of its cycle. The output voltage of the phase detector in this case will be negative because the voltage drop across R_1 is greater than the drop across R_2.

If the oscillator is fast, the sawtooth retrace will pass through its ac axis

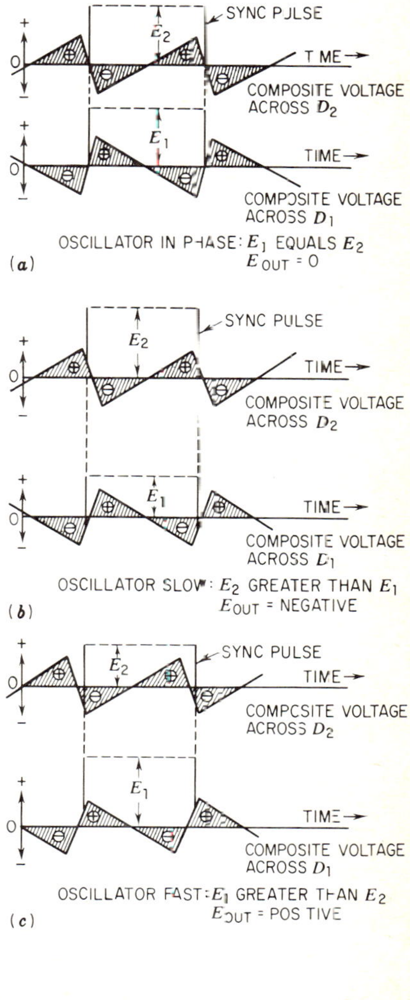

FIGURE 11.37
Operation of the phase detector of Fig. 11.35 when the sync-pulse frequency and sawtooth frequency are (*a*) equal and (*b*) and (*c*) unequal. (*General Electric Co.*)

before the sync pulse occurs (Fig. 11.37*c*). On D_2, therefore, some of the sawtooth voltage will be subtracted from the sync pulse. On D_1, some of the sawtooth voltage will be added to the sync pulse, producing a higher voltage drop across R_2 than across R_1. This will produce a positive output voltage in order to slow down the horizontal oscillator.

Application of the foregoing phase detector to a commercial transistor television receiver is shown in Fig. 11.38. Incoming horizontal sync pulses are brought to diodes D_1 and D_2 by capacitor C_1. At the same time, a sawtooth wave is brought into the circuit by R_{12} and applied to capacitors C_3 and C_4. The two waveforms are compared in frequency by D_1 and D_2, and if any difference exists, a voltage is produced across R_1, and R_2. This voltage is then brought to the base of Q_1 by R_{10}, R_{11}, C_5, C_6, L_1, and C_7. The latter six components serve as filter elements to smooth out instantaneous variations in the control voltage so that smoother control of the oscillator is achieved. The network also tends to prevent the oscillator from shifting back and forth in frequency as it attempts to find the correct operating value.

Incorporated into the phase-detector circuit is the network that permits the set viewer to adjust the operating frequency of the horizontal oscillator. Twelve volts is brought to the emitter of Q_1 from the power line. The same +12

FIGURE 11.38
Commercial application of the phase detector of Fig. 11.35.

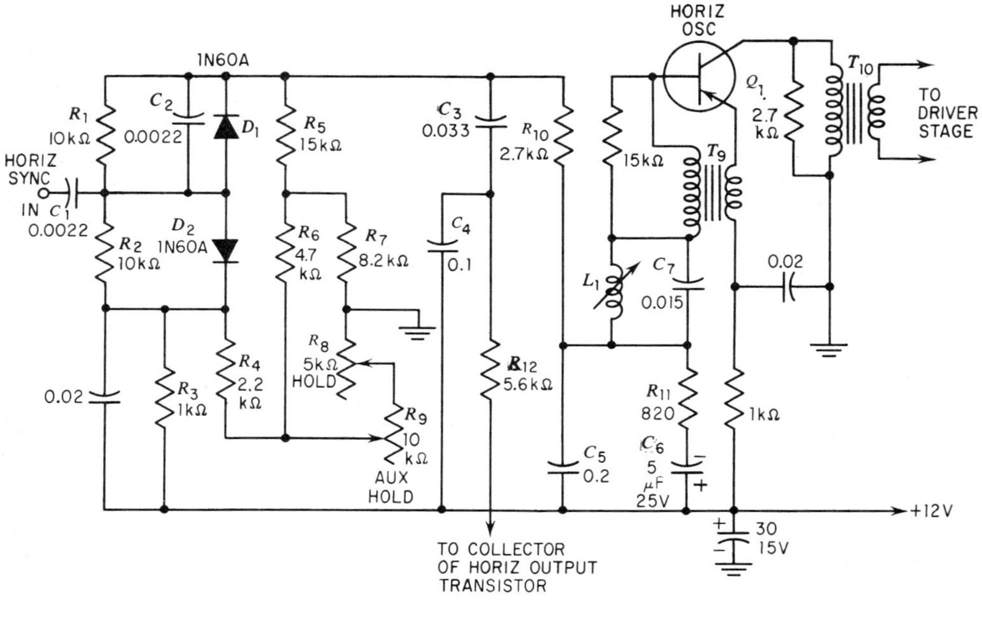

V is also brought to the base of Q_1 through R_3, R_4, R_5, R_6, R_7, R_8, and R_9. Included in this network are two hold controls, R_8 and R_9. By varying either or both of these potentiometers, the base voltage can be altered and thereby the operating frequency of the oscillator. R_8, the hold control, is the front-panel adjustment most frequently adjusted by the set viewer. R_9, the auxiliary hold control, would be mounted on the back panel for occasional adjustment.

Transistor Phase Detector. In a transistor phase-detector circuit, similar operation can be achieved by means of a single transistor in a circuit such as that shown in Fig. 11.39. The transistor, however, is specially constructed so that the collector and emitter junctions are equal in area. (In the transistor designed for general usage, the collector possesses a greater area than the emitter.) This forms a symmetrical transistor. Advantage is taken of the fact that transistors will conduct in either direction to form this phase detector.

FIGURE 11.39
The changing direction of current flow through R_2 when the polarity of sawtooth voltage reverses during sync interval.

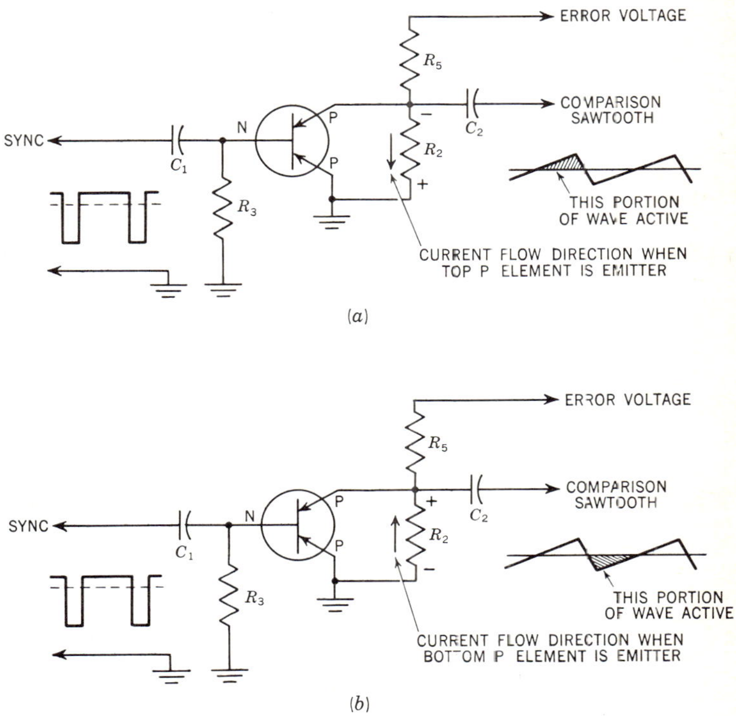

That is, the biasing voltages to emitter and collector can be reversed so that the element that serves as an emitter under one set of conditions becomes a collector under another set of voltages. In short, either element can serve as the emitter or collector, depending upon the applied potentials. It is common, therefore, to refer to either element as a "collector-emitter." This behavior, too, is the reason both emitter and collector in Fig. 11.39 are shown with arrowheads.

Operation of this transistor phase-detector circuit depends on the transistor's conducting only when the sync pulses are present. Thus, if the instantaneous sawtooth voltage across R_2 is positive at the instant the negative sync pulse triggers the base on, then the uppermost element with the arrowhead is serving as the emitter and the lower-arrowed element is the collector. The reason is that this is a PNP transistor and for conduction to occur, the emitter must be positive with respect to the base. Electrons then travel in the direction indicated in Fig. 11.39*a*, producing a negative voltage drop across R_2. The potential represents the error voltage that is transferred via R_5 to the horizontal-control transistor.

Conversely, if the sawtooth voltage is negative when the sync pulses arrive, then the element previously acting as the emitter now becomes the collector and the other P section becomes the emitter. Current flow is now reversed through R_2 (Fig. 11.39*b*) and a positive error voltage is produced. In this way, the transistor acts as a bidirectional switch producing an error (or correction) voltage whose polarity depends on the part of the sawtooth cycle active at the instant the sync pulses arrive.

If the sawtooth wave is passing through zero when the pulses arrive, no voltage appears across R_2.

In order for the circuit to function properly, the transistor must be completely cut off between sync pulses. This is achieved by having the peak-to-peak sync voltage at the base exceed the peak-to-peak sawtooth voltage. When the sync pulses are active, the current that flows in the base circuit causes C_1 to charge to their peak value. In the interval between pulses, this charge decreases very slowly, keeping the base at all times positive enough with respect to the collector-emitter voltage to prevent conduction.

A commercial circuit employing the transistor phase detector just discussed is shown in Fig. 11.40. Positive horizontal-sync pulses are brought to the base of Q_1 by capacitor C_1. At the same time, a portion of the horizontal-deflection voltage is taken from the horizontal-output transformer (not shown), converted into a sawtooth wave by C_6 and R_7, and brought to the collector of Q_1 by C_5. Direction of current flow through Q_1 will depend on whether the sawtooth wave is positive or negative at the moment a horizontal pulse appears at the base of Q_1. This, in turn, will determine the polarity of the control voltage developed across R_2. This voltage will be applied through R_6, R_8, C_3, and C_4 to the base of Q_2, the horizontal blocking oscillator. Note that R_2 is in series with R_5, a potentiometer that supplies a negative dc voltage to the base of Q_2. Rotation of the center arm on R_5 will alter the base-collector voltage of Q_2 and,

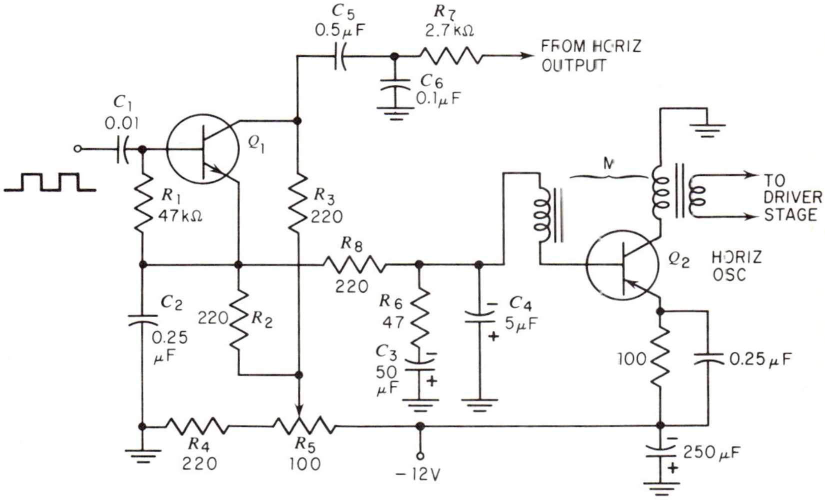

FIGURE 11.40
A transistor phase detector Q_1.

from this, the operating frequency of the oscillator. Thus, R_5 is rightfully the horizontal-hold control.

The positive or negative voltage variations introduced by R_2 will shift the oscillator operating frequency above or below a center frequency established by the setting of R_5. In this way, automatic control of the oscillator frequency is maintained.

Horizontal Oscillator. The horizontal oscillator follows the phase detector, receiving the correction voltage developed by the latter circuit. A frequently employed oscillator is the same blocking oscillator just described for the vertical system. Essentially the same considerations apply, modified only by the higher operating frequency and the need to respond to the afc voltage.

A typical oscillator circuit, integrated into a complete horizontal-deflection system, will be considered presently.

Horizontal-output Stages. The driving signal from the oscillator, if it is powerful enough, may be applied directly to the horizontal-output stage. At the present level of transistor development, however, there is more likely to be a driver stage between the oscillator and the output amplifier. This driver simply takes the output voltage developed by the oscillator, strengthens it, and then applies it to the horizontal-output amplifier.

Whether there is a driver and output amplifier or simply an output stage

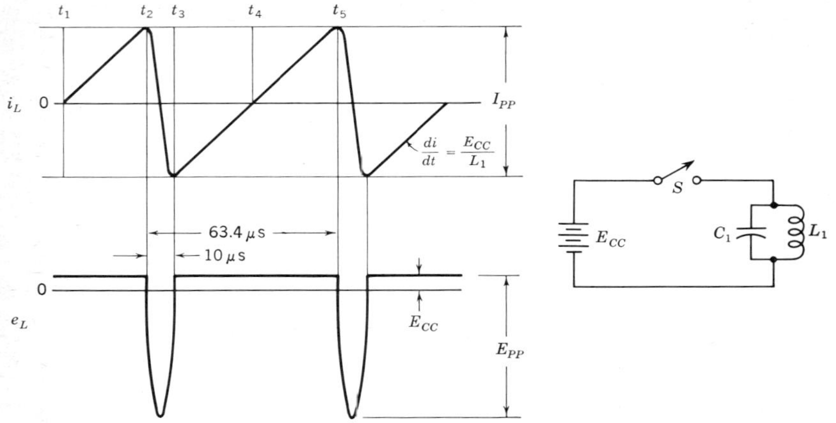

FIGURE 11.41
The basic output circuit of a horizontal-deflection system and its waveforms.

alone, the best place to start is with the output stage, because it is the requirements of this amplifier that will determine, to a large extent, the form of the stages that precede it. The chief purpose of the output transistor is to develop a sawtooth current through a deflection yoke. A secondary goal is to produce a high-voltage pulse that can be rectified and employed as the accelerating dc voltage for the picture tube.

An output stage and driver circuit that has been used successfully is shown in Fig. 11.44. In order to understand fully the operation of this circuit, let us consider the basic circuit from which it was derived. The circuit, shown in Fig. 11.41, consists of a battery E_{CC}, a switch S, a coil L_1, and a capacitor C_1. If switch S is closed at time t_1, the current flowing from the battery through coil L_1 will increase at a linear rate. At time t_2, the switch is opened, interrupting this flow. The interruption shock-excites the capacitor-coil combination and the current starts to oscillate between L_1 and C_1. Since the current had been passing through L_1 when the switch was opened, the inductance will attempt to keep the current flowing, forcing it to flow into C_1, and it is this particular action that starts the oscillatory motion.

For one half-cycle, or from time t_2 to time t_3, the circuit is permitted to oscillate. At t_3, switch S closes again, and now the current will flow back into the battery because the half-cycle of oscillation has caused it to reverse itself. The flow of current into the battery will continue until t_4, at which point it will have decreased to zero. From t_4 to t_5, the current travels from the battery to L_1, repeating the sequence of events just described.

In terms of horizontal-output-circuit operation, the steady rise of current from t_3 to t_5 represents the interval when the electron beam is steadily scanning across the picture screen from extreme left to extreme right. At t_5 (or t_2), the beam is blanked out and swung quickly back to the left-hand side of the tube. This is the retrace interval, and it must be completed in about 10 microseconds (μs). It is during this interval that a sharp pulse of voltage is developed across L_1 by the opening of switch S. This pulse is generally stepped up, rectified, and then applied to the second anode of the picture tube.

Note that if we were to represent the waveform of the voltage obtained from the battery, it would be a square wave. When the switch is closed, the full value of E_{CC} is instantly applied to L_1C_1; when the switch is open, the applied voltage drops instantaneously to zero.

In a practical circuit, the switch can be replaced by a junction transistor and the driving voltage by square waves from either an oscillator or driver. The square waves are required to turn the transistor on or off automatically. The basic transistor circuit and the waveforms in that circuit are shown in Fig. 11.42. From t_1 to t_2, the input signal biases the base-emitter circuit in the forward direction and the transistor conducts, completing the circuit. The input signal is strong enough to place the transistor in saturation, and in this condition the internal resistance (emitter-to-collector) is reduced to a few ohms.

At t_2, the incoming pulse goes sharply positive, which, for a PNP transistor, reverse-biases the base-emitter circuit and cuts off the transistor. The

FIGURE 11.42
The basic transistor output circuit and its waveforms.

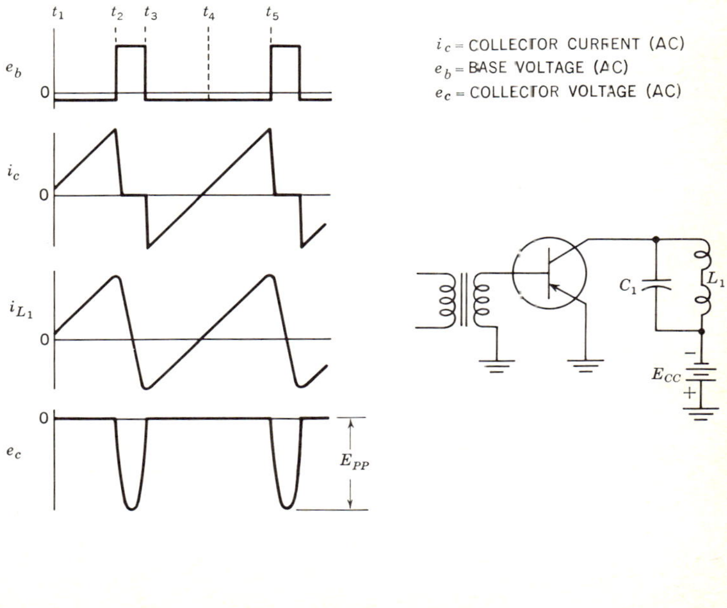

cutoff does not occur as sharply as the pulse change because when a transistor becomes saturated, carriers are accumulated in the base region and a short period is needed to clear this region after the forward bias has been removed.

From t_2 to t_3, the transistor switch is open and L_1 and C_1 are shock-excited into oscillation. For one half-cycle, the circuit oscillates; then the reverse bias is removed and the transistor is once again driven into strong conduction. The current now flows in the reverse direction through the transistor, just as it did in the preceding circuit. At t_4, this reverse current has decreased to zero and the forward flow begins again, repeating the sequence of events.

It is interesting to observe that the same step-by-step process takes place in horizontal-output stages utilizing vacuum tubes. However, since the horizontal-output tube cannot conduct in the reverse direction, a diode (i.e., the damper tube) must be connected across the horizontal-deflection winding. With a transistor, this added item need not be used because the transistor itself will serve the purpose. For best results, the unit should be capable of carrying current equally well in both directions, although some dissymmetry is tolerable. (If a special damper diode is used in transistorized systems, it does lessen the need for the output transistor to be symmetrical and this can be helpful costwise.)

During the interval from t_2 to t_3, when L_1 and C_1 are oscillating freely, a large, sharp pulse is developed across L_1 (Fig. 11.42). Care must be taken to see that the peak voltage of the pulse does not exceed the collector breakdown voltage. The designer of this circuit devised an interesting method of blunting this peak. Another resonant circuit, L_2C_2, tuned to approximately the third harmonic of the resonant frequency of L_1C_1, is connected across L_1 (Fig. 11.43). The voltage developed across L_2C_2 (labeled e_{L2} in Fig. 11.43) during the flyback interval possesses the form shown in Fig. 11.43. This voltage is 180° out of phase with the voltage across C_1 (e_c) at the center of the flyback pulse, and thus reduces the peak collector voltage by about 30 percent.

FIGURE 11.43
The special resonant circuit L_2, C_2, and R_2 helps to reduce the voltage peak at L_1 during the flyback period.

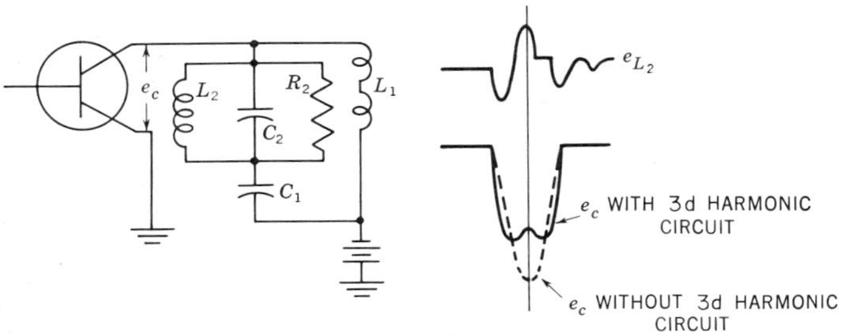

With the foregoing explanation of circuit operation understood, we are ready to consider the actual horizontal-output circuit developed by this designer. The output stage and the driver that precedes it are shown in Fig. 11.44. If we examine the driver stage first, we see that it receives pulses from a preceding oscillator. During the scanning period, the base emitter of the driver is strongly forward-biased and the transistor is operating in a saturated condition. During retrace, the incoming pulse goes sharply positive, cutting off Q_2. Since the transistor is working into a transformer T_3, the sharp cutoff causes a pulse to appear across the primary winding. This pulse, after being stepped down in voltage to reduce the circuit impedance, is employed to cut off the output stage.

Ac coupling between the driver and the output stage results in a dc (i.e., zero) axis about 10 percent above the signal-level scan, as indicated by the base waveform for Q_1. (Because of the ac coupling, the waveform distributes itself about the axis so that there is just as much area above the zero line as below.) The wave below the axis serves as a forward bias, which eliminates the need for an additional bias source. Furthermore, it tends to protect the output stage in case of drive failure, since with no drive both stages are cut off.

In both the driver and the output stage, a bootstrap circuit is used to permit the collector and case to be grounded for best heat conduction. Operation of the two stages is exactly as indicated above. High voltage is obtained from transformer T_2. Whatever flyback pulse appears across the primary winding of T_1 is stepped up by T_2 and then rectified by D_1. The latter is shown here as a tube. However, the same function can be served by several

FIGURE 11.44
Transistorized output stage and driver.

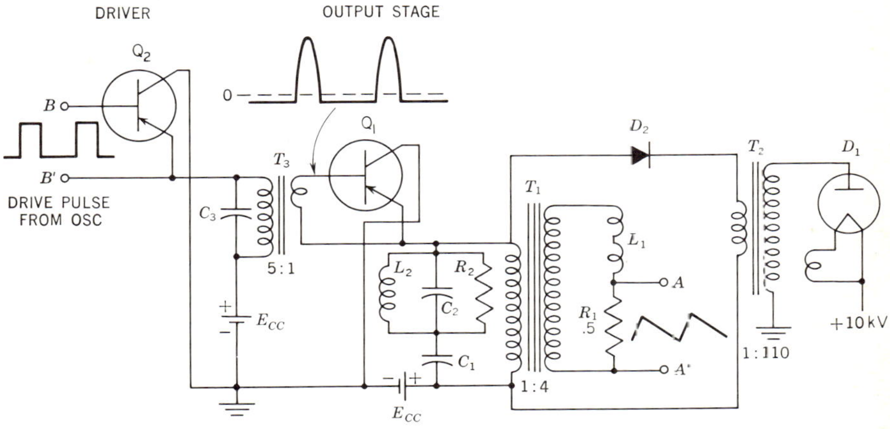

semiconductor diodes connected in series. Diode D_2 is so connected that it is closed during the retrace interval and opened during the scan period. Its purpose is to prevent damped oscillatory voltages in T_2 from causing ripples in the forward-scanning deflection current.

L_1 is the horizontal-deflection winding of the yoke. R_1 is a small resistance, inserted in series with L_1, to obtain a sample of the deflection current. This voltage is applied to a preceding phase detector, where it is compared in frequency with the incoming horizontal-sync pulses. Any difference between the two signals produces a correction voltage for the horizontal oscillator.

It may be of interest to see the oscillator and phase detector that precede the driver and output stages of Fig. 11.44. These are shown in Fig. 11.45. The phase detector is similar to the phase detectors previously discussed. The signal voltage developed across R_1 in Fig. 11.44 is applied between emitter and collector of Q_4. At the same time, negative sync pulses (from a sync

FIGURE 11.45
The transistorized horizontal-sweep oscillator and phase detector that precede the driver and output stages in Fig. 11.44.

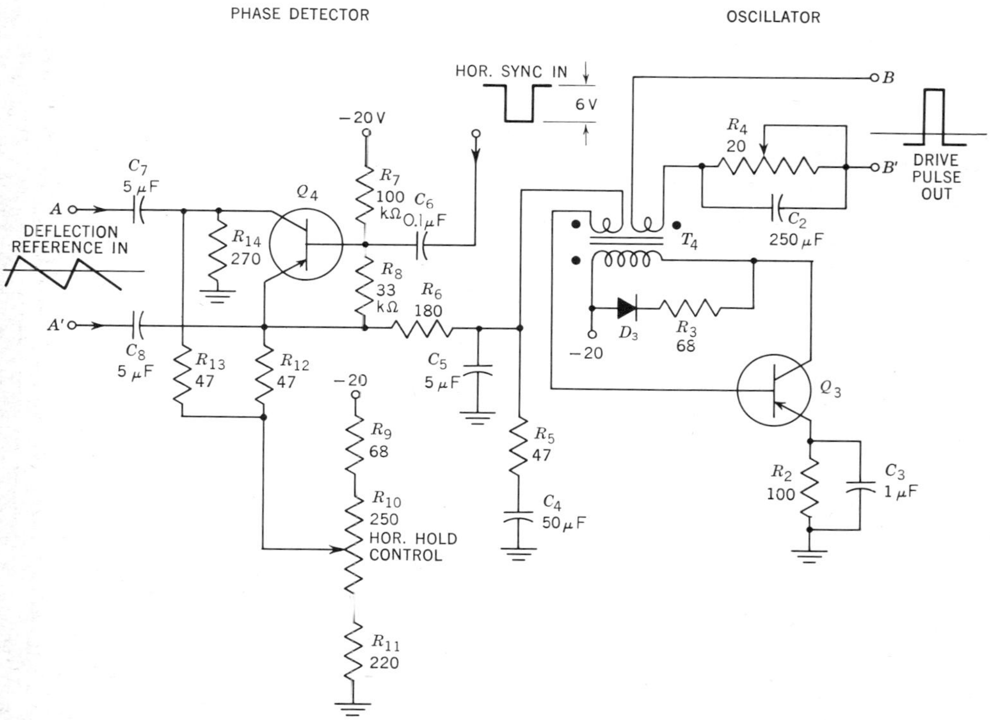

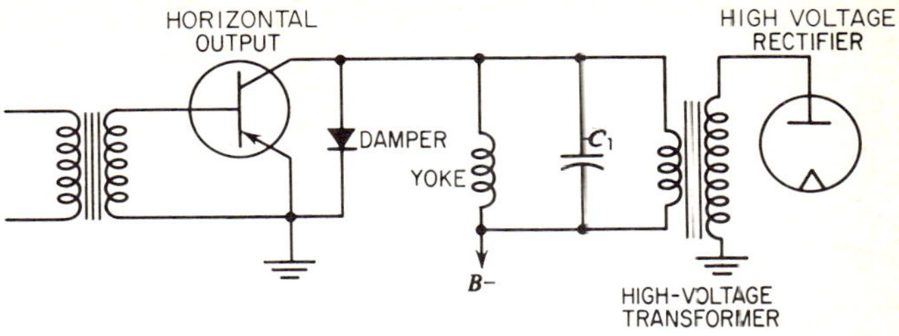

FIGURE 11.46
A horizontal-output circuit in which the transistor is directly coupled to the deflection yoke.

separator) are applied to the base of Q_4. The two voltages are compared in frequency, and any difference produces a voltage across R_{12}. This error signal is applied through the antihunt integrating circuit $R_6C_4C_5R_5$ to the oscillator base winding.

A blocking oscillator is employed for the horizontal oscillator. Oscillator repetition frequency and pulse width depend primarily on R_2, C_3, and the ratio of Q_3 base bias to collector supply voltage. In this circuit, the base bias is varied by the horizontal-hold control. This bias is placed in series with the dc correction voltage from the phase detector.

Diode D_3 serves to limit the transformer inductive overshoot during the time Q_3 is cut off in order that the collector breakdown voltage is not exceeded. The output signal for the driver is obtained from a third winding on the blocking transformer. This method of ac coupling provides forward bias for the output stage. Since the forward bias tends to be higher than necessary, the operating point is shifted in the reverse direction by bias resistor R_4.

A transistorized horizontal-output circuit in which the transistor is directly coupled to the yoke is shown in Fig. 11.46. When this approach is employed, a low-impedance yoke is used, one with an inductance on the order of 100 μH or less. By way of contrast, yoke inductances of 8 to 30 mH are common when a transformer is interposed between the yoke and the output transistor, as in Fig. 11.44.

The transistor in Fig. 11.46 has a damper diode connected in parallel with it. This diode is useful in providing a linear sawtooth current flow through the transistor. It achieves this by conducting current (together with the transistor) during the interval when the output transistor is also carrying the reverse current (during the interval t_3 to t_4 in Fig. 11.41). The sawtooth current wave is made more linear (Fig. 11.47) because a practical transistor will seldom conduct equally in both directions. By paralleling a diode across the transistor, increased current can be passed during the reverse interval to help keep the flow steady.

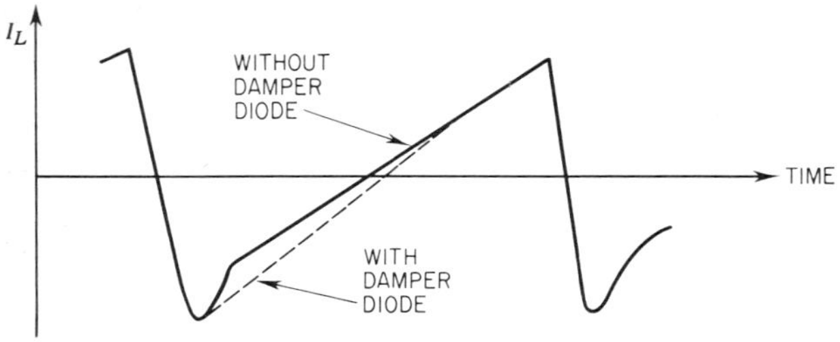

FIGURE 11.47
The shunt damper diode in Fig. 11.46 helps to achieve a linear sawtooth current through the yoke.

Finally, it is interesting to note that the total power dissipated in the output transistor developing 90° deflection at 10,000 V is of the order of 2 watts (W). In the output circuit of a vacuum-tube deflection circuit performing the same function, 20 W would be dissipated.

SCR Horizontal-deflection System.[1] An interesting development in transistorized television receivers, particularly color-television receivers, is a horizontal-output stage that employs silicon controlled rectifiers (SCRs) and diodes in place of a transistor.

A simplified schematic of the basic deflection circuit is shown in Fig. 11.48. It consists of two sets of bidirectional switches, each containing an SCR and diode and each capable of conducting current in both directions. [By this, we do not mean that each SCR and each diode is itself bidirectional but that each set is, because the two components of each set (SCR and diode) are connected in parallel, but with opposing polarity.] One set is formed with SCR_T and D_T and the other set with SCR_C and D_C.

Silicon controlled rectifiers will be covered in Chap. 12, but it can be stated here that they are essentially diodes with an extra element, called a gate, that acts to turn them on (i.e., cause them to conduct). Once they are turned on, they stay on until the voltage across the diode's cathode and anode drops below a minimum value. Once this happens, the SCR becomes nonconducting until pulsed again into operation by the gate.

It is characteristic of an SCR that once it has been turned on, the gate loses all control over it. The SCR cannot be turned off by the gate. Only a lowering of the cathode to anode voltage can do this. SCRs are designed to pass considerable quantities of current, which is, of course, one reason that they are useful in the output stage of the horizontal-deflection system.

In the circuit of Fig. 11.48, SCR_T and diode D_T take care of moving the

[1] Transistor, Thyristor & Diode Manual, Technical Series SC-14, Radio Corporation of America, 1969.

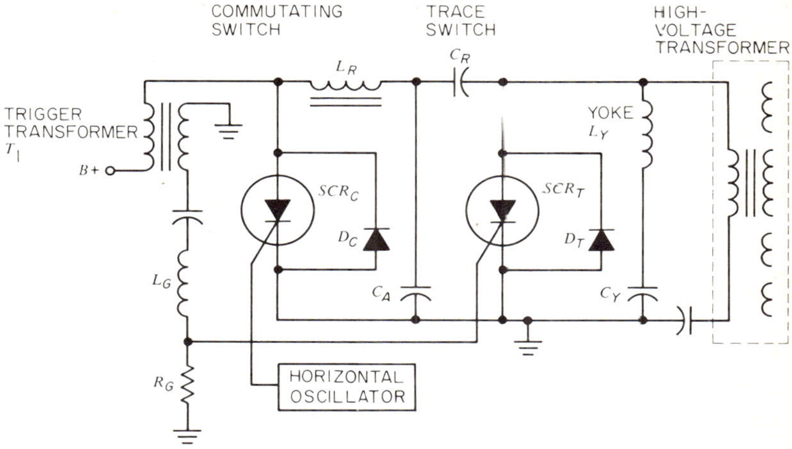

FIGURE 11.48
Basic circuit of a horizontal-deflection system using SCRs in the output stage. (*Courtesy RCA.*)

electron beam across the face of the picture tube. This is the trace T interval. The second set, SCR_C and diode D_C, become active during the beam retrace interval. The C here stands for commutating, and these two components provide the commutating action required for retrace.

To understand fully the operation of this circuit, it is useful to again refer to Fig. 11.41 and to the explanation that accompanied it. From time t_3 to t_4, the electron beam in the picture tube is traveling from the left-hand side of the screen to the right, reaching the center of the screen at time t_4. From t_4 to t_5, the beam is moving from the center of the screen to the right-hand edge, reaching this edge at time t_5. This is now followed by the retrace interval when the beam is brought quickly back to the left-hand side of the screen. Time interval t_2 to t_3 illustrates this retrace action.

Note that within both the trace and retrace intervals, positive and negative current flows exist. The current flow is negative between t_3 and t_4 and positive between t_4 and t_5. A similar variation exists between t_2 and t_3. This is the reason why each SCR and its associated diode in Fig. 11.48 are connected with opposing polarities.

To see in detail how this circuit functions, let us start at the beginning of the trace cycle when the beam is at the left-hand edge of the screen. At this point in time, only the trace-switch diode, D_T in Fig. 11.48, is conducting and sending current through the yoke coil L_y. The diode conducts a linearly decreasing current until the yoke current reaches zero. At this point, the beam has traveled from the left-hand edge of the screen to the center.

Just before the yoke current reaches the zero level, a positive pulse from transformer T_1 is applied to the gate electrode of the trace-switch silicon controlled rectifier SCR_T, thereby placing this component in condition to conduct current. When the yoke current crosses the zero point from negative to positive, the conducting path switches from D_T to SCR_T. This is necessary because a positive current flows in a direction opposite to a negative current. And this, of course, is the reason D_T and SCR_T are connected with opposite polarities.

At this point, capacitor C_y begins to discharge through the yoke winding and SCR_T. Capacitor C_y has sufficient energy so that its voltage remains essentially constant throughout the trace-retrace cycle. And a steady voltage, applied to an inductance, results in a steadily rising current through the inductance. In this instance, the inductance is that of the horizontal-deflection yoke coils, and the steadily rising current moves the electron beam from the center of the screen to the right-hand edge.

Once the beam reaches the right-hand edge, the retrace cycle has to be initiated. To do this, the commutating-switch SCR (i.e., SCR_C) is gated on by the horizontal oscillator. Capacitor C_R then discharges a pulse of current through inductor L_R and both SCR_T and SCR_C. This current pulse, referred to as the commutating pulse, increases until it exceeds the yoke current and thereby causes the trace diode D_T to turn on. This conduction of D_T reverse-biases SCR_T long enough to cause SCR_T to turn off. (SCR_T will now remain off until its gate element again receives a positive pulse from T_1.)

When the commutating pulse current falls to a value less than the yoke current, D_T opens, and the energy in the yoke winding produces a current that charges retrace capacitors C_R and C_A during the first half of retrace (i.e., when the beam swings from the right-hand edge of the screen to the center). This current then rings back into the yoke winding during the second half of retrace. The circuit path for the ringing oscillation during the second half of retrace is completed through diode D_C and allows enough time for SCR_C to turn off. When the yoke current reaches its peak negative value, diode D_T begins to conduct to start the trace interval.

Energy to keep this circuit in operation comes of course from the dc power supply. During the time that current can flow through either SCR_C or D_C, the input transformer T_1 is connected across the B+ supply, and energy is stored in the windings of this transformer. This stored energy charges the retrace capacitors C_A and C_R, replenishing the energy loss in the circuit.

The SCR deflection system exhibits excellent stability, since its critical voltage and current waveforms and timing cycles are determined by passive components in response to the action of two SCR-diode switch networks. Furthermore, only a triggering pulse of nominal value is needed to turn SCR_C on, and hence this deflection system can be driven directly from a pulse developed by the horizontal oscillator.

QUESTIONS

11.1. Why is it more difficult to transistorize a television receiver completely than a radio receiver? Give several specific examples.

11.2. Draw the circuit of an r-f amplifier suitable for use with a television tuner. Indicate what characteristics a transistor for this stage should possess.

11.3. Explain how the circuit drawn for Question 11.2 operates.

11.4. Explain the function of each of the components in the circuit of Fig. 11.4.

11.5. Could the neutralization method employed in Fig. 11.4 be applied to the circuit of Fig. 11.2? Explain your answer.

11.6. How does the circuit of Fig. 11.8 function? Indicate how the transistor can operate even though the collector has zero volts on it.

11.7. Explain the operation of the agc network of Fig. 11.10.

11.8. Draw the diagram of a diode video detector.

11.9. Draw the diagram of a two-stage video-amplifier system. Explain the purpose of each component.

11.10. Indicate what precautions would have to be observed in designing a transistorized video i-f system for a television receiver. (*Note:* Cover such items as β cutoff frequency, temperature stability, and impedance matching.)

11.11. What characteristics of a transistor enable it to be used successfully as a limiter? As a sync separator?

11.12. Explain how the circuit of Fig. 11.23 functions.

11.13. Why is a double-time-constant arrangement, such as used in Fig. 11.25, more desirable than a single-time-constant network?

11.14. Compare the circuits of Figs. 11.23 and 11.25 as to advantages and disadvantages.

11.15. Describe the operation of the sync separator of Fig. 11.25.

11.16. Draw the circuit of a transistor blocking oscillator that can be synchronized by pulses and that develops a sawtooth output voltage.

11.17. Explain the operation of the circuit drawn in response to Question 11.16.

11.18. What is the purpose of each of the controls in Fig. 11.29?

11.19. What is a symmetrical transistor? What counterpart does it have among vacuum tubes?

11.20. How does a double-diode phase detector operate?

11.21. Draw the basic circuit of a transistor phase detector. Explain briefly how it operates.

11.22. Why must the incoming sync pulses in the circuit of Fig. 11.35 possess much larger amplitudes than the sawtooth wave applied to the same circuit?

11.23. Draw the complete circuit of an actual transistor phase detector. Explain the purpose of each component.

11.24. How is bias provided for the two transistors in Fig. 11.44?

11.25. What is the purpose of L_2, R_2, C_2, and C_1 in Fig. 11.44? Explain in detail.

11.26. In what ways does Q_1 in Fig. 11.44 differ in operation from a horizontal-output stage using a vacuum tube?

Chapter Twelve

Additional Solid-state Developments

One might wonder, after becoming aware of the complexity and processing refinements of large-scale integration, if there is much more to be accomplished in solid-state-device development. The answer is that there is no apparent limit to this rapidly advancing technology.

In this chapter we shall discuss some of the "other" solid-state devices, and the mechanisms that were "put to work" to make them practical. It is interesting to note that the majority of these devices are the result of research work done on electronic behavior mechanisms associated with PN junctions and their reaction to external stimuli.

New semiconductor materials, such as gallium arsenide, are beginning to challenge the supremacy of silicon. In all probability, these new semiconductor materials will spur the development of new devices, circuits, and applications. In time, they may even replace contemporary devices, just as silicon is replacing germanium. This, however, is the hallmark of an expanding and developing technology.

GUNN DIODE

The Gunn diode is fabricated from a homogeneous chip of gallium arsenide (GaAs), a compound semiconductor. We shall discuss gallium arsenide in more detail later in the chapter, since it has so many new applications.

This diode, which operates in the microwave region, is the result of research on the mechanisms by which electrons move through semiconductor materials under the influence of an electric field.

In 1963, J. B. Gunn of IBM, while engaged in research of this nature, made the startling discovery that a tiny chip of gallium arsenide could be made to emit microwaves by simply applying a steady voltage across it. In order to understand how a Gunn diode operates, we must know a little about what happens to the electrons in semiconductors, particularly gallium arsenide, at varying applied voltages.

The electrons that carry current in a solid have varying velocities and different thermal energies, which influence their mobility. All the electrons are accelerated by an electric field, and the acceleration increases indefinitely if the electrons do not collide with atoms of the solid, thereby losing some of their energy. The result of this is that the total "extra" energy gained from the electric field is quite small when compared with the electron's random thermal energy. This small increase in mobility is important, since it results in electric current. This type of electron behavior conforms to Ohm's law and prevails in most conductors.

In some semiconductors, however, it is found that increasing the applied electric field can easily result in an increase in electron mobility that will soon challenge the predominant random thermal mobility factor. When this occurs, the distribution of electron energies departs noticeably from its normal distribution, and the electric current no longer increases in proportion to the electric field. In short, Ohm's law is no longer followed.

At the time of his discovery, Gunn was experimenting with a homogeneous (no junctions) N-type chip of gallium arsenide (gallium arsenide may be doped, N or P, in the same way as germanium or silicon). He applied varying voltages across a small N-type chip and studied the current-voltage relationship. At lower voltage levels, Gunn observed nothing unusual. That is, the current increased proportionately with the voltage, and at the higher voltage levels the expected deviations from Ohm's law were encountered.

However, when the field was increased to approximately 3,000 volts per centimeter (V/cm), the current began to fluctuate at an extremely rapid rate. At first, the fluctuations were random, but experimenting further with small chip geometries (0.0005 in), Gunn found that uniform oscillations could be obtained having a frequency of 10^9 hertz (Hz).

Identical microwave oscillations were also achieved with N-type indium phosphide, another compound semiconductor. However, P-type samples of both gallium arsenide and indium phosphide could not be made to produce these oscillations. This was somewhat of a surprise because except for their positive charge, holes behave the same as electrons in a semiconductor. Furthermore, the current-voltage relationships below the 3,000 V/cm level were the same in the N- and P-type samples.

Additional experiments showed that the frequency of oscillation was inversely proportional to the length of the chip; also, the calculated time for an electron to travel the length of the chip, at a specific voltage, was equal to the inverse of the frequency of oscillation. The length of the chip was found to influence also the threshold voltage, or the voltage required to initiate the oscillations.

Magnetic fields, temperature variations, or different types of contacts had little effect on the oscillations. All this data seemed to indicate to Gunn that the oscillation effect was due to a specific type of electronic structure and behavior. Unfortunately, the data he had collected did not fit any known electron behavior mechanism that might explain the oscillations.

Thus, after eliminating the known electron-behavior mechanisms, Gunn resorted to the direct approach. That is, he designed an experiment whereby he would be able to measure variations in the voltage distribution across the surface of the chip, during the current oscillations.

The problems associated with the completion of this experiment were indeed formidable. Remember that the length of the test chip was only 0.0005 of an inch, and in order to measure voltage variations across its surface, extremely small probes had to be fabricated. These probes were built, however, and Gunn was able to detect high-field waves that developed at the cathode and traveled across the chip to the anode.

He found that as the dc voltage applied across the chip was raised, a level was reached (called the threshold level) at which the field at the cathode end of the chip suddenly surged to approximately 3,000 V/cm. When this occurred, the field in the remaining chip area fell below this value, in essence presenting a strong field at the cathode and a low field everywhere else. This pulse, or wave, at the cathode did not remain there but moved rapidly across the chip to the anode. Here it disappeared and another pulse immediately appeared at the cathode, and the sequence repeated itself.

If the voltage across the chip was not fixed, Gunn found that random oscillations were produced. This was so because with varying voltages, the fields on either side of the pulse could be greater than 3,000 V/cm and produce random waves.

Thus, Gunn was able to explain the current oscillations by the existence of high-field waves. That is, the current in the chip drops as the wave is formed at the cathode, since the field around the wave falls below the threshold voltage of 3,000 V/cm, and hence cannot carry as much current. Now, as the wave disappears at the anode the current level in the chip rises, since the field also increases. These current fluctuations, or oscillations, occur in a periodic fashion with the time interval between current spikes being controlled by the transit time of the high-field wave across the sample.

At this point in his experiments, Gunn was approaching the key electron-behavior mechanism that could explain how these waves were formed, and also why they did not dissipate as they traveled from cathode to anode. The answer again involves quantum mechanics and, as Gunn eventually realized, its relationship to the negative-resistance phenomena.

Negative Differential Resistance. Stated briefly, negative resistance occurs in a system when the current initially increases with an applied voltage, then decreases with additional voltage, and finally rises again with further voltage increases (see Fig. 12.1). The tunnel diode, which we shall discuss later, also depends on negative resistance for producing high-frequency oscillations.

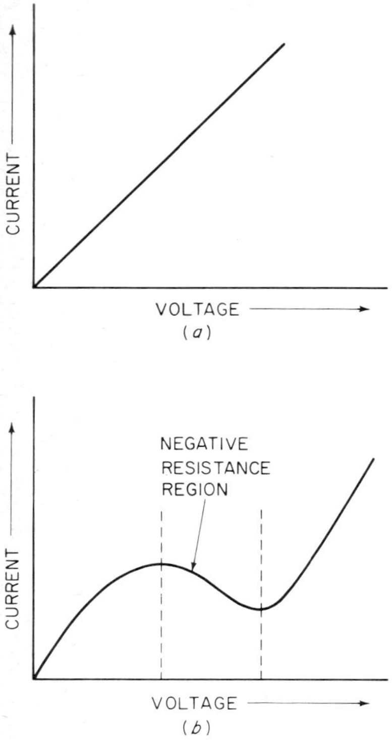

FIGURE 12.1
Current and voltage relationships for positive and negative resistance. (*a*) Positive resistance. (*b*) Negative resistance.

In order to associate the Gunn effect properly with negative resistance, let us discuss for a moment some basic laws of quantum mechanics.

Recall that in Chap. 1 we described how the close proximity of atoms in a solid causes the individual energy levels associated with isolated atoms to "break up" and form bands of energy levels.

We described the conduction band, the valence band, and the forbidden gap, as bands of energy that are either allowed or forbidden to electrons by the laws of quantum mechanics. It was also pointed out that the conduction properties of a solid are governed by the number of electrons in the conduction and valence bands and by the width of the forbidden gap.

Recent research has uncovered many additional details about the energy-band structure of solids. For example, it has been shown that the conduction band can contain more than one energy level. Furthermore, by suitable excitation, "normal" conduction-band electrons may be pushed to a higher conduction-band level. Thus, we must now change our thinking a little and visualize a solid as containing a conduction band that is itself divided into allowed and unallowed energy levels. This also implies that there are additional forbidden-gap regions. The conduction properties of such materials are still governed by the distribution of electrons among the allowed energy levels and the magnitude of the forbidden gaps.

The energy-band structure is now so well known that it is possible to predict the particular conduction behavior of a given solid under various conditions. That is, for any semiconductor, including the compound semiconductors, we know which energy bands are occupied and which are vacant under normal conditions. In addition, we know the precise amount of external energy required to "push" electrons to new (higher) energy levels.

This concept of electron transfer, from the normal conduction-energy-band level to a higher one was first proposed in 1961 by two British scientists, B. K. Ridley and T. B. Watkins. A most important result of this electron-transfer mechanism is that in transferring from the lower to the higher conduction-energy-band level, the effective mass of an electron appears to increase. The result of this is that the transferred electron's mobility and velocity are sharply reduced. The electron-transfer mechanism, with the accompanying curtailed electron mobility, eventually provided the key to the understanding of the Gunn effect.

If we look again at Fig. 12.1*b* it seems quite logical that such a curve could result if electrons initially accelerated by an electric field were somehow removed from the conduction process with further increases in the electric field. In essence, this is what occurs when electrons are pushed into the higher conduction energy band by an appropriate electric field. A semiconductor material, such as gallium arsenide, possesses a conduction-energy-band structure ideal for electron transfer and will therefore exhibit negative resistance.

That is, the current will initially increase with increasing voltage because the electrons in the normal, or lower-condition energy band, are accelerated. However, as the voltage continues to increase, the electric field becomes strong enough to provide some electrons with sufficient energy to transfer to the higher energy-band level. Here, for all practical purposes, these electrons are removed from the conduction process, since they have become essentially immobile. Thus, even though the voltage has been increased, the current decreases, resulting in the characteristic negative-resistance curve of Fig. 12.1*b*.

Let us now attempt to correlate the electron-transfer mechanism, negative resistance, and current oscillations observed by Gunn.

Ridley, continuing Gunn's work with electron-behavior mechanisms, studied the problem of the electric field distribution in a solid when an additional voltage is applied in the negative-resistance region. He was aware at the time of proposals made by William B. Shockley of the Bell Telephone Laboratories and T. Reik of the Phillips Laboratories in Germany. Both men had suggested that if an increase in voltage were applied in the negative-resistance region, the test specimen would become divided into regions of varying electric fields. Specifically, a tiny domain would form in which the electrical field would be extremely high, and the field in the rest of the sample would be low. Ridley then suggested that this high-field domain would travel across the specimen from cathode to anode. A new domain would appear at the cathode when the initial domain, or wave, disappeared at the anode.

Ridley's work was known to Gunn, but he did not immediately accept it as an explanation of the oscillation effect because he initially miscalculated the electron-transfer energy requirements of gallium arsenide.

It was Herbert Kroemer of Varian Associates who asserted that Ridley's moving-domain theory fit the data collected by Gunn and was indeed the correct explanation of the current oscillations. Gunn himself, as we mentioned, verified this by directly measuring the "waves" of high electric field that built at the cathode and traveled across the specimen to the anode.

Let us look briefly now at the electron behavior associated with a moving domain. Figure 12.2 represents a domain, or wave, between the cathode and anode. The electrons in the normal conduction band travel faster than the high-field domain; and once they "catch" the wave, they are forced into the higher energy band and lose a good deal of their mobility. This will cause them to fall behind the wave, where they will quickly lose some energy through scattering and drop back into the lower, fast, energy band and race again after the wave. The slower electrons bunch up at the back of the wave and provide the electron-transfer energy—or in other words, the high field.

The electron-transfer mechanism can also explain the fact that the high-field region carries the same amount of current as the low-field regions. That is, the large number of electrons bunched at the rear of the wave are in the high, immobile energy band, making their net speed about equal to that of the electrons in the rest of the sample.

The electron-transfer-mechanism explanation of the Gunn effect has been neatly verified by an experiment carried out at the Bell Telephone Laboratories. A gallium arsenide Gunn oscillator was subjected to nearly 30,000 atmospheres

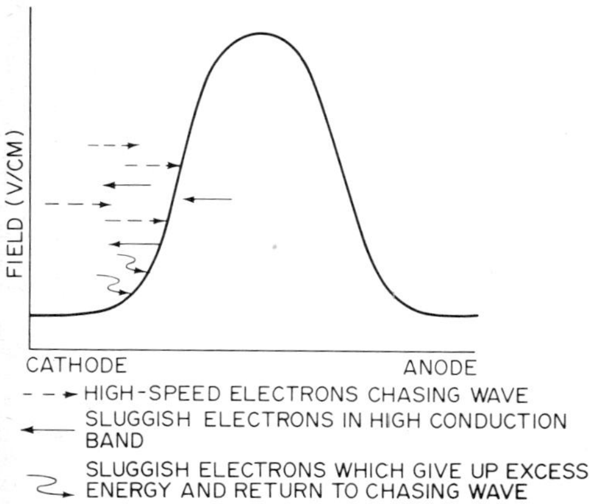

FIGURE 12.2
Sketch of a moving domain.

(atm) of pressure, which decreased the distance between the energy bands. This device was found to require less voltage (i.e., a lower threshold) for the formation of the traveling high-field domains. This, of course, fits in nicely with the electron-transfer mechanism.

Diode Fabrication. The starting material, an N-type gallium arsenide wafer, for Gunn diode fabrication must be extremely uniform and free from imperfections to prevent current crowding and thermal runaway, which would destroy the device. The material parameters are quite critical, in fact, and the Gunn diode is quite difficult and costly to fabricate. However, new material-processing methods are being developed that give promise of mass fabrication of practical, efficient devices.

At present, the most popular fabrication method is the epitaxial planar method. In this process a unique procedure known as the liquid-tin regrowth technique is employed.

An N-type GaAs wafer is first heated, in a quartz tube, to 600°C and then brought into contact with a measured amount of tin saturated with gallium arsenide. After the wafer has made contact with the tin melt, the system is cooled to 550°C at a rate of approximately 12°C/min. This causes the N^{2+} GaAs to precipitate out of the melt and grow epitaxially onto the N-type GaAs wafer surface (Fig. 12.3). After the system reaches 550°C, the tube is flushed with nitrogen gas and the wafer is slowly extracted. Residual tin is etched away with a hydrochloric acid solution and the wafer is then back-lapped.

After this step, the wafer is thoroughly cleaned in hot solvents, such as chloroform, acetone, and ethyl alcohol. The wafer is given a final back-etch, bringing it to its desired thickness.

Contacts play an important role in the formation of microwave oscillations. If one is to obtain uniform oscillations and efficiency, the contact interfaces with the gallium arsenide chip must be well-defined planes, and the anode and cathode must be almost parallel, for uniform field distribution across the device.

One method of contact formation begins with the evaporation of a very thin [100 angstrom (Å)] uniform layer of tin over the epitaxial side of the wafer. The function of this thin layer of tin is to absorb oxygen on the surface of the wafer and thus aid in the formation of a good ohmic contact. A thicker

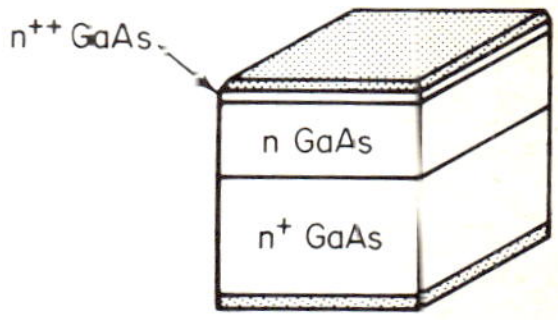

FIGURE 12.3
Basic structure of a Gunn diode.

layer of electroless nickel is then deposited on top of the tin, and is followed by a final layer of tin, which is usually vacuum-evaporated.

Alloying of the contacts is done in a forming gas atmosphere at 400°C for about 30 seconds(s). Other metal systems, such as nickel indium or gold indium, are also used for contact formation. After alloying, the wafer is scribed and cracked just as we described for the planar transistor (Chap. 4).

The Gunn diode is currently being employed to develop signals well into the gigahertz region.

LIGHT-EMITTING DIODES (LEDS)

These devices, as the name suggests, emit light and play a significant role in the expanding new field of optoelectronics. In this section, we shall also discuss light-sensing devices and semiconductor injection lasers, which are additional optoelectronic devices.

Since the LED can be substituted in almost every application where tungsten-filament lamps are presently used, it has to be considered among the most promising semiconductor devices. In addition, together with light-sensing devices, LEDs are fundamental to any optoelectronic system. These systems are finding many new fascinating applications.

Seeing the Light. We have seen that when semiconductors are doped with impurities, majority and minority carriers are present. In a semiconductor doped to produce electrons (N-type dopant), holes are the minority carriers. In a P-type semiconductor, electrons are the minority carriers.

With both types of carriers present, it is inevitable that some recombination will take place. When this occurs in either germanium or silicon, the energy that is released by the recombination appears primarily as heat throughout the crystalline structure of the crystal. With certain other semiconductor materials, such as gallium arsenide, the recombination energy appears in the form of packets or pulses of light called photons.

Since recombination of holes and electrons produces light and we desire as much light emission as possible, it has been found that this is best achieved at a PN junction. When the junction is forward-biased, electrons from the N section are injected into the P section, where recombination occurs. The photons that are produced by this recombination then travel through the crystal structure until they either are absorbed or escape from the crystal into the air as light.

The energy band gap of the material forming the junction determines the wavelength of the emitted light.[1] Specifically, the wavelength of the emitted

[1] Band gap is the energy difference between the conduction band and the valence band in a material.

light is inversely proportional to the energy change occurring at the electron-hole-recombination impact, and this energy change is determined by the energy gap. For example, intrinsic gallium arsenide has a band gap of 1.4 electron-volts (eV) and its emitted-light wavelength is in the infrared region, at about 9,000 Å. Gallium phosphide (GaP), on the other hand, has an energy band gap of 2.25 ev with an output wavelength of 5,500 Å, which is within the visible spectrum.

At the present time most LEDs are made from gallium arsenide phosphide (GaAsP) epitaxial wafers, which have an emitted light wavelength of about 6,700Å. Gallium phosphide (GaP) is a new LED material that can be made to emit both red and green light by "tailoring" the material for a desired wavelength by varying its composition and hence its band gap energies. The processing involved is critical, and the cost of this material (GaP) is not yet competitive with other light sources.

Another interesting processing innovation that is used to improve the light-emitting efficiency of GaAs involves the use of silicon as both the N- and P-type dopants for junction formation. This is possible because group IV elements, such as silicon and germanium, are amphoteric when added to GaAs—that is, they can occupy either gallium or arsenic sites, depending on their concentration. For example, at low concentrations, silicon acts as a donor, displacing arsenic; whereas at higher concentrations, silicon acts as an acceptor, displacing gallium. This type of reaction takes place only in heavily doped N-type regions. This method of PN-junction formation involves changing the temperature of the growing crystal that shifts the silicon-concentration ratio. The technique is similar, in principle, to the rate-growing technique described in Chap. 4. Light emitted from such junctions will have wavelengths ranging from 9,200 to approximately 9,800 Å.

From the previous discussion the reader might well conclude that a LED has a coherent output wavelength that is determined by a specific amount of recombination energy. Unfortunately, this is not the case; commercially available LEDs emit a rather broad wavelength spectrum (Fig. 12.4).

Important concerns are LEDs that emit light in the visible-light spectrum

FIGURE 12.4
Light wavelength spectrum of a GaAs LED. The rated output wavelength is the wavelength at peak power output at 25°C.

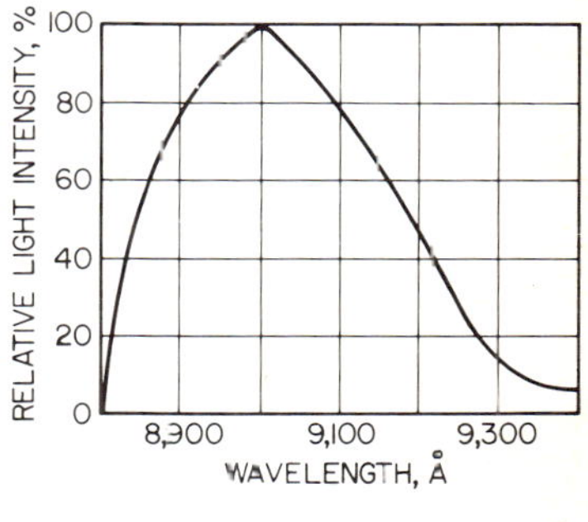

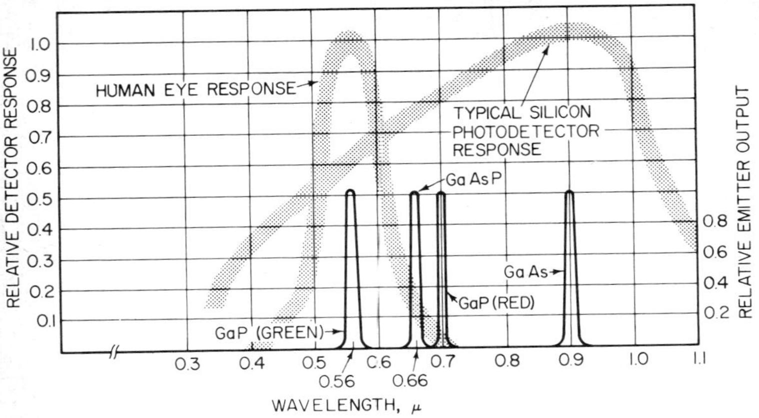

FIGURE 12.5
The spectral output obtained from some common LED materials is plotted against two commonly used detector materials and the human eye. Note how the eye peaks around green. (*Courtesy Electronic Engineer.*)

and the response of the human eye to the light. Such LEDs are most often specified by the manufacturer in terms of brightness and are measured in footlamberts (fL). Under normal room conditions, the eye is most responsive to a wavelength of approximately 5,500 Å (green), which means that a device emitting light outside this range will have to be more efficient if it is to be seen. Figure 12.5 shows the spectral output of some popular LED materials.

The optical power output of a LED device is determined, in general, by the shape of the gallium arsenide pellet and the package in which it is placed. If we utilize a simple, flat N-type gallium arsenide crystal onto which a P-type layer has been diffused, then the light that is developed at the PN junction will travel up through the P-type layer and exit at the chip surface (see Fig. 12.6*a*).

The problem with this approach is that the high index of refraction of gallium arsenide (with respect to air) means that when rays of light deviate more than 15° from the vertical axis, they are reflected back into the crystal and reabsorbed.

This surface reflection can be greatly reduced by forming a hemispherical dome over the junction (Fig. 12.6*b*). The gallium arsenide crystal is ground and polished into the form of a hemisphere and a planar P-type layer is diffused into a section of the base. If the diameter of the hemisphere is properly chosen, virtually all the light that is generated at the PN junction will escape into the air and produce a brightness as much as ten times greater than the unit of Fig. 12.6*a*.

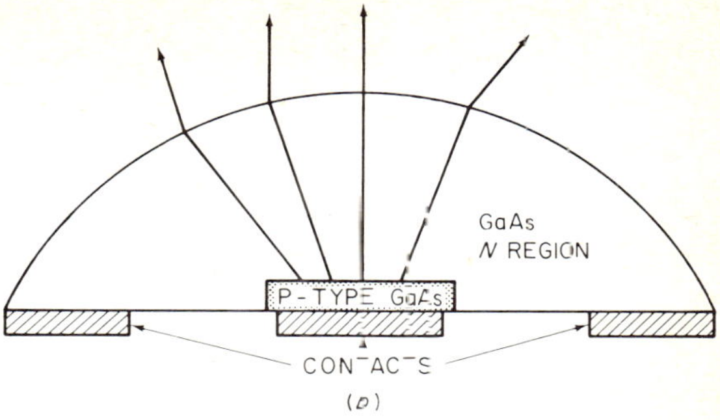

FIGURE 12.6*a*
Flat planar emitter showing internal reflection. Efficiency of approximately .21.5 percent.

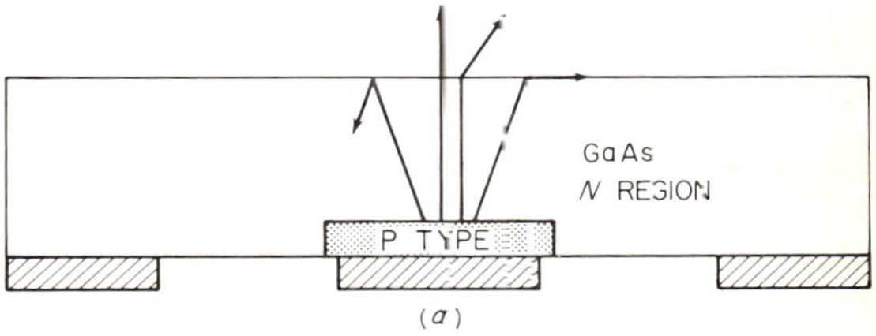

FIGURE 12.6*b*
Hemispherical configuration of LED. Reduces the probability of photon striking interface at more than 15°. Efficiency of approximately 2-10 percent.

The solid-state construction of LEDs makes them particularly sturdy and quite immune to the type of catastrophic failure and mechanical breakdown so common with the conventional pilot lights. Furthermore, they require no sockets or special power sources for operation.

Before we illustrate LED applications, let us look at light-sensing diodes, since they are partners with LEDs in many fascinating optoelectronic applications.

Light-sensing Devices. In 1873, Willoughby Smith observed that when a selenium bar was exposed to light, its resistance decreased. This is photoconductivity, which should not be confused with the photovoltaic effect by means of which a voltage is produced, in certain materials, upon exposure to light.

From these two effects we have two basic types of light-sensitive diodes—photoconductive and photovoltaic. But as we shall see, even though there are many special types of photo diodes, they are all basically diodes and do not deviate from the familiar diode theory. They are, however, optimized for specific light-response applications.

Photo Diodes. In photoconductive diodes, resistivity decreases linearly with increased light intensity. This change in resistivity is utilized in optoelectronic circuits, usually in conjunction with LEDs.

All semiconductor materials are light-sensitive in varying degrees. Not surprisingly, the degree of photoconductivity is governed by the material's energy band gap. That is, if the energy of the impinging light is to be useful in affecting the characteristics of the semiconductor, then this energy must be equal to or greater than the energy band gap of the material. Crystal structure, then, is the most important variable in determining photosensitivity and photoconductivity.

In addition, when the holes and/or electrons are once generated by the impinging light photons, they must have sufficient lifetime and mobility in order to reduce resistivity significantly. An electric field must also exist, together with electrodes, to move these carriers through the material.

Even when not exposed to light, a reverse-biased photoconductive diode will develop a current flow with a bias of only a few tenths of applied voltage. This current—referred to as dark current—is due to the thermal generation of holes and electrons. Now, as the junction is exposed to light, additional electrons and holes are freed, and the current increases. Figure 12.7 illustrates typical VI characteristic curves for a photoconductive diode.

It is important to note that a silicon photo diode operated without any bias whatsoever will respond as a photovoltaic unit, generating voltage in proportion to the light exposure. On the other hand, if the diode is operated in the reverse-biased mode, it will respond as a photoconducting diode with conductivity increasing in proportion to the light intensity.

The photovoltaic diode normally consists of a relatively large PN (silicon) junction, exposed to light. Photons impinging on the junction contain enough energy to break covalent bands, generating electron-hole pairs in the junction. If the load resistance across such a junction is less than approximately 800 ohms (Ω), the device is considered short-circuited, and any current that may be developed would be directly proportional to the light intensity. On the other hand, if the load resistance is in the 10,000-Ω range, open-circuit operation is obtained and the voltage developed across the junction will vary logarithmically with the light intensity.

Photosensitive diodes can be fabricated from any material in which a PN junction, or barrier, can be formed. The most popular materials, at present, are silicon and germanium, and the planar process (Chap. 4) is the

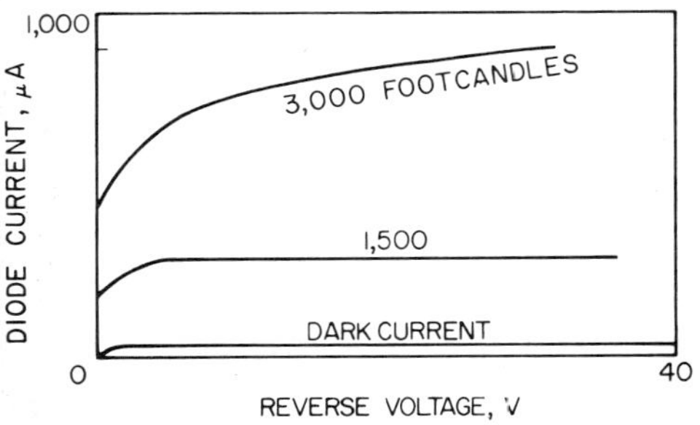

FIGURE 12.7

most common method of fabrication. Device characteristics such as response speed, absorption efficiency, and dark current can be optimized by material properties (crystal structure, resistivity, lifetime, mobility) and by processing techniques.

Phototransistors. Certain optoelectronic applications require more output current than is available from photo diodes. The obvious solution is a phototransistor by means of which current amplification can be obtained.

In the early days of transistor development, it was extremely difficult to measure certain device types because of their sensitivity to light. At that time, experiments were run in which a light spot probe proved that the col-

FIGURE 12.8
A phototransistor. (*a*) Equivalent circuit, (*b*) and (*c*) physical construction of phototransistors.

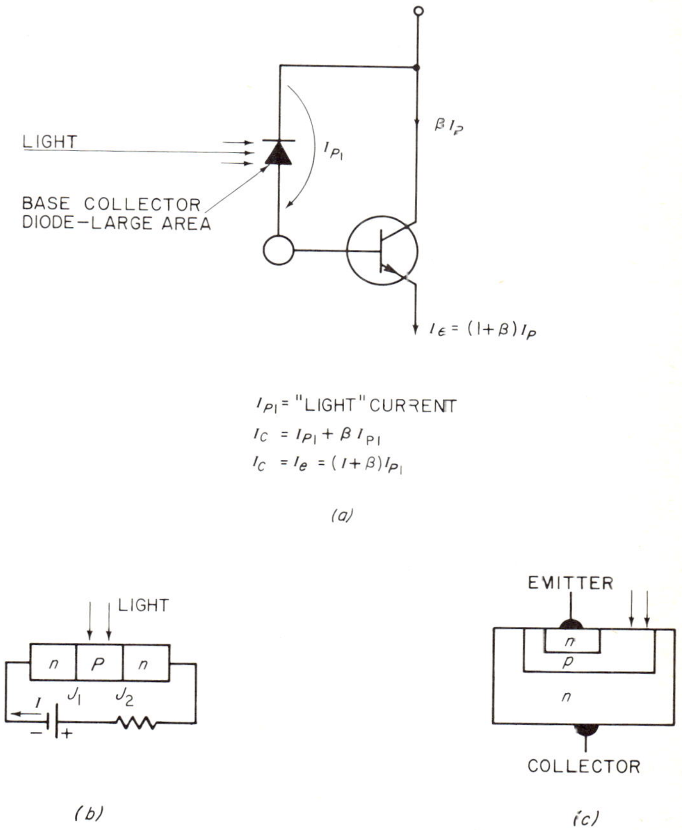

lector-base junction acted as a current generator for the transistor base. Figure 12.8*a* represents the equivalent circuit of a phototransistor. The collector-base junction and the emitter-base junctions are identical (i.e., impurity concentration), except that in practical devices the collector-base-junction area is enlarged for light collection.

To understand the operation of a phototransistor more clearly, consider Fig. 12.8*b*. This is an NPN transistor, in which the two end sections are separated by a wider-than-normal base. Light striking the exposed collector-base junction liberates electron-hole pairs. Because of the applied voltage, the electrons move toward the positive battery terminal. The holes tend to remain in the P region, however, where they form a positive charge that increases the forward bias of junction J_1. This results in an increase in current through the transistor. Figure 12.8*c* shows a planar phototransistor that operates in essentially the same fashion. Note that in neither device is there any connection to the P region (i.e., the base).

It can be seen from Fig. 12.8 that output current will be increased if

1. The collector-base area is made as large as possible
2. The H_{FE} is made as high as possible
3. The light-absorption efficiency (quantum efficiency) of the collector-base diode is optimized by controlled material processing

The operating point of a phototransistor is determined by the light intensity and the load impedance (Fig. 12.9).

FIGURE 12.9
Collector characteristics of a phototransistor.

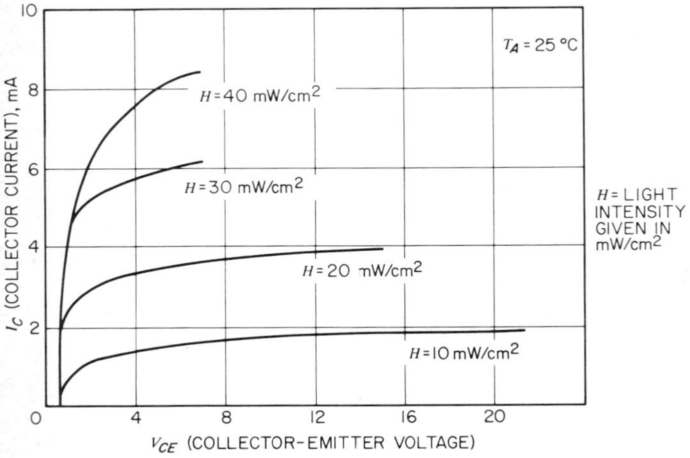

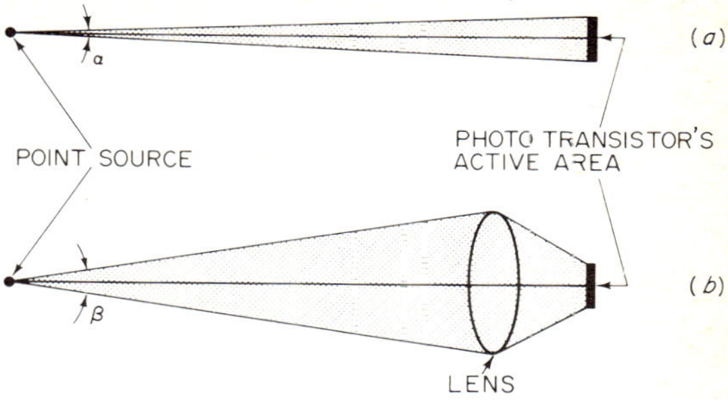

Figure 12.10
Optical gain is possible in a phototransistor by using a lens. In (*a*), the phototransistor intercepts all light in angle α; whereas in (*b*), the same device has a greater collecting angle β because of the lens. (*Courtesy Electronic Engineer Magazine.*)

Lenses are often employed to control the gain of a phototransistor (Fig. 12.10). The lens "gathers" light and guides it to the photo junction of the device. The use of lens gain control has another advantage, in that it permits the device to be somewhat selective in response to light from large off-axis angles. For example, if the data to be sensed is tightly stored, such as on computer cards or tapes, the device selected to sense this data must be selective in its response to light. The use of the lens permits a much smaller collector-base-junction area to be utilized and this, in turn, results in a reduced collector-base-junction capacitance.

A typical sensitivity range for a planar phototransistor would be in the range of 1 (mA) milliampere collector current with a light source (irradiance) value of 1 milliwatt per centimeter squared (mW/cm^2).

Field-effect Phototransistors. The success in using transistors for light-detection applications led to the investigation of other solid-state devices for a similar purpose. It was found that the gate junction of a field-effect transistor can act as a fairly efficient photo diode when light is focused onto it. The drain-to-gate junction is reverse-biased through a gate resistor R_G (see Fig. 12.11). By

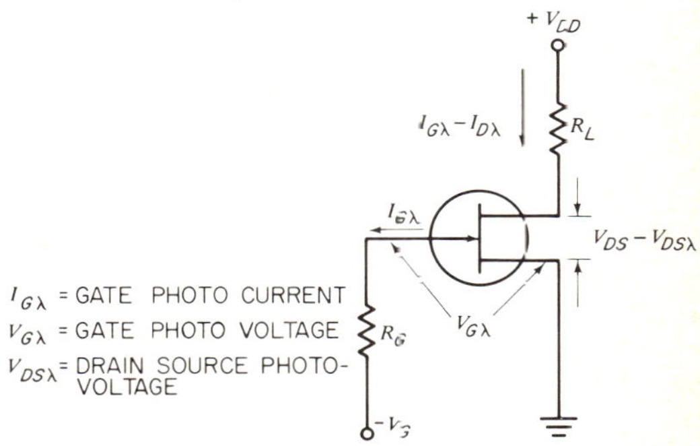

FIGURE 12.11
Photo FET—bias circuit. (*Courtesy Electronic Engineer.*)

controlling the value of R_G and the bias, a wide range of field-effect phototransistor operation is possible. Extra care must be taken in focusing the light source because most of the gate junction is normally covered by contact metallization. Gains of 800 microamperes per foot candle (μA/fc) are attainable with $R_G = 1$ kilohm.

OPTICAL SYSTEMS

A number of variables must be considered in analyzing the operation of an optical system. For example, the light detector and the LED must be precisely aligned, the light-beam spread of the LED must be taken into account, and the level of the ambient light must be considered.

Optoelectronic device manufacturers attempt to supply data that can be used to predict system performance when fixed values are available. We shall discuss a few optoelectronic applications and in so doing, we shall assume that the LED output is at one wavelength, that the emitted light beam has uniform intensity, and that all system components operate at nominal values.

Let us examine an optical system in which there is maximum coupling between the LED and its detector and no lens is employed between them. To predict the performance of such a system, steps similar to the following are taken by the designer.[1]

1. The total power output of the LED is obtained from the data supplied by the manufacturer.
2. The power density on the detector is determined using the following equation: Power density = total power received, mW ÷ detector area, cm²
3. The detector output is obtained from the data supplied by the manufacturer and may have to be adjusted for the wavelength of the LED.

Let us now look at an actual coupling system using an RCA 40598A LED and a Fairchild detector type FPM-100. The LED and detector in this system are to be separated by one-tenth of an inch.

1. The 40598A output power, obtained from Fig. 12.12*a*, is 28 percent $\times$ 1.6 mW = 0.45 mW.
2. The power density on the detector $= 0.45$ mW$/\pi(0.03\text{ m})^2 = 159$ mW/in^2, or 25 mW/cm^2.
3. The detector output current, obtained from Fig. 12.12*b*, is part of the supplied data and is approximately 4 mA at 5 V; adjusted for wavelength, as recommended by the manufacturer, this becomes 12 mA at 5 V.

[1] R. S. Meyers and J. O'Brien, Light-emitting Diodes, *Electronic World,* July, 1969.

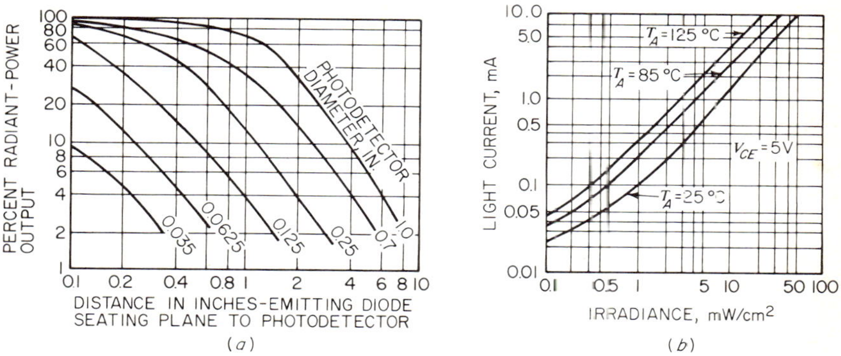

FIGURE 12.12
(*a*) Performance curve for a 40598A LED having a nominal output of 1.6 mW at 50 mA. (*b*) Performance curve for FPM-100 detector. (*Courtesy Electronics World.*)

Optical coupling systems such as this are rapidly replacing electromechanical relays in missiles and rockets because of their relative immunity to the high G environment.

If there is a large separation between the LED and the detector, the ambient light must be taken into consideration, since it can often produce more (detector) light than the LED. This can be overcome by taking advantage of the LEDs high-frequency-modulation capabilities. Such pulsed, or modulated-light, systems employ detectors that "ignore" the ambient light, responding only to the pulse rise time of the light emitted by the LED. This is true even if the ambient light is stronger than the LED light.

The use of a lens between the LED and detector requires precise positioning of the components involved. This can be accomplished. Lenses have greatly increased the output of a system, allowing much greater LED-to-detector separations. Theoretically, this separation distance can be miles, but in practical systems the limit is yards.

A lens placed close to the LED directs divergent light toward the detector, increasing the output of the system. A lens placed close to the detector directs divergent light to the detector and, in effect, increases the detector size to that of the lens.

For many years, engineering efforts have been directed towards finding a way to achieve high-frequency modulation of light. The high-frequency response of the LED makes it possible to transmit multiplexed video and audio signals to detector devices. This work has advanced to the point where computer manufacturers are considering replacing small runs of interconnections with multiplexed LED signals.

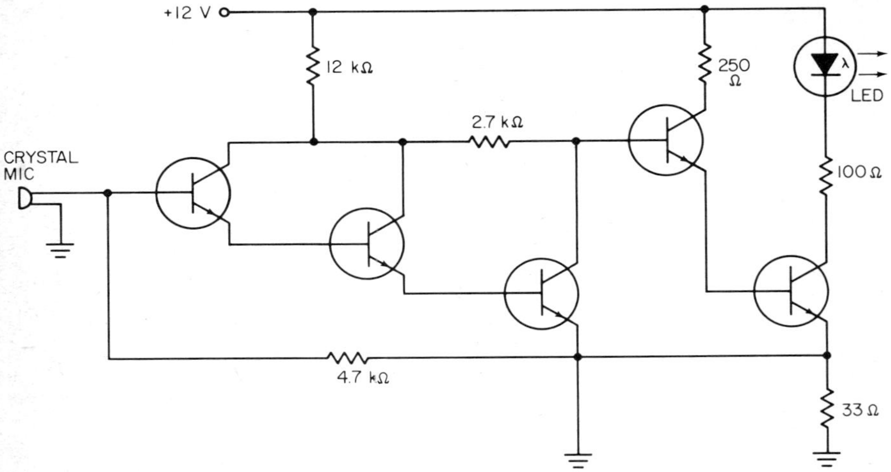

FIGURE 12.13
AM modulation of an LED with an audio signal. (*Courtesy Electronics World.*)

Figure 12.13 is a schematic diagram of an AM light system developed by General Electric, which takes advantage of the LEDs flat audio-frequency response. A modified Darlington amplifier matches the high impedance of the crystal microphone to the low impedance of the LED. This type of circuit is used both for optical recording or for transmitting AM signals through fiber optics or air.

Many new and fascinating optoelectronic applications are in the works. Perhaps one of the most interesting is illustrated in Fig. 12.14. This is a television camera tube that contains a light-sensing semiconductor array designed to image a scene, for use in an all solid-state TV camera.

The RCA SIT camera tube is a sturdy 40-MM Silicon-Intensifier Target (SIT) camera tube designed for use in very low light level TV systems.

At the front end of the tube there is a transparent photo-sensitive plate. In operation, light rays from the scene to be televised are focused by an optical lens system onto this transparent plate. At the inner surface of this plate, electrons are emitted from each point in proportion to the incident light intensity. Note that the light rays must penetrate the transparent plate to reach the photo-sensitive inner surface.

The emitted electron image (in which, at each point, the density of the electrons corresponds to the light at that point) is accelerated toward a silicon diode-array target by a voltage of approximately 12,000 V. This target is a very thin silicon wafer upon which a very tightly spaced matrix of PN junctions

FIGURE 12.14
A camera tube designed for use in very low light level TV systems. The light sensitivity of this tube is achieved by the use of a two-dimensional array of silicon diodes for the target. (*Courtesy RCA.*)

(i.e., diodes) has been formed. The spacing of the diodes is of the order of 14 μm. Gain is achieved when an emitted electron (accelerated to 12,000 V energy) impinges upon the target and causes multiple dissociation of electron-hole pairs. The holes are collected on the opposite or P-side of the diode where the charge is neutralized by a scanning beam that sweeps over the target area in a regular pattern. A resulting signal is produced in the external circuit connected to the target.

An external biasing voltage, applied to each elemental PN diode of the target, reverse biases that diode. That means that the P and N electrodes are separated by the insulating properties of the depletion layer. Hence, each diode can be regarded as a capacitor. When the electrons from the photocathode strike the N-side of the target, this capacitor becomes charged in proportion to the density of the electrons reaching it. The capacitor then discharges when the scanning beam sweeps over the P-side of this capacitor on the reverse side of the target.

It is the succession of pulses produced by the successive diode discharge of the PN matrix that constitutes the signal output of the SIT camera tube.

ZENER DIODES

The term zener diode applies to a semiconductor diode, or PN junction, employed as a voltage or current-controlling device. These diodes are fabricated with precise breakdown voltages, ranging from 3 to 200 V.

We mentioned zeners in Chap. 2 in our discussion of the avalanche-breakdown mechanism of reverse-biased PN junctions. It was pointed out that there are three possible mechanisms that could account for junction breakdown: tunneling, zener or field emission, and the avalanche mechanism. With respect to zener diodes, only the latter two breakdown mechanisms are important.

The reason that these diodes, with their wide range of voltage levels, are called zener diodes is historical, not functional. In 1934 a physicist named Carl Zener, investigating the breakdown phenomena, suggested that the rapid increase in current at breakdown was due to electrons that had obtained sufficient energy from the electric field to transfer from the valence band to the conduction band. Despite the technical misnomer, there are practical characteristics of zener diodes resulting from the variation of breakdown mechanisms.

As the reverse voltage applied to a zener diode (or any PN junction) is increased, the resistance will remain high for some time. At a certain voltage, the resistance then drops very rapidly. The current flow up to this point is low, in the microampere range, but it rapidly increases to the milliampere or ampere range. The change from a low value of current to heavy conduction is very sharp and well defined. It is called the zener knee. The voltage drop across the diode after the "critical" breakdown point remains constant with further increases of applied voltage, resulting in an increase in current (see Fig. 12.15). Zener diodes can be operated through the breakdown sequence indefinitely so long as excessive currents are avoided, which would result in the destruction of the device.

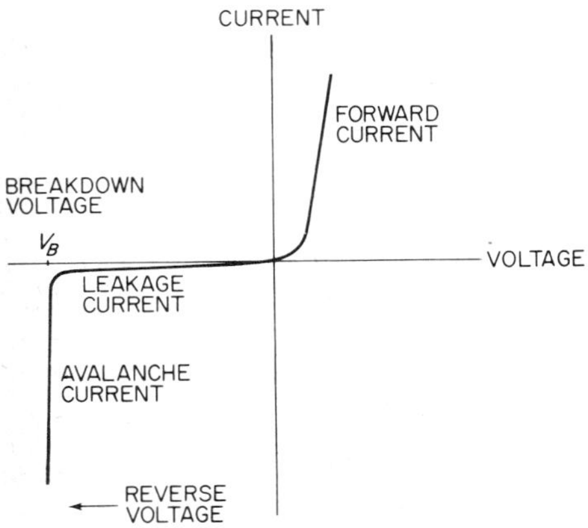

FIGURE 12.15
Current-voltage characteristics of a zener diode.

Zeners are sold in a wide variety of voltage and wattage ratings. The current limitations are based on case temperature and dissipation ratings. Thus, we have the same type of temperature derating curve for zener diodes that we do for other solid-state power-handling devices.

Zener Control Region: Voltage Breakdown Mechanisms.[1] Although all PN junctions exhibit a voltage breakdown, it is important to recall from Chap. 2 that there are two distinct voltage breakdown mechanisms. One is called *zener breakdown* and the other is called *avalanche breakdown.*

Typical diode breakdown characteristics for each category are shown in Fig. 12.16. The factor determining which of the two breakdown mechanisms occurs is the relative concentrations of the impurities in the materials that comprise the junction. If two different resistivity P-type materials are placed against two separate but equally doped low-resistivity pieces of N-type materials, the depletion region spread in the low-resistivity P-type material will be smaller than the depletion region spread in the high-resistivity P-type material. Moreover, in both situations little of the resultant depletion width lies in the N material if its resistivity is low compared with that of the P-type material. In other words, the depletion region always spreads principally into the material having the highest resistivity. Also, the electric field (voltage per unit length) in the less-resistive material is greater than the electric field in the material of greater resistivity because of the presence of more ions per unit volume in the less-resistive material. A junction that results in a narrow depletion

[1] Courtesy of Semiconductor Products Division, Motorola Inc.

FIGURE 12.16
Typical breakdown diode characteristics. Observe effects of temperature for each mechanism.

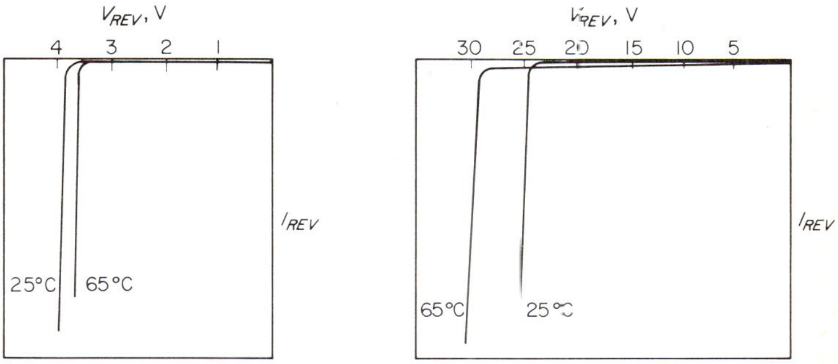

ZENER BREAKDOWN OF A PN FUNCTION
(*a*)

AVALANCHE BREAKDOWN OF A PN FUNCTION
(*b*)

region will therefore develop a high field intensity and breakdown by the zener mechanism. A junction that results in a wider depletion region, and thus a lower field intensity, will break down by the avalanche mechanism.

The zener mechanism can be described qualitatively as follows: because the depletion width is very small, the application of low reverse bias (5 V or less) will cause a field across the depletion region on the order of 3×10^5 V/cm. A field of such high magnitude exerts a large force on the valence electrons of a silicon atom, tending to separate them from their respective nuclei. Actual rupture of the covalent bonds occurs when the field goes beyond 3×10^5 V/cm. Thus, electron-hole pairs are generated in large numbers, and a sudden increase of current is observed. Although we speak of a rupture of the atomic structure, it should be understood that this generation of electron-hole pairs may be carried on continuously so long as an external source supplies additional electrons. If a limiting resistance in the circuit external to the diode junction does not prevent the current from increasing to high values, the device may be destroyed because of overheating.

On most commercially available silicon diodes, the maximum value of voltage breakdown by the zener mechanism is in the region of 5 to 8 V. In order to fabricate devices with higher voltage breakdown characteristics, materials with higher resistivity and, consequently, wider depletion regions are required. These wide depletion regions hold the field strength per centimeter down below the zener breakdown value (3×10^5 V/cm). Consequently, for devices with breakdown voltages lower than 5 V the zener mechanism predominates; between 5 and 8 V both zener and an avalanche mechanism are involved, whereas above 8 V the avalanche mechanism alone takes over.

The foregoing figures are not absolute, but they do indicate the relative voltage magnitudes involved. Also, the name zener diodes is applied commercially to all such diodes, even those that operate by the avalanche breakdown.

Figure 12.16 reveals that in zener breakdown, the value of breakdown voltage decreases as the PN-junction temperature increases. On the other hand, in avalanche breakdown, the value of breakdown voltage increases as the PN-junction temperature increases. The decrease of zener breakdown voltage as junction temperature increases can be explained in terms of energies of the valence electrons. An increase of temperature increases the energies of the valence electrons. This weakens the bonds holding the electrons; consequently, less applied voltage is necessary to pull the valence electrons from their position around the nuclei. Thus, the breakdown voltage decreases as the temperature increases.

The dependence of the avalanche breakdown mechanism on temperature is quite different. Here the depletion region is of sufficient width that the carriers (electrons or holes) can suffer collisions before traveling the region completely; that is, the depletion region is wider than one mean free path (the average distance a carrier can travel before combining with a carrier of opposite conductivity). Therefore, when temperature is increased the increased lattice vibration shortens the distance a carrier travels before colliding and thus requires a higher voltage to get it across the depletion region.

As established earlier, the applied reverse bias causes a small movement of electrons from the P material to the potentially positive N material and holes from the N material to be potentially negative P material (leakage current). As the applied voltage becomes larger, these electrons and holes increasingly accelerate. There are also collisions between these particles and bound-valence electrons of the crystal structure as the particles move through the depletion region.

If the applied voltage is such that the traveling electrons do not have high velocity, then the collisions take some energy away from them, altering their velocity. If the applied voltage is increased, collision with a valence electron will give it enough energy to enable it to break free of its covalent bond. Thus, one electron by collision has created an electron-hole pair. These secondary particles will also be accelerated and will participate in collisions that generate new electron-hole pairs. This phenomenon is called carrier multiplication. Electron-hole pairs are generated so quickly and in such large numbers that there is an apparent avalanche or self-sustained multiplication process (depicted graphically in Fig. 12.17). The junction is said to be in breakdown, and the current is limited only by resistance external to the junction. Zener diodes above 8 V exhibit avalanche breakdown.

As junction temperature increases, the voltage breakdown point for the avalanche mechanism increases. This effect can be explained by considering the vibration of the atoms fixed in their lattice. As temperature increases, the vibrational displacement of atoms in their lattice increases, and this increased displacement corresponds to an increase in the probability that intrinsic

FIGURE 12.17
PN junction in avalanche breakdown. (*Courtesy Motorola.*)

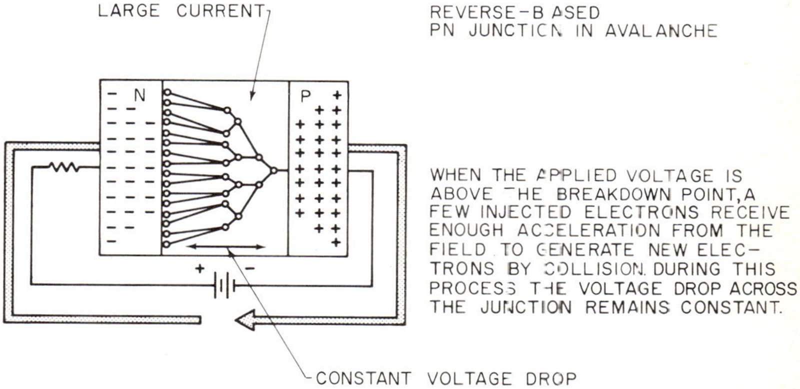

particles in the depletion region will collide with the lattice atoms. If the probability of an intrinsic particle-atom collision increases, then the probability that a given intrinsic particle will obtain high momentum decreases, and it follows that the low-momentum intrinsic particles are less likely to ionize the lattice atoms. Naturally, increased voltage increases the acceleration of the intrinsic particles, providing higher mean momentum and more electron-hole-pairs' production. If the voltage is raised sufficiently, the mean momentum becomes great enough to create electron-hole pairs and carrier multiplication results. Hence, for increasing temperature, the value of the avalanche breakdown voltage increases.

Applications. An obvious zener-diode application is in voltage-regulator circuits. Figure 12.18 shows a basic regulator circuit utilizing a zener diode. The diode is placed in shunt with the load resistor R_1 and in series with R_2. The value of R_2 is chosen so that under no-load conditions, i.e., with R_1 infinite in value (or open-circuited), the current passing through Z_1 will fall within its operating range. This will establish the zener voltage across the diode.

If, now E_{in} increases, the current through the zener diode will increase. However, the diode resistance will decrease, maintaining the original voltage across the diode regulator. If the zener resistance did not decrease, part of the increased current would have gone into R_1 and raised the voltage appearing across this load resistance.

Load-current variations have a similar effect on the diode regulator. As the load increases, or decreases, the zener shunt element will draw less, or more, current, respectively. The net result is substantially a constant output voltage across the diode regulator.

Zener diodes are also suitable for ac regulation (Fig. 12.19). Two closely matched zener diodes, Z_1 and Z_2, are connected back to back across the primary of a power transformer that supplies voltage to a power supply. R_1 is chosen so that the primary voltage will divide equally between it and Z_1 and Z_2, which are selected to operate at approximately half the line voltage. In order to supply a 30-W output, two 60-V 50-W zeners together with an 80-W 70-Ω resistor is required.

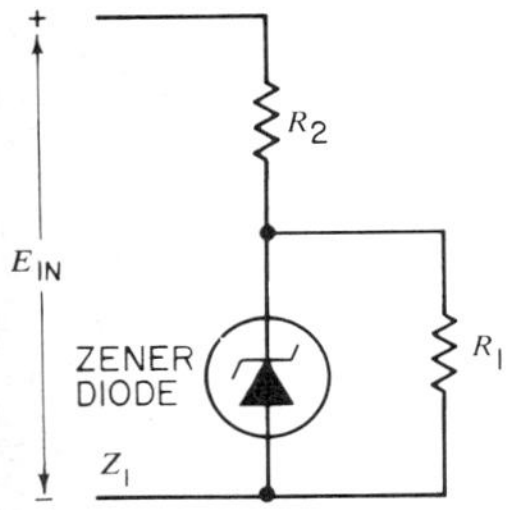

FIGURE 12.18
Basic zener voltage regulator. The symbol for the zener is the standard diagram for these devices.

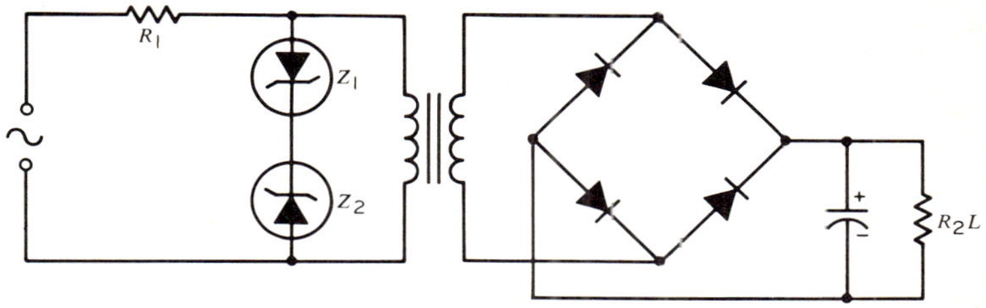

FIGURE 12.19
Ac regulation with zeners.

During the positive half-cycle of the incoming ac wave, Z_2 will serve as the zener regulating element and Z_1 will be forward-biased and hence appear as a short circuit. On the negative half-cycle, Z_2 will be short-circuited and Z_1 will be the zener regulator. In this fashion, both half-cycles of the ac voltage will be kept under control.

The silicon zener diode can be employed to prevent overloading sensitive meter movements used in low-range dc and ac voltmeters, without adversely affecting meter linearity. The zener diode has the advantage over thermal protective devices of instantaneous action and, of course, will function repeatedly for an indefinite time (as opposed to the reset time necessary with thermal devices). Whereas zener protection is presently available for voltages only as low as 2.3 V, forward diode operation can be used for meter protection when the voltage drop is much smaller.

A typical protective circuit is illustrated in Fig. 12.20. Here the meter movement requires 100 μA for full-scale deflection and has 940 Ω resistance. For use in a voltmeter to measure 25 V, approximately 249,000 Ω are required in series.

With an applied voltage of 25 V, the 100 μA current in the circuit produces a drop of 17.9 V across the series resistance of 179,000 Ω. A further increase in voltage causes the zener diode to conduct, and the overload current is shunted away from the meter.

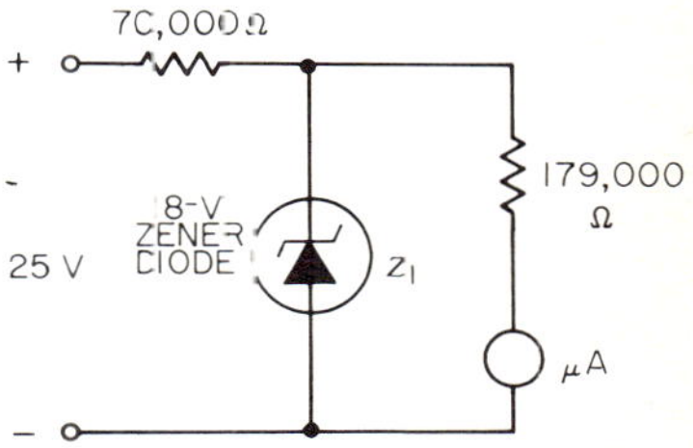

FIGURE 12.20
Meter protection using zener diode.

There is really no limitation to the number of applications of zener diodes; they have displaced many other devices, such as batteries, filament-type regulators, and saturable reactors that were previously employed for voltage regulation.

Fabrication. Zener diodes are comparatively simple to fabricate. There are only two variables: the resistivity of the starting material and the diffusion cycles. Table 12.1 shows the relationship between the resistivity and the breakdown voltages of silicon diodes manufactured from materials of varied resistivities (P type), and diffused for 16 hours (h) at 1,250°C (column 2). The voltages shown in column 3 are the minimum voltages obtainable with that particular resistivity and are achieved using very short diffusion cycles. Thus, breakdown voltages between 5 and 210 V can be obtained by controlling these two parameters.

Voltampere Characteristics. The zener voltampere characteristics for a typical 50-V zener diode are illustrated in Fig. 12.21. It shows that the zener diode conducts current in both directions, the forward current I_F being a function of forward voltage V_F. Note that I_F is small until $V_F \approx$ 0.65 V; then I_F increases very rapidly. For $V_F >$ 0.65 V, I_F is limited primarily by the circuit resistance external to the diode.

The reverse current is a function of the reverse V_R, but for most practical purposes it is zero until the reverse voltage exceeds V_Z, the PN-junction breakdown voltage, at which time the reverse current increases very rapidly. Since the reverse current is small for $V_R < V_Z$ but great for $V_R > V_Z$, each of the current regions is specified by a different symbol. For the leakage current region, i.e., the nonconducting region between zero volts and V_Z, the reverse current is denoted by the symbol I_R; but for the zener control region, $V_R > V_Z$,

TABLE 12.1
Relationship between Resistivity and Breakdown Voltage for Zener Diodes

(1) *Resistivity,* *Ω/cm*	(2) *VB* *(Breakdown Volts)*	(3) *Minimum* *Breakdown Voltage*
0.01	10	5
0.1	28	10
1	75	25
10	210	60

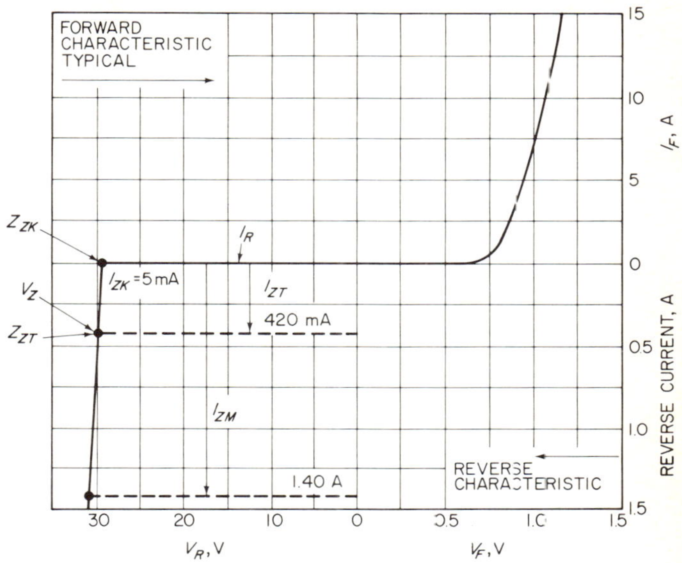

FIGURE 12.21
Zener diode characteristics.

the reverse current is denoted by the symbol I_Z. I_R is usually specified at a reverse voltage $V_R \approx 0.8V_Z$.

The PN-junction breakdown voltage V_Z is usually called the zener voltage, regardless of whether the diode is of the zener or avalanche breakdown type. Commercial zener diodes are available with zener voltages from about 2.4 to 200 V. For most applications the zener diode is operated well into the breakdown region (I_{ZT} to I_{ZM}). Most manufacturers give an additional specification to I_{ZK} (equal to 5.0 mA in Fig. 12.21) to indicate a minimum operating current to assure reasonable regulation.

This minimum current I_{ZK} varies in the various types of zener diodes and is consequently given on the data sheets. The maximum zener current I_{ZM} should be considered the maximum reverse current recommended by the manufacturer. Values of I_{ZM} are also given in the data sheets for all types.

Between the limits of I_{ZK} and I_{ZM}, which are 5 and 1,400 mA (1.40 A) in the example of Fig. 12.21, the voltage across the diode is essentially constant and approximately equal to V_Z. This plateau region, however, has a large positive slope such that the precise value of reverse voltage will change slightly as a function of I_Z. For any point on this plateau region, one may calculate an impedance using the incremental magnitudes of the voltage and current. This impedance is usually called the zener impedance Z_Z and is specified for most zener diodes. Most manufacturers measure the maximum zener impedance at two test points on the plateau region. The first is usually near the knee of the zener plateau, Z_{ZK}, and the latter point is near the midrange of the usable zener current excursion. Two such points are illustrated in Fig. 12.21.

Temperature Compensation of Zener Diodes. As we noted above, the value of breakdown voltage for a zener diode changes with temperature, and this can be a serious limitation when the accuracy of the zener voltage is depended upon to serve as a fixed reference. It is possible, however, to combine a zener diode with another semiconductor device with a response to temperature variation that is opposite to that of a zener and produce a zener reference element exhibiting a very low temperature coefficient.

A common method of developing a temperature-compensated zener is to combine it with a semiconductor diode mounted in the same case as the zener diode but electrically connected in reverse manner. The reason for doing this stems from the fact that the voltage of a forward-biased PN junction, at a specific current, will decrease with increasing temperature. Thus, a device so biased displays a negative temperature coefficient (see Fig. 12.22). If we now combine it with a zener diode operating in the avalanche breakdown mode, where the value of the breakdown voltage increases with temperature, then the overall result will be a voltage drop that remains quite constant.

Thus, by using various combinations of forward-biased and reverse-biased junctions, we can achieve fairly high orders of voltage stability over a considerable temperature range. Typically, temperature-compensated zeners with stability of 100 parts per million per 1,000 h are readily available.

FIGURE 12.22
Temperature compensation employed with zener diodes. (*Courtesy Motorola.*)

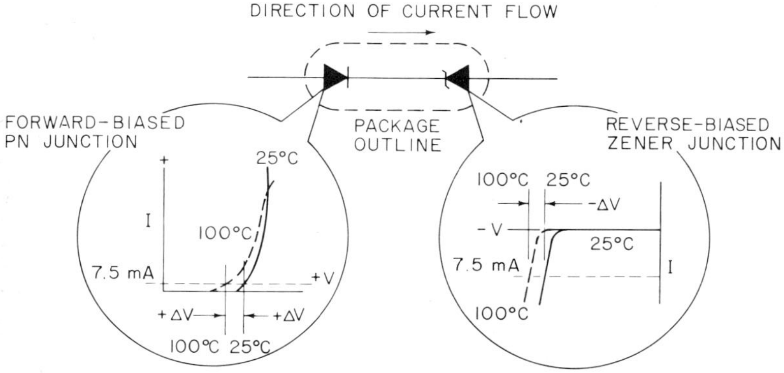

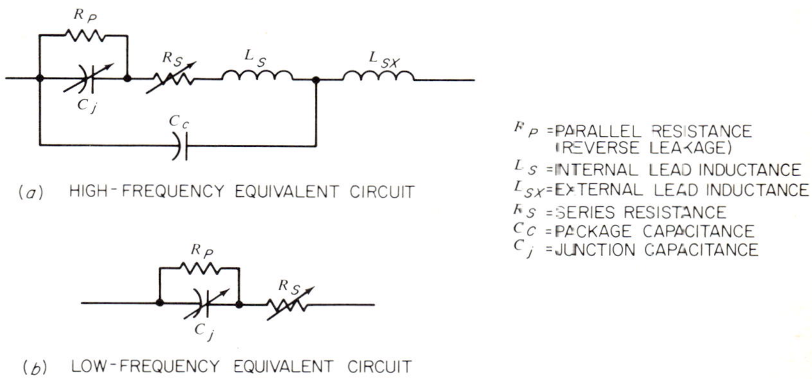

FIGURE 12.23
Equivalent circuits for a varactor diode. (*Courtesy Electronics World.*)

VARACTOR DIODE

The varactor diode is a semiconductor diode that is employed as a variable-reactance circuit element, with the variable reactance being provided by the PN-junction capacitance, which varies as a function of the applied voltage. The varactor operates principally between a very small positive bias and the reverse-breakdown voltage. The junction, under these conditions, can be represented electrically by the equivalent circuit shown in Fig. 12.23a. For low-frequency operation, the circuit simplifies to that of Fig. 12.23b. The fact that the junction capacitance can be varied by an applied voltage, and varied sufficiently to provide a useful capacitance change, enables the varactor diode to be used for circuit tuning and for harmonic generation.

The schematic symbol for a varactor diode is shown in Fig. 12.24.

Controlling Capacitance. When a junction is formed between N-type and P-type semiconductor material, there is a cross-migration of charges across the junction. Electrons from the N region cross the junction to neutralize positive carriers near the junction in the P region. At the same time, holes from the P region cross the junction to neutralize electrons near the junction in the N region. As a result of this migration, all free charged particles are

FIGURE 12.24
Schematic symbol for a varactor diode.

swept out of the immediate vicinity of the junction, creating the depletion layer in the junction area. And in this process, a contact potential or space charge (about 0.6 to 0.7 V for silicon) appears across the junction.

The depletion layer can be thought of as a dielectric. The semiconductor material on either side represents two conductor plates. Hence, we have a slightly charged capacitor. Now, if we reverse-bias the junction, the depletion layer will widen, decreasing the capacitance. Conversely, if the reverse bias across the junction is lowered or if the junction is forward-biased slightly, the depletion layer decreases and the capacitance will increase. If the forward voltage is increased far enough, conduction ensues and the capacitance effect is lost.

Thus, the junction capacitance is controlled by the externally applied voltage, so long as the potential barrier of the junction is not overcome. We can express this relationship in this form:

$$C = \frac{C_0}{(1 + V/\phi)^\gamma} = \frac{\phi^\gamma C_0}{(\phi + V)^\gamma}$$

where C = capacitance at voltage V
C_0 = capacitance at zero bias
V = reverse-bias voltage across the diode
ϕ = contact potential
γ = power law of the junction, determined by the impurity gradient

The capacitance-voltage relationship is nonlinear, which is an extremely important factor in harmonic generation.

The capacitance obtainable from a varactor diode and the diode's range will be determined to a large extent by the manner in which the P and N regions are doped, particularly in the vicinity of the junction itself. Figure 12.25 shows doping profiles of the four most common types encountered. The abrupt and hyperabrupt junctions are used for tuning, whereas the linearly graded and step-recovery junctions are used for harmonic generation.

For tuning application, we require a high tuning ratio, on the order of at least 10:1, which means that the value of the capacitance at (say) zero bias should be 10 times its value at the maximum reverse bias that can be applied without causing breakdown. (Remember that the capacitance value is highest at zero bias when the width of the depletion region is narrowest.) The hyperabrupt junction can provide ratios of 20:1 and higher, whereas the abrupt junction diode has a somewhat lower capacitance ratio change.

For harmonic generation, the step-recovery diode is highly useful. The capacitance change in this diode is quite small, occurring mostly near zero bias. Also, the step-recovery device makes use of forward bias and charge storage. Under these conditions, charge carriers from one region are injected into the other to form minority carriers in that area. If the lifetime of these

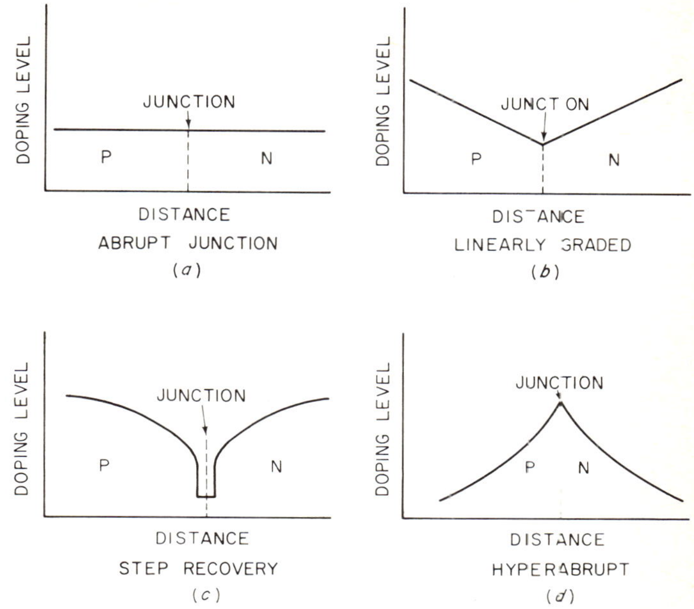

FIGURE 12.25
Doping profiles for various varactor diodes. (*Courtesy Electronics World.*)

carriers is longer than the period of the applied forward voltage, then these carriers are still active (i.e., have not recombined with the oppositely charged majority carriers) when the reverse voltage appears. They are thus swept back to their point of origin and do so in a fairly compact group. While these minority carriers are flowing back, they represent a high reverse current. When they reach their own region, there is an abrupt cessation of current flow. It is these surges of current—first in one direction, then in the reverse direction—that are rich in harmonic frequencies (as, indeed, are all current surges). It is indicated that usable levels of current have been obtained as high as the twenty-fifth harmonic for frequencies well into the microwave region.

Varactor Applications. As previously stated, varactors are employed for tuning and for harmonic generation. A tuning application in the front end of an AM receiver is shown in Fig. 12.26. Here, varactors are utilized to tune the input or antenna tuning coil L_1 as well as the local oscillator tuning coil L_2. As is common in such an arrangement using the more conventional tuning capacitors, both variable sections are ganged, or varied, simultaneously so that only one control is required.

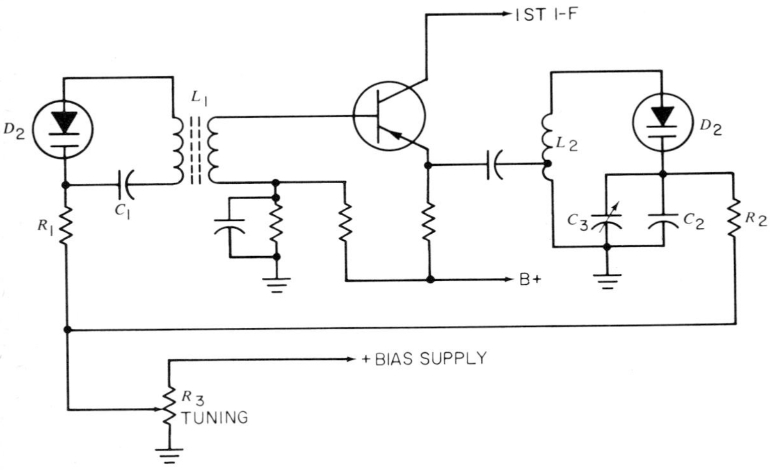

FIGURE 12.26
The front end of an AM radio using varactor-diode tuning. (*Courtesy Electronics World.*)

In the case of varactors, the tuning is accomplished by employing a potentiometer that varies the reverse voltage applied to each varactor. Since a single control is employed, both varactors must track in order that both sets of tuning circuits remain in suitable step with each other. A padding capacitor C_3 in the local oscillator circuit helps to achieve this tracking.

For relatively low-frequency tuning, it is permissible to shunt the varactor across the tuning coil. However, when the frequency becomes high enough, the varactor internal inductance must be taken into account. At such frequencies, the arrangement shown in Fig. 12.27 proves to be more effective, since the diode inductances are placed in series with the circuit inductances and the two are thus simply added together.

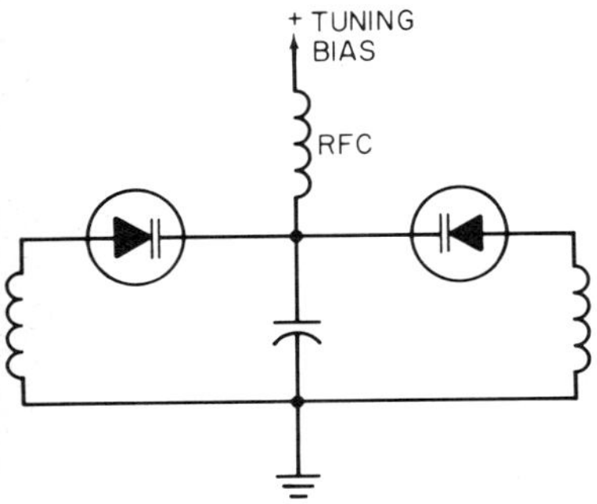

FIGURE 12.27
For high-frequency operation, it is generally more desirable to position the varactor diodes in series with the circuit to be tuned.

An important application of the varactor is as an harmonic generator.

Development of a varactor multiplier circuit is illustrated in Fig. 12.28.[1] The object, as shown in *a*, is the conversion of a signal from a signal source to a harmonic current through the load R_L by means of a varactor. The necessary considerations entail (1) providing for the necessary current paths

[1] The Semiconductor Data Book, 4th ed., Motorola, Inc.

FIGURE 12.28
Development of a varactor multiplier circuit.

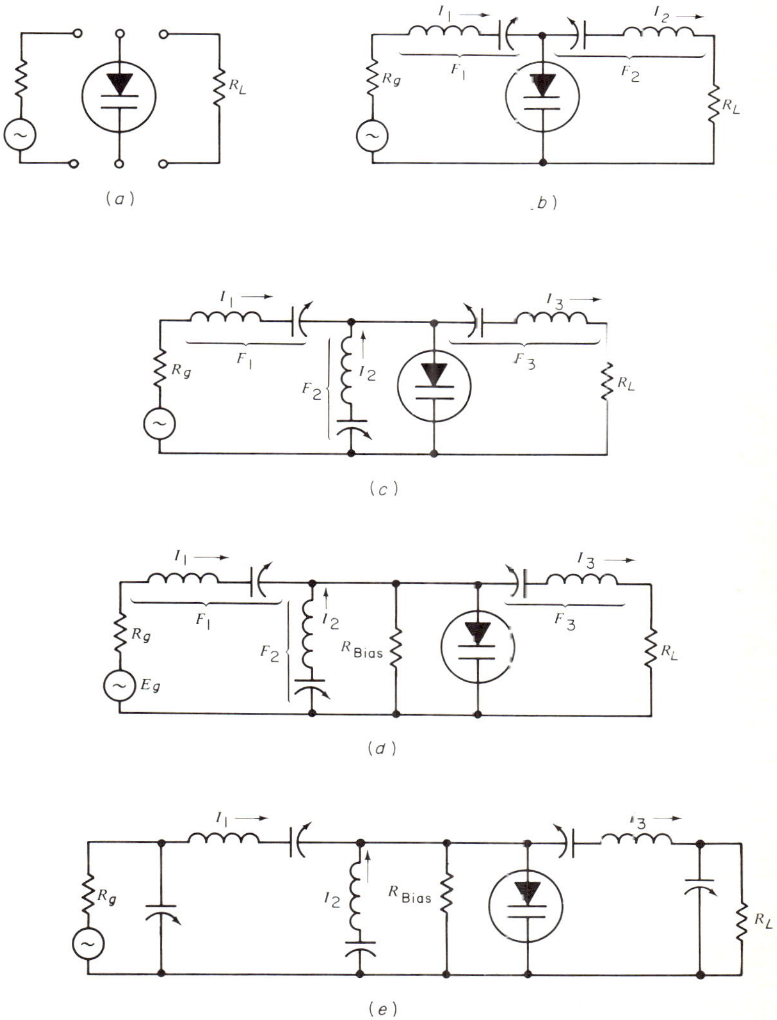

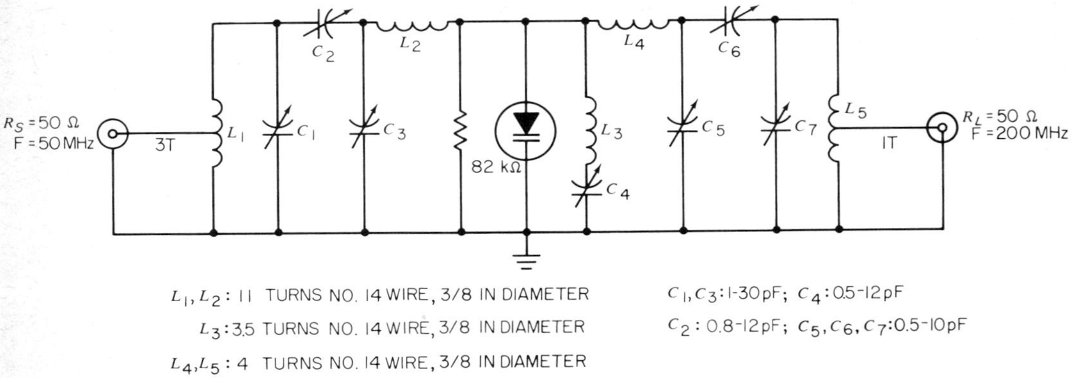

FIGURE 12.29
A 50 to 200-MHz varactor quadrupler. (*Courtesy Motorola.*)

and associated filters, (2) proper matching of source to load, and (3) developing suitable bias voltage for the varactor.

The first step in the design is the addition of suitable current paths, as shown in Fig. 12.28*b*. If the output is to be the second harmonic, filter F_1 is tuned to the fundamental frequency and F_2 is tuned to the second harmonic. In designing the tuned circuits, the capacitance of the varactor must be taken into account. Since this value varies over the applied signal cycle, the "average" varactor capacitance should be used. This can be approximated by the capacitance value at one-third the voltage breakdown rating of the varactor. Since this average capacitance varies with signal power, some circuit detuning occurs if input power is changed appreciably.

If the desired load current is at the third or fourth harmonic, the configuration in *c* may be used. Here, F_1 is again tuned to the fundamental and F_2 is an idler circuit tuned to the second harmonic. This permits fundamental and second harmonic current flow to mix in the varactor to provide a voltage component of F_1, F_2, $F_1 + F_2$, and $2F_2$ across the varactor. (Even if F_2 were omitted, there would be components of higher-order harmonics, such as F_3 and F_4 across the varactor, but it is normally more efficient to employ a suitable addition or multiple of the funamental and second harmonic.) Filter F_3 is then tuned to the desired third or fourth harmonic so that only the desired current will flow through the load.

Bias voltage for the varactor is obtained by shunting the varactor with a high-value (around 100 kΩ) bias resistor, as in Fig. 12.28*d*. Bias current is provided when the varactor is driven slightly into conduction at the peaks of the applied signal.

Proper matching between source and load can be accomplished by adding

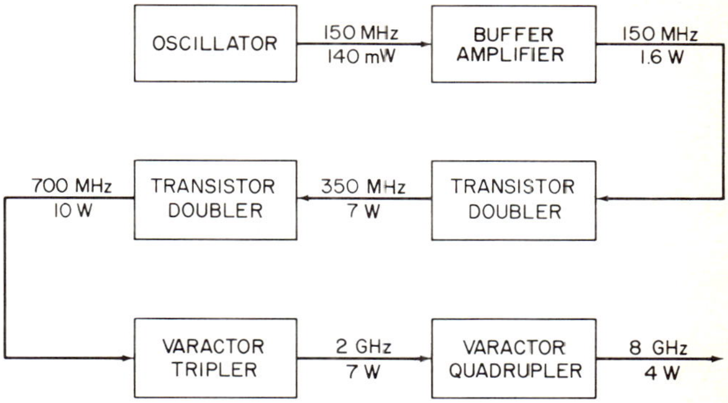

FIGURE 12.30
Solid-state microwave source using varactors. Efficiency is approximately equal to 8.5 percent.

matching capacitors as shown in *e*. Tapped input and output coils could accomplish the same purpose.

Obviously, from the standpoint of performance, the simple circuit developed in Fig. 12.28 can be improved upon. Higher-frequency, distributed-element circuits could be fabricated based on a circuit equivalent to Fig. 12.28. More complex filters, such as double-tuned circuits, could be employed for greater bandwidth and better rejection of spurious signals. In practical applications, the final circuit will almost always be more complex. One such circuit is shown in Fig. 12.29. This is a quadrupler arrangement which accepts an input signal of 50 MHz and provides an output signal of 200 MHz.

The most important use of varactor harmonic generators is in solid-state microwave sources. Figure 12.30 is a block diagram of such a source.

A crystal-controlled transistor oscillator, or mechanically tunable or voltage tunable oscillator furnishes a stable reference frequency for the multiplier circuit. Transistor amplifiers and transistor-amplifier parametric multipliers increase the power level and frequency. The varactor harmonic generators then raise the frequency further.

THYRISTORS AND THE SILICON CONTROLLED RECTIFIER

The term "thyristor"[1] defines any semiconductor switch which possesses two stable states of on and off and whose action depends upon the regenerative

[1] SCR Manual, 4th ed., Semiconductor Products Department, General Electric.

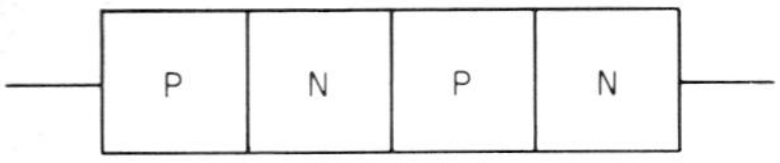

FIGURE 12.31
Basic physical construction of a PNPN thyristor.

feedback in a PNPN structure. Thyristors can have two, three, or four terminals, and can either be unidirectional or bidirectional in their current-conduction capabilities.

The best known of the thyristor devices is the silicon controlled rectifier (SCR), which is the solid-state equivalent of the thyratron electron tube that was used for many years in a variety of control functions in industrial electronic equipment. The SCR has three terminals (an anode, a cathode, and a triggering gate) and is a unidirectional device. It is also known as a reverse-blocking triode thyristor.

In the same reverse-blocking family are also found the silicon unilateral switch, the light-activated silicon controlled rectifier (LASCR), and the gate turn-off switch. There is also a reverse-blocking tetrode thyristor (it has two control gates), as well as the Shockley diode which is a reverse-blocking diode thyristor.

Bidirectional thyristors are classified as PNPN devices that can conduct current in either direction. Commercially available bidirectional thyristors include the triac (for triode ac switch) and the silicon bilateral switch.

Thus, there is quite a family of these solid-state PNPN switches. Because the SCR is by far the most important, it will be discussed here in detail, as will its typical applications. Brief descriptions of the tetrode thyristor and the LASCR will be given at the end of the SCR discussion.

To understand the operation of an SCR, consider the basic physical construction of a PNPN thyristor (shown in Fig. 12.31). It consists of alternate layers of P-type and N-type germanium or silicon, with each P zone having an excess of holes and each N zone having an excess of electrons. In two-terminal PNPN units, connection is made to the end semiconductor sections. In three-terminal PNPN units, a third connection is made to the P section closest to the end N section. This third connection, or terminal, is known as the "gate."

Two-terminal PNPN. The mode of operation of a two-terminal PNPN thyristor can be understood by applying a voltage across the ends of the device as shown in Fig. 12.32. The positive side of the voltage goes to the P end section, and the negative side of the voltage connects to the N end section. This polarity voltage will cause junctions J_1 and J_3 to be forward-biased. The intermediate junction J_2 will be reverse-biased because the voltage on the N side is relatively positive with respect to the voltage on the P side. Thus, junction J_2 will possess a considerably higher impedance than junctions J_1 and J_3. Nearly all the applied voltage will appear across J_2, and the current that flows will be

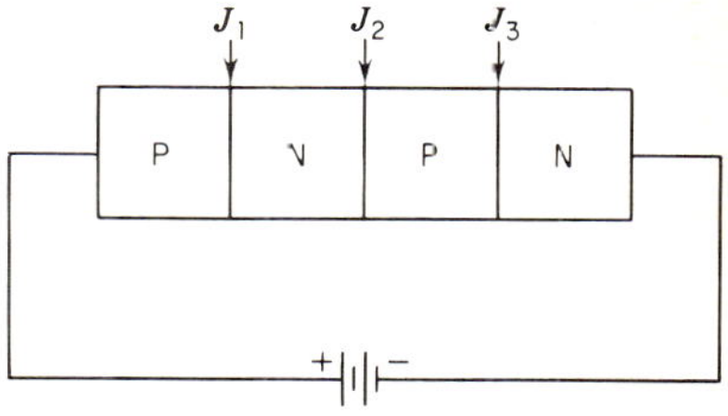

FIGURE 12.32
A PNPN thyristor with voltage applied.

largely that characteristic of a reverse-biased diode. The overall impedance of the unit is very high, on the order of several hundred megohms.

As the voltage across the PNPN thyristor increases, the electric field produced across J_2 will similarly rise. This, in turn, will cause more current to flow through the transistor and its external circuit. The current rise will follow the curve in region I of Fig. 12.33. At a certain value, the voltage will permit carriers from the two end sections to acquire enough energy at J_2 to dislodge additional carriers, producing an avalanche breakdown. The action feeds on itself, quickly producing a current flow whose value is limited only by the resistance of the external circuit. This behavior is depicted by regions I and II of the characteristic curve (Fig. 12.33). The breakdown commences at V_B, at which point, current I_B is flowing. Thereafter, the voltage drop across the PNPN device decreases quite rapidly with current increase. Note that this is a condition of negative resistance because the voltage drop is decreasing while the current is increasing.

FIGURE 12.33
Voltage-current characteristic of a PNPN thyristor. Region I is the high-resistance region where little current flows. Point V_S, I_S is the low-resistance point.

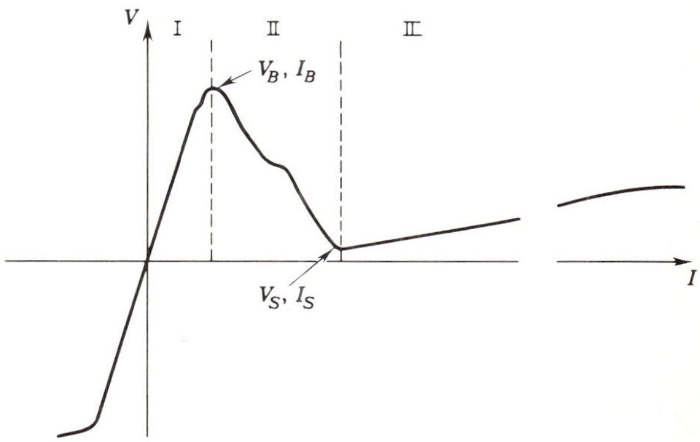

The foregoing action finally results in a condition of zero voltage drop across the center junction, at which time it acts as an emitter as well as a collector. The device is then "closed," the two center sections are flooded with charge, and the overall impedance is exceedingly low. We are now at the lowest point of the curve, between regions II and III. Further increase in current beyond this point will produce a slowly rising resistance (region III).

The PNPN thyristor will remain in this highly conducting condition so long as the current through it remains greater than the value I_S. The latter value is called the holding current. If the current drops below this level, the unit will revert back to the high-impedance condition.

When the PNPN is in its closed, or conducting, condition, each of the three junctions is forward-biased and the two central sections are heavily saturated with holes and electrons. To turn the device off, a reverse voltage must be applied. Upon application of this reverse voltage, the holes and electrons in the vicinity of junctions J_1 and J_3 will diffuse to these junctions and produce a reverse current in the external circuit. So long as the reverse current is appreciable, the voltage drop across the PNPN unit will remain small. After the electrons and holes in the vicinity of J_1 and J_3 have been removed, however, the reverse current ceases and the two junctions become reverse-biased. For all practical purposes, the switch is now open, although some recombinations of holes and electrons must still take place around junction J_2. Once the action is complete, junction J_2 also assumes a blocking condition. The device is now ready to go through the entire process again.

It is comparatively simple to derive an expression for the current flowing through a PNPN. We can depict the device as being composed essentially of a PNP and an NPN transistor interconnected as shown in Fig. 12.34.

Now, when we apply a voltage across the two end sections, with the polarity shown, junctions J_1 and J_3 are forward-biased and J_2 is reverse-biased. This

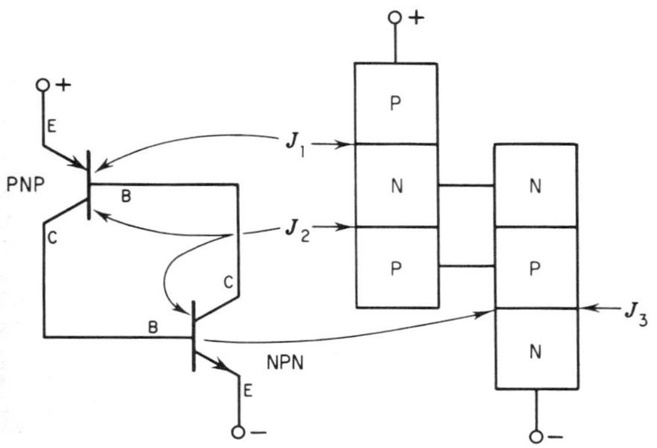

FIGURE 12.34
The PNPN device can be regarded as essentially an NPN transistor and a PNP transistor that are interconnected.

provides each transistor with its conventional biasing arrangement, and each has associated with it a current gain α. Let us call the current gain of the PNP unit α_1 and the current gain of the PNP unit α_2. Since α is, by definition, the fraction of the electron current (for NPN transistors) or hole current (for PNP transistors) injected into the base that reaches the collector, the two structures can be combined to derive the total current flowing.

In Fig. 12.34, J_2 is the common-collector junction, and it is affected by three components of current: $\alpha_1 I$, the hole current from the P end region; α_2I, the electron current from the N end region; and I_{CO}, the leakage current. The total current I flowing in the external circuit must be equal to the current at J_2. Thus

$$I_{J_2} = I = \alpha_1 I + \alpha_2 I + I_{CO}$$

or

$$I - \alpha_1 I - \alpha_2 I = I_{CO}$$

and

$$I(1 - \alpha_1 - \alpha_2) = I_{CO}$$

or

$$I = \frac{I_{CO}}{1 - (\alpha_1 + \alpha_2)} \tag{12.1}$$

From this expression, as $\alpha_1 + \alpha_2$ approaches unity, the current through the device becomes large and is limited only by the resistance of the external circuit. In short, as $\alpha_1 + \alpha_2$ approaches unity, the internal resistance drops to a very low value.

In deriving the above expression, the hole current from the P end region is added to the electron current from the N end region even though they are traveling to opposite directions. The two currents can be added because a hole current, being positive in charge, is equivalent in its electrical effect to an electron current flowing in the opposite direction.

With junctions J_1 and J_3 forward-biased and J_2 reverse-biased, very little current flows. From eq. (12.1), this current is approximately I_{CO}, since α_1 and α_2 are close to zero. This is because at low emitter currents α is also low. This condition corresponds to the off state. To turn the device on, α_1 and α_2 must be increased, which means sending more current through the device. From Eq. (12.1), if $\alpha_1 = 0.49$ and $\alpha_2 = 0.49$, the current through J_2 and the external circuit is

$$I_{J_2} = \frac{I_{CO}}{1 - (0.49 + 0.49)} = \frac{I_{CO}}{0.02}$$

In this case, the current through J_2 is 50 times the normal leakage current. If $\alpha_1 + \alpha_2 = 1$, the current through J_2 is limited only by the resistance of the external circuit.

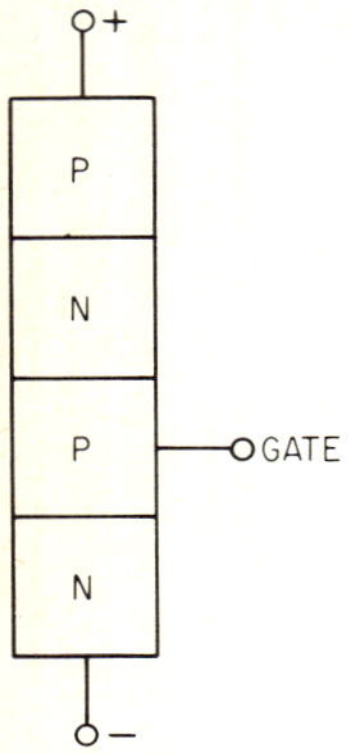

FIGURE 12.35
A three-terminal PNPN.

Three-terminal PNPN. In the three-terminal PNPN, a third connection is made to the P section closest to the end N section (Fig. 12.35). This third terminal is known as the gate. Its purpose is to switch this device from the open to the closed state without the need for exceeding the critical break-over voltage. In essence, what the gate does is inject enough current into the PNPN structure to cause junction J_2 to become forward-biased.

To produce the required gate current, some voltage-switching arrangement in which the gate lead is made positive with respect to the N end section (or cathode, as it is called) is employed. The arrangement may be a simple battery and mechanical switch, or a positive pulse for the gate circuit may be developed electronically. Once conduction through the PNPN device is initiated, the gate loses further control of the current flowing. To return the PNPN to its off condition, either its current must be reduced below the holding level or the voltage between the cathode and the anode (i.e., the other end connection) must be reversed. In this sense, the three-terminal PNPN is equivalent to the thyratron tube.

Turning On SCRs. Figure 12.36 is a typical circuit used for triggering an SCR with a dc gate bias. When the switch is closed, the gate receives a positive voltage sufficient to turn the SCR on. R_1 provides noise suppression and improves the turn-on time.

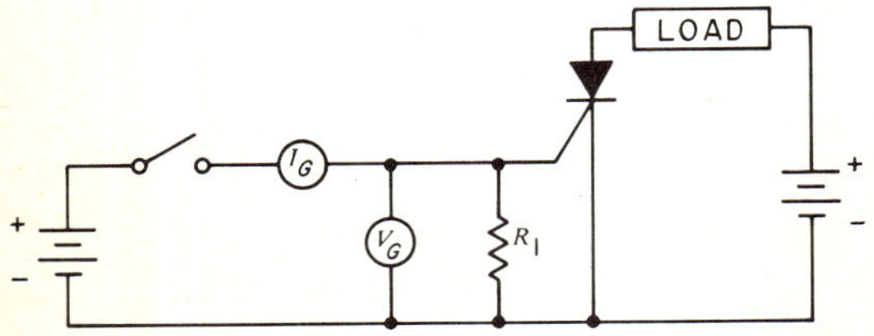

FIGURE 12.36
Typical circuit for turning on SCRs.

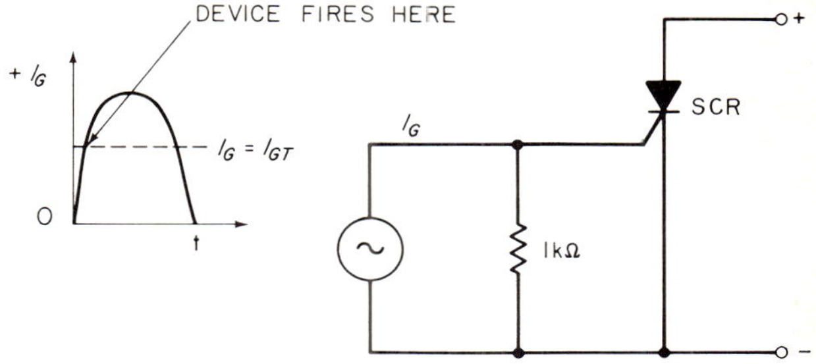

FIGURE 12.37
Ac gate trigger circuit.

Controlled rectifiers may also be turned on with the positive cycle of ac gate current, as Fig. 12.37 illustrates.

Figure 12.38 shows the interrelationships of alternating anode voltage, gate current, and output load current I_L. For the sake of simplicity, the gate current in Fig. 12.38*b* is shown in phase with the anode supply voltage (Fig. 12.38*a*). When the gate current increases to the triggering level I_{GT}, the SCR triggers ON (Fig. 12.38*c*). It remains in the on condition until the current through the SCR drops below the holding current I_H, the minimum amount of

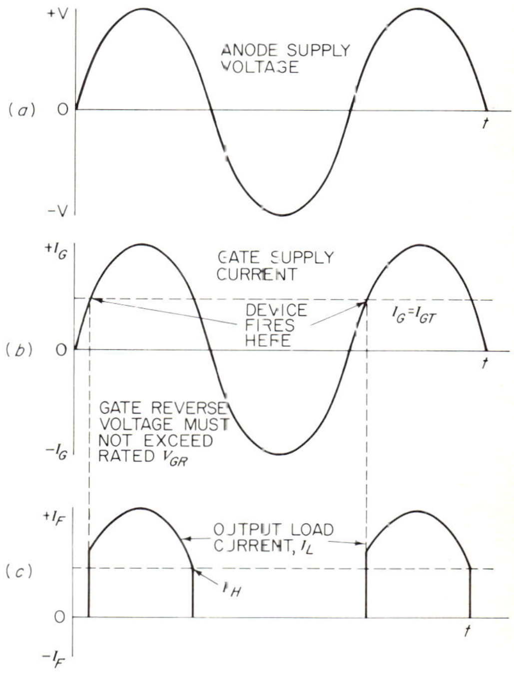

FIGURE 12.38
Ac trigger voltage and current relationships.

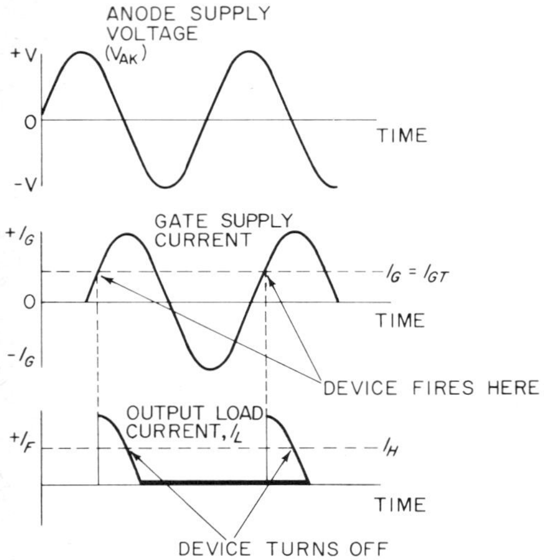

FIGURE 12.39
By altering the phase between the gate and anode ac voltages, the length of conduction time of a SCR can be controlled.

current needed to keep the device on. The holding current is normally low with respect to the load current. In Fig. 12.38*c*, this relationship has been exaggerated for purposes of clarity.

With alternating current on the anode and on the gate, we can vary their relative phase so that the point where the SCR fires can be carefully controlled. Thus, instead of current flowing through the SCR for (say) 120°, as shown in Fig. 12.38, it might only flow for approximately 90°, as shown in Fig. 12.39, or even less, if desired. Such phase shifting can be achieved using the circuit shown in Fig. 12.40. Change of phase is accomplished by rotating resistor R_1.

Under normal circumstances, a positive bias voltage should not be applied to the gate when a negative voltage is being supplied to the anode. Under these conditions transistor action, due to the positive gate current, will result in increased leakage current and excessive power dissipation in the device.

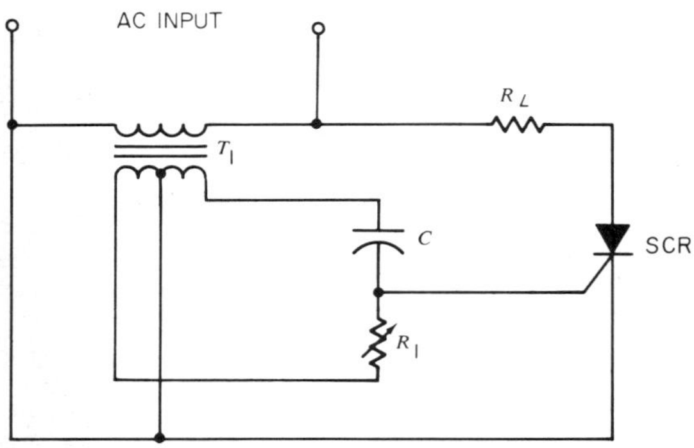

FIGURE 12.40
Controlled firing of an SCR with an *RC* phase-shift network.

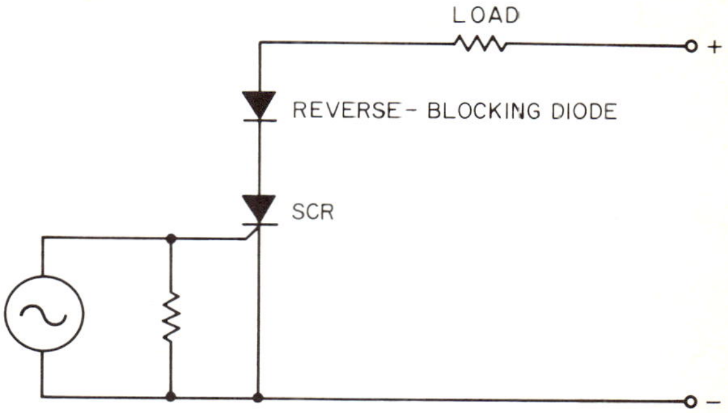

FIGURE 12.41
Ac gate trigger with series blocking diode.

When alternating current is employed at both the gate and the anode, it is possible that this condition will occur. To prevent this, an additional diode can be placed in series with the SCR, as shown in Fig. 12.41.

SCR Switching.[1] The turn-on time (t_{on}) of an SCR, when triggered by a gate signal, occurs in two phases, a delay time t_d and a rise time t_r. The delay time is the time interval between the 50 percent point (midpoint) of the leading edge of the gate-trigger voltage and the 10 percent point of the resultant forward anode current (Fig. 12.42*a*). The rise time is the time interval necessary for the anode current to rise from 10 to 90 percent of its final value. Turn-on time, then, is the time required from the start of the gate signal for the anode current to reach 90 percent of its final value ($t_{\text{on}} = t_d + t_r$).

The turn-on time is influenced primarily by the magnitude of the gate signal. The higher the gate-triggered current level, the shorter the delay time and the total turn-on time.

Current is initially concentrated in the cathode in the vicinity of the gate, during and immediately after the rise time. The finite time it takes for the current to become equally distributed throughout the cathode is the equalization time t_e (Fig. 12.43). An unequal distribution of current may produce hot spots in the cathode; hence, t_e must be minimized. The turn-on losses resulting from the current-equalization-time interval are proportional to the current and the repetition rate of the gate signal, which strongly influences the output capabilities of the device.

The turn-off time of an SCR is also divided into two phases: a reverse recovery time and a gate recovery time (Fig. 12.42*b*). When the forward current is reduced to zero at the end of a condition period, application of reverse

[1] "Transistor Manual," Radio Corporation of America.

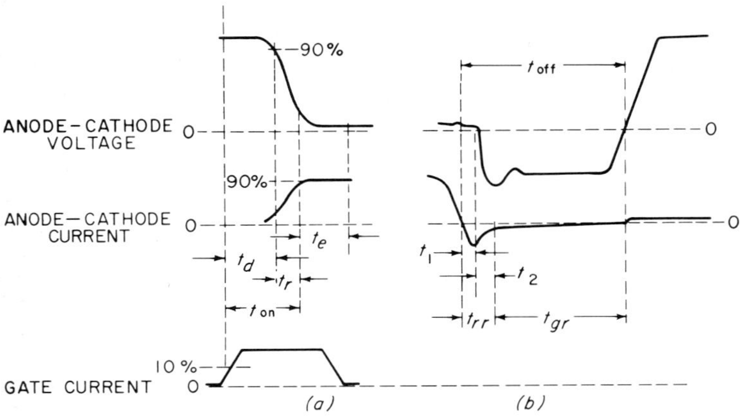

FIGURE 12.42
Switching waveshapes for an SCR. (*Courtesy RCA.*)

anode voltage causes reverse current to flow in the SCR until the reverse blocking junction (between the p-type anode layer and the n-type base or substrate) establishes a depletion region. The time interval between the application of reverse voltage and the time that the reverse current reaches its peak value is called the reverse recovery time t_{rr}. See Fig. 12.42*b*. A second recovery period, called the gate recovery time t_{gr}, must then elapse until the forward blocking junction (between the n-type base or substrate and the p-type gate layer) can establish a forward depletion region so that forward blocking voltage can be reapplied and successfully blocked by the SCR.

The total time, then, from the initial reverse recovery current flow to the initiation of the reapplied blocking voltage is the total turn-off time (t_{off}). The most significant factors affecting the turn-off time are junction temperature and the forward current just before turn-off.

The point at which the forward voltage drops to a comparatively small value and the internal resistance of the device goes from high to low determines the magnitude of the gate trigger pulse. The magnitude of the forward current, at this point, has reached its "latching current" value, which is the region of resistivity transition. Normally, the gate-trigger-pulse width is greater than the turn-on time of the SCR. It is important to note that the total average gate dissipation, in the forward and reverse direction, must not be greater than the unit's rated value.

Tetrode Thyristor. The tetrode thyristor is similar in construction to the SCR except that it has external gate connections made to both bases instead of one (see Fig. 12.43). Hence, either or both bases can be used to initiate

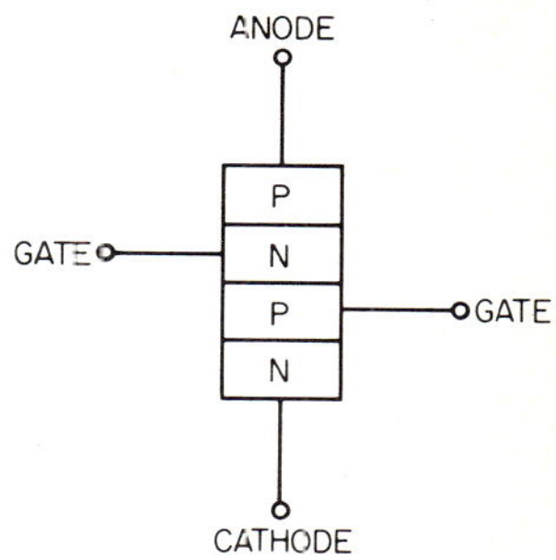

FIGURE 12.43
A tetrode thyristor.

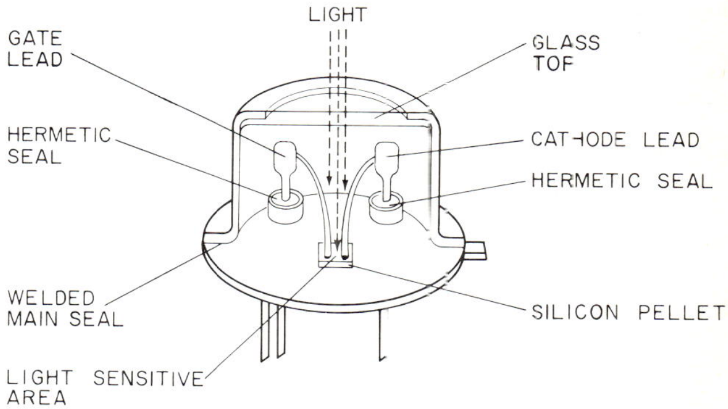

(a) LASCR CONSTRUCTION

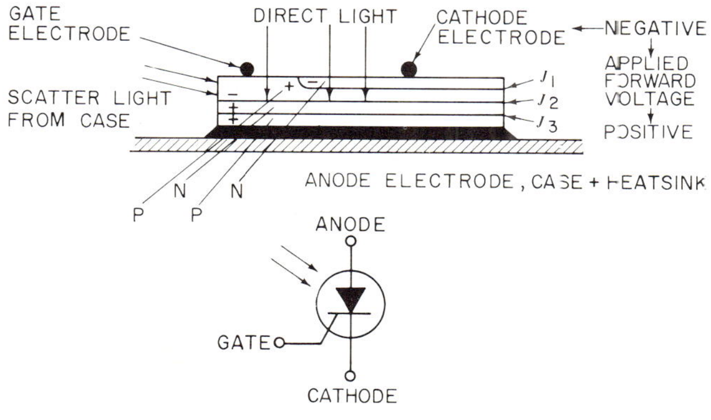

(b) LASCR PELLET AND SYMBOL

FIGURE 12.44
Construction and symbol for a light-actuated SCR. (*Courtesy General Electric Co.*)

switching. It should be noted that because the upper N base is a region of high resistivity, more external gate current is required to trigger the SCR on than when the lower (more normal) gate is used for this purpose.

Light-actuated SCR (LASCR). Silicon controlled rectifiers have been developed that can be triggered into the conduction state when the light falling on the SCR exceeds a given threshold level.

The construction of a light-actuated SCR is shown in Fig. 12.44. A window at the top of the assembly permits the light to enter the SCR and reach the silicon layers. Under normal conditions, junctions J_1 and J_3 are forward-biased by the applied anode-cathode voltage. On the other hand, junction J_2 is reverse-biased and blocks current flow. However, when light enters the case, it creates free hole-and-electron pairs in the vicinity of J_2, and these pairs are swept across J_2, J_1, and J_3 to produce a small current from anode to cathode. As the light is increased, this current increases, and the current gain of the NPN and PNP transistor equivalents in this structure also increase. At some point, the net current gain exceeds unity, and current will increase to a value that is limited only by the external circuit.

The LASCR also has an external gate electrode, and application of a positive voltage to this gate will also trigger the rectifier on.

UNIJUNCTION TRANSISTOR (UJT)

The unijunction transistor cannot be considered a new semiconductor device because the feasibility of such a device was known in 1948 and the first commercial UJTs were available in 1952. It was originally called a double-base diode because of its similarity to a junction diode with two base connections instead of one. The name was changed subsequently to unijunction transistor because the device does possess a control feature; since there is only one junction (i.e., emitter to base), the prefix "uni" was added to the name.

Structurally, a unijunction transistor consists of a uniformly doped N-type silicon bar having ohmic contacts at each end and a PN junction at the center (Fig. 12.45*a*). The two ohmic contacts are called base 1 and base 2, and at room temperature the base-to-base resistance is 5,000 to 10,000 Ω. The PN junction at the center of the bar is formed with an aluminum wire. This junction is called the emitter.

If the two base leads are connected together, the entire unit behaves as a conventional diode, with the aluminum wire representing one terminal and the connected base leads the other. However, the unique properties of the unijunction transistor are obtained when the base leads are separated and a voltage is applied between them (Fig. 12.45*b*). The voltage may have values of 30 V or more, but for this discussion, 10 V will be used. This 10 V, inserted between B_1 (base 1) and B_2 (base 2), produces a uniform drop across the

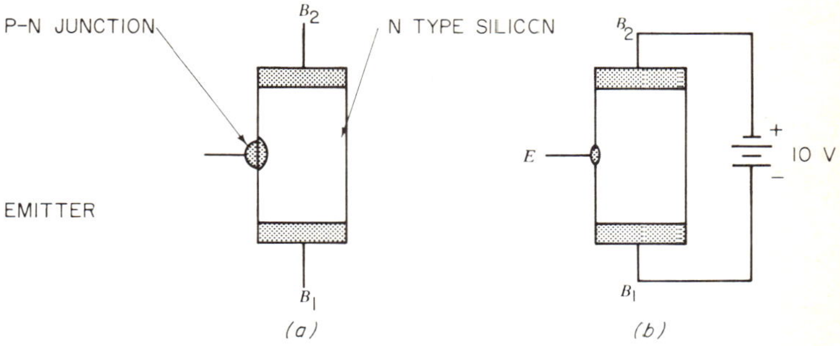

FIGURE 12.45
(*a*) An unijunction transistor. (*b*) Unijunction biasing.

5,000 to 10,000 Ω internal resistance of the silicon bar. Since the emitter is placed halfway along the bar, it sees +5 V (with respect to B_1) opposite it.

If we now apply anything less than +5 V to the emitter lead, the emitter PN junction will be reverse-biased, and very little current will flow in the emitter circuit. However, the instant the emitter voltage rises above +5 V, the junction becomes forward-biased, and holes are injected by the emitter into the silicon bar. The holes are drawn to B_1, increasing the current flow in the emitter-B_1 circuit. The increase is quite steep, because the presence of these holes in the silicon bar attracts a considerable number of electrons to this region and, in consequence, causes the bar resistance between the emitter and B_1 to drop sharply.

Until the control voltage reaches a certain value, determined by the potential between B_1 and B_2 and the silicon-bar temperature, the unit is reverse-biased and essentially cut off. The instant the critical value is exceeded, the emitter PN junction becomes forward-biased, the conductivity between the emitter and B_1 rises sharply (i.e., the resistance of the bar in this region drops to a low value), and the emitter current increases considerably.

A typical characteristic curve of a unijunction transistor is shown in Fig. 12.46. This curve has three distinct regions. Region 1 is the cutoff region, where the emitter is reverse-biased. As the voltage applied between the emitter and B_1 rises, the current through the emitter circuit rises too, although the total current seldom exceeds 10 μA. In essence, this is the leakage current obtained whenever any PN junction is reverse-biased.

Region 1 ends when the applied voltage reaches the point marked "peak." At this point, the unijunction transistor "fires," with a large increase in current and a falling off of the voltage drop between emitter and B_1. This is the negative-resistance section—the feature that gives the unijunction transistor its unique properties, as we shall see presently. Eventually a point is reached, called the minimum (or valley) point, beyond which the device behaves as

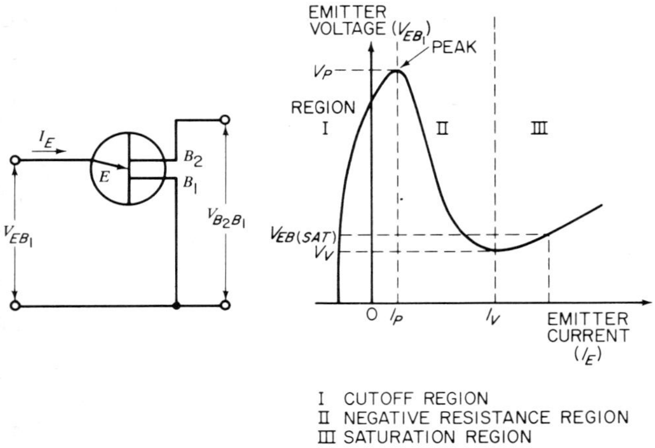

FIGURE 12.46
Typical characteristic curve of a unijunction transistor. The transistor will remain conductive so long as the current flowing through it is at the I_V value or above.

a positive resistor. That is, the current increases slowly with voltage. This region, called the saturation region, does not figure so prominently as the other two regions in the application of the unijunction transistor.

A typical circuit using this transistor is shown in Fig. 12.47*a*. Initially, capacitor C_1 is not charged and the emitter potential is zero. However, because C_1 is connected to the battery through R_1, it will slowly charge until enough voltage develops across C_1 to forward-bias the emitter junction, at which point the unijunction transistor fires. C_1 now discharges rapidly through the low

FIGURE 12.47
(*a*) A sawtooth oscillator using the unijunction transistor.
(*b*) Sawtooth wave developed across C_1.

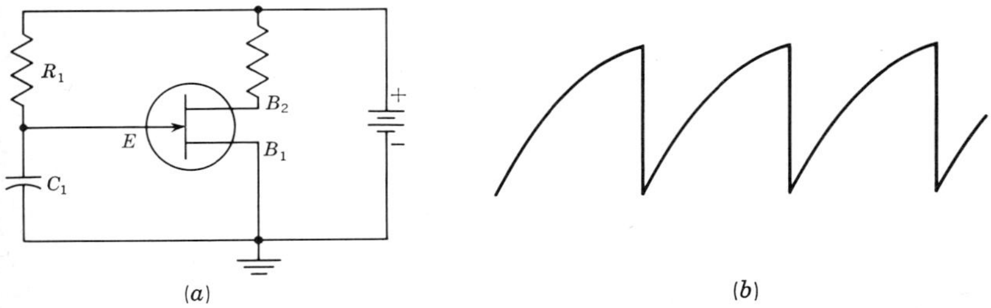

resistance of the emitter-B_1 path. With C_1 discharged, the transistor returns to its nonconducting state and the cycle repeats itself.

The waveforms developed in this circuit are shown in Fig. 12.47*b*. The voltage across C_1 is a sawtooth wave, possessing a slow rise and a rapid descent. At the time of the firing, the current through the entire silicon bar rises; and if a small resistor is placed between B_2 and the battery, a negative pulse is obtained. By the same token, if a positive pulse is desired, a similar resistor can be inserted between B_1 and ground.

Note the extreme simplicity of this circuit, requiring as it does only two resistors and a capacitor. As a matter of fact, if only the sawtooth wave is desired, the resistor can be dispensed with.

Another application of the unijunction transistor is as a multivibrator. However, before we consider its operation in this form, let us return briefly to the characteristic curve shown in Fig. 12.46. From this graph we see that once the transistor is triggered, it will remain conductive as long as the proper amount of current is maintained through the emitter-B_1 path. However, if the current drops to too low a value, the transistor relapses into its nonconductive state—which is what happened in the circuit of Fig. 12.47 when C_1 was unable to maintain the flow of current through the emitter-B_1 path.

Now consider the multivibrator shown in Fig. 12.48*a*. Initially, C_1 is not charged; however, with the power on, it will charge through R_2 and diode D_1. The unijunction transistor is cut off at this time and remains cut off until the potential across C_1 becomes equal to or greater than the peak-point potential of the unit. At this instant, the transistor fires and the potential of point B (the emitter) drops to a value close to ground potential. This causes

FIGURE 12.48
(a) A unijunction multivibrator circuit and (*b*) the waveforms generated at several points in the circuit.

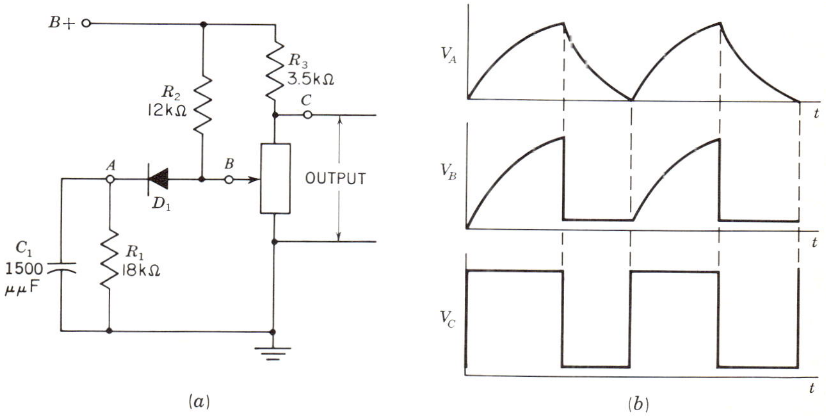

D_1 to cut off because now it is reverse-biased, with the cathode more positive than the anode.

With D_1 nonconducting, points A and B are isolated from each other. The transistor remains triggered because R_2 is passing enough current to keep the transistor activated. At the same time, C_1 discharges through R_1, continuing to do so until the potential at point A is approximately equal to the potential between the emitter and ground. At this point D_1 starts to conduct again. When this happens, enough current is diverted away from the emitter to C_1 to cause the transistor to lapse again into cutoff. The transistor remains in this state until it is retriggered by a suitable voltage rise across C_1.

The waveforms generated at various points in this circuit are shown in Fig. 12.48*b*. The voltage across C_1 follows the usual charge and discharge pattern of RC networks. The waveform at point B follows that of C_1 on charge because D_1 is conducting. However, when D_1 is cut off and the transistor is conducting, the potential at B drops to a value near ground and stays there for as long as D_1 is cut off.

When the transistor is triggered, the current through it rises sharply, producing a square wave across R_3. It is this voltage which can be used as the multivibrator output.

There are a number of additional applications to which the unijunction transistor can be put, including pulse generators, frequency dividers, phase detectors, and one-shot multivibrators. All, however, rely on the basic operation outlined above.

PROGRAMMABLE UNIJUNCTION TRANSISTOR (PUT)

The programmable unijunction transistor, or PUT, was designed to replace the unijunction transistor. It is also referred to as a complementary SCR, which is perhaps more descriptive.

In the conventional SCR, the base of the NPN transistor is the gate, whereas in the PUT, the gate is the base of the PNP transistor (Fig. 12.49). This is the difference between a PUT and an SCR. The operating theories, i.e., turn on, turn off, latching, etc., for the SCR and PUT are the same.

Regeneration in the PUT occurs when the anode is positive with respect to the gate and the current starts flowing across the PN junction between the anode and gate. This constitutes base current for the PNP section, which causes PNP collector current to flow. PNP collector current is NPN base current. Hence, NPN collector current flows and reinforces the initial gate current. Since the NPN and PNP current gains are current- (and voltage-) dependent, the reinforcing action results in increasing gains. Regenerative feedback is soon reached, and the PUT displays a negative resistance region going into a high conduction state.

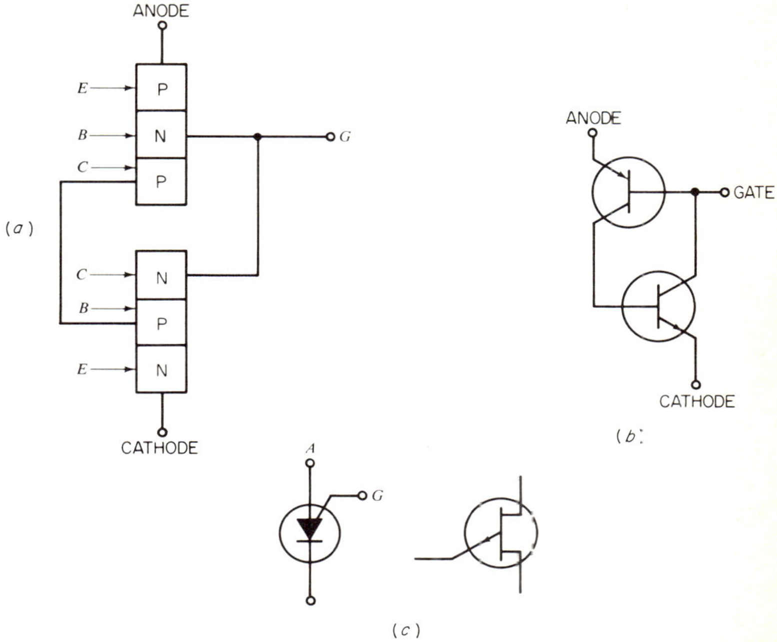

FIGURE 12.49
(*a*) and (*b*) equivalent representation of the PUT. (c) Two schematic diagrams employed for the PUT—one (left) to show its relationship to SCRs, the other (right) to show its ties to the UJT.

The current at which regeneration takes place is defined as the peak-point current (I_p in Fig. 12.50). Once this level is attained, the voltage across the PUT (anode to cathode) drops very sharply in 50 to 100 nanoseconds (10^{-9} s). If there is a charged capacitor externally connected to the PUT, a sharp pulse of current will be produced as the capacitor discharges. Such pulses are excellent for triggering SCRs, and PUTs are widely employed for this purpose.

The turn-on time for a PUT is about 10 times faster than a conventional unijunction transistor, making it ideal for triggering applications.

In the on state, the important parameters are again the holding current I_H and the point at which the PUT turns off, which is a current value less than I_H (or I_V). To turn off a PUT, the holding current need simply be reduced.

The PUT possesses several important advantages over the unijunction transistor. One of the most significant is the stability of the peak point voltage (V_P in Fig. 12.50). This means that the PUT will fire at a specific voltage that possesses a higher stability than the same voltage point in a UJT. It is rela-

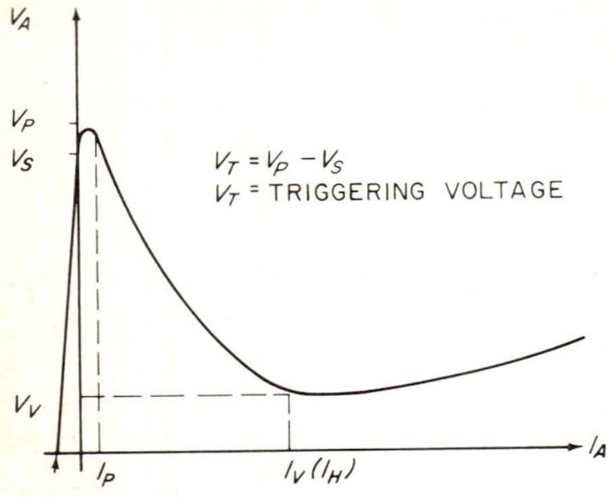

FIGURE 12.50

tively unaffected by temperature changes. Also, when a PUT is triggered on, its impedance decreases very sharply and to a very low value, thereby enabling its external circuitry to develop a high peak pulse output. The dynamic resistance of the PUT in the saturation region is approximately 3 Ω.

Additional features of the PUT include very low leakage when cut off, high breakdown voltages (again, when cut off), excellent performance at low supply voltages, and oscillator frequency stability to ± 1 percent from -55 to $+150°C$.

The PUT can be employed wherever the UJT can.

TUNNEL DIODES

The tunnel diode is a device capable of operation in the 1- to 10-gigahertz (GHz) range. Although it is used in the conventional applications of amplifying, oscillating, and switching, its principle of operation is entirely different from the transistor or the vacuum tube. The name "tunnel diode" has been adopted because the physical mechanism by which the device functions is caused by a complex quantum-mechanical tunneling process. In actual practice, the tunnel diode is basically a very heavily doped PN junction; thus, as one might expect, it possesses many of the properties of a conventional diode.

The tunnel diode has two outstanding properties: (1) extremely high frequency response and (2) very low power consumption. For example, whereas a vacuum tube may operate at 100 MHz and consume 1 watt (W) of power and a conventional transistor may operate at 100 MHz and consume 10 mW of power, a tunnel diode can operate at 500 MHz and consume 1 mW of power.

Operation of the Tunnel Diode. The operation of the tunnel diode is dependent upon a quantum-mechanical principle known as tunneling. Not surprisingly, tunneling is dependent upon special characteristics of the PN junction.

We know that a depletion region amounts to a potential barrier; and to pass current through a diode, sufficient external voltage must be applied, with the proper polarity, to overcome the potential barrier. In the absence of an external voltage, or in the absence of sufficient voltage, very little current will pass across the junction. However, some carriers on both sides of the junction will attain enough energy to surmount the potential barrier of the junction and reach the other side. But this occurs to only a relatively few electrons and holes that can obtain the appropriate energy state.

Classical physical theory states that unless a carrier possesses enough energy to overcome a potential barrier, it will never cross or surmount that barrier. The more recent quantum mechanics, however contradicts classical physics in this instance and states that a carrier can reach the other side of a potential barrier even though it does not have enough energy to surmount the barrier. The carrier does this by tunneling through the potential hill. In fact, quantum mechanics can predict the probability of this occurrence. The probability of a carrier tunneling through the potential hill is dependent on how far the particle must tunnel, i.e., the thickness of the potential hill. The probability of tunneling is nil unless the barrier is extremely narrow, which is proved by the fact that tunneling does not occur in a conventional PN junction.

To make a tunnel diode, the PN junction must be very heavily doped with impurities. The large number of impurities produces a very narrow depletion layer.[1] With this thin layer, electrons can tunnel their way from the N region to the P region. This gives rise to an additional current in the diode at very small forward bias, which disappears when the bias is increased. It is this additional current that produces the negative resistance in a tunnel diode. It has also been found that the electrons travel through the depletion layer at tremendously high velocities. This enables the tunnel diode to operate at far higher frequencies than conventional transistors. The theoretical frequency limit is in the neighborhood of one million megahertz.

In a conventional PN junction diode, the forward voltage required to cause current flow is approximately 0.3 V for germanium diodes and 0.7 V for silicon diodes. The reverse breakdown usually occurs between 20 and 200 V. If we superimpose on this characteristic curve the tunneling current that occurs at a forward bias of about 50 to 100 mV, we obtain the characteristic of Fig. 12.51. The composite characteristic is shown in Fig. 12.52. Note that the tunneling current appears at very small voltages. As we continue to raise the applied voltage, the current dips down and then starts increasing again

[1] In comparison with the conventional diode, the depletion layer of a tunnel diode is 100 times narrower. For example, the width of the depletion region of a tunnel diode is on the order of 0.000001 cm; for a conventional diode, it is 0.0001 cm wide. It is this difference in width that accounts for the major difference in current-flow mechanism between the two devices.

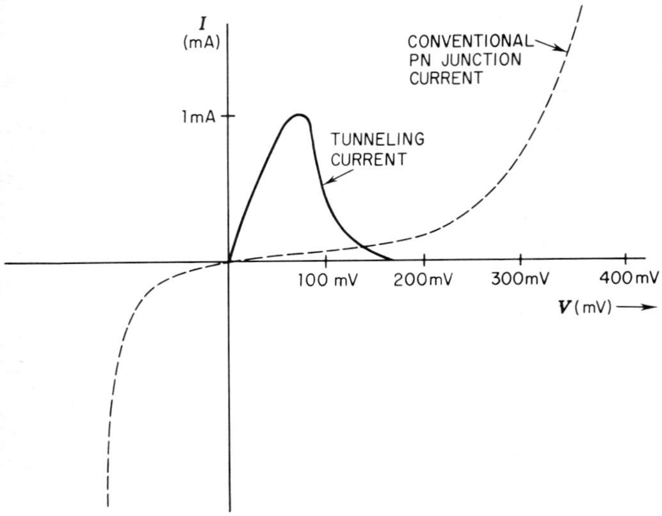

FIGURE 12.51
Combination of tunneling current and conventional PN-junction current.

FIGURE 12.52
Composite characteristic representing overall operational curve of a tunnel diode.

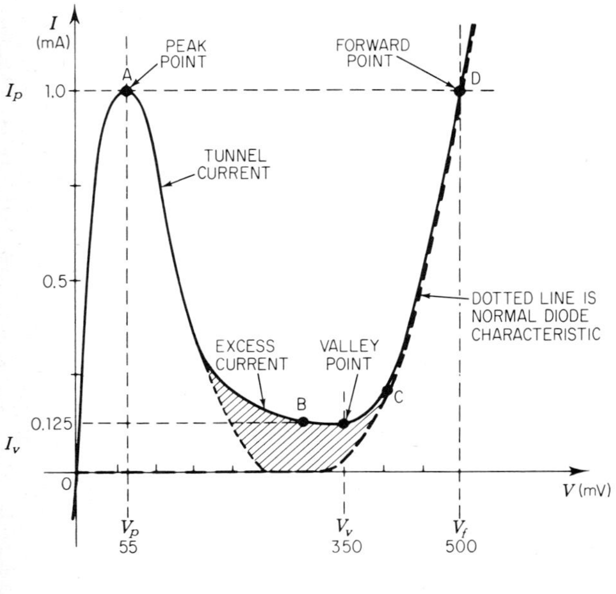

as the tunneling phenomenon disappears and the diode enters the conventional behavior of a PN junction.

The important characteristic to note is that the region between points A and B represents a negative resistance; i.e., as the voltage is increased, the current decreases (just the reverse of the behavior of an ordinary resistor).

Parameters of the Tunnel Diode. Figure 12.52 shows the total voltage-current curve with various dc parameters noted. Typical values of these parameters for a germanium tunnel diode are also given.

Dc Parameters. Let us study these various parameters and note their significance. I_P is the peak current and is dependent upon the junction area and the doping that is used. The actual value of I_P is determined by the intended application. Common values are 1, 5, 10, and 100 mA. Of extreme importance in switching circuits is the variation of I_P from unit to unit. The tolerance of this parameter is usually held to within ± 5 percent or less. V_P, the voltage at which the peak current occurs, is generally in the region of 60 millivolts (mV) for germanium diodes.

I_V, called the valley current, is the lowest current on the voltage-current curve. This current should be as low as possible. Actually, I_V should be small in comparison with I_p, because the largest possible current swing of the device is $I_P - I_V$.

V_V is the voltage at which the valley current occurs. Its value ranges from 200 to 350 mV for germanium diodes. V_V represents the crossover point where the resistance goes from negative to positive.

V_f, called the forward voltage, is measured to a current equal to I_P. Typical values for germanium diodes range from 300 to 500 mV. The difference between V_f and V_P represents the largest possible voltage swing from the tunnel diode. If larger (0.5 to 1 V) voltage swings are required, silicon and gallium arsenide tunnel diodes are used.

All the above information refers only to dc parameters of the tunnel diode; a discussion of the ac characteristics of the tunnel diode follows.

Ac Parameters. In order to form an ac equivalent circuit for the tunnel diode, we must consider what properties of this device would affect its ac operation.

First, the tunnel diode exhibits a negative resistance. This negative resistance is given by the slope of the voltage-current characteristic shown in Fig. 12.52; and for the curve shown in Fig. 12.52, it is about $-150\ \Omega$. A moderate value of negative resistance is most desirable. If the negative resistance is too small, it limits the use of the tunnel diode to low-impedance circuits. If the resistance is too high, the problem of ac instability (oscillation) arises. The value of negative resistance is virtually independent of frequency.

As in the conventional PN junction diode, the tunnel diode also has a

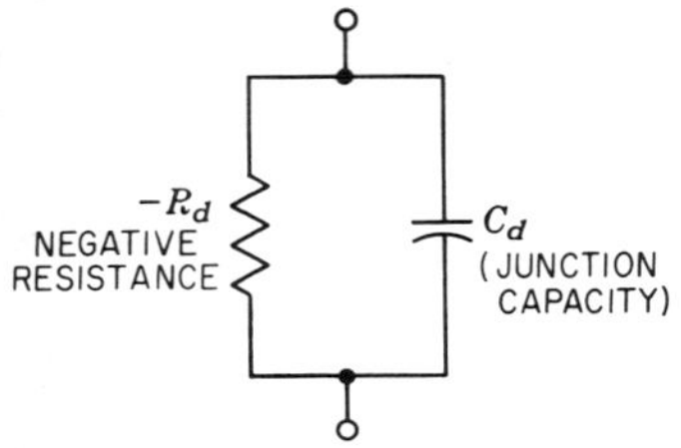

FIGURE 12.53
Simple equivalent circuit of a tunnel diode.

barrier capacitance. This capacitance is analogous to the collector-junction capacitance in a transistor; typical values range from a few picofarads for diodes of very small area to hundreds of picofarads for large-area diodes. This junction capacitance is shown in the simple equivalent circuit of Fig. 12.53 shunting the ac negative resistance.

In addition to the foregoing, the tunnel diode has in series with it two parasitic elements—spreading resistance R_s and lead inductance L_s. The spreading resistance is a combination of the resistance of the bulk semiconductor material and of the leads. Its value is of the order of 0.1 to 10 Ω. The lead inductance is of the order of a few nanohenries; however, even this small inductance can cause problems with parametric oscillations. Both the series spreading resistance and the lead inductance should be kept as low as possible.

Figure 12.54 shows the equivalent circuit obtained when all of the foregoing factors are taken into consideration.

Figures of Merit. The equivalent circuit of Fig. 12.53 may be used to develop the various figures of merit associated with the tunnel diode. Let us consider the figures of merit that pertain to linear, small-signal circuits.

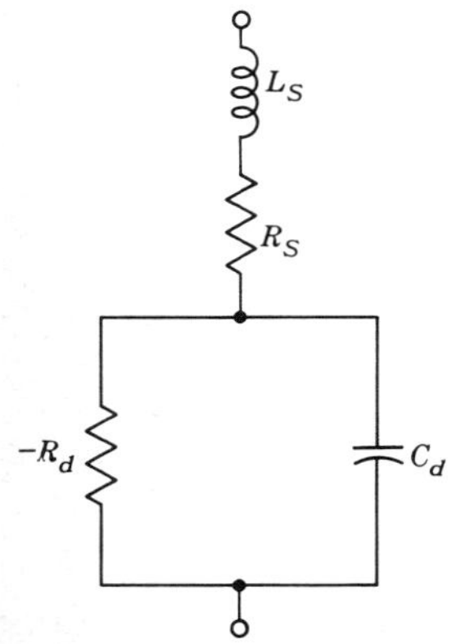

FIGURE 12.54
A more complete equivalent diagram of a tunnel diode.

An important figure of merit for both video and tuned tunnel diode amplifiers is a quantity called the transducer gain-bandwidth product. This is somewhat analogous to the gain-bandwidth product of the transistor f_T discussed previously and is a measure of the frequency response of the diode. It is

$$\sqrt{G_t}\,\mathrm{BW} = \frac{1}{2\pi R_d C_d}$$

where R_d = negative resistance
C_d = junction capacitance
BW = bandwidth

The important point to notice is that the square root of the power gain multiplied by the bandwidth is a constant that is dependent only upon parameters of the diode. It also shows that as the gain G_t increases, the bandwidth decreases, so that at very high gains the bandwidth will be narrow, and vice versa.

Another important figure of merit is the cutoff frequency. Unlike the vacuum tube and transistor, the tunnel diode has two cutoff frequencies. One cutoff frequency, the resistive cutoff frequency, is

$$f_{\mathrm{COR}} = \frac{1}{2\pi R_d C_d}\sqrt{\frac{R_d}{R_s} - 1}$$

This is the frequency at which the resistive part of the diode impedance, measured at its terminals, goes to zero. Above this frequency the tunnel diode has a gain less than 1 and is thus useless as an amplifier. The reactive cutoff frequency is given by

$$f_{\mathrm{COX}} = \frac{1}{2\pi}\sqrt{\frac{1}{L_s C_d} - \left(\frac{1}{R_d C_d}\right)^2}$$

This cutoff frequency is sometimes called the self-resonant frequency because it represents the frequency at which the tunnel diode will oscillate of its own accord. Actually, this frequency represents the point at which the reactive component of the diode impedance goes to zero.

In an actual application, both of the above frequencies are reduced by the effects of the external circuitry. Thus, the highest frequency of operation is very dependent upon the types of circuits, the component values, and the diode package. Lead inductance is especially important, and extreme care should be taken to keep this element as small as possible. The new tunnel diode is packaged in microstrip, pill, and microwave cases in order to reduce lead inductance.

The resistive cutoff frequency is the maximum frequency at which the

FIGURE 12.55
Schematic symbol for a tunnel diode.

tunnel diode may be used as an oscillator. This is a natural consequence of the fact that this is the frequency at which the power gain falls below 1.

Another important figure of merit is the efficiency of the tunnel diode when used as an oscillator. The formula for the efficiency of a class A oscillator is

$$\text{Eff} = 25\left(1 - \frac{R_s}{R_d}\right)\left[1 - \left(\frac{f}{f_{\text{COR}}}\right)^2\right] \qquad \text{percent}$$

where R_s = series diode resistance
f = frequency of oscillation
f_{COR} = resistive cutoff frequency

Note that if $R_d \gg R_s$ a condition that is normally true, the maximum low-frequency efficiency is 25 percent. This is just one-half that attainable when using transistors.

Still another important figure of merit is the noise figure of a tunnel diode. This is expressed by

$$\text{NF} = 10 \log\left(1 + \frac{IR_d}{52}\right) \qquad \text{dB}$$

where NF = noise figure
R_d = the negative diode resistance, Ω
I = the dc bias current, mA

Unlike the noise figure of transistors, the noise figure NF of the tunnel diode is virtually independent of frequency over its entire operable frequency range. The noise figure of the tunnel diode is quite low, usually in the 3- to 5-decibel (dB) range.

A commonly used symbol for the tunnel diode is shown in Fig. 12.55.

TUNNEL-DIODE APPLICATIONS

In view of their good frequency response and low-noise characteristics, the biggest areas of application for tunnel diodes is as oscillators and amplifiers in the gigahertz region.

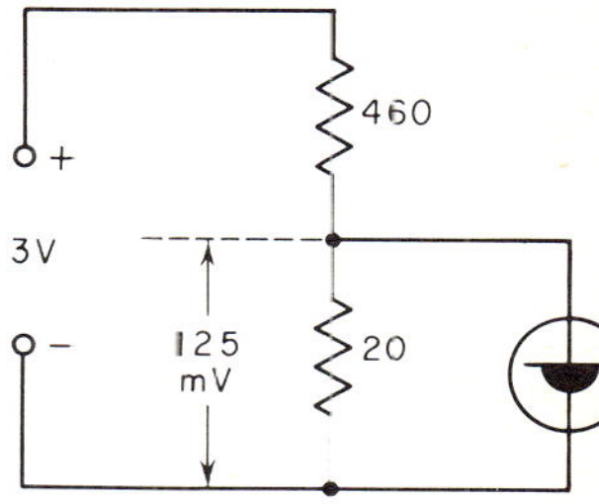

FIGURE 12.56
A method of biasing a tunnel diode.

Oscillators. To achieve oscillation with a tunnel diode, it must be set up so that the negative resistance it provides is greater than the positive resistance of the resonant components in the circuit. A typical tunnel diode has a negative resistance of approximately −150 Ω when it is biased at the center of its negative-resistance region. To bias the diode correctly, approximately 125 mV is required. At this voltage, the current drawn by the diode is approximately 0.5 mA.

There are several ways to set up this bias circuit. We can connect the tunnel diode in series with a resistor and a small battery. If we assume a battery voltage of 3 V, then a series resistor of 6,000 Ω would be required to limit the current flow to 0.5 mA. But the 6,000 Ω would completely overshadow the −150 Ω of the tunnel diode and would prevent us from properly utilizing this negative resistance.

This difficulty can be circumvented by the resistive arrangement shown in Fig. 12.56. The necessary bias voltage is here developed across a 20-Ω resistor, a value considerably less than the −150 Ω of the tunnel diode.

We can now connect a resonant circuit in series with the tunnel diode; and if the circuit possesses a resistance less than 100 Ω, then the tunnel diode will completely balance out the resistive loss of the circuit and enable oscillations to take place. A suitable circuit capable of oscillating into the megahertz range is shown in Fig. 12.57.

A crystal-controlled oscillator using a tunnel diode is shown in Fig. 12.58. R_1 and R_2 are selected so that each has about twice the value it should have

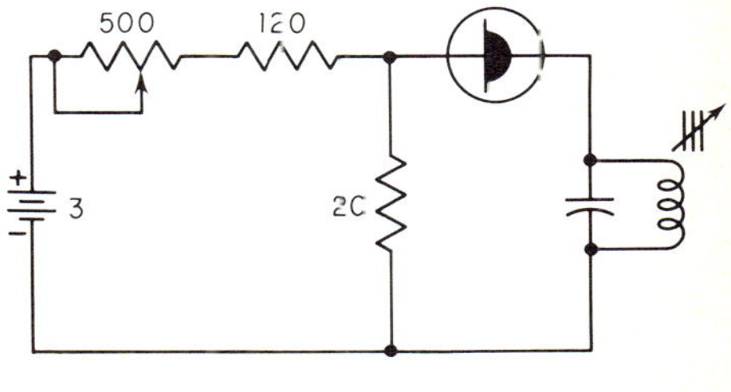

FIGURE 12.57
A tunnel diode oscillator.

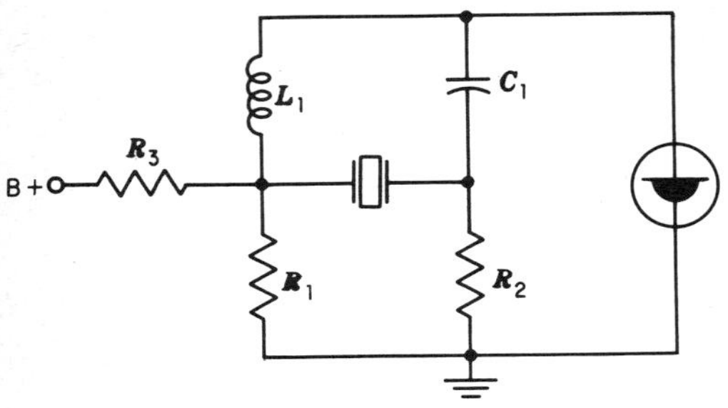

FIGURE 12.58
A tunnel diode crystal-controlled oscillator.

to enable the negative resistance of the tunnel diode to predominate the circuit. A crystal is placed between these resistors. At all frequencies other than the resonance, the crystal impedance is high and the circuit is unable to function. At the resonant frequency, however, the crystal becomes a short circuit and R_1 is placed in parallel with R_2, thus reducing their total resistance value to half their individual values. This new value permits the circuit to oscillate freely as a frequency accurately governed by the crystal.

Amplifiers. If we increase the impedance of the external circuit connected across a tunnel diode until it equals the negative resistance of the tunnel diode, amplification rather than oscillation is obtained. This is done in the 100-MHz amplifier shown in Fig. 12.59. The 1N2939 tunnel diode has a negative resistance that is just counterbalanced by the circuit positive resistance. The latter, in this instance, is equal to 50 Ω from the signal source, 50 Ω from the load, 2 Ω from the internal lead resistance of the 1N2939, and 16 Ω from the series resistor. In this particular configuration, the 16-Ω resistor was added simply to achieve this counterbalancing.

Whereas the actual mathematical justification for the condition indicated above is quite complex, some inkling of how it is arrived at may be seen from

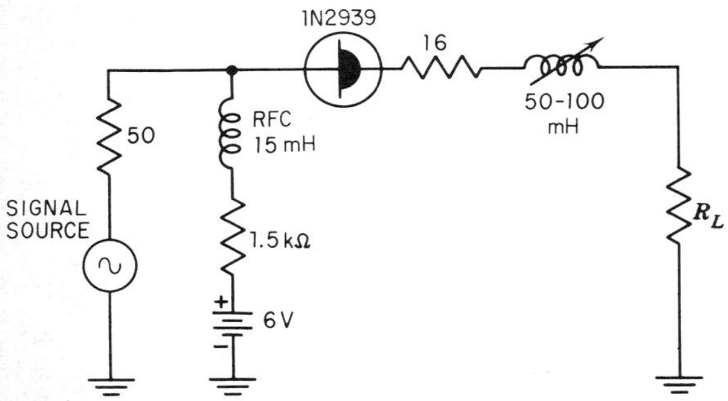

FIGURE 12.59
A tunnel diode amplifier.

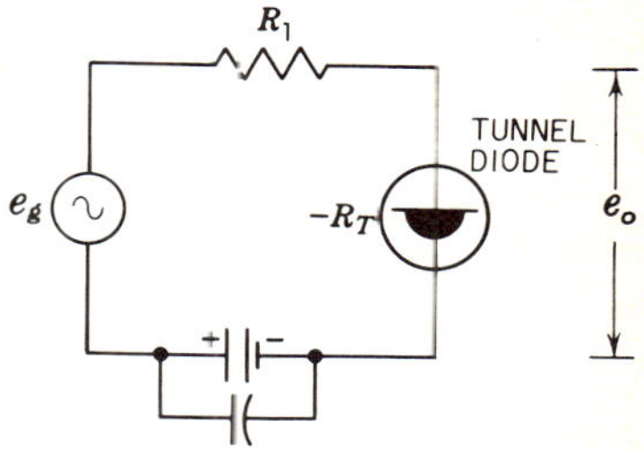

FIGURE 12.60
Simple circuit to demonstrate ability of tunnel diode to amplify. (*Courtesy Radio-Electronics.*)

an examination of the simple circuit shown in Fig. 12.60. Assume that the tunnel diode is biased to the center of its negative-resistance range and that e_g represents an incoming signal and e_o is an output signal. Then the voltage across the tunnel diode, with negative resistance $-R_T$, is

$$e_o = \frac{e_g(-R_T)}{R_1 + (-R_T)}$$

When the circuit positive resistance R_1, exactly equals $-R_T$ in value e_o becomes infinite. This does not happen, of course, in an actual circuit, but the largest output is obtained when $R_1 = R_T$. At this point, the ratio of e_o to e_g, or the voltage gain, is greatest.

Although we have mentioned germanium as the material from which tunnel diodes are manufactured, three other materials are also being employed. These are silicon (Si), gallium arsenide (GaAs), and gallium antimonide (GaSb). Table 12.2 lists some of the parameters of tunnel diodes fabricated from these materials.

Tunnel diodes designed for amplification would generally be manufactured using germanium or gallium arsenide because of their low-noise characteristics.

TABLE 12.2
Comparative Parameters of Tunnel Diodes Manufactured of Different Materials[a]

	Ge	*Si*	*GaAs*	*GaSb*
I_p/I_v	8	3.5	15	12
V_P	0.06	0.07	0.15	≈0.1
V_V	0.35	0.40	0.50	≈0.45
f_{COR}, *GHz*	50	1	20	50

[a] Courtesy *Electronics World.*

Of the two, germanium has a higher noise level, but it is also more stable with respect to temperature. Gallium arsenide exhibits the greatest power-output capabilities; it is hence used for oscillators. Silicon, on the other hand, would be selected for operation at elevated temperatures.

Many tunnel-diode applications are being taken over by the Gunn diode because it can operate at higher frequencies and is rapidly becoming cheaper to produce.

It is interesting to note that both devices utilize negative resistance but different electron-behavior mechanisms, i.e., tunneling and the high-conduction-band and low-mobility mechanisms.

P-I-N DIODES

In some diode applications, it is desirable to have a very low ohmic resistance (in the forward direction) and a very high reverse breakdown voltage. These features are exhibited by the p-i-n diode because it possesses heavily doped P and N regions separated by a layer of nearly intrinsic high-resistivity material. In addition, the r-f resistance of this diode can be varied continuously from high to low values by varying the diode bias. All these properties are useful at microwave frequencies, and it is in this region that the device is employed.

Figure 12.61*a* is a cross section of a typical p-i-n diode, which is fabricated by diffusing boron (P) and phosphorus (N) into high-resistivity silicon. The wafer is about 8 mils thick, with the boron diffused to a depth of about 3.5 mils and the phosphorus to a depth of about 0.6 mil.

In microwave applications, the p-i-n diode would be connected across

FIGURE 12.61*a*
Cross-section of typical p-i-n diode.

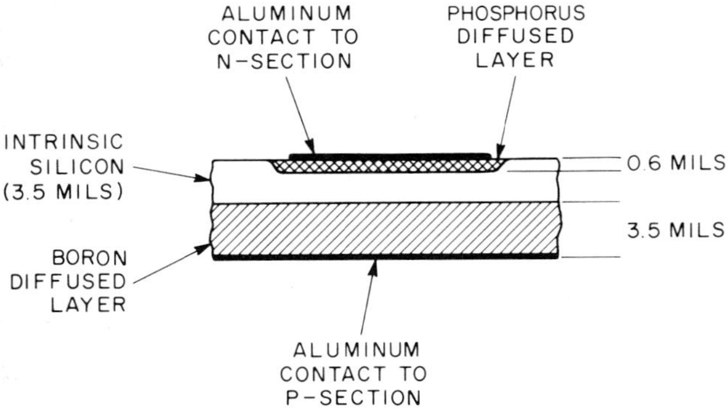

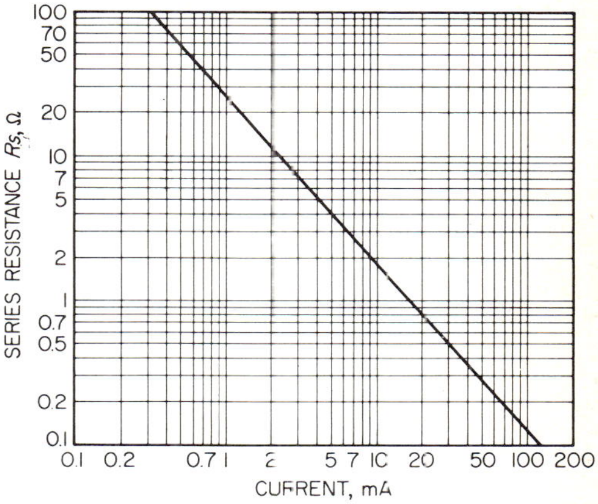

FIGURE 12.61*b*
Resistance vs. forward current for a p-i-n diode. (*Courtesy Electronics World.*)

the transmission line carrying the microwave signal. If the diode is reverse-biased, the diode presents only a small capacitance across the line and the signal passes unimpeded. However, with about 1 to 2 V in the forward direction, the diode presents a very low impedance, effectively short-circuiting the signal and preventing it from traveling further down the transmission line.

By varying the forward voltage across the diode, it can be made to function as a variable attenuator or modulator, presenting more or less impedance to an applied microwave signal. Figure 12.61*b* is a graph of resistance vs. forward current for a p-i-n diode.

When the diode becomes forward-biased, holes (from the P section) and electrons (from the N section) are injected into the intrinsic region. Since conductivity is directly proportional to the number of carriers, the resistance of the intrinsic region decreases as the forward bias is increased. This effect is called conductivity modulation.

The p-i-n diode can thus be represented as a current-variable resistance in the forward direction and as a constant capacitance in the reverse direction.

In microwave applications, where very high frequencies are encountered, it is possible for the p-i-n diode to control peak r-f voltages greater than the breakdown voltage of the diode. This can occur when the time that the applied signal exceeds the breakdown voltage is too short for the avalanche effect to really get going.

HOT-CARRIER DIODE

The hot-carrier diode differs from the conventional semiconductor diode in that the junction consists of a metal and a semiconductor rather than two different semiconductors (see Fig. 12.62). It is because of this altered struc-

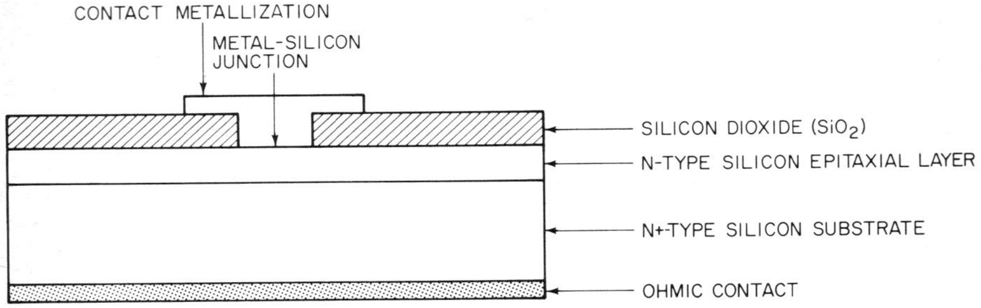

FIGURE 12.62
Internal construction of a hot carrier diode.

ture that the hot-carrier diode provides better performance at higher frequencies for such applications as detectors, mixers, switches, discriminators, limiters, logarithmic converters, gating circuits, sampling circuits, and clamping circuits.

The operation of the hot-carrier diode can best be understood by reference to the energy-level diagram of the metal-semiconductor junction (Fig. 12.63). When the hot-carrier diode is forward-biased, electrons in the conduction band of the semiconductor gain sufficient energy to overcome the junction barrier and are injected into the metal. These electrons, as Fig. 12.63 indicates, have an energy level substantially above the energy of the metal's free electrons, which is why the injected electrons are labeled "hot carriers." In the metal, the hot electrons give up their excess energy very quickly (10^{-15} s), after which they become part of the sea of free electrons in the metal.

Reverse bias (Fig. 12.63*c*) increases the barrier potential, leaving the semiconductor electrons with insufficient energy to be injected into the metal. Barrier potential, as seen from the semiconductor side of the junction, is either increased or decreased depending on external bias.

The electron flow from semiconductor to metal occurs with virtually no flow of minority carriers in the reverse direction. This is in sharp distinction to a conventional PN junction diode where under forward-biasing electrons diffuse to the P section and holes from the P section diffuse to the N section, where both exist as minority carriers. Now, when reverse bias is applied, the minority carriers in each section must either be removed (by flowing back to their original section) or by recombination with an oppositely charged particle. These minority carriers represent a stored charge and their removal is necessary if the junction is to operate normally again.

It is for this reason that PN junction diodes cannot operate at as high a frequency as the hot-carrier diode. Time is required to move them back. In the hot-carrier diode, there are virtually no minority carriers, so that the time lag does not occur.

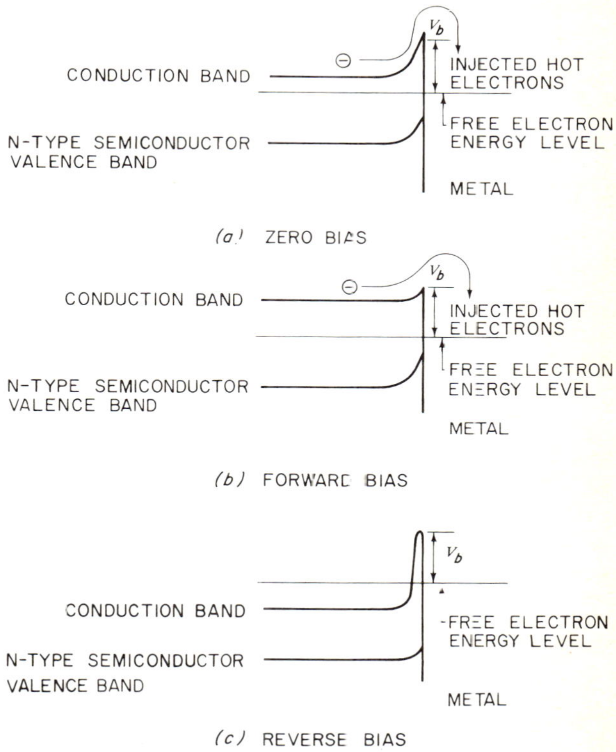

FIGURE 12.63
Energy-level diagrams of the metal-semiconductor junction. V_b is the barrier voltage.

The metal-on-semiconductor concepts on which the hot-carrier diode is based extend back to work previously done by Schottky. Hence, these devices are sometimes referred to as Schottky-barrier diodes.

GALLIUM ARSENIDE

It was mentioned in the introduction to this chapter that gallium arsenide may soon become the most important semiconductor material, replacing silicon. This is true because gallium arsenide has some special features that are just now being optimized for many semiconductor applications.

Gallium arsenide is only one of several intermetallic compounds that have useful semiconductor properties. We shall emphasize GaAs, however, because at present it has many more practical applications than the others.

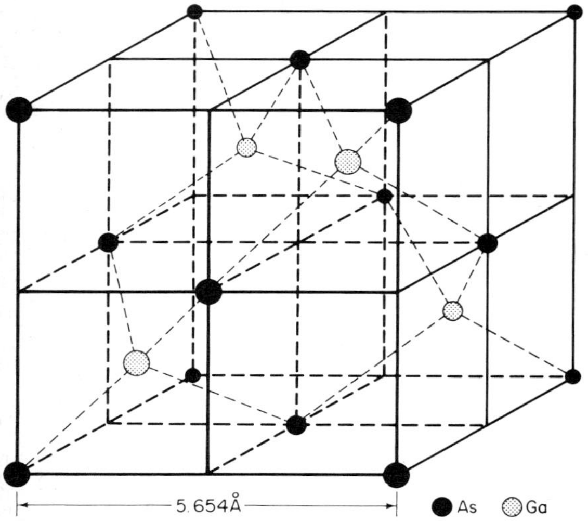

FIGURE 12.64
Crystal structure of gallium arsenide.

All intermetallic compounds differ from semiconductors as silicon or germanium in that they are formed with two pure elements in place of one. Thus, a silicon transistor starts with pure silicon and then has added to it appropriate impurities to form the requisite P and N region. With gallium arsenide, the basic crystal structure consists of two different metallic elements.

TABLE 12.3
Comparative Chemical and Physical Properties of Germanium, Silicon, and Gallium Arsenide[a]

Property	*Ge*	*Si*	*GaAs*
Band gap, eV	0.67	1.106	1.40
Type of band gap	Indirect	Indirect	Direct
Calculated maximum electron mobility at 15°C, $cm^2/(V)(s)$	3,950	1,900	11,000
Calculated maximum hole mobility at 25°C, $cm^2/(V)(s)$	1,900	425	450
Melting point, °C	936	1,420	1,238
Thermal coefficient of expansion, $°C^{-1}$	6.1×10^{-6}	2.33×10^{-6}	5.93×10^{-6}

[a] Courtesy Electronics Magazine.

Suitable impurities are then added to this compound to form the needed junctions.

Silicon and germanium each has four valence electrons in its outer, or chemically active, rings. The crystalline structure is then formed by having the atoms share each other's outer electrons to form bonds. With gallium arsenide, each gallium atom is surrounded by four arsenic atoms at the corners of a tetrahedron, and each arsenic atom is similarly surrounded by four gallium atoms (Fig. 12.64). Table 12.3 compares some chemical and physical properties of gallium arsenide with those of silicon and germanium.

We have seen that a fundamental-electrical property of a semiconductor is the energy needed to free an electron from the bond formed between two atoms. In silicon, more energy is required to liberate electrons, which is the major reason why silicon can be employed at higher temperatures. With the intermetallic compounds, by using different combinations of three valence atoms and five valence atoms, we can achieve a very wide range of energy gaps. This, of course, provides a source of new electron-behavior phenomena and potential new semiconductor devices.

The mobility of electrons and holes in semiconductors (i.e., the speed with which these particles or charges move through a crystal) can also be regulated over a fairly extensive range in intermetallic compounds. Gallium arsenide, with an electron mobility nearly three times greater than germanium and five times greater than silicon, offers excellent high-frequency capabilities.

All these variations provide new materials with the flexibility needed to construct many new semiconductor devices with a wide choice of such properties as current- or power-handling capacity, frequency range, and rectification ratio.

INJECTION LASER

In this chapter we discussed the Gunn diode and light-emitting and light-detecting diodes, all of which are predominantly fabricated from gallium arsenide. Perhaps the most amazing new semiconductor device is the gallium arsenide injection laser. Although the laser chip is just a few thousandths of an inch thick, it produces a beam of coherent light that can be directly modulated with a nanosecond pulsing of input current. This makes the injection laser ideal for portable communication systems.

When a GaAs diode is forward-biased, electrons are injected from the N side to the P side, and in a narrow region less than a micron wide, the electrons recombine with holes in the valence band and give up their energy as radiated light. This is the basis for photo-emitting diodes.

The recombination of electrons and holes can occur spontaneously, where individual events are not correlated, or they can occur by a mechanism proposed by Einstein (in 1918), an emission induced by another photon. It is this latter process, where emission is induced, that is responsible for producing a laser beam. In essence, what we have is a reinforcing action where the initial light that is produced on electron-hole recombination is itself responsible for producing additional light.

The gallium arsenide laser is formed in a resonant cavity structure with parallel reflecting surfaces (Fig. 12.65). With no special precautions, the laser light will emit equally from both the front and back ends at the PN junction of the device. Not all the light (i.e., the photons) produced at the PN junction finds its way out the front or back ends. Some of it is kept within the structure, where it produces additional photons. Some of it is also absorbed and hence cannot reach the ends.

As a matter of fact, to eliminate any emission from the back edge and to strengthen the lasing effect, a mirror reflective surface is placed over the back cleaved edge. This serves to increase front emission by 60 percent.

Thus, we have a PN junction where, when pulsed with current from an external circuit, photons or packets of light of a specific frequency are developed. Some of these photons pass out through the front edge of the semiconductor structure and are seen as light. Many of the other photons move back and forth within the PN junction, producing more photons of light. With the back edge mirrored, light can come out only through the front cleaved edge. Here, too, incidentally, a mirror-smooth finish exists, but it is not as reflective as the back edge so that light is able to pass through. This light falls within a very narrow range of frequencies and hence represents coherent light.

One of the major problems encountered in fabricating solid-state lasers is that of heat dissipation. During operation, the semiconductor chip tends to heat up because of the current passing through and also because much of the stimulated emission is absorbed, rather than escaping as visible light.

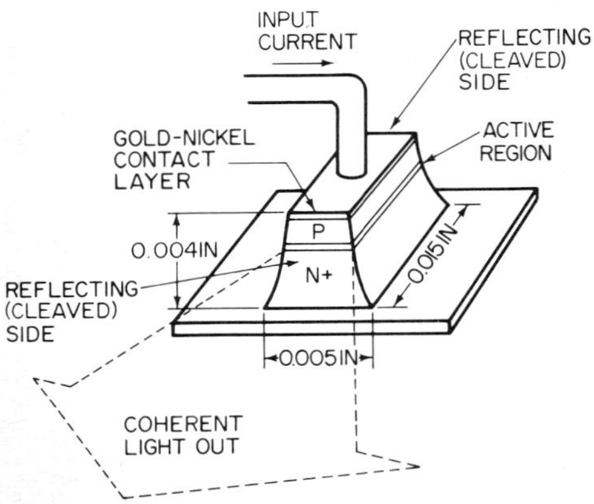

FIGURE 12.65
Gallium arsenide injection laser. (*Courtesy Electronics.*)

To use continuous dc current to power the unit would result in overheating very quickly (at the currents needed to achieve lasing). Hence, only pulses of current are used. Also, to dissipate the heat generated better, the chip is kept as thin as possible, typically on the order of 3 to 6 mils. By mounting a number of chips in arrays, considerable laser power can be developed.

QUESTIONS

12.1. Why is gallium arsenide, and not silicon, suitable material for a Gunn diode?

12.2. Explain, in your own words, the electron-behavior mechanism responsible for the Gunn effect.

12.3. What is negative resistance? How does it occur in GaAs?

12.4. Why is gallium arsenide, and not silicon, suitable material for light-emitting diodes?

12.5. Why is a GaAs light-emitting chip hemispherical?

12.6. What is the relationship between the energy band gap and the wavelength of the emitted light?

12.7. Define amphoteric and explain how it is used in processing light-emitting diodes.

12.8. What is the difference between photoconductive and photovoltaic?

12.9. What is an optical coupling system, and what are its advantages?

12.10. What is dark current in a reverse-biased photoconductive diode?

12.11. Draw a sketch of a phototransistor with its equivalent circuit, and describe how it works.

12.12. What is the distinction between the tunneling and the zener mechanisms?

12.13. Draw the VI characteristic curve for a typical zener diode and explain why the curve has this shape.

12.14. Describe briefly the difference between the zener and avalanche breakdown mechanisms.

12.15. What is a depletion region in a solid-state diode?

12.16. Draw the diagram of a simple voltage-regulated circuit employing a zener.

12.17. How can zener diodes be utilized to provide ac regulation?

12.18. What is a varactor diode?

12.19. Show schematically how a varactor can be used for tuning.

12.20. Explain how the circuit of Fig. 12.29 operates.

12.21. What is a thyristor and what are some of its distinguishing features?

12.22. Draw a diagram of the structure of an SCR and explain briefly how it operates.

12.23. What is the turn-on time and turn-off time of an SCR and what influences each?

12.24. Sketch a typical circuit using a UJT.

12.25. On what one feature does the operation of a tunnel diode depend primarily? Name three other electronic devices that depend upon the same characteristic.

12.26. Describe the operation of a tunnel diode.

12.27. Explain the following tunnel diode characteristics: V_V, I_V, V_f, and I_P.

12.28. Sketch a cross section of a p-i-n diode and label the junctions.

12.29. What is the significance of the intrinsic layer?

12.30. Draw the basic circuit of a p-i-n diode.

12.31. Describe what occurs when a forward bias is suddenly applied to a reverse-biased diode.

12.32. What is the basic difference in the crystal structures of gallium arsenide and silicon?

12.33. Briefly describe a GaAs injecting laser.

12.34. Name several uses for semiconductor lasers.

12.35. In what ways do hot-carrier diodes differ from conventional PN junction diodes?

Chapter Thirteen

Transistor-amplifier Design

In order to be useful as an amplifier, the transistor must be employed in a circuit so designed that the transistor is operated over its most linear range. In this way, minimum distortion will be introduced in the signal passing through the stage. Another precaution that must be observed is that of maintaining the transistor within its heat-dissipation limits. Failure to observe these boundaries will lead to the destruction of the unit just as surely as a tube is destroyed when it is operated with abnormal voltages or currents. In the case of a transistor, there is far less leeway, necessitating even greater care in design.

Selection of a suitable operating point is one aspect of amplifier design, which is governed almost completely by the dc voltages applied to the transistor. The second aspect of amplifier design is the response of the stage to the applied signals. This is governed in part by the electrical characteristics of the amplifying device itself, in this case, the transistor, and in part by the characteristics of the coils, capacitors, and resistors attached to the transistor. It is these items which will frequently limit the bandpass of a circuit, at least for the low and medium frequencies. Beyond this, both the transistor and the circuit share the responsibility of degrading the frequency response.

In this chapter we shall analyze the problem of transistor-amplifier design from each viewpoint separately. Not only is this the conventional approach but it is also an eminently practical one, since it enables the circuit designer to deal with each variable separately. Both the ac gain and the frequency response are functions of the operating point, however. This is possible because we can quite accurately represent the transistor by an equivalent circuit under small-signal conditions. Furthermore, the external circuit components (provided that they are linear) will not affect the distortion produced by the transistor.

SELECTING THE OPERATING POINT

The operating point for the transistor is established by the dc voltages chosen for the base, emitter, and collector elements. In this respect, the transistor is identical to the vacuum tube; as a matter of fact, many of the steps followed in arriving at a bias point are similar to those employed in establishing the operating point for a tube. The biasing methods, however, will frequently be quite different because of the inherent differences between the two devices.

The first items needed to select the proper bias point are a listing of the transistor's characteristics and a graph of its current and voltage behavior for the particular circuit arrangement chosen. In practically all instances, this means with the transistor connected as a grounded emitter, i.e., with the input signal fed into the base and the output signal taken from the collector.

The curves most useful for conventional-design purposes show the variation in collector current for different collector voltages and with different base currents. A representative family of such curves is shown in Fig. 13.1.

Note specifically that these should be for the grounded-, or common-, emitter configuration. Frequently, curves are shown for the common-base arrangement, and the two are very similar in appearance. Both the listed and the graphical characteristics contain important information, and both are re-

FIGURE 13.1
The characteristic curves for a transistor connected in the common-emitter configuration.

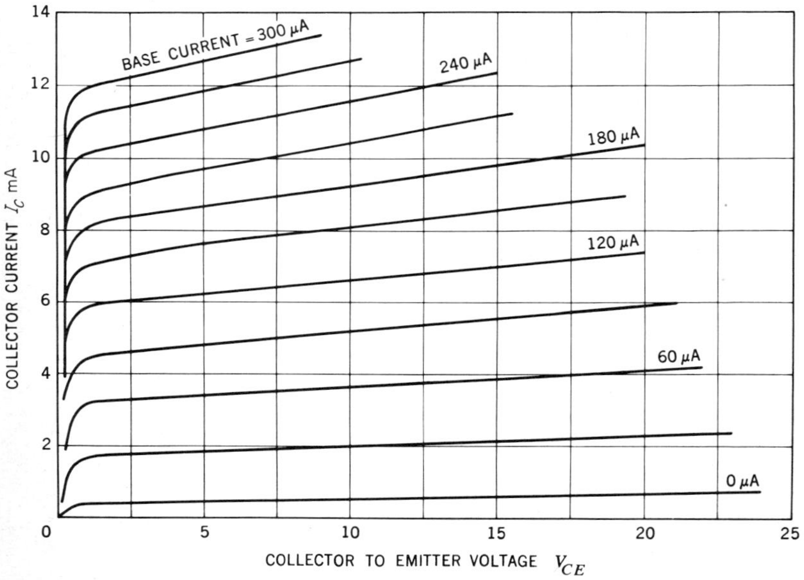

quired to select a dc working point properly. As a start, let us examine the listed characteristics to see what information they can provide that will help us choose a suitable dc operating point for the transistor in question.

At the top of Table 13.1 the absolute maximum ratings are given. These include element voltages, collector current, collector dissipation, and junction temperature. Of specific interest is the collector-to-emitter voltage, because we are dealing here with a common-emitter configuration. The stated limit is —30 V. (The minus sign simply means that the collector voltage is negative with respect to the emitter, this being a PNP transistor. For an NPN unit, the same voltage values would be given as positive.) Whatever dc voltages are applied to the collector and emitter, their *difference* must not exceed 30 V, or a voltage breakdown between these elements may occur. The word "difference" is important, because the individual-element voltages with respect to a common connection point such as ground may be much higher than 30 V. However, it is only the voltage *between* emitter and collector that counts.

Collector current I_C possesses a maximum figure of 10 mA. If, now, we multiply this collector current by the maximum collector-to-emitter voltage (0.01×30), we obtain a power value of 0.3 watt (W). We might suppose that this is the maximum transistor dissipation. Just below the collector-current figure, however, maximum dissipation is given as 60 mW, or 0.060 W. How is this possible?

The answer, of course, is that maximum collector current and maximum collector voltage do not occur at the same time. If they did, the transistor

TABLE 13.1
Maximum Values and Average Characteristics

Ratings—Absolute Maximum Values	
Collector-to-emitter voltage	—30 V
Collector-to-base voltage	—45 V
Collector current	10 milliamperes (mA)
Collector dissipation	60 milliwatts (mW)
Junction temperature (maximum recommended operating temperature)	85°F
Average Characteristics—Design Center	
Collector voltage	—6.0 V
Emitter current	1.0 mA
Current amplification	0.98
Collector-current cutoff I_{CO}	10 microamperes (μA)
Output capacitance	40 picofarads (pF)
Frequency cutoff f_{CO}	1.0 megahertz (MHz)
Maximum power gain	40 decibels (dB)

would be destroyed in short order. What the values do mean is that, within the limits of a transistor dissipation of 60 mW, the collector voltage may go as high as 30 V (with respect to the emitter) or that the collector current may go as high as 10 mA. But the 60-mW value is the figure to watch; it is the limiting factor on the choice of a suitable dc operating point. More on this in a moment.

Also shown in the characteristic listing are recommended design center values. These represent average values that will bias the transistor to a point at which it can be employed gainfully as an amplifier. Note that the recommended collector-to-emitter voltage is -6 V and the recommended current is 1 mA. These values are far from the maximum figures quoted above.

Note, too, that with these values, collector dissipation is

$$6 \times 0.001 = 6.0 \text{ mW}$$

and this is well within the 60 mW given under the maximum ratings. In short, the operating figures provide considerable room for leeway, and this is as it should be.

An interesting way in which the maximum collector dissipation can be kept in front of the circuit designer is shown in Fig. 13.2. The dotted curve,

FIGURE 13.2
The dotted curve indicates the maximum collector dissipation of the transistor represented by this set of characteristic curves.

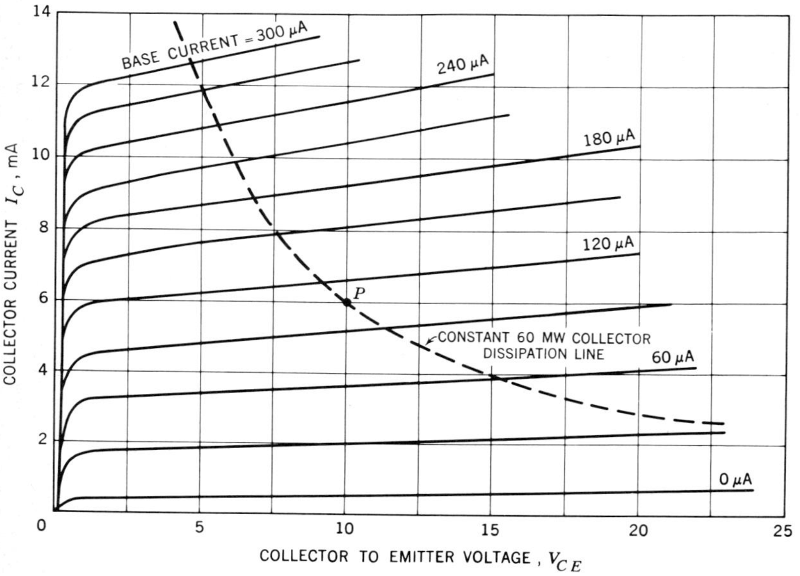

at each point, equals 60 mW; so long as the operating condition of the transistor stays to the left of this line, the unit is within its maximum-dissipation range and no difficulty from burnout should be encountered. On the other hand, if this dotted curve is crossed, the chances are quite good that the useful life of the transistor will be materially shortened.

Some manufacturers provide these maximum-dissipation curves with the transistor characteristic graphs. When this is not done, the designer himself can draw such a curve by choosing, for each collector voltage, a value of collector current such that the product of the two equals the maximum-dissipation figure. For example, in Fig. 13.2 the point P, the collector voltage, is 10 V. The dotted line crosses this point at 6.0 mA; the product of 10 and 6.0 mA is then 60 mW.

As a general rule, to determine the points on the dissipation curve, voltages may be selected along the horizontal axis and corresponding current values calculated ($I_C = P_C/V_C$), or current points may be selected along the vertical axis and corresponding voltage values calculated ($V_C = P_C/I_C$).

LOAD LINES

A collection of curves, such as that shown in Fig. 13.1, represents the behavior of a transistor over a wide range of collector currents and voltages and for a wide variety of base currents. When a transistor is connected with the emitter grounded, the amount of current that will flow from emitter to collector is determined by the base-to-emitter potential. Instead of specifying the latter values, which are very small and exceedingly difficult to measure, the resultant base current is given instead, since for each base-to-emitter potential a certain base current will flow. It is the latter values that are indicated for each of the curves in Fig. 13.1. As a matter of fact, base current values are more significant than voltage values because the transistor is a current-operated device. Furthermore, in the common-emitter configuration, the base is the element to which the incoming signal is applied.

Now, in order to determine how a transistor will operate with specific voltages applied to its elements, we must determine the section of the graph where this operation is to take place, for example, for the transistor represented by the curves in Fig. 13.2, we can place the operating point anywhere to the left of the maximum-dissipation line. The manufacturer recommends −6 V, 1 mA, but we have selected −6 V, 5 mA because it enables us to use a load resistance of 1,200 ohms (Ω), which is what we desire

This point is identified by the letter A in Fig. 13.3. This is the central point of any design we may wish to attempt with the transistor; from this point, the variations of the signal current applied to the base will cause the collector voltage and current to vary above and below the conditions specified by point A.

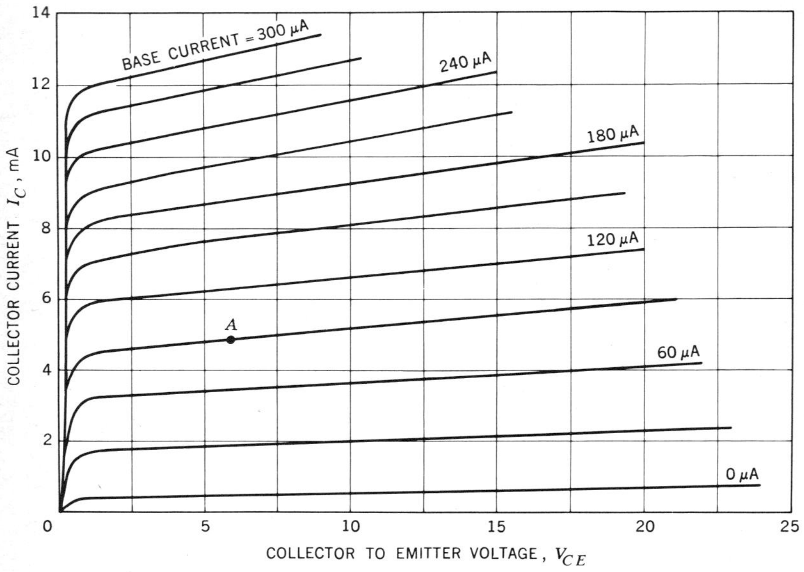

FIGURE 13.3
Point *A* is the selected operating point. A 1,200-Ω load line, drawn through *A*, is shown in Fig. 13.4.

With the location of the operating point, we are ready for the next step, i.e., the drawing of a load line through the operating point. This line reveals graphically what happens when a load resistor is placed in the collector circuit and the input or base current varies. Now the question is, how do we know which line to draw? Several things are already known. First, we know that the load line must pass through point A. Second, it must be so drawn that it does not, at any point, cross the dotted line representing the maximum-dissipation line. These are the two most obvious restrictions on the load line. There are others which will become apparent in the ensuing discussion.

As noted above, the load line represents the load resistance and the value of this resistance will determine the slope of the line. For example, let us assume that the load resistance is 1,200 Ω. Such a line, passing through point A, is shown drawn in Fig. 13.4. To demonstrate that this represents 1,200 Ω, it is necessary to know that the slope of this line is given by the ratio of the change in collector voltage from minimum to maximum current and the current change itself,

$$\frac{\text{Maximum collector voltage} - \text{minimum collector voltage}}{\text{Maximum collector current} - \text{minimum collector current}}$$

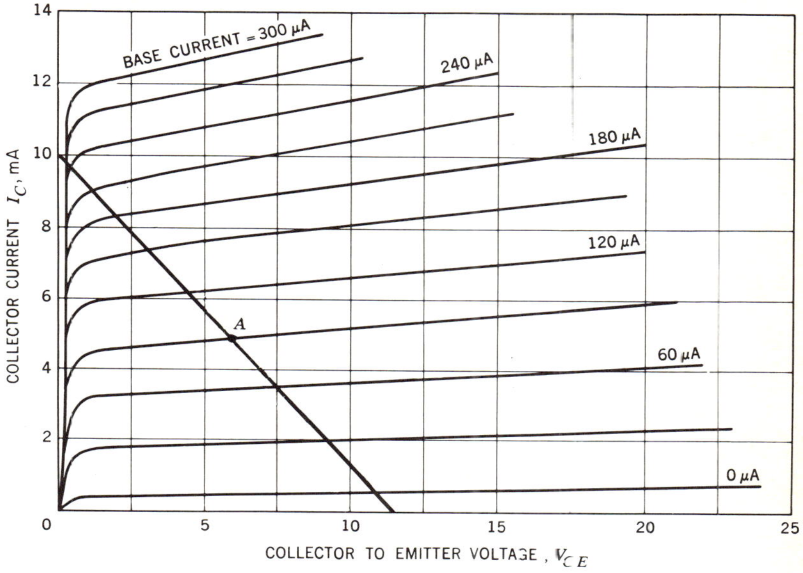

FIGURE 13.4
A 1,200-Ω load line passing through point *A*. The base current, at point *A*, is 90 μA.

with the load resistance in the circuit. From Fig. 13.4, we see that the collector voltage changes from a maximum of 12 V when the collector current is zero to 0 V when the collector current is 10 mA. Substituting these facts in the above ratio, we obtain

$$\frac{12 - 0}{0.010 - 0} = \frac{12}{0.010} = 1{,}200\ \Omega$$

For this value of load resistance, a battery voltage of 12 V is needed. If only 10 V is available and it is desired that the load line pass through point A, then another value of load resistance will be needed. This can be seen by actually drawing a line from 10 V to point A and then extending the line out until it crosses the I_C axis. It does so at 12 mA. From the slope of this second line, the computed value of the required resistor is 833 Ω (Fig. 13.5).

If we examine the load line of Fig. 13.4, we note several things. At the operating point, the base current required is 90 μA. At the bottom end of the load line, it crosses the 0-μA base-current curve; at the upper end of the line, the base-current curve crossed is 240 μA. Thus, the swing below point A extends only for 90 μA, whereas above point A it can go up 150 μA (240 − 90).

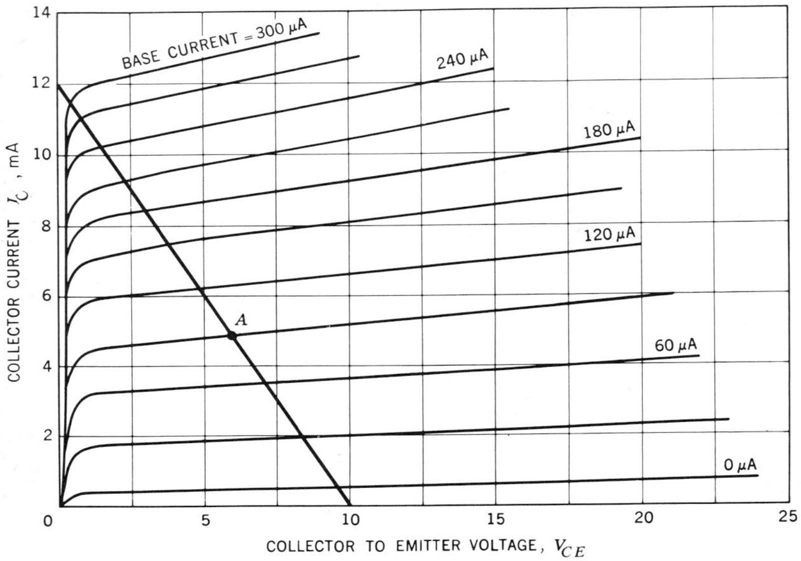

FIGURE 13.5
An 833-Ω load line passing through point *A*.

This occurs because of the progressively closer spacing of the curves at the higher base-current levels because of a reduction in β. Utilization of the entire load line would thus lead to distortion, which is undesirable. In the present instance, it would be better to limit the swing of the signal to the region of more nearly equal base-current curve spacing, particularly when low distortion is more important than maximum output.

Other values of load resistances can be chosen to fit within the area to the left of the maximum-dissipation curve. Several are shown in Fig. 13.6, and the reader can see how the various factors influence the amount of distortion and output obtainable. For example, with the battery voltage fixed, larger and larger resistances permit less and less of an input signal swing between transistor cutoff (when the collector voltage is maximum) and maximum current (when the collector voltage is zero and the characteristic curves are highly distorted). However, with the larger load resistances, most of the swing over the load line is usable, except for the very extreme ends. With low loads, the extent of the curve is greater; but as we saw in Fig. 13.4, the entire curve cannot be used because of the distortion caused by the curves crowding together.

If maximum power output is desired, the load line should be brought as close to the maximum-dissipation line as possible (Fig. 13.7). This enables

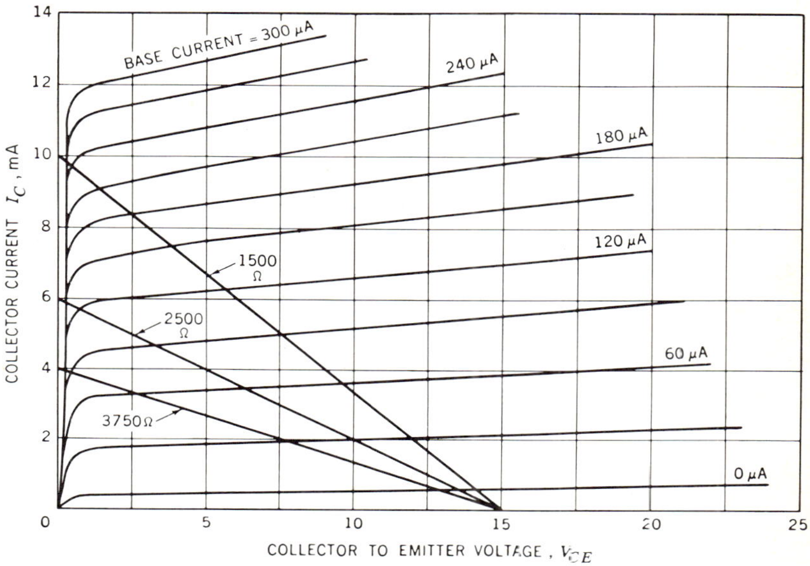

FIGURE 13.6
Three different load lines using the same battery voltage.

the line to encompass as wide an area as possible and this, in turn, means greater output power.

From the foregoing it is seen that there are a number of factors that determine the location of the load line, factors such as the operating point, the maximum collector dissipation permitted, the amount of battery voltage available, and the amount of distortion that can be tolerated. You start with

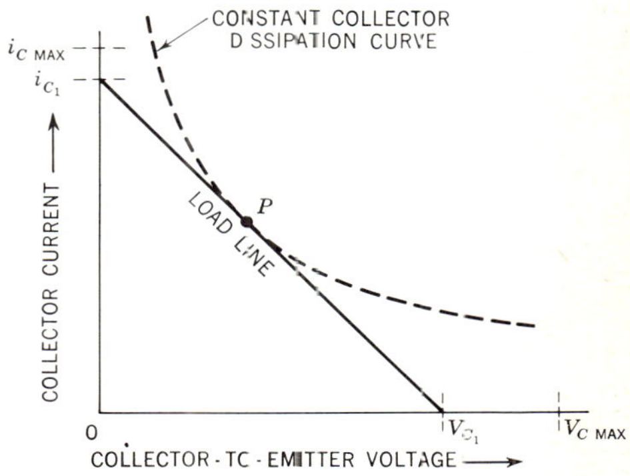

FIGURE 13.7
The condition for maximum power output. The load line is tangent to the maximum-collector-dissipation curve at point *P*. The battery voltage is V_{C1}, and the maximum current that flows with this load line is I_{C1}.

those items that are fixed and adjust the remaining variable factors to fit the specified conditions.

TRANSISTOR BIAS CIRCUITS

Now that we have determined graphically the desired operating conditions for a transistor, let us see what we have to do with respect to the circuit in order to achieve this state. Fig. 13.8*a* is a very simple circuit that will provide the desired conditions. R_L is 1,200 Ω, B_1 is 12 V, B_2 is also 12 V, and R_B is 130,000 Ω. B_2 and R_B are chosen to provide a base current of 90 μA. If we assume that the voltage drop between base and emitter elements in the transistor is negligible compared with the voltage drop across R_B, then

$$R_B = \frac{E}{I_B} = \frac{12}{90 \times 10^{-6}} = 133{,}333\ \Omega$$

The actual value of R_B is 133,333 Ω, but this resistance is close enough to the standard value of 130,000 Ω to permit its use.

A more economical arrangement, using only one battery, is shown in Fig. 13.8*b*. The conditions here are identical to those of Fig. 13.8*a*, and R_B

FIGURE 13.8
Three different biasing methods.

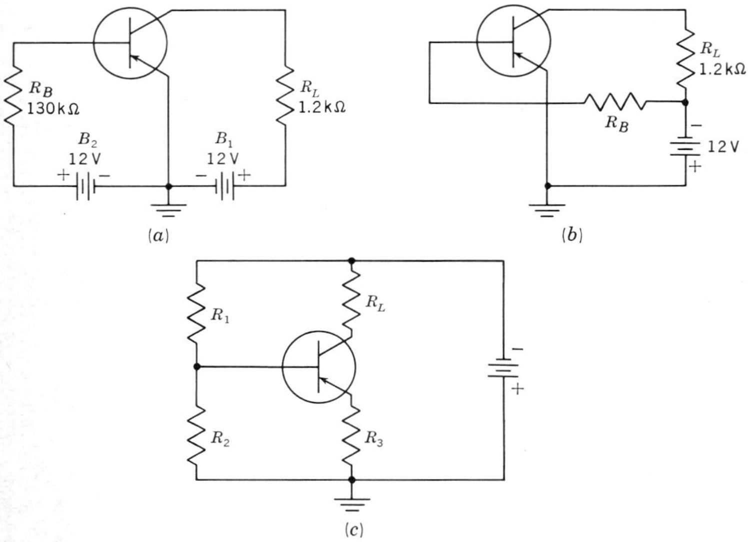

is computed in exactly the same fashion. Still another circuit that is employed more widely than either of the preceding circuits is shown in Fig. 13.8*c*. The base current is supplied by the voltage-divider network of R_1 and R_2. R_L is the collector load resistor and R_3 is an emitter resistor that serves the same purpose as the cathode resistor in a vacuum tube. When this resistor is unbypassed, then it, together with R_L, receives the output signal of the transistor. That is, whatever collector current passes through R_L must also pass through R_3, and the sum of both resistors represents the total load resistance. It is this value which is used to plot the load line.

However, with R_3 unbypassed, degeneration is introduced, and this tends to reduce the gain of the stage. If the degeneration is not desired, then R_3 is bypassed. Now the question is, "Is R_3 still included in the computation of the load line?"

The answer is no if we wish to draw an ac load line in distinction to the dc line. The difference between these lines is this: For various fixed or dc currents, transistor operation will take place along the load line computed using the sum of R_L and R_3. However, if ac signals are applied to the base, then all current variations will be shunted around R_3 by the bypass capacitor, and the corresponding voltage variations will appear only across R_L. Hence, only the value of R_L would be used in computing the ac load-line slope.

Since we are concerned ordinarily with the use of transistors to amplify ac signals, we are more interested in the ac load line than in the dc load line. It is important that this distinction be recognized.

An interesting situation arises when we have an output circuit of the form shown in Fig. 13.9. If we disregard capacitor C_1 and resistor R_2, then the load line will be determined by the value of R_1 alone, in this case, 2,500 Ω. This line, AB, drawn with a 15-V battery, is shown in Fig. 13.10. Q is the quiescent, or operating, point when the input signal is zero.

Now, attach C_1 and R_2. What an ac signal, flowing in the collector circuit, now sees is the parallel combination of R_1 and R_2 (assuming the reactance of C_1 to be negligible with respect to R_2). Hence, a new load line should be drawn representing this new combination. This is CD in Fig. 13.10. Note that this second line is steeper than the preceding load line because the two resistors in parallel possess a total resistance that is less than R_1 alone. Further, the second load line must also pass through point Q because this is the resting position of the transistor when no signal is being received. It is this second

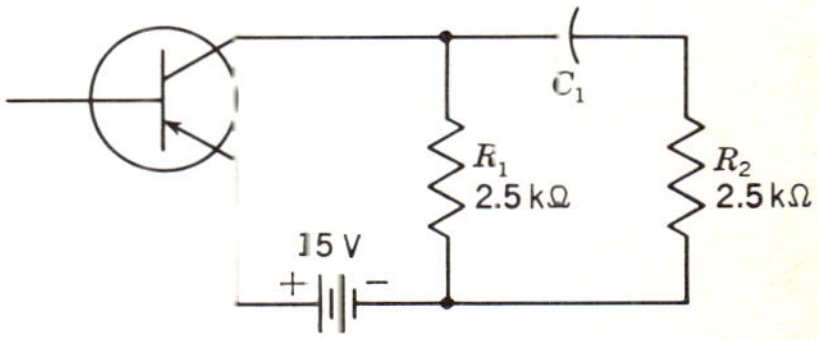

FIGURE 13.9
To draw the ac load lines for this transistor, the effect of C_1 and R_2 must be considered. For the dc load line, only R_1 need be considered.

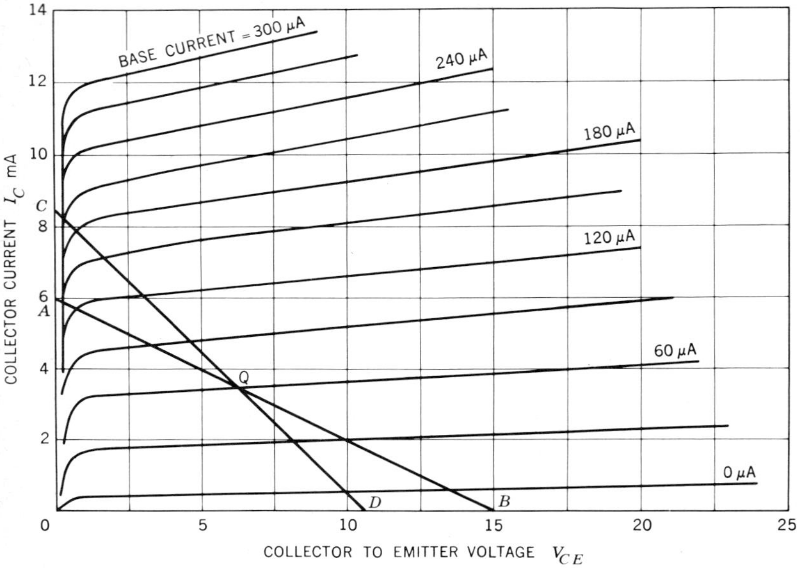

FIGURE 13.10

line, too, that is carefully examined to see if it can accommodate the desired range of input (i.e., base) current without distortion. And this line must not extend into regions where the power (product of collector voltage and current) exceeds the maximum power dissipation of the unit.

DETERMINING THE BEST BIAS ARRANGEMENT

We have just seen several methods of biasing transistor-amplifier circuits in order to obtain a specific operating point and, with it, a desired mode of operation. With so many possible arrangements, it would be logical to ask: Is any one of these better than any other one? And if the answer is yes, how can we determine which circuit is best? It is toward the determination of a suitable answer to these questions that the following discussion is directed.

An important aspect of transistor behavior is its temperature dependence. Increase the temperature of a transistor and it will be found that with all other conditions kept fixed the collector current will increase. This rise stems from two causes: an increase in I_{CO} and a change in the input, or I_BV_{EB}, characteristic. The collector-current rise from the latter source is important when the resistance in the base circuit is low. This occurs in stages that are transformer-coupled, where the base receives its dc bias through the low-resistance winding of the transformer. In RC-coupled stages, the base resistance is fairly high, and here the input characteristic change is not significant in its effect on the collector current. Only changes in I_{CO} need be considered. In the first part of the discussion to follow, we will assume fairly high base resistance and thus consider only the effect that changes in I_{CO} will have on the collector current.

I_{CO} is the current that flows through the collector-base sections when the emitter current is zero. It stems from the presence of minority carriers in the base and collector sections, and it gives rise to a small current when the collector is reverse-biased. I_{CO} is generally below 10 μA, and it is independent of emitter current or collector voltage when the latter is greater than a few tenths of a volt. Because of this it is called a saturation current. The value of I_{CO} is determined chiefly by the particular transistor being used and by the temperature. As a matter of fact, it is extremely sensitive to temperature, actually doubling in value with each 10°C rise in junction temperature. In short order, this can reach a value where it will have a disastrous effect on transistor operation.

To demonstrate this more clearly, let us determine exactly what effect a rise in I_{CO} has on the operation of a common-emitter amplifier. To start, we know that the emitter current is equal to the sum of the base and collector currents. In equation form, this is

$$I_E = I_B + I_C \tag{13.1}$$

The collector current, in turn, is composed of that portion of the emitter current I_E reaching the collector plus the saturation current I_{CO}. Thus,

$$I_C = \alpha I_E + I_{CO} \tag{13.2}$$

Now, if we substitute Eq. (13.1) in (13.2), we obtain

$$\begin{aligned} I_C &= \alpha(I_B + I_C) + I_{CO} \\ &= \alpha I_B + \alpha I_C + I_{CO} \end{aligned}$$

or, rearranging terms,

$$I_C(1 - \alpha) = \alpha I_B + I_{CO}$$

and

$$I_C = \frac{\alpha}{1 - \alpha} I_B + \frac{I_{CO}}{1 - \alpha}$$

$$I_C = \beta I_B + (1 + \beta) I_{CO} \tag{13.3}$$

where

$$\beta = \frac{\alpha}{1 - \alpha}$$

Equation (13.3) tells us a number of things. If we ignore the I_{CO} term, it tells us that for every change in I_B, the effect on the collector current is β times as great. This is the basis for the wide popularity of the common-emitter arrangement. Since β is on the order of 30 or more, a small change in input current applied to the base will cause the collector current to change β times as much, giving us a sizable amplification, or gain, in signal.

Now, consider the I_{CO} factor in Eq. (13.3). As long as I_{CO} is small, even with a multiplying factor of 30 it is still considerably below the level of the normal collector current derived from the emitter. But, with I_{CO} doubling with each 10°C rise in temperature, it can quickly reach a point where the total collector current is appreciably above its normal operating value. This, in turn, will raise the wattage dissipated at the collector and cause the temperature there to increase. With a higher operating temperature, I_{CO} will increase, further raising the collector current and the collector temperature. The process is cumulative, and if no current limiting is provided, the temperature and current at the collector can reach destructive values.

By way of contrast, consider the situation when the common-base arrangement is employed. Under these conditions,

$$I_C = \alpha I_E + I_{CO}$$

A change in emitter current, caused by an input signal, will produce an α change in collector current. Note, however, that I_{CO} stands alone, and even if it doubles with each 10°C rise in temperature, it is still quite small in total value. Consequently, I_C is hardly affected at all. Thus, the common-base arragement is less disturbed by temperature changes than the common-emitter circuit. However, because of the marked gain characteristics of common-emitter circuits, they are used more widely.

Now that the role played by temperature in transistor-circuit operation is understood, the next thing to uncover is which bias arrangement previously shown provides the best defense against such changes. For this determination, a special stability factor S is useful. This is defined as

$$S = \frac{\Delta I_C}{\Delta I_{CO}} = \frac{dI_C}{dI_{CO}} \qquad (13.4)$$

This expression reveals the change in collector current dI_C for a change in collector saturation current dI_{CO}. The lower the value of S, the more stable the arrangement, because changes in I_{CO} have only a small effect on I_C. We can apply Eq. (13.4) to the two situations already discussed, i.e., common base and common emitter. For the former, we saw that

$$I_C = \alpha I_E + I_{CO}$$

from which we obtain

$$dI_C = dI_{CO}$$

or

$$\frac{dI_C}{dI_{CO}} = 1$$

For common-emitter configurations,

$$I_C = \beta I_B + (1 + \beta)I_{CO}$$

and

$$dI_C = (1 + \beta)dI_{CO}$$

or

$$\frac{dI_C}{dI_{CO}} = 1 + \beta \qquad (13.5)$$

Obviously, the latter result possesses a higher value than the former, corroborating the conclusion reached previously, i.e., that the common-base arrangement is less affected by changes in I_{CO} than the common-emitter arrangement.

Note: For those readers who are not familiar with the calculus and differentiation, the following modified procedure will serve equally well. To the original expression

$$I_C = \alpha I_E + I_{CO} \qquad \text{(A)}$$

add a small increment ΔI_C to the left-hand side and a small increment ΔI_{CO} to the right-hand side. This is the mathematical way of stating that increasing I_{CO} by ΔI_{CO} will cause I_C to increase by ΔI_C.

Making the addition just indicated, we obtain

$$I_C + \Delta I_C = \alpha I_E + I_{CO} + \Delta I_{CO} \qquad \text{(B)}$$

Now, substitute Eq. (A) for I_C in Eq. (B):

$$\alpha I_E + I_{CO} + \Delta I_C = \alpha I_E + I_{CO} + \Delta I_{CO}$$

Cancel similar terms on opposite sides of this equation. The result is

$$\Delta I_C = \Delta I_{CO}$$

which is equivalent to

$$dI_C = dI_{CO}$$

as shown above. The same procedure can be followed whenever differentiation is indicated.

To recapitulate the process, whenever you find I_C, substitute $I_C + \Delta I_C$. Whenever I_{CO} appears, substitute $I_{CO} + \Delta I_{CO}$. Then, for I_C, substitute the original equation and cancel out similar terms found on opposite sides of the equals sign. The resulting ratio of ΔI_C to ΔI_{CO} (or dI_C to dI_{CO}) is the stability factor S.

The Stability Factor of Specific Circuits. The foregoing generalized method of determining the thermal stability of a particular network can be applied to specific circuits if we develop the equation for I_C in terms of I_{CO} for the particular amplifier. The object is to obtain as low a stability factor S as possible consistent with the other requirements of the circuit. For example, consider the circuit shown in Fig. 13.11. The expression for I_C in this circuit is

$$I_C = \beta I_B + (1 + \beta)I_{CO}$$

I_B, however, is equal to V divided by R_B, assuming that the voltage drop between base and emitter is close enough to zero to be disregarded. Substituting this into the equation above, we have

$$I_C = \beta \frac{V}{R_B} + (1 + \beta)I_{CO}$$

If, now, we differentiate I_C with respect to I_{CO}, we have

$$S = \frac{dI_C}{dI_{CO}} = 1 + \beta \tag{13.6}$$

The first term after the equals sign, V/R_B, becomes zero, because all its components are fixed and do not change with change in I_{CO}. Thus, the stability factor of Fig. 13.11 is $1 + \beta$, and since β is generally large, the stability of this circuit is quite poor.

An interesting feature of Fig. 13.11 is that the stability (actually, the instability) is governed only by the value of β and not at all by the values of R_B or R_L. While this is generally true, β is somewhat dependent on the choice

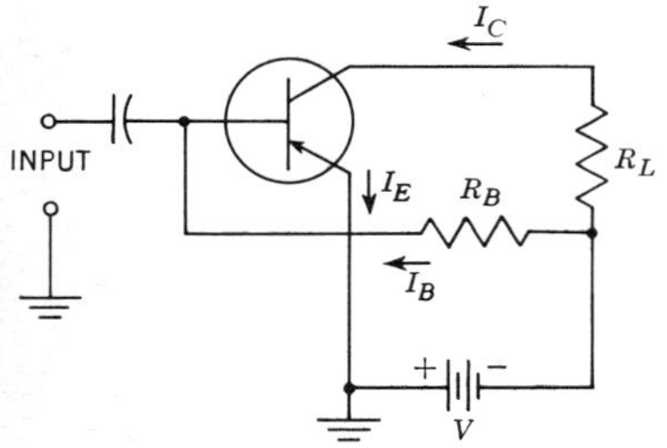

FIGURE 13.11

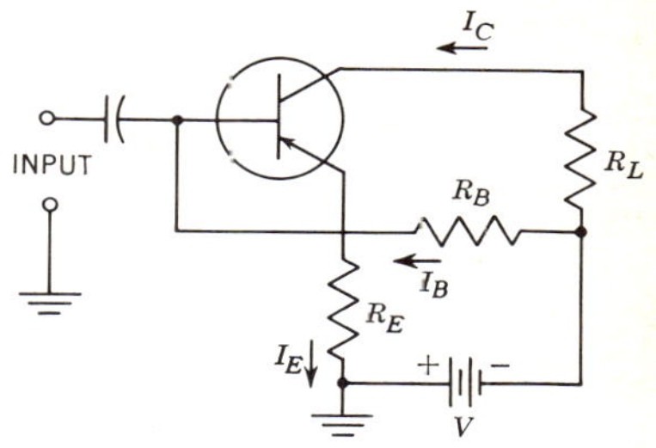

FIGURE 13.12

of the operating point selected for the transistor, so that it will change with different values of R_B and R_L. This is particularly true when the transistor is near cutoff or when the current flow is high.

Let us now obtain the stability factor for the circuit of Fig. 13.12. This is similar to the preceding circuit with the addition of a resistor in the emitter leg. To set up the necessary I_C and I_{CO} expressions, we proceed as follows.

The battery voltage V is equal to the sum of the voltage drops across R_B and R_E (again assuming that the base-emitter voltage drop is zero). Thus,

$$V = I_B R_B + I_E R_E$$

or, transposing,

$$I_B R_B = V - I_E R_E \tag{13.7}$$

The object now is to express I_B in terms of I_C and I_{CO} and to do the same for I_E. This is achieved as follows. I_C, we know, is equal to

$$I_C = \alpha I_E + I_{CO}$$

or

$$I_E = \frac{I_C - I_{CO}}{\alpha} \tag{13.8}$$

This gives us one expression we need. Now, going back to fundamental transistor action,

$$I_E = I_C + I_B$$

Further,

$$I_C = \alpha I_E + I_{CO}$$

We can substitute this expression in the equation just above it. This gives us

$$I_E = \alpha I_E + I_{CO} + I_B$$

or

$$I_B = I_E(1 - \alpha) - I_{CO} \tag{13.9}$$

Now, we substitute expression (13.8) into (13.9):

$$I_B = \frac{I_C - I_{CO}}{\alpha}(1-\alpha) - I_{CO}$$

$$= \frac{I_C}{\beta} - \frac{(1-\alpha)}{\alpha} I_{CO} - I_{CO}$$

$$= \frac{I_C}{\beta} - I_{CO}\left(\frac{1-\alpha}{\alpha} + 1\right)$$

$$= \frac{I_C}{\beta} - \frac{I_{CO}}{\alpha} \qquad (13.10)$$

This is the second equation we need. Now, let us substitute Eqs. (13.8) and (13.10) into Eq. (13.7):

$$\left(\frac{I_C}{\beta} - \frac{I_{CO}}{\alpha}\right) R_B = V - \left(\frac{I_C - I_{CO}}{\alpha}\right) R_E$$

Rearranging terms,

$$I_C\left(\frac{R_B}{\beta} + \frac{R_E}{\alpha}\right) = V + I_{CO}\left(\frac{R_B}{\alpha} + \frac{R_E}{\alpha}\right)$$

Converting β back to its equivalent, $\alpha/(1-\alpha)$, and dividing both sides of the equation by R_B, we obtain

$$I_C\left(\frac{1-\alpha}{\alpha} + \frac{R_E}{\alpha R_B}\right) = \frac{V}{R_B} + I_{CO}\left(\frac{1}{\alpha} + \frac{R_E}{\alpha R_B}\right)$$

$$I_C = \frac{\alpha V/R_B + I_{CO}(1 + R_E/R_B)}{1 - \alpha + R_E/R_B}$$

and

$$S = \frac{dI_C}{dI_{CO}} = \frac{1 + R_E/R_B}{1 - \alpha + R_E/R_B} \qquad (13.11)$$

Here is the stability factor for the circuit of Fig. 13.12. If R_E is set equal to zero, we obtain Eq. (13.6) again, thus verifying the correctness of the derivation. To see what happens as we increase the value of R_E, let us assume a value for α of 0.98 and a value for R_B of 50,000 Ω. R_E, in the first computation, will be 1,000 Ω. Then,

$$S = \frac{1 + 1{,}000/50{,}000}{1 - 0.98 + 1{,}000/50{,}000} = \frac{1.02}{0.04} = 25.5$$

This is already considerably less than $1 + \beta$ for the circuit of Fig. 13.11, assuming that the same transistor is employed. In the latter instance, $1 + \beta = 1 + 48 = 49$.

If we increase R_E to 5,000 Ω in the example above, we obtain

$$S = \frac{1 + 5{,}000/50{,}000}{1 - 0.98 + 5{,}000/50{,}000} = \frac{1.1}{0.12} = 9.1$$

This demonstrates how rapidly S decreases as R_E increases. The reason, of course, is that R_E provides negative feedback. When I_{CO} increases, it produces an amplified current through R_E. This raises the voltage drop across R_E and reduces the voltage across R_B according to Eq. (13.7). Less voltage across R_B means less bias current, and this reduces the collector current. Thus, R_E functions here in much the same way as a cathode resistor in a vacuum-tube circuit.

It may be instructive to examine the stability of one final circuit because of its wide use. This circuit is shown in Fig. 13.13, and it is seen to differ from the circuit of Fig. 13.12 in having a voltage divider, R_1 and R_2, supply the dc voltage (actually, current) to the base. To develop the necessary I_C equation for this arrangement in terms of I_{CO}, let us first note what conditions hold here.

If we assume, as we did before, that there is negligible drop between base and emitter elements of the transistor, then we can write

$$I_E R_E = I_2 R_2$$

Also

$$I_2 R_2 + I_1 R_1 = V$$

$$I_2 + I_B = I_1$$

$$I_1 + I_C = I_3$$

plus the two standbys

$$I_E = I_B + I_C \qquad \text{and} \qquad I_C = \alpha I_E + I_{CO}$$

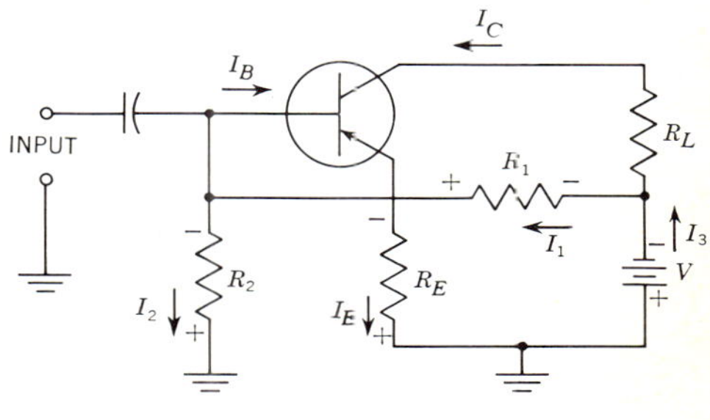

FIGURE 13.13

We have six equations here, and by manipulating them, we can eliminate I_1, I_2, I_3, I_B, and I_E to give us finally the expression

$$I_C\left(\frac{R_E}{\alpha} + \frac{R_E R_1}{\alpha R_2} + \frac{R_1(1-\alpha)}{\alpha}\right) - I_{CO}\left(\frac{R_E}{\alpha} + \frac{R_E R_1}{\alpha R_2} + \frac{R_1}{\alpha}\right) = V$$

Differentiating I_C with respect to I_{CO}, we obtain

$$S = \frac{dI_C}{dI_{CO}} = \frac{R_E/\alpha + R_E R_1/\alpha R_2 + R_1/\alpha}{R_E/\alpha + R_E R_1/\alpha R_2 + (R_1/\alpha)(1-\alpha)}$$

All of the α's in the denominators of the various terms can be canceled out. Furthermore, by dividing numerator and denominator by R_1, we have

$$S = \frac{1 + R_E(1/R_1 + 1/R_2)}{1 - \alpha + R_E(1/R_1 + 1/R_2)} \tag{13.12}$$

If we compare this expression with Eq. (13.11), we see that we have essentially replaced $1/R_B$ in (13.11) by $(1/R_1 + 1/R_2)$. In short, in so far as the circuit of Fig. 13.13 is concerned, R_1 and R_2 act as if they were in parallel with each other.

By substituting appropriate values for R_1 and R_2 in Eq. (13.12), it will be found that S decreases as R_1 and R_2 become less. Offsetting this, of course, is the fact that the lower R_1 and R_2 are made, the greater the current drain on V. Also, R_1 and R_2 are shunted across the input to the stage, and by lowering their overall value, we reduce the input impedance. This could adversely affect the interstage gain.

One further word concerning Eqs. (13.6), (13.11), and (13.12). Each possesses α (or β, which is directly related to α), and hence we can expect a change in stability as α varies. This is an important consideration in present transistor practice because of the fairly wide variation in α among transistors of the same type. It will be found, however, that the better stabilized a circuit is against increases in I_{CO}, the less affected it will be for changes in α or β. This is very fortunate, because variations in α or β present as much difficulty to transistor-amplifier design as instability due to increases in I_{CO}.

The Effect of Input Characteristic Changes. Now that we have seen the effect of I_{CO} variations on the collector current, let us turn our attention to the input circuit. Here we find that a considerable sensitivity to temperature exists, as demonstrated by the curves in Fig. 13.14. Note that if the base-emitter voltage is kept constant, the base current rises quite sharply with temperature. In a common-emitter amplifier, every increase in I_B produces an increase in I_C, which is β times as great. Hence, it would not take a very large increase in I_B

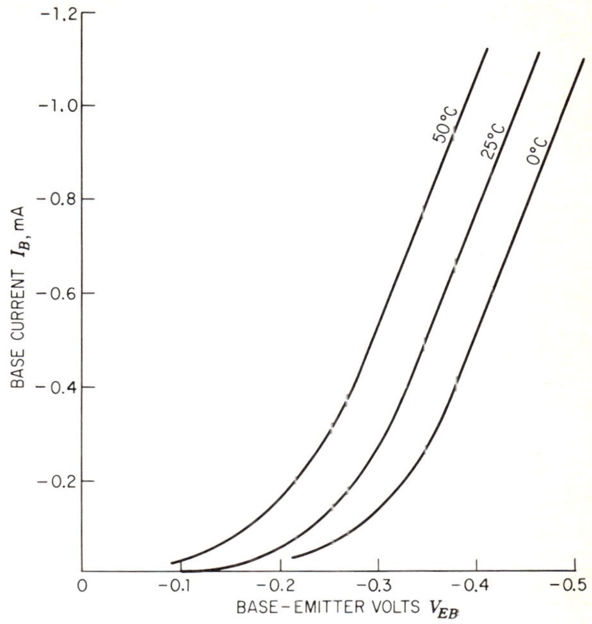

FIGURE 13.14
I_B-V_{EB} curves at different temperatures.

with temperature to alter completely the operating conditions of any transistor circuit.

Now, what the curves in Fig. 13.14 indicate is that as the temperature rises, the base resistance decreases sharply. By maintaining a fixed V_{EB}, we produce a sharply rising I_B. This behavior is not important when a fairly large resistor is employed in the base circuit, because then this external resistor dominates the circuit and maintains a fairly constant dc base current. (Because the base-emitter junction is forward-biased, its resistance is quite low.)

FIGURE 13.15

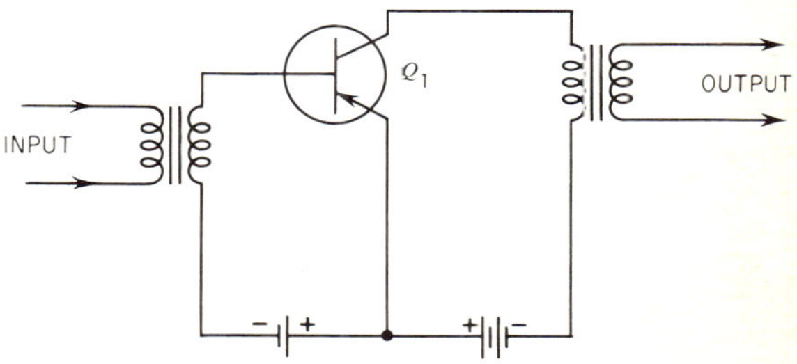

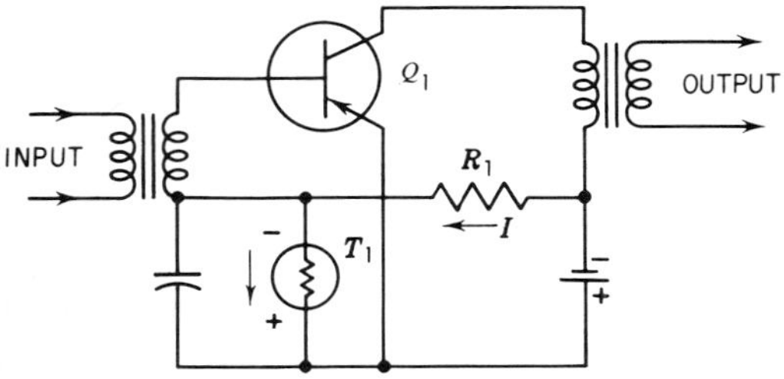

FIGURE 13.16
A thermistor-stabilized transistor amplifier.

Consider, however, the situation when an input transformer is employed (Fig. 13.15). Now, the dc resistance in the external base circuit is low, and changes in the base-emitter resistance with temperature will markedly alter the base current. This, in turn, will affect I_C.

To stabilize I_B under these conditions, either thermistors or compensating diodes have been inserted in the base-emitter circuit. Figures 13.16 and 13.17 demonstrate typical circuits. In Fig. 13.16, the thermistor stabilizes I_B in the following manner. As the temperature rises, the base current also tends to rise. However, the resistance of thermistor T_1 decreases with an increase in temperature, causing more current to flow through it and R_1. The increased current will produce a larger voltage drop across R_1, leaving less battery voltage for T_1. Thus, the available voltage for forward bias, that across T_1, is reduced and this reduces the base current.

In Fig. 13.17, a diode is employed in the same manner, with practically the same results. If the resistance of D_1 can be matched to the base-emitter resistance of Q_1 and if the two vary similarly, I_B can be dc-stabilized over a wide range of temperatures.

TRANSISTOR EQUIVALENT CIRCUITS

In selecting a proper operating point for a transistor, as we just did, we were concerned with the unit solely from a dc point of view. The bias was so

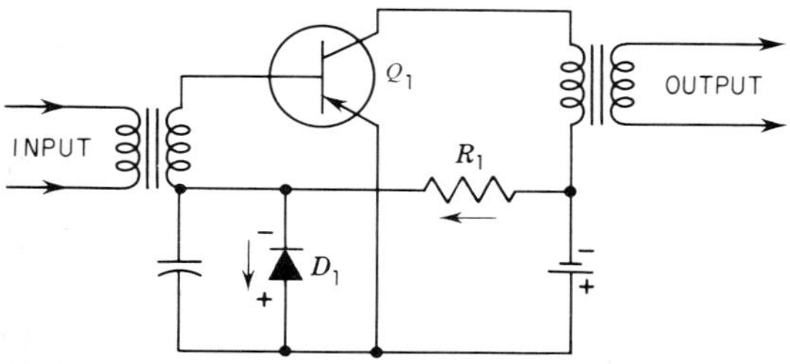

FIGURE 13.17
A diode-stabilized transistor amplifier.

chosen that it would cause the transistor (1) to provide operation as distortion-free as possible, (2) to develop a certain amount of power, or (3) both. We did not know, however, how the circuit would behave over a range of frequencies. Would the gain be the same at 2,000 cycles as it is at 200 cycles? How high could we raise the frequency before the gain became negligible? What effect does the transistor itself have on the frequency response of the circuit? These and other questions show that for a complete picture of transistor-circuit operation at various signal frequencies a more detailed examination is required. Specifically, the information we seek is contained in the equivalent circuit of a transistor. In this section we shall examine such equivalent circuits to see what form they take and how they can be manipulated to provide us with the information we seek.

Vacuum-tube Equivalent Circuits. Prior to the introduction of the transistor, the technique of equivalent circuits was applied most intensively to vacuum-tube circuits, specifically in depicting the vacuum tube as a circuit element. In its most common form, the equivalent circuit for a triode is as shown in Fig. 13.18. The voltage applied to the grid (and cathode) is e_g, which is arbitrarily taken at the instant the grid is positive with respect to the cathode. With this voltage at the grid, an amplified version of the input, μe_g, appears in the plate circuit. Furthermore, μe_g possesses the polarity shown, indicating that when the grid is driven in the positive direction, the plate tends to go in the negative direction. (This, incidentally, is all that the two sets of polarity signs mean.)

In Fig. 13.18, r_p represents the internal plate-cathode resistance of the tube, i_p is the plate current, and e_p is the output voltage, the voltage that would be applied across a load resistor. If we were to write the governing equation for this circuit, it would be

$$e_p + r_p i_p = \mu e_g$$

FIGURE 13.18
The equivalent circuit of a triode; r_p is the internal plate resistance of the tube, and e_g is the ac applied signal.

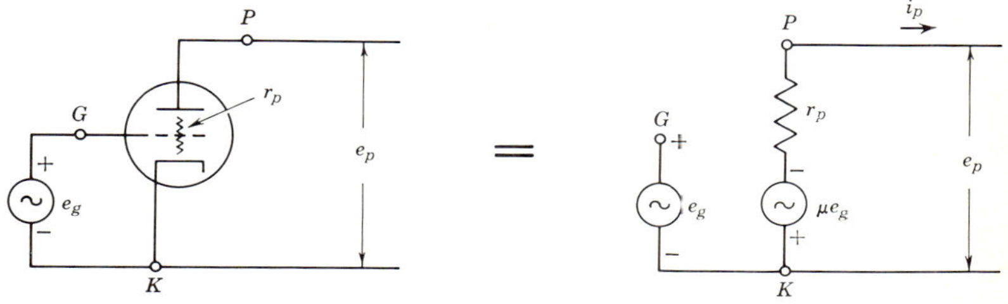

If a load resistor is connected between plate and cathode, e_p will be replaced by i_pR_L, but this is of only incidental importance. The equation is the significant item, since it enables us to deal with a tube in a rather simple manner.

Now, in drawing the equivalent circuit of the tube, as we do in Fig. 13.18, one tacit assumption is made. That is, we assume that while a variable voltage on the control grid produces a voltage in the plate circuit, the converse is not true. That is, if we start with a variable voltage in the plate circuit, no voltage will appear in the grid circuit. In short, there is no internal feedback path in the tube. This agrees with our experience with tubes at low and medium frequencies, but it is not completely true at high frequencies. At high frequencies, some energy does feedback, and we can say that now the tube possesses inverse (or reverse) amplification. It will be found that the value of this inverse amplification is less than 1, and it is generally not desirable. Still, when it exists, it is just as real as the normal forward amplification factor μ, and recognition is taken of this by assigning it the special symbol of μ_r (r stands for reverse).

It is obvious that when μ_r is not zero, the equivalent diagram of Fig. 13.18 no longer holds, since it makes no provision for μ_r. To represent the new conditions, the equivalent circuit of Fig. 13.19 can be drawn. The right-hand side of the circuit remains the same, since nothing has changed there. At the left, however, we now have r_g, representing the grid-to-cathode resistance, and $\mu_r e_p$, representing the voltage that is transferred from the plate to the grid. (r_g is present in the circuit of Fig. 13.18 also and, if shown, would be in parallel with e_g. However, since it does not enter into the governing equation for this equivalent circuit, it was omitted.)

To describe completely this new equivalent circuit mathematically, two equations are needed. One equation, that for the output circuit, remains unaltered, since the circuit is the same. That is,

$$\mu e_g = i_p r_p + e_p$$

The second equation represents the input circuit and its form is

$$e_g = i_g r_g + \mu_r e_p$$

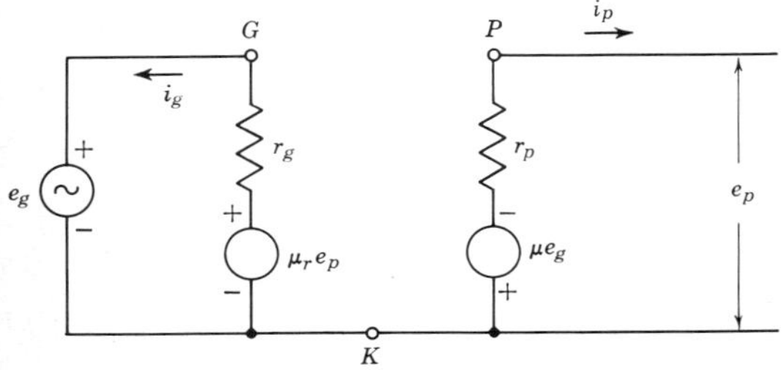

FIGURE 13.19
The equivalent circuit of a triode vacuum tube when signal feedback exists between plate and grid elements.

Thus, we now have two equations where before we had one, and the second equation stems solely from the fact that energy (i.e., a signal) is fed from the plate back to the grid. The tube is no longer a unilateral device. Current can flow in both directions, although not with equal facility.

Before we continue, a word should be said of the fact that dc voltages are never shown in equivalent circuits. In an actual circuit, dc voltages are needed to bias the tube to the desired operating point. There is a dc voltage in the grid-to-cathode circuit and a dc voltage in the plate-to-cathode circuit. However, once the tube is placed at the proper operating point, we are interested only in its behavior to ac signals; thus, only ac voltages are shown in the equivalent circuit. Dc voltages are always understood as being in the background, ready to be utilized once the equivalent circuit has yielded the desired information and the actual circuit is to be designed.

Transistor Equivalent Circuits. Now that we have examined the equivalent circuit of a tube, let us turn our attention to the transistor. We have already learned that a transistor, like a tube, provides gain and that in a transistor there is internal feedback from the collector to the emitter. (This is demonstrated graphically on page 79 for a junction transistor, where it is shown how the input impedance varies as the load impedance changes.) Since such feedback exists, the equivalent circuit best suited for transistors should resemble the circuit of Fig. 13.19 more closely than the circuit of Fig. 13.18. This turns out to be correct, because the equivalent circuit of Fig. 13.20 is eminently suitable for transistors.

In Fig. 13.20 e_e represents the ac input signal applied between emitter and base. e_c is the output voltage, and it would appear across a load resistor connected between collector and base. $r_{12}i_c$ is the voltage (because $e = ir$) transferred from the output circuit to the input circuit. $r_{21}i_e$ is the signal voltage appearing in the output circuit because of a voltage applied to the input circuit.

To set up the equations for the circuit of Fig. 13.20, each branch is considered separately. Thus, for the left-hand branch, we have

$$e_e = r_{11}i_e + r_{12}i_c \qquad (13.13)$$

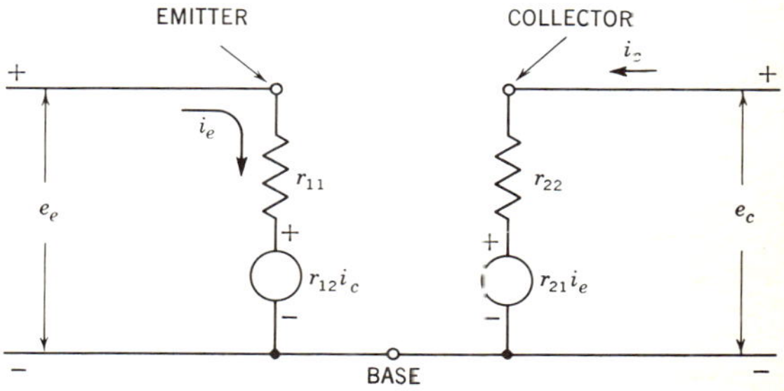

FIGURE 13.20
Equivalent circuit of a transistor. This is one of several forms which will be discussed.

And for the right-hand side, the equation is

$$e_c = r_{21}i_e + r_{22}i_c \tag{13.14}$$

Here, then, are the equations that govern the behavior of the transistor equivalent circuit, and if this combination correctly represents an actual transistor, manipulation of the two equations will tell us what we need to know about the workings of a transistor when a signal is applied. We are considering the operation of a transistor at low and medium frequencies only in this discussion. For high-frequency operation, emitter and collector shunting capacitances have to be added, along with several other components, and the analysis becomes more involved. A discussion of the equivalent circuit of a transistor at high frequencies is given at the end of this chapter.

The next step is to determine what the various items in these two equations represent. i_e, the current in the left-hand circuit, stands for the emitter current. By the same token, i_c is the collector current. If we temporarily set i_c equal to zero in Eq. (13.13), indicating that the output circuit is open, then

$$e_c = r_{11}i_e + 0$$

or

$$r_{11} = \frac{e_e}{i_e} \tag{13.15}$$

If i_e is the input current and e_e the input voltage, then e_e divided by i_e represents the input resistance r_{11}.

If we now permit i_c to flow but prevent i_e from flowing by opening the input circuit, Eq. (13.13) becomes

$$e_e = 0 + r_{12}i_c$$

or

$$r_{12} = \frac{e_e}{i_c} \tag{13.16}$$

This equation tells us that a current i_c flowing in the output circuit produces a voltage in the input circuit. r_{12} is the feedback or reverse impedance common to both input and output circuits, and it is the voltage drop developed by i_c flowing through r_{12} that produces the feedback voltage in the input circuit.

Now, turning to Eq. (13.14), let us consider it when the output circuit is open and i_c is equal to zero. Substituting this value of i_c in (13.14), we obtain

$$e_c = r_{21}i_e + 0$$

or

$$r_{21} = \frac{e_c}{i_e} \tag{13.17}$$

Equation (13.17) reveals that even though the output circuit is open, a voltage

e_c equal to $r_{21}i_e$ is present there. Since i_e is the input current, it means that r_{21} is common to both input and output circuits; and when i_e flows, the voltage drop across r_{21} caused by i_e appears in the output circuit. r_{21} could thus be called the forward-transfer impedance. Note that if r_{21} did not exist, an incoming signal could never reach the output circuit. Thus, r_{21} is the element through which the input circuit affects the output circuit.

For the final step, the input circuit is opened, reducing i_e to zero. When this substitution is made in Eq. (13.14),

$$e_c = 0 + r_{22}i_c$$

or

$$r_{22} = \frac{e_c}{i_c} \tag{13.18}$$

That is, r_{22} is the output impedance or, here, the impedance of the collector circuit. Note that the e_c value in (13.18) would not be the same as the e_c value in (13.17), since the equations are obtained under different conditions. The same is true of e_e in Eqs. (13.15) and (13.16).

As an interesting sidelight, observe that r_{12} and r_{21} are not equal to each other. In a passive network, such as that formed by resistances, inductances, and capacitances only, the impedance in one direction would be the same as the impedance in the opposite direction. In a circuit containing a nonlinear device such as a tube or a transistor, this is not true, a fact we know from our own experience.

Second Transistor Equivalent Circuit. The equivalent circuit of Fig. 13.20 represents the transistor electrically, but it is not the only equivalent circuit. (In similar fashion, Fig. 13.19 is not the only equivalent circuit of a vacuum tube in which feedback from the output to the input circuit is taken into account. For the present discussion, it is the most convenient and, for that reason, it is used.) There are a number of others, one of the most widely used being shown in Fig. 13.21. This circuit is popular because its electrical form ties in closely with the three sections of transistor. For example, r_e is the

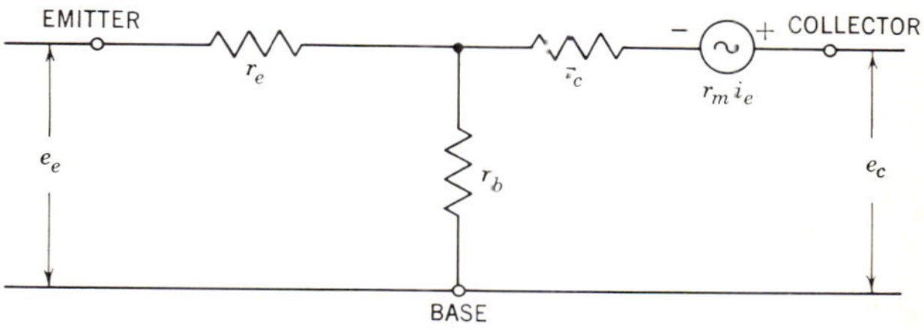

FIGURE 13.21
Another equivalent circuit of a transistor.

resistance of the emitter section, r_b is the resistance of the base section, and r_c is the resistance of the collector section; $r_m i_e$ is a generator placed in the collector arm to represent the effect of the input circuit on the output circuit; and r_m is the mutual resistance between the input and output circuits by which the input current i_e produces a signal in the output. Here this output signal is called $r_m i_e$. It is essentially equivalent to $r_{21} i_e$ of Fig. 13.20, differing from r_{21} only by the value of the base resistance r_b. (r_{21}, however, is so much greater than r_b that the latter may be disregarded in any equation involving r_{21} and r_b.) By the same token, we make provision for the feedback of signal from output to input by the presence of the common resistor r_b. The backward voltage transferred by r_b is equivalent to the generator $r_{21} i_c$ in Fig. 13.20. By separating the voltage that the input delivers to the output from the voltage fed in the opposite direction, we tacitly recognize that they are unequal in size and, furthermore, that they may vary in different fashions. This, of course, again jibes with our experience of tube and transistor circuits. Thus, the form of the equivalent circuit in Fig. 13.21 conforms to the general requirements of a transistor as revealed by what we know about it.

Now, if this equivalent circuit and the preceding one represent the same transistor, there should exist an electrical correspondence between them. To ascertain what this similarity is, let us first set up the equations for the circuit of Fig. 13.21 and then equate them properly to the two equations governing the preceding equivalent circuit.

The second equivalent circuit has been redrawn in Fig. 13.22, this time divided into two loops by the two arrows. The arrows show the assumed directions of flow for current i_e and current i_c. Before we write out the equation for the left-hand and right-hand sides of the circuit, it should be noted that the following rules are being applied.

1. When current flows through a resistor, the end entered first is given a positive sign, while the exit end becomes negative. This is the sign notation associated with the conventional flow of current (i.e., from positive to negative) in distinction to the electron flow. The conventional notation is being employed in the discussion because it is widely used in the literature and will enable the reader to follow other presentations more easily.
2. When there is a generator or a source of voltage in the circuit, the sign to be given this item will depend on its polarity. For example, if the loop arrow encounters the positive side of the generator first, this voltage is added to the loop equation with a plus sign. However, if the arrow meets the negative sign first, the quantity is added to the equation preceded by a negative sign.
3. We will use Kirchhoff's law, which states that the sum of the voltages in a closed loop adds up to zero. Previously, we had applied Kirchhoff's other law, which says that the sum of all the currents flowing toward a point is zero.

4. When two currents flow through the same element, such as r_b, the effect of each current is considered separately. If the currents flow in the same direction through the component, their voltage drops add; if they flow in opposite directions, the voltage drops subtract.

With these rules, we are ready to start the analysis. Consider loop I first. Current i_e flows through resistors r_e and r_b, producing voltage drops across each. i_c also flows through r_b, and since its direction is the same as i_e, their voltage drops add. Finally, i_e meets the negative terminal of e_e first; hence, this voltage is given a negative sign in the equation. Thus

$$r_e i_e + (i_e + i_c)r_b - e_e = 0 \tag{13.19}$$

In the second loop, conditions are somewhat the same except for the presence of $r_m i_e$. This is a voltage generator; furthermore, the arrow encounters the plus sign first. Consequently, $r_m i_e$ is added to the equation bearing a positive sign. Thus

$$r_m i_e + r_c i_c + (i_c + i_e)r_b - e_c = 0 \tag{13.20}$$

Now, let us reshuffle the items in Eqs. (13.19) and (13.20) so that e_c and e_e are each on the other side of the equals sign and all the currents are properly segregated. Doing this, we obtain for Eq. (13.19)

$$e_e = (r_e + r_b)i_e + r_b i_c$$

and for Eq. (13.20)

$$e_c = (r_m + r_b)i_e + (r_c + r_b)i_c$$

Here are the two governing equations for the equivalent circuit of Fig. 13.22. If these are compared with Eqs. (13.13) and (13.14), they are seen to possess the same form with the only difference being the resistance factors of i_e and i_c. And since the circuit of Fig. 13.22 represents the same transistor as the circuit of Fig. 13.21, their governing equations should be equivalent to each other. This will occur when

$$r_{11} = r_e + r_b \tag{13.21}$$

$$r_{12} = r_b \tag{13.22}$$

$$r_{21} = r_m + r_b \tag{13.23}$$

$$r_{22} = r_c + r_b \tag{13.24}$$

These are the relationships between r_{11}, r_{12}, r_{21}, and r_{22} in one equivalent

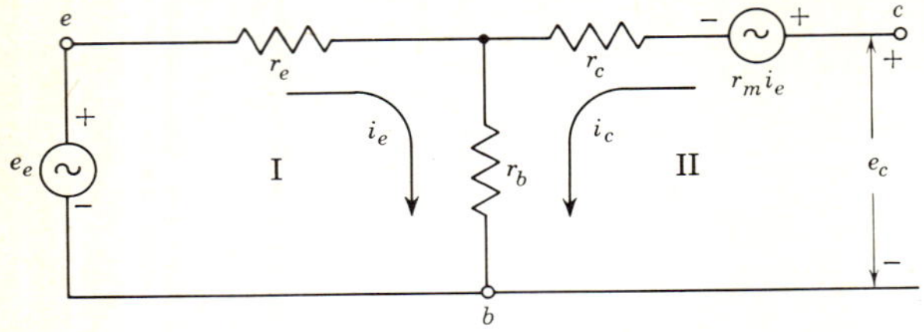

FIGURE 13.22

transistor circuit and r_e, r_b, r_c, and r_m in the other equivalent circuit. Sometimes, the same information is given in the following form:

$$r_e = r_{11} - r_{12} \tag{13.25}$$

$$r_b = r_{12} \tag{13.26}$$

$$r_c = r_{22} - r_{12} \tag{13.27}$$

$$r_m = r_{21} - r_{12} \tag{13.28}$$

These relationships follow directly from the preceding set by a slight manipulation.

Typical values for r_e, r_b, r_c, and r_m in commercial transistors fall within the following ranges:

$$r_e = 20 \text{ to } 30\ \Omega$$

$$r_b = 400 \text{ to } 700\ \Omega$$

$$r_c = 1.0 \text{ to } 2.0 \text{ megohms } (\text{M}\Omega)$$

$$r_m = 0.95 \text{ to } 1.9\ \text{M}\Omega$$

We are now ready to examine the behavior of the equivalent circuit of Fig. 13.22 with the addition of a load resistor across the output and a signal source across the input. We shall do this briefly with the common-base equivalent circuit (Fig. 13.22) and then swing over to an equivalent circuit for a common-, or grounded-, emitter arrangement.

Figure 13.23 is the circuit we shall use. A load resistor R_L has been placed across the output terminals, and a signal generator e_g and its internal resistance R_g have been connected across the input terminals. In accordance with conventional practice, the common base has been grounded, but whether this point is actually so connected in the circuit or not will have no effect on the derivation.

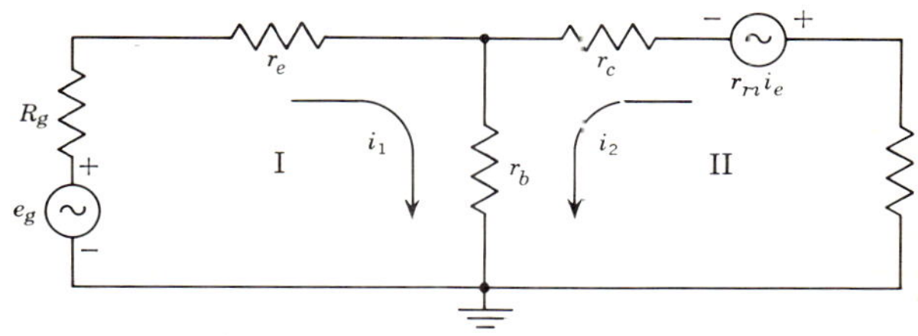

FIGURE 13.23

Proceeding as we did when Eqs. (13.19) and (13.20) were set up, we can write for loop I,

$$-e_g + R_g i_1 + r_e i_1 + r_b(i_1 + i_2) = 0 \tag{13.29}$$

Rearranging terms,

$$e_g = (R_g + r_e + r_b)i_1 + r_b i_2$$

For loop II, we have

$$i_2 R_L + r_m i_e + r_c i_2 + r_b(i_1 + i_2) = 0$$

The term i_e represents the current in the emitter circuit. In the present instance, i_1 is the emitter-circuit current and consequently $i_1 = i_e$. Making this substitution in the last equation and rearranging terms, we have

$$0 = (r_m + r_b)i_1 + (R_L + r_c + r_b)i_2 \tag{13.30}$$

We now have two variables, i_1 and i_2, and two simultaneous linear equations. Hence, we can solve them for i_1 and i_2. Doing this gives us

$$i_1 = \frac{e_g(R_L + r_c + r_b)}{(R_L + r_c + r_b)(R_g + r_e + r_b) - r_b(r_b + r_m)} \tag{13.31}$$

$$i_2 = \frac{-e_g(r_b + r_m)}{(R_L + r_c + r_b)(R_g + r_e + r_b) - r_b(r_b + r_m)} \tag{13.32}$$

These equations can be employed directly to provide the current gain for a grounded-base amplifier. This is denoted by the symbol α and is equal to the ratio of i_2 to i_1. Thus, dividing (13.32) by (13.31), we have

$$\text{Gain} = \frac{i_2}{i_1} = \frac{r_b + r_m}{R_L + r_c + r_b}$$

The gain, then, is dependent upon the load resistance, becoming smaller as R_L becomes larger. The maximum current gain is achieved when R_L is set equal to zero (i.e., a short circuit across the output). Under these conditions

$$\alpha = \frac{i_2}{i_1} = \frac{r_b + r_m}{r_c + r_b} \tag{13.33}$$

Since r_b is numerically much smaller than either r_m or r_c, Eq. (13.33) reduces to

$$\alpha \cong \frac{r_m}{r_c} \tag{13.34}$$

Numerical values of α vary from 0.95 to 0.99 for commercial junction transistors. This information is of course supplied by the manufacturer, although if the values for r_m and r_c are given, α could be computed. Additional manipulations could be performed with i_1 and i_2 to obtain other information concerning the grounded-base amplifier, but rather than spend time on a circuit that is not widely used, let us turn to the equivalent circuit of a common-emitter amplifier and examine it in detail.

COMMON-EMITTER AMPLIFIERS

To obtain the equivalent circuit of the common-emitter arrangement, we employ the circuit of Fig. 13.23, changing only the relative positions of the base and emitter arms (Fig. 13.24). In all other respects, the two circuits are alike.

We would proceed as before to set up the equations for the two loops of Fig. 13.24. Thus, for loop I we have

$$-e_g + R_g i_1 + r_b i_1 + r_e(i_1 + i_2) = 0 \tag{13.35}$$

FIGURE 13.24
Equivalent circuit of a transistor connected with its emitter common to both input and output circuits.

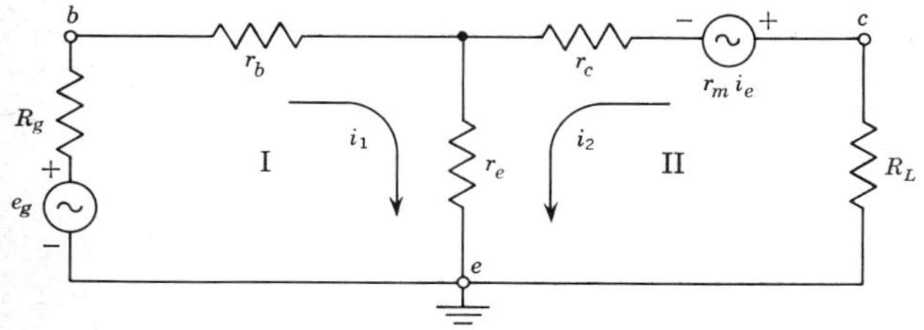

For loop II,

$$R_L i_2 + r_m i_e + r_c i_2 + r_e(i_1 + i_2) = 0 \tag{13.36}$$

Before we consolidate each equation, the question of $r_m i_e$, specifically i_e, must be resolved. In Fig. 13.23, i_e was seen to be equal to i_1 because only i_1 passed through r_e. Furthermore, i_1 passed through r_e going toward the junction of r_e and r_b. The latter point is important because direction must be maintained; otherwise the signs between i_e and i_1 will require reversal.

Now, examine Fig. 13.24. Not only does i_1 pass through r_e but i_2 also passes through r_e. Consequently, i_e is now equal to $i_1 + i_2$. With respect to polarity, it is seen that for i_e to remain unchanged, it should be traveling through r_e toward r_b, as in Fig. 13.22. i_1 and i_2, however, are traveling in the opposite direction. Hence, $i_e = -(i_1 + i_2)$, and this is the substitution to be made for i_e in Eq. (13.36). Thus

$$R_L i_2 - r_m(i_1 + i_2) + r_c i_2 + r_e(i_1 + i_2) = 0 \tag{13.37}$$

Now segregate similar terms in Eqs. (13.35) and (13.37).

$$e_g = (R_g + r_b + r_e)i_1 + r_e i_2 \tag{13.38}$$

and

$$0 = (r_e - r_m)i_1 + (R_L + r_c + r_e - r_m)i_2 \tag{13.39}$$

Solving these equations simultaneously for i_1 and i_2, we obtain

$$i_1 = \frac{e_g(R_L + r_c + r_e - r_m)}{(R_L + r_c + r_e - r_m)(R_g + r_b + r_e) + r_e(r_m - r_e)} \tag{13.40}$$

$$i_2 = \frac{e_g(r_m - r_e)}{(R_L + r_c + r_e - r_m)(R_g + r_b + r_e) + r_e(r_m - r_e)} \tag{13.41}$$

Current Gain. Let us determine next the current gain for the common-emitter arrangement, just as we did previously for the common-base circuit.

$$\text{Current gain} = \frac{i_2}{i_1} = \frac{r_m - r_e}{R_L + r_c + r_e - r_m} \tag{13.42}$$

While this expression differs from that of Eq. (13.34), we see that here, too, current gain will vary with load resistance and in the same manner as before. For maximum current gain, R_L is set equal to zero, giving

$$\text{Current gain} = \frac{r_m - r_e}{r_c + r_e - r_m} \tag{13.43}$$

Divide numerator and denominator by r_c.

$$\text{Current gain} = \frac{r_m/r_e - r_e/r_c}{1 + r_e/r_c - r_m/r_c} \tag{13.44}$$

Numerically, r_e/r_c is very small compared with r_m/r_c and could be disregarded. Hence

$$\text{Current gain} = \frac{r_m/r_c}{1 - r_m/r_c} \tag{13.45}$$

But r_m/r_c is equal to α, as evidenced by Eq. (13.34). Consequently,

$$\text{Current gain} = \frac{\alpha}{1-\alpha} = \beta \tag{13.46}$$

This, of course, agrees with the definition given for the current gain of a common-emitter circuit, and here we see how it is actually derived. Values of β for most commercial transistors fall generally between 30 and 50, but units with higher values are available.

Input Impedance. Another result that we can derive rather easily from Eqs. (13.40) and (13.41) is the input impedance of the circuit of Fig. 13.24. As a first step, it should be recognized that R_g in Fig. 13.24 is not part of the input impedance. Consequently, if we divide e_g by i_1 (giving us the total impedance seen by the generator) and subtract R_g from this, we shall obtain the value we seek. In equation form, this is

$$R_{\text{in}} = \frac{e_g}{i_1} - R_g \tag{13.47}$$

Substituting the expression for i_1 from (13.40) in (13.47), we have

$$\begin{aligned} R_{\text{in}} &= \frac{e_g}{\dfrac{e_g(R_L + r_c + r_e - r_m)}{(R_L + r_c + r_e - r_m)(R_g + r_b + r_e) + r_e(r_m - r_e)}} - R_g \\ &= \frac{(R_L + r_c + r_e - r_m)(R_g + r_b + r_e) + r_e(r_m - r_e)}{(R_L + r_c + r_e - r_m)} - R_g \end{aligned} \tag{13.48}$$

$$R_{\text{in}} = r_b + r_e + \frac{r_e(r_m - r_e)}{R_L + r_c + r_e - r_m} \tag{13.49}$$

If we neglect r_e with respect to r_m in the numerator of the last term, we obtain

$$R_{\text{in}} = r_b + r_e + \frac{r_e r_m}{R_L + r_c + r_e - r_m} \tag{13.50}$$

Again, we see quite clearly that R_L plays a part in determining the input resistance of the transistor. Graphically, this is demonstrated on page 79. Note, too, that the highest input impedance is achieved when R_L is zero. Thereafter, R_{in} decreases as R_L rises, although after R_L reaches 1 MΩ or so, R_{in} remains fairly constant for all practical purposes.

It is interesting to note that when R_L in Eq. (13.49) or (13.50) is small compared with r_c (which is generally the case), the input impedance equation reduces to

$$R_{\text{in}} = r_b + r_e + \beta r_e$$

using Eq. (13.45). Combining terms, we have

$$R_{\text{in}} = r_b + (\beta + 1)r_e$$

or

$$\cong r_b + \beta r_e$$

because β is so much greater than 1. This relationship reveals that the input resistance for the common-emitter configuration is approximately β times as large as the input resistance of the common-base arrangement. The latter is $r_b + r_e$.

Output Impedance. We can derive an expression for the output impedance of the equivalent circuit by utilizing essentially the same techniques as above, modified only for the right-hand side of the circuit instead of the left. To see how this is done, consider the circuit of Fig. 13.25. This is identical to the circuit of Fig. 13.24, except that e_g has been moved from the input to the output circuit In this move, R_g was left untouched so that its effect on the circuit would remain the same. R_L has also been left unchanged, but now, because it is in series with e_g, it assumes the role of the generator impedance. The output impedance of this circuit is the impedance seen by e_g and R_L looking back into the network. Thus, all we need do is find i_A and then apply it in the equation.

$$R_{\text{out}} = \frac{e_g}{i_A} - R_L \tag{13.51}$$

Note that we have altered the labeling of the preceding i_1 and i_2 currents to i_B and i_A. This is because the equations to be set up differ from the preceding equations and consequently the two currents will have other values.

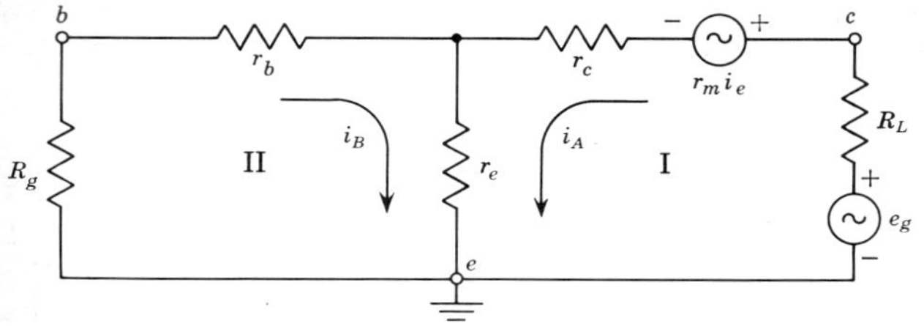

FIGURE 13.25
The equivalent circuit arrangement for determining the output impedance of a transistor.

Following the same setup procedure as we did for Fig. 13.24, the circuit equations obtained for loops I and II are

$$i_B(R_g + r_b + r_e) + i_A r_e = 0 \tag{13.52}$$

$$i_B(r_e - r_m) + i_A(R_L + r_e + r_c - r_m) - e_g = 0 \tag{13.53}$$

Solving these simultaneously for i_A, we obtain

$$i_A = \frac{e_g(R_g + r_e + r_b)}{(R_g + r_b + r_c)(R_L + r_e + r_c - r_m) - r_e(r_e - r_m)} \tag{13.54}$$

Substituting this equation in (13.51),

$$R_{\text{out}} = \frac{e_g}{\dfrac{e_g(R_g + r_e + r_b)}{(R_g + r_b + r_e)(R_L + r_e + r_c - r_m) - r_e(r_e - r_m)}} - R_L \tag{13.55}$$

$$R_{\text{out}} = R_L + r_e + r_c - r_m - \frac{r_c(r_e - r_m)}{R_g + r_e + r_b} - R_L \tag{13.56}$$

$$R_{\text{out}} = r_e + r_c - r_m + \frac{r_e r_m - r_e^2}{R_g + r_e + r_b} \tag{13.57}$$

The presence of R_g in the formula indicates that changes in R_g will affect the output impedance. Thus, a transistor exhibits effective bilateral behavior, with input and output impedances reacting on each other. Incidentally, if we examine the various terms in (13.57), we see that r_e and r_c will always be greater than r_m and that $r_e r_m$ will always be greater than r_e^2, so that the entire expression can never become negative, no matter what value R_g is.

Equation (13.57) also reveals an interesting fact if we assume that R_g is very large compared with r_e and r_b, a situation that is generally true. Under this condition

$$\frac{r_e r_m - r_e^2}{R_g + r_e + r_b}$$

can be ignored, since a large value of R_g reduces the expression to a low value. Equation (13.57) then becomes

$$R_{\text{out}} = r_e + r_c - r_m \tag{13.57a}$$

However, from Eq. (13.34), we have

$$\alpha \cong \frac{r_m}{r_c}$$

Substituting this into (13.57a) gives

$$\begin{aligned} R_{\text{out}} &= r_e + r_c - \alpha r_c \\ &= r_e + r_c(1 - \alpha) \end{aligned} \tag{13.57b}$$

The expression $r_c(1 - \alpha)$ can be readily shown to equal

$$\frac{r_c}{1 + \beta}$$

Substituting this into (13.57b), we have

$$R_{\text{out}} = r_e + \frac{r_c}{\beta + 1}$$

or

$$\cong r_e + \frac{r_c}{\beta} \tag{13.57c}$$

Since it can be shown that the output impedance of a common-base amplifier is r_c, Eq. (13.57c) reveals that the output impedance of a common-emitter stage is approximately β times *lower*.

The foregoing analysis has demonstrated how the equivalent circuit is established and has detailed some of the information that can be derived from it. This does not exhaust the possibilities by any means; for those readers interested in pursuing the subject further, there are a number of engineering texts available, in addition to various technical magazines in which many articles will be found.

HYBRID PARAMETERS AND THEIR EQUATIONS

In the preceding development of the transistor equivalent circuit, the internal structure was represented by resistances r_e, r_b, r_c, and r_m. (Actually, to be more general, impedance symbols Z_e, Z_b, Z_c, and Z_m should have been used. However, for low and medium frequencies, the internal structure can be characterized by resistances. Under this assumption, the bandpass of a transistor amplifier is limited only by the external coupling circuit. This is made evident by the fact that none of the formulas derived above had any frequency-dependent terms in them.) To the circuit designer, the value of these resistances for each different transistor he deals with is important and, ideally, they should be values that can be measured readily with equipment that is not too complex. This is not possible, however, with some of the measurements that have to be made. To see why, let us determine what must be done to measure each parameter. For this discussion, we shall use the equivalent circuit of Fig. 13.20 and its components, r_{11}, r_{12}, r_{21}, and r_{22}. This will cause no difficulty because once these values are obtained, they can be made to yield r_e, r_b, r_c, and r_m by Eqs. (13.21) to (13.24).

In Fig. 13.20 the value of r_{11} is obtained by setting i_c equal to zero in Eq. (13.13) and then measuring the current through the input circuit and the voltage across it. The chief difficulty here lies in making a suitable measurement with $i_c = 0$. This is because all parameter measurements on a transistor should be made with the normal operating dc bias voltages applied. To achieve $i_c = 0$, the collector circuit should be opened for ac signals and closed and complete for dc voltages. This condition can be met by feeding the dc collector voltage to its element through a parallel resonant circuit that would possess negligible dc resistance but a very high ac impedance. Because the collector resistance itself is high, making the output impedance high, it can be seen that the Q of the resonant circuit would have to become impractically high in order to approximate the open circuit needed for the measurement.

The same difficulty arises again for the measurement of r_{21}, because here, too, i_c must be reduced to zero. Obviously, the equivalent circuit, formed by using the resistance parameters, lends itself well to design analysis but not to the measurement of those values on which the design computations would depend.

As a solution to this problem, other forms of equivalent circuit arrangements have been fashioned. For example, there is a configuration that employs conductances and current generators in place of resistances and voltage generators. Such an arrangement is shown in Fig. 13.26, together with the equations that govern it. If this configuration appears strange to the reader, he has only to recall the equivalent circuit that is commonly used for pentodes (Fig. 13.27). Here, $g_m e_g$ is a current generator representing the plate current of the tube and flowing into the parallel combination of r_p and R_L. When the plate resistance is high, the circuit of Fig. 13.27 representing a tube is

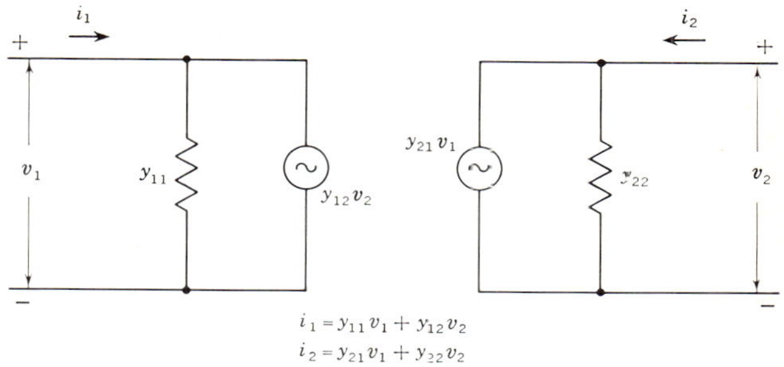

FIGURE 13.26
An equivalent transistor arrangement in which conductances *y*'s and current generators *yv*'s are employed in place of resistance and voltage generators.

easier to deal with than the series equivalent circuit of Fig. 13.19. However, when the internal resistance is low, as it is in the emitter circuit of a transistor, a series arrangement is preferable.

Herein lies the difficulty of using wholly either the resistance or the conductance equivalent circuits to represent the transistor, particularly when measurements are to be made to determine the internal parameter values. The low-resistance input circuit requires one type of approach and the high-resistance output circuit another. What is needed is a combination circuit.

It was in response to this need that the hybrid parameters, labeled h, were developed and are now widely employed by transistor manufacturers to indicate the values of the internal elements of transistors. An equivalent circuit for such hybrid parameters is shown in Fig. 13.28, and its equations are set up in the same manner as they are for Fig. 13.19. That is, for the input loop we have a series circuit, and the equation consists of the voltage drops (and voltage generators), each properly combined with due regard to its polarity. In the present instance,

$$v_b = h_{11}i_b + h_{12}v_c \tag{13.58}$$

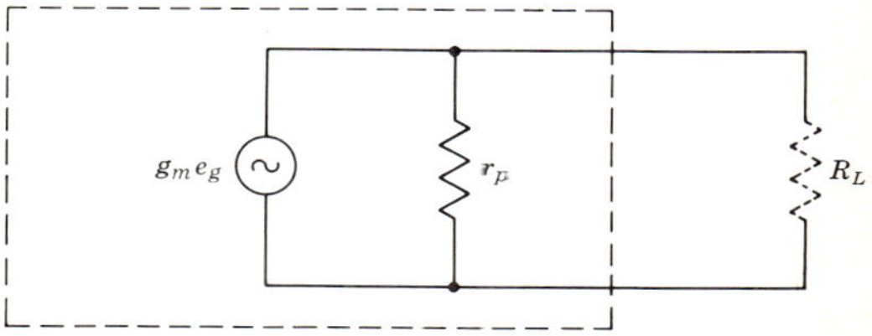

FIGURE 13.27
Equivalent circuit for a pentode vacuum tube.

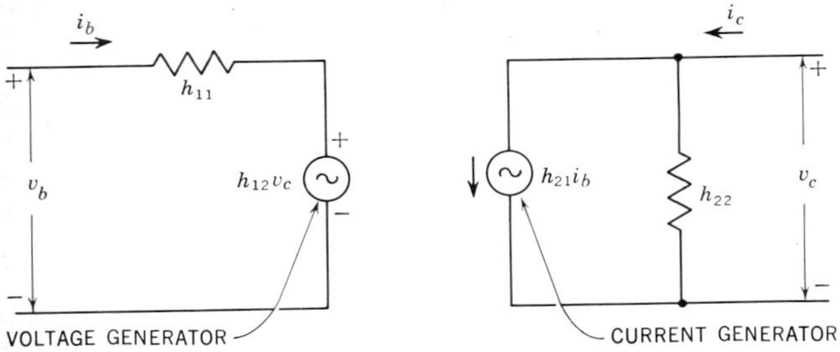

FIGURE 13.28
The hybrid equivalent circuit for a transistor.

In the output loop, we are concerned with currents, and here we assume that the current through h_{22} is in the same direction as the current produced by the current generator $h_{21}i_b$. Hence, the governing equation for this circuit is

$$i_c = h_{21}i_b + h_{22}v_c \tag{13.59}$$

Note that the two equations differ in form from each other because the circuits they are derived from differ. In one instance we are concerned with loop voltages; in the other, loop currents. In both cases, however, the same two independent variables, i_b and v_c, are employed.

The next step is to evaluate the various h parameters such as h_{11}, h_{12}, h_{21}, and h_{22}. This will give us an indication of what they represent.

To obtain h_{11}, v_c in Eq. (13.58) is set equal to zero. That is, the output would be shorted. For this condition,

$$v_b = h_{11}i_b$$

or

$$h_{11} = \frac{v_b}{i_c} \tag{13.60}$$

In this equation h_{11} reveals itself to be the input resistance of the circuit. Consequently, its unit of measurement would be ohms.

h_{12} is determined similarly by setting i_b equal to zero. Equation (13.58) then becomes

$$v_b = h_{12}v_c$$

or

$$h_{12} = \frac{v_b}{v_c} \tag{13.61}$$

Since v_b and v_c have similar units, the units cancel out, leaving h_{12} simply as a number. Now, had v_c been in the numerator and v_b in the denominator, h_{12} would represent the forward voltage gain of the transistor. However, because of the reversal in position, this ratio is sometimes called the feedback voltage ratio.

The next hybrid parameter h_{21} occurs in Eq. (13.59). To find what it represents, let us set v_c to zero. Equation (13.59) now becomes

$$i_c = h_{21}i_b$$

or

$$h_{21} = \frac{i_c}{i_b} \tag{13.62}$$

h_{21} is seen to be the ratio of collector current to base current, or the current amplification of the circuit. Since we are dealing here with a common-emitter arrangement, h_{21} is equal to β.

The final hybrid parameter is h_{22}, and it is obtained by setting i_b in Eq. (13.59) equal to zero. This gives us

$$i_c = h_{22}v_c$$

or

$$h_{22} = \frac{i_c}{v_c} \tag{13.63}$$

This is the equation for the output conductance; hence, h_{22} possesses the unit of mhos.

An alternate designation for the four h parameters that is being used increasingly by manufacturers is

$$h_{11} = h_i \qquad h_{21} = h_f$$

$$h_{12} = h_r \qquad h_{22} = h_o$$

The i of h_i stands for input resistance, the r of h_r for reverse voltage ratio, the f of h_f for forward current ratio, and the o of h_o for output conductance. An additional letter is then added to denote the circuit configuration; for example, h_{ie} denotes the common-emitter arrangement, h_{ib} the common-base arrangement, and h_{ic} the common-collector arrangement.

In arriving at each of the hybrid parameters, we set either $v_c = 0$ or $i_b = 0$. Now, suppose we had an actual transistor whose hybrid parameter values we wished to measure. How could we reduce the ac base current (that is, i_b) or the ac collector voltage v_c to zero without disturbing the dc bias voltages applied to that transistor? (It was on such measurements that the resistance and conductance parameters fell down.)

In the case of $i_b = 0$, the base circuit should be an open circuit for signal currents, yet closed and complete for the dc bias voltage. This can be achieved by feeding the dc voltage to the base via a very high resistor, say 1 to 10 MΩ. Since the transistor input impedance is quite low, constructing such a circuit is easily accomplished by using conventional components.

For $v_c = 0$, we must maintain the correct bias voltage at the collector while bypassing all alternating currents at the collector to ground. This, too, is readily done with a moderate-sized capacitor because the collector impedance is so high that an ac short circuit (for all practical purposes) is attained with a nominal value of shunting capacitance. Thus, use of the hybrid factors provides us with an arrangement whose parameters (such as h_{11}, h_{12}, h_{21}, and h_{22}) are readily measured. This is the outstanding advantage of this configuration.

COMMON-EMITTER HYBRID EQUATIONS

Since the hybrid parameters are employed so extensively, it may be instructive to derive suitable equations governing current gain, voltage gain, power gain, and input and output impedance for a common-emitter amplifier by using hybrid parameters. The basic circuit of Fig. 13.28 will be employed, with the addition of a load resistor R_L, an input signal e_g, and its internal resistance R_g. The corresponding circuit is shown in Fig. 13.29.

The basic relations governing the behavior of this circuit can be written in four equations. The first two equations will relate to the transistor only, the second two will relate to the external circuit. The first two equations are (13.58) and (13.59), previously given:

$$v_1 = h_{11}i_1 + h_{12}v_2 \tag{13.58a}$$

$$i_2 = h_{21}i_1 + h_{22}v_2 \tag{13.59a}$$

Note that the only change we have made is to use v_1 for v_b, i_1 for i_b, v_2 for v_c, and i_2 for i_c. The change was made to enable these equations to represent *all three* possible configurations of a transistor amplifier.

FIGURE 13.29
Hybrid equivalent circuit for a transistor.

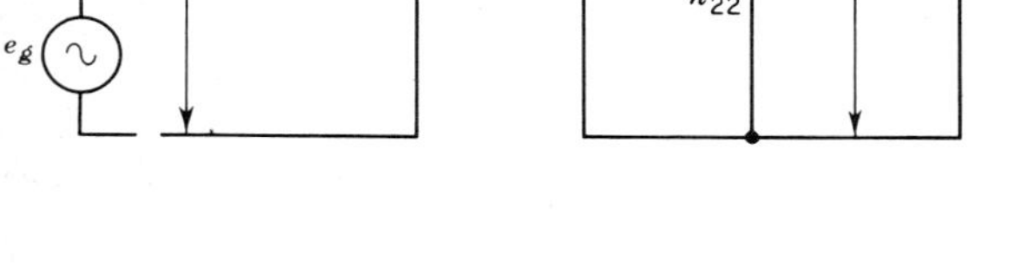

The second two equations needed are

$$v_1 = e_g - i_1 R_g \tag{13.64}$$

$$v_2 = -i_2 R_L \tag{13.65}$$

Equation (13.64) states simply that voltage v_1 is equal to the input signal voltage e_g minus the voltage drop across R_g, the internal resistance of the signal source. Equation (13.65) indicates that v_2 is equal to the voltage developed across R_L. The negative sign is needed before i_2R_L because the current flows up through R_L, while it flows *down* through h_{22} and $h_{21}i_1$. Obviously then, if v_2 possesses the polarity indicated because of the voltage drop across h_{22}, the voltage drop across R_L must be opposite in sign (because of the opposite flow of current). This balancing of the voltages must be made; otherwise, the resulting equations will be incorrect.

Current Gain. To derive the equation for the current gain of this arrangement, we eliminate i_2 between Eqs. (13.59*a*) and (13.65). This gives us

$$0 = h_{21}i_1 + v_2\left(h_{22} + \frac{1}{R_L}\right) \tag{13.66}$$

or

$$v_2 = -\frac{h_{21}i_1}{h_{22} + 1/R_L} \tag{13.67}$$

or

$$v_2 = -i_1\frac{h_{21}R_L}{h_{22}R_L + 1} \tag{13.68}$$

Substituting for v_2 from Eq. (13.65), we obtain

$$-i_2R_L = -i_1\frac{h_{21}R_L}{h_{22}R_L + 1} \tag{13.69}$$

or

$$A_i = \frac{i_2}{i_1} = \frac{h_{21}}{h_{22}R_L + 1} \tag{13.70}$$

Input Resistance. To derive the input resistance, we can substitute Eq. (13.68) into Eq. (13.58*a*). This will remove v_2 and leave only v_1 and i_1 in the equation. The ratio of v_1 to i_1 then is R_{in}. Performing the substitution, we have

$$v_1 = i_1\left(h_{11} - h_{12}\frac{R_Lh_{21}}{R_Lh_{22} + 1}\right) \tag{13.71}$$

$$v_1 = i_1\frac{h_{11}h_{22}R_L + h_{11} - h_{12}R_Lh_{21}}{R_Lh_{22} + 1} \tag{13.72}$$

$$v_1 = i_1\frac{R_L\Delta + h_{11}}{h_{22}R_L + 1} \tag{13.73}$$

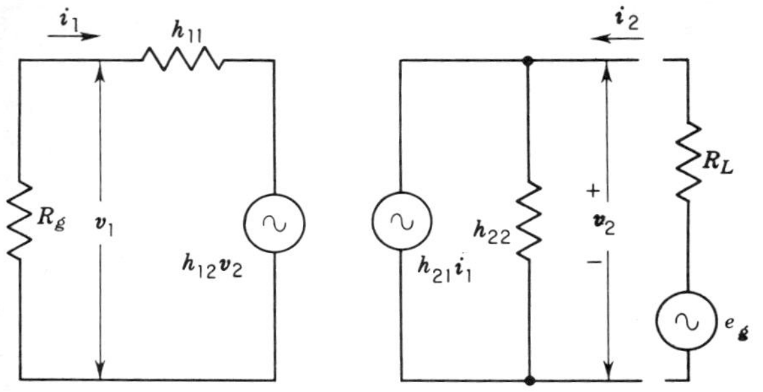

FIGURE 13.30

The input resistance, then, is

$$R_{\text{in}} = \frac{v_1}{i_1} = \frac{R_L\Delta + h_{11}}{h_{22}R_L + 1} \tag{13.74}$$

where Δ is equal to $h_{11}h_{22} - h_{12}h_{21}$

Voltage Gain. We can derive the voltage gain for the circuit by using Eqs. (13.68) and (13.73).

$$\text{Voltage gain} = A_v = \frac{v_2}{v_1} = -\frac{R_L h_{21}}{R_L\Delta + h_{11}} \tag{13.75}$$

Power Gain. Equally easy to derive is the equation for the power gain. This is simply the product of the current and voltage gains.

$$\text{Power gain} = A_i A_v = \left(\frac{h_{21}}{h_{22}R_L + 1}\right)\left(\frac{R_L h_{21}}{R_L\Delta + h_{11}}\right)$$

$$= \frac{h_{21}{}^2 R_L}{(1 + h_{22}R_L)(R_L\Delta + h_{11})} \tag{13.76}$$

Output Resistance. To obtain the governing equation for output resistance, we have to transfer the signal generator to the output terminals. At the same time, we place only R_g across the input terminals (Fig. 13.30). The equations for this arrangement are

$$0 = (R_g + h_{11})i_1 + h_{12}v_2 \tag{13.77}$$

$$i_2 = h_{21}i_1 + h_{22}v_2 \tag{13.78}$$

Note that whereas Eq. (13.78) does not make any overt mention of R_L, this quantity is contained in v_2, as indicated in Eq. (13.65). It is tacitly assumed in Eq. (13.78) that v_2 is developed by generator e_g, which is now in the output circuit.

The required information needed is the ratio of v_2/i_2, since this equals the output impedance seen when looking back into the network from the output terminals.

The first step is to rearrange Eq. (13.77) to the form

$$i_1 = -\frac{h_{12}v_2}{R_g + h_{11}} \tag{13.79}$$

This is then substituted into Eq. (13.78):

$$i_2 = -\frac{h_{12}h_{21}v_2}{R_g + h_{11}} + h_{22}v_2 \tag{13.80}$$

$$i_2 = v_2\left(-\frac{h_{12}h_{21}}{R_g + h_{11}} + h_{22}\right) \tag{13.81}$$

The ratio, then, of v_2 and i_2 is

$$\frac{v_2}{i_2} = \frac{R_g + h_{11}}{(R_g + h_{11})h_{22} - h_{12}h_{21}} \tag{13.82}$$

$$\frac{v_2}{i_2} = \frac{R_g + h_{11}}{R_g h_{22} + \Delta} \tag{13.83}$$

where $\Delta = h_{11}h_{22} - h_{12}h_{21}$.

It is interesting to see that the input impedance, from Eq. (13.74), is dependent upon R_L, while the output impedance, from Eq. (13.83), is dependent on R_g. This, of course, is in keeping with the results previously derived by using resistance parameters.

In deriving Eqs. (13.70), (13.74), (13.75), (13.76), and (13.83), the general circuit of Figs. 13.29 and 13.30 was used. This can represent a common-base, common-emitter, or common-collector arrangement by simply using the proper parameters for each circuit arrangement. This can be better shown by formula, using the notation

$$h_{11} = h_i \qquad h_{21} = h_f$$

$$h_{12} = h_r \qquad h_{22} = h_o$$

Thus, the input resistance from Eq. (13.74) can be expressed as

$$R_{\text{in}} = \frac{R_L\Delta + h_{ie}}{h_{oe}R_L + 1} \tag{13.84}$$

This is the input resistance for the common-emitter circuit. By using h_{ib} and h_{ob} in place of h_{ie} and h_{oe}, the same equation will represent the input resistance for a common-base amplifier. And finally, by substituting h_{ic} and h_{oc}, we obtain the input resistance for a common-collector arrangement. (Δ changes in value, too.)

The remaining equations can be written (for the common-emitter configuration) as

$$A_i = \frac{h_{fe}}{h_{oe}R_L + 1} \tag{13.85}$$

$$A_v = -\frac{R_L h_{fe}}{R_L\Delta + h_{ie}} \tag{13.86}$$

$$\text{Power gain} = \frac{h_{fe}^2 R_L}{(1 + h_{oe}R_L)(R_L\Delta + h_{ie})} \tag{13.87}$$

$$\text{Output resistance} = \frac{R_g + h_{ie}}{R_g h_{oe} + \Delta} \tag{13.88}$$

RELATIONSHIP BETWEEN HYBRID AND RESISTANCE PARAMETERS

Now that we have the hybrid h factors, the next question might be, "How can these be converted into the resistance parameters r_e, r_b, r_c, and r_m?" The answer follows directly if we remember that Eqs. (13.58) and (13.59) can represent a transistor in the common-, or grounded-, emitter configuration. Hence, all we need do is set up the resistance parameters in a similar common-emitter equivalent circuit, obtain the governing equations, and then manipulate (13.58) and (13.59) to the same form. Once this is done, the equivalence between the h and r factors becomes quite evident. The following discussion indicates how this is carried out.

The equivalent circuit of a transistor, in a common-emitter configuration, using resistance parameters, is shown in Fig. 13.31. A signal generator is

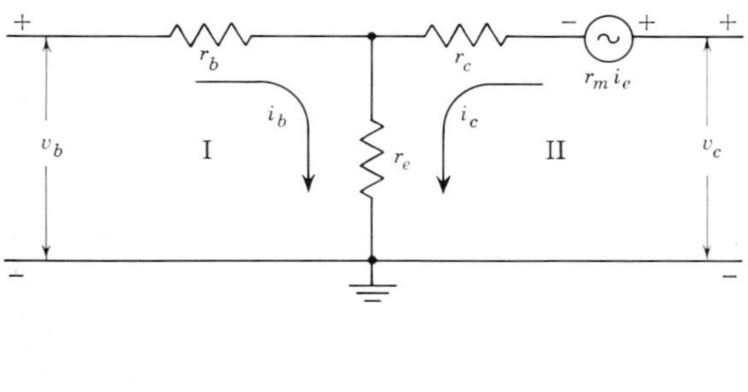

FIGURE 13.31

not being used at the input, nor a load resistor at the output, in order to place this circuit on an equal footing with the circuit of Fig. 13.28. Actually, the generator impedance R_g and the load resistance R_L have no significance so far as the conversion of h to r factors is concerned. Hence, they can be omitted.

The governing equations for Fig. 13.31 are derived by methods identical with those previously explained for Figs. 13.23 and 13.24. Thus, for loop I we have,

$$-v_b + r_b i_b + i_b r_e + i_c r_e = 0$$

or

$$v_b = (r_b + r_e)i_b + r_e i_c \tag{13.89}$$

For loop II the equation is

$$-v_c + r_m i_e + r_c i_c + r_e i_c + r_e i_b = 0$$

But

$$i_e = -(i_b + i_c)$$

Substituting this in the above equation, we obtain

$$v_c = -r_m i_b - r_m i_c + r_c i_c + r_e i_c + r_e i_b$$

or

$$v_c = (r_e - r_m)i_b + (r_c + r_e - r_m)i_c \tag{13.90}$$

Now, if we compare Eqs. (13.58) and (13.59) with (13.89) and (13.90), we see that their forms are not comparable. The next step is to rearrange Eqs. (13.58) and (13.59) so that all the variables, such as v_b, v_c, i_b, and i_c, are assembled in the same order as that possessed by Eqs. (13.89) and (13.90). Here is how this is done. Equation (13.58) is

$$v_b = h_{11} i_b + h_{12} v_c \tag{13.58}$$

Since we are dealing here with a common-emitter arrangement, h_{11} will be converted to h_{ie}, h_{12} will be converted to h_{re}, h_{21} will be changed to h_{fe}, and h_{22} will be written as h_{oe}. Doing this, we have (for the equation just written)

$$v_b = h_{ie} i_b + h_{re} v_c \tag{13.91}$$

This contains v_c, which can be replaced from (13.59).

$$i_c = h_{fe} i_b + h_{oe} v_c \tag{13.59}$$

Rearranging, we have

$$v_c = \frac{i_c}{h_{oe}} - \frac{h_{fe} i_b}{h_{oe}} \tag{13.59b}$$

Now, we substitute this for v_c in Eq. (13.58). Doing this gives us

$$v_b = h_{ie}i_b + \frac{h_{re}}{h_{oe}} i_c - \frac{h_{re}h_{fe}}{h_{oe}} i_b \tag{13.92}$$

Or, collecting similar terms,

$$v_b = \left(h_{ie} - \frac{h_{re}h_{fe}}{h_{oe}}\right) i_b + \frac{h_{re}}{h_{oe}} i_c \tag{13.93}$$

This is one of the equations needed. The other one is (13.59*b*) transposed:

$$v_c = -\frac{h_{fe}}{h_{oe}} i_b + \frac{i_c}{h_{oe}} \tag{13.59c}$$

These are the hybrid equations rearranged to the same form as (13.89) and (13.90), and a direct comparison is now possible. Thus, (13.89) and (13.93) are similar in v_b, i_b, and i_c, and since they deal with the same transistor in the same configuration (here, common emitter), the coefficients of similar terms should be equal. Hence

$$h_{ie} - \frac{h_{re}h_{fe}}{h_{oe}} = r_b + r_e \tag{13.94}$$

and

$$\frac{h_{re}}{h_{oe}} = r_e \tag{13.95}$$

The same comparison can be made between Eqs. (13.90) and (13.59*c*). This gives us

$$-\frac{h_{fe}}{h_{oe}} = r_e - r_m \tag{13.96}$$

and

$$\frac{1}{h_{oe}} = r_e + r_c - r_m \tag{13.97}$$

Further simplification follows from the fact that $r_e = h_{re}/h_{oe}$ by Eq. (13.95). Using this information in (13.94), we can readily determine the hybrid equivalent of r_b.

$$h_{ie} - \frac{h_{re}h_{fe}}{h_{oe}} = r_b + \frac{h_{re}}{h_{oe}}$$

or

$$r_b = \frac{h_{ie}h_{oe} - h_{re}(1 + h_{fe})}{h_{oe}} \tag{13.98}$$

Also, substitution of $r_e = h_{re}/h_{oe}$ in (13.96) will give r_m. Thus

$$-\frac{h_{fe}}{h_{oe}} = \frac{h_{re}}{h_{oc}} - r_m$$

or

$$r_m = \frac{h_{re} + h_{fe}}{h_{oe}} \tag{13.99}$$

Now, with all this information, we can derive the value for r_c alone by suitable substitution in (13.97).

$$\frac{1}{h_{oc}} = \frac{h_{re}}{h_{oe}} + r_c - \frac{h_{re} + h_{fe}}{h_{oe}}$$

$$r_c = \frac{h_{fe} + 1}{h_{oe}} \tag{13.100}$$

Similar comparisons can be made among all the parameters so long as the precautions outlined above are kept in mind.

COUPLING AND BYPASS CAPACITORS

The discussion thus far has dealt with the various aspects of transistor-amplifier design at low frequencies. Before leaving this subject for a consideration of high-frequency transistor-amplifier design, special mention should be made of the problem of selecting the coupling capacitor and the emitter bypass capacitor. Both of these capacitors determine the low-frequency cutoff of the transistor amplifier. The transistor itself has a response that is flat down to and including direct current (zero frequency). However, the coupling and bypass capacitors, acting with the collector and bias resistors and the transistor input resistance, set a lower limit to the attainable frequency response of the overall amplifier.

Figure 13.32 illustrates a two-stage RC-coupled amplifier, together with typical component values. The bypassed emitter resistor stabilizes the operating point, as previously discussed. R_C is the collector resistor of the stage, while R_{B1} and R_{B2} form a voltage divider that establishes the operating point in conjunction with R_E. The coupling capacitor C_K should have such a value that the signal current leaving the collector of the first stage is adequately coupled into the base of the second stage. The emitter bypass capacitor C_E should possess such a value that emitter degeneration caused by R_E is negligible at the lowest frequency of interest; that is, R_E should be adequately bypassed at all frequencies of interest.

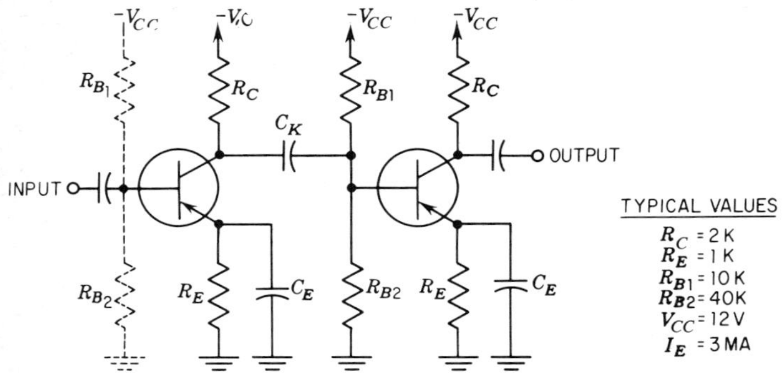

FIGURE 13.32
Two-stage *RC*-coupled amplifier.

It is beyond the scope of this book to derive the formulas for the required values of C_K and C_E; however, a discussion of these formulas is in order. The equation for C_K (in farads) is given by

$$C_K = \frac{1}{2\pi f_{3\text{dB}}\left(R_C + \frac{R_{B_t}R_{\text{in}}}{R_{B_t}R_{\text{in}}}\right)} \tag{13.101}$$

and for C_E by

$$C_E = \frac{\beta_0 + 1}{2\pi f_{3\text{dB}}\left(R_{\text{in}} + \frac{R_C R_{B_t}}{R_C + R_{B_t}}\right)} \tag{13.102}$$

where $f_{3\text{dB}}$ = the desired 3-dB low-frequency cutoff

R_{B_t} = the parallel combination of R_{B_1} and $R_{B_2} = \dfrac{R_{B_1}R_{B_2}}{(R_{B_1} + R_{B_2})}$

R_{in} = the input resistance of the transistor, h_{ie}

β_0 = the low-frequency, small-signal current gain of the transistor h_{fe}.

There are some very important design considerations to be noted here, so let us consider Eqs. (13.101) and (13.102) separately.

The Coupling Capacitor. The value of the coupling capacitor is seen to depend upon the collector resistor of the preceding stage and the bias resistors and input resistance of the following stage. The dominant factor, however, is the collector resistor of the preceding stage. The input resistance is dependent upon the operating point of the transistor; R_{in} will decrease as the emitter bias current increases. Consider the typical component values indicated in Fig. 13.32 and assume a typical input resistance at 3 mA emitter current of $R_{\text{in}} = 500\ \Omega$. In order to obtain a low-frequency 3-dB cutoff of 10 hertz (Hz), Eq. (13.101) predicts that a coupling capacitor of

$$C_K = \frac{1}{2\pi \times 10\left(2{,}000 + \dfrac{8{,}000 \times 500}{8{,}000 + 500}\right)} = \frac{1}{20\pi(2{,}000 + 471)}$$

$$= 6.45 \text{ microfarads } (\mu\text{F})$$

is required. Note that this value depends primarily on the value of R_C (2,000 Ω) of the first stage. This differs considerably from vacuum-tube circuit practice, where the required capacitance depends primarily upon the input resistance of the following stage. Note also that a rather large value of capacitance is required for the transistor amplifier. Whereas coupling capacitors on the order of 0.01 to 0.1 μF are generally quite adequate at this frequency for vacuum-tube amplifiers, transistor amplifiers require values from 1 to 10 μF.

The Emitter Bypass Capacitor. In a similar manner, let us calculate the required value of emitter bypass capacitor C_E. Again using 10 Hz as the lower cutoff frequency, we see from Eq. (13.102) that a transistor having a small-signal current gain β_0 of 50 requires an emitter bypass capacitor of

$$C_E = \frac{50 + 1}{2\pi \times 10\left(\dfrac{2{,}000 \times 8{,}000}{2{,}000 + 8{,}000} + 500\right)} = \frac{51}{20\pi(1{,}600 + 500)} = 387\ \mu\text{F}$$

This is quite large compared with capacitor values required in vacuum-tube circuits where the bypass capacitor is totally dependent upon the size of the cathode resistor. A capacitor of only 25 to 75 μF would be required with a vacuum-tube circuit to obtain the same low cutoff frequency.

The value of C_E is primarily dependent on the current gain of the transistor β_0 and secondarily dependent upon the collector resistance of the preceding stage. It is also influenced by the input resistance of the transistor, which, in turn, is operating-point-dependent. Finally, note that the value of C_E is entirely independent of the size of the emitter resistance R_E, quite the opposite from vacuum-tube theory.

The reader will find it well worth his time to study Eqs. (13.101) and (13.102) in detail, especially from a practical viewpoint. For example, consider the case where stages 1 and 2 are transformer-coupled and stage 2 is a power-output stage with an emitter bias current of 24 mA. Then the transistor input resistance R_{in} will typically be 100 Ω and the value of R_L reflected through the transformer may well be 10 Ω or less. Now, let us calculate the value of C_E required for a low-frequency cutoff of 10 Hz.

$$C_E = \frac{50 + 1}{2\pi \times 10 \left(\dfrac{8{,}000 \times 100}{8{,}000 + 100} + 100\right)} = \frac{51}{20\pi(99 + 100)} = 4{,}080\ \mu\text{F}$$

This value is so large it is impractical.

HIGH-FREQUENCY RESPONSE

Knowledge of Eqs. (13.101) and (13.102) gives one the ability to calculate the low-frequency cutoff of an RC-coupled transistor amplifier. The other frequency response of interest is the high-frequency cutoff. The high-frequency response of a transistor amplifier is determined primarily by the transistor itself and to some extent by the external circuitry. Early transistors were limited in frequency response to the kilohertz range; however, present-day transistors are used in FM and TV tuners in the 100- to 400-MHz range and useful operation into the gigahertz range has been achieved.

In the case of the vacuum tube, the upper-frequency-response limitations are set by the interelectrode capacitances. This is not so with the transistor because whereas the vacuum tube is basically a voltage-operated device, the transistor is essentially a current-operated device. To change the current through a transistor from one value to another requires the movement of charge to or from the base region of the transistor. This movement of charge takes a finite amount of time. For example, consider the illustration of Fig. 13.33. V_{EE}, V_{CC}, R_{E1}, R_{E2}, and R_C are adjusted to provide the desired operating point. V_G and R_G constitute a sine-wave signal source.

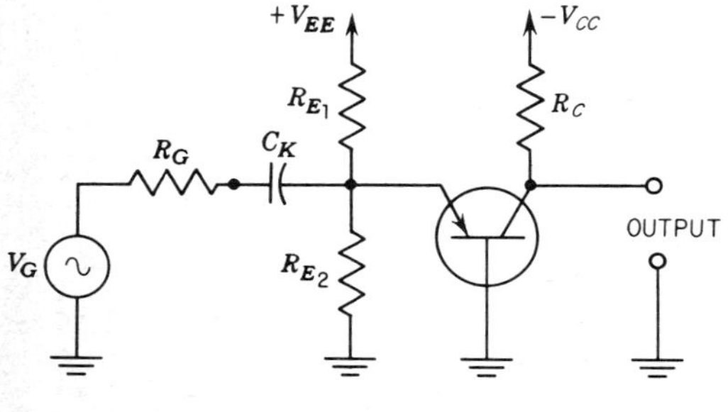

FIGURE 13.33
Common-base amplifier.

If the applied sine-wave signal goes positive the emitter current and the collector current will increase. This means a movement of charge into and through the base region of the transistor. This movement takes time. Another way of regarding this action is to note that the base of the transistor acts like a capacitor. In order to increase the current through the transistor, one must increase the total charge in the base region (capacitor). When the applied sine wave goes negative, the reverse process occurs. This implies the movement of charge out of the base region of the transistor.

At low frequencies, the output sine wave will be an amplified version of the input and will be exactly in phase with the input. However, if the signal frequency is steadily raised, the output amplitude will not only decrease but will also shift in phase with respect to the input signal. In fact, if the input signal is held constant and a plot is made of the output signal vs. frequency, the curve shown in Fig. 13.34 will be obtained.

Let us consider why the gain of the transistor falls off with increasing frequency. First, observe what happens at lower frequencies. As the signal amplitude is increased, i.e., as the sine wave goes positive, charge moves into the base region and the collector current increases. When the signal decreases, i.e., the sine wave goes negative, charge moves out of the base region and the collector current decreases correspondingly. This seems quite simple; however, one must realize that charge flows into and out of the transistor base region by the process of diffusion. Current flow by diffusion, unlike the normal current flow in a wire due to an applied emf, is a random, undirected process. Therefore, the time required to move charges into and out of the base region by diffusion can become important. As the frequency is increased, the time between the positive and negative portions of the sine wave becomes shorter and shorter. Eventually, the frequency reached is such that the positive portion of the sine wave persists for a time that is shorter than that required to move the appropriate amount of charge into the base

FIGURE 13.34
Output amplitude vs. frequency.

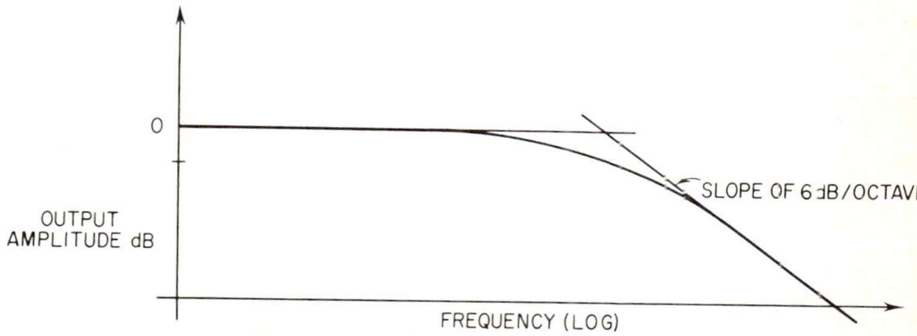

region. Thus, when the sine-wave signal begins to go negative, the charge in the base is still trying to increase. At this point, the transistor is not responding properly to the input signal, resulting in a decrease in transistor gain and a definite phase shift between the input and output signals.

A plot of the gain and phase characteristics of the transistor in the common-base connection is shown in Fig. 13.35. This curve is obtained with an ac open-circuited input (current signal source) and an ac short-circuited output.

The 3-dB upper cutoff frequency in the common-base arrangement is referred to as the α cutoff frequency f_α. The gain above this frequently falls off at 6 dB per octave; i.e., the gain drops 6 dB every time the frequency is doubled. f_α is the frequency at which the small-signal current gain is equal to 0.707 of its value at low frequencies α_0; it is also the frequency at which the phase shift introduced by the transistor is 45°.

A similar set of curves for the common-emitter configuration is shown in Fig. 13.36. Here again, the curves are for an ac open-circuited input and an ac short-circuited output.

The 3-dB upper cutoff frequency in the common-emitter amplifier is called the β cutoff frequency f_β. The gain at frequencies above f_β falls off at 6 dB per octave. Another important parameter shown in Fig. 13.36 is f_T. This is the frequency at which the common-emitter current gain is equal to 1. The value of f_T is referred to as the gain-bandwidth product for the following reason: If the current gain is measured at any frequency along the 6-dB per octave roll-

FIGURE 13.35
Common-base current gain and phase shift vs. frequency.

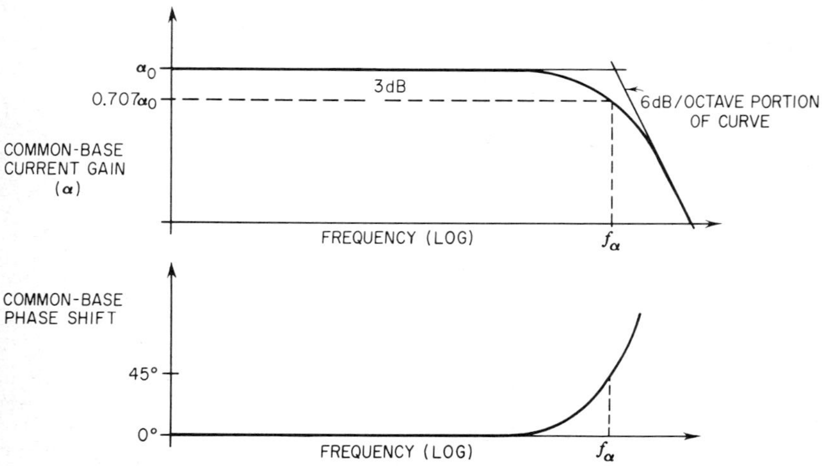

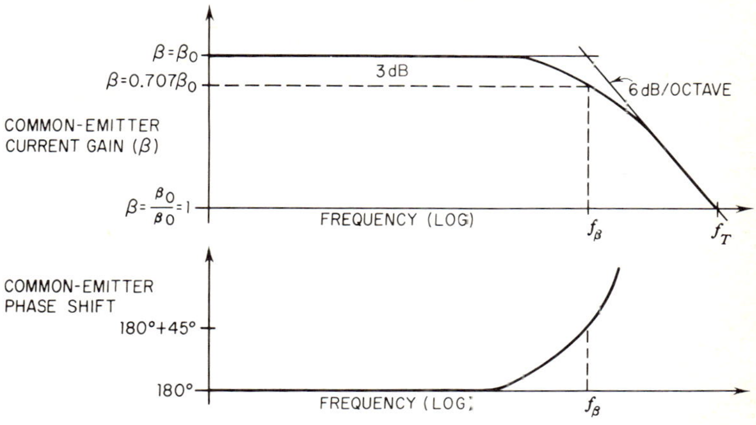

FIGURE 13.36
Common-emitter current gain and phase shift vs. frequency.

off, then the product of the gain and the frequency of measurement will be a constant, namely, f_T. In fact, an expression for the common-emitter current gain at any frequency above f_β is

$$\beta = \frac{\beta_0}{\sqrt{1 + \left(\frac{\beta_0 f}{f_T}\right)^2}}$$

where β_0 = low-frequency current gain, h_{21}
f = frequency of interest
f_T = gain-bandwidth product

Note that the current gain will be 0.707 β_0 when $f = f_T/\beta_0$; thus, this frequency is the value of f_β.

For the common-emitter configuration, the primary criterion in determining the frequency response of the transistor is f_T, the gain-bandwidth product. However, note that the upper cutoff frequency f_β is a function of the low-frequency current gain, as shown in Fig. 13.37.

Thus, if three transistors are obtained, each with the same value of f_T (gain-bandwidth product) but with different values of β_0, and if each unit is inserted into a transistor amplifier, the upper cutoff frequency of the amplifier will be lowest for the high-β unit and highest for the low-β unit. Furthermore, it is interesting to note that in the common-base configuration, the current gain α_0 is less than 1 but the high-frequency cutoff f_α can be high; in the common-emitter configuration, the current gain β is high but the high-frequency

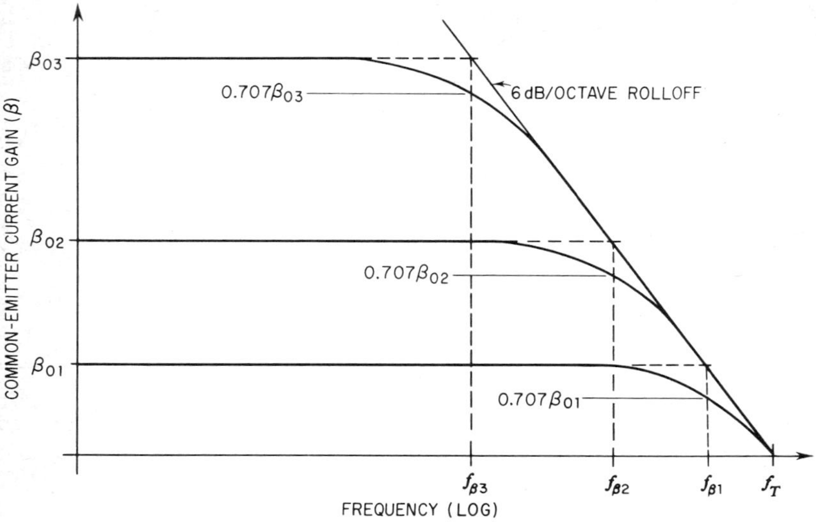

FIGURE 13.37
Common-emitter gain vs. frequency for three low-frequency transistors with the same f_T.

cutoff f_β can be low. For example, consider the 2N1191, for which the following typical values hold:

Common base	$\alpha = 0.99$	$f_\alpha = 2$ MHz
Common emitter	$\beta = 100$	$f_\beta = 19$ kilohertz (kHz)

One additional point is of interest. In transistor-amplifier design, the high-frequency figure of merit is f_α in the common-base connection and f_T in the common-emitter connection. For any given transistor, values of f_α and f_T are not equal. For homogeneous-base transistors such as alloy-junction types, $f_T \approx 0.85\ f_\alpha$, and for the diffused-base transistors such as the mesa types, $f_T \approx 0.5\ f_\alpha$. One must be sure to use f_T when designing a common-emitter amplifier. f_α should be used only when designing a common-base amplifier.

The foregoing facts are very useful in estimating the high-frequency cutoff of a transistor amplifier under the specified conditions, namely, that the input appears open-circuited to the signal and the output is ac short-circuited to the signal. However, in a practical case, the respective input and load resistances need not be infinite and zero, as previously stated. The above estimates will hold so long as the source resistance feeding the base is very large and the load resistance in the collector circuit is small. Let us be more specific in what we mean by "large" and "small." In stating that the source should be large, we mean that it must be large compared with the input resistance of the transistor. Usually, a value of source resistance of 3,000 Ω or greater will suffice. By the same token, the load resistance can be considered small if it is small in comparison with the output resistance of the transistor. Usually a value of 1,000 Ω or less will suffice. For example, the circuit of Fig. 13.32 almost meets these conditions. The source resistance of the stage is the collector resistor of the preceding stage, approximately 2,000 Ω. The load resistor of the stage is the input resistance of the following stage, approximately 500 Ω. Thus, for the transistor-amplifier stage shown in Fig. 13.32, the upper cutoff frequency is approximately $f_\beta = f_T/\beta$.

At this point the reader might well ask what happens to the upper cutoff frequency when the source resistance is not large, the load resistance is not small, or both. This question will be answered in detail presently, but first let us consider other transistor parameters that will affect the upper cutoff frequency of a transistor amplifier.

Collector-junction Capacitance. In the circuit of Fig. 13.38, a capacitance C_c is shown between collector and base. The presence of this capacitance constitutes a feedback path between the output (collector) and the input (base) of the transistor. Since the common-emitter amplifier exhibits a 180° phase shift from input to output, the resulting feedback signal is 180° out of phase with the input signal. The effect of C_c, then, is to introduce negative feedback that subtracts from the input signal and reduces the output signal. Furthermore, since the capacitive reactance of C_c decreases with increasing frequency, the effects of the feedback become more pronounced at the higher frequencies.

There is another important fact concerning the effect of the collector-junction capacitance C_c. The output current that leaves the collector travels to point A and there divides: part goes to the load and part goes through C_c back to the base. The current at point A will divide according to the impedance of the respective current paths. If the load resistance R_c is small com-

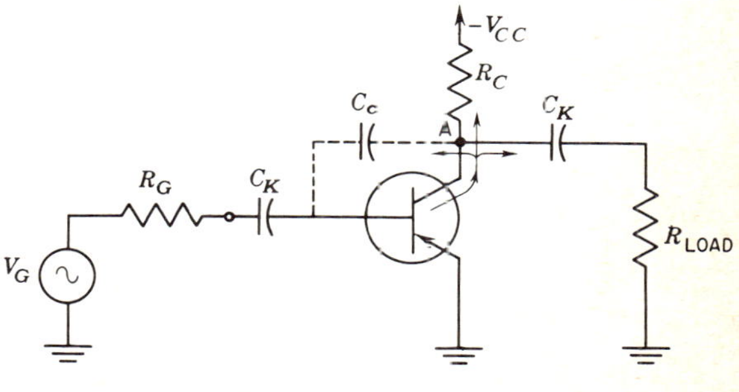

FIGURE 13.38
Circuit showing feedback caused by C_c.

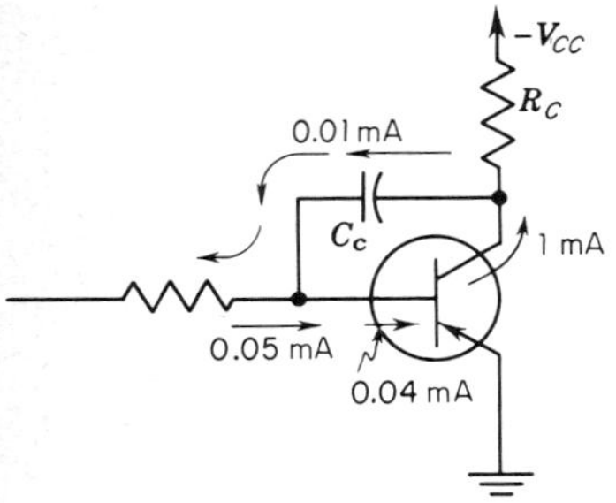

FIGURE 13.39
Effects of C_c on output signal.

pared with the reactance of C_c at the frequency of interest, most of the current will flow into the load, the desirable effect. In this case, the effect of C_c is small and the frequency response is set by f_T, as previously explained. However, if the load resistance is large, much of the output current will be fed back through C_c and a loss in gain will occur. That is why f_T is a valid measure of the high-frequency response only when the output is ac short-circuited or when the load resistance is small. The effect of C_c is quite important because the magnitude of its feedback current is actually multiplied by β, the current gain of the transistor. For example, consider the case of a transistor with a β of 20 and an output signal of 1 mA (Fig. 13.39). If 1 percent of this output signal, 0.01 mA, is fed back to the base (which has a signal current of I_c/β, or 0.05 mA), the net base current after the feedback is subtracted is 0.04 mA. The collector current will then decrease from 1 mA to 0.04β, or 0.8 mA. In this case, a 1 percent feedback factor reduces the output of the amplifier by 20 percent. Thus, the effects of C_c can become quite devastating.

High-frequency Equivalent Circuit. In the preceding section, we discussed the two major parameters f_T and C_c involved in determining the high-frequency response of a transistor. In this section, a more exact method is presented, and it will enable us to calculate the gain and the bandwidth of a transistor amplifier for any values of source and load resistance. In order to do so, a high-frequency equivalent circuit of the transistor is developed. The components of this equivalent circuit are readily lumped with external circuit components. The resulting composite circuit is then useful in obtaining expressions for gain and frequency response.

As a first step in obtaining a useful equivalent circuit for the transistor, let us first isolate all the internal resistances and capacitances associated with the device. Since the transistor is nothing more than two semiconductor diodes connected back to back, an equivalent circuit must certainly contain the resistances and capacitances associated with each diode. However, transistor action is not obtained by simply interconnecting two semiconductor diodes; the very close spacing of the base region is most essential.

One method of acknowledging this mutual coupling between the diodes circuit-wise is to include a current generator in the output, where the amount of current generated depends on the input voltage or current. Another important fact that must be considered is that the base region itself stores charge when the transistor is in operation. This was previously mentioned in our discussion of effects that limit the frequency response of the transistor. Figure 13.40 shows the physical representation of a typical junction transistor. Within this picture are included all the resistances and capacitances affecting the gain and frequency response of the transistor. Although they are shown as lumped components in the model, it must be realized that these resistances and capacitances are actually distributed throughout the emitter, base, and collector regions.

Assuming that the transistor shown in Fig. 13.40 is normally biased for common-emitter operation, let us discuss each of the lumped parameters indicated in both their physical and operational aspects. Starting with the input to the base, the first component that the signal encounters is a resistor $r_{bb'}$. This resistance is referred to as the base-spreading resistance; it is the physical resistance of the bulk semiconductor material that exists between the external base connection and the active base region. The active base region is defined as the region directly between the emitter and collector electrodes. All the carriers that flow between the emitter and the collector must flow through this region. Since we wish to control the flow of these carriers, the signal voltage or current must be applied between the active region and the common terminal (emitter in this case). In order to obtain a high value of β, the resistivity of the base region is usually kept high compared with the resistivity of the material in the emitter and collector regions (commonly 10 to 1,000 times higher). Also, the base region is very thin; a typical value for

FIGURE 13.40
Physical representation of a typical transistor.

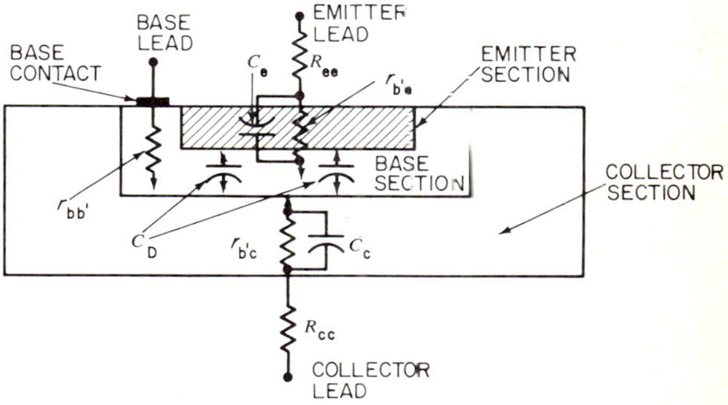

base width is $\frac{1}{4}$ mil to 2 mils. Thus, even though the length of this resistance path is small, its high resistivity in conjunction with its small cross-sectional area results in a value of $r_{bb'}$ that may well exceed 100 Ω. In a homogeneous-base transistor, typical values of $r_{bb'}$ range from 100 to 500 Ω. In a diffused-base transistor, where part of the base is purposely made of low-resistivity material, typical values of $r_{bb'}$ range from 10 to 200 Ω depending upon the intended application. A low value of $r_{bb'}$ is especially essential in transistors that are to be used in oscillator applications. This can be seen from Eq. (13.103), which is the expression for the maximum frequency at which a transistor can oscillate.

$$f_{\max} = \frac{\alpha_0 f_T}{8\pi r_{bb'} C_c} \tag{13.103}$$

where α_0 = common-base low-frequency current gain
f_T = gain-bandwidth product
$r_{bb'}$ = base-spreading resistance
C_c = collector-junction capacitance

If $r_{bb'}$ could be reduced to zero, the transistor could theoretically oscillate at any high frequency. Although this would be true for the isolated transistor, the highest frequency of oscillating would still be limited by the stray capacitances and resistances of the external circuit. In fact, some transistors that are available today require specially constructed circuits and housings in order to attain their high-frequency capabilities. In amplifier applications $r_{bb'}$ is important only when the source resistance is small; generally, if the source resistance is above one or two thousand ohms, $r_{bb'}$ may be neglected.

After the signal passes $r_{bb'}$, it arrives at the active base region where it sees a host of resistances and capacitances. Both the emitter and collector junctions have small-signal ac resistance and capacitance associated with them. Since the emitter-base junction is forward-biased and the collector-base junction is reverse-biased, it is reasonable to expect that the values of these junction resistances and capacitances will be different, and such is the case. Let us consider each junction separately.

At the forward-biased emitter-base junction, the small-signal ac resistance is low. This resistance $r_{b'e}$ exists in the equivalent circuit, as it does in the physical transistor, between the active base region and the grounded emitter. Its value can be determined by finding the slope of the input VI characteristic of the grounded-emitter transistor at the correct operating point. An example of this is shown in Fig. 13.41. Note that the slope of the characteristic curve changes for different values of base current. The slope is given by V_{BE}/I_B, and this is $r_{b'e}$. Generally, the size of this resistance ranges between 100 and 3,000 Ω.

The junction capacitance associated with the emitter-base diode depends upon the area of the junction and the voltage applied to the junction. The

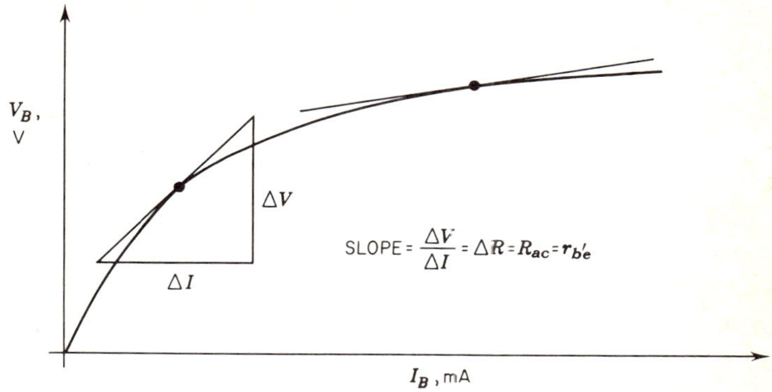

FIGURE 13.41
Input *VI* characteristic. (Common emitter.)

junction capacitance can be compared to a familiar parallel-plate capacitor, whose capacitance is given by

$$C = \frac{\epsilon A}{d} \tag{13.104}$$

where A = area of the plates
d = spacing between the plates
ϵ = dielectric constant of the material between the plates

Capacitance is proportional to the plate area and inversely proportional to the plate spacing. In the single-transistor model, the two plates are the emitter region and the active base region. Between them is the depletion layer whose width is governed by two things: the resistivity of the base in the vicinity of the junction and the applied voltage. A curve of junction capacitance vs. voltage for both a forward- and reverse-biased diode is shown in Fig. 13.42, and it can be seen that the junction capacitance decreases for increasing reverse bias. This is because the width of the depletion layer increases with increasing reverse bias. Since the emitter-base junction is forward-biased its capacitance is larger than that of any reverse-biased diode such as the collector-base junction.

Now consider the collector-base junction. As in the emitter-base junction, there is a small-signal resistance and capacitance. However, since the collector-base junction is reverse-biased, the resistance $r_{b'c}$ is quite large, typically varying from 10,000 Ω to 10 MΩ. In many applications, this is so large it can be ignored.

Because the width of the depletion layer for a reverse-biased junction is large, the collector-base-junction capacitance is small. Even though the area of the collector junction is usually larger than that of the emitter junction,

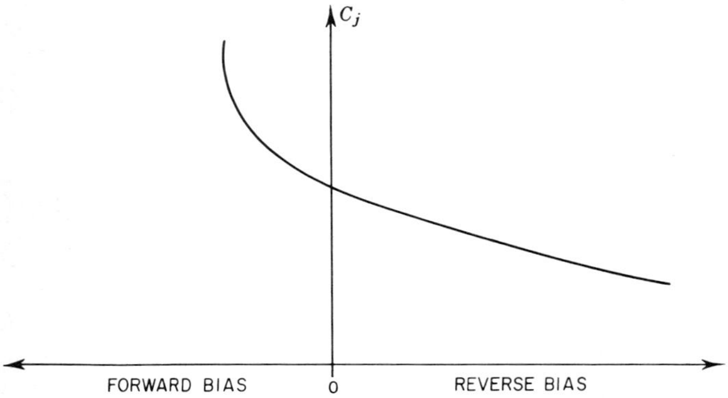

FIGURE 13.42
Diode junction capacitance C_j vs. bias.

the effects of the depletion layers are such that the collector-junction capacitance is generally one-sixteenth to one-half that of the emitter-junction capacitance.

There is still another capacitance associated with the transistor: the diffusion capacitance C_D. C_D is not a physical capacitance in the same sense as a parallel-plate capacitor, but it is an equivalent capacitance representing the charge stored in the active base region. Recall that charge within the active base region must be changed in order to change the collector current of a transistor. As one might expect, the amount of charge stored within the active base region, and thus the equivalent value of diffusion capacitance, is dependent upon the base width of the transistor. As the base width is made smaller, less charge can be stored within the active base region. Therefore, for smaller base widths, the diffusion capacitance will be smaller and the frequency response of the transistor will be higher.

Finally, there are two resistances R_{cc} and R_{ee}. These are spreading resistances, that is, the physical resistance of the bulk material comprising the collector and emitter electrodes. It was previously mentioned that the resistivity of the material within the emitter and collector electrodes was quite low compared with the resistivity in the base region. Therefore, one would expect the value of R_{ee} and R_{cc} to be quite small. In fact, in most practical situations, these resistances are neglected; however, the reader should be aware that they are present even though their value is usually under 10 Ω.

Now let us combine all of the physical resistors and capacitors into a lumped-parameter equivalent circuit. Such a circuit is shown in Fig. 13.43. This circuit is called the hybrid pi because its shape is similar to that of the Greek letter π. It includes all the physical resistors and capacitors shown in

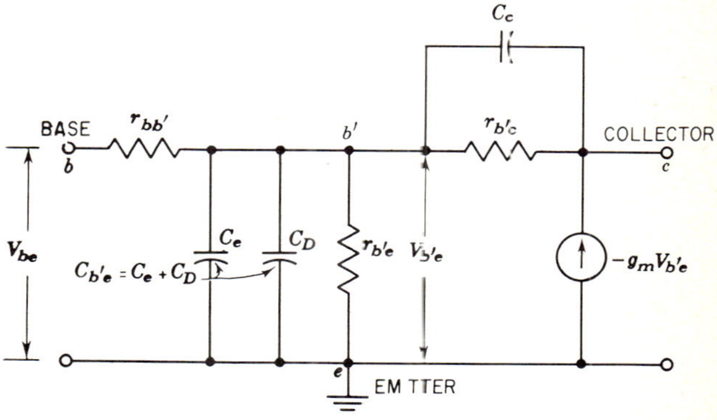

FIGURE 13.43
Lumped small-signal-parameter equivalent circuit for normally biased common-emitter transistor.

Fig. 13.40. In addition, the equivalent circuit includes a current generator at the output between the emitter and collector. The signal from the current generator is equal to the negative of a constant g_m times a voltage $V_{b'e}$.

The minus sign represents the 180° phase reversal between the input signal and the output signal. That is, if the input signal goes positive, the output voltage goes negative, and vice versa. The voltage $V_{b'e}$ is the internal base-to-emitter voltage. This is the voltage between the active base region and the emitter, and it differs from the external base-to-emitter voltage V_{be} by the drop across $r_{bb'}$. The point b' in the equivalent circuit represents the internal base of the transistor: the active base region. It is the voltage across this point, between b' and the grounded emitter, that determines the amount of signal current that will flow out of the collector.

The constant g_m is the transconductance of the transistor. It is quite similar to, and circuit-wise is analogous to, the transconductance of the vacuum tube. g_m relates the small-signal output current of the device to the small-signal input voltage and is given as

$$g_m = \frac{I_E}{27} \qquad \text{mA/V} \tag{13.105}$$

where I_E = emitter current, mA
27 = temperature-dependent constant

At temperatures other than 25°C, the constant, 27, changes slightly. Since the transconductance of the transistor is dependent upon the emitter bias current I_E, it is dependent on the operating point.

Before attempting to derive expressions for amplifier gain and bandwidth by using the equivalent circuit shown in Fig. 13.43, let us review the components of this equivalent circuit and discuss their numerical values. $r_{bb'}$ is the base-spreading resistance. Its value varies with the type of transistor but, in general, is low and on the order of 100 Ω. C_e is the emitter-junction capacitance, and C_D is the diffusion capacitance of the transistor. When the transistor is connected as an amplifier, the input signal to the base sees the two capacitors in parallel and cannot distinguish between them. From a circuit designer's viewpoint, it is their combined value that is of interest; this combination is referred to as $C_{b'e}$, the capacitance between the internal base region and the emitter, and is given by

$$C_{b'e} = \frac{g_m}{2\pi f_T} = C_e + C_D \tag{13.106}$$

where f_T = gain-bandwidth product of the transistor
g_m = transconductance
Since g_m and f_T are operating-point-dependent, $C_{b'e}$ is dependent on the bias point. Since this capacitor is inversely proportional to f_T, the optimum operating point occurs where f_T takes on its highest value.

$r_{b'e}$ is the small-signal resistance of the internal base-emitter diode. Its value is low because the base-emitter diode is forward-biased. It is given by

$$r_{b'e} = \frac{\beta_0}{g_m} \tag{13.107}$$

where β_0 = low-frequency common-emitter current gain
g_m = transconductance
There is a subtle point of interest here. Note that the product of $r_{b'e}$ and $C_{b'e}$ forms a time constant

$$r_{b'e}C_{b'e} = \frac{\beta_0}{g_m}\frac{g_m}{2\pi f_T} = \frac{\beta_0}{2\pi f_T} = \frac{1}{2\pi f_\beta} \tag{13.108}$$

which is the inverse of the frequency response $2\pi f_\beta$ (or f_β). If all other components of the equivalent circuit were neglected, the frequency response would be f_β. It will be shown later that this case holds when the source resistance feeding the stage is very large and the load resistance of the stage is very small; these are the identical conditions mentioned during our definition of f_T and f_β. $r_{b'c}$ is the small-signal resistance of the reverse-biased collector-base diode. This resistor is so large that it is usually neglected in the calculation of gain and bandwidth of a transistor amplifier. C_c is the collector-junction capacitance, which is dependent upon the collector-base voltage. In general, it is wise to minimize this capacitance. This is done in high-frequency transistor amplifiers by using supply voltages above 3 V.

Before proceeding to the problem of obtaining expressions for the gain and bandwidth of a transistor-amplifier stage, let us compare the equivalent circuits of the transistor and the vacuum tube. The primary difference is noted on the input side of the two devices: whereas the input resistance of the tube is high and its input capacitance is low, the transistor has a low input resistance and a high input capacitance. It is interesting to note that the resultant RC time constant may be the same for the two devices.

GAIN AND BANDWIDTH EXPRESSIONS

Now that an understanding of the equivalent circuit of the transistor has been established, it is possible and meaningful to derive the gain and bandwidth expressions for a common-emitter amplifier stage. In deriving these important expressions, we shall completely neglect the effects of $r_{b'c}$ by essentially removing it from the equivalent circuit; this is valid when dealing with load resistances of less than 10,000 Ω, a common condition.

Low-frequency Current and Voltage Gain. To derive the expressions for the current and voltage gain of a single-stage common emitter, we shall first consider the equivalent circuit of Fig. 13.43 at low frequencies. We shall neglect all capacitances by assuming an operating frequency at which neither the high- nor the low-frequency effects come into play.

Let us start first with the current gain. The equivalent circuit under consideration is shown in Fig. 13.44. Current gain is defined as the ratio of the output signal current I_{out} to the input signal current I_{in}. The input current, which may be the signal from another stage or any other signal source, is represented by the current generator I_g and its parallel source resistance R_g. The output current is given by

$$I_{\text{out}} = -g_m V_{b'e} \tag{13.109}$$

FIGURE 13.44
Common-emitter low-frequency equivalent circuit with current source.

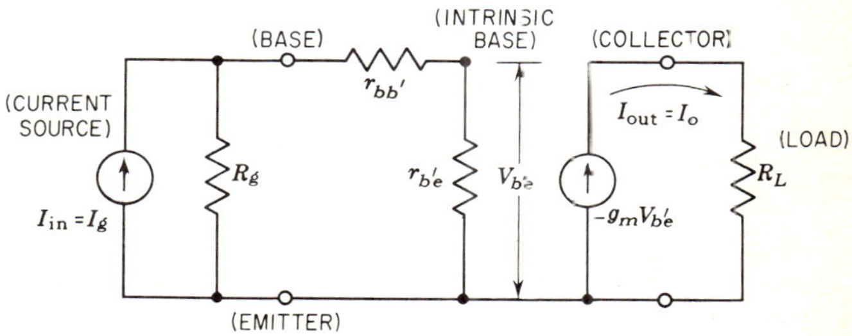

where $V_{b'e}$ may be derived as

$$V_{b'e} = \frac{r_{b'e}}{r_{bb'} + r_{b'e}} \frac{I_g R_g (r_{bb'} + r_{b'e})}{r_{bb'} + r_{b'e} + R_g} = \frac{I_g R_g r_{b'e}}{r_{bb'} + r_{b'e} + R_g} \tag{13.110}$$

In spite of the formidable appearance of Eq. (13.110), it is simply the Ohm's law expression for the voltage developed across $r_{b'e}$ by the input current I_g. The circuit is complicated by the fact that R_g is in parallel with the series combinations of $r_{bb'}$ and $r_{b'e}$.

The current gain is

$$G_i = \frac{I_{out}}{I_g} = \frac{-g_m R_g r_{b'e}}{r_{bb'} + r_{b'e} + R_g} \tag{13.111}$$

The minus sign indicates that there is a 180° phase shift from input to output. If R_g becomes very large, this expression reduces to

$$G_i = -g_m r_{b'e} = \beta_0 \tag{13.112}$$

which has previously been shown to be the small-signal low-frequency current gain under conditions of large source resistance.

To derive the expression for the low-frequency voltage gain, consider the equivalent circuit shown in Fig. 13.45. Voltage gain is defined as the ratio of output signal voltage V_{out} to input signal voltage V_{in}. The output voltage is developed across the load R_L, and the input signal, which may be the signal from a preceding stage or any other source, is represented by the voltage generator V_g and its series resistance R_g. The output voltage is given by

$$V_{out} = I_{out} R_L = -g_m V_{b'e} R_L \tag{13.113}$$

FIGURE 13.45
Common-emitter low-frequency equivalent circuit with voltage source.

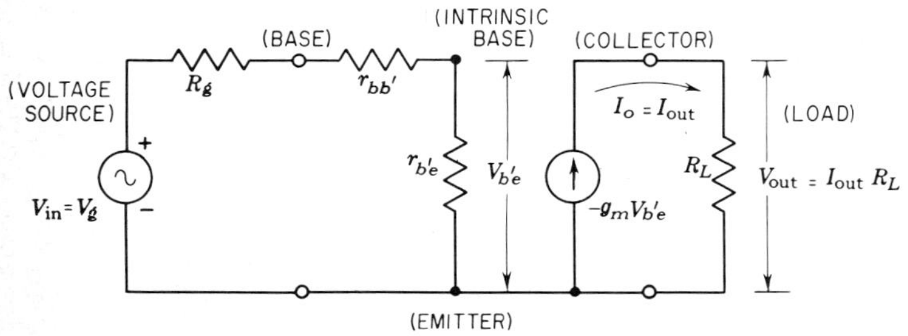

where $V_{b'e}$ may be derived as

$$V_{b'e} = \frac{r_{b'e}V_g}{r_{bb'} + r_{b'e} + R_g} \tag{13.114}$$

Thus, the voltage gain is found to be

$$G_v = \frac{V_{\text{out}}}{V_g} = \frac{-g_m r_{b'e} R_L}{r_{bb'} + r_{b'e} + R_g} \tag{13.115}$$

Notice that the expressions for current and voltage gain differ only by the factor R_L/R_g; thus, the following relationship may be seen to exist as

$$G_v = G_i\left(\frac{R_L}{R_g}\right) \tag{13.116}$$

Upper Cutoff Frequency. It was previously explained that two effects control the upper cutoff frequency of the transistor: the gain-bandwidth product f_T, whose effect is shown by the diffusion capacitance C_D, and the collector-junction capacitance C_c. We wish to derive an expression for the upper cutoff frequency of the amplifier that will include the effect of both C_D and C_c. In order to accomplish this, a rather clever circuit trick is employed whereby the collector capacitance C_c is reflected back to the input side of the transistor equivalent circuit. This reflection of the feedback capacitance into the input is called the Miller-effect transformation, and it is similar to the transformation employed with the grid-to-plate capacitance in a vacuum tube. In order to visualize the transformation and its simplicity, consider the circuit of Fig. 13.46*a*.

If one breaks the equivalent circuit at points b' and e and looks to the right, he will see the circuit of Fig. 13.46*b*. The following equations may readily be written. The output voltage is

$$V_{\text{out}} = I_{\text{out}}R_L = -g_m V_{b'e} R_L \tag{13.117}$$

This may be rewritten as

$$\frac{V_{\text{out}}}{V_{b'e}} = -g_m R_L \tag{13.118}$$

Now, let us hold Eq. (13.118) in abeyance while we obtain an expression for the input impedance of the circuit of Fig. 13.46*b*. The input impedance

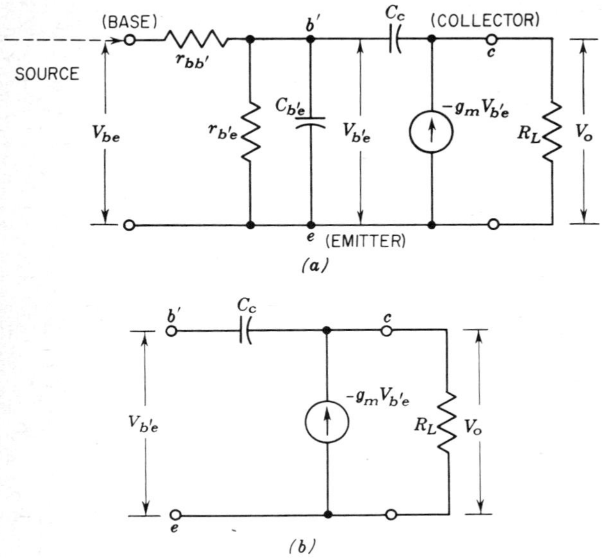

FIGURE 13.46
(*a*) Equivalent circuit including capacitances. (*b*) Circuit to right of points *b′* and *e*.

would be Z_{in}, and this value divided into $V_{b'e}$ would provide the current flowing into the circuit. Thus,

$$I = \frac{V_{b'e}}{Z_{in}} \tag{13.119a}$$

This same current must also flow through C_c. To obtain an expression for this capacitor current, note that C_c has voltage $V_{b'e}$ on one side and voltage V_{out} on the other. It is the difference between these voltages that will produce a current through C_c. Hence, we can write

$$I_c = \frac{V_{b'e} - V_{out}}{X_{c_c}} \tag{13.119b}$$

Since the currents of Eqs. (13.119a) and (13.119b) are equal, we may equate them:

$$\frac{V_{b'e}}{Z_{in}} = \frac{V_{b'e} - V_{out}}{X_{c_c}} \tag{13.119c}$$

or

$$Z_{in} = \frac{X_{c_c} V_{b'e}}{V_{b'e} - V_{out}} = \frac{X_{c_c}}{1 - V_{out}/V_{b'e}} = \frac{X_{c_c}}{1 + g_m R_L} \tag{13.120}$$

This may finally be written as

$$Z_{\text{in}} = \frac{1}{j\omega C_c} \frac{1}{1 + g_m R_L} \tag{13.121}$$

which implies an equivalent, or transferred, capacitance of

$$C = C_c(1 + g_m R_L) \tag{13.122}$$

From this it is seen that the effect of the feedback of C_c is to increase its size by $1 + g_m R_L$ when reflected across the input at points b' and e. The result of this transformation is illustrated in Fig. 13.47, where the new equivalent circuit is shown. This circuit shows the effect of C_c transformed to the input; at the same time, the feedback path between input and output has been removed. The transformed equivalent circuit has many advantages, the most important of which is that the input circuit now takes the form of a simple parallel RC circuit. This gives rise to a simple expression for the upper cutoff frequency. Since the output current of the transistor is directly proportional to the internal base-to-emitter voltage $V_{b'e}$, the value of the output current will be down 3 dB (0.707 of its low-frequency value) when the voltage $V_{b'e}$ is also down 3 dB. Let us consider the simple circuit shown in Fig. 13.48. It is easily shown that the output voltage of this circuit will be down 3 dB at an upper frequency that is given by

$$f_{3\text{dB}} = \frac{1}{2\pi RC} \tag{13.123}$$

The input of the equivalent circuit of Fig. 13.47 is almost of this simple form. If one combines the effect of $r_{bb'}$ and R_g, the source resistance, an equivalent circuit that is a condensed form of Fig. 13.47 is obtained. This is shown

FIGURE 13.47
Equivalent circuit with C_c reflected to input.

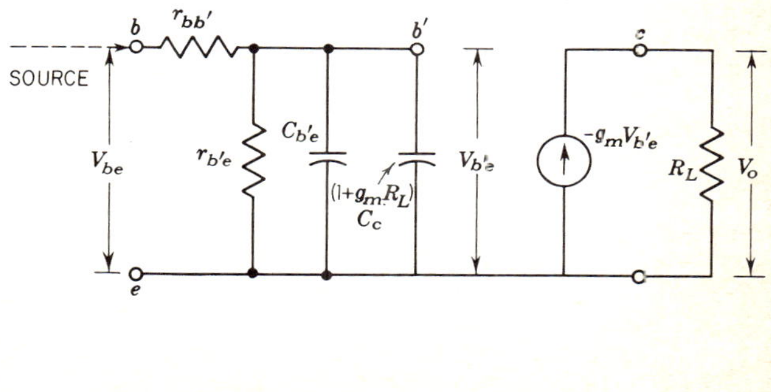

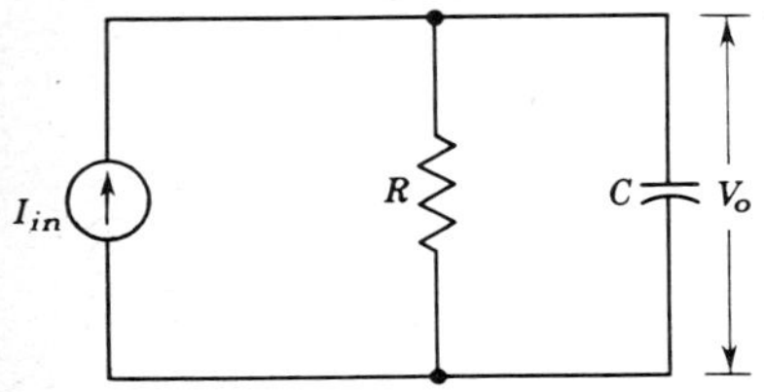

FIGURE 13.48
Single *RC* network.

in Fig. 13.49. The input to this equivalent circuit is of the same form as the simple circuit shown in Fig. 13.48; thus, the upper cutoff frequency of this circuit is given by the expression

$$f_{3\text{dB}} = \frac{1}{2\pi \dfrac{(r_{bb'} + R_g)(r_{b'e})}{R_g + r_{bb'} + r_{b'e}} [C_{b'e} + C_c(1 + g_m R_L)]} \tag{13.124}$$

which may be rewritten as

$$f_{3\text{dB}} = \frac{r_{bb'} + r_{b'e} + R_g}{2\pi(r_{bb'} + R_g)(r_{b'e})[C_{b'e} + C_c(1 + g_m R_L)]} \tag{13.125}$$

Notice that the upper cutoff frequency of the transistor-amplifier stage is dependent upon the transistor parameters and upon the source and load resistance. Consider the ideal case when R_g becomes very large and R_L very small. Under these conditions.

$$f_{3\text{dB}} = \frac{1}{2\pi r_{b'e}(C_{b'e} + C_c)} \tag{13.126}$$

FIGURE 13.49
Condensed equivalent circuit.

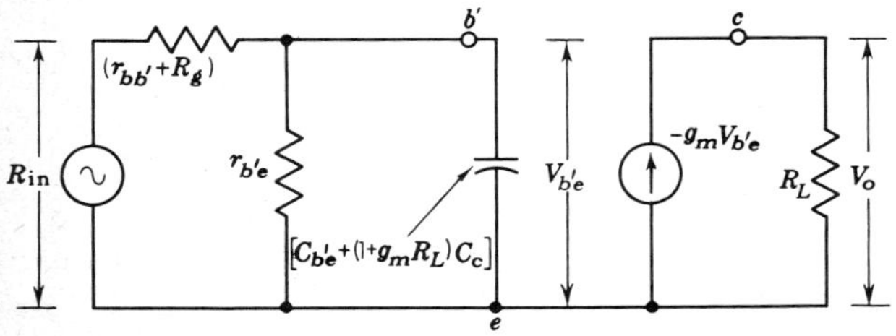

and since $C_{b'e}$ is much larger than C_c, this reduces (recalling preceding equalities) to

$$f_{3\text{dB}} \approx \frac{1}{2\pi r_{b'e} C_{b'e}} = \frac{1}{2\pi \dfrac{\beta_0}{g_m} \dfrac{g_m}{2\pi f_T}}$$

$$\approx \frac{f_T}{\beta_0} = f_\beta \tag{13.127}$$

This is the result one would expect, since we previously demonstrated that a large source resistance and a small load resistance were required for the upper cutoff frequency to be defined as it is in Eq. (13.127) [see discussion concerning Eqs. (13.106) to (13.108)].

We might pause and note that now we have examined, in varying detail, four different equivalent circuits of transistors. Actually, many more equivalent circuits have been suggested by numerous engineers; and each of these undoubtedly had some reason for favoring their particular arrangement. Equivalent circuits are designed to represent a transistor electrically, and since a transistor can perform a multiplicity of functions, it is logical to expect a number of different arrangements. The most desirable circuit is the one that is simplest in form yet represents the transistor for the widest possible range of application.

COMPLETE DESIGN OF A TRANSISTOR AMPLIFIER

In the preceding sections, we derived the formulas for the gain and the upper cutoff frequency of a transistor-amplifier stage. These equations, together with those developed for the coupling and emitter bypass capacitors and for the stability factor, provide the necessary expressions to design a complete transistor amplifier. In order to illustrate the utility of these expressions, let us perform an actual design of a typical amplifier.

The information necessary to complete the design and the desired performance of the amplifier are listed as follows (keep in mind that this is a typical application and that the requirements will vary from one application to another):

1. Lower cutoff frequency, $f_{3\text{dB}}$ low, less than 100 Hz
2. Upper cutoff frequency, $f_{3\text{dB}}$ high, greater than 2 MHz
3. Stability factor S less than 10
4. Voltage gain G_v greater than 500
5. Source resistance R_g and load resistance R_L equal to 1,000
6. Collector supply voltage V_{cc}, 32 V

The first step is selection of a suitable transistor. Since the upper cutoff frequency must be above 2 MHz, the choice is limited to transistors having an f_T greater than 100 MHz. For our example, we will select the 2N1742. This is a high-frequency PNP transistor with an f_T on the order of 200 MHz. Since the frequency-determining characteristics f_T and C_c are dependent upon the device and the operating point, choice of the operating point is generally concurrent with the choice of transistor. The 2N1742 data sheet indicates a typical f_T of 200 MHz at a collector-emitter voltage V_{ce} of -10 V and an emitter current I_E of 2 mA. The maximum C_c is specified as 1 pF at $V_{ce} = -10$ V. In choosing the operating point, the maximum power dissipation, 60 mW for the 2N1742, must not be exceeded; in our case $P_c = V_{ce}I_c = 20$ mW.

With the operating point determined, it is necessary to determine the additional parameters of interest; these are usually included on the data sheet, or they may be measured or calculated from other available data. The available data are listed as follows (at $I_c = 2$ mA, $V_{ce} = -10$ V):

$$f_T = 200 \text{ MHz (degraded for worst-case transistor)}$$
$$\beta_0 = 40 \text{ (degraded for aging)}$$
$$C_c = 1 \text{ pF}$$

$$r_{bb'}C_c = 60 \text{ picoseconds (ps)} \qquad \therefore\ r_{bb'} = 60\ \Omega$$

From these parameters, we may calculate the following:

$$\alpha_0 = \frac{\beta_0}{\beta_0 + 1} = \frac{40}{41} = 0.975$$

$$g_m = \frac{I_E}{27} = \frac{2 \text{ mA}}{27} = 0.074\ \mho$$

$$C_{b'e} = \frac{g_m}{2\pi f_T} = \frac{0.074}{2\pi \times 200} = 58.9 \text{ pF}$$

$$r_{b'c} = \frac{\beta_0}{g_m} = \frac{40}{0.074} = 540\ \Omega$$

The next step in the design is to devise a suitable circuit that will provide the desired operating point and the desired stability factor. The general form of our circuit is shown in Fig. 13.50. The two base biasing resistors form a voltage divider that in conjunction with the emitter resistor establishes the required bias. The emitter resistor serves to aid stability but cannot be too large or gain will be lowered; the collector resistor, on the other hand, may be large to avoid interstage loss and to fix the operating point. Before obtaining design values for these components, consider the collector characteristics for the 2N1742 shown in Fig. 13.51. The supply voltage is 32 V and the

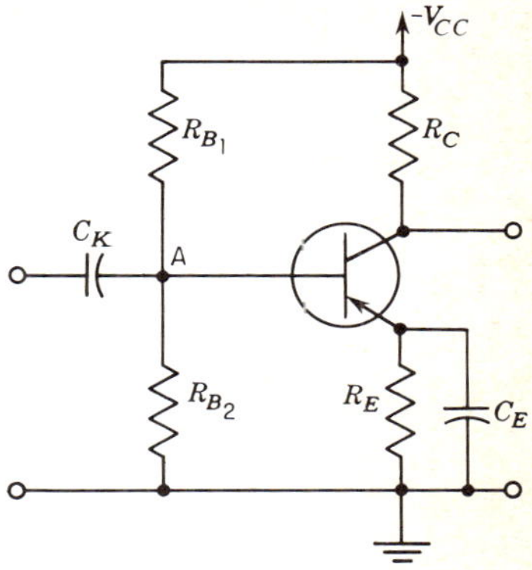

FIGURE 13.50
General amplifier circuit.

operating point is 2 mA, 10 V. This point correspcnds to a base current of 30 μA. A load line may be constructed througn the two points to obtain the total series resistance $R_C + R_E$ as

$$\frac{32}{3\text{ mA}} = 10{,}700\ \Omega$$

Thus, we shall choose $R_E = 2{,}200\ \Omega$ and $R_C = 8{,}700\ \Omega$.

FIGURE 13.51
Typical collector characteristic curves for a 2N1742 transistor.

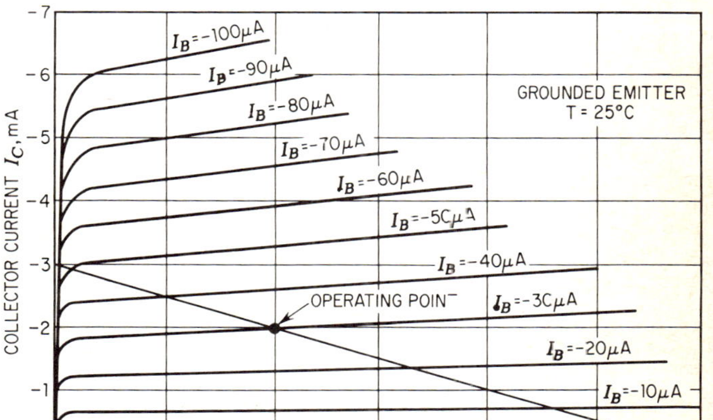

Experience is probably the best design tool in the matter of selecting the values of R_{B_1} and R_{B_2}. Lack of experience may be compensated for by making an operating circuit and experimentally adjusting the values of R_{B_1} and R_{B_2} until the desired operating point is achieved. However, we can "zero in" on the approximate values by a few approximations. Referring to Fig. 13.50, we see that the current I_1 that travels down through R_{B1} divides at point A into two currents. One segment of this current, I_2, passes through R_{B_2}, while the other segment flows into the base, becoming I_B for the transistor. In equation form, this relationship is expressed as follows:

$$I_1 = I_2 + I_B \tag{13.128}$$

From a design standpoint, it is more convenient to express Eq. (13.128) in terms of the resistances and voltages in the circuit. I_{CO} is neglected and I_B, since it is a design parameter, remains. Doing this gives us

$$\frac{V_{CC} - (V_{BE} + I_E R_E)}{R_{B_1}} = \frac{V_{BE} + I_E R_E}{R_{B_2}} + I_B \tag{13.128a}$$

If we neglect V_{BE}, which will be the order of 0.3 V, this becomes

$$\frac{V_{CC} - I_E R_E}{R_{B_1}} = \frac{I_E R_E}{R_{B_2}} + I_B$$

Substituting the assumed value for I_E, R_E, V_{CC}, and I_B, we have

$$\frac{32 - (2\text{ mA})(2.2\text{ k}\Omega)}{R_{B_1}} = \frac{(2\text{ mA})(2.2\text{ k}\Omega)}{R_{B_2}} + 0.03\text{ mA}$$

or

$$R_{B_1} = \frac{27.6}{4.4/R_{B_2} + 0.03\text{ mA}}$$

If we assume a value of 10,000 Ω for R_{B_2}, then

$$R_{B_1} = 59{,}000\ \Omega,\ \text{approx}$$

Allowing for I_{CO} and V_{BE}, R_{B_1} finally comes down to a value of 52,000 Ω.

We are now in a position to calculate the stability factor, which is given by

$$S = \frac{1 + R_E\left(\dfrac{R_{B_1} + R_{B_2}}{R_{B_1}R_{B_2}}\right)}{1 - \alpha_0 + R_E\left(\dfrac{R_{B_1} + R_{B_2}}{R_{B_1}R_{B_2}}\right)} \tag{13.129}$$

Substituting the appropriate values given,

$$S = \frac{1 + 2.2(62/520)}{1 - 0.975 + 2.2(62/520)} = 4.4$$

which indeed meets the requirements of $S \leq 10$.

The next step is to calculate the voltage gain available from our single stage. From this, we may calculate the number of stages required.

$$G_v = \frac{g_m r_{b'e} R_L}{r_{bb'} + r_{b'e} + R_g} = \frac{0.074 \times 540 \times 1{,}000}{60 + 540 + 1{,}000} \tag{13.130}$$

$$= 25$$

This indicates that two stages will be necessary. The bandwidth for a single stage is [using Eq. (13.125)]

$$\text{BW} = \frac{r_{bb'} + r_{b'e} + R_g}{2\pi(r_{bb'} + R_g)(r_{b'e})[C_{b'e} + C_c(1 + g_m R_L)]} \tag{13.131}$$

$$= \frac{60 + 540 + 1{,}000}{2\pi(60 + 1{,}000)(540)[58.9 + 1(1 + 0.074 \times 1{,}000)]}$$

$$= 3.32 \text{ MHz} \tag{13.106}$$

We shall design the second stage in the same fashion as the first stage. However, we must calculate not only the individual gains and bandwidths but also the overall gain and bandwidth of the two-stage amplifier.

Before we make these calculations, bear in mind that the load on the first stage is the input resistance of the second stage and the source for the second stage is the output resistance of the first stage. In finding the load on the first stage, the biasing network may be neglected, since the resistors are large compared with $r_{b'e}$; however, the biasing network does become important when determining the source for the second stage.

Thus, the load that stage 1 works into is the input resistance of stage 2. This is

$$\begin{aligned} R_{\text{in}} \text{ of stage 2} &= r_{bb'} + r_{b'e} \\ &= 60 + 540 \\ &= 600\ \Omega \end{aligned}$$

The source resistance for stage 1 is the series resistance of the input signal, be it a generator or a preceding stage. In Fig. 13.52, this is indicated to be 1,000 Ω.

The input or source resistance for stage 2 consists of the 8,700-Ω collector load resistor of stage 1 in parallel with the 52,000-Ω resistor and the

10,000-Ω resistor of the base biasing network of stage 2. Hence R_{g_2} possesses a value of 4,260 Ω, obtained by performing the indicated calculation. Finally, the load resistance of stage 2 is 1,000 Ω, as revealed by Fig. 13.52.

We must use the expressions for current gain and then convert into voltage gain because two stages, not one, are involved. The current gains are calculated as

$$G_{i_1} = \frac{g_m r_{b'e} R_g}{r_{bb'} + r_{b'e} + R_g} = \frac{(0.074)(540)(1{,}000)}{60 + 540 + 1{,}000} = 25$$

$$G_{i_2} = \frac{g_m r_{b'e} R_g}{r_{bb'} + r_{b'e} + R_g} = \frac{(0.974)((540)(4{,}260)}{60 + 540 + 4{,}260} = 35$$

Recalling that voltage gain is equal to $(R_L/R_g)G_i$, we obtain

$$G_v = G_{i_1} G_{i_2} \times \frac{R_{L_2}}{R_{g_1}} = 25 \times 35 \times \frac{1{,}000}{1{,}000} = 875$$

which satisfies our requirement of $G \geq 500$.

The individual bandwidths may be calculated as in Eq. (13.131). The results are

$$\begin{aligned} \text{BW}_1 &= \frac{60 + 540 + 1{,}000}{2\pi(60 + 1{,}000)(540)[58.9 + 1(1 + 0.074)(600)]} \\ &= 4.26 \text{ MHz} \end{aligned}$$

$$\begin{aligned} \text{BW}_2 &= \frac{60 + 540 + 4{,}260}{2\pi(60 + 4{,}260)(540)[58.9 + 1(1 + 0.074 \times 1.000)]} \\ &= 2.48 \text{ MHz} \end{aligned}$$

FIGURE 13.52
Final circuit design.

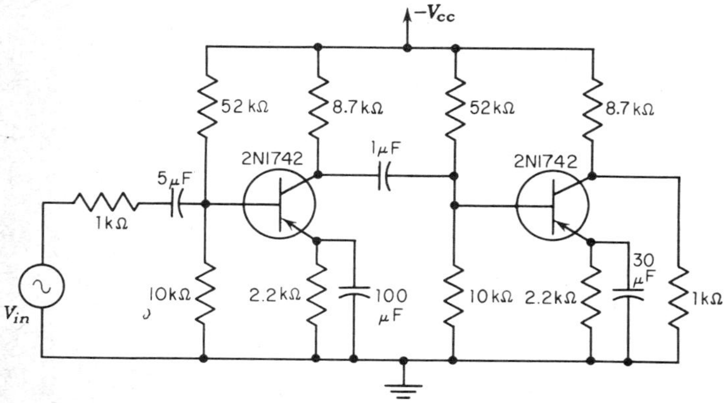

From vacuum-tube theory, one recalls that the bandwidth of a two-stage amplifier is smaller than the smallest bandwidth of either stage. The calculation of the overall bandwidth is easily accomplished with the aid of the bandwidth-degradation curve shown in Fig. 13.53. To use this curve, we must find the ratio of the larger individual bandwidth to the smaller individual bandwidth as $K = f_L/f_s$. This ratio is used to obtain the degradation factor D from Fig. 13.53. The overall bandwidth is then given by the product of the degradation factor and the smaller individual bandwidth as $\text{BW} = Df_s$. In our case $f_L = f_1 = 4.26$ MHz and $f_s = f_2 = 2.48$ MHz. Thus the ratio is

$$K = \frac{4.26\text{ MHz}}{2.48\text{ MHz}} = 1.72$$

From Fig. 13.53 we obtain the value of the degradation factor as $D = 0.81$. Thus the overall bandwidth is

$$\text{BW} = (0.81)(2.48\text{ MHz}) = 2.01\text{ MHz}$$

which satisfies the requirements of $\text{BW} \geq 2$ MHz.

The only calculations that remain are those required to find the values of the coupling capacitors and the emitter bypass capacitors. These capacitors

FIGURE 13.53
Bandwidth-degeneration curve for a two-stage amplifier.

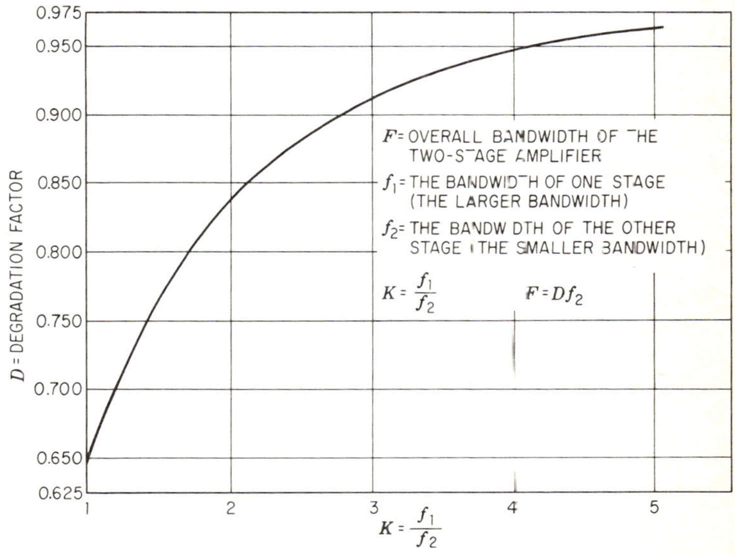

will be different for each stage because of the different source impedances involved. The coupling capacitor is given by

$$C_K = \frac{1}{2\pi f_{3\text{dB}}\left(R_C + \dfrac{R_{B_t}R_{\text{in}}}{R_{B_t} + R_{\text{in}}}\right)}$$

where
$$R_{B_t} = \frac{R_{B_1}R_{B_2}}{R_{B_1} + R_{B_2}}$$

We may calculate R_{B_1} as

$$R_{B_t} = \frac{(10\ \text{k}\Omega)(52\ \text{k}\Omega)}{10\ \text{k}\Omega + 52\ \text{k}\Omega} = 8.4\ \text{k}\Omega$$

and it is the same for both stages. Similarly for both stages, $R_{\text{in}} = r_{bb'} + r_{b'e} = 60 + 540 = 600$. Thus, $R_{Bt}R_{\text{in}}/R_{Bt} + R_{\text{in}} = 560$. For the first stage the R_C term is the source resistance $R_{g_1} = 1$ kΩ, and for the second stage it is the collector resistor $R_C = 8.7$ kΩ. We may now calculate the coupling capacitors as

$$\text{First-stage } C_{K_1} = \frac{1}{2\pi(100\ \text{Hz})(1{,}000 + 560)} = 1\ \mu\text{F}$$

$$\text{Second-stage } C_{K_2} = \frac{1}{2\pi(100\ \text{Hz})(8{,}700 + 560)} = 0.17\ \mu\text{F}$$

The emitter bypass capacitor is given by

$$C_E = \frac{\beta_0 + 1}{2\pi f_{3\text{dB}}\left(R_{\text{in}} + \dfrac{R_C R_{B_t}}{R_C + R_{B_t}}\right)}$$

We may calculate $R_{Bt}R_C/R_{Bt} + R_C$ for the first stage as

$$\frac{(8.4\ \text{k}\Omega)(1{,}000)}{8.4\ \text{k}\Omega + 1{,}000} = 895\ \text{k}\Omega$$

and for the second stage as

$$\frac{(8.4\ \text{k}\Omega)(8.7\ \text{k}\Omega)}{8.4\ \text{k}\Omega + 8.7\ \text{k}\Omega} = 4.28\ \text{k}\Omega$$

We can now determine the emitter bypass capacitor values

$$\text{First-stage } C_{E_1} = \frac{40 + 1}{2\pi(100 \text{ Hz})(600 + 895)} = 43.8 \ \mu\text{F}$$

$$\text{Second-stage } C_{E_2} = \frac{40 + 1}{2\pi(100 \text{ Hz})(600 + 4{,}280)} = 13.4 \ \mu\text{F}$$

In order to further ensure the low-frequency response, let us increase each of these capacitors to the following values:

$$C_{K_1} = 5 \ \mu\text{F} \qquad C_{E_1} = 100 \ \mu\text{F} \qquad C_{K_2} = 1 \ \mu\text{F} \qquad \text{and} \qquad C_{E_2} = 30 \ \mu\text{F}$$

The capacitors required for the second stage are smaller than those for the first stage; this is because the source resistance feeding the second stage is larger than that feeding the first.

The completed design circuit is given in Fig. 13.52. We have so designed this circuit that it offers a stability factor of 4.4, a voltage gain of 875, a bandwidth of 2.01 MHz, and a low-frequency cutoff of less than 100 Hz.

As a summary of the foregoing design, let us again list the steps involved and the measures that have to be taken if the design requirements cannot be met in all instances. The first step is to select the operating point. Perhaps the only critical limitation we might encounter here would be transistor power dissipation or voltage rating limits. The remedial action would be to go to a less desirable operating point or to change the supply voltage. The next step is to obtain the derived stability factor. If that were not possible, we would have to reduce the base biasing resistor or would have to increase the emitter resistor. In the former case, we would sacrifice gain. Gain may be increased, within limits dictated by device linearity, by adding more stages. However, as indicated by the bandwidth-degradation chart, an increase in gain is accompanied by a decrease in bandwidth. Unique and complex feedback methods are available for increasing the bandwidth with less sacrifice in gain, but such techniques are beyond the scope of this text. An example of one such feedback technique is shown in Fig. 13.54. R_e and R_f are feedback resistors that increase the bandwidth at the expense of gain. This two-stage amplifier employing the 2N1742 yields a voltage gain of 10 with a bandwidth of 50 MHz if $R_g = 50$ and $R_L = 50$. A voltage gain of 40 is possible with a bandwidth of 15 MHz if $R_g = 1{,}000\ \Omega$ and $R_L = 1{,}000\ \Omega$. Finally, the low-frequency cutoff is attained by using large coupling and emitter bypass capacitors. There are limits on how large these capacitors may be both electrically and physically, however.

As a final word, keep in mind that there may be design requirements that are unattainable. For instance, it would be a difficult problem indeed to design our amplifier for a voltage gain of 500 and a bandwidth of 20.0 MHz.

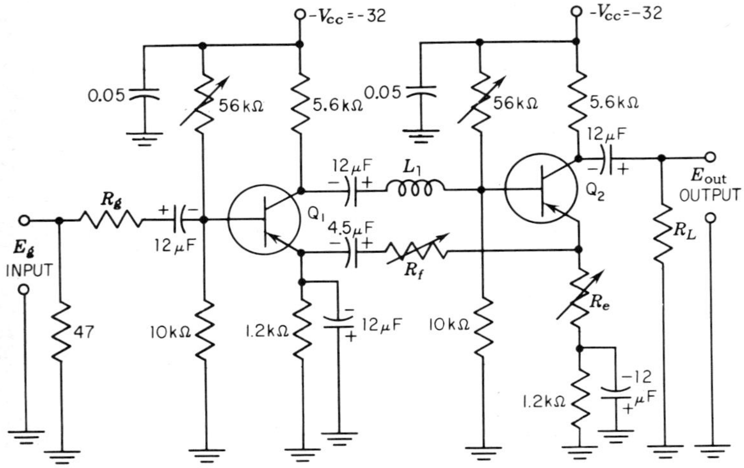

FIGURE 13.54
Band amplifier employing feedback.

Such a requirement might be met, but only after resorting to very complex feedback techniques such as those shown in Fig. 13.54.

QUESTIONS

13.1. What factors must be considered when selecting a dc operating point for a transistor amplifier?

13.2. Obtain the collector-voltage–collector-current characteristics for a transistor to be operated as a grounded emitter. Plot the maximum collector dissipation curve for that transistor on the characteristic chart.

13.3. How is the position for a load line determined?

13.4. Draw a suitable load line for the transistor represented by the characteristics selected for Question 13.2. Explain why the line drawn is suitable.

13.5. What is the difference between dc and ac load lines? Which is more important to the transistor in its role as an amplifier?

13.6. Why is the behavior of the saturation current I_{CO} with temperature more critical in grounded-emitter amplifiers than in grounded-base circuits?

13.7. Derive the relationship between β and α of a transistor.

13.8. What do we mean by the stability factor S?

13.9. Derive the stability factor S for the circuit of Fig. 13.11.

13.10. Explain, without resorting to S, why the circuit of Fig. 13.11 is more sensitive to temperature changes than the circuit of Fig. 13.12.

13.11. Derive Eq. (13.12).

13.12. When would the equivalent circuit of Fig. 13.18 be valid in representing a triode and when would the circuit of Fig. 13.19 be required?

13.13. What equations govern the circuit of Fig. 13.19? Explain each term put down.

13.14. Explain the significance of each item in Fig. 13.20.

13.15. What advantage does the equivalent circuit of Fig. 13.21 possess over the equivalent circuit of Fig. 13.20 in so far as they both represent the same transistor?

13.16. Explain how the equations governing the circuit of Fig. 13.24 are set up.

13.17. Derive the equation for the input impedance of the circuit in Fig. 13.24.

13.18. What assumption is made in deriving Eq. (13.45)? Using a commercial transistor, show from its characteristic (as given by the manufacturer) that this assumption is valid.

13.19. Explain the procedure to follow when deriving the output impedance of a transistor by using its equivalent circuit.

13.20. In what ways are hybrid parameters more useful than the other parameters discussed?

13.21. Draw the equivalent circuit of a transistor using hybrid parameters. Explain the significance of each item in this circuit.

13.22. Outline briefly how the relationship between the hybrid and resistance parameters of the same transistor circuit is obtained.

13.23. Draw a high-frequency equivalent circuit of a transistor.

13.24. Explain each component appearing in the circuit drawn for Question 13.23.

13.25. What can be done, in the fabrication of a transistor, to improve its high-frequency response? What compromises must be made?

Chapter Fourteen

Servicing Transistor Circuits

The transistor exhibits a curious combination of ruggedness and fragility. For example, it is far more physically rugged than even the most powerfully built vacuum tube; it is capable of withstanding centrifugal forces with accelerations as high as 31,000 times the force of gravity and impact tests as great as 1,900 times. These are far in excess of the forces that will completely shatter any vacuum tube.

On the other hand, the transistor is much more fragile than a vacuum tube with respect to the application of excess voltages, particularly with regard to voltages possessing the wrong polarity.

It is important, then, to be familiar with the physical-handling limitations of transistors, so that transistor equipment can be built or serviced with a minimum adverse effect either on the transistors themselves or on the miniature components with which they are often employed.

TOOLS

Probably the first step to take in preparing yourself for transistor work is the acquisition of the proper tools. Since transistors and their associated components are extremely small, conventional-sized tools are frequently unsuitable for effective use. In their place, the technician requires tools that, because of their own reduced size, are better able to cope with the limited space encountered in compact, miniaturized equipment. In addition to the smallest cutting pliers that can be obtained, it is suggested that two or three shapes and sizes of tweezers be acquired. These will come in handy when fine wires must be soldered (or unsoldered) in the circuit. Another useful device is a soldering aid one end of which has a notch for gripping wires while the other end comes to a fine point for probing or cleaning away solder from small openings.

Servicemen have also found a large reading or magnifying glass to be useful, particularly one mounted on a holding stand so that both hands are left free. Other tools that should be available include needle-nose pliers and small- and long-shank screwdrivers having narrow blades.

Another change required by the transistor is the use of a small, low-wattage soldering iron (or pencil) possessing a narrow point or wedge. Wattage ratings on the order of 35 to 40 watts (W) are satisfactory; anything larger than this could damage the transistor while it is being soldered into the circuit. (The same low-wattage iron is also required for the printed-circuit wiring of a transistor receiver.)

To provide the transistor with the maximum protection while it is being soldered or unsoldered, it is good practice to grasp the terminal lead tightly with long-nose pliers positioned between the transistor body and the lead end. With this arrangement, any heat traveling along the wire will be shunted away from the transistor housing. It is desirable to retain the pliers on the wire for a short time after the iron has been removed to make certain that all the heat has been dissipated. It is also good practice to provide such a heat shield when other wires are being soldered to any terminal lugs to which a transistor lead is attached.

Two helpful rules to follow are to keep the transistor leads as long as possible, consistent with the space available and the application, and to get whatever soldering that has to be done over with as quickly as possible. Helpful in this respect is 60/40 low-temperature rosin-core solder.

In some instances, transistors are constructed with leads stiff enough to permit plugging the transistor into a specially constructed socket (Fig. 14.1). In such cases, of course, soldering is no problem, and the only precaution to observe is to remove the transistor from the socket before the soldering iron is brought into contact with any of the socket terminal lugs.

As a final word concerning the use of any tools on transistors and their associated miniaturized components, always remember that because these units

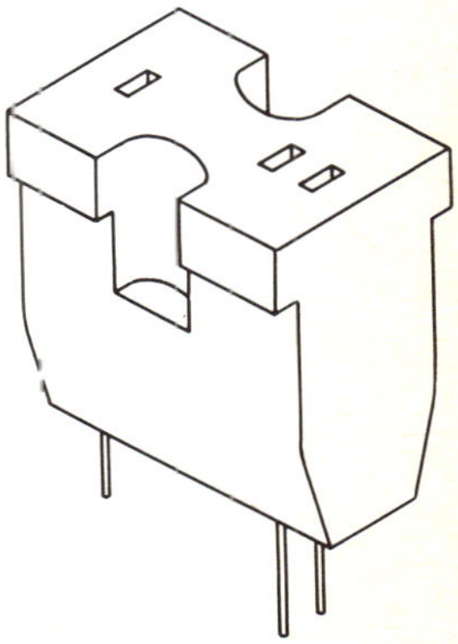

FIGURE 14.1
A representative transistor socket.

are small, their connecting wires are quite fragile. Handle these wires carefully and gently, both when the part is being installed and when it is being removed.

BATTERY POTENTIALS

Two factors combine to make transistors particularly sensitive to applied bias voltages. First, there is the fact that the emitter-base junction is biased in the forward, or low-resistance, direction and the impedance of this circuit, under these conditions, is extremely low. Any voltage in excess of the required value could easily result in so large a current that the resultant heat would permanently damage the transistor. The correct operation of a transistor—any transistor—is intimately tied in with the maintenance of its crystal lattice structure and the distribution of certain impurity atoms throughout that structure. If enough heat is generated to disrupt the crystal structure, the effectiveness of the transistor to function as desired is seriously undermined. This is the reason for the oft-repeated warnings against applying too much heat or permitting the unit to become too warm during operation.

The second factor that makes transistors sensitive to applied bias voltages is the extremely minute dimension of the several elements and their very limited heat-dissipating ability. Collector current is important in this respect because this current, passing through the relatively high collector resistance, develops a certain amount of heat. If this heat, added to the ambient heat at which the transistor is operating, exceeds the maximum limits of the transistor, behavior becomes erratic. That is why the maximum collector dissipation is always specified at a definite ambient temperature. If the surrounding temperature is higher than specified, the collector-dissipation rating must be reduced proportionately. This is called derating and was discussed in Chap. 3. The maximum safe value of collector voltage is important also, since too high a value will lead to a reverse voltage breakdown.

Thus, because of the foregoing limitations, the value and the polarity of any voltages applied to the circuit must be scrutinized carefully. Make certain first that you have the right voltage, then check polarity before making final connection to the circuit. If you are at all in doubt about the latter point, check the type of transistors being employed. PNP units require negative collector voltages and positive emitter voltages, both taken with respect to the base. In NPN transistors, the reverse situation holds.

Before the battery is connected to the circuit, the various transistors should be firmly in place. Never insert or remove a transistor when voltages are present. This is designed to prevent the development of surge currents that can damage a transistor. Always remove the voltage first. If you are experimenting with a new circuit or building a transistor kit, double-check all wiring before applying bias voltages. If you are doubtful about the outcome, insert a current meter in series with the collector circuit and then use a potentiometer arrangement to apply the collector voltage gradually. If the collector current begins to exceed the specified maximum, you know something is at fault.

To men who have gained all their radio and television experience on vacuum-tube circuits, all these precautions may appear somewhat excessive. However, experience has revealed that they are most certainly required. Transistors are highly sensitive to heat, and anything that develops heat, such as current flow, must be watched with a wary eye.

Another source of potential danger lies in the signal generators that the technician uses to service radio and television systems. When a signal is injected into a transistor circuit, start with a very low amplitude signal and gradually increase the generator output until the desired indication is obtained. Never inject strong signals into a transistor circuit, particularly when it is a low-level stage. Frequently, indirect- rather than direct-coupling methods of signal injection are advisable. For example, clip the "hot" output lead from the generator across the insulated body of a nearby resistor or capacitor. The signal will then enter the circuit by radiation and capacitive coupling. This approach is widely practiced in television-receiver alignment when a marker signal must be brought into the system without swamping the sweeping signal.

It has also been suggested that signal injection can be achieved by connecting the output of the signal generator to a suitable coil and then aligning the axis of the coil with that of the input of the circuit under test. This will bring the signal into the circuit by inductive coupling. In using this method, the radiating coil should be geared to the signal frequency, i.e., a high-inductance coil for low frequencies and a low-inductance coil for high frequencies.

The sensitivity of a transistor to surge currents should be borne in mind when a voltmeter is being used to check voltages at various points in a transistor receiver. Because of the closeness with which components are placed, it is easy for the probe tip to touch two closely spaced terminals accidentally if the technician is not exceptionally careful. This simple slip may result in battery burnout or be responsible for a current surge through the transistor as, for example, when the probe makes simultaneous contact with the collector and base electrodes. Special emphasis is placed on this precaution because of the ease with which the mistake can be made. In vacuum-tube circuits, similar slips may occasionally cause a component to burn out, although they rarely affect tubes. In a transistor circuit, the transistor is usually the weakest link, and it becomes the victim.

Along these lines, here are some meter precautions that are issued by the manufacturer of the pocket radio receiver shown in Fig. 9.3.

Some volt-ohmmeters (VOMs) and vacuum-tube voltmeters (VTVMs) have their red leads positive, whereas others have a negative potential on the red lead. The technician should investigate his meter to determine the polarity of its leads. This can be done easily by connecting a voltmeter across the ohmmeter test prods. When measuring circuits that are critical with regard to polarity (such as those containing electrolytic capacitors), the technician should keep in mind the polarity of the meter leads and should connect them accordingly. The positive lead, whether it is red or black, should be connected to the positive lead of the electrolytic capacitor. The transistors in this receiver would not be ruined if an ohmmeter were to be connected into the circuit in the reversed polarity, but the electrolytic capacitors would give incorrect readings because they would be measured backwards. It is also imperative not to use an ohms range that utilizes a battery of more than 3 volts (V), because the transistors can be damaged if too much voltage is applied to them.

Before we leave the subject of heat and its effect on transistors, one word might be said about the precautions to observe when positioning transistors in electronic equipment. Keep transistors clear of any component, be it tube, resistor, or transformer, that passes enough current to develop a noticeable amount of heat. The ratings specified for a transistor are always given at a certain ambient temperature, generally 25°C. For every degree above this figure, a corresponding lowering of the transistor ratings must be made, thereby effectively reducing the operating range of the unit. It might be useful to remember this when you find that transistor equipment is not operating as it should and no component is apparently at fault. Measure the ambient temperature of the enclosure where the transistor is contained. Make this measurement under the same conditions that prevail when the equipment is functioning normally, that is, with the chassis in the cabinet and all removable sections or panels in place. If you possess a Fahrenheit thermometer, the equation for conversion to centigrade is

$$C = \frac{5}{9}(F - 32)$$

where C = temperature, °C
F = temperature, °F

SAFETY PRECAUTIONS WITH THE MOSFET

The MOSFET (metal oxide semiconductor field-effect transistor) is a very delicate device when it is out of the circuit, far more so than a bipolar transistor. In circuit, it can be just as rugged as the junction-type FET (field-effect transistor). However, out of the circuit, the MOSFET is subject to damage from static charges when handled. This stems from the fact that in the construction of a MOSFET, the gate is electrically isolated from the main body of the transistor. Since there is an insulating film between the gate and the MOSFET channel or body beneath it, not much of a static voltage charge between the gate and the body of the transistor is required to puncture this insulating film and cause a permanent short circuit that ruins the MOSFET. Note that this condition does not appear in an ordinary junction FET because here the gate makes a direct ohmic connection to the body of the transistor and thus no static charge will be built up between the gate and the other terminals of the junction FET; the same precaution consequently need not be observed.

The MOSFET is generally shipped with all of its leads shorted together to prevent damage in shipment and handling. To test a MOSFET out of circuit, the leads from the tester should be connected to the MOSFET before the short is removed from the MOSFET leads. The following points should be closely followed when checking the MOSFET.

1. When a MOSFET is to be unplugged from a unit for testing, your body should be at the same potential as the unit. This can be accomplished by placing one hand on the chassis before the MOSFET is unplugged. Before the MOSFET is connected to a tester, put the hand holding the MOSFET in contact with any metal surface of the tester, generally its front panel. Now, connect the various leads from the tester to the appropriate MOSFET leads. When a large number of MOSFETs are to be tested, a lead clipped from a metal surface on the tester to your watchband or ring is very helpful.
2. When handling the MOSFET, the leads must be shorted together. This is generally done in shipment by a shorting ring or piece of wire. Connect the test leads from the tester to the MOSFET, with the source lead being first. Now, connect the other leads from the tester to the appropriate MOSFET leads. The shorting ring or wire may then be removed without fear of damage.
3. When soldering or unsoldering the MOSFET, the iron tip must be at ground potential. Connect a clip lead from the barrel of the soldering iron to the chassis in which the MOSFET is to be soldered or unsoldered. Do not use a soldering gun.
4. Turn off the power to the circuit before inserting or removing a MOSFET. The voltage transients that are generated may damage the transistor.

All these precautions may seem somewhat excessive, but it is a fact that a MOSFET can be readily damaged by static charges, and every precaution must be taken to avoid such a buildup between the gate and the body of a MOSFET transistor.

TRANSISTOR TESTING

The transistor may be checked, in roundabout fashion, by substituting another unit known to be good. If the circuit operation returns to normal, the previous unit was defective; if the trouble persists, the transistor was not at fault.

A more direct approach is to test the suspected transistor in a suitable tester. A number of such units are commercially available. Many of these testers are fairly simple in design, checking only the leakage current and current gain of a transistor. If the results of both checks are satisfactory, the unit is probably functioning normally.

In most testers, leakage is checked first because it is the more sensitive indicator of the condition of a transistor. Leakage will almost always drag the gain down with it, and a transistor that possesses more than the normal amount of leakage will have a greater distorting effect on the circuit than one with low leakage and gain. If a transistor passes the leakage test, it should then be checked for gain.

Leakage Measurement. In nearly all commercial transistor testers, leakage tests are performed in one of two ways. The most widely employed method is the I_{CEO} test in Fig. 14.2*a*, in which a meter, resistor, and battery are connected in series. If a PNP transistor is being checked, the positive terminal of the battery is connected to the emitter and the negative terminal is connected to the end of the circuit terminating at the collector. The current through the circuit is a function of temperature, the resistivity of the germanium or silicon, and the applied voltage. Any contamination on the surface of the transistor or a short circuit within the device will produce a high reading on the meter.

The I_{CEO} notation of this leakage test indicates that current is flowing between the collector (C) and emitter (E) with the third element or base open (O). Thus, the first two letters after the I indicate the circuit in which the current is being measured. The disposition of the third element is then identified by the third letter, O, here standing for open circuit.

Perhaps a more sensitive indicator of the leakage condition of a transistor is the I_{CBO} current (frequently shortened to I_{CO}). As the notation indicates, current is measured between the collector and base with the emitter circuit open.

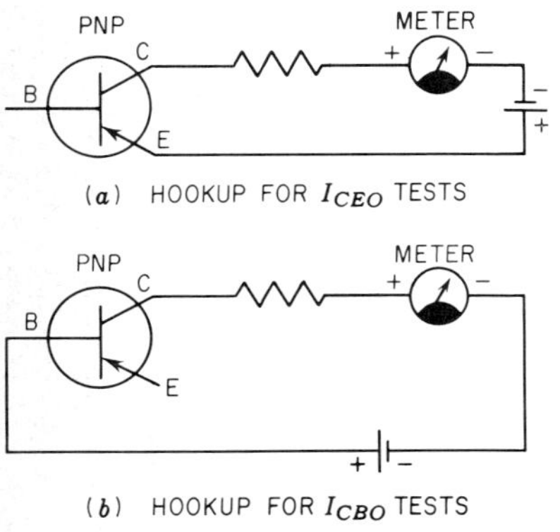

FIGURE 14.2
Two leakage-current tests that are performed on transistors.

FIGURE 14.3
A comparison of the leakage currents for various types of germanium and silicon transistors.

TYPE OF LEAKAGE	GERMANIUM		SILICON	
	SIGNAL	POWER	SIGNAL	POWER
I_{CBO}	1 μA	10 μA	0.001 μA	0.01 μA
I_{CES}	5 μA	50 μA	0.005 μA	0.05 μA
I_{CEO}	50 μA	300 μA	0.01 μA	0.5 μA

This circuit is shown in Fig. 14.2*b*. The battery is connected between the collector and base, the negative battery terminal going to the collector of a PNP transistor and the positive terminal to the base. In essence, the transistor is reverse-biased, and only a very small current should flow (in the low microampere range). Since the transistor is reverse-biased, the reader may wonder why any current at all flows. The reason is the energy that the ambient, or surrounding, temperature supplies to the internal atoms of the transistor structure. This energy causes some of the atomic electrons to vibrate strongly enough to enable them to break away from their parent nucleus and leave behind an equivalent positive charge called a hole. The positive charges are attracted to the negative terminal of the battery, while the electrons are drawn to the positive terminal. It is the flow of these two opposite charges that produces a current flow. If a transistor is in good operating condition, however, this current should be very low.

Just what constitutes a very low reverse current for a transistor depends upon its power rating and whether it is made of germanium or silicon. In milliwatt transistors (the type found in most signal stages except the output), 1 μA for a germanium transistor and 0.001 μA for a silicon transistor would be considered an acceptable I_{CO}. In a power transistor, the I_{CO} for germanium can be 10 μA or more, whereas for silicon it can be 1μA or more. As a general rule, the larger the power capabilities of a transistor, the greater its acceptable I_{CO}.

It is interesting to note, as demonstrated by the figures shown in Fig. 14.3, that the three types of leakage associated with a bipolar transistor, I_{CBO}, I_{CES}, and I_{CEO}, all have different values for an operating unit. Note, too, that I_{CES} falls between I_{CBO} and I_{CEO}. If any of the two of these three leakages are approximately equal in value, the device is generally defective. Furthermore, observe the wide difference in leakage currents between germanium and silicon transistors. As a matter of fact, *any significant* leakage for a silicon device indicates that the unit is probably defective.

Leakage current can increase with transistor age, especially if some slight impurities were left in the transistor when it was manufactured. Leakage will cause the most problems in circuits that operate at both high and low temperatures, because leakage current will increase with heat. The current will approximately double for each 10°C increase for germanium devices and for each 6°C increase for silicon devices.

Some transistor testers check I_{CES} using a circuit such as that shown in Fig. 14.4. The diagram indicates that the current to be measured is the reverse current that flows between the collector and the emitter, with the base shorted to the emitter.

Gain Measurement. Once it has been established that the transistor leakage falls within a normally acceptable limit, the dc β gain of the transistor is measured next. This gain is a measurement of the change in collector current for a small change in base current. A small current is introduced into the base circuit and its effect on the collector circuit is noted. There are several ways to do this, and they will be brought out as we examine the various commercial testers.

The most desirable condition for a low-frequency transistor is low leakage and high β. As the leakage current rises and the β value falls, the transistor becomes increasingly less desirable until its usefulness in the circuit is negligible. (Keep this in mind when checking any transistor.) If several similar transistors are available, it is possible by the foregoing tests to select the one that gives the best indication and to grade the other in terms of their indications.

For high-frequency transistors [say those above 10 megahertz (MHz)], low leakage is still desirable. The β value measured by most checkers, however, is not too indicative of the high-frequency behavior of the transistor. A β reading of 5 or more is still desired to ensure that the unit will amplify, but it gives no direct clue of what the transistor will do in the circuit. The latter operating condition can be determined only by an actual test at the frequency in question, and such a facility is not generally available to the technician.

Remember that transistors have dc and ac β values, which in high-frequency units do not necessarily have any direct and simple relationship to each other. In low-frequency transistors, on the other hand, it is more likely that such dependence will exist. In general, high-frequency transistors have lower dc and ac β than low-frequency ones have.

One final word concerning the behavior of defective transistors in circuits. Transistors will short, they will develop leakage, and they may become open. However, the chances of a transistor's becoming intermittent is considerably less than is the case with vacuum tubes. Occasionally, it will be found that when a transistor does exhibit intermittent qualities, one of the connecting wires within the transistor case itself is no longer making a solid connection,

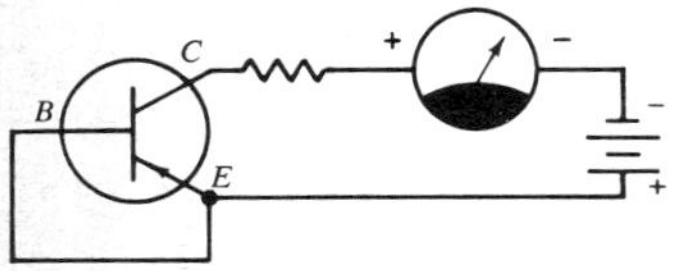

FIGURE 14.4
A test circuit to measure I_{CES}.

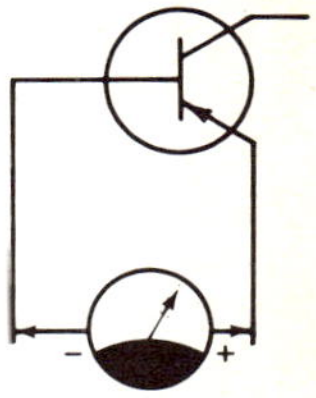

FIGURE 14.5

which can result in intermittent operation under conditions of vibration. However, the number of times that a transistor will be found intermittent will be quite low.

Before we consider transistor testers, it should be noted that transistors can usually be checked rather quickly using a simple volt-ohm meter. This method of testing depends upon the fact that any two terminals of a bipolar transistor can be considered as a diode and can actually be made to function as a diode. If this diode is forward-biased, its resistance will be quite low. On the other hand, if the terminals of this diode are reverse-biased, the resistance reading will be high.

As an illustration, consider the PNP transistor shown in Fig. 14.5. If we connect the positive lead of an ohmmeter to the emitter terminal and the negative lead of the ohmmeter to the base terminal, we will be forward-biasing the emitter-base diode and consequently the ohmmeter will indicate a very low resistance reading between these two elements. If we now reverse the leads to these same elements, we will be reverse-biasing this diode and a high reading will be obtained.

We can readily check the forward and reverse resistances between emitter and base, emitter and collector, and collector and base. If the transistor is in good operating condition, there will be a significant resistance difference between any two terminals when measured with forward and reverse biasing. If it is discovered that the reading is the same for both forward and reverse biasing, it can be assumed that the transistor is not operating normally and should be replaced.

In using an ohmmeter for these tests, it is generally best to use the $R \times 100$ or $R \times 1{,}000$ range. On any of the lower or higher ranges we may obtain voltage or current conditions that could injure the transistor.

COMMERCIAL TRANSISTOR TESTERS

Commercial transistor testers come in a variety of forms and with a fairly broad range of testing capabilities. For the very simple and economical testers, leakage in the form of I_{CO} or I_{CEO} is measured, followed by a very simple check of the dc beta for the transistor.

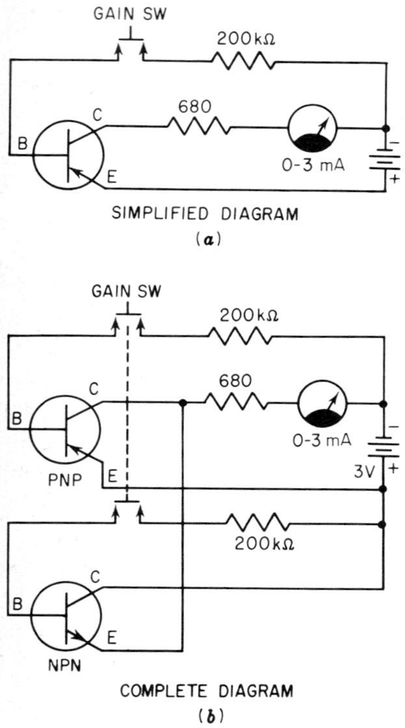

FIGURE 14.6
Simplified and complete diagrams of an economical transistor tester.

A simple circuit to measure leakage is shown in Fig. 14.6*a*. This setup is for a PNP transistor. This unit performs the I_{CEO} leakage test in which the base is left floating while a voltage is applied between the collector and emitter. A 680-ohm (Ω) resistor is connected together with a 0- to 3-milliampere (mA) meter. When the transistor is inserted into the socket, the meter will indicate the leakage current flow; generally there are also markings on the scale itself to indicate when the transistor leakage can be assumed to be beyond normal limits.

After this has been established, the gain switch is depressed. It connects a 200,000-Ω resistor to the base element, through which a current of 15 μA flows. With this current in the base circuit, a larger current should flow in the collector circuit. This larger current will be indicated by the meter.

Note that this circuit arrangement does not directly give the value of β for the transistor under test. Rather, a relative reading is taken. So long as the needle moves a sufficient number of divisions for the gain test and the leakage does not exceed a certain value, the transistor is presumed to be satisfactory.

A complete diagram of the tester is shown in Fig. 14.6*b*. The connections for the NPN socket complement those of the PNP socket to provide the proper voltages for each.

A second inexpensive transistor test circuit is shown in Fig. 14.7. The transistor is inserted into the tester and the switch labeled "Leakage-Gain" is shifted to "Gain." Calibration control R_2 is then rotated until the meter reads "1," or full scale. The switch is then returned to the leakage position and the meter reading is again noted. If the leakage current is less than the gain current (here, arbitrarily set at 1), the transistor can be presumed good. Actually, the greater the difference between the gain and leakage readings, the better the unit. Thus, although this instrument has the same basic circuitry as the preceding transistor checker has, the method of checking is somewhat different. The gain for every transistor is arbitrarily set at 1 by the calibration control while the switch is in the gain position. When the switch is returned to the leakage position, the current through the transistor should be considerably less than 1. (The leakage current being measured here is I_{CEO}.)

The foregoing arrangement permits a wide range of transistors to be checked, because no matter what the actual β of a specific transistor is, the calibration control always establishes it at 1. Thus, this checker does not directly measure β either; instead, it measures the ratio between β and leakage. The higher this ratio, the better the transistor.

RCA Transistor Tester. The RCA model WT-501A transistor tester (Fig. 14.8) is designed to test transistors for collector-to-base leakage I_{CBO}, collector-to-emitter leakage I_{CEO}, and dc beta. Collector current I_C is continuously adjustable from 20 μA to 1 ampere (A) in four ranges. The instrument can also test the in-circuit dc current gain of a transistor.

A 100-μA meter movement is used in the measuring circuits for the var-

FIGURE 14.7
The schematic diagram of a second inexpensive transistor tester.

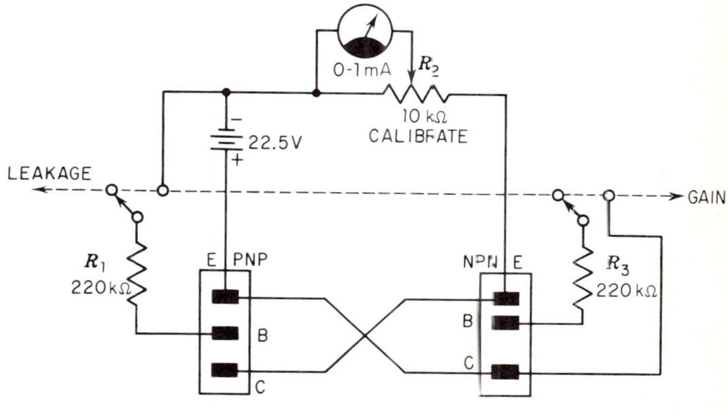

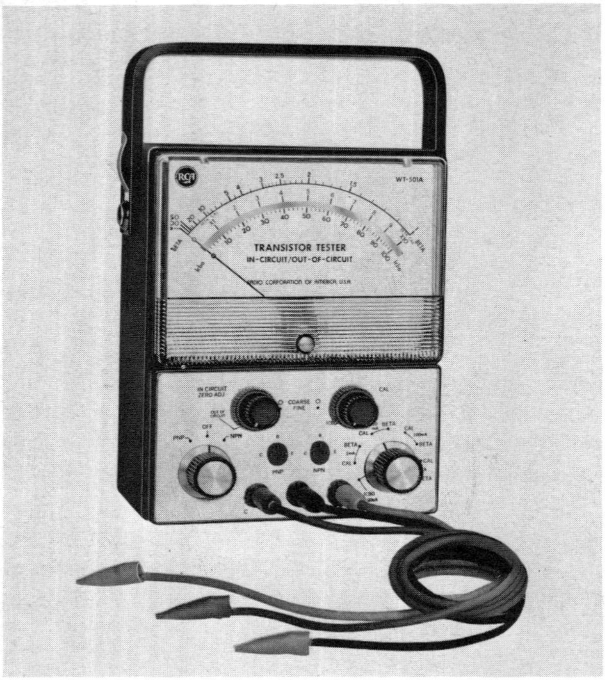

FIGURE 14.8
RCA model WT-501A transistor tester.

ious test functions. Precision resistors are used to ensure accurate test results.

An NPN/PNP switch provides the proper bias polarity to the transistor. Two dual potentiometers provide coarse and fine adjustment of collector current (CAL) and in-circuit zero.

The instrument has two internal 1.5 volt D-size batteries. One battery is used in NPN test and the other is used in PNP test. The batteries are also used during in-circuit tests to provide voltage in reverse polarity to cancel the effect of circuit leakage.

A simplified diagram of the dc beta test circuit of the WT-501A tester is shown in Fig. 14.9. Resistors R_b and R_c serve to establish the collector current and to shunt the meter to the required sensitivity. R_b and R_c values are shown below.

Range	R_b	R_c
1 mA	1,000 Ω	110 Ω
10 mA	110 Ω	10 Ω
100 mA	10 Ω	1 Ω
1 A	1 Ω	0.1 Ω

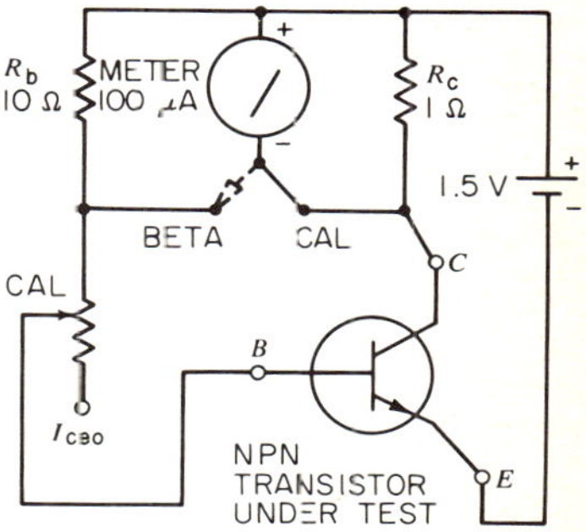

FIGURE 14.9
Simplified beta measuring circuit employed in RCA transistor tester.

When the range switch is set to the CAL function, the meter is in the collector circuit. Collector current is determined by the value of the collector resistor for the particular range, and by the setting of the CAL control.

In the beta function, the meter is switched to the base circuit. Dc beta is defined as the ratio of the dc collector current to the dc base current. Since the collector current is established at a known value by the CAL adjustment, the base current meter reading can be interpreted in terms of dc beta for the transistor.

The I_{CBO} measuring circuit is shown in Fig. 14.10. There are 1.5 V applied to the collector and base of the transistor, and the meter is connected in the collector circuit. Collector-to-base leakage is indicated directly in microamperes.

For I_{CEO} measurements, a voltage of 1.5 V is applied to the transistor, and the meter is connected in the collector circuit (see Fig. 14.11). The resistor shunting the meter reduces the meter sensitivity to 1 mA.

Measurement of I_{CEO} is normally made on the 1-mA range. However, if I_{CEO} exceeds 1 mA, the range switch can be set to the 10- or 100-mA range as necessary. Collector-to-emitter leakage is indicated in milliamperes, depending on the current range that is used.

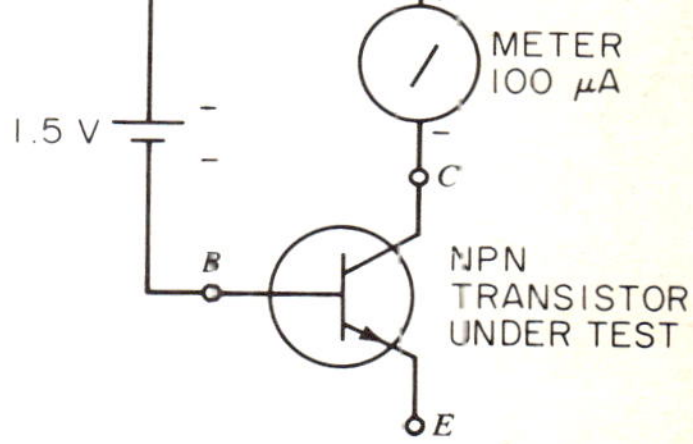

FIGURE 14.10
I_{CBO} test circuit used in RCA transistor tester.

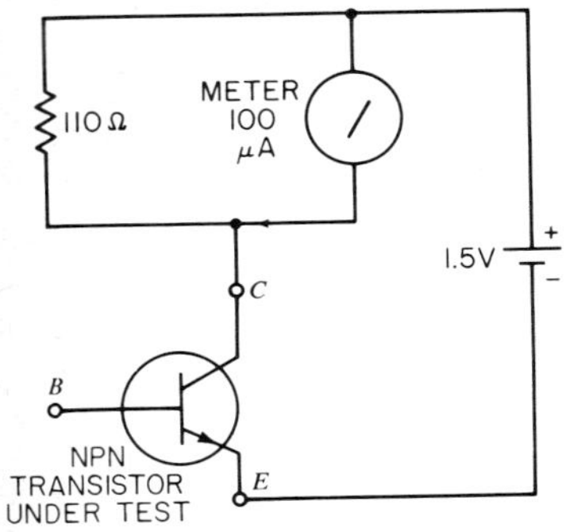

FIGURE 14.11
***I_{CEO}* test circuit, 1 mA range, in RCA transistor tester.**

IN-CIRCUIT TESTING

A very useful convenience feature that a number of transistor testers are offering is the ability to check the condition of a transistor while it is still in the circuit. If such a test can be carried out successfully, it saves the technician the time and trouble of unsoldering the transistor from the circuit and testing it, and soldering it back into the circuit if it is found to be good. In many cases, in-circuit tests can be performed, but there are limitations to this method that we will discuss shortly.

In-circuit testing is usually confined to the measurement of the current gain in essentially the same manner as that used for out-of-circuit beta measurement. The in-circuit test circuit for the RCA model WT-501A tester is shown in Fig. 14.12. The only essential difference is an in-circuit zero-adjust control that applies a voltage of reverse polarity to the collector metering circuit. This voltage compensates for the collector-to-emitter leakage through the components in the circuit under test, and permits the meter to be set to zero.

The CAL adjustment and the metering circuit are the same as for out-of-circuit measurement.

The resistance of the measuring circuit is designed so that no significant loading effect occurs from the circuit being tested.

While in-circuit testing provides a considerable convenience in checking through a nonoperating circuit, the technician must recognize that there are some very definite limitations on this method of checking transistors and field-effect transistors. It will be found, for example, that an in-circuit tester will generally uncover a defective transistor, but it will not necessarily indicate or uncover all the transistors that are good. Because of the associated circuitry, some good transistors may give a defective indication on the tester, in which case additional testing is generally desirable. That is, when a device is believed to be defective by the test indication that it gives, it should be removed from the circuit and a further out-of-circuit check should be made.

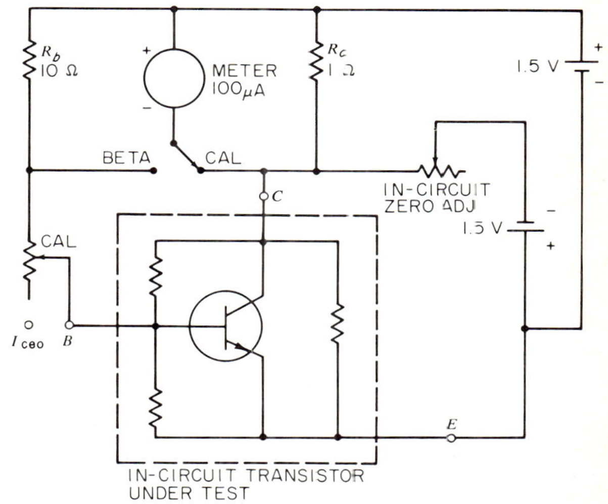

FIGURE 14.12
A simplified diagram of the in-circuit beta test employed in the RCA model WT-501A tester.

However, when a transistor gives a good indication on the tester while it is in circuit, then that decision should be accepted.

Another limitation of in-circuit testing is its inability to test in-circuit leakage. When a transistor develops a leakage current beyond its normal limit, the gain is affected, and distortion may also be introduced because the operating point of that transistor is shifted by the leakage current. Hence, it is perfectly possible for all of the transistors in a receiver to check well with an in-circuit tester and still have the output distorted. In situations in which leakage is suspected, out-of-circuit tests will have to be made.

Despite these few limitations, in-circuit testing does offer a very decided convenience, and more and more commercial testers are offering this facility.

FET TESTERS

With the increasing use of junction FETs and MOSFETs, a number of transistor testers are providing the facility to check these units. Unlike bipolar transistors, FETs (and MOSFETs) are not tested for current gain or leakage but for a quantity known as mutual conductance, or G_m. This quantity is defined as the ratio of the current coming out of the drain electrode to the voltage applied between the gate and source electrodes.

In equation form, G_m is

$$\frac{\text{Current out}}{\text{Voltage in}} = G_m$$

If we examine this equation, we see that it is essentially I/V or $1/R$ (by Ohm's law). Thus, if resistance is measured in ohms, $1/R$ is measured in mhos. Actually, because the resistance value is quite high, G_m is given in micromhos.

In the transistor/FET tester shown in Fig. 14.13, advantage is taken of the fact that theoretically the G_m of a field-effect transistor can be measured by measuring the channel resistance at a very low voltage. A shunt-type ohmmeter is used in this tester to measure the channel resistance. The voltage impressed across the channel during the test is less than 0.1 V. The G_m scale is then basically one over R_{DS} (resistance between drain and source).

The ohmmeter measures the G_m of the device; however, it does not check the gate. The gates of a FET are checked by reverse-biasing the gates, one at a time, and noting that the G_m decreases. The G_m reading should go down as the gate is reverse-biased. Two leakage tests are provided for FET testing.

The first of these is the gate leakage. Any nonzero reading of gate leakage indicates a bad device, since I_{GSS} is typically in the nanoampere region. The second test actually measures the channel current I_{DSS} at $1\frac{1}{2}$ V. This is not a leakage inasmuch as most FETs display some current in the 2- to 4-mA region. This I_{DSS} test verifies that the channel is conducting and is useful in matching devices.

A second transistor and FET tester is shown in Fig. 14.14. A bridge circuit is employed here, to measure the G_m of an FET or MOSFET. Both of

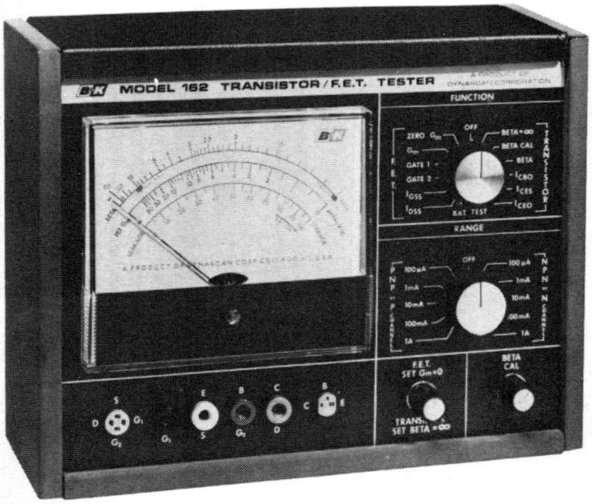

FIGURE 14.13
The B & K model 162 transistor/FET tester.

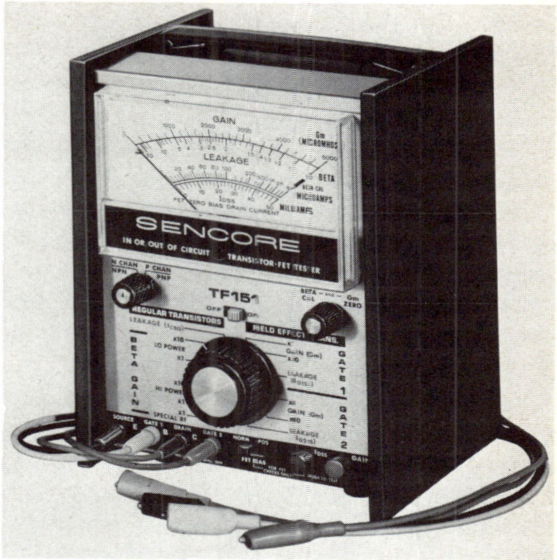

FIGURE 14.14
Sencore model TF151 transistor/FET tester.

these latter testers can check bipolar and FETs in circuit as well as out of circuit.

SERVICING TRANSISTOR CIRCUITS

In the servicing of transistor devices, much can be learned from a measurement of the voltages found in the circuits. This voltage-analysis method has long been employed with considerable success in vacuum-tube circuits; that it can be applied equally well to transistor circuits is indeed fortunate. There is this very important difference, however: whereas electrons always flow through a tube in one direction, their path through a transistor is dependent on the type of transistor, PNP or NPN. Also, the relationship between the various electrode voltages is significantly different in transistors as compared with tubes. But once these obstacles have been overcome, voltage analysis of defective transistor circuits will pay off handsomely as a servicing tool.

The following are some general statements that apply to both PNP and NPN circuits. Since both have currents flowing through them, both will develop voltage drops across resistors in the base, emitter, and collector circuits. A typical audio-amplifier stage using a PNP transistor is shown in Fig. 14.15. Resistors R_1 and R_2 provide the base with the proper voltage (or current). R_3 is an emitter-stabilizing resistor; R_4 is the collector load resistor. Since the battery potential applied to the circuit is negative, electrons will flow in the

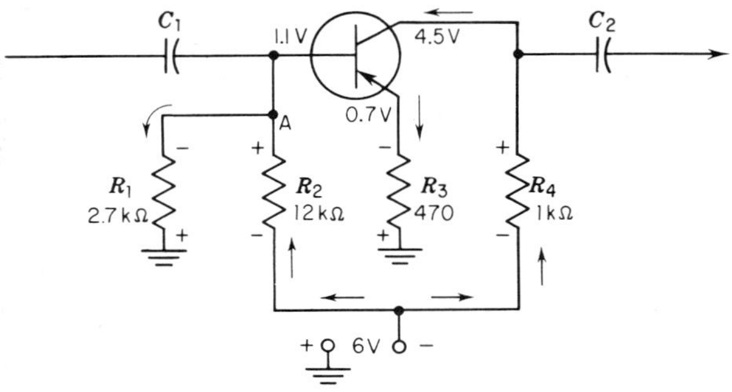

FIGURE 14.15
Current flow in a PNP transistor circuit.

directions indicated by the arrows. In PNP transistors, electrons travel into the collector and base and out of the emitter element. In NPN transistor circuits, Fig. 14.16, electrons flow into the emitter and out of the collector and base. Thus, the NPN transistor comes closest to duplicating the electron flow of vacuum tubes.

In examining the voltages at the transistor elements, it will be found that the differential between emitter and base is generally less than 1 V. This is because the emitter-base diode is forward-biased; hence, the internal resistance across this PN junction is very low and so is the voltage drop there. On the other hand, the voltage difference between collector and base or collector and emitter is considerably greater.

The one exception to the above statements occurs when a transistor is normally cut off, conducting only when it is pulsed. In this case, the difference between base and emitter voltages can be greater than the range indicated.

FIGURE 14.16
Current flow in an NPN transistor circuit.

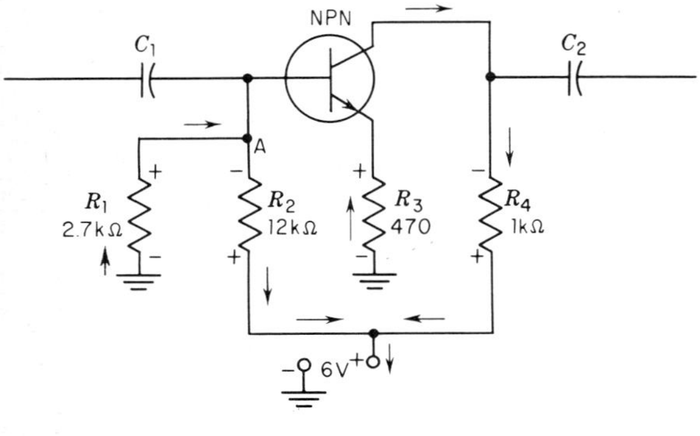

This situation does not occur in radio-receiver circuits, but it is found in television circuits and industrial equipment.

When the transistor is conducting, every resistor in its immediate circuit will have a voltage drop. The drops should be measured when a circuit is being investigated to establish the fact that current is flowing. If there is no voltage drop across one or more resistors but voltages are present, then an open transistor is a possibility. On the other hand, if the voltage drops are higher than normal, the transistor may be leaky.

Of the voltages associated with a transistor, those at the base and emitter are the most critical. When both of these elements have the same voltage, the transistor is at cutoff. This will bring the collector voltage to the same value as the applied battery voltage to that stage.

Now, normally, cutoff will be difficult to achieve in a single stage if no outside influences enter the circuit. That is, in Fig. 14.15, R_1, R_2, and R_3 can normally change values within a wide range and the transistor will still conduct (although, perhaps, at a different value from its normal condition). This is because the emitter voltage follows the base and as the base voltage changes, so will the emitter voltage. But if an outside voltage is brought into the circuit, such as a leaky C_1 might bring, then cutoff can occur.

Note, too, that the collector voltage can vary over a fairly wide range and still provide normal or close-to-normal operation. This is evident from the characteristic curves of transistors.

Voltage variations of 10 to 20 percent from those stated by the manufacturer for the circuit or system are ordinarily permissible. Furthermore, since low voltages are the norm for transistor circuits, 10 or 20 percent of these voltages will likewise be small. Thus, where we might be inclined to dismiss a variation of 5 to 10 V (from a stated value) in a vacuum-tube circuit, in a transistor circuit this could represent a change considerably more than the permissible 10 to 20 percent. So it is necessary to be more alert to voltage variations generally and always try to evaluate them from the percentage standpoint.

A circuit arrangement that is fairly common in power stages is shown in Fig. 14.17. The collector of this PNP transistor is dc-grounded, while a large positive voltage is applied to the emitter. This has the same effect as applying a similar negative voltage to the collector and grounding the end of the 470-Ω emitter resistor. The arrangement in Fig. 14.17, however, is advantageous with power transistors because it necessitates less stringent insulating precautions from the chassis on which the transistor case is mounted. In power transistors, the collector is usually connected directly to the transistor case.

Note that a similarly large positive voltage must also be applied to the base to establish the proper base-to-emitter bias. This is done in Fig. 14.17 by tying R_1 to the +9-V line. The voltage-divider arrangement of R_1 and R_2 then provides the base with the necessary voltage.

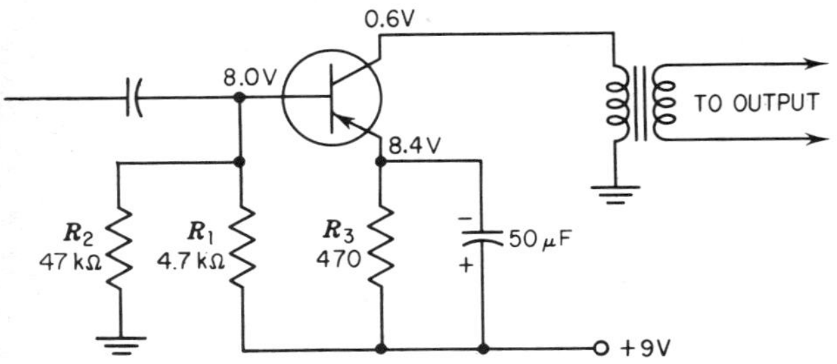

FIGURE 14.17

Particular attention is called to the circuit setup of Fig. 14.17 not only because it enjoys widespread use but also because a zero collector voltage reading (or one close to it) will often be taken as an indication of a defect.

The Effect of Open Elements and Open Leads. When an open element develops in a transistor or when an open circuit appears in one of the components (or connections) attached to a transistor, the voltages that are then measured at the transistor can easily prove confusing unless the situation is carefully analyzed. For example, in Fig. 14.17 the emitter voltage is ordinarily 8.4 V, the base is 8.0 V, and the collector is 0.6 V. If, now, R_1 should open up, conduction through the transistor would effectively stop. In making a voltage check of the stage, however, it would be found that a meter would show 9 V on the emitter *and* the base. That is, the base voltage would not show up as zero. The reason for this stems from the very low internal resistance between base and emitter. With this low resistance, the base will show, to the voltmeter, the same voltage as the emitter. This could easily lead the serviceman to conclude that the base was receiving its voltage through its voltage-divider network and pass up the defect that exists in R_1.

Exactly the same result will be obtained when the open circuit develops in the emitter lead but the base is intact. Thus, if R_3 should open up, the base voltage (because of R_1 and R_2) would remain about 9.0 V. Measuring the voltage at the emitter would reveal the same value, again because of the low resistance between the two elements.

On the other hand, if the emitter opened up internally so that no direct path then existed between base and emitter, the emitter voltage would be 9 V (Fig. 14.17), whereas the base would be 8.0 V. This difference of +1.0 V between emitter and base in a PNP transistor would ordinarily cause excessive current flow through the circuit. Since it does not here and, further, because it reveals no voltage drop across R_3, suspicion is directed at the transistor. An open internal base lead will give precisely the same voltage readings.

One interesting effect appears when an open circuit develops in the collector external circuit. With the removal of the battery voltage, we would expect the collector potential to drop to zero. Instead, however, a voltage equal to the base voltage appears at the collector. To see why this occurs, let us consider the internal conditions in the transistor. Figure 14.15 will serve as an illustration. With the collector voltage gone, we find −1.1 V on the base (which has not been affected by the change in the collector) and nothing on the collector. Since the base is an N-type semiconductor and the collector a P-type semiconductor, the foregoing voltages will forward-bias the base-collector diode and produce a low-resistance condition there. Hence, when a voltmeter probe is touched to the collector electrode, the voltage at the base will be obtained. As a matter of fact, with the collector circuit open, the potentials at all three transistor elements will be fairly close to each other.

Finally, if the collector opens internally, no current will flow in the collector circuit. The collector voltage will then be at the full battery voltage.

Resistance measurements in a transistor circuit may pose a problem because of the presence of the transistors. To see how this can occur, consider the partial input circuit shown in Fig. 14.18. All power in the circuit has been turned off, and one of the resistance checks we wish to make is that of R_2. Now, depending on what ohmmeter lead we connect to the top end of R_2 and what lead we connect to the bottom end, the value indicated by the ohmmeter will be either very low or 15,000 Ω, assuming the resistor to be good.

The reason for this behavior stems from the nature of transistors and the manner in which ohmmeters operate. An ohmmeter applies a voltage across the resistance to be measured, and from the resulting current flow the resistance value is determined. Let us suppose that the ohmmeter has an internal battery of 3 V, this being a fairly common value. This voltage must make one lead positive and one lead negative. Now, if the positive lead goes to the top of R_2 and the negative lead to the bottom, the true value of R_2 will be indicated on the meter, because under these conditions

FIGURE 14.18
The resistance value indicated by the ohmmeter will depend on the way the internal battery of the instrument is connected.

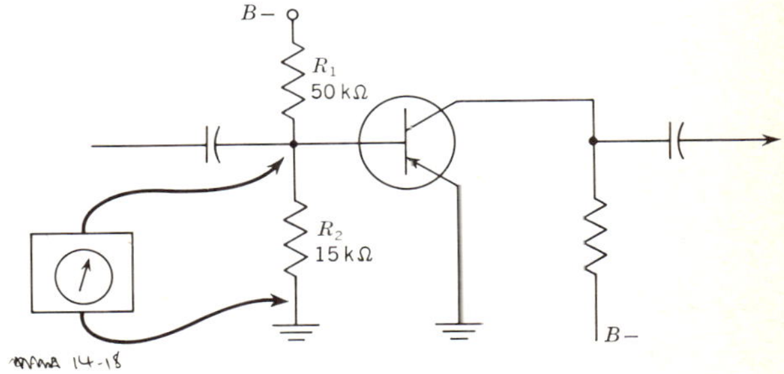

the base-emitter circuit of the transistor is reverse-biased. That is, the positive meter voltage is being applied to an N base here while the negative meter voltage goes to a P emitter (through the ground connection).

From this, it is readily apparent that if we reverse the ohmmeter leads, a negative voltage will be applied to the base and a positive voltage to the emitter. This will forward-bias the unit and throw an extremely small impedance across R_2. The measured result of this combination will also be quite small.

Another possible side effect of the second reading is damage to the transistor because of the excessive current flow. To avoid damage, either the transistor should be removed prior to such measurements or care should be taken to see that the proper ohmmeter lead goes to the top end of R_2. If the polarity of the ohmmeter leads is not known it can be determined readily by connecting the leads to a dc voltmeter and noting whether the meter needle moves to the right or left.

What has been stated for resistance measurements in the emitter circuit is just as true in the collector circuit. Remember that a forward-biased collector has practically the same low impedance as a forward-biased emitter.

It is generally good practice to inspect a transistor circuit visually as part of the servicing procedure. Look for such things as breaks in the printed circuitry or even in the printed board itself. A sudden twist, a sharp jar, or an inadvertent fall of a transistor device can readily damage any one of a number of miniature components common to transistor circuitry, including the mounting board containing the printed wiring. A careful visual inspection will frequently bring these defects to light and shorten what could otherwise be a lengthy service job.

Because of the wide variety of battery voltage values found in portable receivers, it is helpful to have a low-voltage power supply available for transistor-receiver servicing. Since battery voltages used in transistor sets seldom exceed $22\frac{1}{2}$ V, the power supply need not provide a higher voltage than that. Current requirements are in the milliampere range and are generally below 100 mA. The voltage output of the supply should be variable, preferably with a front-panel indicating meter to reveal the exact output voltage. Then, when the batteries in a unit are suspected of being low or dead, the power supply can be substituted directly for the batteries.

In selecting a power supply for testing purposes, make certain that a good dc output is obtainable. Because of the low value of the dc voltages required by a transistor circuit, even minute amounts of ac ripple can produce annoying hum from the loudspeaker.

PRINTED CIRCUITS (PC)

Transistors are used extensively with printed circuits, so that it behooves the technician to become familiar with the proper methods of removing or adding components to a printed-wiring chassis. The following discussion, from information furnished by the Admiral Corporation, will be helpful in this respect.

A printed circuit begins as a laminated plastic board with a sheet of thin copper foil bonded to one side. To form the necessary wiring, some of the copper foil is removed by a photographic and etching process. Holes through which various component leads are inserted are punched in the board. The leads of the various components are cut and bent over the copper-foil wiring. The wiring side of the board is then dipped in molten solder to make all solder connections at once. The copper-foil wiring also picks up solder, thus increasing its ability to carry current. Finally, a coat of silicone-resin varnish is applied to the wiring side of the board. This prevents dust or moisture from causing short circuits. The result is a circuit with uniformity of wiring, compactness, and freedom from wiring errors.

The foregoing method of producing printed wiring boards is known as the etched wiring method. It is more widely used today than any other form of printed wiring, largely because of the reliability, great flexibility, and low setup cost of the method. There are, however, other methods of manufacture, such as embossed wiring and stamped wiring. Since we are interested primarily in the end result, none of the other methods will be described here.

Circuit tracing of a printed-circuit board is usually simpler than that of conventional wiring owing to the uniform layout of the wiring. Also, many boards are translucent, and a 60-W light bulb placed underneath the side being traced will facilitate location of connections. Test points can frequently be located rather easily in this manner without the necessity of viewing both sides of the board.

Resistance or continuity measurements of coils, resistors, and some capacitors can be made from the component side of the board. In some cases a magnifying glass will assist in locating very small breaks in the wiring. Voltage measurements can be made on either side of the board. However, on the wiring side of the board (some printed circuit boards have component parts on both sides), a needle-point probe for circuit checking should be used, since the varnish coating must be pierced to make contact.

Be careful when removing components from the board. However, if the copper-foil wiring is damaged, a piece of wire can be used to replace the damaged foil. Small breaks can be "jumped" with molten solder. Larger breaks can be repaired with ordinary hook-up wire. It is seldom necessary to replace an entire board because of foil breakage.

Do not apply excessive pressure to the printed-circuit board or components. Although the board is sturdy in construction and mounting, it may crack or break if proper care is not taken when servicing. On extremely rare occasions, access to components on the board may be difficult. In that case the board may be removed from the chassis by removing the mounting screws around the edges and unsoldering a few leads between the board and the chassis. If this is done, a vise with protected jaws should be used to hold the board while servicing and care should be taken not to exert excessive pressure against the board.

In some areas on the printed board, the wiring is very closely spaced. When resoldering a new component, avoid excessive deposits of solder. Excessive solder may cause a short or intermittent trouble to occur later that will be difficult to locate.

When using the soldering iron (35 W or less), do not overheat the component terminals or the copper foil. Excessive heat (applying the soldering iron longer than necessary, using a higher-wattage iron than recommended, or using a soldering gun) may cause the bond between the board and foil to break. This will necessitate the replacement or repair of the foil connection.

REPLACING CAPACITORS, RESISTORS, COUPLATES, AND PEAKING COILS

Defective resistors, couplates, and ceramic-disk and wax-encased capacitors can be replaced by either of the following two methods:

1. If the leads extending from the defective component are long enough for a replacement component to be soldered to it, cut the leads where they enter the defective component (Fig. 14.19).

FIGURE 14.19
Cutting a defective resistor free of the printed-circuit board. (*Admiral Corp.*)

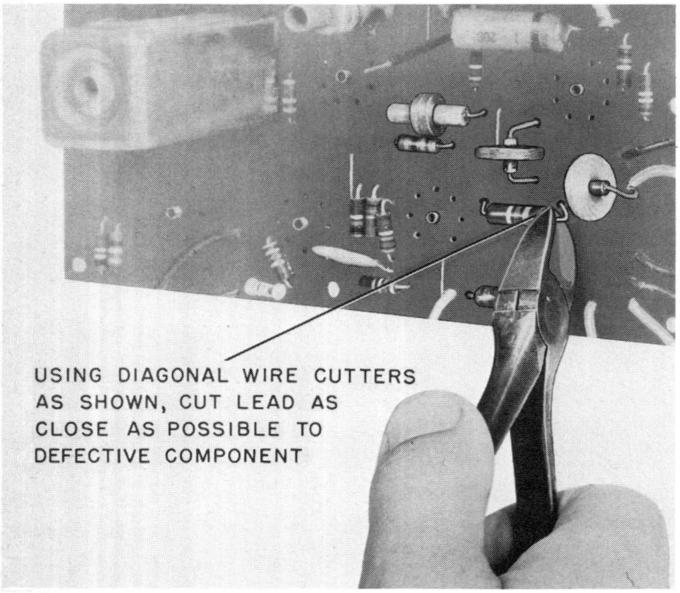

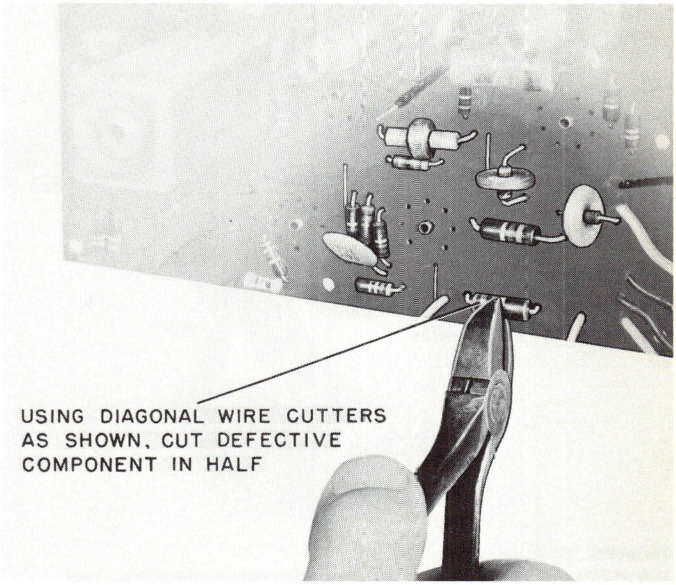

FIGURE 14.20
Cutting a defective resistor apart so as to have maximum lead length left. (*Admiral Corp.*)

2. If there is not enough length in the leads extending from the defective component to use the method described in 1, cut the defective component in half. Then cut through each half of the component until it is broken away from its lead. If this procedure is performed carefully, enough extra lead inside the component will be gained to permit soldering the replacement component to it (Figs. 14.20 and 14.21).

Clean off the ends of the remaining leads, leaving as much of the leads as possible. Make a small loop in each lead of the replacement component and slide the loops over the remaining leads of the old component (Fig. 14.22). Caution should be observed not to overheat the connection, since the copper foil may peel or the original component lead may fall out of the board. This is possible because of heat transfer through the leads. The lead length of the replacement part should be kept reasonably short to provide some mechanical rigidity.

In some cases, components are mounted in such a manner that neither of the above methods can be used. In that event it is necessary to unsolder the defective component completely and replace it. The following procedure should be used whenever it is necessary to unsolder any connections to replace defective components.

FIGURE 14.21
Cleaning remaining leads of component that has been cut apart. (*Admiral Corp.*)

FIGURE 14.22
Soldering replacement resistor in place. (*Admiral Corp.*)

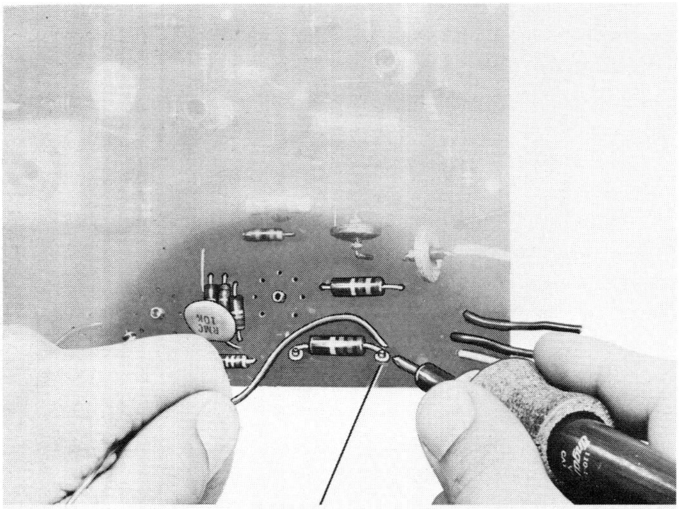

FIGURE 14.23
All excess solder should be carefully brushed away. (*Admiral Corp.*)

1. Heat the connection on the wiring side of the board with a small soldering iron. When the solder becomes molten, brush away the solder (Fig. 14.23). A 60-W bulb placed over the component side of the board will facilitate location of the connections on the wiring side if the board is translucent (and many boards are). In the process of removing the solder, caution is needed to prevent excessive heating. Therefore, do not leave the iron on the connection while brushing away the solder. Melt the solder, remove the iron, and quickly brush away the molten solder. (For this purpose, a small wire brush is suitable.) More than one heating and brushing process may be required to remove the solder completely.
2. Insert a knife blade between the wiring foil and the bent-over component lead, and bend the lead perpendicular to the board. (It may be necessary to apply the soldering iron to the connection while performing this step, because it is sometimes difficult to break the connection completely by brushing.) Do not overheat the connection.
3. While applying the soldering iron to the connections, wiggle the component until it is removed.
4. Remove any small particles of solder embedded in the silicone resin (if such a coating is employed) by using a clean cloth dipped in solvent.

5. A thin film of solder may remain over the hole through the board after removing the component. Pierce the film with the lead from the new component after heating the film with the soldering iron.
6. Insert the leads of the new component through the holes provided, cut to desired length, and bend over the ends against the copper foil. Resolder the connection with 60/40 low-temperature solder.
7. It is recommended that the cleaned area be recoated with clear lacquer or sprayed with Krylon for protection against shorts.

REPLACING COILS

The terminal lugs of these components are not bent over against the foil in most cases; therefore, brushing is not necessary. Heat one connection until the solder becomes molten, and wiggle the coil back and forth until the connection is broken. Continue to wiggle and apply the soldering iron to the other connections, and lift the coil from the board while the solder is still molten. Insert the replacement coil in the exact same position and solder the connections. Cover the connection points with a coat of lacquer or Krylon.

SOLDER-EXTRACTION METHOD

A recently developed method for solder-joint removal is called solder extraction. This method provides controlled combinations of heat, pressure, vacuum,

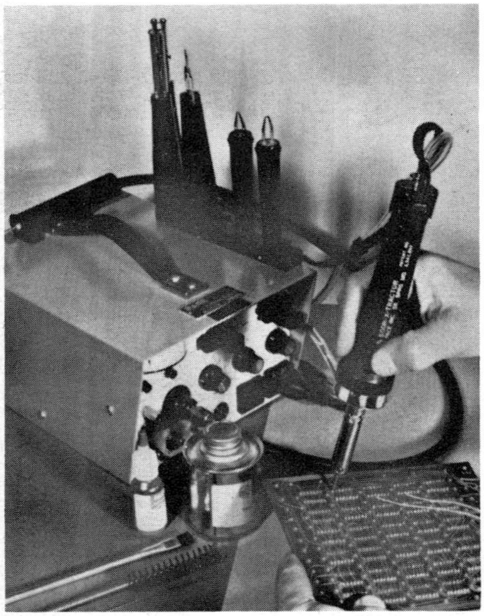

FIGURE 14.24
The Pace solder-extraction device.

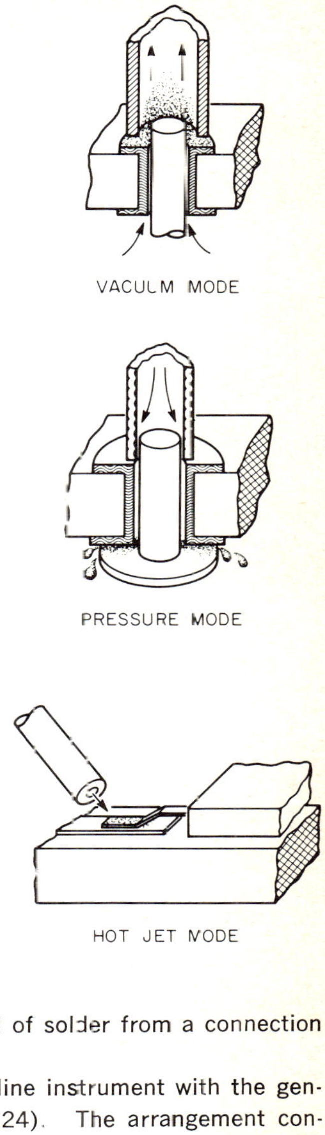

FIGURE 14.25
Solder-extraction methods. (*Courtesy Pace Incorporated.*)

air flow, and manipulative qualities for the removal of solder from a connection or joint where it is desired to undo the connection.

The solder-extraction device[1] is a coaxial, in-line instrument with the general shape of a small soldering iron (see Fig. 14.24). The arrangement consists of a hollow-tip heating element, transfer tube, and collecting chamber located within the handle that collects and solidifies the waste solder and clipped leads. The unit is operated by a power source that provides controlled vacuum pressure and electrical supply.

The advantage of this type of unit is that it is easily manipulated and fully controllable. It provides three basic modes of operation: (1) heat and vacuum, (2) heat and pressure, and (3) hot-air jet. See Fig. 14.25. The

[1] Manufactured by Pace Incorporated, Silver Springs, Md.

power source provides variable control over pressure and vacuum levels and flow rates, as well as temperature control over the heated tubular tip. It permits selection of power combinations and controls as required.

General Operation and Application of Extractor. The solder extractor is provided with various sizes of tubular tips, which may be modified to suit joint configurations and which perform three basic functions. First, these tips are the primary means of transferring heat to melt the solder joint in the pressure and vacuum modes. Second, it initiates the path up which the molten solder will flow in the vacuum mode. Third, it provides the path and pattern for hot-jet operations.

Considerations for selecting the proper tip size are based on a minimum inside diameter to fit over the lead or pin while permitting molten solder to easily pass up through it. The maximum outside diameter of the tip selected should not cover the pad completely or even touch the circuit-board base material itself nor extend over the flanges of an eyelet or funnelet (Fig. 14.26*a*). For terminal desoldering, the tip OD is selected to permit entry into the desired area (Fig. 14.26*b*).

In the *vacuum mode,* the heated tip is applied to the solder joint and when a melt is noted, the vacuum is activated, causing the solder to be withdrawn from the joint and deposited into the chamber. If the lead is preclipped, it will also be drawn into the chamber. To overcome the critical problem of the lead resweating to the side walls of a plated through-hole-type joint in a PC board, the lead is oscillated with the tip while the vacuum is being

FIGURE 14.26
(*a*) Tip diameter should be smaller than terminal-pad diameter. (*b*) Tip should enter soldered area. (*Courtesy Pace Incorporated.*)

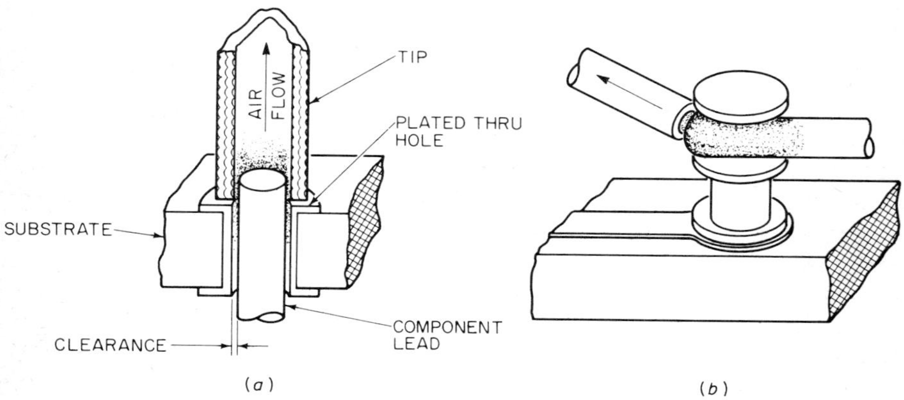

applied. This permits cool air to flow into and around the lead and hole side walls, causing them to cool down and prevent resweat. Solder extraction with continuous vacuum does help overcome this resweat problem for either dual or multileaded devices terminating into through-hole solder joints.

In the *pressure mode*, the tip can apply heat to a pin for melting a back-side sweat joint, without contacting the delicate pad, and the air pressure is forced down the hold to break up back-side sweat joints.

The *hot-air-jet mode* uses a controlled flow of heated air that melts the solder joint without physical contact. This permits the reflow of very delicate joints, while minimizing the possibilities of mechanical damage. The hot-jet mode is particularly useful in unsoldering IC flat packs or in removing conformal coatings so common on PC boards.

TRANSISTOR-RECEIVER SERVICING

Transistor-receiver servicing does not differ appreciably from the servicing of vacuum-tube-operated receivers. There are certain differences of initial approach, however, because of the use of batteries and transistors, and it is these differences (rather than the more familiar similarities) which will be considered here.

For example, when the output of a transistor receiver is distorted, weak, or completely dead, the prime suspect is the battery. The measurement is made with a vacuum-tube voltmeter or high-resistance voltmeter and is best taken with the battery in the receiver and the latter turned on. If the voltage reading is at or near the correct value, the battery can be presumed to be good. If the voltage reading is off by 20 percent or more, then the receiver output may be weak or distorted, but it should not be dead. Since transistor characteristics are linear to very low voltages and currents, chances are that distortion will not occur until the battery voltage drops more than 20 percent. There is, however, no set rule regarding this, and it is best to try a new battery when the voltage of the existing battery has decreased by this amount. If the distortion of weakness still persists, then some other defect is indicated.

Whenever a weak battery is found, it may be advisable to check the resistance of the circuit across the battery clips before a new unit is inserted. For example, in the receiver shown in Fig. 9.3, the manufacturer indicates that the resistance between the battery clips (with the battery removed and the receiver turned on) should be between 6,000 and 15,000 Ω as read by an ohmmeter with an internal battery of not more than 3 V. A reading lower than 6,000 Ω will usually indicate a defective component somewhere in the receiver.

Some manufacturers indicate what the current drain on the battery should be instead of quoting the circuit resistance across the battery clips. In that case, a milliammeter must be inserted in series with the battery. For example, with the negative terminal of the battery touching the negative clip, a wire is connected from the negative terminal of the milliammeter to the positive battery clip. Then one end of another wire is connected to the positive terminal of the milliammeter while the other end of this wire is touched to the positive end of the battery. The value of current indicated on the meter should fall in the range specified by the manufacturer.

If the battery proves to be good, then the rest of the trouble-shooting procedure follows established practice. As an example of this, the method of attack for the receiver of Fig. 9.3 is given below. Study this in conjunction with the schematic diagram of the set.

The alignment procedure for the same receiver is given also following the trouble-shooting outline.

TROUBLE-SHOOTING PROCEDURE FOR A TRANSISTOR RADIO RECEIVER

This section is reprinted from *PF Reporter* by permission of Howard W. Sams & Co., Inc.

Dead Receiver—Absolutely No Output

1. Remove the battery and turn on the switch. Measure the resistance between the battery clips. (Make sure the positive meter lead is on the positive clip.) If the resistance is
   a. Approximately 10,000 Ω, the B+ circuit is normal.
   b. Less than 2,500 Ω, check the leads of capacitors C_{17} and C_{21}, and make sure they are not touching the battery clips or the frame of the output transformer. Check for a shorted condition in either C_{17} or C_{21}. Measure the resistance from the top ends of R_3 and R_7 to ground. These should be 2,200 Ω more than the reading across the battery clips.
   c. Infinity, check for an open switch.
2. Turn the volume control to maximum and insert the battery. If a click or noise is heard from the speaker, check Q_4 by shorting its base to the frame of the output transformer. The audio stage is operating if a click is heard. If no click or noise is heard, proceed as follows:
   a. Check for an open or shorted jack. Indicative readings can be obtained by measuring the resistance from the fixed contact of the jack to ground. These readings are
      0 Ω—shorted jack
      2 Ω—normal
      15 Ω—jack is open, or the ground between the wiring board and chassis is open
   b. Check for an open condition in the speaker or in the output transformer.
   c. Voltage at the base of Q_4 (about +2 V normal).

   d. Voltage at the emitter of Q_4 (should measure approximately 0.15 V less than the base voltage).

3. Check capacitor C_{19} by paralleling it with a capacitor known to be good.
4. Measure the voltage at the output of the diode D_1 (should be approximately $+0.1$ V).
   a. If voltage is zero, check the resistance to ground with the positive meter lead on the output. This resistance should measure between 20 and 100 Ω. If the resistance is zero, check for a shorted condition in the diode circuit. If the resistance is 200 Ω or greater, check for an open in the diode circuit or for an open diode.
   b. If voltage is negative when the receiver is tuned to a station, move the tuning dial so that no station is received. The negative voltage should decrease.
   c. If voltage is negative by 1 V or more and does not drop when the receiver is tuned off the station, the receiver is oscillating. Proceed to the section entitled Oscillating Receiver.
5. Make voltage and resistance measurements in the i-f stages.

Dead Receiver—Noise But No Signal. Check the local oscillator in the receiver as follows: Tune another receiver to any station above 850 kilohertz (kHz). On the receiver being serviced, rock the dial above and below a setting that is approximately 262 kHz below the frequency of the station being received by the other receiver. If the local oscillator in the receiver being serviced is operating, a whistle will be heard from the other receiver as the radiation from the oscillator beats with the station frequency.

If the oscillator is dead, proceed as follows:

1. Check the voltage at the base of Q_1. This should be between 3 and 10 V.
2. Check the voltage at the emitter of Q_1. This voltage should be within 0.1 V of the base voltage.
3. Check the voltage at the top end of R_3. This should be measured from the B+ line, and it should be between 0.6 and 2 V.
4. If any of the voltages measured in the three preceding steps are incorrect, check for an open oscillator-coil primary or an open first i-f transformer.
5. Check resistances of
   a. The high side of the antenna coupling coil to ground (should be less than 1 Ω).
   b. The secondary of the oscillator coil (should be approximately 10 Ω).
   c. Stator of oscillator section of the tuning capacitor to ground (should be infinity).

Weak or Distorted Output

1. Turn volume control to maximum. Check capacitors C_{19} and C_{21} by paralleling a good capacitor across each.

2. Perform step 5 under section Dead Receiver—Absolutely No Output.
3. Measure voltages at
   a. Base of Q_4 (should be approximately $+2$ V).
   b. Emitter of Q_4 (should be approximately 0.15 V less than the base voltage).
   c. Base of Q_3 (should be approximately 0.15 V less than the voltage at the emitter of Q_3).
   d. Top end of R_{11} (should be approximately -0.5 V when receiving a signal of average strength).
   e. Avc line (should be from approximately 0V with signal to 0.5 V with no signal).
   f. Emitter of Q_2 (should be approximately 0.15 V less than the avc line).
4. Check the alignment of the receiver.

Oscillating Receiver

1. Measure the battery voltage. If it is below 1.5 V, the battery should be replaced.
2. Check the local oscillator as in step 1 under the section entitled Dead Receiver—Noise But No Signal.
3. Check capacitors C_{17} and C_9 by paralleling a good capacitor across each.
4. Check ground connection between wiring board and chassis. This connection is the twisted lug near the negative battery clip and is the only lug which has been soldered to the board. Measure between an i-f transformer can and the metal chassis. These readings are

   0 Ω—normal

   15 Ω—ground lead is open

ALIGNMENT PROCEDURE

The alignment of this receiver is quite simple. Signal injection is accomplished by connecting the signal generator to a loop formed of several turns of wire and situated close to the antenna coil of the receiver. Set the generator to 262 kHz with 400-cycle modulation and reduce the output to as low a value as is usable. Connect an output meter (with a 0.1-V scale) across the voice-coil terminals. (The high side of the voice coil is easily accessible at the spring of the phone jack in this set.) Set the volume control in the receiver to maximum. Adjust each of the cores of the i-f transformers for maximum indication on the output meter. Set the receiver dial to its maximum counterclockwise position, tune the generator to 535 kHz, and adjust the core of the oscillator coil for maximum output. Tune the generator to 1,630 kHz, set the receiver dial to its maximum clockwise position, and adjust the oscillator trimmer capacitor for maximum output. Repeat these last two adjustments alternately until no further improvements can be made. Then tune the generator to 1,500 kHz, tune in this signal with the receiver dial, and adjust the antenna trimmer capacitor for maximum output. Turn the receiver dial to the high-frequency end and determine whether or not the range extends to 1,630 kHz. If not, the oscillator trimmer capacitor must be readjusted and the alignment at 1,500 kHz must be repeated.

QUESTIONS

14.1. In what respects are transistors sturdy? In what respects fragile?

14.2. What changes in working tools are necessary when dealing with transistors and their associated components?

14.3. Indicate several precautions to follow when soldering transistors into a circuit.

14.4. Why are transistors especially sensitive to applied voltages?

14.5. Outline the precautions to observe when injecting signals into a transistor circuit. Indicate a suitable safe method of bringing such signals into a circuit.

14.6. How could one determine whether a transistor was good or bad?

14.7. Draw a simple test circuit to measure I_{CBO}.

14.8. Indicate briefly how transistor gain is checked.

14.9. Indicate generally the precautions to observe when removing components from printed circuit boards.

14.10. In what respects does the servicing of transistor receivers differ from the servicing of comparable vacuum-tube sets?

14.11. In what respects is an in-circuit transistor and FET tester useful and in what respects is it limited in application?

14.12. For any given transistor, what is the relationship between the values for I_{CBO}, I_{CES}, and I_{CEO}?

14.13. Indicate some of the safety precautions that must be observed when handling MOSFETS.

14.14. Is it better to check a battery (using a vacuum-tube voltmeter) while it is in the circuit or after it has been removed completely from the set? Give reasons for your answer.

14.15. How would you align the receiver shown in Fig. 9.3?

14.16. If it were determined that the oscillator of this receiver was not functioning, how would you proceed to localize the defect?

14.17. What might cause a weak or distorted output from this receiver (of Fig. 9.3)?

14.18. Under what conditions would the battery in the receiver not be the first component checked? (Assume that the receiver is not functioning properly.)

Chapter Fifteen

Experiments with Transistors

In the preceding chapters, the theory and application of transistors were covered in detail. The information contained there represents the first step toward the acquisition of a basic understanding of transistor operation. The next step for one who is going to work with these units is actual physical contact so that he may become practically proficient in handling transistors and learning first hand of their characteristics and peculiarities. Toward this end, a series of experiments are presented, and all readers are urged to perform them prior to any work on transistors in commercial equipment.

The circuitry involved in these experiments has been kept as simple as possible. This serves the twofold purpose of making each experiment easy to perform and keeping component cost down. Furthermore, the same basic components are used over and over again. Because of this, caution should be exercised when leads are trimmed prior to soldering lest the amount of wire removed be so much that the unit will not be usable for as many times as required.

Another very important precaution to observe, one that was mentioned in Chap. 14 on transistor servicing, is the use of a low-wattage soldering iron when soldering transistor leads into the circuit. Keep the leads as long as possible, and grip the lead being soldered with a pair of long-nose pliers. Since the lead is held between the point where the heat is applied and the body of the transistor, any heat traveling along the lead wire will be shunted away from the transistor.

It is also important to observe battery polarity when connection is made to the circuit. If the collector terminal receives a forward-biasing voltage in place of a reverse-biasing voltage, the transistor can be damaged permanently. Double-check wiring before connecting any batteries. Do not use voltages higher than those indicated in the experiments. Also, make certain that the battery is disconnected before any wiring connections are altered. Finally, be especially careful to avoid circuit shorts between various wires. If necessary, cover all bare wires (or exposed ends of wires) with protective spaghetti.

No specific chassis form or size is recommended for the ensuing experiments. They may be performed on a breadboard or on any of the small metallic chassis that are obtainable at a parts jobber. In the latter instance, there are a number of standard base sizes ranging generally from 2 × 6 × 4 in (height, width, depth) to 5 × 17 × 13 in. A recommended size of 3 × 7 × 5 in was found to be entirely adequate for the experiments; however, any suitable dimensions may be used.

It would also be desirable to use terminal strips on which the components may be mounted. Choice of such strips is left to the reader.

EXPERIMENT 1. ADJUSTING TRANSISTOR VOLTAGES AND CURRENTS

Note: NPN transistors can be used in place of the PNP units specified. If they are, the battery voltage as well as all electrolytic capacitors must be reversed. Otherwise, identical results should be obtained.

Object. To adjust the voltages on a transistor and to establish the principle of phase reversal in a grounded-emitter amplifier.

Material Required

1 5-microfarad (μF) electrolytic capacitor
1 0.01-μF capacitor
1 1-megohm (MΩ) potentiometer
1 2N404 transistor (or equivalent). *Note:* Any low-frequency PNP transistor that operates with the voltages noted would be suitable in place of the recommended 2N404. Other high-frequency or higher-power transistors could also be employed, but these units generally cost more, and they would not provide any more useful information (in these experiments) than the inexpensive transistor.
1 5,600-Ω resistor (¼ W)
1 1,000-Ω resistor (½ W)
1 4½-V battery)

Test Equipment

1 vacuum-tube voltmeter or a good multimeter (preferably 20,000 Ω/V)

Procedure

1. Wire the circuit of Fig. 15.1.
2. Check all connections before attaching the battery.
3. Connect the battery and adjust P_1 until the voltage between collector and emitter is approximately 3 V.
4. Leave the voltmeter connected to the collector. Take a 10,000-Ω resistor and touch it between base lead and ground. Notice how the collector potential becomes more negative. When the 10,000-Ω resistor was touched

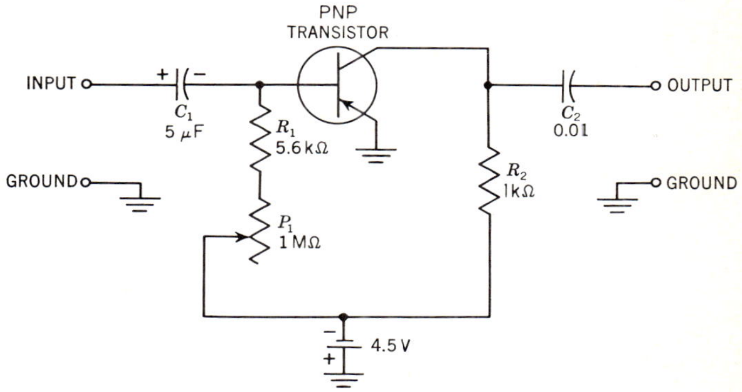

FIGURE 15.1

from the base to ground, it caused the base to become less negative, or more positive. At the same time the collector became more negative.

Conclusions

1. The principal conclusion that can be drawn from the foregoing behavior is that a grounded-emitter amplifier reverses the phase of an applied signal.
2. As a secondary consideration, it was noted that as P_1 was rotated, it varied the base and collector voltages. We could also conclude that if a resistor had been inserted in the emitter lead, the voltage drop across the resistor would have varied also. Ninety-five percent of the emitter current passes through the collector, and if we vary the collector current, we must also vary the emitter current.

EXPERIMENT 2. DISTORTION AND TEMPERATURE EFFECTS

Object. To observe the operation of a single-stage transistor with signal input and to demonstrate the effect of temperature change.

Material Required

1 5-μF electrolytic capacitor
1 0.01-μF capacitors
1 1-MΩ potentiometer
2 2N404 transistors (or equivalent)
1 5,600-Ω resistor (¼ W)
1 1,000-Ω resistor (½ W)
1 4½-V battery

Test Equipment

1 audio signal generator or a filament transformer arrangement as shown in Fig. 15.2.

1 oscilloscope

Procedure

1. Apply a very low voltage signal, about 0.01 V rms, between C_1 and ground in Fig. 15.1. An audio oscillator is desirable for this purpose. (Be careful not to use too strong a signal, because the transistor can be damaged by overdrive.) The signal lead goes to C_1, while the other generator lead goes to ground. If an audio oscillator is not available, a filament transformer can be used with the circuit shown in Fig. 15.2.
2. Connect the vertical terminals of an oscilloscope between the output of C_2, Fig. 15.1, and ground.
3. Starting from zero, adjust the output of the audio generator until an undistorted sine-wave signal is just seen on the oscilloscope.
4. By varying P_1, the least distorted signal may be obtained. When adjusting P_1, note how the output signal reaches a peak and then starts clipping. The clipping occurs on the negative half-cycle of the input signal. While this portion of the signal is active, the collector voltage is actually becoming less negative because the collector current is increasing. At the negative input peak, the voltage drop across load resistor R_2 is almost equal to the battery voltage. Further increase in input signal cannot further lower the collector voltage, and the output signal flattens out or clips. This is similar to plate clipping in a vacuum tube.

   If the input signal is increased beyond this point, clipping of the other half-cycle also results. This occurs when the positive half-cycle of the applied signal causes the emitter-to-base voltage to reach the reverse bias or cutoff region. Collector current is cut off too, at the same time.
5. Signal leakage may be observed by adjusting P_1 until an undistorted signal is observed and then disconnecting the battery. Notice how the input

FIGURE 15.2

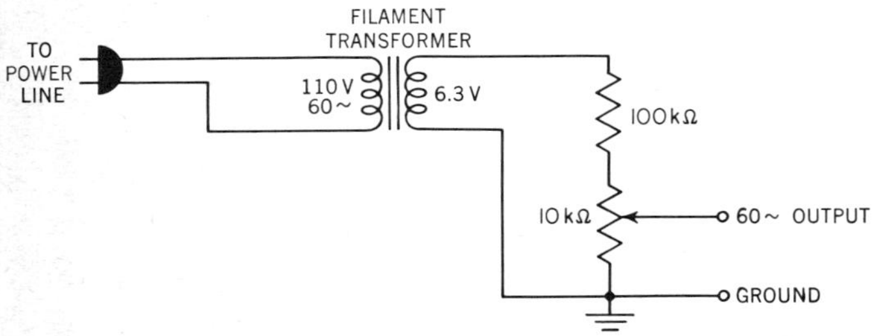

signal goes through the transistor and appears at the collector. It possesses the reverse phase from that normally seen on the collector; it is usually distorted; and, of course, there is no gain.

6. Reconnect the battery. With a vacuum-tube voltmeter or an oscilloscope, measure the input and output signals. The ratio of the output to input signals is the signal gain. After a numerical value is obtained, convert it to an equivalent decibel figure.
7. Readjust P_1 for the best possible gain with a fixed input signal. Without changing any of the instrument settings, note the height and waveform of the output signal. Now bring a hot soldering iron in the general vicinity of the transistor, thereby warming it. Be careful not to bring the iron too close. Note how the output signal changes as the transistor cools down. The amplitude of the input signal should be such that the output signal is just below distortion on top and bottom.
8. With the signal returned to normal and P_1 adjusted for maximum undistorted output, disconnect the battery and change transistors. Do not touch any of the other settings. Reconnect the battery and notice how the amplitude of the signal has changed. It may be even quite distorted. Readjust P_1 for the least distorted signal.

Conclusions

1. Clipping occurs when the signal is too strong.
2. Transistors are very sensitive to temperature variations.
3. The base bias current may frequently have to be adjusted for each transistor.
4. The characteristics of transistors of the same type will often vary considerably. (In time it is expected that the variations among individual units will become less and less.)
5. A very definite voltage gain is obtained with grounded-emitter amplifiers.

EXPERIMENT 3. COMPENSATION, INPUT IMPEDANCE, AND BIAS

Object. To establish temperature compensation, to note the effects of emitter degeneration on input impedance, and to note the effects of collector voltage on gain.

Material Required

2 2N404 transistor (or equivalent)
2 5-μF electrolytic capacitors
1 0.01-μF capacitor
1 4,700-Ω resistor (¼ W)
1 10,000-Ω resistor (¼ W)
2 1,000-Ω resistor (¼ W)

1 100,000-Ω potentiometer
2 4½-V batteries
1 6-V battery

Test Equipment

1 audio signal generator or filament transformer arrangement as mentioned in Experiment 2
1 oscilloscope

Procedure

1. Note the gain, or output signal level, in Experiment 2.
2. Change the circuit to that shown in Fig. 15.3. Adjust P_1 until the voltage drop across R_2 is 1½ V. (This occurs when the collector current is 1½ mA.)
3. Note how the gain has fallen. This may be explained as follows: Voltage measurements will show that the base and emitter voltages are higher than before but the base-to-emitter voltage is still about −0.4 V. The collector voltage, on the other hand, is still the same. Consequently, the collector-to-emitter voltage is lower, and this accounts for the lower gain.
4. Try changing transistors. Note that transistors may now be changed without further adjustment. The circuit is much more stable because any variation in I_{CO} (collector saturation current) or β is minimized by the degeneration introduced by R_3. This is also true for temperature changes.
5. Increase the collector battery voltage (B_2) to −6 V. Notice how the gain increases. This is because of the high collector-to-emitter voltage. The transistor can also handle greater power with this increase. Do not go any higher than −6 V.
6. Measure the gain of the stage by dividing the output signal voltage by the input signal voltage.
7. Note the amplitude of the output voltage and then remove the emitter bypass capacitor C_3. Notice how the gain drops markedly.

FIGURE 15.3

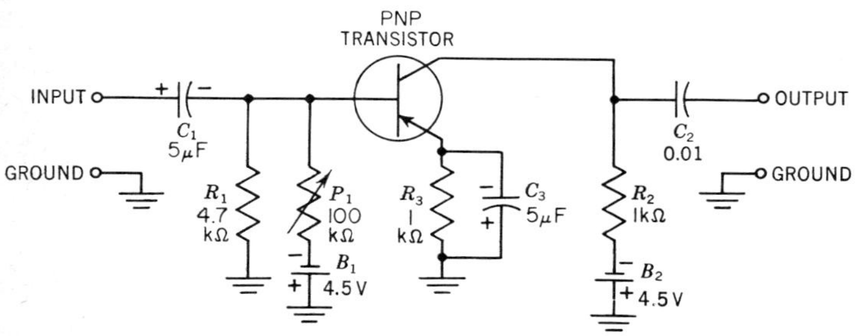

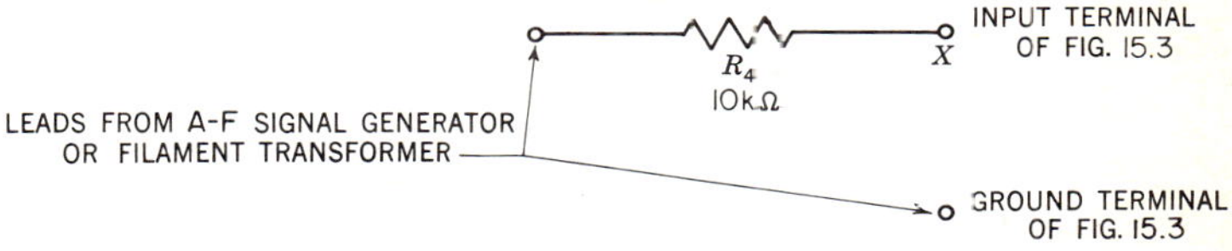

FIGURE 15.4

8. The input impedance will also increase if the bypass capacitor is removed. This can be shown as follows:

a. Reconnect the bypass capacitor C_3.

b. Place a 10,000-Ω resistor in series with the signal input generator and the input capacitor C_1 (Fig. 15.4).

c. Adjust the output of the signal generator for maximum undistorted output from C_2.

d. Note the height of the input signal at the junction of C_1 and R_4 (point X in Fig. 15.4).

Remove C_3 and note the increase in input signal at point X. This action can be explained as follows: With the bypass capacitor in the circuit, the base loading caused a voltage drop across R_4. When the bypass capacitor was removed, the loading was reduced and consequently more of the input signal became available to the transistor. Changes in input impedance of 5:1 or even 10:1 are common.

Conclusions

1. Emitter compensation improves transistor temperature stability.
2. Emitter compensation permits different transistors to be employed without separately adjusting the circuit for each.
3. The emitter bypass capacitor eliminates signal degeneration.
4. Removing the bypass capacitor increases the input impedance.
5. The higher the collector-to-base voltage, the greater the gain.

EXPERIMENT 4. TESTING TRANSISTORS DC PARAMETERS[1]

Object. Four basic parameters are of prime importance in the operation of a transistor. These are I_{CBO}, I_{EBO}, h_{FE}, and I_{CEO}. The purpose of this experi-

[1] From Milton S. Kiver and Bernard Van Emden, "Transistor Laboratory Manual," McGraw-Hill Book Company, 1962. This manual contains a wide variety of experiments designed to provide familiarity with transistor circuit operation.

ment is to familiarize the reader with these parameters and their methods of measurement. The characteristics of a number of transistors will be measured to obtain from the results an idea of the order of magnitude of these parameters.

Material Required

1 5,000-Ω ½-W resistor
1 0-to-10-V adjustable dc power supply
1 2N404 PNP or equivalent
1 2N3702 PNP or equivalent
1 2N1707 PNP or equivalent
1 2N525 PNP or equivalent
1 2N5225 NPN or equivalent
1 2N718 NPN or equivalent
1 100,000-Ω 2-W potentiometer
1 1,000-Ω ½-W ± 10 percent resistor
1 4,700-Ω ½-W ± 10 percent resistor
1 2,200-Ω ½-W ± 10 percent resistor
1 10,000-Ω ½-W ± 10 percent resistor

Test Equipment

1 microammeter (0 to 100 μA)
1 milliameter (0 to 10 mA)
1 0-to-10-V dc voltmeter

Background Information. For determining the overall condition or usability of a transistor, the magnitude of the cutoff or saturation current I_{CBO} and the dc current gain h_{FE} or beta (β) are of prime importance. These two quantities are also important to the circuit designer when establishing the bias for the desired operating point. In addition, I_{CEO} will be found useful because it is related to both current gain and I_{CBO} (and, also, because it is more easily measured than I_{CBO}). In general, nearly all transistor parameters are functions of collector current, voltage, or temperature. I_{CBO}, as we have noted doubles when the temperature increases approximately 10°C; it is multiplied by 10 for a 33°C temperature rise. (This is true for germanium transistors; for silicon transistors, I_{CBO} doubles for a somewhat lesser change in temperature.) The current gain of the transistor is not a constant. It is dependent on the collector current and the ambient temperature.

I_{CBO} or I_{CO}. I_{CBO}, the collector cutoff current, is measured with the collector-base junction reverse-biased and the emitter open. The I_{CBO} notation is frequently shortened to I_{CO}; the element connections, however, remain the same. A typical test circuit to measure I_{CO} is shown in Fig. 15.5. Since I_{CO}, even for a germanium transistor where it is greatest, is of the order of microamperes,

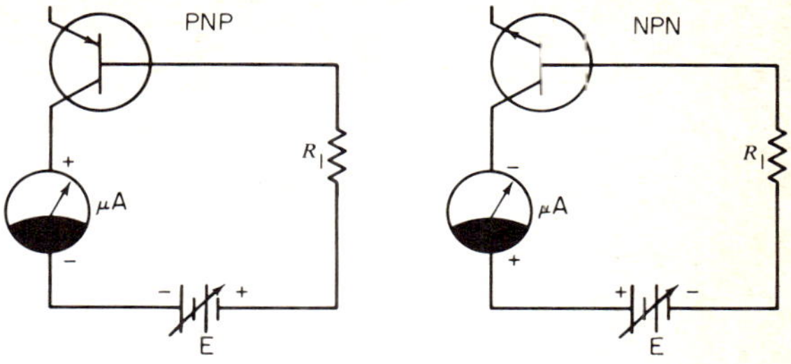

FIGURE 15.5
Test circuits for I_{CO} measurements.

a microammeter is required. (In low-power transistors, I_{CO} seldom exceeds 10 μA; in high-power transistors, I_{CO} may be in the milliampere range.) R_1, in Fig. 15.5, serves as a current limiter in case the transistor is connected improperly or is defective. I_{CO} is of considerable importance when the transistor is to be biased. Since I_{CO} varies with temperature, care must be taken to see that the operating point of the collector current (i.e., the collector-current value when no signal is being received) does not go beyond the limits of proper circuit operation. (*Note:* I_{CO} in silicon transistors is far lower than that of germanium transistors.)

I_{EBO} *or* I_{EO}. I_{EBO} is the current that flows when the emitter-base junction is reverse-biased and the collector is open-circuited. Sometimes I_{EBO} is shortened to I_{EO}. A typical test circuit for its masurement is shown in Fig. 15.6. Again, resistor R_2 acts as a current limiter in the event that the transistor is improperly connected or defective. I_{EO} is important when operating the transistor as a switch because under switching conditions, the emitter-base junction is back-biased in order to turn off the transistor completely.

h_{FE}. h_{FE} or β is the dc current gain of a transistor operating with the emitter common to the input and output circuits. This parameter can be measured

FIGURE 15.6
Test circuits for I_{EO} measurements.

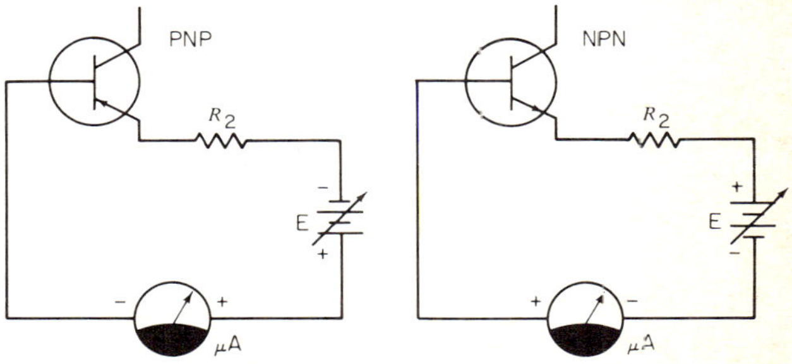

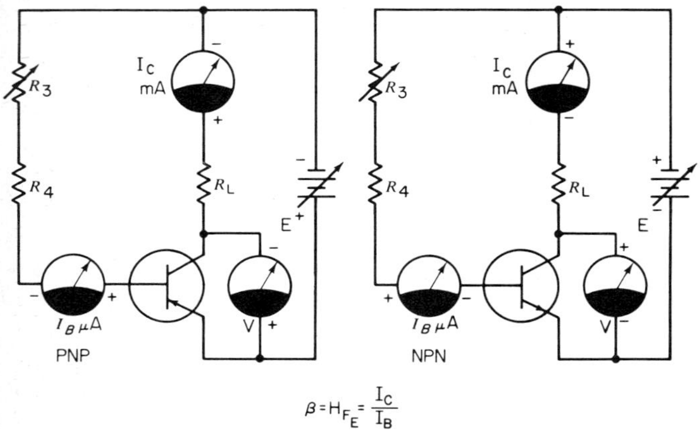

FIGURE 15.7
Test circuits for h_{FE} (β) measurement.

by the circuit shown in Fig. 15.7. β or h_{FE} is equal to the ratio of the collector current (output current) and the base current (input current). Numerically, h_{FE} is greater than 1 and is basically a property of the transistor itself. It is not constant but varies with the magnitude of collector current and with temperature. In general, as temperature goes up, current gain also increases, with the amount of increase obtained depending on the particular transistor type in question. To measure β, the collector and the base currents are measured and their ratio is computed; or the transistor can be driven with a constant current input and the resultant output collector current measured. From these values, the ratio can be computed.

It will be noted that transistor circuits actually have not one but three types of current gains. These are associated with the three ways transistors may be connected: common base, common emitter, and common collector. These can be compared, respectively, with grounded-grid, grounded-cathode, and grounded-plate (or cathode-follower) vacuum-tube amplifiers. Each transistor configuration has a different current gain. The common base has a current gain of alpha (α); the common emitter has a current gain of β; and the common collector has a current gain of $1 + \beta$. In this experiment, we are concerned solely with the current gain of a common-emitter amplifier because this is by far the most widely used configuration.

I_{CEO}. I_{CEO} is the collector-to-emitter current of a transistor when measured with the base open and the collector junction reverse-biased, as shown in Fig. 15.8. I_{CEO} is approximately equal to I_{CO} $(1 + h_{FE})$ or I_{CO} $(1 + \beta)$. Since h_{FE} is much greater than 1, $(I_{CO}\beta)$ is a good approximation of this formula. I_{CEO} is more frequently checked in transistor testers than is I_{CO} (to evaluate the condition of a transistor) because it is a much larger quantity and therefore measurable with a milliammeter instead of a microammeter.

Procedure

1. I_{CO} *Test.* Connect the transistor into the I_{CO} test circuit shown in Fig. 15.5. Be sure to use the proper circuit for the type of transistor you have, i.e., a PNP circuit for a PNP transistor and an NPN circuit for an NPN transistor. Resistor R_1 is a 5,000-Ω ½-W resistor. Power supply E is a 0-to-10-V adjustable dc power supply. The microammeter shown in these circuits reads the values of I_{CO} directly. Be sure the proper polarity is used on this meter. The microammeter may be part of a multiscale combination milliammeter and microammeter.
   a. Connect the transistor in the proper test circuit with E equal to 6 V. Measure I_{CO} and note its value.
   b. Repeat step *a* with E equal to 3 V.
   c. With E equal to 3 V, carefully hold a match so that the flame is approximately 1 in away from the transistor for a period of 10 seconds (s). At the end of this period, remove the match and note the peak value of I_{CO}. (Be careful not to bring the match flame too close to the transistor case. Just bring it close enough so that the heat of the flame is felt at the transistor.)
   d. Allow the transistor to cool, and with E equal to 3 V, measure and note the value of I_{CO}.
2. I_{EO} *Test.* Connect the transistor to the appropriate circuit shown in Fig. 15.6 R_2 equals 4,700 Ω. With E set at 6 V, the microammeter will indicate the appropriate value of I_{EO}. Note this value.
3. h_{FE} *Test.* Connect the transistor into the appropriate circuit shown in Fig. 15.7. R_3 is a 100,000-Ω 2-W potentiometer. A series of ½-W ± 10 percent resistors (or a resistance decade box) can be used for R_L. The desired values will be given as required. R_4 is a 1,000-Ω ½-W ± 10 percent resistor. E is equal to 6 V.

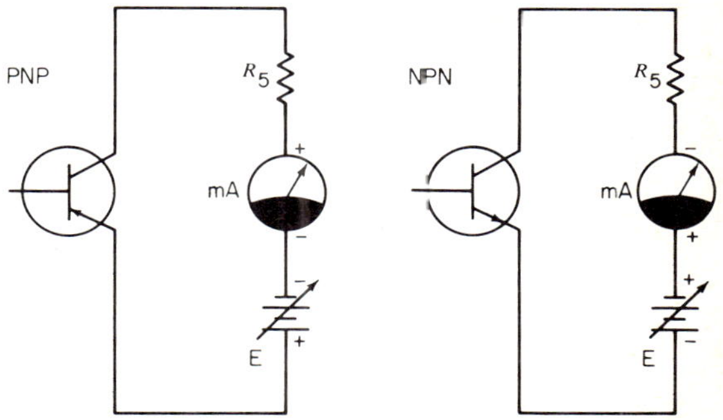

FIGURE 15.8
Test circuits to measure I_{CEO}.

a. After making R_L equal to 10,000 Ω, turn on the power supply and adjust potentiometer R_3 until voltmeter V first reaches a minimum. Read I_B and I_C and note their appropriate values. Compute the value of β, or I_C/I_B.

b. Repeat step a with R_L equal to 4,700 Ω.

c. Repeat step a with R_L equal to 2,200 Ω.

4. I_{CEO} *Test.* Connect the transistor to the appropriate circuit shown in Fig. 15.8, with R_5 equal to 4,700 Ω and E equal to 6 V. The meter shown will indicate I_{CEO}. Note this value and compare it with I_{CO} obtained above.

   Change the value of R_5 to 10,000 Ω, and with E equal to 6 V, note the value of I_{EO} obtained.

   With the circuit unchanged, carefully hold a match so that the flame is approximately 1 in away from the transistor for a period of 10 s. At the end of this period, remove the match and note the peak value of I_{CEO}.

ADDITIONAL EXPERIMENTS

Repeat the above experimental procedures for each of four or five different transistor types.

Conclusions

1. I_{CO} changes with the base-collector potential, tending to rise as the voltage between these two elements rises.
2. I_{CO} is temperature-sensitive.
3. h_{FE} will vary with I_C. The latter current, in turn, will vary with the load resistor in the collector circuit.
4. I_{CO} and I_{CEO} tend to act in a similar manner and to be related by the beta of their transistor.
5. I_{CEO} is more temperature-sensitive than I_{CO}.

EXPERIMENT 5. GROUNDED-BASE, GROUNDED-EMITTER, AND GROUNDED-COLLECTOR AMPLIFIERS

Object. To determine the relative differences among the grounded-emitter, the grounded-base, and the grounded-collector amplifiers.

Material Required

1 2N404 transistor (or equivalent)
1 0.01-μF capacitor
2 5-μF electrolytic capacitors
1 4,700-Ω resistor (¼ W)

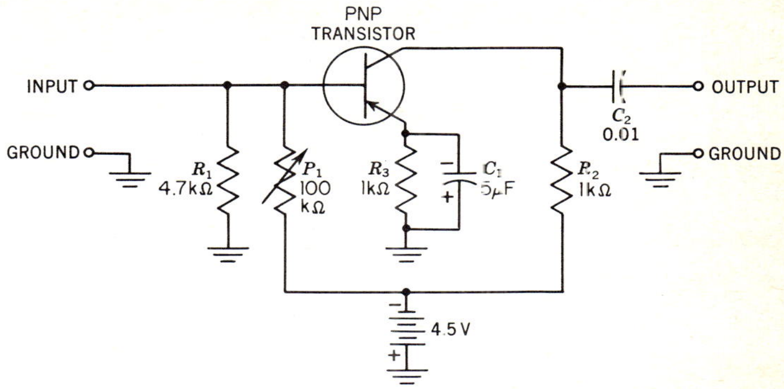

FIGURE 15.9

1 10,000-Ω resistor (¼ W)
2 1,000-Ω resistors (¼ W)
1 100,000-Ω potentiometer
1 4½-V battery

Test Equipment

1 oscilloscope
1 audio signal generator or filament transformer arrangement

Procedure

1. Wire the circuit of Fig. 15.9. Adjust P_1 until the voltage drop across R_2 is 1½ V.
2. Connect the vertical input terminals of an oscilloscope between the output (at C_2) and ground.
3. Feed an input signal [through a 5-μF electrolytic capacitor and a 10,000-Ω resistor in series (Fig. 15.10)] to the input (which is the base lead).
4. Increase the signal input until maximum undistorted output is obtained. Check the magnitude of the output signal and also check the gain. The latter measurement should be taken as output voltage divided by the input

FIGURE 15.10

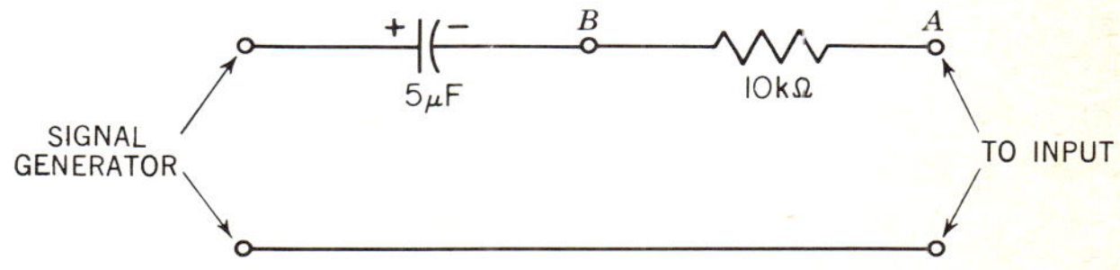

voltage at point A (Fig. 15.10) and then computed again as output voltage divided by the generator voltage at point B.

There will be a difference between these two values because of the voltage drop across the 10,000-Ω resistor through which the signal is fed to the amplifier. This drop is caused by the input impedance of the transistor amplifier.

The true gain figure is the value obtained when the output voltage is divided by the input voltage at point A.

5. Connect a 0.01-μF capacitor from the collector of the transistor amplifier to the *external sync* terminal of an oscilloscope. When the instrument is properly synchronized, check the phase of the signal at the output and input terminals of the amplifier. The two waveforms should be 180° out of phase.
6. Repeat steps 1 to 5 using the circuit of Fig. 15.11, which is a grounded-base amplifier. Although the base does not go to ground directly, the bypass capacitor essentially brings it to ground so far as alternating current is concerned. (In setting up the circuit of Fig. 15.11, adjust P_1 for a voltage drop across R_2 of 1½ V.)
7. Repeat steps 1 to 5 using the circuit of Fig. 15.12. There is a grounded-collector amplifier. The 5-μF capacitor from collector to ground places this element at ac ground potential. (Adjust P_1 until the voltage drop across the 1,000-Ω emitter resistor is 1½ V.)

Conclusions

1. The input and output signals of a grounded-emitter stage are 180° out of phase.
2. The input and output signals of a grounded-base stage are in phase. The same is true of a grounded-collector amplifier.

FIGURE 15.11

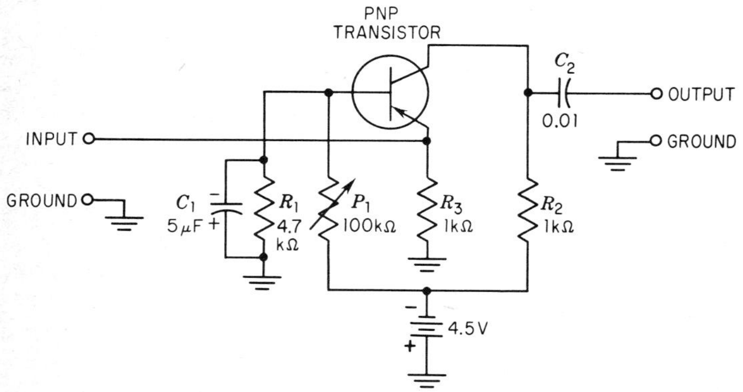

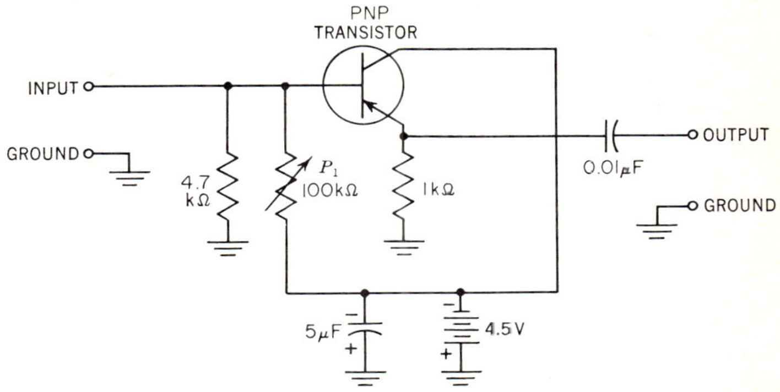

FIGURE 15.12

3. The grounded-base amplifier possesses a higher voltage gain than the grounded-emitter amplifier. However, the input impedance of a grounded-base amplifier is lower than it is in a grounded-emitter stage. This was shown by the greater signal drop across the series 10,000-Ω resistor. The grounded-collector stage possesses a voltage gain less than 1; input impedance is the highest of all three arrangements.

Actually, to measure the input impedance of any of these amplifiers, substitute a 100,000-Ω potentiometer for the 10,000-Ω resistor of Fig. 15.10. As an initial step, adjust this potentiometer for zero resistance (i.e., short it out of the circuit). Now feed in just enough signal for maximum undistorted output on the oscilloscope connected across the output terminals.

Note the amplitude of the signal on the scope screen. (The easiest way of doing this is by having the pattern cover a specific number of squares on a calibrated screen overlay mask.) Then gradually increase the resistance of the potentiometer until the amplitude of the output signal has been reduced to half. The resistance of the potentiometer at this point equals the input impedance of the amplifier circuit.

EXPERIMENT 6. RESISTANCE-CAPACITANCE AND IMPEDANCE-COUPLING

Object. To observe a two-stage audio amplifier with RC and impedance coupling.

Material Required

1 0.01-μF capacitor.
4 5-μF electrolytic capacitors.

1 50-μF electrolytic capacitor (25 V).
2 4,700-Ω resistors ($\frac{1}{4}$ W).
4 1,000-Ω resistors ($\frac{1}{4}$ W).
2 100,000-Ω potentiometers.
2 2N404 transistors (or equivalent).
1 interstage transformer. *Note:* Miniature interstage audio transformers designed especially for transistors are available at local parts jobbers. A suitable unit would possess a primary impedance of 20,000 Ω and a secondary impedance of 500 Ω.
1 $4\frac{1}{2}$-V battery.

Test Equipment

1 audio signal generator or filament transformer arrangement as described in Experiment 2
1 vacuum-tube voltmeter

Procedure

1. Wire the circuit of Fig. 15.13. This is a two-stage RC-coupled transistor audio amplifier. Adjust P_1 for a voltage drop across R_3 of $1\frac{1}{2}$ V. Adjust P_2 for the same drop across R_6.
2. Feed a very low level signal into C_1. The output is taken from C_4 and applied to the vertical input of an oscilloscope. Compute the overall gain of this system. Also compute the individual gain of each stage. This is done by measuring the amplitude of the signal at the base and collector of each transistor and then taking the ratio of collector signal to base signal.
3. Increasing the signal input will show clipping of the positive and negative cycles. Do not increase too much, since it is possible to damage the transistor with too strong a signal input.
4. If a microphone is available, feed its output into the amplifier instead of

FIGURE 15.13

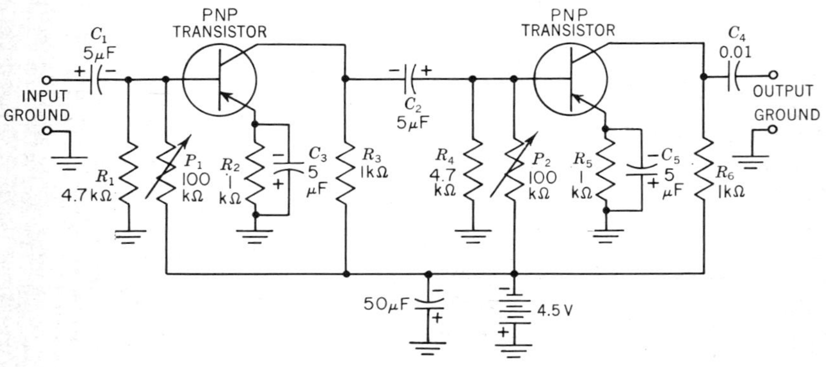

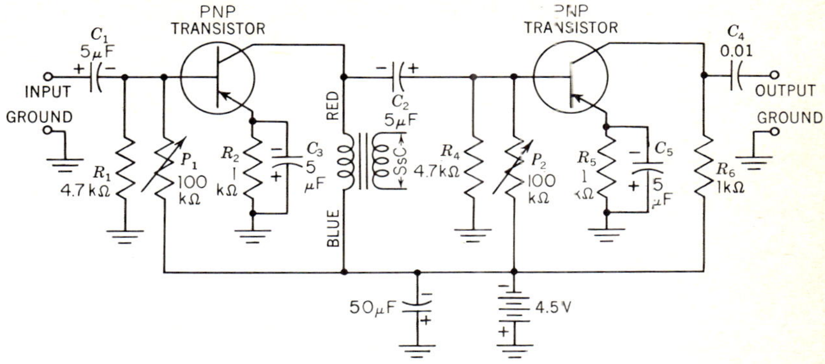

FIGURE 15.14

a signal generator. A considerably amplified version will be seen at the output.

5. Feed the signal generator into the input and note the maximum undistorted output. Normally this is about 1 V rms, but it will vary with the type of transistor used.
6. Change to the circuit of Fig. 15.14 by removing R_3 and substituting the interstage transfomer primary. This is impedance coupling. The blue lead goes to B− and the red lead to the collector. Do not connect the secondary. Notice how the signal clips about the same level. The gain is also the same. Adjust P_2 for a voltage drop of $1\frac{1}{2}$ V across R_6. Adjust P_1 for the same drop across R_2. (What we are doing here is setting each transistor to an operating point of $1\frac{1}{2}$ mA.)
7. Remove the transformer, put back R_3, and remove R_6. Connect the primary of the transformer in place of R_6. The blue lead of the transformer connects to B− and the red lead to the collector (Fig. 15.15). Readjust P_1 and P_2 for $1\frac{1}{2}$ V across R_2 and R_5. Lower the input signal until maximum undistorted output is obtained. The gain will be found to have increased by 5 to 10 times.

What we have just seen may be summarized as follows: In Fig. 15.13, we have conventional resistance coupling. In Fig. 15.14, although the collector voltage of the first stage was increased, the output could not handle a larger signal than the resistance coupling of Fig. 15.13. In Fig. 15.15, however, the output voltage was much greater, since the collector voltage of the second stage was higher. It could, therefore, handle a larger signal from the first stage and also develop a greater output voltage. This indicates that higher outputs require higher collector voltages—a simple achievement with impedance coupling, since there is very little voltage drop through the low dc resistance of the transformer primary. The ac resistance, however, is quite high.

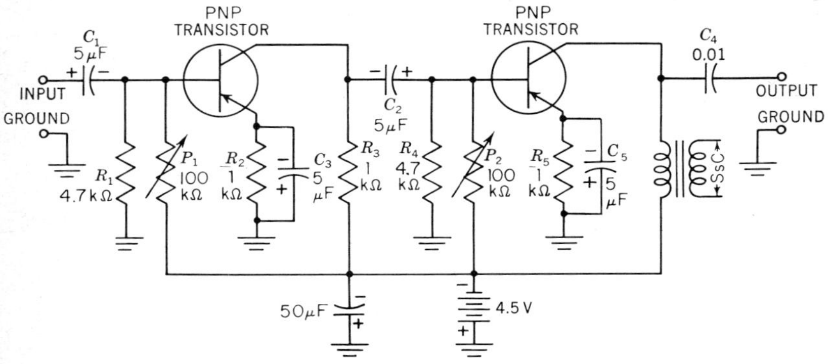

FIGURE 15.15

Conclusions

1. An RC amplifier has considerable gain.
2. Impedance coupling increases the collector voltage and therefore increases the gain and voltage-handling capacity of the stage. It is most valuable in the final stage where it is not hampered by any succeeding amplifiers.

EXPERIMENT 7. AUDIO-FREQUENCY AMPLIFIER WITH TRANSFORMER COUPLING

Object. To observe the performance of a transformer-coupled amplifier.

Material Required

2 2N404 transistors (or equivalent)
4 5-μF electrolytic capacitors
1 50-μF capacitor
1 0.01-μF capacitor
2 4,700-Ω resistors (¼ W)
4 1,000-Ω resistors (¼ W)
2 100,000-Ω potentiometers
1 interstage transformer
1 4½-V battery

Test Equipment

1 audio signal generator or filament transformer arrangement as described in Experiment 2
1 oscilloscope
1 vacuum-tube voltmeter

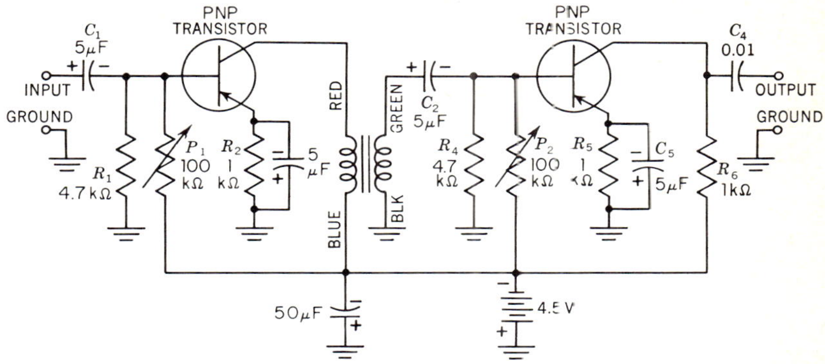

FIGURE 15.16

Procedure

1. Wire the circuit of Fig. 15.13 again.
2. Compute the overall gain.
3. Change to the circuit of Fig. 15.16. Keep each collector current at 1½ mA, as discussed previously.
4. Compute the overall gain.
5. An excellent example of the current-amplifying ability of the transistor can be observed by disconnecting C_2 from the transformer secondary and connecting it to the collector of the first transistor. (Be sure to reverse C_2 polarity when doing this step.) The circuit is now the same as that in Fig. 15.14. Note how the gain decreases when using impedance coupling. This is in spite of the fact that the signal voltage across the secondary winding is very much lower than it is across the primary winding. The gain, however, increases when the input of the second stage is obtained from the secondary rather than from the larger primary. This is opposite to the action of interstage transformers in vacuum-tube amplifiers. The vacuum tube, of course, operates primarily on a voltage drive, while the transistor operates on a current drive. And the secondary of the transformer possesses a greater current than the primary.

Conclusions

1. Transformer coupling increases gain considerably by impedance matching.
2. The transistor is a current-amplifying device.
3. The transistor in the grounded-emitter connection has a much higher output impedance than input impedance. This can be more firmly established by rewiring the transformer in the circuit so that the secondary now goes to the collector and B— of the preceding stage while the primary connects to the base and ground of the following stage. Note the effect on distortion and overall gain.

If the reader possesses an audio generator, he might try checking the frequency responses of the RC-coupled, impedance-coupled, and transformer-coupled amplifiers. Look particularly for the points where the low-frequency response begins to fall off and where the high-frequency response begins to drop, and also note if the curve is flat between these two end frequencies. If desired, the reader might plot these responses on semilog paper.

EXPERIMENT 8. COMPLETE A-F AMPLIFIER AND SUPPLEMENTARY EXPERIMENTS

Object. To analyze audio amplifiers further.

Material Required

3 2N404 transistors (or equivalent)
4 5-μF electrolytic capacitors
1 50-μF electrolytic capacitor
1 interstage transformer
2 1,000-Ω resistors ($\frac{1}{4}$ W)
2 4,700-Ω resistors ($\frac{1}{4}$ W)
2 100,000-Ω potentiometers
1 1-MΩ potentiometer
1 speaker and output transformer
1 phonograph input
1 4$\frac{1}{2}$-V battery
1 15-V battery

FIGURE 15.17

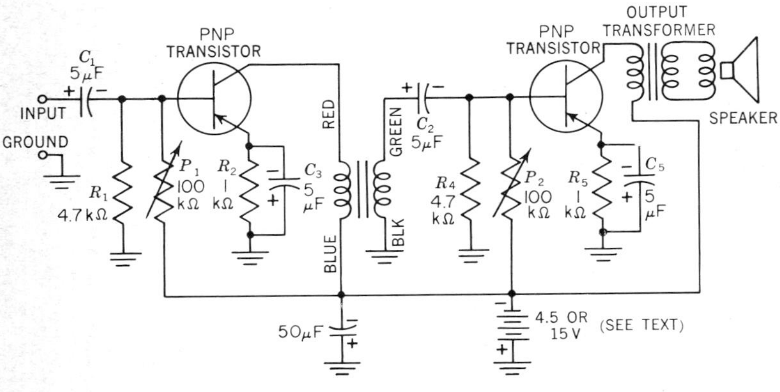

Test Equipment. None.

Procedure

1. Wire the circuit of Fig. 15.17. Adjust P_1 and P_2 for collector currents of $1\frac{1}{2}$ mA through each transistor. These will produce voltage drops of $1\frac{1}{2}$ V across each emitter resistor. This is the same as Fig. 15.16 except that a loudspeaker and its output transformer have been added. High-impedance magnetic earphones (phones with low dc resistance) may be used in place of the loudspeaker and output transformer.
2. A phonograph pickup is fed into the input. If possible, use a pickup with a high output voltage (i.e., a crystal pickup).
3. By changing to a 15-V battery, a much greater output can be obtained. If a 15-V battery is employed in the circuit shown in Fig. 15.17, adjust P_1 and P_2 for collector-emitter voltages of $7\frac{1}{2}$ V. The operating point will now be in the middle of the load line and the output signal is able to swing equally above and below this point.
4. Further output power may be developed by decreasing the values of emitter resistors R_2 and R_5 to 100 Ω each. Additional power output can be obtained by removing all base resistors and connecting a potentiometer (1-MΩ) from the base of each transistor to B−. The collector current can then be adjusted for maximum output.
5. It is startling to measure the total current used with 15 V as a supply. The drain is only several milliamperes, yet enough volume is produced to serve an average living room. If more gain is desired, a phonograph step-down transformer may be used to match the relatively high phonograph impedance to the low input impedance of the transistor. The circuit is shown in Fig. 15.18.
6. An alternate circuit, Fig. 15.19, contains a class A push-pull output arrange-

FIGURE 15.18

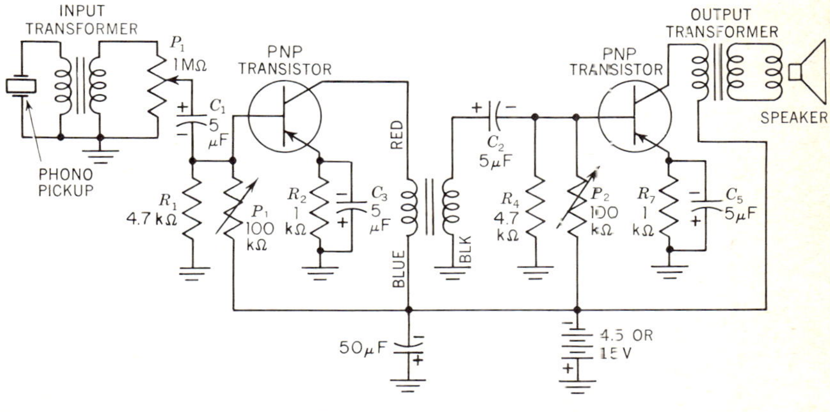

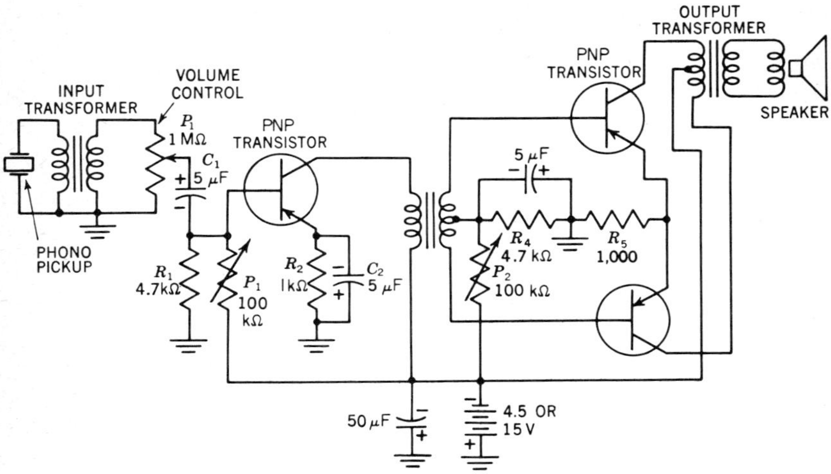

FIGURE 15.19

ment. The second-stage emitter bypass capacitor is omitted because the emitter currents are 180° out of phase, eliminating the need for a capacitor. (*Note:* This could not be done for class B push-pull output stages.) This circuit has the conventional advantage of push-pull operation. A 1-MΩ potentiometer controls volume.

EXPERIMENT 9. DIFFERENTIAL AMPLIFIER

Object. To study the operation of a differential amplifier.

Material Required

2 2N404 transistors (or equivalent)
1 100-Ω 2-W potentiometer
2 1,000-Ω 2-W potentiometers
2 1,500-Ω ½-W ± 10 percent resistors
1 560-Ω ½-W ± 20 percent resistor
1 0 to 6-V dc power supply

Test Equipment

1 VTVM

Procedure

1. Connect the differential amplifier circuit shown in Fig. 15.20. Initially, set potentiometers R_1 and R_2 so that they are at midrange. Then connect the power supply and carefully adjust it to 5 V.

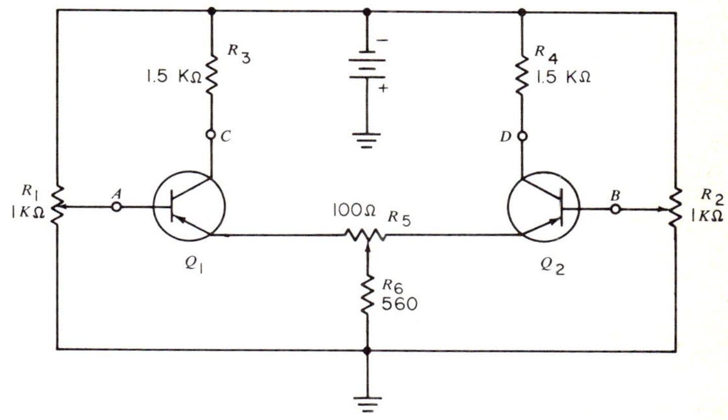

FIGURE 15.20
Schematic diagram of a differential amplifier.

2. The next step is to bring the two stages into balance with each other. To achieve this, terminals A and B are connected together with a jumper wire. Rotate potentiometer R_5 until the voltage between points C and D is equal to zero.
3. Now, remove the jumper wire from A, B. Adjust potentiometer R_1 until the voltage from point A and ground (V_A) is 1 V. Adjust R_2 until V_B is equal to 1 V. Measure the voltage now present between points C, D. On a separate sheet of paper, draw up a chart such as that shown in Table 15.1.

TABLE 15.1

V_A	V_B	V_{A-B}	V_{CD}
1	1	0	
1.3	1		
1.5	1		
1.8	1		
2.0	1		
2.5	1		
3	1		
1	1.3		
1	1.5		
1	1.8		
1	2.0		
1	2.5		
1	3		

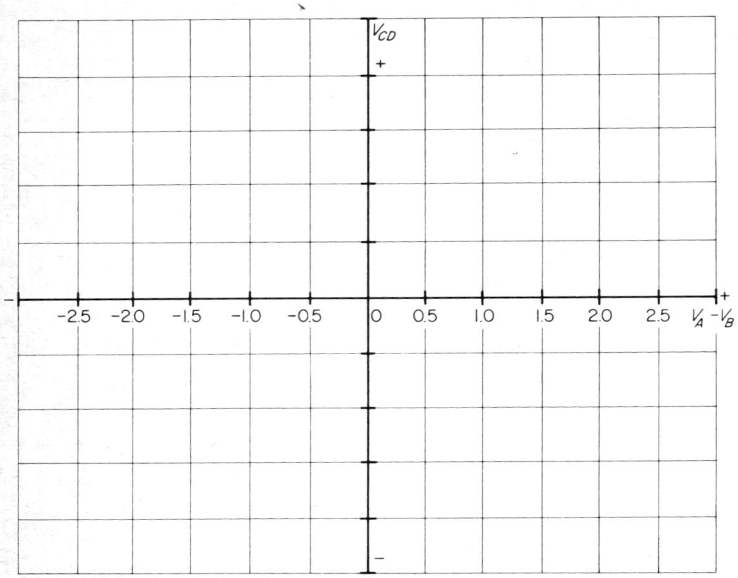

FIGURE 15.21

Record the measured value of voltage CD and write this value down in the first line of the fourth column.

4. Change the voltage at V_A and V_B (where necessary) to the values indicated on the second line of Table 15.1. Note and record the resulting voltage V_{CD}. Calculate the difference between V_A and V_B and record this value in the second line, third column. Be careful that you use the proper sign for this difference value. (*Note:* when V_B is larger than V_A, the result is a negative value.)
5. Repeat this same procedure for the remaining values shown in Table 15.1. Then, plot V_{CD} versus $(V_A - V_B)$ on a chart such as that shown in Fig. 15.21.
6. After step 5 has been completed, adjust R_2 for a 1-V potential at point B. Adjust R_1 for a similar 1-V potential at point A. Measure the potential of point C with respect to ground. Do the same for point D and ground.
7. Now, rotate R_1 to make point A more negative. Measure the voltage between point C and ground and compare this with the value of point C obtained in step 6. Point C will now be found to be less negative or more positive. This indicates a 180° phase shift in the change (representing a signal) that has passed through Q_1.
8. Measure the potential of point D. This will be found to be more negative than step 6.

Conclusions

1. Equal voltages at the inputs of a differential amplifier produce zero output. This feature, called common-mode rejection, is a valuable characteristic of differential amplifiers.
2. Difference input voltages produce an amplified output at terminals CD.
3. Difference voltages at the input can be measured to a high degree of accuracy by the amplified effect they produce at the output.
4. A signal applied to point A will appear at point C with its phase shifted 180°.
5. A signal at point A will appear at point D with zero phase shift.

EXPERIMENT 10. TRANSISTOR VOLTAGE REGULATORS[1]

Object. Transistors are being used increasingly in power systems. One excellent application of the transistor and related semiconductor devices is in the regulation of voltage. Transistor voltage regulators have been designed to achieve 0.005 percent regulation with excellent long-term stability. The purpose of this experiment is to explain the operation of three of the simpler types of transistor regulators and to provide some practical experience in the measurement of their circuit characteristics. The three types of regulators will be discussed, constructed, and tested.

Material Required

1 47-Ω 1-W ± 10 percent resistor
1 270-Ω ½-W ± 20 percent resistor
1 6-V zener diode
1 Variable-load resistor (decade box)
1 2N696 transistor
1 Variable-voltage power supply (0 to 12 V)
1 2N404 transistor
1 500-Ω 2-W potentiometer
1 1.5-V battery

Test Equipment

2 voltmeters (0 to 15 V)
1 milliammeter (0 to 10, 100 mA)

Background Information. Three of the most common regulators in use are zener, series, and shunt. Each operates in an entirely different manner, embodying unique features that enable it to function best under certain conditions.

[1] From Milton S. Kiver and Bernard Van Emden, "Transistor Laboratory Manual," McGraw-Hill Book Company, 1962.

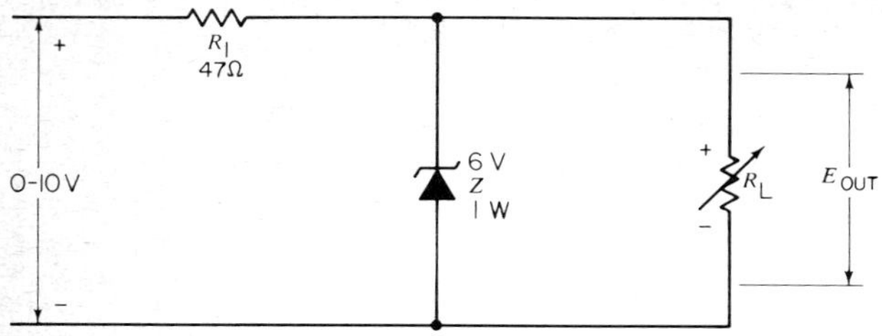

FIGURE 15.22
A zener-diode regulator circuit.

Zener Regulators. The zener regulator utilizes a zener diode. An example of the simplest form of zener-diode regulator is shown in Fig. 15.22. The diode Z exhibits the characteristic shown in Fig. 15.23. In the forward region, it functions as a conventional rectifier. In the reverse region, it exhibits a characteristic similar to that of a vacuum-tube gaseous regulator. That is, in the breakdown region, the voltage drop across the diode remains fairly constant over a wire range of currents. This behavior is the outstanding feature of zener diodes.

There are other characteristics, however, that make this device attractive for regulatory application. One of these is the zener voltage. Zener diodes are presently available with breakdown voltages between 2 and 200 V. Zener voltage is normally specified as the voltage drop across a reverse-biased diode at a specific current level. A particular value of zener voltage is indicated in Fig. 15.23 by he notation V_Z. This voltage is normally specified to within 20, 10, or 5 percent of a nominal value, and almost any value of zener voltage may be obtained. The characteristic provides a wide range of reference, or regulated, voltages.

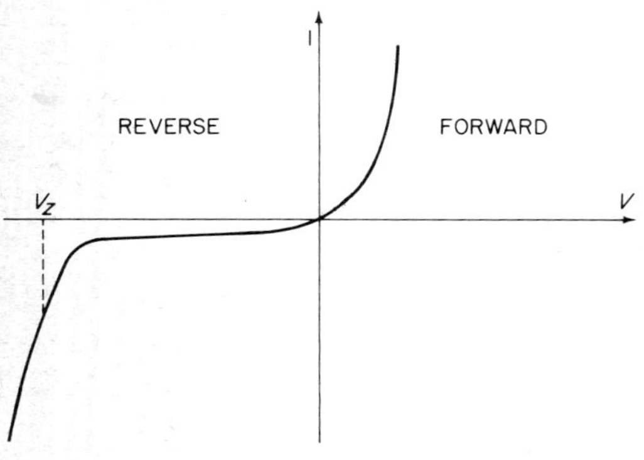

FIGURE 15.23
The characteristic behavior of a zener diode. V_Z represents the zener-voltage value.

Another zener-diode characteristic of importance is its power dissipation. Zener diodes are available in a variety of power-dissipation values, namely, 250, 500, and 750 mW; 1, 1.5, 3.5, 10, and 50 W. The rating is governed by the size of the device, its thermal resistance, and the ambient temperature. It can be made higher by the use of suitable heat sinks that will more efficiently remove the heat generated by the current flow. Power-dissipation ratings are given for specific ambient temperatures, usually 25°C. If the diode is to be operated at a higher temperature, the permissible dissipation should be lowered or "derated." For example, the data sheet for a zener diode may state, "derate 200 mW per degree centigrade above 25°C." This means that for each degree centigrade above 25°C, the rating given must be lowered by 200 mW. Thus, if the operating temperature is 35°C, the derating figure is 200 × 10, or 2,000 mW. The power-dissipation value is obtained from the product of the maximum safe diode current and the zener voltage. Thus, to determine the maximum safe diode current, divide the power-dissipation rating of the diode by its zener-voltage rating.

The zener voltage is temperature-dependent and varies normally at a predictable rate. Figure 15.24 is a typical curve of the temperature coefficient of zener voltage in percent per degree centigrade versus zener voltage. The curve shows that a zener diode with a zener voltage of approximately 5 to 6 V would have a zero temperature coefficient. Above this value, the temperature coefficient is positive; below this value it is negative. The curve also indicates that the increase in the zener-voltage temperature coefficient is not constant but tends to level off around 30 V. Various methods of compensation for this change in zener voltage have been devised. Some of these methods include using additional semiconductor devices or designing the zener diode into composite semiconductor devices. Resistors have also been employed as compensating elements.

Another important zener-diode characteristic is the zener impedance. As shown in Fig. 15.23, the zener voltage increases slightly with increase in current. Consequently, a measure of the rate of voltage increase is desired. The zener impedance may be defined as the ratio of the zener-voltage change resulting from a slight change in current. The zener impedance is normally given by the manufacturer on the data sheet.

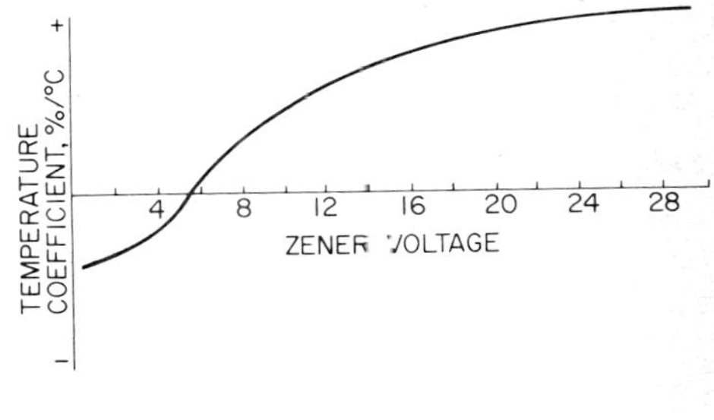

FIGURE 15.24
A typical curve of the temperature coefficient of zener voltage in percent per degree centigrade versus voltage.

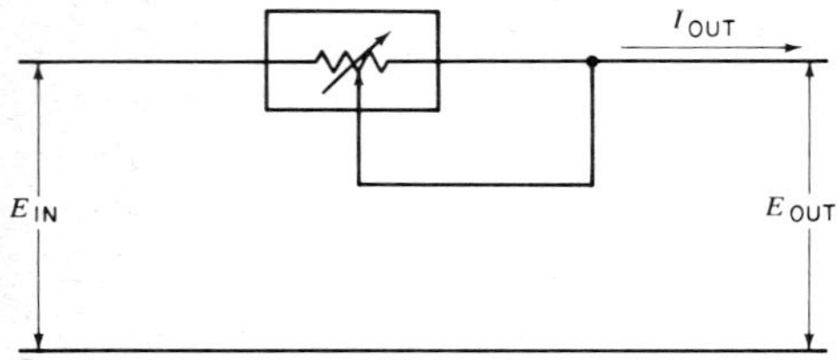

FIGURE 15.25
The basic circuit of a series voltage regulator.

Series Regulator. The series regulator is one of the most common transistor dc voltage regulators. Its mode of operation can be seen in Fig. 15.25. A variable impedance is placed in series with a dc line. By means of sensing and amplification circuits, the value of the impedance is adjusted until a desired output voltage is obtained. Under these conditions, E_{out} is equal to E_{in} minus the voltage drop across the series impedance. Thus, E_{in} must be greater than E_{out} in order for the circuit to function properly. The power dissipation in such a regulator is equal to the product of the voltage across the series impedance and the output current I_{out}.

A transistor circuit based on this principle is shown in Fig. 15.26. The emitter Q_2 is held at a constant potential with respect to the positive output terminal. When the output voltage tends to increase, the base voltage rises, but not as much as the emitter voltage. Since the emitter voltage remains fixed with respect to the positive output voltage, the net effect of this action is to reduce the base-to-emitter voltage. This lessens the collector current of Q_2 and the base current of Q_1. The collector current of Q_1 is thereby reduced, lowering the current flowing through the output load resistance. Thus, the output-voltage rise is counteracted.

FIGURE 15.26
An operational transistor series voltage regulator.

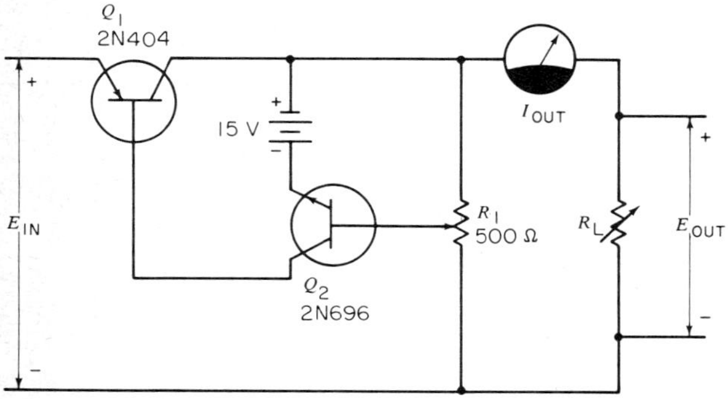

When it is stated that the emitter of Q_2 is held at a constant potential with respect to the positive output terminal, it means that when the output voltage rises, the emitter potential increases by exactly the same amount (in order to keep their difference constant). On the other hand, the base-voltage rise, under the same conditions, is governed by the setting of the center arm of R_1. If this arm is at the midpoint of the resistor, the base receives only one-half of the rise. Thus, it is seen that the emitter voltage will rise more than the base voltage. With a decrease in positive output voltage, the emitter voltage will drop more than the base voltage. That is, the difference between the positive output terminal and the emitter of Q_2 remains constant, no matter what the output voltage may do. The setting of the arm on R_1 will determine the value of the regulated output voltage.

The series regulator shown in Fig. 15.26 is a simplified version of this class of regulators. Series regulators using several transistors for the series element and a multistage transistor amplifier for Q_2 have been constructed and used in much high-performance transistorized equipment. Usually a zener diode is used as a voltage reference rather than the battery shown in Fig. 15.26.

Shunt Regulator. The basic circuit of a shunt regulator is shown in Fig. 15.27. It consists of a fixed impedance in series with the dc line and a variable impedance across the output. Should the input voltage increase, the variable impedance is reduced, causing the voltage drop across the series fixed resistance to increase, thereby maintaining the output voltage at the desired value. The power dissipation in such a regulator is equal to the output voltage times the current through the shunt regulator. This current, by Kirchhoff's law, is equal to the difference between E_{in} and E_{out} divided by the series resistance minus any current that flows in a load, such as R_L shown in Fig. 15.27. In this type of regulator, the power dissipation in the regulator element is at its maximum when the load resistance is infinite. Conversely, in the series regulator previously mentioned, the power dissipation of the regulator is at its maximum when the load resistance is at its minimum value.

FIGURE 15.27
Basic circuit of a shunt voltage regulator.

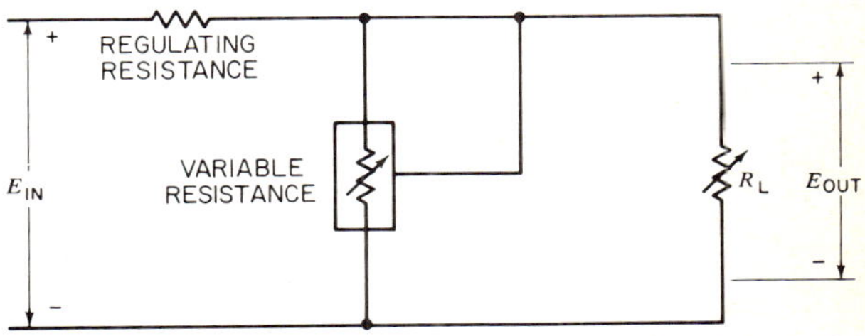

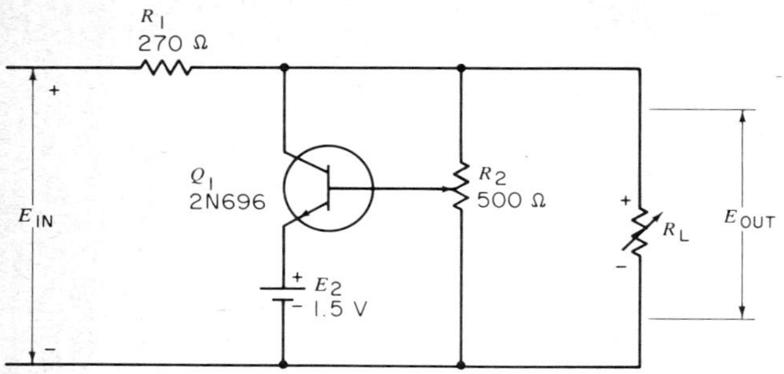

FIGURE 15.28
A simple transistor shunt voltage regulator.

A circuit that embodies the shunt regulator is shown in Fig. 15.28. The battery in the circuit is the reference for the system. If the regulated output voltage increases, the base will become more positive than it was before. Since the emitter voltage is fixed by the reference-battery voltage, the net effect will be to increase the base-to-emitter potential. This will increase the current through the transistor, which, in turn, will increase the voltage drop across the regulating resistor R_1 and decrease the output voltage. The correcting process continues until the output voltage is brought back to its initial value.

The process is essentially the same for a decrease in output voltage, except that now the base-to-emitter voltage decreases, the collector current drops, less voltage is developed across R_1, and the output voltage is brought back to normal again. All these changes take place in less than a second.

The value of the regulated output voltage is determined by the potentiometer setting. This, in turn, is limited by the operating range of the transistor as a function of base-to-emitter voltages. There can be no more voltage difference between the base and emitter than that which produces the safe maximum collector current.

As in the case of the series regulator, the circuit shown in Fig. 15.28 is one of the simplest of the shunt-regulator class. More complicated circuits using many more transistors as amplifiers have been constructed. Shunt regulators are normally used in circuits where the output voltage is low and the load resistance remains relatively constant. A zener diode may again be used as a reference instead of the battery shown in Fig. 15.28.

Procedure. The procedure for this experiment will be divided into three sections covering each of the three regulators discussed.

1. *Zener Diode.* Connect the circuit shown in Fig. 15.22. Set the input voltage at 2 V. Set R_L to infinity (i.e., completely open). Now draw up

a table such as Table 15.2. Record the output voltage in the space provided in this table. Repeat this test for the different values of input voltage and load resistance shown in Table 15.2.

2. Now, set up a graph like Table 15.3 and plot output voltage versus input voltage for the four load-resistance values given in Table 15.2. (Your graph, when completed, should contain four separate plots of output voltage versus input voltage.)
3. *Series Regulator.* Connect the circuit shown in Fig. 15.26. With the input voltage set at 10 V and the load resistance infinite (open circuit), set potentiometer R_1 so that the output voltage is equal to 4 V. Record this value in a chart such as that of Table 15.4. Change the load resistance to 1,000 Ω and measure the output voltage with the input voltage at 10 V. Record the value of voltage measured. Repeat this test for the values of load resistance and input voltage given in Table 15.4. *Note:* Be sure to readjust the input voltage to the desired value after each change in load resistance. Plot output voltage versus input voltage for the four values of load resistance in Table 15.4 on a graph such as that of Table 15.5.
4. *Shunt Regulator.* Connect the circuit of Fig. 15.28. With the input voltage at 10 V and infinite load resistance, set potentiometer R_2 so that the output voltage is equal to 4 V. Record this value in a chart such as that of

TABLE 15.2

E_{IN}	E_{OUT}			
	$R_L = 5\ k\Omega$	$R_L = 1\ k\Omega$	$R_L = 100\ \Omega$	$R_L = \infty$
2				
4				
6				
8				
10				

TABLE 15.3

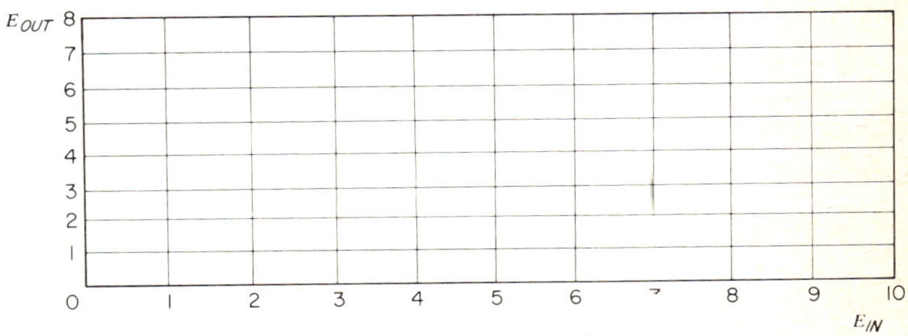

Table 15.6. Adjust the load resistance to 5,000 Ω, keeping the input voltage at 10 V. Measure and record the output voltage. Repeat this test for the values of input voltage and load resistance given in Table 15.6.

5. The output resistance of a regulator can be defined as the ratio $\Delta E_{out}/\Delta I_{out}$, where ΔE_{out} is the change in output voltage for a change of output current ΔI_{out}. From your data, calculate the output resistance for each of the three regulators studied. Use as small a change in E and I as possible.

TABLE 15.4

E_{IN}	E_{OUT}			
	$R_L = \infty$	$R_L = 1000\,\Omega$	$R_L = 500\,\Omega$	$R_L = 220\,\Omega$
1				
3				
5				
6				
7				
10				

TABLE 15.5

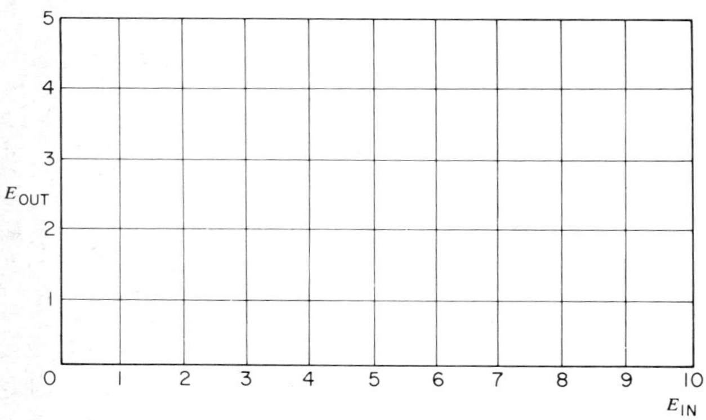

TABLE 15.6

E_{IN}	E_{OUT}			
	$R_L = \infty$	$R_L = 5\,k\Omega$	$R_L = 1\,k\Omega$	$R_L = 100\,\Omega$
2				
4				
6				
8				
10				

Conclusions

1. The output voltage of a regulator circuit is always less than the input voltage.
2. The maximum output current of a regulator is determined by the safe maximum current that can be handled by the regulating element.
3. Of the three regulator circuits studied in this experiment, the shunt regulator would be selected when a low output voltage is desired and the load resistance is fairly constant.
4. Of the three circuits, the series regulator operates best for widely varying load resistances.
5. The zener-diode regulator possesses a simple circuit arrangement but does not provide as wide a range of output voltages as the series or shunt transistor regulators.

EXPERIMENT 11. TRANSISTOR CURRENT REGULATORS[1]

Object. In many transistor-circuit applications it is desirable to have a power supply that maintains constant output current rather than constant output voltage. In this experiment, the operation of a current regulator or limiter will be examined.

Material Required

1 2N404 transistor
1 100-Ω, ½-W ± 5 percent resistor
1 500-Ω, 2-W potentiometer
1 1.5-V power supply
1 0 to 10-V power supply

Test Equipment

1 0-to-20-mA meter
1 0-to-10-V voltmeter

Background Information. It has previously been shown that the collector-voltage-vs.-collector-current characteristics of a common-emitter or common-base amplifier tends to become a straight horizontal line beyond a collector voltage of a volt or so. In other words, the collector current remained constant and was not dependent on load resistance or supply voltage.

The circuit of Fig. 15.29 takes advantage of this behavior. Voltage V_R is applied as shown. This voltage, in conjunction with the 100-Ω resistance R_1 and the transistor, forms a current regulator capable of supplying a specific

[1] From Milton S. Kiver and Bernard Van Emden, "Transistor Laboratory Manual," McGraw-Hill Book Company.

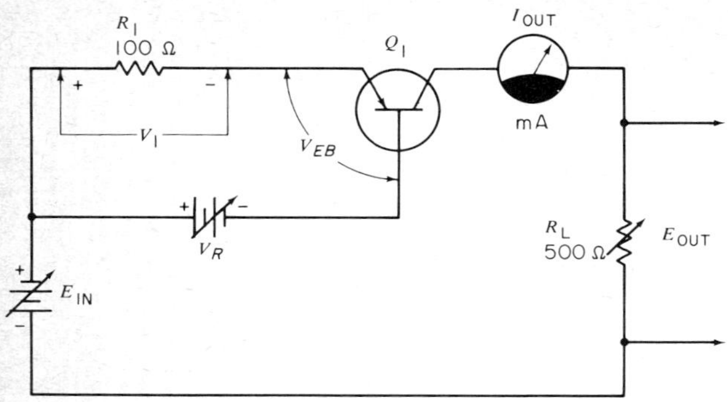

FIGURE 15.29
Basic circuit of a transistor current regulator.

value of emitter current. So long as the voltage drop across the emitter is either negligible or constant, the emitter current will remain more or less constant. If the emitter current is held at a constant value, the collector current will be equal to h_{FB} (α) times this emitter current. This statement is true so long as E_{in} minus V_R is sufficiently large to allow the desired current to flow through resistance R_L. The circuit then acts as a constant-current regulator or current limiter and maintains a specific value of current flowing through the load.

From an operational standpoint, all the load current must pass through transistor Q_1; hence, the objective of the circuit is to have changes in the output current alter the internal resistance of Q_1 so that these current changes are counteracted. In Fig. 15.29 this is achieved by having the output, or load, current pass through R_1. A voltage drop is produced here with the polarity indicated. At the same time, V_R establishes a fixed, or reference, voltage with a similar polarity.

The voltage applied between base and emitter V_{EB} is then equal to

$$V_{EB} = V_R - V_1$$

A typical value of V_{EB} is 0.3 to 0.4 V for germanium transistors.

Now, as the load current varies, it varies the voltage across R_1. This alters the base-to-emitter voltage because V_R remains fixed. A variation in V_{EB} causes the collector current to vary, and thus regulation is achieved.

Procedure

1. In order to study the characteristics of the regulator in Fig. 15.29, a three-part procedure will be followed. The first part will be devoted to the initial

calibration of the circuit. The second part will investigate the effects of load-resistance changes. The third part will examire what happens when the input voltage varies.

2. *Initial Calibration.* To calibrate the constant-current regulator, set resistor R_L equal to approximately 250 Ω. Set voltage E_{in} equal to 6 V. Adjust voltage V_R until current I_{out} is 10 mA. Do not change voltage V_R after this initial setting has been made.
3. *Effect of Load Resistance.* To determine the effect of load resistance on the constant-current-regulator circuit shown in Fig. 15.29, set voltage E_{in} equal to 5 V. Set the load resistance equal to 0 Ω. Measure the output voltage and output current. Draw a chart such as that of Table 15.7 and record the current just measured. Readjust the load resistance until an output voltage of 1 V appears. Record the value of output current. From Ohm's law, compute the value of load resistance and record the results of your computation in the space provided in the table. Repeat this procedure for the voltages listed in Table 15.7.

   Note that E_{in} has not been changed throughout the foregoing procedure. Now, draw a chart such as that shown in Fig. 15.30 and plot output current versus output voltage for a constant input voltage, in this case 5 V. Note over what range the current remains steady.
4. *Effect of Input Voltage.* Set the load resistance R_L equal to 100 Ω. Set voltage E_{in} equal to 0 V. Record the value of output voltage and output current in a table such as that of Table 15.8. Repeat this procedure for the values of input voltage E_{in} given in Table 15.8. Now, plot the output current versus output voltage for a constant output resistance.

Conclusions

1. The current regulator examined in this experiment will yield constant output current over a wide range of load resistances.
2. A well-designed transistor current regulator will protect a power supply against short circuits in the load.

E_{OUT}, V	I_{OUT}, mA	R_L, Ω
0		
1		
2		
3		
4		
4.2		
4.5		
4.7		
5		

TABLE 15.7

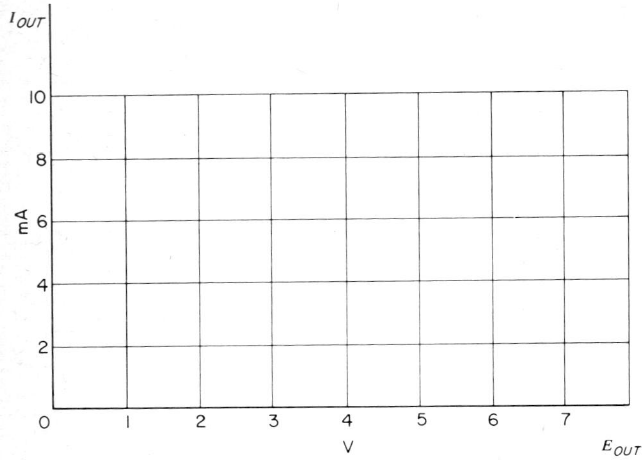

FIGURE 15.30

3. A current regulator will act to maintain a constant output current over a range of input voltages.
4. A transistor with a high output resistance functions well as a constant-current regulator. The common-base configuration, possessing the highest output resistance of the three amplifier arrangements, functions best as a constant-current regulator.

E_{IN}, V	E_{OUT}, V	I_{OUT}, mA
0		
0.2		
0.5		
0.7		
1		
1.2		
1.5		
2		
4		
6		
7		

TABLE 15.8

EXPERIMENT 12. TRANSISTOR SIGNAL TRACER AND ELEMENTARY RADIO RECEIVER

Object. To construct a signal tracer and then tune it, thereby converting it to a tuned signal tracer or a trf radio receiver.

Material Required

2 2N404 transistors (or equivalent)
1 loop antenna (with ferrite core)
4 5-μF electrolytic capacitors
2 50-μF electrolytic capacitors
1 0.01-μF capacitor
2 1,000-Ω resistors ($\frac{1}{4}$ W)
2 4,700-Ω resistors ($\frac{1}{4}$ W)
2 10,000-Ω resistors ($\frac{1}{4}$ W)
2 100,000-Ω potentiometers
1 interstage transformer
1 0.0015-μF capacitor
1 crystal diode
1 $4\frac{1}{2}$-V battery
1 15-V battery
1 speaker and output transformer or earphone
50 ft of antenna wire
1 365-pF midget variable tuning capacitor

Test Equipment. None.

Procedure

1. Wire the circuit of Fig. 15.31. Adjust P_1 and P_2 for collector currents of $1\frac{1}{2}$ mA through each transistor. Earphones may be substituted for the loudspeaker and its output transformer.
2. Connect a short length of wire to the input capacitor and another length of wire to ground. This system may then be employed as a signal tracer in practically any circuit from audio through TV by connecting the ground lead to the chassis of the device being tested and probing with the other lead. This instrument can be used in any application where the well-known vacuum-tube signal tracer can be employed.

   Louder signals can be obtained from the transistor tracer by using a 15-V battery.
3. The signal tracer may be tuned by the addition of a loop antenna and a variable capacitor as shown in Fig. 15.32. An antenna 50 ft in length should be used. The chassis of the transistor receiver should be grounded to a radiator or water pipe. The importance of a good antenna and ground, particularly in an area somewhat remote from high-power broadcast stations, cannot be overemphasized.

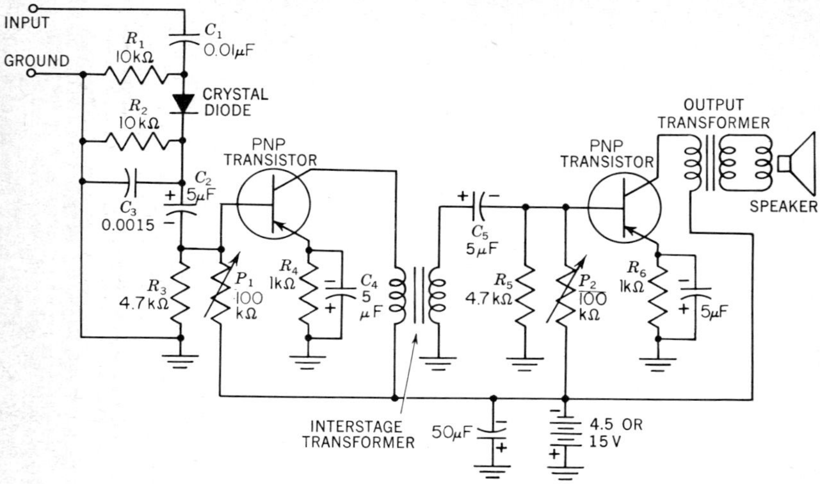

FIGURE 15.31

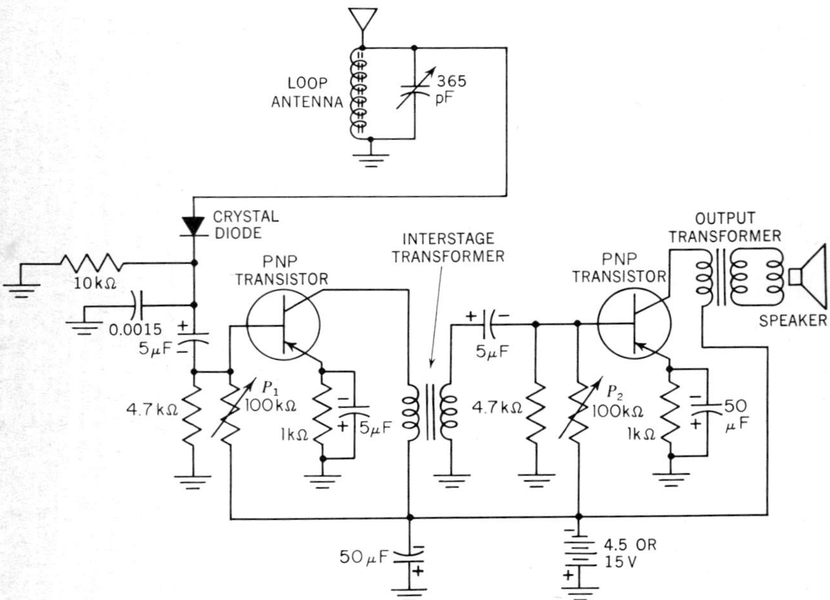

FIGURE 15.32

Conclusions

1. The transistor audio amplifier may be converted to a signal tracer by the addition of a solid-state diode.
2. Tuning the signal tracer converts it to a trf receiver.

EXPERIMENT 13. RELAXATION OSCILLATOR

Object. To analyze the mode of operation and the waveforms in a relaxation oscillator.

Material Required

2 2N404 transistors (or equivalent)
2 10,000-Ω resistors (¼ W)
2 1,000-Ω resistors (¼ W)
2 0.01-μF capacitors
2 100,000-Ω potentiometers
3 5-μF electrolytic capacitors
1 0.0015-μF capacitor
1 4½-V battery

Test Equipment

1 oscilloscope

Procedure

1. Wire the circuit of Fig. 15.33. Adjust P_1 and P_2 for a collector-to-emitter voltage of 2 V in each transistor.

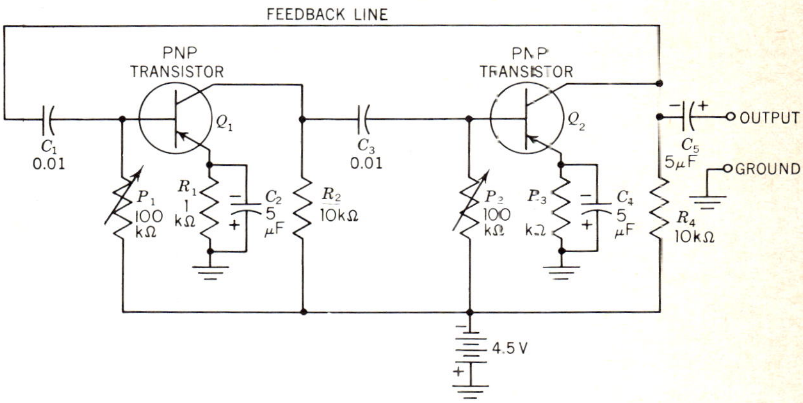

FIGURE 15.33

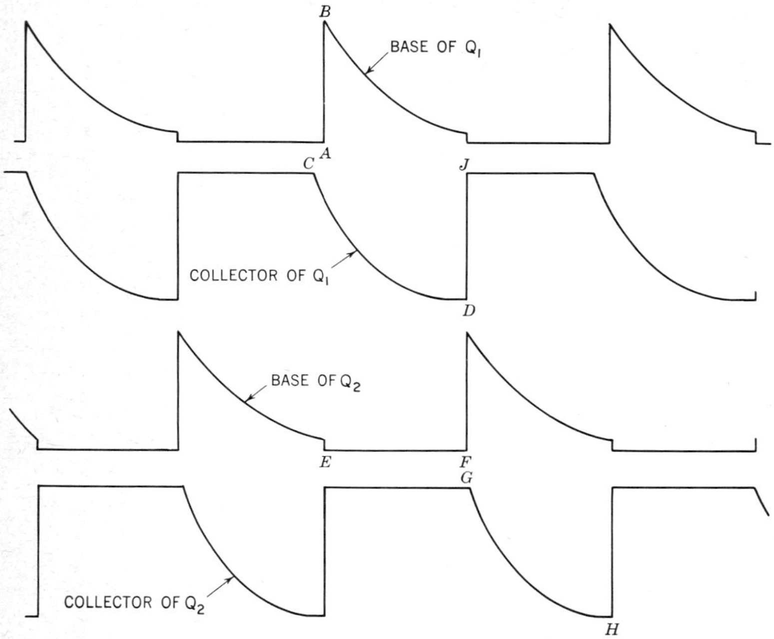

FIGURE 15.34

2. When connected, the circuit will oscillate, producing the nonsymmetrical waveforms shown in Fig. 15.34.
3. When observing waveforms with an oscilloscope, it is suggested that external synchronization be used. This may be accomplished by attaching a 0.0015-μF capacitor in series with a lead from the external sync terminal of the oscilloscope. The other end of this lead should be connected to the collector of transistor 2. The vertical input probe of the oscilloscope may be used for observing the voltage waveform at various points in the circuit. If external synchronization is not used, the waveform will shift as the vertical lead is moved from one point to another.
4. To analyze the operation of the circuit, let us assume that the base of Q_1 goes positive momentarily. This decreases the collector current in Q_1, causing the voltage at the collector (and the top end of R_2) to become more negative than it was before. This applies a negative voltage to the base of Q_2. Since this acts to increase the forward bias between base and emitter of Q_2, the collector current of Q_2 will increase and the voltage at the collector will become less negative or more positive. This signal makes the base of Q_1 go more positive, thereby aiding the initial positive signal. This

is the fundamental requirement of oscillation—positive feedback with a gain greater than 1.

The buildup will continue with the current through Q_1 decreasing and the current through Q_2 increasing until Q_1 is cut off. When the latter condition is reached, the base of Q_2 no longer receives any driving signal from the collector of Q_1 and the current though Q_2 starts decreasing. This brings a negative signal to the base of Q_1, gradually bringing this transistor out of cutoff. When Q_1 starts conducting, the base of Q_2 receives a positive-going signal from Q_1, and its current is further decreased. Through this process, the current is built up in Q_1 and reduced in Q_2 until Q_2 becomes nonconducting. In this fashion, control is passed back and forth between Q_1 and Q_2.

5. Waveform analysis. As the base of Q_1 goes positive (points A and B of Fig. 15.34), it cuts off its collector current. Capacitor C_3 therefore starts charging through R_2. The charging is shown by points C and D. Since capacitor C_3 is charging at a constant rate, the voltage at the base of Q_2 is maintained fairly constant (points E and F). When the charging of capacitor C_3 starts to round off slightly (approaching time-constant value), the base voltage of Q_2 starts to decrease (go less negative, point F). This causes the collector of Q_2 to go more negative (points G and H). When the collector of Q_2 goes more negative, it drives the base of Q_1 negative, which in turn makes the collector of Q_1 go positive (see points D and J). The action is now the same as described previously for the base of Q_1.
6. The frequency of oscillation may be changed by varying the coupling capacitors. The frequency may also be altered by varying the base resistors, but this may prove to be somewhat difficult because it will result in a change in the dc base voltages.
7. By placing a potentiometer in the lead between the collector of Q_2 and C_1, a square-wave output may be obtained.

Conclusions

1. A transistor multivibrator is feasible.
2. The frequency of oscillation is determined by the time constant of the RC circuits. In the circuit shown, it is easier to vary the frequency by varying the coupling capacitors than the resistors.
3. Square and sawtooth waves can be developed by this oscillator.

EXPERIMENT 14. BLOCKING OSCILLATOR AND SAWTOOTH GENERATOR

Object. To study a blocking oscillator and sawtooth generator.

Material Required

1 2N404 transistor (or equivalent)
1 1,000-Ω resistor ($\frac{1}{4}$ W)

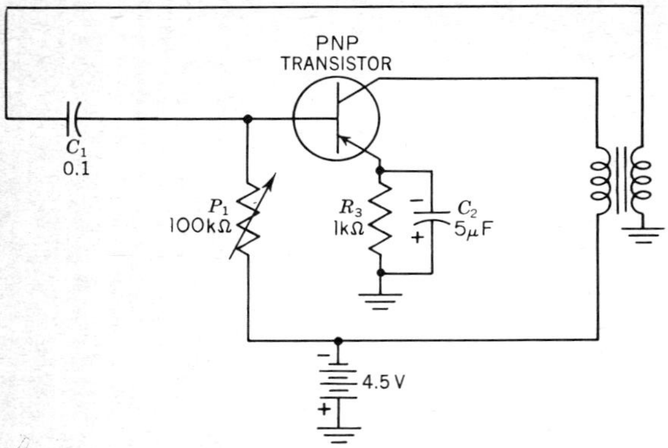

FIGURE 15.35

1 100,000-Ω potentiometer
1 0.1-μF capacitor
1 5-μF electrolytic capacitor
1 transformer (use interstage transformer)
1 4½-V battery

Test Equipment

1 oscilloscope

Procedure

1. Wire the circuit of Fig. 15.35. Adjust P_1 for a 0.5-V drop across R_3.
2. Observe the waveforms at the base, collector, and secondary of the transformer. (Use external synchronization from the secondary of the transformer as described in Experiment 13.)
3. To analyze the operation, at point A, Fig. 15.36, the base starts to go negative, and this drives the collector from point B to point C. At point C the collector current can no longer increase. This removes the feedback that the transformer has been supplying to the base. The base, however, does

FIGURE 15.36

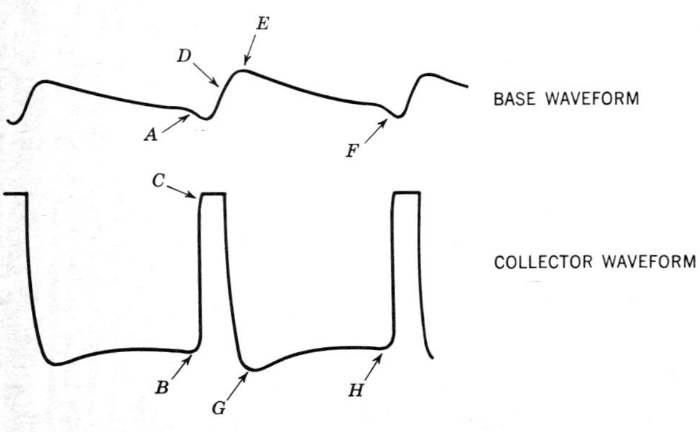

not reach its most negative point until capacitor C_1 has been fully charged. Once the charging rate of C_1 ceases, the voltage across the base potentiometer P_1 starts to drop, becoming less negative, or more positive. This drives the collector more negative, which, because of the phase reversal of the transformer, causes the base to go more positive (point D). The gradual slope in the base and collector curves (points E and F and G and H, respectively) is the time constant of the capacitor and resistor.

4. If a capacitor is now placed from the collector to ground, a sawtooth wave will be obtained. Note the similarity between this and the blocking oscillator used in a television set. The frequency may be changed by varying C_1 or R_3.
5. It is suggested that two 10,000-Ω resistors be placed in parallel across the green and black leads of the transformer in waveform experiments. This will eliminate any transformer ringing, if present. By changing the value of this loading resistor you will see how various waveforms may be developed.

Conclusions

1. A transistor may be employed as a blocking oscillator.
2. The addition of a capacitor from the collector to ground converts the blocking oscillator to a sawtooth wave generator.
3. The frequency of oscillation is determined by the time constant of the circuit if the transistor has a high enough cutoff frequency. If it does not, then the transistor will determine the oscillation frequency.

EXPERIMENT 15. A TRANSISTOR COLPITTS OSCILLATOR

Object. To construct a variable-frequency Colpitts r-f oscillator

Material Required

1 2N404 transistor (or equivalent)
2 0.01-μF capacitors
2 47-pF capacitors
1 1,000-Ω resistor (½ W)
1 coil, which may be the antenna coil of Experiment 12
1 100,000-Ω potentiometer
1 4½-V battery

Test Equipment

Oscilloscope or radio receiver

Procedure

1. Wire the circuit of Fig. 15.37. Adjust P_1 for a collector-to-emitter voltage of 2 V.

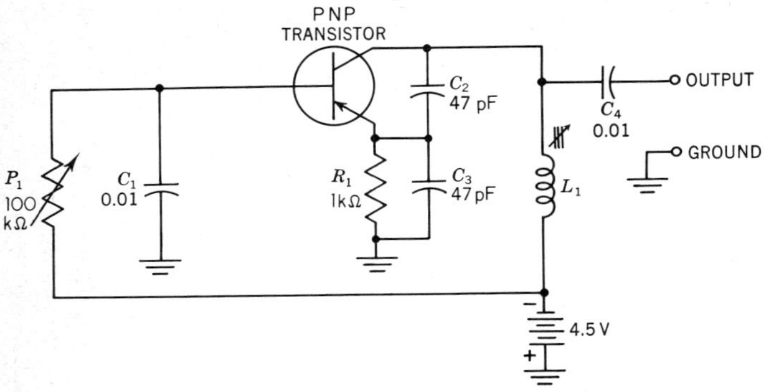

FIGURE 15.37

2. Oscillation should be observed by connecting the vertical lead from the oscilloscope to the collector output capacitor C_4. The other lead from the oscilloscope goes to ground.
3. Vary the frequency of oscillation by varying the slug in the coil. The frequency can also be varied by varying capacitor C_2, C_3, or both.
4. Touch the vertical lead of the oscilloscope and note the change in oscillator frequency. This is because there is no intervening buffer stage.
5. The frequency of oscillation will probably not go much beyond several megahertz, with the transistor specified. However, with a high-frequency transistor, operation to 100 MHz could be obtained.
6. Check the amplitude of oscillations as the frequency is raised.
7. Try to reach the highest operating frequency by continually reducing the inductance of the tuning coil. Check the value of this frequency with the transistors you have.

Conclusions

1. A Colpitts oscillator can be readily formed with a transistor.
2. The circuit is quite sensitive to "hand-capacitance" effects.
3. Oscillations tend to become weaker as the frequency is raised.
4. The highest operating frequency will tend to vary among similar transistors.

EXPERIMENT 16. COLPITTS OSCILLATOR AND BUFFER USING A GROUNDED-COLLECTOR CONNECTION

Object. To construct a buffer Colpitts oscillator arrangement.

Material Required

2 2N404 transistor (or equivalent)
2 0.01-μF capacitors

2 47-pF capacitors
1 coil as in Experiment 15
1 3,300-Ω resistor
1 1,000-Ω resistor
2 100,000-Ω potentiometers
1 4½-V battery

Test Equipment

1 oscilloscope or radio receiver

Procedure

1. Wire the circuit shown in Fig. 15.38. Adjust P_1 for a collector-to-emitter voltage of 2 V.
2. The output is taken here from the emitter of the second transistor. To observe the waveform of the generated signal, connect the vertical input lead of an oscilloscope to the output terminal. Use a capacitor in series with the oscilloscope lead.
3. Observe the frequency of the waveform and then touch the output lead. The level of the signal may decrease, but its frequency remains essentially unchanged.
4. The grounded-collector stage acts very much as a vacuum-tube- grounded-plate (i.e. cathode-follower) circuit. There is no voltage gain through the grounded-collector circuit, but there is a definite buffer action. This stems from the change in resistance levels. The output is now a very low impedance and thus hand-capacitance effects will not be noted.

Conclusions

1. It is desirable to employ a buffer stage when operating transistor oscillators that may be subject to varying load conditions. The buffer here is a

FIGURE 15.38

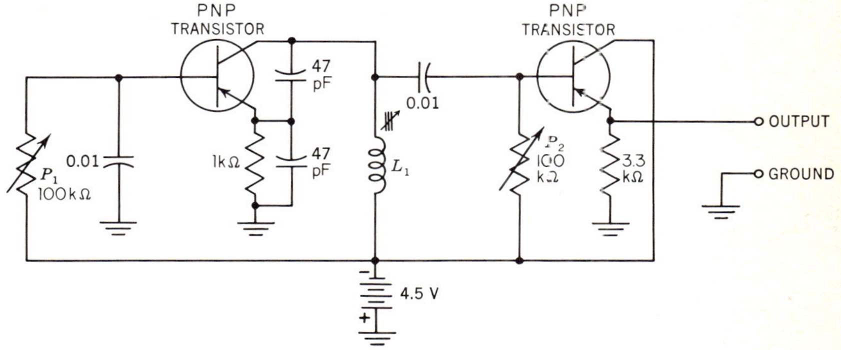

grounded-collector circuit. This is a good buffer circuit because load changes have less effect on the buffer input.

EXPERIMENT 17. TRANSISTOR CHARACTERISTIC CURVES

Object. To obtain the characteristic curves of transistors on the screen of an oscilloscope.

Material Required

1 2N404 transistor (or equivalent)
1 100-Ω resistor (½ W)
1 25-mA silicon rectifier
1 12-V filament transformer (1-A rating)
1 1½-V battery
1 50,000-Ω potentiometer

Test Equipment

1 oscilloscope
1 vacuum-tube voltmeter or VOM

Procedure

1. Wire the circuit shown in Fig. 15.39. The leads to the horizontal terminals of the oscilloscope provide a horizontal sweep that is directly proportional to variations in collector voltage. The voltage across R_2 is applied to the vertical input terminals of the oscilloscope, giving a vertical deflection

FIGURE 15.39

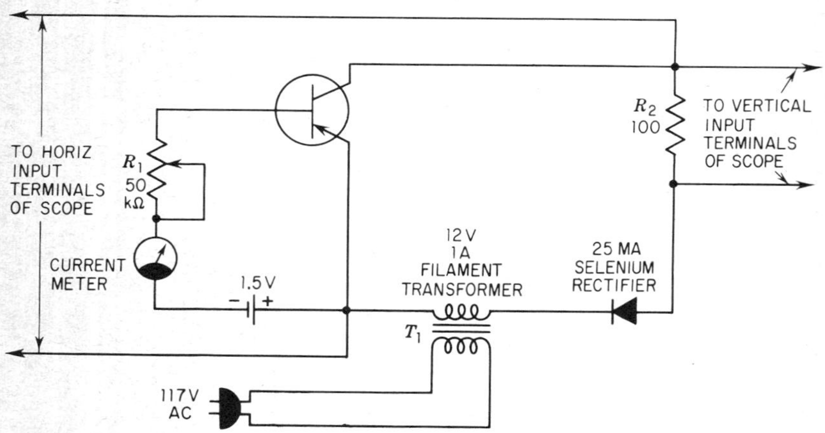

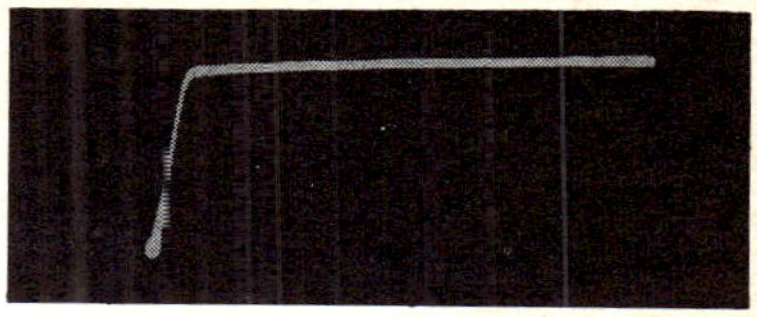

FIGURE 15.40
Form of a single characteristic curve of a transistor.

that is directly proportional to the variations in collector current. In this way, we obtain a collector-voltage–collector-current characteristic curve. Note that the internal sweep of the oscilloscope is not used; the horizontal amplifier is switched to "horizontal input."

2. The transistor is connected in Fig. 15.39 as a grounded-emitter amplifier. With this arrangement, a series of curves are desired for different base currents. (See Chap. 3, where typical transistor characteristic curves are shown.) To start, adjust R_1 to its maximum value. Then, the vertical-gain and horizontal-gain controls of the oscilloscope are adjusted until the desired image size is obtained, Fig. 15.40. (Do not advance the vertical-gain control too far. The curve obtained when the base current is lowest possesses the lowest vertical amplitude. If this curve is made too high, some of the curves to follow will fall beyond the top of the screen.)

3. For each different value of base current, a different characteristic curve is obtained (Fig. 15.41). The base current is varied by adjusting potentiometer R_1.

If desired, a transparent scale can be placed over the scope screen and the vertical-gain and horizontal-gain controls then adjusted for a known amount of gain. Here is how to do this. Apply a known ac voltage first to the vertical-input and then to the horizontal-input terminals of the scope and adjust the respective vertical-gain and horizontal-gain controls for a given deflection. For example, an ac voltage having a peak-to-peak amplitude of 6 V is applied to the horizontal-input terminals of the scope and the horizontal-gain control is adjusted for a deflection of 3 in. The horizontal sensitivity then becomes equal to 2 V/in. This enables you to

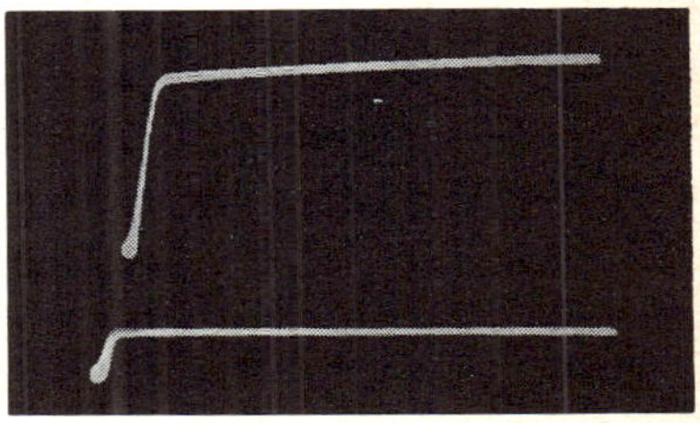

FIGURE 15.41
Characteristic curves obtained with base current of 100 μA (above) and of 20 μA (below).

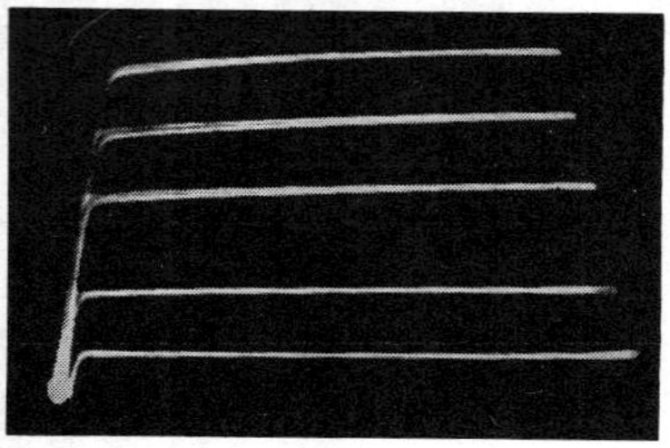

FIGURE 15.42
Family of collector-current-vs.-collector-voltage curves for a 2N404 transistor at base currents of 20, 50, 100, 150, and 200 μA (bottom to top).

measure the collector voltage at any point on a traced-out curve by measuring its distance horizontally along the screen.

For the vertical-input terminals, a smaller voltage, say 0.6 V peak to peak, is recommended. Then, if the vertical-gain control is adjusted for a deflection of 3 in, the vertical sensitivity becomes 0.2 V/in. Since the vertical axis of the curves represents current, these voltage values must be converted into equivalent current values. This is readily accomplished, since whatever voltage is applied to the vertical-input terminals of the scope must come from the 100-Ω resistor of Fig. 15.39. Thus, if the transparent scale over the scope screen indicates a vertical amplitude of 0.4-V, then

$$E = IR$$

$$0.4 = I \times 100$$

$$I = \frac{0.4}{100} = 0.004 \text{ A} = 4 \text{ mA}$$

This is the collector current at the point where the above 0.4-V amplitude was measured.

Note that once the oscilloscope gain controls are set, they are not touched again; otherwise, the calibration is disturbed.

4. The circuit shown in Fig. 15.39 is for a PNP transistor. To obtain the characteristic curve for an NPN transistor, reverse the connections to the battery and to the selenium rectifier.
5. To obtain the characteristic curves for a grounded-base arrangement, the connections to the base and emitter of Fig. 15.39 would be interchanged. Battery polarity, too, would be altered accordingly. Then the curve for a number of emitter current values would be obtained.
6. One curve of a family will be obtained for one value of base (or emitter) current. To obtain a composite or family group of curves, a series of successive photos would have to be taken (Fig. 15.42).
7. If greater changes in collector voltage are desired, substitute another transformer for T_1 that will provide a greater secondary voltage. For example, a 25-V transformer may be used here.

Index